C000231010

1,000,000 Books

are available to read at

www.ForgottenBooks.com

Read online
Download PDF
Purchase in print

ISBN 978-1-5278-8479-3
PIBN 10898178

This book is a reproduction of an important historical work. Forgotten Books uses
state-of-the-art technology to digitally reconstruct the work, preserving the original format
whilst repairing imperfections present in the aged copy. In rare cases, an imperfection in
the original, such as a blemish or missing page, may be replicated in our edition. We do,
however, repair the vast majority of imperfections successfully; any imperfections that
remain are intentionally left to preserve the state of such historical works.

Forgotten Books is a registered trademark of FB &c Ltd.
Copyright © 2018 FB &c Ltd.
FB &c Ltd, Dalton House, 60 Windsor Avenue, London, SW19 2RR.
Company number 08720141. Registered in England and Wales.

For support please visit www.forgottenbooks.com

1 MONTH OF
FREE
READING

at

www.ForgottenBooks.com

By purchasing this book you are
eligible for one month membership to
ForgottenBooks.com, giving you
unlimited access to our entire
collection of over 1,000,000 titles via
our web site and mobile apps.

To claim your free month visit:
www.forgottenbooks.com/free898178

* Offer is valid for 45 days from date of purchase. Terms and conditions apply.

English
Français
Deutsche
Italiano
Español
Português

www.forgottenbooks.com

Mythology Photography **Fiction**
Fishing Christianity **Art** Cooking
Essays Buddhism Freemasonry
Medicine **Biology** Music **Ancient**
Egypt Evolution Carpentry Physics
Dance Geology **Mathematics** Fitness
Shakespeare **Folklore** Yoga Marketing
Confidence Immortality Biographies
Poetry **Psychology** Witchcraft
Electronics Chemistry History **Law**
Accounting **Philosophy** Anthropology
Alchemy Drama Quantum Mechanics
Atheism Sexual Health **Ancient History**
Entrepreneurship Languages Sport
Paleontology Needlework Islam
Metaphysics Investment Archaeology
Parenting Statistics Criminology
Motivational

received personally May 25ᵈ 1871

THE

JOURNAL

OF THE

ROYAL GEOGRAPHICAL SOCIETY.

VOLUME THE FORTIETH.

1870.

EDITED BY THE ASSISTANT-SECRETARY.

LONDON:

910.6
R 87
Vol. 40
212762

LONDON:

PRINTED BY WILLIAM CLOWES AND SONS, STAMFORD STREET,
AND CHARING CROSS.

STANFORD LIBRARY

CONTENTS OF VOL. XL.

[N.B. The Authors are alone responsible for the contents of their respective papers.]

ILLUSTRATIONS.

₊ The MAP of TURKISTAN (by Arrowsmith), before mentioned as in preparation,
is not yet ready. It was intended in this volume to insert it as illustration to Severtsof's
paper, p. 343.

Royal Geographical Society.

1870.

REPORT OF THE COUNCIL,

READ AT THE ANNIVERSARY MEETING ON THE 23RD MAY.

THE Council have to lay before the Fellows the following Report of the financial and general condition of the Society.

Members.—The number of new Fellows added to the roll of the Society during the year, is 191—namely, two Honorary Corresponding, and 189 Ordinary Fellows, of whom 25 have paid their Life Compositions. In the previous year, the elections amounted to 175; in 1868, to 190; and in 1867, to 147. The losses the Society has sustained are 49 by death, and 27 by resignation; besides these, 14 have been struck off the list for arrears of subscription, making a total of 90. The net increase for the year is 101, in 1869 the increase was 87; in 1868, 79; and in 1867, 26.

Finances.—The balance-sheet for the financial year 1869 (Appendix A), as will be seen on examination, shows an income of 6859*l.* 16*s.*; which is an increase of 868*l.* 12*s.* over the previous year, when the amount was 5991*l.* 4*s.*; in 1867 the income was 5462*l.* 7*s.* 11*d.*; and in 1866, 5085*l.* 8*s.* 3*d.* The increase over 1868 is more than accounted for by the much larger amount of extraordinary receipts, the difference on this head being 1215*l.* 14*s.* 6*d.* The amount received by subscriptions was less by 420*l.* 18*s.* in 1869 than in 1868.

The ordinary expenditure (including Expeditions) was

4454*l.* 12*s.* 1*d.*, an increase of 297*l.* 14*s.* 3*d.* over that of the previous year. After adding the item comprising the legal expenses and legacy duty paid on the termination of the Oliveira law-suit, the total expenditure will be found to be 4606*l.* 8*s.* 8*d.* This leaves the excess of income over expenditure 2214*l.* 15*s.* 4*d.*, nearly the whole of which sum has been added to the invested property of the Society. The amount of funded capital at the end of December, 1869 (the termination of the financial year), was 18,250*l.* Since then 1000*l.* more stock has been purchased, making the present amount 19,250*l.*

The sum accruing to the Society, out of the legacy of 4000*l.* bequeathed by the late Mr. Benjamin Oliveira, on the termination of the law-suit mentioned in several of the previous Reports, has been received during the past year. The gross amount was 1506*l.* 17*s.* 1*d.*, from which must be deducted 191*l.* 8*s.* 7*d.* for legacy duty and law charges, leaving a net addition to the property of the Society, by this bequest, of 1315*l.* 8*s.* 6*d.*

The Finance Committee of Council have continued, as in previous years, to hold their monthly Meetings to supervise the accounts, and all bills due by the Society have been paid by them at the next Meeting after their presentation and examination.

The annual Audit was held on the 6th of April last, the gentlemen acting as Auditors being Mr. A. G. Findlay and Mr. Charles White on behalf of the Council, and General G. Balfour and Mr. H. Jones Williams on behalf of the Fellows. The Council have already returned their thanks, in their own name and that of the Society at large, to these gentlemen, for giving their valuable time and the benefit of their great experience to this arduous but necessary duty.

STATEMENT showing the RECEIPTS and EXPENDITURE of the Society from the Year 1848 to the 31st Dec. 1869.

Year.	Cash Receipts within the Year.			Cash Amounts invested in Funds.			Deducting Amounts invested in Funds; actual Expenditure.		
	£.	s.	d.	£.	s.	d.	£.	s.	d.
1848	606	10	5	..	..		755	6	1
1849	778	3	0	..	..		1098	7	6
1850	1096	10	5	..	..		877	2	10
1851	1056	11	8	..	..		906	14	7
1852	1220	3	4	..	..		995	13	1
1853	1917	2	6	..	..		1675	6	0
1854	2565	7	8	..	..		2197	19	3
1855	2594	7	0	..	..		2636	3	1
1856	3372	5	1	533	10	0	2814	8	1
1857	3142	13	4	378	0	0	3469	19	9
1858	3069	15	1	..	..		2944	13	6
1859	3471	11	8	950	0	0	3423	3	9
1860	6449	12	1	466	17	6	5406	8	7
1861	4782	12	9	1356	2	6	3074	7	4
1862	4659	7	9	1369	7	6	3095	19	4
1863	5856	9	3	1637	10	0	3655	4	0
1864	4977	8	6	1796	5	0	3647	7	10
1865	4665	8	8	1041	5	0	4807	4	5
1866	5095	8	8	1093	15	0	4052	15	0
1867	5462	7	11	1029	0	6	3943	17	4
1868	5991	4	0	1837	3	9	4156	17	10
1869	6650	16	0	2181	5	0	4646	0	8

In 1856 a Treasury Grant of 1000*l*. for the East African Expedition received.

In 1860 a Treasury Grant of 2500*l*. for the East African Expedition received.

STATEMENT showing the Progress of the INVESTMENTS of the Society from the Year 1832 to the 31st Dec. 1869.

End of the Year, Dec. 31.	Cash invested.			Amount of Stock purchased.		
	£.	s.	d.	£.	s.	d.
1832	3657	10	0	4000	0	0
1833	4130	0	0	4500	0	0
1834	4426	0	0	4800	0	0
1835	4426	0	0	4800	0	0
1836	4426	0	0	4800	0	0
1837	4426	0	0	4800	0	0
1838	4426	0	0	4800	0	0
1839	4129	15	0	4500	0	0
1840	3768	10	0	4150	0	0
1841	2801	0	0	3150	0	0
1842	2801	0	0	3150	0	0
1843	2219	18	6	2578	4	4
1844	2219	18	6	2578	4	4
1845	2219	18	6	2578	4	4
1846	1933	1	0	2278	4	4
1847	2133	1	0	2502	6	2
1848	1886	16	8	2224	1	10
1849	1886	16	8	2224	1	10
1850	1886	16	8	2224	1	10
1851	1886	16	8	2224	1	10
1852	1886	16	8	2224	1	10
1853	1662	14	10	2000	0	0
1854	1662	14	10	2000	0	0
1855	1662	14	10	2000	0	0
1856	2216	4	10	2600	0	0
1857	2594	4	10	3000	0	0
1858	2594	4	10	3000	0	0
1859	3544	4	10	4000	0	0
1860	4011	2	4	4500	0	0
1861	5369	4	10	6000	0	0
1862	6758	12	4	7500	0	0
1863	8596	2	4	9500	0	0
1864	10365	7	4	11500	0	0
1865	11406	12	4	12500	0	0
1866	12435	7	4	13500	0	0
1867	13464	7	10	14500	0	0
1868	15321	11	7	16250	0	0
1869	17452	16	7	18250	0	0*

Publications.—The 39th volume of the 'Journal' was published on the 14th of March last, for the first time as a bound volume, with cut leaves; the Council having decided on this improvement, originally suggested by Mr. Francis Galton, as

* Of which 4000*l*. is India 5 per Cents., and 1000*l*. India 5 per Cent. Debentures, and 1000*l*. India 4 per Cent. Certificates.

likely to be most acceptable to the Fellows. The increased cost
of the issue of the annual volume in this form is about 40*l.*
The 13th volume of the 'Proceedings' has also been completed since the last Report, and one part of Volume 14 published and sent out.

Library.—1120 volumes of books and scientific pamphlets
have been added to the library since the last anniversary,
163 of which were by purchase, and the remainder by donation
or exchange.

Among the additions may be mentioned, Siebold's 'Nippon,
in 6 quarto volumes, and Von der Decken's 'Reise in Ost-
Afrika,' 4 volumes, both by purchase; Solvyn's 'Hindus,' in
6 folio volumes, presented by C. H. Bracebridge, Esq.; 'The
Balearic Islands,' in one volume folio, presented by the
anonymous author; 'Mémoires sur la Chine,' by D'Escayrac
de Lauture, presented by Sir Roderick Murchison, and many
other important works.

As in previous years, the management of the library has
been superintended by a Committee of Council, who have held
numerous meetings, and selected books for purchase out of the
fund granted by the Council for this purpose, viz., 100*l.* They
have entrusted the compilation of the classified catalogue of
the library (which was mentioned in the Reports of the two
previous years as being in the hands of the Librarian) to Mr.
Evans of the British Museum, who has completed the manuscript, as far as concerns the alphabetical catalogue printed in
1865, and is now engaged on the works added since that time.
The Council hope before the next anniversary to be able to
announce the publication of this most useful work. The sum of
99*l.* 12*s.* 10*d.* has been expended during the year in the purchase
of books, and 46*l.* 10*s.* 8*d.* in binding.

Map-Collection.—The accessions to the Map Department
during the past year consist of 919 sheets of Maps and
Charts, 4 Diagrams, and 90 Views. All have been mounted,
catalogued, and incorporated into the classified collection.

The following are the most important:—

506 Sheets of the Ordnance Survey of Great Britain and Ire-
    land. Presented by the Topographical Office, through
    Sir Henry James, Director.

83 Admiralty Charts. Presented by the Admiralty, through Capt. G. H. Richards, R.N., Hydrographer.

4 Sheets of the Topographical Map of Sweden, and 34 Sheets of the Government Map of Norway. Presented by the Swedish Government.

A Chart of the Surface Temperatures in the Atlantic Ocean between Shetland and Greenland. By Admiral C. Irminger.

62 Sheets of India, comprising Governmental and District Maps of the Revenue Survey. By Purchase.

Chart of the North Atlantic Ocean, showing the track of all the Telegraph Cables between Europe and America. Presented by Captain S. Osborn, C.B., R.N.

Maps of the various Railways of the United States of America. Presented by the Foreign Secretary, C. C. Graham, Esq.

Plans of the Ordnance Survey of Jerusalem. By Captain C. W. Wilson, R.E. Presented by Sir H. James, Director.

Plans of the Suez Canal. By D. A. Lange, Esq., Director of the Company in London.

Russian Map of the Province of Khokan and Kirghiz Steppe. Presented by Lady Strangford.

78 Photographs, taken during the Abyssinian War, from Zoulla to Magdala. Presented by the Topographical Office.

7 Photographs of Relief-Models of Mountain Ranges in France and Italy. By M. Bardin.

A Bust of Richard Lander, gold medallist. By the sculptor, N. N. Burnard, Esq.

Grants to Travellers.—A donation of 100*l.* has been granted to Mr. St. Vincent Erskine, of Pietermaritzburg, Natal, as a contribution towards the expenses of his adventurous journey in search of the mouth of the Limpopo, the account of which appears in the Thirty-ninth volume of the 'Journal.' A second grant of 300*l.* has also been made to Mr. G. W. Hayward, in aid of the exploration in which he is now engaged in Central Asia, the Society having had, in return for the first grant of a similar amount, the Report and Map of this most able and courageous traveller, relating to his journey across the Kuen Lun to Yarkand and Kashgar. To Sir Samuel Baker have been granted instruments for his present White Nile expedition, to the amount of 114*l.* 6*s.*; and to Mr. Palmer instruments to the value of 4*l.* 1*s.*, for his exploration of the northern part of the Sinaitic Peninsula. The total amount thus expended during the year is 518*l.* 7*s.*

The sum expended in Medals and other awards has this year been 140*l.*; a larger amount than in other years, chiefly caused by the addition of the prizes of gold and bronze medals offered to the Public Schools.

APPENDIX A.

BALANCE-SHEET FOR THE YEAR 1869.

Receipts.

	£ s. d.	£ s. d.
1869.		
Balance in Banker's hands 31st Dec, 1868 ..	566 19 9	
Ditto Accountant's Ditto	2 15 10	569 15 7
Subscriptions of 1444 Fellows ..	2887 10 0	
Entrance Fees of 133 Ditto ..	399 0 0	
Life Compositions of 16 Ditto..	400 0 0	
Arrears of Subscriptions ..	390 0 0	4076 10 0
Parliamentary Grant..		500 0 0
Royal Premium..		52 10 0
Sale of Publications ..		122 9 1
Advertisements ..		18 18 3
Half Year's Dividend on 3000l. India 5 per Cents.	73 2 6	
Half Year's Dividend on 1000l. ditto Debentures	24 7 6	
Half Year's Dividend on 12,250l. New 3 per Cents.	179 3 2	
Half Year's Dividend on 4000l. India 5 per Cents.	97 18 4	
Half Year's Dividend on 1000l. India Debentures	24 9 7	
Half Year's Dividend on 12,250l. New 3 per Cents.	179 18 6	578 19 7
Amount received on a/c of the Bequest of 4000l. made by the late B. Oliveira, Esq. ..		1506 17 1
Miscellaneous ..		3 12 0
		£7429 11 7

Expenditure.

	£ s. d.	£ s. d.
1869.		
Rent, Taxes, and House Expenses ..		530 18 8
Salaries and Wages..		1338 9 8
Library and Map-Room ..		184 18 3
Subscriptions, &c., returned ..		21 0 0
Gold Medals, School Prizes, and other awards ..		140 0 0
Postages, &c. ..		131 9 6
Office Expenses ..		244 4 10
Publications, Illustrations, &c.		1155 9 1
Furniture and Repairs, ..		125 14 7
Expeditions ..		518 7 0
Miscellaneous ..		64 0 6
		4454 12 1
Legacy Duty and charges in connexion with the Bequest of the late Benjamin Oliveira, Esq.		191 8 7
Addition to Funded Property (1000l. India 5 per Cents.; 1000l. India 4 per Cent. Certificates) ..		2131 5 0
Balance in Banker's hands, 31st December, 1869 ..	637 1 10	
Ditto Accountant's ditto ..	15 4 1	652 5 11
		£7429 11 7

REGINALD T. COCKS, *Treasurer, R.G.S.*

Audited 6th April, 1870.

G. BALFOUR,
CHARLES WHITE,
H. JONES WILLIAMS,
A. G. FINDLAY,
} *Auditors.*

APPENDIX B.

ESTIMATE FOR THE YEAR 1870.

Receipts.

	£	s.	d.
Cash Balance, 1st Jan., 1870	652	5	11
Annual Subscriptions	3050	0	0
Life Compositions	500	0	0
Entrance Fees	450	0	0
Arrears of Subscriptions	200	0	0
Royal Premium for 1870	52	10	0
Parliamentary Grant	500	0	0
Sale of Publications, Advertisements, &c.	125	0	0
Dividends and Small Receipts	650	0	0
	£6179	15	11

Expenditure.

	£	s.	d.
Rent, Taxes, and House Expenses	550	0	0
Salaries and Wages	1400	0	0
Library and Map-Rooms	300	0	0
Gold Medals and other Awards	150	0	0
Postages, &c.	150	0	0
Office Expenses	300	0	0
Journal, with Maps, and Proceedings	1200	0	0
Library Catalogue (classified)	600	0	0
	4650	0	0
Amount available for investment or advancement of Geographical Science	1000	0	0
Cash Balance	529	15	11
	£6179	15	11

Library Regulations.

I. The Library will be open every day in the week (Sundays excepted) from 10·30 in the morning to 4·30 in the afternoon,* except on New-Year's Day, Good Friday to Easter Monday inclusive, and Christmas week; and it will be closed one month in the year, in order to be thoroughly cleaned, viz. from the first to the last day of September.

II. Every Fellow of the Society is entitled (*subject to the Rules*) to borrow as many as four volumes at one time.

Exceptions :—

1. Dictionaries, Encyclopædias, and other works of reference and cost, Minute Books, Manuscripts, Atlases, Books and Illustrations in loose sheets, Drawings, Prints, and unbound Numbers of Periodical Works, *unless with the special written order of the President.*
2. Maps or Charts, *unless by special sanction of the President and Council.*
3. New Works before the expiration of a month after reception.

III. The title of every Book, Pamphlet, Map, or Work of any kind lent, shall first be entered in the Library-register, with the borrower's signature, or accompanied by a separate note in his hand.

IV. No work of any kind can be retained longer than one month: but at the expiration of that period, or sooner, the same must be returned free of expense, and may then, upon *re-entry*, be again borrowed, provided that no application shall have been made in the mean time by any other Fellow.

V. In all cases a list of the Books, &c., or other property of the Society, in the possession of any Fellow, shall be sent in to the Secretary *on or before the 1st of July in each year.*

VI. In every case of loss or damage to any volume, or other property of the Society, the borrower shall make good the same.

VII. No stranger can be admitted to the Library except by the introduction of a Fellow, whose name, together with that of the Visitor, shall be inserted in a book kept for that purpose.

VIII. Fellows transgressing any of the above Regulations will be reported by the Secretary to the Council, who will take such steps as the case may require.

By Order of the Council.

---

* On Saturday the Library is closed at 2·30 P.M.

ROYAL GEOGRAPHICAL SOCIETY.

Patron.
HER MAJESTY THE QUEEN.

Vice-Patron.
H.R.H. THE PRINCE OF WALES.

COUNCIL.
(ELECTED 23RD MAY, 1870.)

President.

MURCHISON, Sir Roderick I., Bart., K.C.B., G.C.ST.A., M.A., D.C.L., V.P.R.S., G.S., and L.S., Grand Officer of the Order of the Crown of Italy, Director-General of the Geological Survey of Great Britain and Ireland, Trust. Brit. Mus., Hon. Mem. R.S. of Ed., R.I.A., Foreign Member of the Academy of Sciences, Paris. Mem. Acad. St. Petersburg, Berlin, Stockholm, Brussels, and Copenhagen, Corr. Ins. Fr., &c. &c.

Vice-Presidents.

FRERE, Sir H. Bartle, K.C.B., G.C.S.I. | RAWLINSON, Maj.-Gen. Sir H. C., K.C.B.
GALTON, Francis, Esq., M.A., F.R.S., &c. | WAUGH, Gen. Sir A. Scott, F.R.S.

Treasurer.
COCKS, Reginald T., Esq.

Trustees.
HOUGHTON, Lord.
TREVELYAN, Sir Walter C., Bart., &c.

Secretaries.
MARKHAM, Clements R., Esq., F.S.A. | MAJOR, Richard Henry, Esq., F.S.A.

Foreign Secretary.
GRAHAM, Cyril C., Esq.

Members of Council.

BACK, Admiral Sir G., D.C.L., F.R.S., &c. | INGLEFIELD, Admiral E. A., C.B., F.R.S.
BRODRICK, Hon. Geo. C. | McCLINTOCK, Capt. Sir F. L., R.N.
CAMPBELL, George, Esq., D.C.L. | NICHOLSON, Sir Charles, Bart., D.C.L.
COLLINSON, Admiral R., C.B. | RAE, John, Esq., M.D.
FERGUSSON, James, Esq., F.R.S., D.C.L. | RICHARDS, Admiral G. H., R.N., F.R.S.
FINDLAY, A. G., Esq. | RIGBY, Major-Gen. C. P.
GRANT, Lieut.-Col. J. A., C.B. | RUSSELL, A. J. E., Esq., M.P.
GRANT-DUFF, Rt. Hon. M. E., M.P. | SILVER, S. W., Esq.
HALL, Vice-Admiral Sir W. H., K.C.B., F.R.S. | STRATFORD DE REDCLIFFE, Visc., K.G.
HUXLEY, Prof. T. H., F.R.S., &c. | WELLINGTON, His Grace the Duke of.
 | WHITE, Charles, Esq., J.P.

Bankers.
Messrs. COCKS, BIDDULPH, and Co., 43, Charing-cross.

Assistant Secretary and Editor of Transactions,
H. W. BATES, Esq.

HONORARY AND HONORARY CORRESPONDING MEMBERS.

21st January, 1871.

HONORARY.

His Imp. Majesty Dom Pedro II., Emperor of Brazil.

H.M. Charles XV., Louis Eugene, King of Sweden and Norway.

H.M. Leopold II., King of the Belgians.

H.I.H. the Grand Duke Constantine, President of the Imperial Geographical Society of St. Petersburg.

H.R.H. the Duke of Edinburgh.

H.I.H. Ismail Pasha, Viceroy of Egypt.

HONORARY CORRESPONDING.

ABICH, Dr. William Hermann, St. Petersburg

ALMEIDA, Dr. Candido Mendes de

BAER, Chev. de K. E., Mem. Imp. Acad. of Science .. .. .. St. Petersburg

BALBI, M. Eugène de .. .. .. Milan

BASTIAN, Dr. Adolph .. .. .. Bremen

BERGHAUS, Prof. Heinrich .. .. Berlin

BURMEISTER, Dr. Hermann, Buenos Ayres

CHAIX, Prof. Paul .. .. .. Geneva

COELLO, Don Francisco .. .. Madrid

DANA, Professor James D., New Haven, Connecticut

D'AVEZAC, M... .. .. Paris

DUFOUR, Gen., Director of the Topo. Depart., Switzerland .. .. Geneva

DUVEYRIER, M. Henri .. .. Paris

EHRENBERG, C. P., For. M.R. and L.S., Berlin

ERMAN, Prof. Adolph .. .. .. Berlin

FAIDHERBE, Général L., Général Commandant à Bone .. .. .. Algérie

FIGANIÈRE, Command. Jorge César, Lisbon

FORCHHAMMER, Prof. P. W. .. .. Kiel

FREMONT, General .. .. .. New York

GRINNELL, Henry, Esq. V.P. Geogr. Soc. of .. .. New York

GUYOT, Prof., LL.D., Princeton, New Jersey

HANSTEIN, Prof., For. M.R S. Christiania

HAZELIUS, M.-Gen. J. A., Chief of the Topo. Corps of Sweden .. Stockholm

HELMERSEN, Col. P. .. St. Petersburg

HOCHSTETTER, Dr. Ferdinand von, Pres. Imp. Geograph. Soc. of .. .. Vienna

IRMINGER, Rear-Admiral C. L. C., R.D.N. Copenhagen

JANSEN, Captain M. H., D.R.N., Delft, Holland

JOCHMUS, Field Marshal Lieutenant Baron Vienna

KENNELLY, D. J. Esq., F.R.A.S.

KHANIKOF, M. .. .. Paris

KIEPERT, Dr. H. .. .. Berlin

LEAL, His Exc. Senhor Fernando da Costa, Governor of Mozambique

LEAL, José da Silva Mendes, Minister of the Colonies .. .. Lisbon

LINANT Pasha .. .. .. Alexandria

LIVINGSTONE, David, Esq., M.D., LL.D.

LÜTKE, Admiral F. B., Pres. of the Imp. Academy of Sciences .. St. Petersburg

MACEDO, J. J. da Costa de .. .. Lisbon

MADOZ, Don Pascual .. .. .. Madrid

MALTE-BRUN, M. V. A., Hon. Sec. Geogr. Soc. of .. .. Paris

MAURY, Commodore M. F.

MUNZINGER, Werner, Esq.

NARDI, Monsignor Francesco .. Rome

NEGRI, Chevalier Cristoforo .. .. Turin

OSTEN SACKEN, Baron, Sec. to the Imp. Geogr. Soc. of .. .. St. Petersburgh

PETERMANN, Dr. Augustus .. .. Gotha

PHILIPPI, Dr. Rodulfo Armando .. Chili

PLATEN, His Excellency Count.

RAIMONDY, Don Antonio.. .. .. Lima

RANUZZI, Count Annibale .. .. Bologna

RÜPPELL, Dr. E., For. M.L.S... Frankfort

SALAS, Don Saturnino, Pres. Topo. Depart., Argentine Repub. .. .. Buenos Ayres

SCHEDA, Herr von, Director of the Imp. Inst. of Military Geogr. .. .. Vienna

SCHERZER, Dr. Karl von .. .. Vienna

SOLDAN, Don Marino Felipe Paz Lima.

SONKLAR, Lieut.-Col. the Chev. de, Wiener Neustadt, Vienna

STRUVE, Prof. Otto, Imp. Observ. of Pulkowa .. .. .. St. Petersburg

SYDOW, Lt.-Col., Emil von (Chief of the Geog. Dep. of the Staff of the Prussian Army), Behrenstrasse, 66. Berlin

TCHIHATCHEF, M. Pierre de .. Paris

TSCHUDI, Herr T. T. von .. .. Vienna

VANDER MAELEN, Mr. Ph. .. Brussels

VASCONCELLOS É SILVA, Dr. Alfredo Casimiro de .. .. .. Rio de Janeiro

VERNEUIL, M. E. de .. .. .. Paris

VILLAVICENCIO, Don Manuel Guayaquil

WRANGELL, Adm. Baron.. St. Petersburg

ZIEGLER, M. J. M. .. .. Winterthur

FELLOWS.

(To 21st January, 1871.)

N.B.—*Those having * preceding their names have compounded for life.*

Year of Election.	
1869	Abbott, Keith E., H.M. Consul. *Odessa.*
1868	*Abbott, Wm., S. D., Esq. 28, *Pembridge-crescent, W.*
1863	Abdy, Rev. Albert, M.A. *Linslade-vicarage, Leighton Buzzard.*
1851	Abinger, W. F. Scarlett, Lord. *Guards' Club, S.W.*
1865	Acheson, Frederick, Esq., C.E. 7, *College-hill, Highbury-park, North, N.*
1861	Acland, J. Barton Arundel, Esq. *Mount Peel, Canterbury, New Zealand.* Care of *A. Mills, Esq., 34, Hyde-park-gardens, W.*
1853	Acland, Sir Peregrine Palmer F. P., Bart. *Fairfield, Somerset.*
1830	*Acland, Sir Thomas Dyke, Bart., F.R.S. 34, *Hyde-park-gardens, W.;* and *Killerton, Exeter, Devon.*
1867	Adair, Col. Alex. Shafto. 7, *Audley-square, W.*
1867	10 Adare, Viscount. *Clearwell-court, Coleford, Gloucestershire.*
1862	Addison, Col. Thomas, C.B.
1859	Ainslie, Col. H. Francis. *Burlington-chambers,* 180, *Piccadilly, W.;* and *United Service Club, S.W.*
1830	*Ainsworth, W. F., Esq., F.S.A. *Ravenscourt-villa, New-road, Hammersmith, W.*
1859	Airlie, David Graham, Earl of. *Holly-lodge, Kensington, W.*
1860	Aitchison, David, Esq. 180, *Piccadilly, W.*
1830	*Albemarle, George Thomas, Earl of. 11, *Grosvenor-square, W.; Quiddenham-hall, Larlingford, Norfolk;* and *Elvedon-hall, Suffolk.*
1862	Alcock, Sir Rutherford, K.C.B. *Athenæum Club, S.W.*
1838	*Aldam, William, Esq. *Frickley-hall, near Doncaster.*
1865	Aldom, Joseph R. Esq., M.A., PH. DR. *Salway-house, Leyton, Essex.*
1857	20 Aldrich, Captain Robert D., R.N. *Windmill-road, Croydon, Surrey, S.*
1830	Alexander, Colonel Sir Jas. Ed., K.C.L.S., F.R.A.S., F.R.S.E., &c., 14th Regt. *United Service Club, S.W.;* and *Westerton-house, Bridge of Allan, N.B.*
1870	Alford, Lewis, Esq. 2, *Little Love-lane, E.C.*
1864	Allan, C. H., Esq. *Lloyd's, E.C.;* and 31, *Park-street, Stoke Newington, N.*
1857	Allan, G. W., Esq. *Moss Park, Toronto, Canada.* Care of *Gen. J. H. Lefroy, R.A.,* 82, *Queen's-gate, W.*
1858	Allan, Jas., Esq. 122, *Leadenhall-street, E.C.*

VOL. XL. *b*

Year of Election.	
1871	*Allcroft, John D., Esq. 55, *Porchester-terrace, W.; Hartington, Middlesex; and Stokesay, Shropshire.*
1865	Allen, James Pearce, Esq. 13, *Waterloo-place, S.W.*
1854	Ancona, J. S., Esq. 8, *John-street, Adelphi, W.C.*
1860	Anderdon, John Edmund, Esq. *Henlade-house, Taunton, Somerset.*
1867	30 Anderson, Sir Henry L., K.C.B. *India-office, S.W.*
1862	Anderson, James, Esq. 1, *Billiter-court, City, E.C.*
1861	Anderson, John, Esq. Messrs. W. R. Adamson and Co. *Shanghai. Care of Messrs. Jno. Burd and Co., Hong Kong. Per Messrs. Adam, Thomson, and Co., 48, Lime-street, E.C.*
1868	Anderson, John, Esq. *Conservative Club, S.W.*
1868	Anderson, Joseph, Esq. 7, *Cleveland-square, Hyde-park, W.*
1870	Anderson, William Jas., Esq. *Sans Souci, Newlands, near Cape Town, Cape of Good Hope.*
1856	*Andrew, William P., Esq.
1867	Andrews, G. H., Esq. *The Cedars, New Brentford.*
1866	Andrews, John R., Esq. *East-hill-house, Wimbledon, S.W.*
1868	Angas, George F., Esq. 72, *Portland-road, Notting-hill, W.*
1861	40 Annesley, Col. the Hon. Hugh, M.P. 25, *Norfolk-street, Park-lane, W.*
1860	*Anson, Sir John William Hamilton, Bart. 55, *Portland-place, S.W.; and Sherley-house, Croydon.*
1853	Ansted, Prof. D. T., M.A., F.R.S., &c. 33, *Brunswick-square, W.C.; Athenæum Club, S.W.; and Chateau Vieux, St. Léonard, Boulogne-sur-Mer.*
1868	*Anstey, Chisholm, Esq. *Bombay. Care of Messrs. King and Co.*
1857	Anstruther, M.-Gen. Philip, C.B., Madras Artil. *Airth-castle, by Falkirk, N.B.*
1864	Anstruther, Lieut. R. L., Rifle Brigade. *Staff-college, Farnboro'-station, Hants.*
1863	Arber, Edward, Esq., A.K.C. 5, *Queen's-square, Bloomsbury, W.C.*
1858	Arbuthnot, George, Esq. 23, *Hyde-park-gardens, W.*
1862	Arbuthnot, Major George, R.H.A. *Cowarth, Sunningdale.*
1861	Archer, Graves Thos., Esq. 1, *Ennismore-place, Prince's-gate, S.W.*
1866	50 Arconati, The Marquis Giammartino. *Casa Prini, Pisa. Care of Messrs. Bocca, Bros., Turin. Per Messrs. Barthes and Lowell, Great Marlborough-street, W.*
1870	Ardagh, John C., Esq., Lieut. R.E. *Junior United Service Club, S.W.*
1855	*Arden, Richard Edward, Esq. *Sunbury-park, Middlesex, S.W.*
1858	*Armistead, Rev. Charles John, M.A., F.S.A. *United University Club, S.W.; and National Club, S.W.*
1863	Armitage, Edward, Esq. 3, *Hall-road, St. John's-wood, N.W.*
1867	Armitstead, Geo., Esq., M.P. *Errol-park, Errol, N.B.*
1857	Armstrong, Alexander, Esq., M.D., R.N., F.R.C.P.. Director-General of the Navy Medical Department. *Admiralty, Somerset-house, W.C.; and Junior United Service Club, S.W.*
1850	*Arrowsmith, John, Esq., F.R.A.S. 35, *Hereford-square, Old Brompton, S.W.*
1863	Arthur, Captain William, R.N. *The Priory, Leatherhead.*

Year of Election	
1868	Ashbee, Edmund Wm., Esq., F.G.S. 17, *Mornington-crescent, Regent's-park, N.W.*
1870	60°Ashton, Charles, Esq. *New University Club, S.W.*
1864	*Ashton, R. J., Esq. *Hatton-court, Threadneedle-street, E.C.*
1863	*Ashwell, James, Esq., M.A., F.G.S.
1836	*Atkins, John Pelly, Esq., F.S.A. *Halsted-place, near Sevenoaks.*
1870	Atkinson, Wm., Esq., F.L.S. &c. 47, *Gordon-square, W.C.*
1868	Atlee, Charles, Esq. *The Park, Ealing, W.*
1860	Attwell, Professor Henry. *Barnes, S.W.*
1869	Auld, Thomas Reid, Esq. 36, *Portland-place, W.*
1859	Austen, Major Henry H. Godwin, 24th Foot, Trig. Survey, Punjaub. *Junior United Service Club, S.W.; and Chilworth-manor, Guildford, Surrey.*
1868	Austin, John G., Esq. *Care of the Colonial Company,* 16, *Leadenhall-street, E.C*
1854	70 Ayrton, Acton S., Esq., M.P. 11, *Bolton-street, Piccadilly; and Office of Works, Whitehall.*
1846	*Ayrton, Frederick, Esq.
1866	*Babington, William, Esq., *St. Kilda, Buckhurst Hill, Essex; and Bonny River, West Coast of Africa.*
1826	*Back, Admiral Sir Geo., D.C.L., F.R.S. 109, *Gloucester-place, Portman-sq., W.*
1866	Bacon, Geo. Washington, Esq. 127, *Strand, W.C.*
1864	Badger, Rev. Geo. P. 21, *Leamington-road-villas, Westbourne-park, W.*
1863	Bagot, Christopher N., Esq. *Oriental Club, W.*
1862	Bagot, Capt. L. H. *Care of C. S. Bagot, Esq.,* 40, *Chancery-lane, W.C.*
1869	Bailey, L. C., Esq., Staff Commander, R.N. *Topographical Department, New-street, Spring-gardens, S.W.*
1857	Baillie, Lt.-Colonel John, Bengal Staff Corps. 22, *Palace-gardens-terrace, Kensington, W.*
1861	80 Baillie, William Henry, Esq. 61, *Jermyn-street, S.W.*
1857	Baines, Thomas, Esq. *Care of E. L. King, Esq.,* 35, *Austin-street, King's Lynn, Norfolk.*
1861	*Baker, John, Esq.
1863	Baker, Capt. Robert B. *Oriental Club, Hanover-square, W.*
1867	Baker, Sir Samuel White, Pasha. *Hedenham-hall, Bungay, Norfolk; and* 118, *Belgrave-road, S.W.*
1865	Baker, Major Wm. T., 85th Regt. *Junior United Service Club, S.W.; and* 31, *Grosvenor-place, Bath.*
1861	Baldwin, William Charles, Esq. *Leyland-vicarage, Preston.*
1861	Balfour, David, Esq. *Balfour-castle, Kirkwall, N.B.*
1847	Balfour, Maj.-General Sir George, K.C.B., R.A. 6, *Cleveland-gardens, Hyde-park, W.; and Oriental Club, Hanover-square, W.*
1870	Balfour, Capt. George M., R.N. 3, *Surrey-villas, Upper Norwood.*
1858	90 Balfour, John, Esq. *New South Wales; and Colinton, Queensland;* 39, *St. James's-street, S.W.*

Year of Election.	
1863	Balfour, William, Esq. 1, *Gladstone-villas, Walmer-road, Deal, Kent.*
1860	Ball, John, Esq. 24, *St. George's-road, Eccleston-square, S.W.*
1863	Bamforth, Rev. J.,
1852	Bancroft, Capt. W. C., 16th Regt. *Aide de Camp and Military Sec., King's House, Jamaica; M'Gregor and Co., Charles-street, S.W.*
1862	Banks, George F., Esq., Surgeon R.N.
1858	Bannerman, Sir Alexander, Bart. 46, *Grosvenor-place, S.W.*
1869	Barchard, Francis, Esq. *Horsted-place, Uckfield.*
1870	Barclay, Wm. L., Esq., B.A. *Leyton, Essex.*
1863	Barford, A. H., Esq., M.A. 1, *Cornwall-terrace, Regent's-park, N.W.*
1870	100 Baring, Evelyn, Esq. Lieut. R.A. 11, *Berkeley-square, W.*
1835	*Baring, John, Esq. *Oakwood, Chichester.*
1844	*Baring, Thomas, Esq., M.P. 41, *Upper Grosvenor-street, W.*
1862	Barlee, Frederick Palgrave, Esq. *Perth, Western Australia. Care of G. Lawrence, Esq., 12, Marlboro'-road, Lee, S.E.*
1868	Barlow, Frederick Thomas Pratt, Esq. 26, *Rutland-gate, S.W.*
1864	Barnett, H. C., Esq., J.P. *York, West Australia.*
1867	*Barns, John W., Esq. *Bhawulpore, Punjaub, India; care of Messrs. Grindlay.*
1870	Barr, Edward G., Esq. 76, *Holland-park, W.; and 36, Mark-lane, E.C.*
1859	Barrington, Lord, M.P. 19, *Hertford-street, Mayfair, W.*
1867	Barrington Ward, Marcus J., Esq., B.A., B.C.L. OXON., F.L.S., &c. *Clifton College, Bristol; and 14, Alfred-street, Belfast.*
1833	110 Barrow, John, Esq., F.R.S., F.S.A. 17, *Hanover-terrace, Regent's-park, N.W.*
1863	Barry, Alfred, Esq.
1869	Barry, Dr. Alfred, M.A., &c. *King's College, Strand, W.C.*
1857	Bartholomew, John, Esq. 17, *Brown-square, Edinburgh.*
1861	Bartlett, Herbert Lewis, Esq. *Union Club, S.W.*
1862	Barton, Alfred, Esq., M.D. *Hampton-court.*
1864	Basevi, Capt. J. P., R.E. *Messrs. Grindlay & Co., Parliament-street, S.W.*
1837	*Bateman, James, Esq., F.R.S., L.S. 9, *Hyde-park-gate South, W.*
1859	Bateman, John F., Esq., C.E. 16, *Great George-street, Westminster, S.W.*
1866	Bates, Henry Walter, Esq., F.Z.S. 15, *Whitehall-place, S.W.*
1866	120 Bateson, George, Esq. *Heslington-hall, York.*
1866	Batten, John H., Esq. *Alverne-hill, Penzance; and Oriental Club, Hanover-square, W.*
1864	Bax, Capt. Henry G. 2, *Sussex-place, Hyde-park-gardens, W.*
1867	*Baxter, Sir David, Bart. *Dundee; 5, Moray-place, Edinburgh; and Kilmaron-castle, Cupar, Fife.*
1867	Baxter, Richard, Esq., Barrister-at-Law. 19, *Leinster-gardens, Bayswater, W.*
1858	Baxendale, Joseph H., Esq. *Worplesdon, Guildford.*
1867	Bayley, Chas. Jno., Esq., C.B., M.A. 51, *Victoria-road, Kensington, W.*
1863	Bayley, H. Esq. *Blackheath-park, Kent.*
1862	Bayly, Lieut.-Col. John, R.E. *Ordnance Survey Office; 131, St. George's-road, Pimlico, S.W.*

Year of Election	
1862	Baynes, Lieut.-Col. R. Stuart. *Army and Navy Club, S.W.; and 38, Jermyn-street, S.W.*
1868	130 Bayston, Capt. Edward. *Trafalgar-lodge, Shirley, Southampton.*
1866	Beamish, Captain H. H., R.N. *Care of Messrs. Hallett and Co., 7, St. Martin's-place, W.C.*
1852	Beardmore, Nathaniel, Esq., C.E. *30, Great George-street, Westminster, S.W.*
1854	Beaufort, William Morris, Esq., Bengal Civil Service. *Oriental Club, Hanover-square, W.*
1856	Beaumont, John Aug., Esq. *Wimbledon-park-house, Wimbledon, S.W.; and 50, Regent-street, W.*
1870	Beaumont, Somerset, Esq., M.P. *23, Park-street, Park-lane, W.*
1851	*Beaumont, Wentworth B., Esq., M.P. *144, Piccadilly, W.; Bywell-hall, Newcastle-upon-Tyne; and Bretton-park, Wakefield.*
1867	*Beazeley, Michael, Esq., M.I.C.E. *Trinity Works, Penzance, Cornwall.*
1865	Bebb, Horatio, Esq. *13, Gloucester-place, W.; and Leamington.*
1861	Beckett, James F., Esq., Staff Commander R.N., F.R.A.A. *6, Boyne-terrace, Notting-hill, W. Care of Captain George.*
1838	140 Beckford, Francis Love, Esq. *Sunningdale Vicarage, Surrey.*
1859	Bedford, Rear-Admiral G. Augustus, R.N. *South-view, Widmore-road, Bromley, Kent.*
1870	*Beer, Julius, Esq. *23, Park-crescent, Portland-place, W.*
1868	Bedingfeld, Felix, Esq., C.M.G. *Reform Club, S.W.; and 41, Clarges-street, W.*
1861	*Begbie, James, Esq. *17, Trinity-square, Tower-hill, E.C.*
1860	*Begbie, Thomas Stirling, Esq. *4, Mansion-house-place, E.C.*
1846	*Beke, Charles Tilstone, Esq., PH. DR., F.S.A., &c. *Bekesbourne-house, near Canterbury.*
1853	*Belcher, Rev. Brymer. *St. Gabriel's, Pimlico, S.W.*
1830	*Belcher, Vice-Adm. Sir Edward, K.C.B., F.R.A.S. *22A, Connaught-square, W.*
1858	Beldam, Edw., Esq. *Royston, Herts.*
1863	150 Belmore, The Earl of. *Governor of New South Wales.*
1858	*Bell, C. Davidson, Esq., Surveyor-General, Cape of Good Hope. *Cape Town. Care of the S. A. Pub. Library, Cape Town. Per Messrs. H. S. King and Co.*
1830	Bell, James Christian C., Esq. *42, Westbourne-terrace, W.; and 15, Angel-court, Throgmorton-street, E.C.*
1868	Bell, Wm. A., Esq. B.A., M.D., *18, Hertford-street, Mayfair, W.*
1864	Bellamy, Edward, Esq. *10, Duke-street, St. James's, S.W.*
1870	Benjamin, Joseph, Esq. *22, Glasshouse-street, Regent-street, W.*
1830	*Bennett, John Joseph, Esq., F.R.S. *British Museum, W.C.*
1857	Bennett, J. Risdon, Esq., M.D. *15, Finsbury-square, E.C.*
1858	*Benson, Robert, Esq. *16, Craven-hill-gardens, Bayswater, W.*
1856	*Benson, William, Esq., Barrister-at-Law. *16, Craven-hill-gardens, Bayswater, W.*
1830	160 Bentham, George, Esq., Pres. L.S. *25, Wilton-place, S.W.*
1808	Bentley, George, Esq. *Upton-park, Slough.*
1863	Bentley, Richard, Esq. *New Burlington-street, W.*

Year of Election	
1870	Benyon, Wm. H., Esq. *Stainley-hall, Ripon, Yorkshire.*
1859	Berens, H. Hulse, Esq. *Sidcross, Foot's Cray, Kent.*
1865	Bernard, P. N., Esq. 16, *Leadenhall-street, E.C.*
1866	Berridge, F., Esq. *Winchester-house, Winchester-road, Adelaide-road, N.W.*
1856	Berry, Josiah, Esq. 16, *Regent-square, W.C.*
1863	Best, William, Esq. 3, *Belle-vue-place, Southampton.*
1867	Best, William John, Esq. 104, *Franklin-street, New York.*
1867	170 Bethune, Alexander M., Esq. *Otterburn, Hamlet-road, Upper Norwood; and* 122, *Leadenhall-street, E.C.*
1842	*Bethune, R.-Adm. C. R. Drinkwater, C.B. 4, *Cromwell-rd., South Kensington, W.*
1864	*Betts, E. L., Esq. *Preston-hall, Maidstone, Kent.*
1836	Betts, John, Esq. 115, *Strand, W.C.*
1868	Bevan, George Phillips, Esq. *Junior Athenæum Club; and 4, Suffolk-square, Cheltenham.*
1866	Bevan, William, Esq. 8, *Cedars-road, Clapham-common, S.*
1862	Bicker-Caarten, Peter, Esq. 30, *Northumberland-place, Bayswater, W.*
1868	*Bickmore, A. S., Esq., M.A. 252, *Pearl-street, New York.*
1866	Bicknell, Algernon S., Esq. 37, *Onslow-square, S.W.*
1869	Bidie, Geo., Esq., M.D., &c. Madras Establishment, Madras. *Care of Messrs. H. S. King and Co.*
1865	180 Bidwell, Charles Toll, Esq. *Garrick Club, 35, King-st., Covent-garden, W.C.; and* 28, *Grosvenor-st., Eaton-sq., S. W. Care of John Bidwell, Esq., Foreign-office.*
1860	Bidder, G. Parker, Esq., C.E. 24, *Gt. George-st., S. W.; and Mitcham, Surrey, S.*
1859	Biggs, Frederick W., Esq. *Debden-hall, Saffron Walden.*
1868	Biggs, C. H. Walker, Esq. 2, *Alexandra-terrace, Reading.*
1850	Bigsby, John J., Esq., M.D. 89, *Gloucester-place, Portman-square, W.*
1860	Birch, H. W., Esq. 46, *Welbeck-street, Cavendish-square, W.*
1858	Birch, John William, Esq. 9*, *New Broad-st., E.C.; and 27, Cavendish-sq., W.*
1862	*Birchill, Capt. B. H. H. *Old-lodge, Hartfield, Tunbridge-wells.*
1867	*Bischoffsheim, Henri Louis, Esq. 7, *Grafton-street, New Bond-street, W.*
1858	Bishop, George, Esq., F.R.A.S. *Union Club, S.W.; and The Meadows, Twickenham, S.W.*
1861	190 Bishop, James, Esq. 11, *Portland-place, W.*
1870	Bishop, Wm. Henry, Esq. 7, *Sumner-terrace, Onslow-square, S.W.*
1867	Bisson, Fredk. S. de Carteret, Esq., Lieut. R.I.M. 70, *Berners-street, W.*
1836	*Blaauw, William H., Esq., M.A., F.S.A., F.Z.S. *Beechlands, near Uckfield, Sussex.*
1870	Black, Andrew H., Esq. 23, *Royal-crescent, Glasgow.*
1860	*Black, Francis, Esq. 6, *North-bridge, Edinburgh.*
1867	Black, Thomas, Esq., Superintendent P. and O. Steam Navigation Company's Dockyard. *Oriental-place, Southampton.*
1869	Blacker, Louis, Esq. *Flowermead, Wimbledon-park, S.W.*
1858	Blackett, Henry, Esq. 13, *Great Marlborough-street, W.*
1870	Blackie, Thos. M., Esq. *Chipping-hill School, Witham, Essex.*
1849	200 Blackie, W. Graham, Esq., PH. DR. 36, *Frederick-street, Glasgow.*

Year of Election	
1862	*Blackstone, Frederick Elliot, Esq., B.C.L. *British Museum, W.C.*
1854	Blaine, D. Roberton, Esq., Barrister-at-Law. 3, *Paper-buildings, Temple, E.C.*; and 8, *Southwick-place, Hyde-park-square, W.*
1869	Blaine, Henry, Esq. 2, *Cleveland-road, Castle-hill, Ealing, W.*
1868	Blair, William Edward, Esq. *Windham Club, S.W.*
1865	Blake, Brig.-Gen. H. W.
1857	*Blake, Wollaston, Esq. 8, *Devonshire-place, W.*
1868	Blakiston, Matthew, Esq. *Mobberley, Knutsford, Cheshire.*
1857	Blakiston, Captain Thomas, R.A. 28, *Wellington-street, Woolwich, S.E.*
1861	*Blakeney, William, Esq., R.N. *Hydrographic-office, S.W.*
1868	210 Blanc, Henry, Esq., M.D., &c. *Care of Messrs. H. S. King and Co., 45, Pall-mall, S.W.*
1830	*Blanshard, Henry, Esq., F.R.A.S.
1857	Blanshard, Richard, Esq. *Fairfield, Lymington, Hants.*
1865	Blaxall, Fras. H., Esq., M.D. *Tendring, near Colchester.*
1854	Blencowe, W. Robert, Esq. *The Hook, Lewes.*
1861	Blenkin, William, Esq. *Addlestone, Surrey.*
1839	*Blewitt, Octavian, Esq. 4, *Adelphi-terrace, Strand, W.C.*
1864	Blore, Edward, Esq., D.C.L., F.R.S., F.S.A., &c. 4, *Manchester-square, W.*
1866	Blow, William Wootton, Esq. 22, *Wigmore-street, W.*
1861	Bloxsome, Oswald, jun., Esq. *Berrington-hall, Leominster.*
1868	220 Blumberg, George F., Esq. 4, *Ladbroke-square, Kensington-park, W.*
1837	*Blunt, Jos., Esq.
1863	*Blunt, Wilfred S., Esq. *Worth, Crawley, Sussex.*
1868	Blyth, Philip P., Esq. (J.P. for Middlesex). 53, *Wimpole-street, W.*
1858	Bohn, Henry G., Esq. *York-street, Covent-garden, W.C.*; and *North-end-house, Twickenham, S.W.*
1850	Bollaert, William, Esq. 21A, *Hanover-square, W.*
1862	Bolton, Major Francis John, 12th Regt. 2, *Westminster-chambers, S.W.*
1861	Bompas, George Cox, Esq. 15, *Stanley-gardens, Kensington-park, W.*
1864	Bone, John William, Esq., B.A., F.R.S.L., F.S.S. 26, *Bedford-place, Russell-square, W.C.*
1861	Bonney, Charles, Esq. *Adelaide, Australia.*
1858	230 Bonnor, George, Esq. 49, *Pall-mall, S.W.*; and 2, *Bayswater-terr., Kensington-square, W.*
1865	Bonwick, James, Esq. *St. Kilda, Melbourne. Care of W. Beddow, Esq., 22, South Audley-street, W.*
1866	Booker, Wm. Lane, Esq. *Care of F. B. Alston, Esq., Foreign-office.*
1859	Borough, Sir Edward, Bart. 4, *Nassau-street, Dublin.*
1845	*Borrer, Dawson, Esq. *Altmont Ballon, Co. Carlow, Ireland.*
1856	*Botcherby, Blackett, Esq., M.A. 174, *Brompton-road, S.W.*
1858	*Botterill, John, Esq. *Flower-bank, Burley-road, Leeds.*
1860	Boustead, John, Esq. 34, *Craven-street, Strand, W.C.*
1866	*Boutcher, Emanuel, Esq. 12, *Oxford-square, Hyde-park, W.*

Year of Election.	
1865	Bouverie, P. P., Esq. 32, *Hill-street, Berkeley-square,* W.
1867	240 Bowell, Wm., Esq. *Chandos-house, Hereford; and Gate-house Grammar-school, Hereford.*
1861	*Bowen, Charles Christopher, Esq. *Christchurch, Canterbury, New Zealand. Care of A. O. Ottywell, Esq.,* 16, *Charing-cross,* S.W.
1854	*Bowen, Sir George Ferguson, K.C.M.G., M.A., *Governor of New Zealand.*
1866	Bower, Anthony Maw, Esq.
1862	Bowie, John, Esq. *Conservative Club,* S.W.
1869	Bowker, James Henry, Esq. *Basutuland, South Africa. Care of Messrs. King and Co., Cornhill,* E.C.
1868	Bowly, William, Esq. *Cirencester.*
1856	Bowman, John, Esq. 9, *King William-street,* E.C.
1869	Bowra, E. C., Esq., Commissioner of Maritime Customs. *Ningpo, China.*
1865	Bowring, John Charles, Esq. *Larkbeare, Exeter.*
1866	250 Bowring, Samuel, Esq. 1, *Westbourne-park,* W.
1868	Bowser, Alfred T., Esq. *Cromwell-house, Hackney, N.E.*
1862	Boyce, Rev. W. B., Secretary to Wesleyan Missionary Society. *Wesleyan Mission House, Bishopsgate-street,* E.C.
1845	*Boyd, Edward Lennox, Esq., F.S.A. 35, *Cleveland-square, Hyde-park,* W.
1869	Boyle, Richard Vicars, Esq., C.S.I., &c. 9, *Stanhope-place, Hyde-park,* W.
1856	Boyne, G. Hamilton-Russell, Viscount. 22, *Belgrave-square,* S.W.; *Brancepeth-castle, Durham; and Burwarton-hall, Ludlow, Salop.*
1851	Bracebridge, Charles Holte, Esq. *Atherstone, Warwick.*
1870	*Bragge, Wm., Esq., C.E. *Shirle's-hill, Sheffield.*
1870	Braim, the Ven. the Archdeacon. *The Rectory, Bishop's-caundle, Sherborne, Dorset.*
1862	Braithwaite, Isaac, Esq. 27, *Austin-friars,* E.C.
1863	260*Bramley-Moore, John, Esq. *Langley-lodge, Gerrard's-cross, Bucks.*
1859	*Brand, James, Esq. 109, *Fenchurch-street,* E.C.
1868	Brand, Jas. Ainsworth, Esq. 100, *Cannon-street,* E.C.
1867	Brandis, Dr. D., F.L.S. *Director of Forests, Calcutta. Care of W. H. Allen, Esq.,* 13, *Waterloo-place,* S.W.
1859	Braybrooke, Philip Watson. *Assistant Colonial Secretary, Ceylon. Messrs. Price and Co., Craven-street.*
1861	*Brenchley, Julius, Esq. *Oxford and Cambridge Club,* S.W.; *and Milgate, near Maidstone, Kent.*
1833	*Brereton, Rev. John, LL.D., F.S.A. *Bedford.*
1834	*Breton, Wm. Henry, Esq., Commd. R.N., M.R.I. 26, *Queen-square, Bath.*
1862	Brett, Charles, Esq.
1867	Bridge, John, Esq. *Altrincham, Cheshire.*
1858	270 Bridges, Nathaniel, Esq.
1852	*Brierly, Oswald W., Esq. 8, *Lidlington-place, Harrington-square, Hampstead-road,* N.W.
1865	Briggs, Colonel, J. P. *Lantern Tower, Jedburgh.*

Year of Election	
1861	*Bright, Sir Charles T., F.R.A.S. 6, *Westminster-chambers, Victoria-street, S.W.;* and 69, *Lancaster-gate,* W.
1868	Bright, Henry Arthur, Esq. *Ashfield, Knotty Ash, Liverpool.*
1860	Bright, James, Esq., M.D. 12, *Wellington-square, Cheltenham.*
1854	Brine, Major Frederic, R.E. K.T.S. A.I.C.E., Executive Engineer, Punjaub. *Athenæum Club, S.W.;* Army and Navy Club, S.W.; *Garrick Club, W.C.*
1856	Brine, Captain Lindesay, R.N. *Army and Navy Club, S.W.;* All Saints' *Rectory, Axminster. Care of Messrs. Woodhead.*
1861	Bristowe, Henry Fox, Esq.
1861	Broadwater, Robert, Esq. 3, *Billiter-square, Fenchurch-street, E.C.*
1861	280 Brodie, Walter, Esq. 13, *Delamere-terrace, Hyde-park,* W.
1861	Brodie, William, Esq. *Eastbourne, Sussex.*
1863	*Brodrick, Hon. George C. 32A, *Mount-street,* W.
1864	*Brooke, Sir Victor A., Bart. *Colebrooke-park, Co. Fermanagh, Ireland.*
1862	Brooke, Thomas, Esq. *Mattock-lane, Ealing,* W.
1856	*Brooking, George Thomas, Esq. 25, *Sussex-gardens, Hyde-park,* W.
1856	*Brooking, Marmaduke Hart, Esq. 11, *Montagu-place, Bryanston-square,* W.
1870	*Brooks, Wm. Cunliffe, Esq., M.P., M.A., F.S.A., &c. 5, *Grosvenor-square, S.W.;* Barlow-hall, near Manchester; and Forest of Glen-Tanar, Aboyne, Aberdeenshire.*
1863	*Broughall, William, Esq. *Broadwater, Down, Tunbridge-wells.*
1866	*Brown, Col. David (Madras Staff Corps). *India.*
1856	290*Brown, Daniel, Esq. *The Elms, Larkhall-rise, Clapham, S.*
1864	Brown, Edwin, Esq., F.G.S. *Burton-on-Trent.*
1867	Brown, Geo. H. Wilson, Esq. *Victoria, Vancouver Island, British Columbia. Care of H. C. Beeton, Esq., 5, Bow-churchyard, E.C.*
1860	Brown, James, Esq. *Rossington, Yorkshire.*
1868	Brown, James P., Esq. *Cocaes, Brazil. Care of Mr. C. Williams, 25, Poultry, E.C.*
1865	*Brown, James R., Esq., F.R.S.N.A. Copenhagen. 5, *Langham-chambers, Langham-place,* W.
1861	*Brown, John Allen, Esq. *The Laurels, The Haven, Ealing,* W.
1867	Brown, Richard, Esq., C.E. 115, *Lansdowne-road, Notting-hill,* W.
1866	Brown, Robert, Esq. 4, *Gladstone-terrace, Hope-park, Edinburgh.*
1868	*Brown, Samuel, Esq. 11, *Lombard-st., E.C.;* and The Elms, Larkhall-rise, Clapham, S.*
1862	300*Brown, Thomas, Esq. 8, *Hyde-park-terrace, Hyde-park,* W.
1868	Brown, William, Esq. *Loat's-road, Clapham-park, S.*
1872	Browne, H. H., Esq. *Moor-close, Binfield, Bracknell.*
1869	Browne, John Comber, Esq., Superintendent and Inspector of Government Schools. *Port Louis, Mauritius.*
1866	*Browne, John H., Esq. *Montpellier-lawn, Cheltenham.*
1862	Browne, Samuel Woolcott, Esq. 58, *Porchester-terrace, Hyde-park,* W.
1864	*Browning, Capt. Wade. 85, *Charles-street, Berkeley-square,* W.

Year of Election	
1858	Browne, William J., Esq. *Marston-lodge, Pitville, Cheltenham.*
1870	Browne, Wm. A. Morgan, Esq. *Junior Athenæum Club, S.W.*
1869	Browning, George Frederick, Esq. *Weston School, Bath.*
1852	310 Browning, H., Esq. *73, Grosvenor-street, Grosvenor-square, W.; and Old Warden-park, Biggleswade.*
1856	*Browning, Thomas, Esq. *6, Whitehall, S.W.*
1859	Bruce, Henry Austin, Esq. *Dufryn, Aberdare, Glamorganshire.*
1863	Brunton, John, Esq., M.I.C.E., F.G.S. *Care of Messrs. Willis and Sotheran, Charing-cross, S.W.*
1856	Bryant, Walter, Esq., M.D., F.R.C.S. *7, Bathurst-street, Hyde-park-gardens, W.*
1843	*Buchan, John Hitchcock, Esq. *The Grove, Hanwell, W.*
1867	*Buccleuch, his Grace the Duke of, K.G., F.R.S. *Dalkeith Palace, near Edinburgh; and Montagu-house, Whitehall, S.W.*
1869	Buckley, John, Esq. *Care of Messrs. Dalgety, Du Cros, and Co., 52, Lombard-street, E.C.*
1863	Budd, J. Palmer, Esq. *Care of J. J. Milford, Esq., 18, Austin-friars, E.C.*
1867	*Bulger, Maj. George Ernest, F.L.S., &c. *Care of Mr. Booth, 307, Regent-st., W.*
1868	320*Bull, William, Esq., F.L.S. *King's-road, Chelsea, S.W.*
1865	Buller, Sir Edward M., Bart., M.P. *Old Palace-yard, S.W.; and Dilhorn-hall, Cheadle, Staffordshire.*
1869	Buller, Walter L., Esq., F.L.S. *Wanganui, New Zealand. Care of Mr. J. Van Voorst, 1, Paternoster-row, E.C.*
1863	Bullock, Commander Charles J., R.N. *Hydrographic-office, S.W.*
1830	*Bullock, Rear-Admiral Frederick. *Woolwich, S.E.*
1864	Bullock, W. H., Esq. *Grosvenor-hill, Wimbledon, S.W.*
1860	*Bunbury, Sir Charles James Fox, Bart.; F.R.S. *Barton-hall, Bury St. Edmund's.*
1839	Bunbury, E. H., Esq., M.A. *35, St. James's-street, S.W.*
1863	Bundock, F., Esq. *Windham Club, S.W.*
1861	Burges, William, Esq. *Fethard, Co. Tipperary.*
1866	330*Burgess, James, Esq., M.R.A.S., Hon. Sec. Bombay Geogr. Soc., Principal of Sir J. Jejeebhoy's Parsee B. Institution. *Hornby-row, Bombay. Care of J. McGlashan, Esq., 31, Essex-st., Strand, W.C.*
1864	Burn, Robert, Esq. *5, Clifton-place, Sussex-square, W.*
1863	*Burns, John, Esq. *1, Park-gardens, Glasgow; and Castle Wemyss, by Greenock, N.B.*
1861	*Burr, Higford, Esq. *23, Eaton-place, S.W.; and Aldermaston-court, Berkshire.*
1857	Burstal, Capt. E., R.N. *6, Park-villas, Lower Norwood, S.*
1830	*Burton, Alfred, Esq. *64, Marina, St. Leonard's.*
1833	*Burton, Decimus, Esq., F.R.S. *37, Gloucester-gardens, Hyde-park, W.*
1850	*Burton, Capt. Richd. Fras., 18th Regt. Bombay N.I., H.B.M. Consul at Damascus. *14, St. James's-square, S.W. Care of R. Arundell, Esq., Admiralty, Somerset-house, W.C.*
1869	Burton, William Samuel, Esq. *South-villa, Regent's-park, N.W.*
1858	Bury, William Coutts, Viscount. *48, Rutland-gate, S.W.*
1861	340 Bush, Rev. Robert Wheler, M.A. *1 Milner-square, Islington, N.*

Year of Election.	
1868	Busk, William, Esq., M.C.P., &c. 28, *Bessborough-gardens, S.W.*
1861	Butler, Charles, Esq. 3, *Connaught-place, Hyde-park, W.*
1867	Butler, E. Dundas, Esq. *Geographical Department, British Museum, W.C.*
1860	*Butler, Rev. Thomas. *Rector of Langar, Nottinghamshire.*
1870	Butter, Donald, Esq., M.D., &c. *Hazelwood-church-road, Upper Norwood, S.*
1862	*Buxton, Chas., Esq., M.P. 7, *Grosvenor-crescent, S.W.; and Fox-warren, Surrey.*
1870	Buxton, Fras. W., Esq., B.A. 23, *Upper Brook-street, W.*
1869	Buxton, Henry Edmund, Esq., B.A. *Bank-house, Great Yarmouth, Norfolk.*
1858	*Buxton, Sir Thomas Fowell, Bart. *Brick-lane, N.E.*
1866	350 Byass, Robert B., Esq. *Melville-park, Tunbridge-wells.*
1864	Bythesea, Capt. J., R.N., V.C. 20, *Grosvenor-place, Bath.*
1830	*Cabbell, B. B., Esq., M.A., F.R.S., F.S.A. 1, *Brick-court, Temple, E.C.; 52, Portland-place, W.; and Aldwick, Sussex.*
1866	Caldbeck, Capt. J. B. (P. and O. Sup. at Aden). 17, *West Mall, Clifton; and* 122, *Leadenhall-street, E.C. Care of Mrs. Caldbeck, Brownswood-house, Ennis-road, Hornsey, N.*
1863	Callaghan, Thos. F., Esq. *Garrick Club, W.C.*
1861	Calthorpe, the Hon. Augustus Gough. 33, *Grosvenor-square, W.*
1855	*Calthorpe, F. H. Gough, Lord. 33, *Grosvenor-square, W.*
1854	Calvert, Frederic, Esq., Q.C. 38, *Upper Grosvenor-street, W.*
1861	Cameron, Donald, Esq., M.P. *Auchnacarry, Invernesshire.*
1858	Cameron, Major-General Sir Duncan Alexander, R.E., C.B. *New Zealand.*
1864	360 Cameron, J., Esq. *Singapore.*
1866	Cameron, R. W., Esq. *Staten Island, New York. Care of Messrs. Brooks and Co., St. Peter's-chambers, Cornhill, E.C.*
1871	Campbell, Allan, Esq. 35, *St. James's-place, S.W.*
1861	Campbell, Capt. Frederick, R.N. 12, *Connaught-place, Hyde-park, W.*
1866	Campbell, George, Esq., D.C.L. Lieut.-Governor of Bengal; *and Athenæum Club, S.W.*
1844	*Campbell, James, Esq. *Grove-house, Hendon, Middlesex; and* 37, *Seymour-street, W.*
1857	Campbell, James, Esq., Surgeon R.N. *Bangkok, Siam. Care of Messrs. H. S King and Co.*
1834	*Campbell, James, Esq., jun. *Hampton-court-green, S.W.*
1863	*Campbell, James Duncan, Esq. *Peking. Care of H. C. Batchelor, Esq.,* 155, *Cannon-street, E.C.*
1869	Campbell, Robert, Esq., J.P. 31, *Lowndes-square, S.W.; and Buscot-park, Lechlade, Gloucestershire.*
1857	370 Camps, William, Esq., M.D.
1866	Canning, Sir Samuel, C.E. *The Manor-house, Abbots Langley, near Watford, Herts.*

Year of Election.	
1864	Cannon, John Wm., Esq. *Castle-grove, Tuam.*
1857	Cannon, Lieut.-General R. 10, *Kensington-gardens-terrace, W.*
1853	*Cardwell, Right Hon. Edward, M.P. 74, *Eaton-square, S.W.*
1863	*Carew, R. Russell, Esq., J.P. *Carpenders-park, Watford, Herts; and Oriental Club, W.*
1869	Carey, Rev. Tupper. *Fifield, Bavant, Salisbury; and* 15, *Hyde-park-gardens, W.*
1862	Cargill, John, Esq., Member of the Legislative Assembly of New Zealand and Legislative Council of Otago. *Dunedin, Otago, New Zealand.*
1863	*Cargill, Wm. W., Esq. 4, *Connaught-place, Hyde-park, W.*
1870	Carleton, Col. Dudley. 42, *Berkeley-square, W.*
1864	380 *Carmichael, L. M., Esq., M.A., 5th Lancers. *Oxford and Cambridge Club, S.W.*
1865	*Carnegie, David, Esq. *Eastbury, by Watford, Herts.*
1863	Carnegie, Commander the Hon. J., R.N. 26, *Pall-mall, S.W.*
1869	Carr, William, Esq. *Dene-park, near Tunbridge.*
1864	Carrington, R. C., Esq. *Admiralty, S.W.*
1861	Carter, Captain Hugh Bonham, Coldstream Guards. *Guards' Club, S.W.; and* 1, *Carlisle-place, Victoria-street, S.W.*
1868	Carter, Thomas Tupper, Esq., Lt. R.E. *Care of Messrs. H. S. King and Co.,* 45, *Pall-mall.*
1860	Cartwright, Capt. Edmund Henry, F.S.A. *Magherafelt-manor, Londonderry, Ireland.*
1857	Cartwright, Col. Henry, Grenadier Guards, M.P. 1, *Tilney-street, Park-street, Grosvenor-square, W.*
1860	*Carver, the Rev. Alfred J., D.D., Master of Dulwich College. *Dulwich, S.E.*
1869	390 Casberd-Boteler, Lieut. W. J., R.N. *Higham-hill, Walthamstow.*
1858	Casella, Louis P., Esq. 23, *Hatton-garden, E.C.; and South-grove, Highgate, N.*
1860	Cave, Amos, Esq. 109, *New-road, Kennington-park, S.; and Rathbone-place, Oxford-street, W.*
1857	Cave, Capt. Laurence Trent. 75, *Chester-square, W.*
1858	Cave, Right Hon. Stephen, M.P. 35, *Wilton-place, S.W.*
1869	Cayley, Dr. Henry.
1863	Challis, John Henry, Esq. *Reform Club, S.W.*
1865	Chambers, Charles Harcourt, Esq., M.A. 2, *Chesham-place, S.W.*
1864	Chambers, David Noble, Esq. *Paternoster-row, E.C.*
1858	Champion, John Francis, Esq. *High-street, Shrewsbury.*
1866	400 *Chandless, Wm., Esq., B.A. 1, *Gloucester-place, Portman-square, W.*
1870	Chapman, E. F., Esq., Lt. R.A. *The Grove, Tunbridge.*
1867	Chapman, James, Esq. *Cape Town, Cape of Good Hope.*
1863	*Chapman, Spencer, Esq. *Roehampton, S.W.*
1860	Charlemont, Lord. *Charlemont-house, Dublin.*
1870	Charles, Rev. David, M.A. *Aberystwith, South Wales.*
• 1861	Charnock, Richard Stephen, Esq., PH.DR., F.S.A. 8, *Gray's-inn-square, W.C.; and The Grove, Hammersmith.*
1864	Cheadle, Walter, Esq., B.A., M.D. Camb. 2, *Hyde-park-place, Cumberland-gate, W.*

1861		Cheetham, John Frederick, Esq. · *Eastwood, Stalaybridge.*
1855		Cheshire, Edward, Esq. *Conservative Club, S.W.*
1828	450	Chesney, Major-General Francis Rawdon, R.A., D.C.L., F.R.S. *Athenæum Club, S.W.; and Ballyardle, Down, Ireland.*
1856		Chetwode, Augustus L., Esq. 7, *Suffolk-street, Pall-mall-east, S.W.; and Chilton-house, Thame, Oxfordshire.*
1870		Chichester, Sir Bruce, Bart. *Arlington-court, Barnstaple.*
1858		Childers, Right Hon. Hugh C. E., M.P. 17, *Prince's-gardens, W.; and Australia.*
1854		Childers, John Walbanke, Esq. *Cantley-hall, near Doncaster.*
1867		*Chimmo, Commr. William, R.N. *H.M.S. ' Nassau.' Care of the Hydrographic-office, S.W.*
1868		Chinnock, Frederick George, Esq. 37, *Portland-place, W.*
1830		*Church, W. H., Esq.
1849		Churchill, Lord Alfred Spencer. 16, *Rutland-gate, S.W.*
1856		Churchill, Charles, Esq. *Weybridge-park, Surrey.*
1869	420	Churchill, Henry A., Esq., H.M. Consul, Zanzibar. *Care of Messrs. King and Co., 45, Pall-mall, S.W.*
1870		Clapton, Edward, Esq., M.D., &c., *St. Thomas's-street, Southwark, S.E.*
1863		Clark, Lieut. Alex. J. 14, *St. James's-square, S.W.; and Evewell-house, Maindee, Newport, Monmouthshire.*
1870		Clark, Charles, Esq. 20, *Belmont-park, Lee, Kent, S.E.*
1866		Clark, J. Howarth, Esq. *Cheetham Collegiate-school, Manchester.*
1868		Clark, John Gilchrist, Esq. *Speddock, Dumfries, Dumfriesshire.*
1862		Clark, Latimer, Esq. 1, *Victoria-street, Westminster, S.W.; and Cairo.*
1870		Clark, Robert, Esq. 46, *Chepstow-villas, Bayswater, W.*
1868		Clark, William, Esq. *The Cedars, South Norwood.*
1859		Clark, Rev. W. Geo., M.A. *Trinity College, Cambridge.*
1865	430	Clark, W. H., Esq. 6, *Leinster-terrace, Hyde-park, W.*
1859		Clarke, Capt. A., R.E. *Army and Navy Club, S.W.*
1855		*Clarke, Rev. W. B., M.A. *St. Leonard's, Sydney, New South Wales. Care of Messrs. Richardson, Cornhill.*
1868		Clarke, W., Esq. 44, *Ladbroke-grove, W.*
1862		Claude, Eugène, Esq. *Villa Helvetia, Carlton-road, Tufnell-park, N.*
1848		*Clavering, Sir William Aloysius, Bart., M.A. *United University Club, S.W.; Axwell-park, near Gateshead; and Greencroft, Durham.*
1863		Clayton, Capt. John W., late 15th Hussars. 14, *Portman-square, W.*
1866		Clayton, Sir W. R. *Harleyford, Great Marlow, Bucks.*
1859		*Cleghorn, Hugh, Esq., M.D., Conservator of Forests, Madras. *Stravithy, St. Andrew's.*
1858		Clements, Rev. H. G. *United University Club, S.W.*
1870	440	Clements, Robert Geo., Esq. 97, *Victoria-park-road, N.E.*
1862		Clerk, Capt. Claude. *Military Prison, Aldershott, Hants.*
1852		Clermont, Thomas, Lord. *Ravensdale-park, Newry, Ireland.*
1855		*Cleveland, His Grace the Duke of. *Cleveland-house, 17, St. James's-square, S.W.*

1861	Clifford, Sir Charles. *Campden-house, Broadway, Worcestershire.*
1858	Clifford, Charles Cavendish, Esq. *House of Lords, S.W.*
1871	Clifford, Henry, Esq., C.E. *Hamilton-villas, Hyde-vale, Greenwich, S.E.*
1866	Clinton, Lord Edward. *Army and Navy Club, S.W.*
1865	Clipperton, Robert Charles, Esq., H.B.M. Consul, Nantes. *Care of T. G. Staveley, Esq., Foreign-office, S.W.*
1856	Clive, Rev. Archer. *Whitfield, Hereford.*
1863	450 Clowes, E., Esq. *Salisbury-square, Fleet-street, E.C.*
1854	Clowes, George, Esq. *Duke-street, Stamford-street, Blackfriars, S.E.; Charing-cross, S.W.; and Surbiton, Surrey.*
1864	Clowes, Rev. George, M.A. *Surbiton, Surrey.*
1854	Clowes, William, Esq. *Duke-street, Stamford-street, Blackfriars, S.E.; Charing-cross, S.W.; and 51, Gloucester-terrace, Hyde-park, W.*
1861	Clowes, William Charles Knight, Esq., M.A. *Duke-street, Stamford-street, Blackfriars, S.E.; and Surbiton, Surrey.*
1852	Cobbold, John Chevallier, Esq. *Athenæum Club, S.W.; and Ipswich, Suffolk.*
1859	*Cochrane, Rear-Admiral the Hon. A., C.B. *Junior United Service Club, S.W.*
1869	*Cockburn, Captain James George. *Rawul Pindee, Bengal. Care of Colonel Cockburn, Bracon Ash, Norwich.*
1862	Cockerton, Richard, Esq. *Cornwall-gardens, South Kensington, W.*
1862	Cockle, Captain George. *9, Bolton-gardens, South Kensington, W.*
1859	460 Cocks, Colonel C. Lygon, Coldstream Guards. *Crediton, Devon.*
1865	Cocks, Major Octavius Yorke. *180, Piccadilly, W.*
1841	*Cocks, Reginald Thistlethwayte, Esq. *43, Charing-cross, S.W.; and 22, Hertford-street, Mayfair, W.*
1857	Coghlan, Edward, Esq. *Training-institution, Gray's-inn-road, W.C.*
1861	Coghlan, J., Esq., Engr.-in-Chief to the Government. *Buenos Ayres. Care of Messrs. J. Fair and Co., 4, East India-avenue, Leadenhall-street, E.C.*
1862	Coghlan, Major Gen. Sir William M., K.C.B., R.A. *Rumsgate, Kent.*
1865	Colchester, Reginald Charles Edward, Lord. *All Souls' College, Oxford.*
1868	Cole, William H., Esq. *23, Portland-place, W.*
1867	Colebrook, John, Esq. *15, Hans-place, Chelsea, S.W.*
1841	*Colebrooke, Sir Thomas Edward, Bart., V.R.A.S. *37, South-st., Park-lane, W.*
1854	470 Coleman, Everard Home, Esq., F.R.A.S. *Registry and Record Office, Adelaide-place, London-bridge, E.C.*
1848	Coles, Charles, jun., Esq. *86, Great Tower-street, E.C.*
1835	*Collett, William Rickford, Esq. *Carnarvon; and Carlton Club, S.W.*
1867	Collier, C. T., Esq., Barrister of the Middle Temple. *Cedar-villa, Sutton, Surrey; and Oriental Club, W.*
1858	Collinson, Henry, Esq. *7, Devonshire-place, Portland-place, W.*
1866	Collinson, John, Esq., C.E. *9, Clarendon-gardens, Maida-hill, W.*
1855	Collinson, Vice-Admiral Richard, C.B. *Haven-lodge, Ealing, W.; and United Service Club, S.W.*
1866	Collison, Francis, Esq. *Herne-hill, Surrey, S.*
1864	Colnaghi, Dominic E., Esq. *Care of F. B. Alston, Esq., Foreign-office, S.W.*

Year of Election.	
1862	Colquhoun, Sir Patrick, M.A.
1869	480 Colvill, William H., Esq., Surg. H.M. Ind. Army. *Lawn-bank, Hampstead; and Baghdad.*
1861	*Colville, Charles John, Lord. 42, *Eaton-place, S.W.*
1865	Colvin, Binney J., Esq. 71, *Old Broad-street, E.C.*
1868	Colvin, Captain W. B., Royal Fusiliers. *Care of Messrs. Cox and Co., Craig's-court, S.W.*
1868	Combe, Lieut. B. A.
1861	Combe, Thomas, Esq., M.A. *University Press, Oxford.*
1864	Commerell, Commr. J. E., R.N., V.C. *Alverbank, near Gosport.*
1864	Conder, John, Esq. *Hallbrooke-house, New Wandsworth, S.W.*
1868	Coney, Rev. T., M.A. *Chaplain to the Forces, Chatham.*
1861	Constable, Capt. Chas. Golding, I.N. 68, *Hamilton-ter., St. John's-wood, N.W.*
1868	490 Cook, Edward, Esq. *Kingston-on-Thames.*
1868	Cook, H. Esq., M.D., &c. *Care of Messrs. Forbes and Co., 12, Leadenhall-st., E.C.*
1859	Cooke, Lt.-Colonel A. C., R.E. *Bermuda.*
1863	*Cooke, E. W., Esq., A.R.A., F.R.S., F.L.S., F.Z.S., F.G.S., Accad. Bell. Art. Venet. et Holm Socius. *Glen-Andred, Groombridge, Sussex; and Athenæum Club, S.W.*
1856	Cooke, John George, Esq. 18, *Pelham-place, Brompton, S.W.*
1866	Cooke, Rev. J. Hunt. *Victoria-house, Southsea, Hants.*
1860	Cooke, Nathaniel, Esq. 5, *Ladbroke-terrace, Notting-hill, W.*
1852	Cooke, Robt. F., Esq. 50, *Albemarle-street, W.*
1860	Cooke, William Henry, Esq., Barrister-at-Law. 4, *Elm-court, Temple, E.C.*
1830	Cooley, William Desborough, Esq. 13, *College-place, Camden-town, N.W.*
1862	500 Cooper, Sir Daniel. 20, *Prince's-gardens, South Kensington, S.W.*
1856	Cooper, Lt.-Col. Edward, Grenadier Guards. 5, *Bryanston-square, W.*
1860	Cooper, Lt.-Col. Joshua H., 7th Fusiliers. *Dunboden, Mullingar.*
1857	*Coote, Captain Robert, R.N. *Shales, Bittern, Southampton.*
1862	Cope, Walter, late H.M.'s Chargé d'Affaires at the Equador. 14, *The Terrace, Camberwell, S.*
1853	Copley, Sir Joseph William, Bart. *Sprotborough, Doncaster.*
1864	Cork and Orrery, Earl of. 1, *Grafton-street, W.*
1868	Cork, Nathaniel, Esq. *Ivy-lodge, 9, Warwick-road, Upper Clapton, N.E.*
1868	Corner, William M., Esq. *Cobden-house, Leytonstone; and 104, Leadenhall-street, E.C.*
1868	*Cornish-Brown, Charles, Esq. 7, *Lansdowne-place, Clifton, Bristol.*
1865	510 Cornthwaite, Rev. T., M.A. *Forest, Walthamstow.*
1860	Cornwell, James, Esq., PH. DR. *Trev'er-byn, Dartmouth-park, Forest-hill.*
1839	*Corrance, Frederick, Esq. *Parkham-hall, Wickham Market, Suffolk.*
1870	Corrie, John, Esq. 42, *Lancaster-gate, Hyde-park, W.*
1868	Cory, Frederic C., Esq., M.D. *Portland-villa, Buckhurst-hill, Essex; and Nassau-place, Commercial-road, E.*
1869	Coster, Guillaume F., Esq. 11, *Park-crescent, Regent's-park, N.W.*

Year of Election	
1853	*Cosway, William Halliday, Esq. *Oxford and Cambridge Club, S.W.*
1848	Courtenay, L. W., Esq. *British Post-office, Constantinople. Care of R. Wood, Esq., 139, Fleet-street.*
1865	Cowan, John E., Esq. *58, Denbigh-street, S.W.*
1862	Coward, William, Esq. *7, Southsea-terrace, Southsea, Portsmouth.*
1857	520*Cowell, Major Sir J. C., K.C.B., R.E. *Buckingham-palace, S.W.*
1854	Cowley, Norman, Esq. *4, Montagu-place, Montagu-square, W.*
1862	*Cowper, Sedgwick S., Esq. *Windale, Westwood, Rockhampton, Queensland.*
1862	Cox, Mr. Sergeant, Barrister-at-Law, Recorder of Falmouth. *1, Essex-court, Temple, E.C.; and Moat-mount, High-wood, Middlesex.*
1865	Coysh, John S., Esq. *Levant-house, St. Helen's-place, E.C.*
1870	*Cracroft, Bernard, Esq. *Oxford and Cambridge Club, S.W.; and 4, Austin-friars, E.C.*
1867	Crane, Leonard, Esq., M.D. *7, Albemarle-street, W.*
1857	Craufurd, Lieut.-General James Robertson, Grenadier Guards. *Travellers' Club, S.W.; and 36, Prince's-gardens, W.*
1848	Crawford, Robert Wigram, Esq., M.P. *71, Old Broad-street, E.C.*
1866	Crawfurd, O. J., Esq. *Athenæum Club, S.W.*
1861	530 Creswell, Rev. Samuel Francis, M.A., F.R.A.S. *Principal of the High School, Dublin.*
1859	*Creyke, Capt. Richard Boynton, R.N. *Ulverstone, Lancashire.*
1856	Croker, T. F. Dillon, Esq. *19, Pelham-place, Brompton, S.W.*
1864	Croll, A. A., Esq., C.E. *Southwood, Southwood-lane, Highgate.*
1868	Croll, Alex., Esq. *16, The Boltons, Brompton, S.W.*
1860	*Croskey, J. Rodney, Esq.
1860	Cross, the Rev. Thomas, D.C.L., M.R.A.S. *Hastings.*
1862	Crossman, James Hiscutt, Esq. *Rolls-park, Chigwell, Essex.*
1863	*Crowder, Thos. Mosley, Esq., M.A. *Thornton-hall, Bedale, Yorkshire.*
1852	Crowdy, James, Esq. *17, Serjeant's-inn, E.C.*
1859	540 Cull, Richard, Esq., F.S.A. *13, Tavistock-street, Bedford-square, W.C.*
1857	Cumming, William Fullarton, Esq., M.D. *Athenæum Club, S.W.; and Kinellan, Edinburgh.*
1860	Cunliffe, Roger, Esq. *24, Lombard-street, E.C.; and 10, Queen's-gate, South Kensington, W.*
1864	Cunningham, H. Esq.
1853	Cunningham, John Wm., Esq., Sec. King's College. *Somerset-house, W.C.; and Harrow, N.W.*
1862	*Cunynghame, Major-Gen. A. T., C.B. *Mount Dillon, Dundrum, Co. Dublin.*
1870	Cunynghame, Sir Edward A., Bart. *Army and Navy Club, S.W.*
1865	Cure, Capel, Esq. *51, Grosvenor-street, W.*
1868	Currie, A. A. Hay, Esq., C.E. *The Manor-house, Tunbridge-wells.*
1843	*Cursetjee, Manockjee, Esq., F.R.S.N.A. *Villa-Byculla, Bombay.*
1839	550*Curtis, Timothy, Esq.

Year of Election	
1865	Caxton, Hon. R. 24, *Arlington-street, W.; and Parham-park, Steyning, Sussex*
1867	Cuttmoe, John Fras. J., Esq. *Cleveland-house, Greville-road, Kilburn, N.W.*
1868	*Dalgety, Fred. G., Esq. 8, *Hyde-park-terrace, W.*
1864	Butler, A. G., Esq. 36, *Beaufort-gardens, W.*
1870	Dallas, Geo. E., Esq. *Foreign-office, S.W.*
1865	D'Almeida, W. B., Esq. 19, *Green-park, Bath.*
1868	Dalrymple, Donald, Esq. *Thorpe-lodge, Norwich.*
1867	Dalrymple, Geo. Elphinstone, Esq. *Logie, Elphinstone, Aberdeenshire.*
1857	Dalton, D. Foster Grant, Esq. *Shanks-house, near Wincanton, Somerset.*
1839	560 Dalyell, Sir Robt. Alex. Oxborn, Bart. *H.M.'s Consul at Rustchuk, Bulgaria and 126, Belgrave-road, S.W.*
1868	Difield, William R., Esq. 5, *Gresham-park; Brixton, S.*
1866	Damer, Lt.-Col. Lionel S. Dawson. 2, *Chapel-street, Grosvenor-square, W.*
1863	Darvall, John Bayly, Esq.
1838	*Darwin, Charles, Esq., M.A., F.R.S. 6, *Queen Anne-street, Cavendish-square, W*
1860	Dasent, John Bury, Esq. 22, *Warwick-road, Maida-hill, W.*
1868	Davies, R. H., Esq. Chief Commissioner of Oudh, Lucknow. *Care of Messrs Twining, 215, Strand, W.C.*
1869	*Davies, Robert E., Esq., J.P. *Crescent-villa, Kington, near Portsmouth.*
	Davis, Edmund F., Esq. 6, *Cork-street, Bond-street, W.*
	Davis, Frederick E., Esq. 20, *Blandford-square, N.W.*
	570 Davis, Richard, Esq. 9, *St. Helen's-place, E.C.*
	Davis, Staff-Commander John Edward, R.N. *Hydrographic-office, Admiralty, S.W*
1846	Davis, Sir John Francis, Bart., K.C.B., F.R.S., F.R.S.N.A. *Athenæum Club, S.W. and Hollywood, near Bristol.*
1840	*Dawnay, the Hon. Payan. *Beningborough-hall, Newton-upon-Ouse, Yorkshire.*
1870	Dawson, John Edward, Esq. *Oak Lodge, Watford, Herts.*
1865	Denny, Rev. Thomas, M.A. 85, *Mount-street, W.*
1866	Debenham, William, Esq. 3, *Porchester-square, Hyde-park, W.*
1858	De Bourghe, T. J., Esq. 6, *Charing-cross, S.W.*
1856	De Crespigny, Lieut. C., R.N.
1856	De Gex, William Francis, Esq. 25, *Throgmorton-street, E.C.*
1833	580 De Grey and Ripon, George Frederick Samuel, Earl. 1, *Carlton-gardens, S.W., and Studley Royal, Ripon.*
1865	De Ladd, A., Esq.
1869	De Leon, Dr. Hananel. 26, *Redcliffe-gardens, West Brompton, S.W.*
1858	Dell, William, Esq. *Messrs. Combe, Delafield, and Co., Long Acre.*
1862	Denham, Adm. Sir Henry Mangles, K.C.B. 21, *Carlton-road, Maida-vale, W.*
1860	Denison, Alfred, Esq. 6 *Albemarle-street, W.*

Year of Election.	
1836	Denman, Rear-Admiral the Hon. Joseph. 17, *Eaton-terrace, S. W.*
1870	Denniss, Col. Shuckburgh. 30, *Duke-street, St. James's, S. W.*
1870	Dentry, J., Esq. *Crescent School, Margate.*
1853	*Derby, Edward Henry, Earl of, P.C., LL.D., D.C.L. 23, *St. James's-square, S. W.; and Knowsley-park, Prescott, Lancashire.*
1867	590 De Salis, Col. Rodolph, C.B. 123, *Pall-mall, S. W.*
1853	De Wesselow, Lieut. Fras. G. Simpkinson.
1854	*Devaux, Alexander, Esq. 2, *Avenue-road, Regent's-park, N. W.*
1837	*Devonshire, William Cavendish, Duke of, LL.D., D.C.L., M.A., F.R.S. *Devonshire-house, Piccadilly, W.; and Hardwicke-hall, Derbyshire.*
1870	Dibdin, Charles, Esq. 62, *Torrington-square, W.C.*
1870	Dibdin, Robert W., Esq. 62, *Torrington-square, W.C.*
1864	Dick, A. H. Esq., M.A., LL.D. *Free Church Normal College, Glasgow.*
1862	Dick, Capt. Charles Cramond. *Exeter, Devon.*
1866	*Dick, Fitzwilliam, Esq., M.P. 20, *Curzon-street, Mayfair, W.*
1861	Dick, Robert Kerr, Esq., Bengal Civil Service. *Oriental Club, W.*
1866	600 Dick, William Græme, Esq. 29, *Leinster-square, W.*
1854	*Dickenson, Sebastian Stewart, Esq., M.P., Barrister-at-Law. *Brown's-hill, Stroud, Gloucestershire.*
1830	*Dickinson, Francis Henry, Esq., F.S.A. 119, *St. George's-square, Pimlico, W.; and Kingweston-park, Somerset.*
1852	Dickinson, John, Esq , jun. *Athenæum Club, S. W.*
1859	Dickson, A. Benson, Esq. 4, *New-square, Lincoln's-inn, W.C.*
1860	Dietz, Bernard, Esq., of Algoa Bay. 3, *Dorset-square, W.*
1859	Digby, G. Wingfield, Esq. *Sherborne-castle, Dorset.*
1860	Digby, Lieut.-Col. John Almerus. *Chalmington-house, Cattstock, Dorchester.*
1869	Digby, Kenelm T. Esq., M.P. *Shaftesbury-house, Kensington, S. W.*
1859	*Dilke, Sir Charles Wentworth, Bart., M.P. 76, *Sloane-street, S. W.*
1856	610 Dillon, the Hon. Arthur. 17, *Clarges-street, W.*
1864	Dimsdale, J. C., Esq. 50, *Cornhill, E.C.; and 52, Cleveland-square, S. W.*
1867	Dix, Thomas, Esq. 10, *Amwell-street, W.C.*
1861	Dixon, Lieut.-Colonel John. 18, *Seymour-street, Portman-square.*
1854	Dixon, W. Hepworth, Esq., F.S.A. 6, *St. James's-terrace, Regent's-park, N. W.*
1857	Dobie, Robert, Esq., M.D., R.N. 7, *Houghton-pl., Ampthill-sq., Hampstead-road, N. W.*
1854	Dodson, John George, Esq., M.P. 6, *Seamore-place, Mayfair, W.*
1854	Domville, William T., Esq., M.D., R.N. *Army and Navy Club, S. W.; and Naval Dockyard, Malta.*
1867	Donald, James, Esq. 5, *Duke-street, York-place, Edinburgh.*
1858	Donne, John, Esq. *Instow, North Devon.*
1864	620 Doran, Dr. John, F.S.A. 33, *Lansdowne-road, Notting-hill, W.*
1868	Douglas, James A., Esq. *The Grange, Coulsdon, near Caterham.*
1868	Douglas, John, Esq. *Angas-lodge, Portsea.*
1870	Douglas, John, Esq. 32, *Charing-cross, S. W.*

Year of Election.	
1868	Douglas, Capt. N. D. C. F. *Guards' Club, S.W.*
1850	Dover, John William, Esq. 132, *Stanley-street, Belgravia, S.W.*
1853	Doyle, Sir Francis Hastings C., Bart. *Custom-house, E.C.*
1845	*Drach, Solomon Moses, Esq., F.R.A.S. 39, *Howland-street, Fitzroy-square, W.*
1869	Drake, Francis, Esq., F.G.S. *Leicester.*
1869	Drummond, Alfred, Esq. *Charing-cross, S.W.*
1865	630 Drummond, E. A., Esq. 2, *Bryanston-square, W.*
1846	Drummond, Lieut.-General John. *The Boyce, Dymock, Gloucestershire.*
1846	Drury, Capt. Byron, R.N. *The United Service Club, S.W.*
1851	*Du Cane, Major Francis, R.E. 64, *Lowndes-square, S.W.*
1851	*Dacie, Henry John, Earl, F.R.S. 30, *Prince's-gate, S.W.*
1859	Duckworth, Henry, Esq. *Holmfield-house, Aigburth, near Liverpool.*
1860	*Duff, Right Hon. Mountstuart Elphinstone Grant, M.P. 4, *Queen's-gate-gardens, South Kensington, W.*
1868	Duff, Wm. Pirie, Esq. *Calcutta. Care of Messrs. John Watson, and Co., 34, Fenchurch-street, E.C.*
1857	*Dufferin, Right Hon. Lord, K.P., K.C.B. *Dufferin-lodge, Fitzroy-park, Highgate, N.*
1866	*Dugdale, Captain Henry Charles G. *Merevale-hall, Atherstone, Warwick.*
1867	640 *Dugdale, John, Esq. 1, *Hyde-park-gardens; and Llwyn, Llanfyllin, Oswestry.*
1868	Dunbar, John Samuel A., Esq. 28, *Pembridge-crescent, Bayswater, W.; and 4, Barnard's Inn, Holborn.*
1863	Duncan, Capt. Francis, R.A., M.A., F.R.S. 32, *The Common, Woolwich, S.E.*
1861	*Duncan, George, Esq. 45, *Gordon-square, W.C.*
1840	*Dundas, Right Hon. Sir David, Q.C. 13, *King's-Bench-walk, Temple, E.C.; and Ochtertyre, Stirling.*
1860	Dunell, Henry James, Esq. 12, *Hyde-park-square, W.*
1859	*Dunlop, R. H. Wallace, Esq., C.B., Indian Civil Service. *Northwood, near Rickmansworth, Herts.*
1860	*Dunmore, Charles Adolphus Murray, Earl of. 24, *Carlton-house-terrace, S.W.*
1868	Dunn, Capt. F. J. A. *Portillon, Tours, France; and 4, Cambrian-grove, Gravesend, E.C.*
1837	*Dunraven, Edwin Richard, Earl of, F.R.S. *Adare-manor, Limerick; and Dunraven-castle, Glamorganshire.*
1856	650 Duprat, Le Vicomte. *Consul-Général de Portugal, 9, Upper Porchester-street, Cambridge-square, S.W.*
1869	Durham, Edward, Esq. *Beauchamp-house, Kibworth, near Leicester.*
1852	D'Urban, M.-Gen. W. J. *U.S. Club, S.W.; and Newport, near Exeter.*
1865	Dutton, F. S., Esq. *Reform Club, S.W.; and Adelaide, Australia.*
1868	Dutton, Frederick H., Esq. 45, *Dover-street, W.*
1870	Dymes, Daniel David, Esq. *Windham Club, S.W.; and 9, Mincing-lane, E.C.*
1867	Eadie, Robert, Esq. *Blaydon-on-Tyne, Durham.*
1854	Eardley-Wilmot, Capt. A. P., R.N., C.B. *Deptford Dockyard, E.*
1858	Eardley-Wilmot, Major-Gen. F., M.R.A. 22, *Victoria-rd., Clapham-common, S.W.*
1869	Eastwick, Edward B., Esq., F.R.S., M.P. 38, *Thurlow-square, Brompton, S.W.*

Year of Election.		
1857	660	Eastwick, Captain W. J. 12, *Leinster-terrace, Hyde-park, W.*
1863		Eaton, F. A., Esq. *New University Club, St. James's-street, S.W.*
1862		*Eaton, H., Esq. 16, *Prince's-gate, Hyde-park, W.*
1862		*Eaton, Henry William, Esq., M.P. 16, *Prince's-gate, Hyde-park, W.*
1864		*Eaton, William Meriton, Esq., 16, *Prince's-gate, Hyde-park, W.*
1866		Eatwell, Surgeon-Major W. C. B., M.D. *Oriental Club, Hanover-square, W.*
1861		Eber, General F. 33, *St. James's-square, S.W.*
1862		Ebury, Lord. 107, *Park-street, Grosvenor-square, W.; and Moor-park, Herts.*
1862		Eden, Vice-Adm. Charles, C.B. 20, *Wilton-place, S.W.*
1858		Edge, Rev. W. J., M.A. *Benenden-vicarage, near Staplehurst, Kent.*
1863	670	Edgeworth, M. P., Esq., BENG. C.S. *Mastrim-house, Anerly, S.*
1867		*Edward, James, Esq. *Balruddery, by Dundee, N.B.*
1866		*Edwardes, Thomas Dyer, Esq. 5, *Hyde-park-gate, Kensington, W.*
1868		Edwards, Rev. A. T., M.A. 39, *Upper Kennington-lane, S.*
1865		Edwards, G. T., Esq., M.A.* *Devon-lodge, Alexandra-road, London, N.W.*
1861		*Edwards, Henry, Esq. 53, *Berkeley-square, W.*
1860		Edwards, Major J. B., R.E. *United Service Club, S.W. Care of Messrs. H. S. King and Co., 65, Cornhill, E.C.*
1858		Egerton, Captain the Hon. Francis, R.N., M.P. *Bridgewater-house, S.W.; and H.M.S. 'St. George.'*
1868		Elder, A. L., Esq. *Carlisle-house, Hampstead.*
1863		*Elder, George, Esq. *Knock-castle, Ayrshire.*
1857	680	Eley, Charles John, Esq. *Old Brompton, S.W.*
1865		Elias, Ney, Jun., Esq. 64, *Inverness-terrace, Bayswater, W.*
1845		Ellenborough, Edward, Earl of, G.C.B. *Southam-house, near Cheltenham.*
1863		Ellerton, John L., Esq. 6, *Connaught-place, Hyde-park, W.*
1860		Elliot, G., Esq., M.P., C.E. *The Hall, Houghton-le-Spring, near Fence Houses, Durham.*
1857		*Elliot, Capt. L. R. *La Mailleraye-sur-Seine, Seine Inférieure. Care of J. L. Elliot, Esq., 10, Connaught-place, W.*
1830		*Elliott, Rev. Charles Boileau, M.A., F.R.S. *Tattingstone, Suffolk.*
1868		Ellis, C. H. Fairfax, Esq., Lieut. R.A. *Shoeburyness, Essex.*
1865		Ellis, W. E. H., Esq. *Hasfield-rectory, Gloucester; Oriental Club, W.; and Byculla Club, Bombay.*
1858		Elphinstone, Major Howard C., R.E. *Buckingham-palace, S.W.*
1869	690	Elsey, Colonel William. *West-lodge, Ealing, W.*
1857		Elton, Sir A. H., Bart. *Athenæum Club, S.W.; und Clevedon-court, Somersetshire.*
1868		Ely, John Henry Wellington Graham Loftus, Marquis of. 9, *Prince's-gate, W.*
1862		*Emanuel, Harry, Esq. 8, *Clarence-terrace, Regent's-park, N.W.*
1866		Emanuel, Joel, Esq., F.A.S. *Norfolk-villa, Lansdowne-road, Notting-hill, W.*
1863		Emslie, John, Esq. 47, *Gray's-inn-road, W.C.*
1830		Enderby, Charles, Esq., F.R.S., F.L.S. 13, *Great St. Helen's, E.C.*
1860		Enfield, Edward, Esq., F.S.A. 19, *Chester-terrace, Regent's-park, N.W.*

Year of Election	
1863	Engleheart, Gardner D., Esq. *Gatton-cottage, Reigate.*
1870	Erskine, Claude J , Esq., Bombay Civil Service. 87, *Harley-street, W. ; and Athenæum Club, S. W.*
1852	700 Erskine, Admiral John Elphinstone, M.P., C.B. *H.M.S.* 'Edgar;' 1 L, *Albany, W.; and Cardross, Stirling, N.B.*
1857	*Esmeade, G. M. M., Esq. 29, *Park-street, Grosvenor-square, W.*
1865	Evans, Colonel William Edwyn. 24, *Great Cumberland-place, Hyde-park, W.*
1870	*Evans, Edward Bickerton, Esq. *Whitbourne-hall, near Worcester.*
1857	Evans, F. J., Esq., Staff Captain R.N., F.R.S., F.R.A.S. 4, *Wellington-terrace, Charlton, Blackheath, S.E.*
1830	*Evans, Vice-Admiral George. 1, *New-street, Spring-gardens, S. W.; and Englefield-green, Staines.*
1870	Evans, Lieut.-Colonel Henry Lloyd. 14, *St. James's-square, S.W.*
1857	Evans, Thos. Wm., Esq. 1, *Dartmouth-street, Westminster, S. W.; and Allestree-wall, Derby.*
1830	*Evans, W., Esq.
1867	Evans, W. Herbert, Esq. 32, *Hertford-street, Mayfair, W.*
1861	710 Evelyn, Lieut.-Colonel George P. 34, *Onslow-gardens, Brompton, S.W.*
1851	*Evelyn, William J., Esq., F.S.A. *Evelyn Estate-office, Evelyn-street, Deptford.*
1830	*Everett, James, Esq., F.S.A.
1865	Everitt, George A., Esq. *Knowle-hall, Warwickshire.*
1856	Ewing, J. D. Crum, Esq. 3, *Lime-street-square, E.C.*
1869	Ewing, John Orr, Esq. *Levenfield-house, Alexandria, Dumbartonshire.*
1857	Eyre, Edward J., Esq.
1861	Eyre, George E., Esq. 59, *Lowndes-square, Brompton, S.W.*
1856	Eyre, Major-Gen. Sir Vincent, C.B. *Athenæum Club, S.W.; and 33, Thurloe-square, S. W.*
1870	Fairbridge, Charles, Esq., Queen's Proctor. *Court of Vice-Admiralty, Cape Town.*
1861	720 Fairbairn, Sir William, Bart., C.E., F.R.S. *Manchester.*
1869	Fairfax, Captain Henry, R.N. *Army and Navy Club, S.W.*
1856	Fairholme, George Knight, Esq.
1866	Fairman, Edward St. John, Esq., F.G.S., &c. 874, *Via Santa Maria, Pisa. Care of H. Fairman, Esq.*
1838	Falconer, Thomas, Esq. *Usk, Monmouthshire.*
1868	Falconer, William, Esq. 23, *Leadenhall-street, E.C.; and 42, Hilldrop-road, Camden-new town, N.*
1857	Falkland, Lucius Bentinck, Viscount. *Skutterskelfe, Yorkshire.*
1855	*Fanshawe, Admiral E. G. 63, *Eaton-square, S. W.*
1868	*Farquharson, Lieut.-Col. G. McB. *Care of Messrs. King and Co., Cornhill, E.C.*
1863	*Farrer, W. Jas., Esq. 24, *Bolton-street, Piccadilly, W.*

Year of Election.	
1864	730 Faulkner, Charles, Esq., F.S.A., F.G.S. *Deddington, Oxon.*
1863	*Faunthorpe, Rev. J. P., M.A. *Training-college, Battersea.*
1869	Fawcett, Captain Edward Boyd, M.A. *Dunolly, Torquay.*
1869	Fawcett, Henry, Esq. *Wainsford, Lymington.*
1853	*Fayrer, Joseph, Esq., M.D. *Calcutta.* Care of General Spens, 4, Rosslyn-street, Pilrieg, Edinburgh.*
1858	Fazakerley, J. N., Esq. 17, *Montagu-street, Portman-square,* W.
1866	Felkin, Wm., Esq., Jun., F.Z.S. *Beeston, near Nottingham.*
1840	*Fergusson, James, Esq., F.R.S., D.C.L. 20, *Langham-place.* W.
1863	Ferreira, Baron De. 12, *Gloucester-place, Portman-square,* W.
1860	Ferro, Don Ramon de Silva.
1871	740 Festing, Capt. Robert, R.E. *South Kensington Museum, S.W.*
1865	Field, Hamilton, Esq. *Thornton-lodge, Thornton-road, Clapham-park.*
1844	Findlay, Alex. George, Esq. 53, *Fleet-street, E.C.*; and *Dulwich-wood-park,* S
1862	Finnis, Thomas Quested, Esq., Alderman. *Wanstead, Essex, N.E.*
1870	Firth, John, Esq., J.P. *Care of J. W. Firth, Esq., 2, Gresham-place, Lombard-street, E.C.*
1863	Fisher, John, Esq. 60, *St. James's-street, S.W.*
1869	Fitch, Frederick, Esq., F.R.M.S. *Hadleigh-house, Highbury-new-park, N.*
1857	*Fitzclarence, Commander the Hon. George, R.N. 1, *Warwick-square, S.W.*
1863	Fitzgerald, J. F. V., Esq. 11, *Chester-square, S.W.*
1861	Fitzgerald, Captain Keane. 2, *Portland-place,* W.
1864	750 Fitzpatrick, Lieut. Francis Skelton, 42nd Regt. Madras Army.
1857	Fitzwilliam, the Hon. C. W., M.P. *Brooks' Club, St. James's-street, S.W.*
1837	*Fitzwilliam, William Thomas, Earl. 4, *Grosvenor-square, W.; and Wentworth-house, Rotherham, Yorkshire.*
1865	*Fitzwilliam, Wm. S., Esq. 28, *Ovington-square, Brompton, S.W.*
1863	Fleming, G., Esq. *Brompton Barracks, Chatham.*
1861	*Fleming, John, Esq. 18, *Leadenhall-street, E.C.*
1865	Fleming, Rev. T. S. *Roscoe-place, Chapeltown-road, Leeds.*
1853	*Flemyng, Rev. Francis P. *The Rectory, Laurence Kirk, Scotland.*
1862	Fletcher, John Charles, Esq. *Dale-park, Arundel; and Eaton-place, S.W.*
1868	Fletcher, John Thompson, Esq. 15, *Upper Hamilton-ter., St. John's-wood, N.W.*
1857	760 Fletcher, Thomas Keddey, Esq. *Union-dock, Limehouse, E.*
1866	Flood, John Edwin, Esq. *Alexandra-house, Carshalton, Surrey.*
1864	Flower, Capt. L. *Banstead, Surrey; and Queen's United Service Club, S.W.*
1863	Foley, Col. the Hon. St. George, C.B. 24, *Bolton-street,* W.
1861	Foord, John Bromley, Esq. 52, *Old Broad-street, E.C.*
1860	Forbes, Commander Charles S., R.N. *Army and Navy Club, S.W. Care of Messrs. Woodhead.*
1863	Forbes, Capt. C. J. F. Smith. 5, *Hatch-street, Dublin.*
1867	Forbes, Geo. Edward, Esq. *Colinton, Ipswich, Queensland; care of Messrs. Edenborough and Co., 54, Moorgate-street, E.C.; Union Club, S.W.; 11, Melville-street, Edinburgh; New Club, Edinburgh.*

Year of Election	
1860	Forbes, Lord, **M.A.** *Castle Forbes, Aberdeenshire.*
1869	Ford, Col. Barnett, Governor of the Andaman Islands. 24, *Upper-park-road, Hampstead.*
1868	770 Forster, Hon. Anthony. *Newsham-grange, Winston, Darlington, Durham.*
1845	Forster, Rev. Charles, B.D. *Stisted-rectory, Essex.*
1839	*Forster, William Edward, Esq. *Burley, near Otley.*
1867	Forsyth, T. Douglas, Esq., C.B. (B.C S.), Commissioner, Jullundhur, Punjaub. *Care of Messrs. H. S. King and Co., 65, Cornhill, E.C.*
1861	Forsyth, William, Esq., M.P., Q.C. 61, *Rutland-gate, S. W.*
1858	Fortescue, Right Hon. Chichester S., M.P. 7, *Carlton-gardens, S. W.*
1861	*Fortescue, Hon. Dudley F., M.P. 9, *Hertford-street, Mayfair, W.*
1869	Foster, Ebenezer, Esq. 19, *St. James's-place, St. James's, S. W.*
1866	Foster, Edmond, Esq., Jun. 79, *Portsdown-road, Maida-vale, W.*
1864	Foster, H. J., Esq.
1863	780*Fowler, J. T., Esq. *Government Inspector of Schools, Adyar, Madras, India. Care of Rev. A. Wilson, National Society's Office, Sanctuary, Westminster.*
1850	*Fowler, Robert N., Esq., M.P., M A. 50, *Cornhill, E.C.; and Tottenham, N.*
1859	Fox, Lieut.-Colonel A. Lane. 10, *Upper Phillimore-gardens, Kensington, W.*
1830	*Fox, Lieut.-General C. R. *Travellers' Club, S. W.; and 1, Addison-road, Kensington, W. *
1866	Fox, D. M., Esq., Chief Engineer of the Santos and St. Paulo Railway. *St. Paulo, Brazil.*
1864	*Fox, Francis E., Esq., B.A. *Falmouth.*
1865	Fox, Samuel Crane, Esq. 31, *Cambridge-gardens, Notting-hill, W.*
1865	*Franks, Aug. W., Esq. 103, *Victoria-street, S. W.*
1860	Franks, Charles W., Esq. 2, *Victoria-street, S. W.*
1867	Fraser, Edward John, Esq.. Solicitor. 1, *Percy-villas, Campden-hill, Kensington, W.*
1862	790 Fraser, Capt. H. A., I.N.
1860	Fraser, Thos., Esq.
1866	Fraser, Capt. T. *Otago, New Zealand.*
1868	Frater, Alexander, Esq. *Canton. Care of Thomas Frater, Esq., National Provincial Bank of England, Brecon, Wales.*
1869	Freke, Thomas George, Esq. 1, *Cromwell-houses, Kensington, W.*
1860	Freeman, Daniel Alex., Esq., Barrister-at-Law. *Plowden-buildings, Temple, E.C.*
1868	Freeman, Henry W., Esq. *Junior Athenæum Club, S. W.*
1864	Fremantle, Lieut.-Col. Arthur. *Guards' Club, S. W.*
1863	Fremantle, Captain Edmund Robert, R.N. 4, *Upper Eccleston-street, S. W.*
1856	Fremantle, Rt. Hon. Sir Thomas F., Bart. 4, *Upper Eccleston-street, Belgrave-square, S. W.*
1864	800 Freme, Major James H. *Wrentnall-house, Shropshire; and Army and Navy Club, S. W.*
1850	Frere, Bartle John Laurie, Esq. 45, *Bedford-square, W.C.*
1839	*Frere, George, Esq. *Cape of Good Hope. Care of the Foreign-office, S. W.*

1859		Fryer, William, Esq.
1863		Fuidge, William, Esq. *5, Park-row, Bristol.*
1865		Fuller, Thomas, Esq. *United University Club, S.W.*
1860		Fussell, Rev. J. G. Curry. *16, Cadogan-place, S.W.*
1868	810	Fyfe, Andrew, Esq., M.D. *112, Brompton-road, S.W.*
1861		Fynes Clinton, Rev. Charles J., M.A. *3, Montagu-place, Russell-square, W.C.; and Cromwell, Notts.*
1866		Fytche, Colonel Albert. *Reform Club, S.W.*

1863		*Gabrielli, Antoine, Esq. *6, Queen's-gate-terrace, Kensington, W.*
1858		Gaisford, Thomas, Esq. *Travellers' Club, S.W.*
1855		*Galloway, John James, Esq.
1869		Galsworthy, Frederick Thomas, Esq. *8, Queen's-gate, Hyde-park, W.*
1848		*Galton, Capt. Douglas, R.E. *12, Chester-street, Grosvenor-place, S.W.*
1850		*Galton, Francis, Esq., M.A., F.R.S. *42, Rutland-gate, S.W.; and 5, Bertie-terrace, Leamington.*
1871		Galton, Theodore Howard, Esq. *78, Queen's-gate; and Hadzor-ho., Droitwich.*
1854	820	*Gammell, Major Andrew. *Drumtochty, Kincardineshire, N.B.*
1861		Garden, Robert Jones, Esq. *30, Cathcart-road, South Kensington, S.W.*
1869		Gardner, Christopher T., Esq. *3, St. James's-terrace, Paddington, W.*
1865		Gardner, Capt., G. H., R.N. *7, James-street, Westbourne-terrace, W.*
1866		Gardner, John Dunn, Esq. *19, Park-street, Park-lane, W.*
1863		Gascoigne, Frederic, Esq. *Parlington, Yorkshire.*
1859		*Gassiot, John P., Jun., Esq. *6, Sussex-place, Regent's-park, N.W.*
1866		Gastrell, Lieut.-Col. James E. (B. Staff Corps). *Surveyor-General's Office, Calcutta. Care of H. T. Gastrell, Esq., 36, Lincoln's-inn-fields, W.C.*
1866		*Gatty, Charles H., Esq., M.A., *Felbridge-park, East Grinstead, Sussex.*
1870		*Gellatly, Edward, Esq. *Uplands, Sydenham.*
1865	830	George, Rev. H. B. *New College, Oxford.*
1859		Gerstenberg, Isidore, Esq. *Stockley-house, North-gate, Regent's-park, N.W.*
1866		*Gibb, George Henderson, Esq., *13, Victoria-street, Westminster, S.W.*
1865		*Gibbons, Sills John, Esq., Alderman. *13, Upper Bedford-place, Russell-square, W.C.*
1859		*Gibbs, H. Hucks, Esq. *St. Dunstan's, Regent's-park, N.W.*
1870		Gibson, James Y., Esq. *Edinburgh. Care of Messrs. Williams and Norgate,*
1855		Gibraltar, Right Rev. and Hon. C. A. Harris, Bishop of. *Gibraltar Palace, Malta.*
1855		Gillespie, Alexander, Esq. *Heathfield, Walton-on-Thames, Surrey.*

Year of Election	
1866	*Gillespie, William, Esq. (of Torbane-hill). 46, Melville-street, Edinburgh.
1857	Gillespy, Thomas, Esq. Brabant-court, Philpot-lane, E.C.
1868	840*Gillett, Alfred, Esq. 113, Piccadilly, W.; and Banbury, Oxon.
1863	*Gillett, William, Esq. 6L, Albany, W.
1861	Gilliat, Alfred, Esq. Postern-house, Tunbridge.
1868	Gilliat, Algernon, Esq. Fernhill, near Windsor; and 7, Norfolk-crescent, W.
1863	Gillies, Robert, Esq., C.E. Dunedin, Otago, New Zealand.
1867	Gisborne, Fred. N., Esq., Engineer and Electrician. 445, West Strand, W.C.
1836	Gladdish, Col. William. Bycliffes, Gravesend.
1864	Gladstone, George, Esq. Clapham-common, S.
1863	Gladstone, J. H., Esq., PH.DR. 17, Pembridge-square, W.
1862	*Gladstone, Robert Stuart, Esq.
1846	850*Gladstone, William, Esq. 57½, Old Broad-street, E.C.
1864	*Gladstone, W. K., Esq. 39a, Old Bond-street, W.; and Fitzroy-park, Highgate, N.
1860	Glascott, Commander Adam Giffard, R.N.
1867	Glass, H. A., Esq. 4, Gray's-inn-square, W.C.
1857	Gleig, Rev. G. R., M.A. Chaplain-General, Chelsea-hospital, S.W.
1854	Glen, Joseph, Esq., Mem. Geogr. Soc. of Bombay. Oriental Club, W.
1857	Glover, Commr. John H., R.N. Lagos; and Army and Navy Club, S.W.
1866	Glover, Robert Reaveley, Esq. 30, Great St. Helen's, E.C.
1870	Glover, Col. T. G., R.E. Barwood, Hersham, near Esher, Surrey.
1868	Glyn, Richard H., Esq. 10, King's-arms-yard, E.C.; and Oriental Club, S.W.
1864	860 Glyn, Sir Richard George, Bart. Army and Navy Club, S.W.
1869	Goldney, G. Esq., M.P. 40, Hill-street, Berkeley-square, W.
1868	Goldsmid, Sir Francis, Bart., M.P. Inner-circle, Regent's-park, N.W.
1863	Goldsmid, Lt.-Colonel Frederick John. Harrow-on-the-hill; Southborough, Kent; and United Service Club, S.W.
1861	Goldsmid, Julian, Esq. 49, Grosvenor-street. S.W.
1868	Goldsworthy, Major W. T. British Service Club, 4, Park-place, St. James's, S.W.
1860	Gooch, Thomas Longridge, Esq. Team-lodge, Saltwell, Gateshead-on-Tyne.
1864	Goodall, George, Esq. Messrs. Cox and Co., Craig's-court; and Junior Carlton Club, W.
1863	*Goodenough, Capt. J. G., R.N. U. S. Club, S.W. Care of Messrs. Stilwell, 22, Arundel-street, Strand, W.C.
1864	*Goodenough, Lieut.-Col., R.A. F. Battery, 9th Brigade, Royal Artillery, Ahmedabad, Bombay. Care of Messrs. Cox and Co., Craig's-court, S.W.
1861	870 Gooldin, Joseph, Esq. 18, Lancaster-gate, W.
1865	*Goolden, Charles, Esq. United University Club, S.W.
1856	*Gordon, Colonel the Hon. Alexander H., C.B. 4, Warrior-square, St. Leonard's on Sea.
1854	Gordon, Harry George, Esq. 1, Clifton-place, Hyde-park-gardens, W.; and Killiechassi, Dunkeld, Perthshire.

Year of Election.	
1856	Gordon, Admiral the Honourable John.
1870	Gordon, Russell Manners, Esq. 38, *Alpha-road, St. John's-wood, N.W.*
1866	Gore, Augustus F., Esq., Colonial Secretary. *Demerara.*
1853	Gore, Richard Thomas, Esq. 6, *Queen-square, Bath.*
1859	Gosling, Fred. Solly, Esq. 18, *New-street, Spring-gardens, S.W.*
1862	Goss, Samuel Day, Esq., M.D. 111, *Kennington-park-road, S.*
1870	880 Gottlieb, Felix Henry, Esq., J.P. *Garden-court, Temple; and 16, Colcherne-road, West Brompton, W.*
1868	Gough, Hugh, Viscount, F.L.S. *Lough Cutra Castle, Gort, Co. Galway.*
1835	Gould, Lieut.-Colonel Francis A. *Buntingford, Herts.*
1846	Gould, John, Esq., F.R.S., F.L.S. 26, *Charlotte-street, Bedford-square, W.C.*
1870	Gould, Rev. Robert John (Vicar of Wadham College, Oxford). *Stratfield Mortimer, near Reading.*
1865	Gowen, Colonel J. E.
1867	Grabham, Michael, Esq., M.D. *Madeira.* Care of C. R. Blandy, Esq., 25, *Crutched-friars, E.C.*
1868	Graeme, H. M. S., Esq.
1869	Graham, Andrew, Esq., Staff Surg., R.N. *Army and Navy Club, S.W.*
1868	Graham, Cyril C., Esq. 9, *Cleveland-row, St. James's, S.W.; and Debroe-house, Watford, Herts.*
1868	890*Graham, Thomas Cuninghame, Esq. *Carlton Club, S.W.; and Dunlop-house, Ayrshire.*
1861	Grant, Alexander, Esq. *Oakfield-house, Hornsey, N.*
1870	*Grant, Andrew, Esq. *Oriental Club, Hanover-square, W.*
1861	Grant, Daniel, Esq. 11, *Warwick-road, Upper Clapton, N.*
1865	*Grant, Francis W., Esq. *Army and Navy Club, S.W.*
1860	Grant, Lieut.-Col. James A., C.B., C.S.I. *E. India U. S. Club, S.W.; and 7, Park-square, Regent's-park, N.W.*
1862	Grant, Lieut. J. M. (late 25th Reg.) *Elands Port, Cape of Good Hope.* Care of Messrs. Ridgway and Sons, 2, *Waterloo-place, S.W.*
1860	Grantham, Capt. James, R.E.
1867	Graves, Rev. John. *Underbarrow-parsonage, Milnthorpe, Westmoreland.*
1870	Gray, Charles W., Esq. 19, *Regent's-park-road, N.W.*
1830	900*Gray, John Edw., Esq., PH. DR., F.R.S., Z.S. and L.S. *British Museum, W.C.*
1868	Gray, Lieut.-Col. William, M.P. 26, *Prince's-gardens, W.; and Darcy Lever-hall, near Bolton.*
1862	Greathed, Lieut.-Colonel Wilberforce, W. H., C.B.
1863	Grenves, Rev. Richard W. 1, *Whitehall-gardens, W.*
1861	Green, Capt. Francis, 58th Regt. 89, *Eccleston-square, S.W.*
1871	Green, John Henry, Esq. 11, *Green-bank-terrace, Falmouth.*
1868	Green, Rev. W., M.A. *Chaplain to the Tower of London.*
1869	Green, Sir W. H. R., K.C.S.I., C.B. 36, *St. George's-road, Eccleston-square, S.W.*

Year of Election.	
1830	Greene, Thomas, Esq. *Whittington-hall, near Burton, Westmoreland.*
1857	*Greenfield, W. B., Esq. 59, *Porchester-terrace, Hyde-park, W.; and Union Club, S. W.*
1870	910 Greenup, W. Thomas, Esq. *Woodhead-road, London-road, Sheffield.*
1865	Greg, W. R., Esq., Comptroller of H.M.S. Stationery Office. *Wimbledon, S. W.*
1858	*Gregory, Augustus Charles, Esq. *Surveyor-General, Brisbane, Queensland, Australia.*
1858	Gregory, Charles Hutton, Esq., C.E. 1, *Delahay-street, Westminster, S. W.*
1860	*Gregory, Francis Thomas, Esq. *Queensland.*
1858	*Gregory, Isaac, Esq. *Merchants'-college, Blackpool.*
1857	*Grellet, Henry Robert, Esq. *Care of L. Mentzendorff, Esq., 87, Great Tower-street, E.C.*
1865	Grenfell, Henry R., Esq., M.P. 15, *St. James's-place, S. W.*
1858	Grenfell, Pascoe St. Leger, Esq. *Maesteg-house, Swansea.*
1853	Grenfell, Riversdale W., Esq. 27, *Upper Thames-street, E.C.*
1830	920 *Greswell, Rev. Richard, M.A., F.R.S. 39, *St. Giles, Oxford.*
1866	Grey, Charles, Esq. 13, *Carlton-house-terrace, S. W.*
1837	*Grey, Sir George, K.C.B. *Colonial Office; and Grosvenor-mansions, S. W.*
1864	Grierson, Charles, Esq. *Alexandria. Care of Rev. W. Grierson Smith, Ashkirk-by-Hawick, Scotland.*
1868	Griffin, Daniel, Esq. *Guildford-place, W.C.*
1862	Griffin, James, Esq. 2, *Eastern-parade, Southsea; and The Hard, Portsea, Hants.*
1861	*Griffith, Daniel Clewin, Esq. 20, *Gower-street, W.C.*
1839	Griffith, John, Esq. 16, *Finsbury-place-south, E.C.*
1863	Griffith, Sir Richard. 20, *Eccleston-square, S. W.*
1836	Griffith, Richard Clewin, Esq. 20, *Gower-street, W.C.*
1867	930 Griffiths, Captain A. G. F., 63rd Reg. (Major of Brigade, Gibraltar). *St. Mary's-vale, Chatham.*
1869	Griffiths, William, Esq., J.P. 24, *Great Cumberland-place, W.; and The Welkin, Lindfield, Sussex.*
1855	Grindrod, R. B., Esq., M.D., LL.D., F.L.S., &c. *Townsend-house, Malvern.*
1861	Grosvenor, Lord Richard, M.P. 33, *Upper Grosvenor-street, W.*
1858	Grote, George, Esq. 12, *Savile-row, W.*
1857	Gruneisen, Charles Lewis, Esq. 16, *Surrey-street, Strand, W.C.*
1861	Gunnell, Captain Edmund H., R.N. *Army and Navy Club, S. W.; and 21, Argyll-road, Campden-hill, W.*
1859	*Gurney, John H., Esq. *Marldon, Totnes.*
1857	Gurney, Samuel, Esq. 20, *Hanover-terrace, Regent's-park, W.*
1862	Guthrie, James Alexander, Esq. 30, *Portland-place, W.*
1865	940 Gwyther, John H., Esq. *Meadowcroft, Lower Sydenham, S.E.*

Year of Election	
1870	Habicht, Claudius Edward, Esq. *Garrick Club, W.C.*
1863	Hadfield, Wm., Esq. 11, *Inverness-road, W.*
1865	Hadley, Henry, Esq., M.D. *Needwood-lodge, Bay's-hill, Cheltenham.*
1863	Hadow, P. D., Esq. *Sudbury-priory, Middlesex.*
1865	Halcombe, Rev. J. J. *Charter-house, E.C.*
1868	Hale, Rev. Edward, M.A. *Eton College; and United University Club, S.W.*
·1835	Hale, Warren S., Esq., Alderman. 71, *Queen-street, Cheapside, E.C.*
1860	Haliday, Lieut.-Colonel William Robert. *United Service Club, S.W.*
1853	Halifax, Viscount, G.C.B. 10, *Belgrave-sq., S.W.; and Hickleton, Yorkshire.*
1853	950*Halkett, Rev. Dunbar S. *Little Bookham, Surrey.*
·1853	*Halkett, Lieut. Peter A., R.N.
1861	Hall, Charles Hall, Esq. *Watergate-house, Emsworth.*
1863	Hall, Henry, Esq. 109, *Victoria-street, S.W.*
1862	Hall, James Febbutt, Esq. *Fore-street, Limehouse, E.*
1869	*Hall, James MacAlester, Esq. 15, *Woodside-crescent, Glasgow.*
1863	Hall, Thomas F., Esq., F.C.S. 29, *Warwick-square, S.W.*
1853	Hall, Admiral Sir William Hutcheson, K.C.B., F.R.S. *United Service Club, S.W.; and 48, Phillimore-gardens, Kensington, W.*
1865	Hallett, Lieut. Francis C. H., R.H.A.
1858	Halloran, Arthur B., Esq. 3, *Albert-terrace, St. Leonard's, Exeter.*
1871	960*Hamilton, Andrew, Esq., Lieut. 102nd Regiment *The House of Falkland, Fyfe; and Naval and Military Club, W.*
1862	Hamilton, Archibald, Esq. *South Barrow, Bromley, Kent, S.E.*
1866	Hamilton, Rear-Admiral C. Baillie. *Care of Messrs. Walford, Bros., 320, Strand, W.C.; and 50, Warwick-square, S.W.*
1861	Hamilton, Lord Claude, M.P. 19, *Eaton-square, S.W.; and Barons-court, County Tyrone.*
1830	*Hamilton, Capt. Henry G., R.N. 71, *Eccleston-square, S.W.*
1869	Hamilton, Capt. Richard Vesey, R.N. *H.M.S. 'Achilles,' Portland.*
1861	Hamilton, Col. Robert William, Grenadier Guards. 18, *Eccleston-square, S.W.*
1863	Hamilton, R., Esq. *Care of J. Forster Hamilton, Esq., Oriental Club, W.*
1830	Hamilton, Terrick, Esq. 121, *Park-street, Grosvenor-square, W.*
1846	Hamilton, Rear-Admiral W. A. Baillie. *Macartney-house, Blackheath, S.E.*
1853	970*Hand, Admiral George S., C.B. *U. S. Club, S.W.; and H.M.S. 'Victory.'*
1860	*Handley, Benjamin, Esq. *Chandos-road, Stratford; and Grafton Club, Grafton-street, W.*
1866	Hanham, Commr. T. B., R.N. *Manston-house, near Blandford, Dorset.*
1861	*Hankey, Blake Alexander, Esq.
1870	*Hankey, Rodolph Alexander, Esq. 9, *Suffolk-place, Pall-mall, S.W.*
1857	Hankey, Thomson, Esq. 45, *Portland-place, W.*
1837	*Hanmer, Sir J., Bart., M.P., F.R.S. *Hanmer-hall and Bettisfield-park, Flintshire.*
1839	*Hansard, Henry, Esq. 13, *Great Queen-street, W.C.*
1870	Harbord, John B. Esq., M.A., Chaplain R.N. 69, *Victoria-park-road, E.*

Year of Election	
1840	*Harcourt, Egerton V., Esq. *Whitwell-hall, York.*
1864	980*Hardie, Gavin, Esq. 113, *Piccadilly, W.*
1864	Harding, Captain Charles, F.R.S.L., F.S.S., F.A.S.L. *Grafton Club,* 10, *Grafton-street, Piccadilly, W.*
1864	Harding, J. J., Esq. 1, *Barnsbury-park, Islington, N.*
1864	Hardinge, Capt. E., R.N. 32, *Hyde-park-square, W.*
1861	Hardinge, Henry, Esq., M.D. 18, *Grafton-street, Bond-street, W.*
1862	Hardman, William, Esq., M.A. *Norbiton-hall, Kingston-on-Thames.*
1864	Hardwick, B. Esq. 157, *Fenchurch-street, E.C.*
1865	Harper, J. A. W., Esq. 23, *Grosvenor-road, Pimlico, S.W.; and Lloyd's, E.C.*
1853	Harris, Admiral the Hon. E. A. J., C.B. *H.B.M.'s Envoy Extraordinary and Minister Plenipotentiary, Legation Britannique, Berne. Messrs. Woodhead.*
1869	Harris, Lieut. G. F., 20th Regiment. *Care of Colonel Harris,* 28, *Leinster-road, Dublin.*
1850	990 Harris, Capt. Henry, H.C.S. 35, *Gloucester-terrace, Hyde-park, W.*
1866	Harris, John, Esq. 31, *Belsize-park, Hampstead, N.W.*
1865	Harris, John M., Esq.
1863	Harrison, Chas., Esq. 3, *Great Tower-st., E.C.*
1870	Harrison, Charles, Esq. 10, *Lancaster-gate, W.*
1860	· Harrison, Capt. Thos. A. John, R.A. *Lansdowne-road, Charlton, S.E.*
1865	*Harrison, William, Esq., F.S.A., F.G.S., &c. *Conservative Club, S.W.; Royal Thames Yacht Club,* 7, *Albemarle-street, W.; and Samlesbury-hall, near Preston, Lancashire.*
1868	Harrowby, Dudley, Earl of. *Sandon-ho., Lichfield; and Norton, Gloucestershire.*
1868	*Hart, J. L., Esq. 20, *Pembridge-square, W.*
1864	*Hartland, F. Dixon, Esq., F.S.A., &c. 14, *Chesham-place, S.W.; and the Oak-lands, near Cheltenham.*
1868	1000Harvey, Charles, Esq. *Rathgar-cottage, Streatham, S.*
1864	Harvey, C. H., Esq., M.D. 17, *Whitehall-place, S.W.*
1865	Harvey, Edward N., Esq. *Springfield, near Ryde, Isle of Wight; and Carlton Club, S.W.*
1862	Harvey, James, Esq. (Solicitor). *Esk-street, Invercargill, Southland, New Zealand. Care of the Bank of Otago, Old Broad-street, E.C.*
1864	Harvey, John, Esq. *Ickwell Bury, Biggleswade.*
1864	Harvey, John, Esq. 7, *Mincing-lane, E.C.*
1862	Harvey, John, Esq., LL.D. *Château Deslyons, Boulogne-sur-Mer. Care of Capt. Felix Jones, Fernside, Church-road, Westow-hill, Upper Norwood, S.*
1858	Harvey, Richard M., Esq. 13, *Devonshire-street, Portland-place, W.*
1864	Harvey, W. D., Esq. *Holbrooke-house, Richmond.*
1868	Hawker, Edward J., Esq. 37, *Cadogan-place, S.W.*
1854	1010Hamilton, Francis Bisset, Esq., M.D., F.R.S. 146, *Upper Harley-street, W.; and Lytlet-lodge, Dorchester.*

	hill, N.W.
1863	*Hay, Lord John, M.P. 15, *Cromwell-road, South Kensington, W.*
1865	Hay, Lord William. 3, *Cleveland-row, S.W.*
1859	Hay, Major W. E. 7, *Westminster-chambers, Victoria-road, S.W.; and Garrick Club, Garrick-street, W.C.*
1870	Haynes, Stanley L., Esq., M.D. 6, *Trinity-crescent, Edinburgh.*
1868	Haysman, David, Esq. *Portway-house, Weston, Bath.*
1864	1020 Haysman, James, Esq.
1862	Head, Alfred, Esq. 13, *Craven-hill-gardens, Bayswater, W.*
1863	Headlam, Right Hon. Thomas E., M.P. 27, *Ashley-place, Victoria-street, S.W.*
1856	Heath, J. Benj., Esq., F.R.S., F.S.A. 31, *Old Jewry, E.C.*
1863	Heathfield, W. E., Esq. *Arthur's Club, S.W.*
1861	Hector, Alexander, Esq. 6, *Stanley-gardens, Bayswater, W.*
1861	Hector, James, Esq., M.D. *Care of E. Stanford, Esq.*
1862	Hemans, Geo. Willoughby, Esq., C.E. *Westminster-chambers, Victoria-street, S.W.*
1870	Henderson, David Mitchell, Esq. 1, *Carden-place, Aberdeen; and Old Calabar, W. Africa.*
1853	Henderson, John, Esq. 2, *Arlington-street, Piccadilly, W.*
1866	1030 Henderson, Patrick, Esq. *Care of George Reid, Esq., 21, Abchurch-lane, E.C.*
1864	Henderson, R., Esq. 7, *Mincing-lane, E.C.*
1832	Henderson, William, Esq. 5, *Stanhope-street, Hyde-park-gardens, W.*
1844	*Heneage, Edward, Esq. *Stag's-end, Hemel Hempstead.*
1861	Henn, Rev. J., B.A., Head Master of the Manchester Commercial Schools. *Old Trafford, Manchester.*
1860	Hennessey, J. B. N., Esq. 1st Asst. Trig. Survey of India, Dehra in the Dhoon, N.W. Provinces, India. *Care of Messrs. H. S. King and Co.*
1838	*Henry, Wm. Chas., Esq., M.D., F.R.S. *Haffield, near Ledbury, Herefordshire.*
1861	*Henty, Douglas, Esq. *Chichester.*
1870	Hepworth, Campbell, Esq. 2, *St. James's-square, Cheltenham.*
1857	Herd, Captain D. J. 2, *Norway-house, Limehouse, E.*
1858	1040 Hertslet, Edward, Esq. *Librarian, Foreign-office, S.W.; and Belle-vue-house, Richmond, S.W.*
1861	Hough, John, Esq. *Tunbridge-wells.*
1840	*Heywood, James, Esq., F.R.S. *Athenæum Club, S.W.; and 26, Kensington-palace-gardens, W.*
1869	Heywood, Samuel, Esq. 171, *Stanhope-street, Hampstead-road, N.W.*
1860	Heyworth, Capt. Lawrence, 4th Royal Lancashire. *Junior United Service Club, S.W.*

Year of Election	
1853	Hickey, Edwin A., Esq. 116, *Piccadilly, W.*
1867	Higgins, Edmund Thomas, Esq., M.R.C.S. 122, *King Henry's-road, Haverstock-hill, N.*
1868	Hiley, Rev. W., M.A. 3, *Cambridge-gardens, Richmond-hill, S.W.*
1856	Hill, Arthur Bowdler, Esq. *South-road, Clapham-park, Surrey, S.*
1866	Hill, Berkeley, Esq. 14, *Weymouth-street, Portland-place, W.*
1867	1050 Hill, O'Dell Travers, Esq. *Clarence-cottage, Surrey-lane-south, Battersea, S.W.*
1854	Hill, Lieut.-Colonel Stephen J., Governor of Antigua. *Army and Navy Club, S.W. Care of Capt. E. Barnett, R.N.,* 14, *Woburn-square, W.*
1861	Hilliard, Lieut.-Colonel George Towers, Madras Staff Corps. 10, *Harbury-crescent, Notting-hill, W.*
1870	Hilliard, R. Harvey, Esq., M.D. 258, *Kingsland-road, N.; and Kinsembo, W. Coast of Africa.*
1856	Hinchliff, T. Woodbine, Esq., Barrister-at-Law. 64, *Lincoln's-inn-fields, W.C.*
1852	*Hinde, Samuel Henry, Esq. *Windham Club, S.W.*
1846	*Hindmarsh, Frederick, Esq. 4, *New-inn, Strand, W.C.*
1870	Hitchins, T. M., Esq., Lieut. R.A. *Sandown, Isle of Wight.*
1861	Hoare, Deane John, Esq. *Royal Thames Yacht Club, Albemarle-street, W.*
1866	Hoare, Samuel, Esq., M.A. 1, *Upper Hyde-park-street, W.*
1866	1060 Hobson, Stephen James, Esq. 32, *Nicholas-lane, Lombard-street; and* 10, *Regent's-park-road, N.W.*
1866	Hobbs, Wm. Geo. Ed., Esq. *Beulah-cottage, London-road, Enfield, N.*
1869	Hodgson, Henry, Esq. *Bondesbury-lodge Collegiate-school, Kilburn.*
1846	*Hodgson, Arthur, Esq., Superintendent of the Australian Agricultural Company.*
1861	*Hodgson, James Stewart, Esq. 8, *Bishopsgate-street, E.C.*
1857	Hodgson, Kirkman Daniel, Esq. 8, *Bishopsgate-street, E.C.*
1852	Hodgson, William H., Esq. *Treasury-chambers; and* 1, *Whitehall-gardens, S.W.*
1866	Hogg, James, Esq. 217, *Piccadilly, W.*
1866	Holdich, Thos. Hungerford, Esq., Lt. R.E.
1866	Hale, Rev. Charles. *Loughborough-house-school, East Brixton, S.*
1869	1070 *Holford, Robert S., Esq. *Dorchester-house, Park-lane, W.*
1867	Holland, Rev. Fred. Whitmore. 38, *Bryanstone-street, W.*
1866	Holland, Sir Henry, Bart., M.D., F.R.S. 25, *Lower Brook-street, W.*
1853	Holland, Colonel James. *Southside, The Park, Upper Norwood, S.E.*
1866	Holland, Leton, Esq. 6, *Queen's-villas, Windsor.*
1866	Holland, Robert, Esq. *Stanmore-hall, Great Stanmore, Middlesex.*
1866	Holland, Major T. J., C.B. *Topo. Dept., War-office; The Park, Upper Norwood; East India U.S. Club; and Club of Western India, Poona.*
1861	*Hollingsworth, John, Esq., M.R.C.S. *Maidenstone-house, Greenwich, S.E.*
1853	Holmes, J. Wilson, Esq., M.A. *Downwood, Beckenham, Kent, S.E.*
1866	Holmes, James, Esq. 4, *New Ormond-street, Queen-square, W.C.*
1869	*Holroyd, Arthur Todd, Esq., M.D., F.L.S. *Master's-office, Sydney, New South. Care of Edgar Howell, Esq.,* 3, *St. Paul's-churchyard, E.C.*
1867	Holroyd, Edward, Esq., Barrister-at-Law. 2, *Elm-court, Temple, E.C.*

Year of Election	
1867	*Holstein, The Marques de Souza. *Lisbon. Care of Messrs. Kraentler and Mieville, 12, Angel-court, E.C.*
1869	Holt, George, Esq. *Union-street, Willenhall.*
1864	Holt, Vesey, Esq. *17, Whitehall-place, S.W.*
1857	Homfray, William Henry, Esq. *6, Storey's-gate, S.W.*
1865	Honywood, Robert, Esq. *Manor-house, Wethersfield, Braintree: and Windham Club, S.W.*
1864	*Hood, Sir Alex. Acland, Bart. *St. Andrie's-park, Bridgewater, Somerset.*
1862	Hood, Henry Schuback, Esq. *War-office, S.W.; and 10, Kensington-park-gardens, W.*
1861	Hood, T. H. Cockburn, Esq. *Stoneridge, Berwickshire.*
1859	1090*Hood, William Charles, Esq., M.D. *Bethlehem-hospital, S.*
1866	*Hooker, Joseph, Esq., M.D., F.R.S., F.L.S., &c. *Director of the Royal Gardens, Kew.*
1868	Hooper, Alf., Esq. *City of London Club, Old Broad-street, E.C.*
1870	Hooper, George Norgate, Esq. *139, King Henry's-road, Adelaide-road, N.W.*
1870	Hooper, Rev. Robert Poole. *29, Cambridge-street, Brighton.*
1861	Hopcraft, George, Esq. *3, Billiter-square, E.C.*
1846	*Hope, Alex. James Beresford, Esq. *Arklow-house, Connaught-place, Hyde-park, W.; and Bedgebury-park, Hurst-green, Kent.*
1862	Hope, Capt. C. Webley, R.N. *H.M.S. 'Brisk,' Australia; Messrs. Hallett & Co.*
1869	Hopkins, Capt. David. *New Calabar, near Bonny, W. Africa. Care of the Company of African Merchants, 6, Water-street, Liverpool.*
1870	*Hopkins, Edward M., Esq. *66, Great Cumberland-place, Hyde-park, W.*
1869	1100Horne, Charles, Esq., H.M. Ind. Civ. Serv. *"Innisfail," Beulah-hill, Upper Norwood.*
1869	Horrex, Theophilus, Esq. *18, Connaught-square, Hyde-park, W.*
1868	Horton, James Africanus B., Esq., M.D., &c. *Care of Sir John Kirkland, 17, Whitehall-place, S.W.*
1870	Hosenson, Capt. John C., R.N. *United Service Club, S.W.*
1861	Hoskins, Capt. A. H., R.N. *Army and Navy Club, S.W. Care of Messrs. Woodhead.*
1859	Hoskyns, Chandos Wren, Esq. *Wraxhall-abbey, Warwickshire.*
1853	Houghton, Lord. *Travellers' Club, S.W.; The Hall, Bawtry; and Fryston-hall, Ferrybridge, Yorkshire.*
1856	Hovell, William Hilton, Esq. *Goulburn, New South Wales. Care of Mr. W. Chamberlin, 74, Fleet-street, E.C.*
1869	Howard, John, Esq., C.E. *Exmouth, Devon.*
1853	Howard, Sir Ralph, Bart. *17, Belgrave-sq., S.W.; and Bushy-park, Wicklow.*
1857	Howard, Samuel Lloyd, Esq. *Goldings, Loughton, Essex.*
1864	1110Howell, W. G., Esq.
1842	*Hubbard, J. Gellibrand, Esq. *24, Prince's-gate, Hyde-park, S.W.*
1867	*Hubbard, William Egerton, Esq. *St. Leonard's-lodge, Horsham.*
1867	*Hubbard, William Egerton, Esq., jun., R.A. *St. Leonard's-lodge, Horsham.*
1870	Hudson, George B., Esq. *Imperial Club, Chancery-lane, W.C.*

Date of Election	
1857	Hughes, Capt. Sir Frederic. *Ely-house, Wexford.*
1836	Hughes, William, Esq. 4, *Lawford-road, Kentish-town, N.W.*
1828	*Hume, Edmund Kent, Esq.
1860	*Hume, Hamilton, Esq. *Cooma Yass, New South Wales. Care of Rev. A. Hume, 24, Fitzclarence-street, Liverpool.*
1861	1120Hunt, George S. Lennox, Esq., H.B.M. Consul, *Rio de Janeiro.*
1868	Hunt, John Percival, Esq., M.D. *Great Ouseburn, near York, Yorkshire.*
1866	Hunt, Joseph, Esq. *Cave-house, Uxbridge, Middlesex.*
1865	Hunt, Capt. Thomas, R.H.A. *The Barracks, Maidstone.*
1857	Hunt, Zacharias Daniel, Esq. *Aylesbury.*
1866	Hunter, Major Edward. *Junior United Service Club, S.W.*
1862	Hunter, Henry Lannoy, Esq. *Beech-hill, Reading.*
1870	Hutchins, Edward, Esq. 10, *Portland-place, W.*
1864	Hutchinson, Capt. R. R. *Parkville-house, Forest-hill.*
1868	Hutchinson, Thomas J., Esq., F.R.S.L., F.R.S., F.A.S.L., H.B.M. Consul, *Rosario, Argentine Republic. Care of Fras. O'Brien, Esq., 43, Parliament-st., S.W.*
1870	1130*Hutton, Charles W. C., Esq. *Belair, Dulwich, S.*
1869	Huxley, Thomas H., Esq., F.R.S., &c. 26, *Abbey-place, St. John's-wood, N.W.*
1860	*Hyde, Captain Samuel. 8, *Billiter-square, E.C.*
1866	Illingworth, Rev. Edward A. 3, *Mecklenburg-street, W.C.*
1853	Illingworth, Richard Stonhewer, Esq. 9, *Norfolk-crescent, Hyde-park, W.*
1830	*Imray, James Frederick, Esq. 89, *Minories, E.; and Beckenham, Kent, S.E.*
1867	Ince, Joseph, Esq., F.L.S., &c., &c. 26, *St. George's-place, Hyde-park-corner, W.*
1862	*Ingall, Samuel, Esq. *Forest-hill, Kent, S.E.*
1851	Inglefield, Admiral Edward A., C.B., F.R.S. *United Service Club, S.W.; and* 10, *Grove-end-road, St. John's-wood, N.W.*
1866	Ingram, Hughes Francis, Esq. *University Club, S.W.*
1869	1140Inman, Robert Matthew, Esq., M.D. *Edinburgh-house, West-street, Brighton.*
1860	*Inskip, Staff Commander G. H., R.N. *H.M. Surveying Vessel 'Porcupine;' and* 6, *Park-place-west, Sunderland.*
1852	*Inskip, Rev. Robert Mills. 8, *Boon's-place, Plymouth.*
1868	*Irby, Frederick W., Esq. *Athenæum Club, S.W.*
1870	Irvine, James, Esq. 18, *Devonshire-road, Claughton, Cheshire.*
1864	*Irving, John, Esq.
1868	Irving, Thomas, Esq.
1861	Irwin, James V. H., Esq. 10, *Nottingham-place, Euston-road, N.*

Year of Election	
1866	Jackson, Robert Ward, Esq. 28, *Inverness-road, Hyde-park, W.*
1855	Jackson, William, Esq. 44, *Portland-place, W.*
1862	1150 Jacomb, Thomas, jun., Esq. 23, *Old Broad-street, Gresham-house, E.C.*
1870	James, William Morris, Esq. 8, *Lyndhurst-road, Hampstead, N.W.*
1857	James, Colonel Sir Henry, R.E., F.R.S. *Director of the Ordnance Survey, Southampton.*
1861	James, William Bosville, Esq. 13, *Blomfield-road, Maida-hill, W.*
1868	Jamieson, Robert Alexander, Esq., M.A. *Shanghai. Care of J. P. Watson, Esq., 85, Gracechurch-street, E.C.*
1868	Jamieson, Hugh, Esq. *Junior Carlton Club, S.W.*
1862	*Jaques, Leonard, Esq. *Wentbridge-house, Pontefract, Yorkshire.*
1863	*Jardine, Andrew, Esq. *Lanrick-castle, Stirling.*
1863	*Jardine, Robert, Esq., M.P. *Castlemilk, Lockerby, N.B.*
1857	Jefferson, Richard, Esq. A4, *The Albany, W.*
1865	1160 Jeffreys, J. G., Esq. 25, *Devonshire-place, W.*
1850	*Jejeebhoy, Sir Jamsetjee, Bart. *Bombay.*
1854	Jellicoe, Charles, Esq. 12, *Cavendish-place, W.*
1859	Jencken, H. Diedrich, Esq. *Goldsmith's-buildings, Temple, E.C.; and 2, York-terrace, Upper Sydenham, S.E.*
1854	Jenkins, Capt. Griffith, I.N., C.B. *East India Club, St. James's-square, S.W.; and Derwen, Welch Pool, Montgomeryshire.*
1837	*Jenkins, R. Castle, Esq. *Beachley, near Chepstow.*
1854	*Jennings, William, Esq., M.A. 13, *Victoria-street, Westminster, S.W.*
1860	Jermyn, Rowland Formby, Esq. *War-office, S.W.*
1850	Jessopp, Rev. Augustus, M.A., Head Master, King Edward VI. School. *Norwich.*
1870	Jessop, Captain Thomas. 37, *Clarges-street, Piccadilly, W.*
1864	1170 *Jeula, Henry, Esq. *Lloyd's, E.C.*
1847	Johnson, Edmund Chas., Esq. 4, *Eaton-place, Belgrave-square, S.W.*
1859	*Johnson, Henry, Esq. *Messrs. Johnson, 7, Bedford-row, Worthing, Sussex.*
1854	Johnson, John Hugh, Esq.
1870	Johnson, T. Scarboro, Esq. 42, *Gloucester-place, Hyde-park, W.*
1866	Johnson, W. H., Esq., Civil Assistant G. T. S. India. *Sealkote, Punjaub.*
1843	Johnston, Alex. Keith, Esq., F.R.S.E., Hon. Mem. Berl. Geog. Soc., &c. *Marchhall-park; and 4, St. Andrew-square, Edinburgh.*
1868	*Johnston, Alexander Keith, Esq., jun. 74, *Strand, W.C.*
1856	Johnston, A. R., Esq., F.R.S. *Heatherley, Sandhurst, near Wokingham, Berks.*
1857	Johnston, J. Brookes, Esq. 29, *Lombard-street, E.C.*
1868	1180 Johnston, Thomas, Esq. 12, *Belvedere, Bath; and King Edward VI. Grammar-school, Bath.*
1866	Johnstone, Colonel H. C. *Murree, Punjaub, India. Care of Messrs. H. S. King and Co., Cornhill, E.C.*
1867	*Johnstone, John, Esq. *Castlenau-house, Mortlake, S.W.*
1858	Jones, Capt. Edward Monckton, 20th Regt. 16, *The Terrace, York-town, Farnborough-station.*

1864	Jones, Captain Felix, late I.N. *Fernside, Church-road, Westow-hill, Upper Norwood, S.*
1868	Jones, Capt. H. M., V.C. *Care of the Foreign-office, S.W.*
1857	Jones, Lt.-Colonel Jenkin, Royal Engineers. 1, *Lennard-place, Circus-road, St. John's-wood, N.W.; and India.*
1862	Jones, John, Esq. 338, *Strand, W.C.*
1861	Jones, John Pryce, Esq. *Grove-park-school, Wrexham.*
1861	Jones, Sir Willoughby, Bart. *Cranmer-hall, Fakenham, Norfolk.*
1862	1190 Jones, William S., Esq. 2, *Verulam-buildings, Gray's-inn, W.C.*
1867	*Jordan, Wm. Leighton, Esq. 1, *Powis-square, Notting-hill, W.*
1868	Joshua, Moss, Esq. *Melbourne.*
1868	Kantzow, Capt. H. P. de, R.N. *United Service Club, S.W.*
1858	Kay, David, Esq. 19, *Upper Phillimore-place, Kensington, W.*
1865	Kaye, J. W., Esq. *India-office, S.W.*
1860	Keate, R. W., Esq., Lieutenant-Governor, *Trinidad.*
1857	Keating, Sir Henry Singer, Q.C., one of the Judges of the Court of Common Pleas. 11, *Prince's-gardens, S.W.*
1857	Keene, Rev. C. E. Ruck. *Swynscombe-park, Henley-upon-Thames.*
1863	Keir, Simon, Esq. *Conservative Club, S.W.*
1845	1200 *Kellett, Rr.-Adm. Sir Henry, K.C.B. *Clonmel, Ireland.*
1861	Kelly, William, Esq. *Royal Thames Yacht Club, 7, Albemarle-street, W.*
1866	*Kemball, Col. Sir Arnold Burrowes, C.B., Indian Army. *United Service Club, S.W.*
1868	Kempster, J., Esq. 1, *Portsmouth-place, Kennington-lane, Surrey, S.*
1870	*Kenlis, Lord. 85, *Dover-street, W.; and Underley-hall, Kirkby Lonsdale, Westmoreland.*
1861	Kennard, Adam Steinmetz, Esq. 7, *Fenchurch-street, E.C.*
1860	Kennard, Coleridge J., Esq. 44, *Lombard-street, E.C.*
1861	Kennedy, Edward Shirley, Esq.
1864	Kennedy, Rev. John, M.A. 4, *Stepney-green, E.*
1866	Kerr, J. H., Esq., Staff-Commr. R.N. *Hydrographic-office, S.W.*
1864	1220 Kerr, Lord Schomberg. 15, *Bruton-street, W.*
1858	Kershaw, Wm., Esq. 16, *St. Mary Axe, E.C.; and Suffolk-lodge, Brixton-road, S.*
1862	Key, J. Binney, Esq. *Oriental Club, W.*
1867	Keysall, Francis P., Esq. *Sycamore-villa, 35, Carlton-hill, St. John's-wood, N.W.*
1864	*Kiddle, W. W., Esq. *Linton-villa, Clarendon-road, Southsea.*
1864	Kimber, Dr. E. *Murchison-house, Dulwich, S.E.*
1845	King, Lieut.-Colonel Edward R., 36th Regt. *Junior United Service Club, S.W.*
1859	King, Henry S., Esq. J.P. 65, *Cornhill, E.C.; 45, Pall Mall, S.W.; Manor House, Chigwell, Essex; and Junior Carlton Club, S.W.*
1863	King, John, Esq. *Compton-field-place, Guildford, Surrey.*

Year of Election.	
1861	King, Lieut.-Col. W. Ross, Unatt., F.S.A. Scot. *Tertowie, Kinellar, Aberdeenshire; and Army and Navy Club, S.W.*
1868	1220Kingsley, Henry, Esq. *Wargrave, Henley-on-Thames, Berks; and Garrick Club, W.C.*
1857	*Kinnaird, Hon. Arthur F., M.P. 2, *Pall-mall-east, S.W.*
1867	Kinnaird, George William Fox, Lord, K.G. *Rossie-priory, Inchture, N.B.; and 33, Grosvenor-street, W.*
1860	Kinns, Samuel, Esq., PH. DR., F.R.A.S. *Highbury-new-park College, N.*
1858	Kirk, John, Esq., M.D. *Care of J. F. Rogers, Esq., 25, South-castle-street, Edinburgh.*
1863	Kirke, John, Esq., Barrister. *C. Thorold, Esq., Welham, Retford, Notts.*
1870	Kirkland, Major-Gen. John A. Vesey. *Beckenham-hill, Kent; and 17, Whitehall-place, S.W.*
1868	Kisch, Daniel Montagu, Esq. *1, Devonshire-place, Seven Sisters'-road, Upper Holloway, N.*
1866	*Kitson, James, jun., Esq. *Hanover-square, Leeds.*
1868	Kitto, Richard L. Middleton, Esq. *Church-hill-villa, Fryerstown, Victoria, Australia.*
1835	1230*Kjaer, Thomas Andreas, Esq. *Hjornet af Kongins Nystow og Gatheragaden, No. 26, 3d Sahl, Copenhagen.*
1867	Knight, Andrew Halley, Esq. *Care of R. Philpott, Esq., 3, Abchurch-lane, E.C.*
1862	Knollys, Lieut.-General Sir William T., K.C.B., V.-Pres. Council of Military Education. *Eaton-square, S.W.*
1867	Knox, Alex. A., Esq. 91, *Victoria-street, Westminster, S.W.*
1861	Knox, Thomas G., Esq. *India. Care of Messrs. H. S. King and Co., 45 Pall-mall, S.W.*
1866	Kopsch, Henry, Esq. *Custom-house, Shanghai. Care of H. C. Balchelor, Esq., 155, Cannon-street, E.C.*
1861	Kyd, Hayes, Esq., M.R.C.S. *Wadebridge, Cornwall.*
1859	Labrow, Lieut.-Colonel Valentine H., F.S.A., F.G.S. *Mitre-court-chambers, Temple, E.C.; and Club-chambers, S.W.*
1870	Lackersteen, Mark Henry, Esq., M.D., &c. 28, *St. Stephen's-rd., Westbourne-pk., W.*
1849	*Laffan, Capt. Robert Michael, R.E. *Army and Navy Club, S.W.; and Otham-lodge, Kent.*
1870	1240Laing, Arthur, Esq. 18, *Kensington-gardens-square, Hyde-park, W.*
1869	Lamb, Hon. Edward William. *Brisbane, Queensland, Australia.*
1859	Lamb, Lieut. Henry, I.N. *H.M. India Store Department, Belvedere-road, Lambeth, S.*
1863	*Lambert, Alan, Esq. *Heath-lodge, Putney-heath, S.W.*
1864	Lambert, Charles, Esq. 2, *Queen-street-place, Upper Thames-street, E.C.*
1867	Lambert, Wm. Blake, Esq., C.E. 3. *Morden-road. Blackheath, S.E.*

Year of election	
1861	Lamont, James, Esq. *Knockdow, Greenock, N.B.*
1870	Lamplough, Charles Edward, Esq. *City of London Club, E.C.*
1866	Lamprey, John, Esq. 16, *Camden-square, N.W.*
1867	1250 Lamprey, Jones, Esq., M.B., Surgeon-Major 67th Regt.
1864	Lampson, Sir C. M., Bt. 64, *Queen-street, Cheapside, E.C.*
1866	*Lance, John Henry, Esq., F.L.S. *The Holmwood, Dorking.*
1868	Lane, Rev. W. W., B.A. 13, *Merchant-street, Bow.*
1861	*Lang, Andrew, Esq. *Dunmore, Hunter-river, New South Wales; and Dunmore, Teignmouth, Devon.*
1859	*Lange, Sir Daniel A. 21, *Regent-street, W.*
1867	Langlands, John, Esq., Engineer. *Melbourne, Australia.*
1865	Langley, Edward, Esq. *Well-hall, Eltham, Kent.*
1866	*Langler, John R., Esq., B.A. *Wesleyan Training College, Westminster; and Gothic-villas, 2, Bridge-road-west, Battersea, S.W.*
1870	Lanyon, Charles, Esq. 3, *Paper-buildings, Temple, E.C.*
1856	1260 *Larcom, Maj.-General Sir Thomas Aiskew, R.R., K.C.B., F.R.S. *Heathfield, Fareham, Hants.*
1861	Lardner, Col. John. *United Service Club, S.W.*
1860	Larnach, Donald, Esq. 21, *Kensington-palace-gardens, W.*
1870	Lassetter, Frederic, Esq. *Sydney, New South Wales; and 3, Belsize-park, N.W.*
1864	Latrobe, Ch. J., Esq. *Clapham-house, Lewes, Sussex.*
1870	Laughton, Lieut.-Col. George Arnold (Bombay Staff Corps). *Superintendent Bombay Survey, Bombay.*
1869	Laughton, J. K., Esq. *Denton-house, Victoria-road, Southsea; and Royal Naval College, Portsmouth.*
1870	Law, Colonel Charles Edmund. 16, *Chester-street, Belgrave-square, S.W.; and Holly Spring, Bracknell, Reading.*
1866	*Law, Hon. H. Spencer, M.A. 40, *Eaton-place, S.W.*
1870	Lawrence, Alexander, Esq. *Clyde-house, Thurlow-road, Hampstead; and Windsor-chambers, Great St. Helen's, E.C.*
1861L	1270 Lawrence, Edward, Esq. *Beechmont, Aigburth, Liverpool.*
1860	*Lawrence, Philip Henry, Esq. 12, *Whitehall-place, S.W.*
1870	Lawrence, Lord, G.C.B. 26, *Queen's-gate, W.*
1866	Lawrie, James Esq., 63, *Old Broad-street, E.C.*
1867	Lawson, Wm., Esq. 21, *Walham-grove, Fulham, S.W.*
1869	*Lay, Horatio, N., Esq.
1857	Layard, Right Hon. Austen H., D.C.L. 130, *Piccadilly, W.*
1866	*Layard, Lieutenant Brownlow Villiers (3rd W. India Regt.). 38, *Upper Mount-street, Dublin; and Lane's-hotel, 1, St. Alban's-place, S.W.*
1866	Laybourne, Augustine, Esq. 9, *King-street, Finsbury-square, E.C.; and Loughton, Essex.*
1865	*Leaf, Chas. J., Esq. *Old-change, E.C.; and The Rylands, Norwood, S.*
1861	*Leach, T. H., Esq. *Burlington-lodge, Streatham-common, S.W.*

Year of Election	
1861	*Learmonth, Dr. John. 11, *Gloucester-gardens, Hyde-park*, W.
1866	Lebour, G. A., Esq. 28, *Jermyn-street*, S.W.
1853	*Le Breton, Francis, Esq. 21, *Sussex-place, Regent's-park*, N.W.
1868	Le Couteur, Lt.-Col. J. Halkett. 17, *Chapel-street, Belgrave-square*, S.W.
1861	Leckie, Patrick C., Esq. 7, *Palace-road, Roupell-park, Streatham*, S.
1870	Lecky, Squire Thornton Stratford, Esq., Lt. Royal Naval Reserve. 47, *Aubrey-street, Everton, Liverpool*.
1868	Lee, John, Esq. 62, *Loughborough-park*, S.
1839	Lee, Thomas, Esq. *Royal Institution, Albemarle-street*, W.
1869	*Lees, Lieutenant-Colonel Nassau, D.C.L. *Athenæum Club*, S.W.
1865	1290 Le Feuvre, W. H., Esq., C.E. 68, *Bedford-gardens, Kensington*, W.
1833	*Lefevre, Sir John George Shaw, M.A., D.C.L., F.R.S., Vice-Chancellor of the University of London. 18, *Spring-gardens*, S.W.
1853	Lefroy, General John Henry, R.A., F.R.S. 82, *Queen's-gate*, W.
1862	Leggatt, Clement Davidson, Esq. 43, *Inverness-terrace*, W.
1861	Legh, Wm. John, Esq. 37, *Lowndes-square*, S.W.; *and Lyme-park, Cheshire*.
1861	*Lehmann, Frederick, Esq. 139, *Westbourne-terrace*, W.
1845	Leigh, John Studdy, Esq., F.G.S. 8, *Old Jewry*, E.C.
1869	Leigh, Roger, Esq. *Barham-court; and Hindley-hall, Hindley*.
1863	Le Mesurier, Henry P., Esq., C.E. *St. Martin's, Guernsey*.
1863	Le Mesurier, M.-Gen. A. P. 2, *Stanhope-terrace, Hyde-park*, W.
1856	1300 Leslie, the Hon. G. W. 4, *Harley-street*, W.
1867	L'Estrange, Carleton, Esq. *Carlton Club*, S.W.
1840	*Letts, Thomas, Esq. 8, *Royal Exchange*, E.C.
1863	Leveaux, E. H., Esq. 25, *The Cedars, Putney*, S.W.
1857	Leverson, George B. C., Esq. 73, *Gloucester-terrace, Hyde-park*, W.
1869	Leveson, Edward J., Esq. *Cluny; Crescent-wood-road, Sydenham-hill*, S.E.
1862	Levick, Joseph, Esq. 8, *Great Winchester-street, Old Broad-street*, E.C.
1866	Levinge-Swift, Richard, Esq. *Levinge-lodge, Richmond, Surrey*.
1859	Levinsohn, Louis, Esq. *Vernon-house, Clarendon-gardens, Maida-hill*, W.
1865	Levy, William Hanks, Esq. *Institution of the Association for the Welfare of the Blind*, 210, *Oxford-street*, W.
1869	1310 *Lewin, Capt. Thomas (Beng. Staff Corps). *East India United Service Club*, S.W.
1852	Leycester, Captain Edmund M., R.N. 18, *Castlenau-villas, Barnes, Surrey*.
1861	Leyland, Luke Swallow, Esq. *The Leylands, Hatfield, Doncaster*.
1859	Lichfield, Thomas George, Earl of. *Shugborough, Staffordshire*.
1869	Ligar, C. W., Esq., Surveyor General of Victoria. 4, *Royal Exchange-avenue*, E.C.; *and Melbourne, Australia*.
1870	Light, Rev. John. 13, *Notting-hill-terrace*, W.
1856	Lilford, Thomas Lyttleton Powys, Lord. 10, *Grosvenor-place*, W.
1860	Lindsay, H. Hamilton, Esq.
1857	Lindsay, Major-General the Hon. J., Grenadier Guards, M.P. 20, *Portman-square*, W.

Year of Election	
1867	*Lindsay, Col. Robert J. L., M.P., V.C. Lockinge-house, Wantage, Berks; and 2, Carlton-gardens, S.W.*
1855	1320*Lindsay, Wm. S., Esq. Manor-house, Shepperton, Middlesex.*
1869	Lindsay, Mark John, Esq. 33, Ludgate-hill, E.C.; and Burnt-ash-lane, Lee, Kent.
1868	Linton, Robert P., Esq., Surg.-Major. 14, St. James's-square, S.W.
1856	Lister, John, Esq.
1866	Little, Archibald J., Esq. 71, Brook-street, Grosvenor-square, W.
1870	Littleton, the Hon. Henry S. Teddesley, Penkridge, Staffordshire.
1857	*Lloyd, George A., Esq. George-yard, Lombard-street, E.C.
1863	Lloyd, Sir Thomas Davis, Bart. United University Club, S.W.; and Bronwydd, Carmarthen.
1854	Lloyd, W., Esq. Moor-hall, near Sutton Coldfield.
1867	Lloyd, Rev. Wm. V., M.A. 16, Lancaster-gate, W.
1861	1330Llnellyn, Capt. Richard. 20, Montagu-square, W.
1869	Llnellyn, Captain William R., R.A. Army and Navy Club, S.W.
1868	Lobley, James L., Esq. 50, Lansdowne-road, Kensington-park, W.
1863	Loch, George, Esq. 12, Albemarle-street, W.
1850	Loch, Henry Brougham, Esq. Government-house, Isle of Man.
1861	Loch, John Charles, Esq. 12, Albemarle-street, W.; and Hong-Kong.
1857	Loch, William Adam, Esq. 8, Great George-street, Westminster, S.W.
1844	Locke, John, Esq. 83, Addison-road, Kensington, W.
1856	Lockhart, William, Esq., F.R.C.S. Park-villas, Granville-park, Blackheath, S.E.; and China.
1860	Lockwood, James Alfred. United Arts Club, Hanover-square, W.
1856	1340*Logan, Sir William Edmond, F.R.S. Montreal, Canada.
1860	Londesborough, Wm. Henry Forester, Lord. 3, Grosvenor-square, W.
1850	*Long, George, Esq., M.A. 32, Buckingham-street, Brighton.
1857	*Long, W. Beeston, Esq.
1868	Longden, Morrell D., Esq. 4, Ennismore-place, Hyde-park, S.W.
1865	*Langley, Major George, R.E. 60, Prince's-gate, W.
1847	Longman, Thos., Esq. Paternoster-row, E.C.; and 8, Sussex-sq., Hyde-park, W.
1859	Longman, William, Esq. 36, Hyde-park-square, W.
1870	*Longstaff, Capt. Llewellyn Wood. 12, Albion-street, Hull.
1851	Lonsdale, Arthur Pemberton, Esq.
1849	1350Looker, William Robert, Esq. Melbourne, Australia. Care of Mr. Ashhurst, 16, Bishopsgate-street-within, E.C.
1853	Lovett, Phillips Cosby, Esq. Liscombe-ho., Liscombe, Leighton Buzzard, Bucks.
1862	Low, Alex. F., Esq. 84, Westbourne-terrace, W.
1866	Low, S. P., Esq. 55, Parliament-street, S.W.
1858	Lowndes, Rev. George Rouse. Brent-villa, Hanwell, Middlesex.
1859	Lowe, Capt. W. Drury. Myria, Bettws-y-Coed, Llanrwst, North Wales.
1865	Lowndes, E. C., Esq. 84, Eaton-place, S.W.

Year of Election	
1830	Lowry, Joseph Wilson, Esq. · 45, *Robert-street, Hampstead-road, N.W.*
1860	Loyd, Col. W. K. *Union Club, S.W.*
1870	Luard, Capt. Charles Edward, R.E. *Gibraltar.*
1866	1360 Luard, Wm. Charles, Esq. *Llanduff-house, Cardiff; and Athenæum Club, S.W.*
1860	Lumsden, Rev. Robert Comyn, M.A. *Cheadle, Manchester.*
1860	Lush, Robert, Esq., Q.C. *Balmoral-house, Avenue-road, Regent's-park, N.W.*
1870	Lyall, George, Esq. 73, *Eaton-place, S.W.; and Hedley, near Epsom.*
1866	Lydall, J. H., Esq. 12, *Southampton-buildings, Chancery-lane, W.C.*
1869	Lye, John Gaunt, Esq. 18, *Prince of Wales-terrace, Kensington, W.*
1830	*Lyell, Sir Charles, Bart., M.A., LL.D., F.R.S. 73, *Harley-st., Cavendish-sq., W.*
1837	*Lynch, Capt. H. Blosse, I.N., C.B., F.R.A.S. *Athenæum Club, S.W.*
1861	*Lynch, Thomas Kerr, Esq. 31, *Cleveland-square, Hyde-park, W.*
1858	Lyne, Francis, Esq.
1862	1370 *Macarthur, Major-Gen. Sir Edward, K.C.B. 27, *Prince's-gardens, W.*
1863	Macbraire, James, Esq. *Broadmeadows, Berwick-on-Tweed.*
1862	Macdonald, Chessborough C., Esq. 32, *Belsize-park, Hampstead, N.W.*
1843	Macdonnell, Sir Richard Graves, C.B.
1865	Macfarlan, John G., Esq. *Locksley, Victoria-road, Gipsy-hill, S.E.*
1865	Macfie, Rev. M. *Moseley-road, Birmingham.*
1868	MacGregor, Capt. C. M. *Simla. Care of Messrs. Grindlay.*
1861	Mackintosh, Alexander Brodie, Esq. *Oriental Club, W.; and Dunoon, Scotland.*
1845	*Macintyre, Patrick, Esq., F.S.A., Off. Assoc. Inst. Act. 1, *Maida-hill, W.*
1868	Mackay, Dr. A. E., R.N. *Admiralty, Somerset-house, W.C.*
1859	1380 Mackay, Rev. Alexander, LL.D. 1, *Hatton-place, Grange, Edinburgh.*
1870	Mackay, Nevile F., Esq. 2, *Elm-court, Temple, E.C.*
1859	*Mackean, Thos. W. L., Esq. *Bank of British Columbia, 5, East India-avenue.*
1845	Mackenzie, Right Hon. Holt, F.R.A.S. *Athenæum Club, S.W.; and 28, Wimpole-street, W.*
1860	*Mackenzie, James T., Esq. 69, *Lombard-street, E.C.*
1863	Mackenzie, John H., Esq. *Wallington, Carshalton, Surrey.*
1864	*Mackeson, Edward, Esq. 59, *Lincoln's-inn-fields, W.C.*
1862	Mackinlay, D., Esq. *Oriental Club, W.*
1867	Mackinlay, John, Esq., J.P., M.I.C.E., Chief Engineer and Inspector of Machinery, H.M. Dockyard, and Surveyor to the Port, Bombay. *Care of Charles Bannerman, Esq., 193, Camberwell-new-road, Kennington, S.*
1864	Mackinnon, C. D., Esq. *Care of Messrs. J. Clinch and Sons, 31, Abchurch-lane, E.C.*
1861	1390 Mackinnon, Lachlan, Esq. *Reform Club, S.W.*
1855	*Mackinnon, Wm. Alex., Esq., M.P., F.R.S. 4, *Hyde-park-place, W.*
1865	*Mackinnon, W., Esq. *Balinakiel-by-Harbert, Argyleshire.*
1860	Mackirdy, M.-Gen. Elliot, 69th Rgt. *U.S. Club, S.W.*

Year of Election	
1860	Maclean, William Crighton, Esq., F.G.S. 31, *Camperdown-pl., Great Yarmouth.*
1858	MacLeay, George, Esq. *Pendell-court, Bletchingly.*
1870	MacLeod, Lieut. Angus, R.N. *Briardale, Birkenhead: and H.M.S. ' Excellent.'*
1865	Maclure, Andrew, Esq. *Maclure, Macdonald, and Macgregor,* 37, *Walbrook, E.C.*
1861	Maclure, John William, Esq. *Fallowfield, near Manchester.*
1860	Macmillan, Alex., Esq. 16, *Bedford-street, Covent-garden, W.C.*
1855	1400 Macnab, John, Esq. *Findlater-lodge, Trinity, near Edinburgh.*
1868	Macnair, Geo. Esq. *Oriental Club, Hanover-square, W.*
1861	Macpherson, William, Esq. 32, *Lancaster-gate, W.*
1865	Mactaggart, Malcolm, Esq. *Sydney, New South Wales.*
1870	Macturk, John, Esq. *Tillicoultry.*
1863	McArthur, Alex., Esq. *Raleigh-hall, Brixton-rise, Brixton, S.*
1867	McArthur, William, Esq. 1, *Guyder-houses, Brixton-rise, S.*
1860	McClintock, Capt. Sir Francis Leopold, R.N. *United Service Club, S.W.*
1861	*McConnell, W. R., Esq., Barrister-at-Law. 12, *King's-Bench-walk, E.C. ; and Charleville, Belfast.*
1862	McCosh, John, Esq., M.D. *Junior United Service Club, S.W.*
1855	1410 *M'Clure, Admiral Sir Robert J. le M., C.B. *Chipperfield, Herts ; and Athenæum Club, S.W.*
1865	M cDonald, James, Esq. *Oriental Club, Hanover-square, W.*
1865	McEuen, D. P., Esq. 24, *Pembridge-square, Bayswater, W.*
1855	McGregor, Duncan, Esq. *Board of Trade, S.W. ; and Athenæum Club, S.W.*
1867	McGregor, Duncan, Esq. *Clyde-place, Glasgow.*
1869	McGrigor, Alexander Bennett, Esq. 19, *Woodside-terrace, Glasgow.*
1866	*McIvor, W. G., Esq., *Superintendent of Chinchona Plantations, Ootacamund, Madras.*
1858	McKerrell, Robert, Esq. 45, *Inverness-terrace, W. ; and Mauritius.*
1869	McLaren, Robert, Esq. 6, *Räcknitz-platz, Dresden. Care of Messrs. McLaren and Co., 5, South Hanover-street, Glasgow.*
1868	McLean, Frank, Esq., M.A., C.E. 23, *Great George-street, Westminster, S.W.*
1867	1430 McLean, Hon. John. *Oamaru, New Zealand. Care of Messrs. Redfern, Alexander, and Co., 3, Great Winchester-street-buildings, E.C.*
1870	McLeod, Major-Gen. W. C. 14, *St. James's-square, S.W.*
1852	M'Leod, Walter, Esq. *Head Master of the Royal Military Asylum, Chelsea, S.W.*
1866	McNair, Capt. John F. A., R.A.
1855	M'Neil, The Right Hon. Sir John, G.C.B. *Granton, near Edinburgh.*
1855	Maitland, Geo. Gammie, Esq. *Shotover-house, Wheatley, Oxon.*
1845	*Major, Richard Henry, Esq., F.S.A. *British Museum, W.C.*
1865	*Makins, Henry F., Esq. 19, *Prince of Wales-terrace, Kensington-palace, W., and Reform Club, S.W.*
1858	Malby, John Walter, Esq. 15, *Richmond-villas, Seven-sisters'-rd., Holloway, N.*
1853	*Malby, Thomas, Esq. 2, *Park-villas, Seven-sisters'-road, Holloway, N.*
1848	1440 *Malcolm, Capt. Edward Donald, R.E. *Chatham.*

Year of Election	
1863	Malcolm, Jas., Esq. 22, *Prince's-gate, Knightsbridge,* W.
1843	*Malcolm, W. E., Esq. *Burnfoot, Langholme, near Carlisle.*
1853	*Mallet, Charles, Esq. *Audit-office, W.C.;* and 7, *Queensborough-terrace, Bayswater,* W.
1870	Man, J. Alexander, Esq. (Commissioner of Customs for Formosa, &c.) *Care of Messrs. King and Co.,* 65, *Cornhill, E.C.;* and 16, *Stanley-crescent, Kensington-park,* W.
1860	Mann, James Alexander, Esq., M.R.A.S. *Shock-villa, Brecon; and Kensington Palace-avenue,* W.
1866	Mann, Robert James, Esq., M.D. 4, *Belmont-villas, Surbiton-hill, and 6, Duke-street, Adelphi,* W.C.
1866	Manners, Geo., Esq., F.S.A. *Lansdowne-road, Croydon.*
1868	Manners-Sutton, Graham, Esq., 7, *Gloucester-terrace, Hyde-park,* W.
1856	Manning, Frederick, Esq. *Byron-lodge, Leamington; and 8, Dover-street,* W.
1864	1440*Mansell, Capt. A. L. *Hydrographic-office, Admiralty,* S.W.
1869	Mantell, Sir John Iles. *Swinton-park, Manchester; and Windham Club,* S.W.
1859	Mantell, Walter Baldock Durant, Esq. *Wellington, New Zealand. Care of E. Stanford, Esq.*
1869	March, Edward Bernard, Esq., H.M. Consul, Fiji Islands. 13, *Buckingham-street, Strand,* W.C.
1860	Mariette, Prof. Alphonse, M.A. 27, *St. Stephen's-square, Bayswater,* W.
1854	Markham, Clements Robert, Esq. *India-office, S.W.; and 21, Eccleston-sq., S.W.*
1870	Markham, John, Esq. (consul at Chifu). *Care of Messrs. King & Co., Cornhill.*
1857	Marlborough, George, Duke of. *Blenheim, Woodstock. Care of E. Stanford, Esq.*
1864	Marsden, Rev. Canon J. H. *Higher Broughton, Manchester.*
1857	Marsh, Matthew Henry, Esq. *Oxford and Cambridge Club, S.W.; and 41, Rutland-gate, S.W.*
1870	1450*Marsh, Rev. W. R. Tilson, M.A. *Oxford and Cambridge Club, S.W.; Conservative Club; and Stretham-manor, Isle of Ely.*
1862	Marshall, Capt. J. G. Don. *Downton, Wilts.*
1854	Marshall, James Garth, Esq. *Headingley, near Leeds; and Monk Coniston, Ambleside.*
1862	Marshall, William, Esq. 4, *Paper-buildings, Inner Temple,* E.C.
1859	*Marsham, the Hon. Robert. 5, *Chesterfield-street, Mayfair,* W.
1857	Marshman, J. C., Esq. 7, *Kensington-palace-gardens,* W.
1867	Marthin, Guillermo E. de, Consul-General United States of Columbia.
1857	Martin, Francis P. B., Esq.
1861	Martin, Henry, Esq. *Sussex-house, Highbury-new-park,* N.
1860	*Martin, Richard Biddulph, Esq. *Clarewood, Bickley,* S.E.
1862	1460Martin, Thomas, Esq. 5, *Compton-terrace,* N.
1867	Martin, Wm., Esq.
1870	Martin, Wm. Coleman, Esq. *Shireoaks, Worksop, Notts.*
1870	*Martindale, William, Esq. 66, *Upper Thames-street, E.C.; and Woodford, Essex.*

Date of Election	
1865	Mamroon, Wm. R., Esq. *Elm-villa, South-bank, Notting-hill, W.*
1870	Masterman, Edward, Esq. 30, *Threadneedle-street, E.C.; and 27, Clement's-lane, Lombard-street, E.C.*
1870	Masterman, Edward, Esq., Jun. 57½, *Old Broad-street, E.C.; and Walthamstow.*
1859	*Matheson, Alexander, Esq., M.P. 33, *South-street, Park-lane, W.; and Ardross Castle, Ross-shire, N.B.*
1845	*Matheson, Sir James, Bart., F.R.S. 13, *Cleveland-row, S.W.; and Achany, Bonar-bridge, Sutherlandshire, &c.*
1858	Mathieson, James Ewing, Esq. 77, *Lombard-street, E.C.; and 16, Queen's-gardens, Bayswater, W.*
1869	1470 Maude, Col. Francis Cornwallis, R.A., V.C., &c. *Army and Navy Club, S.W.*
1868	Mavrogordato, M. Lucas. *Belgrave-mansions, Grosvenor-gardens, S.W.; and Messrs. Ralli, Brothers, 25, Finsbury-circus, E.C.*
1868	*Maxwell, Sir William Stirling, Bart. 128, *Park-street, Grosvenor-square, W.*
1855	May, Daniel John, Esq., R.N., Staff-Commr. *Care of Case and Loudensack.*
1856	Mayer, Joseph, Esq., F.S.A. 68, *Lord-street, Liverpool.*
1861	Mayers, William S. F., Esq., Interpreter to H.M. Consulate. *Shanghai. Care of F. J. Angier, Esq., 12, George-yard, Lombard-street, E.C.*
1867	Mayhew, Rev. Samuel Martin. 158, *New Kent-road, S.*
1862	Mayne, Captain Richard Charles, R.N., C.B. 80, *Chester-square, S.W.*
1858	Mayo, Capt. John Pole. *Army and Navy Club, S.W.*
1867	Mayson, John S., Esq., J.P. *Oakhill, Fallowfield, near Manchester.*
1863	1480 Meade, the Hon. Robert Henry. *Foreign-office, S.W.; and 8, Belgrave-sq., S.W.*
1862	*Medlycott, Lieut. Mervyn B., R.N. *Care of Messrs. Woodhead.*
1854	Melvill, Major-Gen. Sir Peter Melvill, Mil. Sec. to the Bombay Gov. 27, *Palmeira-square, Brighton.*
1858	Melvill, Philip, Esq., F.R.A.S. *Ethy-house, Lostwithiel, Cornwall.*
1868	Merewether, Col. Sir William Lockyer, C.B. *Kurrachee.*
1842	*Merivale, Herman, Esq., C.B., Under Sec. of State for India. *India-office, S.W.; and 26, Westbourne-terrace, W.*
1866	Messiter, Charles A., Esq. *Barwick, near Yeovil, Somerset.*
1867	Metcalfe, Frederic Morehouse, Esq. *Wisbech, Cambridgeshire.*
1837	*Mexborough, John Chas. Geo., Earl of. 33, *Dover-street, W.; and Methley-park, near Leeds.*
1868	*Michell, Lieut.-Colonel J. E., R.H.A.
1868	1490 Michell, Robert, Esq. 17, *King-street, St. James's, S.W.*
1863	*Michie, A., Esq. 26, *Austin-friars, E.C.*
1870	*Midwinter, William Colpoys, Esq. *St. Michael's Rectory, Winchester; and Akyab, British Burmah.*
1848	Middleton, Rear-Admiral Sir G.N. Broke, Bart., H.M.S. 'Hero,' Sheerness; and Broke-hall, Suffolk.*
1868	*Miers, John William, Esq., C.E. 74, *Addison-road, Kensington, W.*
1859	Miland, John, Esq. *Clairville, Lansdown-road, Wimbledon.*
1856	Mildmay, Capt. Herbert St. John (Rifle Brigade). 19, *Charles-street, Berkeley-square, W.*

Year of Election.	
1860	Miles, Rev. R. *Bingham, Notts.*
1861	*Miller, Commander Henry Matthew, R.N. *The Grove, Exeter; and Junior United Service Club, S.W.*
1868	Miller, Robert Montgomerie, Esq. *Culverden-grove, Tunbridge-wells.*
1853	1500*Miller, Capt. Thomas, R.N. *H.M.S. 'Royal George;' and U.S. Club, S.W.*
1861	Milligan, Joseph, Esq. 15, *Northumberland-street, W.C.*
1857	Mills, Arthur, Esq. 34, *Hyde-park-gardens, W.*
1863	*Mills, John R., Esq. *Kingswood-lodge, Tunbridge-wells.*
1864	Mills, Rev. John. 40, *Lonsdale-square, N.*
1863	*Milton, Viscount, M.P. 17, *Grosvenor-street, W.*
1860	Milman, Capt. Everard, Royal Horse Artillery. *Care of Mrs. Milman, 9, Berkeley-square, W.*
1866	Milne, Vice-Admiral Sir Alex., K.C.B. *United Service Club, S.W.*
1867	Milner, Rev. John, B.A. *Chaplain of H.M.S. 'Galatea.'*
1860	Mitchell, Capt. Alexander. 6, *Great Stanhope-street, Park-lane, W.*
1862	1510*Mitchell, George, Esq. 22, *Bolton-street, Piccadilly, W.*
1864	Mitchell, Thomas, Esq., C.E. *Oldham.*
1859	Mitchell, Sir William. 6, *Hyde-park-gate, Kensington-gore, W.*
1865	Mitchell, Wm. H., Esq. *Junior Carlton Club, S.W.*
1851	*Mocatta, Frederick D., Esq. 6, *Connaught-place, W.*
1853	Moffatt, George, Esq. 103, *Eaton-square, S.W.*
1868	Moffitt, John, Esq. 5, *Canning-place, South Kensington, W.*
1861	Mollison, Alexander Fullarton, Esq. *Woodcote, Tunbridge-wells.*
1870	Moneta, Don Pompeio (Chief Engineer, Argent. Repub.). *Buenos Ayres.*
1842	*Montagu, Major Willoughby. *Clapham-common, S.*
1862	1520*Montagu, Capt. Horace. 24, *Chapel-street, Park-lane, W.*
1830	*Montefiore, Sir Moses, Bart., F.R.S., F.R.S.N.A. 7, *Grosvenor-gate, Park-lane, W.; and East-cliff-lodge, Ramsgate.*
1859	Montgomerie, Major T. G., Engrs., 1st Assist. Trig. Survey. *Care of Messrs. Alexander Fletcher & Co., 10, King's-arms-yard, Moorgate-street, E.C.*
1860	Montgomery, Robert Mortimer, Esq.
1865	Montgomery, Sir Robert, K.C.B. 7, *Cornwall-gardens, Queen's-gate, W.*
1839	Moody, General R. C., R.E. *Caynham-house, near Ludlow, Shropshire.*
1857	*Moor, Rev. Allen P., M.A., F.R.A.S. *Sub-Warden St. Augustine College, Canterbury.*
1870	Moran, Benjamin, Esq. 20, *Norfolk-terrace, Bayswater, W.; and 17, Arlington-street, W.*
1863	Moore, H. Byron, Esq. *Survey-office, Melbourne, Australia. Care of Mr. Wadeson, 100, St. Martin's-lane.*
1861	Moore, John Carrick, Esq. *Corswall, Wigtonshire; Geological Society, W.C.; and 23, Bolton-street, W.*
1870	1530Moore, John, Esq. 36, *Mark-lane. E.C.*
1870	*Moore, Joseph, Esq. *Brockwell-house, Dulwich.*

Year of Election.	
1857	Moore, Major-General W. Y. *United Service Club, S.W.*
1870	Moran, Benjamin, Esq. 20, *Norfolk-terrace, Bayswater, W.; and* 17, *Arlington-street, W.*
1863	More, R. Jasper, Esq. *Linley-hall, Salop.*
1869	*Morgan, Delmar, Esq. 26, *Ryder-street, St. James's, S.W.*
1864	Morgan, D. L., Esq. *H.M.S. ' Euryalus.'*
1861	Morgan, Junius Spencer, Esq. 13, *Prince's-gate, Hyde-park, S.W.*
1861	Morgan, William, Esq., R.N. 1, *Sussex-place, Southsea, Hants.*
1866	Morland, Lieut. Henry, late I.N. *Assistant Dockmaster, &c., Bombay.*
1839	1540*Morris, Charles, Esq. *University Club, S.W.*
1868	Morris, Eugene, Esq. *Birchwood, Sydenham, Kent.*
1863	Morrison, Col. J. D. *United Service Club, S.W.*
1867	Morrison, Pearson, Esq. *Care of William Fletcher, Esq.,* 35A, *Moorgate-street, E.C.*
1865	Morson, T., Esq. 124, *Southampton-row, Russell-square, W.C.*
1869	Moser, Robert James, Esq. 45, *Bedford-square, W.C.*
1869	Mott, F. T., Esq. 1, *De Montfort-street, Leicester.*
1861	*Mount, Frederick J., Esq., M.D., Surgeon-Major and Inspector-General of Prisons, Bengal Army, &c. *Athenæum Club, S.W.; and* 45, *Arundel-gardens, Notting-hill, W. Care of Messrs. A. C. Lepage & Co.,* 1, *Whitefriars-street, Fleet-street, E.C.*
1868	*Mounsey, Aug. Henry, Esq. *British Legation, Florence. Care of F. B. Alston, Esq., Foreign-office, S.W.*
1858	Mudie, Charles Edward, Esq.
1858	1550Mueller, Ferdinand, Esq., M.D., PH. DR. *Director of the Botanical Gardens, Melbourne. Care of Messrs. Dulau and Co.,* 37, *Soho-square, W.*
1862	Muir, Francis, Esq., LL.D.
1855	Muir, Thomas, Esq. 24, *York-terrace, Regent's-park, N.W.*
1867	*Muir, Thomas, Esq., Jun. *Madeira; and* 24, *York-terrace, Regent's-park, N.W.*
1869	Müller, Albert, Esq. *Eaton-cottage, South Norwood, S.*
1866	Mundella, A. J., Esq. *Nottingham.*
1869	Munton, Francis Kerridge, Esq. 21, *Montagu-street, Russell-square, W.C.*
1866	*Murchison, John H., Esq. *Surbiton-hill, Kingston-on-Thames; and Junior Carlton Club, S.W.*
1859	Murchison, Kenneth R., Esq. *Junior United Service Club.*
1830	*Murchison, Sir Roderick Impey, Bt., K.C.B., G.C.ST.A., M.A., D.C.L., V.P.R.S., G.S., and L.S., Grand Officer of the Order of the Crown of Italy, Director-General of the Geological Survey of Great Britain and Ireland, Trust. Brit. Mus., Hon. Mem. R.S. of Ed., R.I.A., Foreign Mem. of the Academy of Sciences, Paris, Mem. Acad. St. Petersburg, Berlin, Stockholm, Brussels, and Copenhagen, Corr. Ins. Fr., &c., &c. 16, *Belgrave-square, S.W.; and* 28, *Jermyn-street, S.W.*
1864	1560Murchison, Capt. R. M. *Bath and County Club, Queen-square, Bath.*
1830	*Murdock, Thomas W. C., Esq. 8, *Park-street, Westminster, S.W.; and River-bank, Putney, S.W.*

1860	Murray, George J., Esq. *Purbrook-house, Cosham, Hants; and Junior Carlton Club, S. W.*
1868	*Murray, Henry, Esq. *Hong Kong. Care of Messrs. Jardine, Matheson, and Co., 3, Lombard-street, E.C.*
1844	*Murray, James, Esq. *Foreign-office, S. W.*
1830	Murray, John, Esq. 50, *Albemarle-street, W.; and Newstead, Wimbledon, S. W.*
1860	*Murray, Lt. W., 68th Beng. N. Inf., Topo. Assist. G. Trig. Survey. *Mussoorie, India. Messrs. H. S. King and Co.*
1870	Murray, T. Douglas, Esq. 13, *Great Cumberland-place, W.*
1870	Murray, William Vaughan, Esq., M.R.I., &c. 4, *Westbourne-crescent, Hyde-park, W.*
1865	Mussy, H. G. de, Esq., M.D. 4, *Cavendish-place, W.*
1865	1570Nairne, P. A., Esq. 2, *Grove-hill, Camberwell, S.*
1858	Napier, Maj.-General Geo. Thomas Conolly, C.B. *Jun. United Service Club, S. W. Care of Sir J. Kirkland.*
1868	Napier, of Magdala, Lord, G.C.B., F.R.S.
1861	Napier, William, Esq.
1870	Napier, Wm. Jno. Geo., Esq. (Master of Napier). *Thatched-house Club, St. James's-street; and Thirlestane-castle, Selkirkshire.*
1870	Nash, Samuel, Esq., B.A., &c. 44, *Renshaw-street, Liverpool.*
1859	*Nasmyth, Capt. David J., 1st Assist. Trigonometrical Survey. 5, *Charlotte-street, Edinburgh.*
1857	*Nesbitt, Henry, Esq. 12, *Victoria-villas, Kilburn, N. W.*
1869	Neville, Lieut.-Col. Edward. 30, *Clarges-street, Piccadilly, W.*
1870	Newall, Wm. Johnstone, Esq. 33, *South-street, Park-lane, W.*
1868	1580Newbatt, Benjamin, Esq., F.R.S., &c. 7, *Vicarage-gardens, Campden-hill, W.*
1867	Newdigate, Lieut.-Col. Francis W. (Coldstream Guards). *Byrkley-lodge, Needwood Forest, Burton-upon-Trent.*
1856	Newman, Thomas Holdsworth, Esq. 43, *Green-street, Grosvenor-square, W.*
1868	Nicol, Geo. William, Esq. *Care of Messrs. Glyn, Mills, and Co., 67, Lombard-street, E.C.*
1866	Nicol, James D., Esq., M.P. 13, *Hyde-park-terrace, Cumberland-gate, W.*
1869	*Nicol, Robert, Esq. *Reform Club, S. W.; and Westminster-palace-hotel, S. W.*
1868	Nicol, William, Esq. 41, *Victoria-street, S. W.; and Fawsyde, Kenneff, Kincardine.*
1870	Nicholl, Henry Jno., Esq. 16, *Hyde-park-gate, W.*
1870	Nicholas, W., Esq. 31, *Lansdowne-road, Dalston, E.*
1870	Nichols, James, Esq. 22, *Laurence Pountney-lane, E.C.; and The " Mount," Kenley, Surrey.*
1865	1590*Nichols, Robert C., Esq. 5, *Sussex-place, W.*
1856	Nicholson, Sir Charles, Bart., D.C.L., Chancellor of the University, Sydney. 26, *Devonshire-place, Portland-place, W.*

Year of Election	
1856	Erskine, Rear-Admiral Sir Frederick Wm. Erskine, Bart. 15, *William-street, Lowndes-square, S.W.*
1864	Nissen, H. A., Esq. *Mark-lane, E.C.*
1858	Nix, John H., Esq. 77, *Lombard-street, E.C.*
1861	Noel, the Hon. Roden. 11, *Chandos-street, Cavendish-square, W.;* and *Exton-hall, Oakham, Rutlandshire.*
1857	•Nolloth, Captain Matthew S., R.N. 13, *North-terrace, Camberwell, S.E.;* and *United Service Club, S.W.*
1865	Norman, H. J., Esq. 106, *Fenchurch-street, E.C.*
1860	North, Henry, Esq. *Colonial-office, S.W.;* and 4, *Little St. James's-street, S.W.*
1861	North, Alfred, Esq. *Birthwaite-lodge, Windermere.*
1855	1500 Northumberland, Algernon George, Duke of. *Northumberland-house, S.W.*
1862	Notman, Henry Wilkes, Esq. 7, *Great Marlborough-street, W.*
1862	Nourse, Henry, Esq. *Conservative Club, S.W.*
1868	•Oakeley, R. Banner, Esq. *Kilmaronaig, Inverary, Argyllshire, N.B.*
1867	O'Brien, James, Esq. 109, *Belgrave-road, Pimlico, S.W.;* and *Clare, Ireland.*
1856	O'Connor, Major-General Luke Smyth, C.B. *U.S. Club, S.W.*
1858	Ogilvie, Edward D., Esq. *Yulgillar, Clarence-river, New South Wales. Care of Messrs. Marryat and Sons, Laurence Pountney-lane, E.C.*
1863	Ogilvy, Col. Thos., 23, *Grafton-st., Piccadilly, W.;* and *Ruthven, Forfarshire, N.B.*
1864	Ogilvy, Thos., Esq. 62, *Prince's-gate, Hyde-park, W.*
1870	Oldham, Henry, Esq., M.D. 26, *Finsbury-square, E.C.*
1870	1510 Oldham, Robert, W., Esq. *Lloyds', E.C.*
1861	Oldershaw, Capt. Robert Piggott. 74, *Warwick-square, Belgrave-road, S.W.*
1858	Oliphant, Laurence, Esq. *Athenæum Club, S.W.*
1866	Oliver, Lieut. S. P., 12th Brigade R.A. 1, *Buckingham-villas, Brockhurst-road, Gosport, Hants.*
1845	•Ommanney, Adml. Erasmus, C.B., F.R.A.S. 6, *Talbot-square, Hyde-park, W.;* and *United Service Club, S.W.*
1858	•Ommanney, H. M., Esq. *Blackheath, S.E.*
1857	Ormathwaite, John Benn-Walsh, Lord. 28, *Berkeley-square, W.*
1868	Osborn, Sir George R., Bart. *Travellers' Club, S.W.;* and *Chicksand-priory, Beds.*
1855	Osborn, Capt. Sherard, R.N., C.B., Officier de Légion d'Honneur, &c. *Athenæum Club, S.W.;* and 119, *Gloucester-terrace, W.*
1861	Osborne, Lieut.-Col. Willoughby. *Political Agent, Bhopal, Schira, India.*
1866	1520 O'Shaughnessy, Richard, Esq. 12, *Cornwall-gardens, South Kensington, W.*
1861	Oswell, William Cotton, Esq.
1855	Otway, Arthur John, Esq., M.P. *Army and Navy Club, S.W.*
1866	•Overy-North, the Rev. J. *East Acton, Middlesex, W.*
1870	•Overbeck, M. the Chev. G. de. *Hong Kong. Care of Messrs. King and Co., 65, Cornhill, E.C.*

Year of Election	
1844	*Overstone, Samuel, Lord, M.A., M.R.I. 2, *Carlton-gardens, S.W.; and Wickham-park, Surrey.*
1868	Owden, Thomas S., Esq. *Mount-pleasant, Philip-lane, Tottenham.*
1867	Owen, Capt. Chas. Lanyon (Adj. R. M. Light Inf., Portsmouth Division), *Glendowan-lodge, Bury-road, Gosport.*
1861	Page, Thomas, Esq., C.E., F.G.S. 3, *Adelphi-terrace, W.C.; and Tower Cressy, Aubrey-road, Bayswater, W.*
1853	Pakington, Right Hon. Sir John Somerset, Bart., M.P. 41, *Eaton-square, S.W.; and Westwood-park, Droitwich, Worcestershire.*
1868	1630 Palæologus, William Thomas, Esq. *Care of Messrs. McGregor and Co., Charles-street, S.W.*
1855	Palmer, Major Edm., R.A. *Boxhill, Pennycross, Plymouth.*
1870	Palmer, F. J., Esq., R.N. 4, *Furnival's-inn, W.C.*
1865	*Palmer, Commander George, R.N. *H.M.S. 'Rosario,' Australia; and Cavers, Hawick, Roxburghshire, N.B.*
1870	*Palmer, John Linton, Esq., Surg. R.N. 19, *Royal Parade, Cheltenham.*
1862	Palmer, Rev. Jordan, M.A., F.S.A., Chaplain to St. Ann's Royal Society, *Streatham, S.*
1838	*Palmer, Samuel, Esq.
1870	Pannel, Charles S., Esq. *Walton-lodge, Torquay.*
1855	*Papengouth, Oswald C., Esq., C.E. *Care of W. Hornibrook, Esq., 6, Regent's-square, W.C.*
1870	Parfitt, W. S., Esq., C.E. *Montevideo. Care of Mrs. Parfitt, Devizes, Wilts.*
1863	1640 *Paris, H.R.H. Le Comte de. *Claremont.*
1864	Parish, Capt. A. *Chislehurst, Kent.*
1849	*Parish, Capt. John E., R.N. *Army and Navy Club, S.W. Care of Messrs. Stilwell.*
1833	*Parish, Sir Woodbine, K.C.H., F.R.S., &c. *Quarry-house, St. Leonard's-on-Sea.*
1866	Parker, Capt. Francis G. S., 54th Regt., F.G.S., A.I.C.E. *Curragh, Ireland.*
1862	Parker, Robert Deane, Esq. *Barham, Canterbury.*
1850	Parkes, Sir Harry S., C.B., &c. *Oriental Club, W.; and Athenæum Club, S.W.*
1850	*Parkyns, Mansfield, Esq., F.Z.S. *Arthur's Club, St. James's-street, S.W.; and Woodborough-hall, Southwell.*
1859	Pasteur, Marc Henry, Esq. 38, *Mincing-lane, E.C.*
1867	Paterson, John, Esq. 19A *Coleman-street, City, E.C.*
1857	1650 Paton, Andrew A., Esq. *H.B.M.'s Vice-Consul, Missolonghi, Greece.*
1863	Pattinson, J., Esq. 21, *Bread-street, E.C.*
1868	Paul, J. H., Esq., M.D. *Camberwell-house, Camberwell, S.*
1858	Paul, Joseph, Esq. *Ormonde-house, Ryde, Isle of Wight.*
1865	Payne, Captain J. Bertrand, M.R.I., F.R.S.L., Mem. Geograph. Soc. of France. *Conservative Club, S.W.; Royal Thames Yacht Club, W.; and Tempsford-house, Grange-terrace, Brompton, W.*

Year of Election	
1847	*Paynter, William, Esq., F.R.A.S. 21, *Belgrave-square, S.W.; and Camborne-house, Richmond, Surrey, S.W.*
1853	Peacock, George, Esq. *Starcross, near Exeter.*
1865	Pearse, Capt. R. B., R.N. 13, *Hyde-park-street, W.; Arthur's Club.*
1853	Pearson, Fred., Esq.
1853	*Peckover, Alexander, Esq. *Wisbeach.*
1860	1660*Peek, Henry William, Esq., M.P. *Care of G. Thorpe, Esq., 21, Eastcheap, E.C.*
1861	Peel, Archibald, Esq. *The Gerwyn, Wrexham, N. Wales.*
1858	Peel, Sir Robert, Bart., M.P. 4, *Whitehall-gardens, S.W.; and Drayton-manor, Tamworth.*
1868	*Pender, John, Esq. 18, *Arlington-street, W.*
1863	*Pennant, Col. S. S. Douglas. *Penrhyn-castle, Bangor, N.B.*
1859	*Penrhyn, Lord. *Penrhyn-castle, Bangor.*
1853	Percy, Major-General the Hon. Lord Henry M. (Guards). 40, *Eaton-square, S.W.*
1865	Pereira, Francisco E., Esq. *Care of Messrs. Richardson, 13, Pall-mall.*
1860	Perkins, Frederick, Esq. *Mayor of Southampton.*
1865	Perkins, William, Esq. *Rosario, Argentine Republic. Care of W. Bollaert, Esq.*
1850	1670Perry, Sir Erskine, Member Indian Council. 36, *Eaton-place, S.W.*
1865	Perry, Gerald R., Esq., British Consulate, Stockholm. *Care of E. Hertslet, Esq., Foreign-office, S.W.*
1850	Perry, William, Esq., H.B.M.'s Consul, Panama. *Athenæum Club, S.W.*
1863	*Perry, William, Esq. 9, *Warwick-road, Upper Clapton, N.E.*
1862	Peter, John, Esq.
1857	*Peters, William, Esq. 85, *Nicholas-lane, Lombard-street, E.C.*
1860	*Petherick, John, Esq. *Henley-on-Thames.*
1856	Peto, Sir S. Morton, Bart. 12, *Kensington-palace-gardens, W.*
1861	Petrie, Alexander S., Esq. 4, *St. Mark's-square, N.W.*
1866	Petrie, Major Martin, 97th Regiment. *Hanover-lodge, Kensington-park, W.*
1866	1680Pharasyn, Robert, Esq. *Wellington, New Zealand. Care of Messrs. Scales and Rogers, 24, Mark-lane, E.C.*
1867	Phayre, Col. Sir Arthur.
1854	Phelps, William, Esq. 18, *Montagu-place, Russell-square, W.C.*
1862	Phené, John Samuel, Esq., F.G.S. 5, *Carlton-terrace, Oakley-street, S.W.*
1860	Philip, George, Esq. 32, *Fleet-street, E.C.*
1865	Philipps, Edward B., Esq. 105, *Onslow-square, S.W.*
1857	Phillimore, Capt. Augustus, R.N. 25, *Upper Berkeley-st., W.; and U.S. Club, S.W.*
1859	Phillimore, Chas. Bagot, Esq. *India-office, S.W.; and 25, Upper Berkeley-st., W.*
1859	Phillimore, Wm. Brough, Esq., late Capt. Grenadier Guards. 5, *John-street, Berkeley-square, W.*
1854	Phillips, Major-General Sir B. Travell. *United Service Club, S.W.*
1866	1690Phillips, Edward Augustus, Esq. *Chantry-lodge, 34, Abbey-road, St. John's-wood, W.*

VOL. XL.

1871	Pierce, Josiah, Esq. 19, *Harley-street, W.*
1869	Piggot, John, jun., Esq , F.S.A., &c. *The Elms, Ulting, Maldon, Essex.*
1870	Pigott, Robt. Turtle, Esq. *Torrington-villas, Lee, Kent; and 36, Southampton-street, Strand, W.C.*
1864	*Pigou, F. A. P., Esq. *Dartford, Kent.*
1865	Pigou, Rev. F., M.A. *The Rectory, Doncaster.*
1861	Pike, Frederick, Esq. *Co-operative Stores, Haymarket.*
1852	1700*Pike, Captain John W., R.N. *United Service Club, S.W.*
1855	Pilkington, James, Esq. *Blackburn.*
1865	Pilkington, William, Esq. *War-office.*
1870	Pimblett, James, Esq. *Tutenhill, Burton-on-Trent.*
1852	*Pim, Capt. Bedford C. T., R.N. *Belsise-square, Hampstead, N.W.; and Senior and Junior United Service Club, S.W.*
1858	Pincott, James, Esq. *Tolham-house-school, Brixton-hill, S.*
1859	Pinney, Colonel William. 30, *Berkeley-square, W.*
1867	Plant, Nathaniel, Esq. *Hotel Exchange, Rio de Janeiro; and De Montfort-house, Leicester.*
1865	Player, John, Esq. 24, *Duchess-road, Edgbaston, Birmingham.*
1860	Playfair, Lieut.-Col. Robert Lambert. H.B.M. Consul-General, Algiers. *Care of E. Hertslet, Esq., Foreign-office.*
1866	1710Plowden, Charles, C., Esq. *Belgrave-mansions, Grosvenor-gardens, S.W.*
1856	*Plowes, John Henry, Esq. 39, *York-terrace, Regent's-park, N.W.*
1870	Plunkett, Maj.-Gen. Hon. Chas. Dawson. *United Service Club, S.W.*
1855	*Pollexfen, Capt. J. J. *India.*
1860	*Pollington, Jno. Horace, Viscount. 33, *Dover-street, W.*
1853	Pollock, Field-Marshal Sir George, G.C.B. *Clapham-common, Surrey, S.*
1835	*Ponsonby, Hon. Frederick G. B. 3, *Mount-street, Grosvenor-square, W.*
1860	Pook, Captain John. 6, *Colfe's-villas, Lewisham-hill, S.E.*
1870	Poole, C. M., Esq., C.E. 5, *Cambridge-terrace, Notting-hill, W.*
1857	Pope, Captain Wm. Agnew. 12, *Stanhope-place, Hyde-park, W.*
1863	1720*Porcher, Captain Edwin A., R.N. 3, *Montagu-square, W.*
1853	Porter, Edwd., Esq. *Athenæum Club, S.W.; and 26, Suffolk-street, Pall-mall, S.W.*
1864	Portugal, Chev. Joaquim de.
1868	Potter, Archibald Gilchrist, Esq. *Woodham-lodge, Lavender-hill, Wandsworth, S.W.*
1867	Potter, Wm. Henry, Esq. *Dunsden-lodge, Sonning, near Reading.*
1861	*Pounden, Captain Lonsdale. *Junior United Service Club, S.W.; and Browns-wood, Co. Wexford.*
1862	Povah, Rev. John V., M.A. 11, *Endsleigh-street, W.C.*

Date of Election		
1864		*Powell, F. S., Esq. 1, *Cambridge-square, Hyde-park*, W.
1859		Power, E. Rawdon, Esq. Retired List, Ceylon Civil Service. *Heywood-lodge, Tenby, South Wales; and Thatched House Club, S. W.*
1854		Power, John, Esq. 3, *College-terrace, Cambridge-road, Hammersmith*, W.
1868	1730	Pownall, John Fish, Esq. 63, *Russell-square*, W.C.
1864		Powys, the Hon. C. J. F.
1864		Powys, the Hon. E. R.
1864		Powys, Hon. Leopold. 17, *Montagu-street, Portman-square*, W.
1870		*Prance, Reginald H., Esq. *Frognal, Hampstead.*
1868		Price, Charles S., Esq. *Bryn Derwen, Neath.*
1869		Price, F. G. H., Esq. 1, *Fleet-street*, E.C.
1852		Price, James Glenie, Esq., Barrister-at-Law. 14, *Clement's-inn*, W.C.
1868		Pritchard, Iltudus Thomas, Esq. 29, *Granville-park, Blackheath*, S.E.
1860		*Prickett, Rev. Thomas William, M.A., F.S.A.
1868	1740	Prideaux, W., F. Esq., Bombay Staff Corps. 13, *Avenue-road, Regent's-pk.*, N.W.
1865		*Pringle, A. Esq. *Yair, Selkirk*, N.B.
1855		*Pringle, Thomas Young, Esq. *Reform Club*, S.W.
1866		*Prinsep, Edw. Aug., Esq., B.C.S., Commissioner of Settlements in the Punjaub. *Umritsur.* Care of Messrs. H. S. King and Co., 65, *Cornhill*, E.C.
1868		Pritchard, Lieut.-Col. Gordon Douglas. *Chatham.*
1868		Pryce, James E. Coulthurst, Esq. *Conservator of the Port of Bombay.*
1861		*Prodgers, Rev. Edwin. *The Rectory, Ayott St. Peter's, Herts.*
1852		Prout, John William, Esq., M.A., Barrister-at-Law. *Athenæum Club, S. W.; and Neasdon, Middlesex, N.W.*
1862		*Puget, Major J., 8th Hussars. 71, *Queen's-gate, Hyde-park, South Kensington*, W.
1860		Puller, Arthur Giles, Esq. *Athenæum Club, S. W.; Arthur's Club, S.W.; and Youngsbury, Ware.*
1857	1750	Purcell, Edward, Esq., LL.D.
1869		Purdon, Lieut. George Frederic, R.N. *Woodlands, Bracknell, Berks.*
1865		*Pusey, Sidney E. Bouverie, Esq. 7, *Green-street, Grosvenor-square*, W.
1870		Pycroft, Sir Thos., K.C.S.I. 10, *Kensington-gardens-terrace, Hyde-park*, W.
1866		*Quin, Francis Beaufort Wyndham, Esq. *Wistanswick-house, near Market Drayton, Salop.*
1861		Quin, Lord George. 15, *Belgrave-square*, S.W.
1865		Quin, John Thos., Esq. *Care of Mr. Lambeon, Epsom.*
1868		Quin, T. Francis, Esq. *Bathurst-house*, 418, *Clapham-road, Clapham*, S.

Year of Election.	
1858	*Radstock, Graville Augustus, Lord. 30, *Bryanston-square, W.*
1869	Rae, Edward, Esq. *Claughton, near Birkenhead.*
1862	1760*Rae, James, Esq. 32, *Phillimore-gardens, Kensington, W.*
1853	Rae, John, Esq., M.D. 2, *Addison-gardens-south, Holland-villas-road, Kensington, W.*
1870	Raikes, Lieut.-Col. Geo. W. *Albany, Piccadilly, W.*
1870	Raikes, Fras. Wm. Esq. *St. Peter's College, Cambridge.*
1867	Raleigh, Rev. A., D.D. *Arran-house, Highbury-new-park.*
1870	Ralston, W. R. Shedden, Esq., M.A. *British Museum, W.C.*
1866	Ramsay, Alex., Jun., Esq. 45, *Norland-square, Notting-hill, W.*
1866	*Ramsay, Admiral G. *United Service Club, S.W.*
1867	Ramsay, John, Esq. *Islay, N.B.*
1851	*Ramsay, Rear-Admiral Wm., C.B., F.R.A.S. *Junior United Service Club, S.W.; and 23, Ainslie-place, Edinburgh.*
1867	1770*Ramsden, Richard, Esq., B.A. *Camp-hill, Nuneaton, Warwickshire.*
1869	Randell, Thomas, Esq. 1, *Redcliff-parade, Bristol.*
1868	Rankin, William, Esq. *Tiernaleague, Carndonagh, Donegal.*
1866	Ransom, Edwin, Esq. *Kempstone, near Bedford.*
1869	Rassam, Hormusd, Esq., Assistant Political Resident, Aden. *Ailsa-park-lodge, Twickenham, S.W.*
1859	Ratcliff, Charles, Esq., F.S.A. *National Club, S.W.; Edgbaston, Birmingham; and Downing College, Cambridge.*
1870	Ratcliffe, Rev. Thomas, B.D., &c. 74, *Belgrave-road, Belgrave-square, S.W.*
1861	Rate, Lachlan Macintosh, Esq. 9, *South Audley-street, W.*
1846	Ravenshaw, E. C., Esq., M.R.A.S. *Oriental Club, W.; and 36, Eaton-sq., W.*
1859	Ravenstein, Ernest G., Esq. *Topographical-depôt, Spring-gardens, S.W.*
1861	1780Rawlinson, Sir Christopher. *Manydown-park, Basingstoke; and United University Club, S.W.*
1844	*Rawlinson, Maj.-General Sir Henry C., K.C.B., D.C.L., F.R.S. *Athenæum Club, S.W.; and 21, Charles-street, Berkeley-square, W.*
1838	Rawson, His Excellency Rawson Wm., C.B., Colonial Secretary. *Barbadoes.*
1869	Ray, Capt. Alfred William. *The Lodge, Brixton-oval, S.*
1866	Ray, W. H., Esq. *Thorn-house, Ealing.*
1869	Read, Col. William Fitzwilliam. *Junior United Service Club, S.W.*
1863	Reade, W. Winwood, Esq. 11, *St. Mary Abbot's-terrace, Kensington, W.; Conservative Club.*
1865	Redhead, R. Milne, Esq. *Springfield, Seedley, Manchester; Conservative Club, S.W.; and Junior Carlton Club, S.W.*
1868	*Redman, John, B., Esq., C.E. 6, *Westminster-chambers, Victoria-street, S.W.*
1861	*Reid, David, Esq. 95, *Piccadilly, W.*
1858	1790Rees, L. E. R., Esq. 7, *Park-road, New Wandsworth, S.W.*
1859	Reeve, John, Esq. *Conservative Club, S.W.*
1866	*Rehden, George, Esq. 9, *Great Tower-street, E.C.*
1856	Reid, Henry Stewart, Esq., Bengal Civil Service.

Year of Election	
1857	Reid, Lestock R., Esq. *Athenæum Club, S.W.; and 122, Westbourne-ter.., W.*
1861	Reilly, Anthony Adams, Esq. *Belmont, Mullingar.*
1869	*Reiss, James, Esq. 7, Cromwell-road-houses, South Kensington, W.*
1826	*Rennie, Sir John, C.E., F.R.S., F.S.A. 7, Lowndes-square, S.W.*
1868	*Rennie, John Keith, Esq., M.A. Camb. 56, Gloucester-terrace, Hyde-park, W.*
1854	*Rennie, M. B., Esq., C.E. Care of James Rennie, Esq., 9, Motcombe-street, Belgrave-square, S.W.*
1854	1800 Rennie, W., Esq. 14, *Hyde-park-square, W.*
1830	*Renwick, Lieutenant, R.E.*
1861	Reuter, Julius, Esq. 1, *Royal Exchange-buildings, E.C.*
1858	Reynardson, Henry Birch, Esq. *Adwell, near Tetsworth, Oxfordshire.*
1857	Rhodes, Arthur John, Esq. 38, *Ordnance-road, St. John's-wood, N.W.*
1870	Rice, Joseph Marcus, Esq., M.D. 17, *Pleasant-street, Worcester City, Mass., U.S. Care of Messrs. Haseltine, Lake, and Co., 8, Southampton-buildings, W.C.*
1870	Rice, Wm., Esq. 2, *Albert-villas, Evelyn-road, Richmond, S.W.; and Stanford's, Geograph. Estab., Charing-cross, S.W.*
1868	Richards, Alfred, Esq. *Tewkesbury-lodge, Forest-hill.*
1857	Richards, Admiral George H., F.R.S. *Admiralty, Whitehall, S.W.; and 12, Westbourne-terrace-road, W.*
1860	Richards, the Rev. George, D.D. *Spring Mount, St. Leonard's-on-Sea.*
1864	1810 Richardson, F., Esq. *Juniper-hall, Mickleham, Dorking.*
1859	Richards, Edward Henry, Esq. 4, *Connaught-place, Hyde-park, W.*
1865	*Rideout, W. J., Esq. 51, Charles-street, Berkeley-square, W.*
1854	Ridley, F. H., Esq. 11, *Mortimer-road, Kilburn, W.*
1864	Ridley, George, Esq. 2, *Charles-street, Berkeley-square, W.*
1862	*Rigby, Major-General Christopher Palmer. Oriental Club, W.; and 14, Mansfield-street, W.*
1862	Rigby, Joseph D., Esq. *Esher, Surrey; and Kew-green, Surrey, W.*
1868	Riley, Capt. Charles Henry. *Junior United Service Club, S.W.*
1849	Ristoul, Robert, Esq. *Windham Club, S.W.*
1850	*Robe, Maj.-General Fred. Holt, C.B. United Service Club, S.W.; and 10, Palace-gardens-terrace, Kensington, W.*
1866	Roberts, Charles W., Esq. *Penrith-house, Effra-road, Brixton, S.*
1861	Roberts, Capt. E. Wyane. *Junior Carlton Club, S.W.; and 18, Great Cumberland-street, Hyde-park, W.*
1855	Robertson, A. Stuart, Esq., M.D. *Horwich, near Bolton.*
1860	Robertson, D. Brooke, Esq., H.B.M.'s Consul. *Canton. H. S. King and Co.*
1861	*Robertson, Graham Moore, Esq. 21, Cleveland-square, Hyde-park, W.*
1870	*Robertson, Jas. Nisbet, Esq. 23, Porchester-square.*
1869	Robertson, Rev. J. S. S., M.A., F.R.A.S. *Duncrub-castle, Duncrub-park, Dunning, Perthshire, N.B.*
1866	Robertson, R. B., Esq. *H.M.'s Legation, Yokahama, Japan. Care of Capt. Bryson, R.N., 10, Portland-terrace, Southsea, Hants.*
1870	Robinson, Alfred, Esq. *Mountjoy-house, Huddersfield.*

Year of Election.	
1830	*Robinson, Rear-Admiral Charles G. 30, *Blomfield-ter., Upper Westbourne-ter.*, W.
1859	1830Robinson, Lieut.-Col. D. G., R.E., Director-Gen. of Telegraphs in India. *Calcutta.*
1864	Robinson, H. O., Esq.
1859	Robinson, Sir Hercules G. R, *Governor of Ceylon. Messrs. Burnett,* 17, *Surrey-street*, S.W.
1865	Robinson, J. R., Esq., LL.D., F.S.A. Scot., LL.D., F.R.S.A. du Nord, Copenhagen, F.G.S., Edin. Membre Société Asiatique de Paris, &c. *South-terrace, Dewsbury.*
1860	Robinson, Mr. Serjeant. 8, *King's-Bench-walk, Temple, E.C.; and* 43, *Mecklenburgh-square*, W.C.
1862	Robinson, Lieut.-Col. Sir John Stephen, Bart. *Arthur's Club, S.W.; and* 16A, *Park-lane*, W.
1864	Robinson, John, Esq. *Care of E. Street, Esq.,* 30, *Cornhill, E.C.*
1855	Robinson, Thomas F., Esq., F.L.S. 9, *Derwent-road, South Penge-park, Anerley, S.*
1850	*Robinson, Capt. Walter F., R.N. 15, *Montpellier-villas, Brighton.*
1830	*Rodd, James Rennell, Esq. 29, *Beaufort-gardens*, W.
1860	1840Roe, Capt. Jno. Septimus, Surveyor-General, W. Australia. *Care of Mrs. Ellie Jervoise,* 7, *Euston-place, Leamington.*
1863	Rogers, John T., Esq. 38, *Eccleston-square, S.W.*
1861	Rollo, Lord. *Dumcrieff-castle, Moffat, N.B.*
1863	Rönn, M. Herman von. 21, *Kensington-park-gardens*, W.
1866	Rooke, Capt. W., R.A. *Formosa, Lymington, Hants.*
1868	Roos, Gustaf, Esq.
1834	*Rose, the Right Hon. Sir George, F.R.S., LL.D. 4, *Hyde-park-gardens, W., and* 25, *Southampton-buildings, Chancery-lane, W.C.*
1868	Rose, Henry, Esq. 8, *Porchester-square, Hyde-park, N.W.*
1861	Rose, Jas. Anderson, Esq. *Wandsworth, Surrey, S.W.; and* 11, *Salisbury-street, W.C.*
1857	*Rose, Col. Sir Wm. Anderson, Alderman, F.R.S.L. *Carlton Club, S.W.;* 63, *Upper Thames-street, E.C.; *and Upper Tooting, S.*
1870	1850Rose, the Right. Hon. Sir John. 18, *Queen's-gate, Hyde-park*, W.
1864	Ross, B. R., Esq. *Care of the Hudson-bay Company, Hudson-bay-house,* 1, *Lime-street, E.C.*
1870	Ross, Capt. Geo. Ernest Augustus (King's Own Light Inf. Militia). *Bryn-Ellen, Clapham-park, Surrey.*
1863	Ross, Wm. Andrew, Esq.
1867	Rossiter, Wm., Esq., F.R.A.S. *South London Working Men's College,* 91, *Black-friars-road, S.E.*
1868	Ross-Johnson, H. C., Esq. 7, *Albemarle-street*, W.
1864	*Roundell, C. S., Esq. 44, *Piccadilly*, W.
1862	Roupell, Robert Priolo, Esq., M.A., Q.C. J 5, *Albany*, W.
1839	*Rous, Vice-Admiral the Hon. Henry John. 13, *Berkeley-square*, W.
1864	Routh, E. J., Esq. *St. Peter's College, Cambridge.*
1862	1860Rowe, Sir Joshua, C.B., late Chief Justice of Jamaica. 10, *Queen Anne-street, Cavendish-square*, W.

Rowlands, Percy J., Esq. *India-office, S.W.*

Rowley, Capt. C., R.N. '33, *Cadogan-place, S.W.*

Rucker, J. Anthony, Esq. *Blackheath, S.E.*

Rumbold, Charles James Augustus, Esq. *Downing College, Cambridge ; and 5, Percival-terrace, Brighton.*

Rumbold, Thomas Henry, Esq.

Rumley, Major-General Randall, Vice-President Council of Military Education. 12, *Cadogan-place, S.W.*

*Russell, Arthur John Edward, Esq., M.P. 10, *South Audley-street, W.*

Russell, George, Esq., M.A. *Viewfield, Southfields, Wandsworth; and 16, Old Change, St. Paul's, E.C.*

*Russell, Jesse Watts, Esq., D.C.L., F.R.S.

1870 Russell, John, Earl, F.R.S. 37, *Chesham-place, S.W.; Pembroke-lodge, Richmond, S.W.; Endsleigh-ho., Devon; and Gart-ho., near Callandar, N.B.*

Russell, Wm. Howard, Esq., LL.D. *Carlton Club, S.W.; and 18, Sumner-place, Onslow-square, W.*

Rutherford, John, Esq. 2, *Cavendish-place, Cavendish-square, W.*

*Ryder, Admiral, Alfred P. *U.S. Club, S.W.; and Lownds-abbey, Uppingham.*

Ryder, G., Esq. 10, *King's-Bench-walk, Temple, E.C.*

Rylands, Peter, Esq. *Bewsey-house, Warrington.*

Sabben, J. T., Esq., M.D., *Northumberland-house, Stoke Newington, N.*

Sabine, Lieut.-General Sir Edw., K.C.B., R.A., Pres. R.S., F.R.A.S., &c. &c. 13, *Ashley-place, Victoria-street, Westminster, S.W.; and Woolwich, S.E.*

St. Clair, Alexander Bower, Esq., H.B.M. Consul, Jassy, Moldavia. *Care of R. Herbert, Esq., Foreign-office, S.W.*

St. David's, Counop Thirlwall, Bishop of. *Abergwilly-palace, Carmarthen.*

St. John, Lieut. Oliver Beauchamp Coventry, R.E. *National Club, S.W.*

St. John, R. H. St. Andrew, Esq., 60th Rifles.

St. John, Spenser, Esq., Chargé d'Affaires, Port-au-Prince, Haiti. 25, *Grove-end-road, St. John's-wood, N.W.*

1863   Sale, Lieut. M. T., R.E. *The Crescent, Rugby; and Cherrapoonjee, Bengal.*

Saibold, Colonel J. C. (H.M.I. Forces). 29, *St. James's-street, S.W.*

Fallex, J. de, Esq. *Belgrave-mansions, Grosvenor-gardens.*

*Salmond, Robert, Esq. *Reform Club, S.W.; 14, Woodside-crescent, Glasgow; and Rankinston, Patna, Ayr.*

1845   *Salomons, Alderman Sir David, Bart., M.P., F.R.S., F.R.A.S. 26, *Great Cumberland-place, Hyde-park, W.; and Broom-hill, near Tunbridge-wells.*

1863   *Salt, Henry, Esq. 29, *Gordon-square, W.C.*

Salting, William Severin, Esq. 60, *St. James's-street, S.W.*

*Sandbach, Wm. Robertson, Esq. 10, *Prince's-gate, Hyde-park, S.W.*

Year of Election.	
1867	Sandeman, David George, Esq., *Cambridge-house, Piccadilly, W.*
1862	Sanford, Major Henry Ayshford. 29, *Chester-street, Grosvenor-place, W.; and Nynehead-court, Wellington, Somerset.*
1870	Sanford, W. Ayshford, Esq. *Nynehead-court, Wellington, Somerset.*
1860	Sarel, Lieut.-Colonel H. A., 17th Lancers. *Army and Navy Club, S.W.; and Shanghae.*
1862	Sargood, F. J., Esq. *Moorgate-street-buildings, E.C.*
1869	Sarll, John, Esq. *Englefield-house, De Beauvoir-town, N.*
1860	Sartoris, Alfred, Esq. *Abbottswood, Stow-on-the-Wold.*
1852	Saumarez, Captain Thomas, R.N. *The Firs, Jersey.*
1866	Saunders, James Ebenezer, Esq., F.L.S., F.G.S., F.R.A.S. 9, *Finsbury-circus; and Granville-park, Blackheath, S.E.*
1864	1900 Saurin, Admiral E. *Prince's-gate, S.W.*
1863	Sawyer, Col. Charles, 6th Dragoon Guards. 50, *Sussex-square, Kemp-town, Brighton.*
1838	Scarlett, Lieut.-General the Hon. Sir J. Yorke, K.C.B. *Portsmouth.*
1861	Schenley, Edward W. H., Esq. 14, *Prince's-gate, S.W.*
1870	Scobell, Sandford Geo. T., Esq. *Kingwell-hall, near Bath.*
1866	Scott, Adam, Esq. 10, *South-street, Finsbury, E.C.*
1866	Scott, Arthur, Esq. *Rotherfield-park, Alton, Hants; Travellers' Club, S.W.*
1859	Scott, Lord Henry. 3, *Tilney-street, Park-lane, W.*
1861	*Scott, Hercules, Esq. *Brotherton, near Montrose, N.B.*
1855	Scott, Admiral Sir James, K.C.B. *United Service Club, S.W.*
1866	1910 Scott, John, Esq., M.D.
1868	Scott, William Cumin, Esq. *Mayfield-house, Blackheath-park, S.E.*
1863	Scovell, George, Esq. 34, *Grosvenor-place, S.W.*
1861	Searight, James, Esq. 80, *Lancaster-gate, W.*
1869	Searle, Frank Furlong, Esq., M.R.C.S., &c. 26, *Cathedral-yard, Exeter.*
1867	Seaton, Col. the Right Hon. Lord. D 3, *Albany, W.*
1868	Seaton, Joseph, Esq., M.D. *Halliford-house, Sunbury, Middlesex, S.W.*
1830	*Sedgwick, the Rev. A., Woodwardian Lecturer, M.A., F.R.S. *Athenæum Club, S.W.; and Cambridge.*
1869	Sedgwick, John Bell, Esq. 1, *St. Andrew's-place, Regent's-park, N.W.*
1862	Seemann, Berthold, Esq., PH. DR., F.L.S. 57, *Windsor-road, Holloway, N.*
1866	1930 Sendall, Walter T., Esq., Inspector of Schools in Ceylon. *Colombo.*
1865	Sercombe, Edwin, Esq. 49, *Brook-street, Grosvenor-square, W.*
1858	*Serocold, Charles P., Esq. *Brewery, Liquorpond-street, E.C.*
1853	Sevin, Charles, Esq. 155, *Fenchurch-street, E.C.*
1870	Sewell, Edward, Esq., M.A. *Ilkley, Yorkshire.*
1867	Seymour, Alfred, Esq., M.P. 47, *Eaton-square, S.W.*
1858	Seymour, George, Esq. 54, *Lime-street, E.C.*
1858	*Seymour, Henry Danby, Esq. *Athenæum Club, S.W.; Knoley-Hindon, Wilts; and Glastonbury, Somersetshire.*

Date of Election	
1854	*Shadwell, Admiral Charles F. A., C.B. *Meadow-bank, Melksham, Wilts.*
1860	*Shadwell, Lieut.-Colonel Lawrence. 9, Queensberry-place, Cromwell-road, Kensington, W.*
1856	1930*Share, Staff Commander James Masters, R.N. *The Willows, Wyke Regis, Weymouth, Dorset.*
1866	Sharp, Henry T., Esq. 102, Piccadilly, W.
1861	Sharp, Peter, Esq. *Oakfield, Ealing, W.*
1861	*Sharpe, William John, Esq. 1, Victoria-street, Westminster, S.W.; and Norwood, Surrey, S.
1869	Shaw, James V., Esq. *The Elms, Twickenham, S.W.*
1862	*Shaw, John, Esq. *Finegand, Otago, New Zealand.* Care of John Morrison, Esq., New Zealand Government Agency, 3, Adelaide-place, E.C.*
1861	Shaw, John Ralph, Esq. *Arrowe-park, Birkenhead.*
1870	*Shaw, Robert B., Esq. (Manager of Tea Plantations, Kangra.)
1870	Shearme, Edward, Esq. *Junior Athenaeum Club, W.*
1846	Sheffield, George A. F. C., Earl of. 20, Portland-place, W.; and Sheffield-park, Sussex.
1857	1940Shell, Lieut.-Gen. Sir Justin, K.C.B. 13, Eaton-place, Belgrave-square, S.W.
1868	*Shelley, Capt. G. Ernest. 32, Chesham-place, W.
1861	Shephard, Chas. Douglas, Esq., Surg. R.N. *Plantation House, near Cheadle, Staffordshire.*
1867	Shepherd, Chas. Wm., Esq., M.A., F.Z.S. *Trottescliffe, Maidstone.*
1860	Sheridan, H. Brinsley, Esq. *Bellefield-house, Parson's-green, Fulham, S.W.*
1868	Sheridan, Richd. B., Esq., M.P. *Oaklands, St. Peter's, Thanet.*
1857	Sherrin, Joseph Samuel, Esq., LL.D., PH. DR. *Leyton-house, Leyton-crescent, Kentish-town, N.W.*
1850	*Sherwill, Lieut.-Col. W. S., F.G.S. *Perth, N.B.*
1850	*Shipley, Conway M., Esq. *Twyford Moors, Winchester;* and Army and Navy Club, S.W.
1868	Shirley, Lionel H., Esq., C.E., &c. *Raleigh Club; and The Lypiatts, Cheltenham.*
1864	1950Short, the Rev. Thos. Vowler.
1856	Shuttleworth, Sir J. P. Kay, Bart. 3, Victoria-street, S.W.; and Gawthorp-hall, Burnley, Lancashire.
1869	Silk, George Chas., Esq. *The Vicarage, Kensington, W.*
1870	*Silva, Samuel, Esq. 8, Sheen Villas, Park-road, Richmond.
1865	*Silva, Frederic, Esq. 12, Cleveland-square, Bayswater, W.
1850	Silver, the Rev. Fred., M.A., F.R.A.S. *Norton-rectory, Market Drayton, Salop.*
1850	*Silver, Stephen Wm., Esq. 66, Cornhill, E.C.; and Norwood-lodge, Lower Norwood, S.
1860	Sim, John Coysgarne, Esq. 13, James-street, Buckingham-gate, S.W.
1868	Simmons, Edward R., Esq., Barrister-at-Law. 4, Hyde-park-gate, S.W.
1846	*Simmons, Major-General Sir John L. A., R.E., K.C.B. *Lieut.-Governor Royal Military Academy, Woolwich, S.E.*
1868	1950Simmons, Henry M., Esq. *Tyersall-crescent, Wood-road, Sydenham-hill, S.E.*

Year of Election.	
1864	Simpson, Frank, Esq. 17, *Whitehall-place, S.W.*
1862	Simpson, Henry Bridgeman, Esq. 44, *Upper Grosvenor-street, W.*
1863	*Simpson, Wm., Esq. 64, *Lincoln's-inn-fields, W.C.*
1866	· *Sims, Richard Proctor, Esq., c.e. *Malabar-hill, Bombay. Care of Messrs. King and Co.*
1858	Skelmersdale, Edward, Lord. *Lathom-park, Ormskirk, Lancashire.*
1866	Skinner, John E. H., Esq. 3, *Dr. Johnson's-buildings, Temple, E.C.*
1863	Skrine, Hy. D., Esq. *Warleigh-manor, near Bath.*
1870	Sladen, Major E. B. (Polit. Agent at the Court of H.M. the King of Burmah). *Oriental Club, W.*
1861	Sladen, Rev. Edward Henry Mainwaring. *Alton, near Marlborough, Wilts.*
1861	1970Sligo, G. J. Browne, Marquis of. 14, *Mansfield-street, W.; and Westport, County Mayo.*
1865	Smedley, Joseph V., Esq., m.a. *Oxford and Cambridge Club, S.W.*
1860	*Smith, Augustus Henry, Esq. *Flexford-house, Guildford.*
1857	*Smith-Bosanquet, Horace, Esq. *Broxbourne-borough, Hoddesdon.*
1866	Smith, Drummond Spencer-, Esq. 7, *Mount-street, Berkeley-square, W.*
1859	Smith, Edward, Esq. *Windham Club, S.W.*
1867	Smith, Frederick, Esq. *The Priory, Dudley.*
1865	Smith, Guildford, Esq. 63, *Charing-cross, S.W.*
1861	Smith, Jervoise, Esq. 47, *Belgrave-square, S.W*
1853	Smith, John Harrison, Esq. 55, *Chepstow-place, Bayswater, W.*
1853	1980Smith, John Henry, Esq. 1, *Lombard-street, E.C.; and Purley, Croydon, Surrey.*
1861	*Smith, Joseph Travers, Esq. 25, *Throgmorton-street, E.C.*
1838	*Smith, Octavius Henry, Esq. 28, *Prince's-gate, Hyde-park, W.*
1868	*Smith, Major Robert M., r.e., Director of the Telegraphic Establishment in Persia, Teheran. *Care of the Foreign-office, S.W.*
1857	Smith, Captain Philip, Grenadier Guards.
1841	*Smith, Thomas, Esq.
1859	*Smith, W. Castle, Esq. 1, *Gloucester-terrace, Regent's-park, N.W.*
1857	Smith, Wm. Gregory, Esq. *Hudson-bay Company, 1, Lime-street, E.C.*
1859	Smith, William Henry, Esq., m.p. 1, *Hyde-park-street, W.*
1869	Smyth, Colonel Edmund. *Elkington-hall, Wotton-hill, Lincolnshire.*
1869	1990*Smyth, Warington, Esq., f.r.s. 92, *Inverness-terrace, W.*
1837	*Smyth, Rear-Adm. William. *Care of Messrs. Child and Co., Temple-bar.*
1850	*Smythe, Colonel William J., r.a.
1863	Snowden, Francis, Esq., m.a. 1, *Dr. Johnson's-buildings, Temple, E.C.*
1865	Solomons, Hon. Geo. *Jamaica.*
1839	*Somers, Charles, Earl. 33, *Prince's-gate, S.W.; Eastnor-castle, Herefordshire; and The Priory, Reigate, Surrey.*
1862	Somerset, Capt. Leveson E. H., r.n. *Care of Messrs. Chard, 3, Clifford's-inn Fleet-street, E.C.*
1858	*Somes, Joseph, Esq. *Burntwood-lodge, Wandsworth-common, S.W.*

Year of Election.	
1855	Sopwith, Thos., Esq., M.A., C.E., F.R.S. 103, *Victoria-street, Westminster, S.W*
1861	South, John Flint, Esq. *Blackheath-park, S.E.*
1860	2000°Southey, Jas. Lowther, Esq. *Care of Messrs. Stilwell.*
1850	Southesk, the Right Hon. Jas. Carnegie, Earl of. 38, *Portland-place, W.*
1869	Southwell, Thomas Arthur Joseph, Viscount. *Windham Club, S.W.*
1865	Spalding, Samuel, Esq. 7, *Upper Park-road, South Hampstead.*
1870	Spurin, J. Hyde, Esq. *Conservative Club, S.W.*
1850	*Spencer-Bell, James, Esq. 1, *Devonshire-place, Portland-place, W.*
1857	Spicer, Edward, Esq. *Woodside, Muswell-hill, N.*
1863	Spickernell, Dr. Geo. E., Principal of Eastman's Royal Naval Establishment *Eastern-parade, Southsea.*
1855	*Spottiswoode, William, Esq., F.R.S. 50, *Grosvenor-place, S.W.*
1850	*Spratt, Capt. Thos. A. B., R.N., C.B. *Clare-lodge, Nevill-park, Tunbridge wells, Kent.*
1865	2010Spruce, Richard, Esq., PH. DR. *Welburn, Castle Howard, York.*
1869	Stafford, Edward W., Esq. *Colonial Secretary of New Zealand. Care o Mr. J. S. Tytler, 19, Castle-street, Edinburgh.*
1868	Staley, the Right Rev. Bishop, D.D. 5, *Park-place, St. James's, S.W.*
1853	Stanford, Edward, Esq. 6, *Charing-cross, S.W.*
1855	Stanhope, Philip Henry, Earl, Pres. Soc. of Antiquaries. 3, *Grosvenor-place houses, Grosvenor-place, S.W.; and Chevening, Sevenoaks, Kent.*
1869	*Stanhope, Walter Spencer, Esq. *Cannon-hall, Barnsley, Yorkshire.*
1856	Stanley, Edmund Hill, Esq. *Leicester-house, Gipsy-hill, Norwood.*
1865	Stanton, Charles Holbrow, Esq. 1, *Mitre-court-buildings, Inner Temple, E.C.*
1863	Stanton, Geo., Esq. *Coton-hill, Shrewsbury; and Conservative Club, S.W.*
1867	Stanton, Henry, Esq. 1, *River-street, Myddelton-square, W.C.*
1870	2020Starling, Joseph, Esq. *Chichester-lodge, Dyke-road, Brighton.*
1855	Statham, John Lee, Esq. 60, *Wimpole-street, W.*
1863	*Staveley, Miles, Esq. *Old Sleningford-hall, Ripon.*
1868	Staveley, Major-Gen. Sir Charles, K.C.B. *Government-house, Devonport; an United Service Club, S.W.*
1859	Stebbing, Edward Charles, Esq. *National Debt Office, 19, Old Jewry, E.C.*
1867	Steel, J. P., Esq., Lieut. R.E. *Junior United Service Club, S.W. Care of Messrs Grindlay and Co., 55, Parliament-street, S.W.*
1868	Steel, William Strang, Esq. 13, *Leadenhall-street, E.C.*
1870	Stensing, Charles, Esq. 3, *Upper Hamilton-terrace, N.W.*
1864	*Stephen, Sir George. *Melbourne. Care of Mr. H. W. Ravenscroft, 7, Gray's-inn-square, W.C.*
1870	Stephens, Thos. Wall, Esq. *North-villa, Regent's-park, N.W.*
1866	2030Stephenson, B. Charles, Esq. 12, *Bolton-row, Mayfair, W.*
1857	Stephenson, Sir R. Macdonald, C.E. 72, *Lancaster-gate, W.; and East-cottage Worthing.*
1868	Stephenson, Henry P., Esq. 8, *St. Mary-axe, E.C.*
1868	Stepney, A. K. Cowell, Esq. 6, *St. George's-terrace, Knightsbridge, W.*

Year of Election.	
1860	Sterling, Col. Sir Anthony. *South-lodge, South-place, Knightsbridge, W.*
1862	Sterry, Henry, Esq. *7, Paragon, Southwark, S.E.*
1869	Steuart, Col. T. R., Bombay Army. *Esgair, Machynlleth, Wales.*
1855	Stevens, Henry, Esq., F.S.A. *4, Trafalgar-square, W.C.*
1841	Stevenson, Thomas, Esq., F.S.A. *37, Upper Grosvenor-street, W.*
1866	Stewart, Rev. Dr. James. *Lovedale, Alice, South Africa. Care of Robert Young, Esq., Offices of the Free Church of Scotland, Edinburgh.*
1860	2040*Stewart, Major J. H. M. Shaw, Royal Madras Engineers.
1869	Stewart, J. L., Esq., M.D., Forest Department, India. *Kew, W.*
1870	Stilwell, Henry, Esq., M.D. *Moorcroft, Hillington, Uxbridge.*
1868	Stirling, the Hon. Edward. *34, Queen's-gardens, Hyde-park, W.*
1860	Stirling, Capt. Frederick H., R.N. *H.M.S. 'Hero;' and United Service Club, S. W.*
1863	Stirling, Sir Walter, Bart. *36, Portman-square, W.*
1868	Stock, Thomas Osborne, Esq., M.P. *51, Portsdown-road, W.*
1860	Stooker, John Palmer, Esq. *93, Oxford-terrace, Hyde-park, W.*
1845	*Stokes, Rear-Admiral John Lort. *United Service Club, S. W.; and Scotchwell, Haverfordwest, Wales.*
1868	Stone, David H., Esq., Alderman. *Sydenham-hill, S.*
1867	2050*Story, Edwin, Esq., M.A. *3, King Edward's-terrace, Liverpool-road, Islington, N.*
1876	Stoton, Wm. O., Esq. *Junior Carlton Club, S. W.*
1868	Stovin, Rev. Charles F. *8, Grosvenor-mansions, Victoria-street, S.W.*
1866	Strachey, Colonel Richard, R.E., F.R.S.
1861	Strange, Lieut.-Col. Alexander. *India Store Department, Belvedere-road, Lambeth, S.*
1858	Stratford de Redcliffe, Stratford Canning, Viscount. *29, Grosvenor-square, W.*
1864	Straton, Rev. N. D. J. *Kirkby-wharf, Tadcaster.*
1860	Strickland, Edward, Esq., C.B., Commissary-General. *Halifax, Nova Scotia.*
1868	*Strode, Alf. Rowland Chetham, Esq. *Dunedin, Otago, New Zealand. Care of J. G. Cooke, Esq.*
1865	Strong, F. K., Esq., K.H. *Hamburg, Germany; and 8, St. Martin's-place, S. W.*
1853	2060Strousberg, Dr. Bethel Henry. *70, Wilhelm-strasse, Berlin. Care of Messrs. Asher & Co., Bedford-street, Covent-garden, W.C.*
1853	Strutt, George H., Esq., F.R.A.S. *Bridge-hill, Belper.*
1858	Strutt, Captain Hammel Ingold, F.R.A.S. *Royal Mail Steam Packet Company, Southampton.*
1853	*Strzelecki, Count P. E. de, C.B., F.R.S. *23, Savile-row, W.*
1859	Stuart, Lieut.-Col. J. F. D. Crichton. *25, Wilton-crescent, Belgrave-sq., S.W.*
1861	Stuart, Vice-Chancellor Sir John. *11 and 12, Old-buildings, Lincoln's-inn, W.C.; 5, Queen's-gate, Hyde-park, W.; and Grushernish, Isle of Skye, Invernesshire.*
1866	Stuart, Major Robert. *Janina, Albania.*
1858	Sudeley, Charles G. Hanbury Tracy, Lord. *5, Bolton-row, W.; and Toddington,*

Year of Election	
1869	2070 Summerhayes, William, Esq., M.D. 18, *Sandringham-gardens, Ealing, W.*
1862	Surridge, Rev. Henry Arthur Dillon, M.A. 21, *Berners-street, W.*
1862	Surtees, Capt. Charles Freville. *Chalcott-house, Long Ditton, Surrey.*
1861	*Sutherland, George Granville William, Duke of. *Stafford-house, St. James's Palace, S.W.*
1869	Sutherland, Robert, Esq. *Egham-rise, Surrey.*
1869	Sutherland, Thomas, Esq. *H 3, Albany, Piccadilly, W.*
1857	Swansy, Andrew, Esq. *Sevenoaks, Kent.*
1836	*Swinburne, Rear-Admiral Charles H. *Capheaton, near Newcastle-upon-Tyne.*
1862	*Swinburne, Com. Sir John, Bart., R.N. *Capheaton, Newcastle-on-Tyne.*
1863	Swinhoe, R., Esq., H.B.M. Consul, Taiwan. 33, *Oakley-square, S.W.*
1851	2080 Sykes, Colonel William Henry, M.P., F.R.S., Hon. M.R.I.A. *Athenæum Club, S.W.; and* 47, *Albion-street, Hyde-park, W.*
1864	Symonds, F., Esq., M.D. *Beaumont-street, Oxford.*
1852	*Synge, Col. Millington H., R.E. *Alverclif, Alverstoke, Hants.*
1852	Tagart, Courtenay, Esq. *Rockleaze Point, Durdham Down, near Bristol.*
1860	Tagart, Francis, Esq. 31, *Craven-hill-gardens, Hyde-park, W.*
1866	Taintor, Edward C., Esq. (Impl. Chinese Customs). *Tientsin, China. Care of H. C. Batchelor, Esq.,* 155, *Cannon-street, E.C.*
1864	Tait, P. M., Esq. 38, *Belsize-park, N.W.; and Oriental Club, W.*
1857	*Tait, Robert, Esq. 14, *Queen Anne-street, W.*
1867	Talbot, Right Hon. Richard Gilbert. *Ballinclea, Kingstown, County Dublin.*
1861	Talbot de Malahide, James Talbot, Lord. *Malahide Castle, Co. Dublin.*
1861	2090 Taylor, Commander A. Dundas, I.N. 2, *Gloucester-villas, Upper Eglinton-road, Shooter's-hill, S.E.*
1869	Taylor, George N., Esq. 15, *Elvaston-place, Queen's-gate, W.*
1863	Taylor, H. L., Esq. *Reform Club, S.W.; and* 23, *Phillimore-gardens, Kensington, W.*
1866	Taylor, Rev. Jas. Hudson. *Ningpo, China. Care of Mr. Berger, Saint-hill, East Grinstead.*
1865	Taylor, John, Esq. *The National Bank,* 13, *Old Broad-street, E.C.*
1869	Taylor, John Fenton, Esq. 5, *Horbury-crescent, Notting-hill; and* 12, *New-street, Spring-gardens, S.W.*
1867	*Taylor, John George, Esq. *H.B.M. Consul in Kurdistan, Diarbekr. Care of Messrs. O'Brien and Co.,* 43, *Parliament-street, S.W.*
1854	*Taylor, John Stopford, Esq., M.D. 1, *Springfield, St. Anne-street, Liverpool.*
1863	Taylor, Col. R. C. H. 16, *Eaton-place, S.W.; and Carlton Club, S.W.*
1864	Taylor, W. R., Esq.

Year of Election	
1857	2100 Teesdale, John M., Esq. *Eltham-house, Eltham, S.E.*
1863	Tegg, Wm., Esq. 13, *Doughty-street, Mecklenburg-square*, W.C.
1865	Temple, Sir Richard, B.C.S.I.
1860	Templeton, John, Esq. 24, *Budge-row*, E.C.
1857	Tennant, Professor James. 149, *Strand*, W.C.
1870	Teschemacher, Edward Fred., Esq. 1, *Highbury-park North*, N.
1830	*Thatcher, Colonel E.I.C.
1865	Theed, William S., Esq.
1863	Thomas, G., Esq. 6, *Queen's-gate-terrace, Hyde-park*, W.
1854	Thomas, Henry Harrington, Esq. 8, *Camden-crescent, Bath.*
1864	2110 Thomas, J. R., Esq., Staff Assist. Surg. *Castle-hill, Fishguard, Pembrokeshire.*
1865	Thomas, John Henwood, Esq. *East India Dept., Custom-house, E.C.*
1869	Thompson, Lieut.-Col. George, C.E., Cordoba, Argentine Confederation. *Care of Messrs. Parlane, Graham, and Co., Buenos Ayres; Messrs. Lumb, Wanklyn, and Co., 10, Angel-court, Throgmorton-street, E.C.*
1869	Thompson, Henry Yates, Esq. *Vice-regal Lodge, Dublin; 2, Cleveland-row, St. James's, S.W.; and Thingwall-park, near Liverpool.*
1854	Thompson, William C., Esq.
1863	Thomson, James, Esq. *Dunstable-house, Richmond.*
1863	Thomson, James Duncan, Esq., Portuguese Consul. *St. Peter's-chambers, Cornhill, E.C.*
1848	*Thomson, J. Turnbull, Esq., Chief Surveyor. *Otago, New Zealand.*
1866	Thomson, John, Esq. *Care of John Little, Esq., 21, Cannon-street, E.C.*
1861	*Thomson, Ronald Ferguson, Esq., 1st Attaché to the Persian Mission. *Care of F. B. Alston, Esq., Foreign-office, S.W.*
1854	2120 *Thomson, Thomas, Esq., M.D., F.R.S. *Hope-house, Kew, W.*
1865	Thomson, W. T., Esq. *Arlary-house, Kinross.*
1862	*Thorne, Augustus, Esq. 4, *Cullum-street, City, E.C.*
1867	Thornton, Edward, Esq., C.B. *Harrow.*
1847	Thornton, Rev. Thomas Cooke, M.A., M.R.I. *Brock-hall, near Weedon, Northamptonshire.*
1858	Thorold, Rev. A. W. 31, *Gordon-square*, W.C.
1868	Thorold, Alexander W. T. Grant, Esq. *Medsley, Great Grimsby, Lincolnshire.*
1854	Thorold, Henry, Esq. *Cuxwold, Lincolnshire.*
1861	Thrupp, John, Esq.
1859	Thuillier, Lt.-Col. H. L., Surveyor-General of India. *Calcutta; Messrs. Grindlay, and Co. Care of J. Walker, Esq., India Office.*
1865	2130 *Thurburn, C. A., Esq. 29, *Queensborough-terrace, Kensington-gardens*, W.
1864	Thurburn, Hugh, Esq. 108, *Westbourne-terrace*, W.
1861	Thurlow, the Hon. Thos. J. Hovell. *Dumphail, Forres, N.B.*
1868	Tilley, Henry Arthur, Esq. *Hanwell, Middlesex*, W.
1859	*Tinne, John A., Esq. *Briarley, Aigburth, near Liverpool.*
1862	Todd, John, Esq. *Eastcote-lodge, St. John's-park, Blackheath, S.E.*

Year of Election.	
1865	Todd, Rev. John W. *Tudor-hall, Forest-hill, Sydenham, S.*
1853	*Tomlin, George Taddy, Esq., F.S.A. *Combe-house, Bartonfields, Canterbury.*
1853	Tomline, George, Esq. 1, *Carlton-house-terrace, S. W.*
1835	*Tooke, Arthur Wm., Esq., M.A. *Pinner-hill-house, near Watford, Middlesex.*
1856	2140 Torrance, John, Esq. 5, *Chester-place, Hyde-park-square, W.*
1866	Torrens, Robert Richard, Esq. 2, *Gloucester-place, Hyde-park, W.; and The Cott, Holm, near Ashburton, South Devon.*
1859	Townsend, Commander John, R.N. *Lona, Weston-super-Mare.*
1866	Townson, Wm. Parker, Esq., B.A. Cantab. *Care of Miss Townson, Ash-house, Caton, near Lancaster.*
1846	*Towry, George Edward, Esq.
1858	Towson, J. Thomas, Esq. *Secretary Local Marine Board, Liverpool.*
1864	*Toynbee, Capt. Hy. 25, *Inverness-road, Kensington-gardens, W.*
1863	*Tozer, Rev. H. F., M.A. *Exeter College, Oxford.*
1864	Tracy, the Hon. C. H. 11, *George's-street, W.*
1863	*Travers, Arch., Esq. *Addison-road (opposite the Napier-road), Kensington, W.*
1867	2150 Tremenheere, Col. C.W., R.E. *Bombay.*
1859	Tremlett, Rev. Francis W., M.A., D.C.L., HON. PH.DR. of Jena. *Belsize-park, Hampstead, N.W.*
1869	Trench, Capt. Frederic. *Naval and Military Club, Piccadilly, W.*
1865	*Trench, Capt. the Hon. Le Poer, R.E. 32, *Hyde-park-gardens, W.; and Ordnance-survey-office, Pimlico, S.W.*
1863	Trestrail, Rev. Frederick. *Stanmore-villa, Beulah-hill, Upper Norwood, S.*
1862	Trevelyan, Sir Charles Edward, K.C.B. 8, *Grosvenor-crescent, S.W.*
1830	Trevelyan, Sir Walter Calverly, Bart., M.A., F.S.A., F.L.S., F.R.S.N.A., &c. *Athenæum Club, S.W.; Wallington, Northumberland; and Nettlecombe, Somerset.*
1864	Trimmer, Edmund, Esq. *Care of Messrs. Trimmer and Co., New City-chambers, Bishopsgate-street, E.C.*
1867	Tritton, Joseph Herbert, Esq. 54, *Lombard-street, E.C.*
1869	Trotter, Lieut. Henry, R.E. 11, *Hertford-street, Mayfair, W.*
1867	2160 Tryon, Capt. George, R.N., C.B. *Army and Navy Club, S.W.*
1862	Tuckett, Frns. Fox, Esq. *Frenchay, near Bristol.*
1835	*Tuckett, Frederick, Esq. 4, *Mortimer-street, Cavendish-square, W.*
1865	Tuckett, Philip D., Esq. 28, *Cleveland-gardens, Hyde-park, W.*
1852	Tudor, Edward Owen, Esq., F.S.A. 80, *Portland-place, W.*
1857	Tudor, Henry, Esq. 80, *Portland-place, W.*
1870	Tupper, Lieut.-Col. D. W. *Army and Navy Club, S.W.*
1864	Turnbull, George, Esq., C.E., F.R.A.S. 23, *Cornwall-gardens, South Kensington, W.*
1834	*Turnbull, Rev. Thomas Smith, F.R.S. *University Club, S.W.; and Blofield, Norfolk.*
1870	Turner, Maj.-Gen. Henry Blois, Bomb. Eng. 131, *Harley-street, W.*
1853	2170 Turner, Thos., Esq. *Guy's Hospital, Southwark, S.*
1867	Tweedie, Capt. Michael, H.A. *Care of R. W. Tweedie, Esq., 5, Lincoln's-inn-fields.*

Year of Election	
1864	*Twentyman, A. C., Esq. *Tettenhall-wood, near Wolverhampton.*
1863	Twentyman, Wm. H., Esq. *Ravensworth, St. John's-wood-park, N.W.*
1863	*Twiselton, Hon. E. F. *Rutland-gate, S.W.*
1849	Twiss, Sir Travers, D.C.L., F.R.S. 19, *Park-lane, W.*
1858	Twyford, Capt. A. W., 21st Hussars. Deputy-Governor, H.M.'s Convict Service, Brixton. *Reform Club, S.W.*
1865	Tyer, Edward, Esq., C.E., F.R.A.S. 15, *Old-jewry-chambers, E.C.*
1862	*Tyler, George, Esq. 24, *Holloway-place, Holloway-road, N.*
1859	Tytler, Capt. W. Fraser. *Aldourie, Inverness.*
1869	2180Underdown, E. M. Esq., 3, *King's-Bench-walk, Temple, E.C.*
1862	Underhill, Edward Benn, Esq., LL.D. *Derwent-lodge, Thurlow-road, Hampstead, N.W.*
1868	Unwin, Howard, Esq., C.E. 24, *Bucklersbury, E.C.*
1861	Ussher, John, Esq. *Arthur's Club, St. James's-street, S.W.*
1844	*Vacher, George, Esq. *Manor-house, Teddington.*
1862	*Vander Byl, P. G., Esq., M.P. *Care of Mr. H. Blyth,* 17, *Gracechurch-st., E.C.*
1865	Vane, G., Esq. *Ceylon. Messrs. Price and Boustead.*
1856	*Vaughan, James, Esq., F.R.C.S. *Builth, Breconshire.*
1861	Vaughan, J. D., Esq.,
1849	Vaux, William S. W., Esq., M.A., F.S.A. 4, *St. Martin's-place, W.C.*
1852	2190*Vavasour, Sir Henry M., Bart. 8, *Upper Grosvenor-street, W.*
1855	Vavasseur, James, Esq. *Knockholt, near Sevenoaks, Kent.*
1867	Venner, Capt. Francis John S. *Dilston-house, Upper Norwood, S.; and Elmbank, near Worcester.*
1863	*Vereker, the Hon. H. P., LL.D., H.M. Consul at Charante. 1, *Portmansquare, W.*
1862	Verner, Edward Wingfield, Esq., M.P. *The Ashe, Bray, County Wicklow.*
1862	*Verney, Edmond H., Commr. R.N. 32, *South-street, Grosvenor-square, W.*
1837	*Verney, Major Sir Harry C., Bart., M.P., F.R.A.S. *Travellers' Club, S.W.; 32, South-street, Grosvenor-square, W.; and Claydon-house, Bucks.*
1857	Verrey, Charles, Esq.
1852	Verulam, James Walter, Earl of. *Gorhambury, near St. Alban's; Barry-Mill, Surrey; and Messing-hall, Essex.*
1865	Vile, Thomas, Esq. 75, *Oxford-terrace, W.*
1865	2200*Vincent, M. C., Esq., Professor of Economic Geology and Metallurgy; Inspector of Mines, &c. *Cincinnati, U.S.; and* 127, *Strand, London.*
1857	Vincent, John, Esq. 4, *Granville-park, Blackheath, S.E.*

Year of Election.	
1856	Vines, William Reynolds, Esq., F.R.A.S. *Care of R. P. Hollyer, Esq., Ivy Garth, Shrewsbury-lane, Shooter's-hill, Plumstead, S.E.; and 4, Thavies-inn, Holborn-hill, E.C.*
1863	Vivian, Major Quintus, late 8th Hussars. 17, *Chesham-street, S.W.*
1838	*Vyvyan, Sir Richard Rawlinson, Bart., F.R.S. Trelowarren, Cornwall.*
1852	Wade, Mitchell B., Esq. 66, *South John-street, Liverpool.*
1864	Wade, R. B., Esq. 13, *Seymour-street, Portman-square, W.*
1863	Wade, Thos. F., Esq., C.B., H.B.M. Secretary of Legation. *Pekin, China.*
1858	*Wagstaff, William Racster, Esq., M.D., M.A.*
1860	Waite, Charles, Esq., LL.D., Principal of St. John's College. *Weighton-road, South Penge-park, S.E.*
1863	2210 Waite, Henry, Esq. 3, *Victoria-street, Pimlico, S.W.*
1867	*Waite, Rev. John.*
1870	*Walker, Albert, Esq. Auckland Club, New Zealand. Care of L. C. Walker, Esq. 3, Hartley-villas, Lansdowne-road, Croydon.*
1862	Walker, Col. C. P. Beauchamp, C.B. 97, *Onslow-square, S.W.; and United Service Club, S.W.*
1861	Walker, Edward Henry, Esq., Consul at Cagliari. *Newton-bank, Chester.*
1863	*Walker, Frederick John, Esq. Alltyr Odyn, Llandyssil, Carmarthen, Wales.*
1863	Walker, James, Esq., Managing Director of Madras Railway. 23, *Cambridge-square, Hyde-park, W.*
1850	*Walker, Lt.-Col. James, Bombay Engineers. Murree, near Rawul Pindd, Punjab. Care of Messrs. H. S. King and Co., Pall-mall, S.W.*
1820	Walker, John, Esq., Hydrog. India Office. 31, *Keppel-street, Russell-square, W.C.*
1861	*Walker, John, Esq. 60, Porchester-terrace, W.*
1858	2220 *Walker, Captain John, H.M.'s 66th Foot. Broom-hill, Colchester.*
1864	Walker, R. B. N., Esq. *Care of Mr. Blissett, 38, South Castle-st., Liverpool.*
1863	*Walker, T. F. W., Esq. 6, Brock-street, Bath; and Athenæum Club, S.W.*
1866	Walker, Captain William Harrison, H.C.S. 3, *Gloucester-terrace, W.; and Board of Trade, S.W.*
1861	Walker, Rev. William. *Grammar-school, Hanley-castle, Upton-on-Severn.*
1868	Walker, William, Esq., F.S.A. 48, *Hilldrop-road, Tufnell-park, N.*
1866	Walkinshaw, William, Esq. 74, *Lancaster-gate, Hyde-park, W.*
1854	Wallace, Alfred Russell, Esq. *Holly-house, Tanner-street, Barking, E.*
1867	Wallace, Rev. Charles Hill, M.A. 3, *Harley-place, Clifton, Bristol.*
1868	Waller, Major George Henry. 16, *Eaton-square, S.W.*
	Waller, Rev. Horace. *The Vicarage, Leytonstone.*

Year of Election.	
1865	Waller, Sir Thos. Wathen, Bart. 16, *Eaton-square, S.W.*
1863	Wallich, George C., Esq., M.D. 11, *Earl's-terrace, Kensington, W.*
1864	Walmsley, Joshua, Government Resident Agent. *Natal.*
1860	Walpole, Capt. the Hon. F., M.P. *Travellers' Club, S.W.; and Rainthorpe-hall, Long Stratton, Norfolk.*
1863	Walpole, Rt. Hon. Spencer, M.P. 109, *Eaton-square, S.W.*
1853	Walter, Henry Fraser, Esq. *Papplewick-hall, near Nottingham.*
1865	Walton, H. C., Esq., C.E. 26, *Savile-row, W.*
1863	Walton, J. W., Esq. 26, *Savile-row, W.*
1864	Walton, R. G., Esq., C.E. *Bombay.*
1853	2240*Ward, George, Esq.
1860	Ward, Admiral J. Hamilton. *Oakfield, Wimbledon-park, S.W.*
1865	Ward, Swinburne, Esq., Civil Commissioner. *Seychelles Islands.*
1868	Ward, Capt. the Hon. Wm. John, R.N. *H.M. Legation, Washington, Care of the American Department, Foreign-office, S.W.*
1869	Ward, William Robert, Esq. *Nea-house, Christchurch, Hants.*
1862	Wardlaw, John, Esq. 44, *Princes-gardens, Hyde-park, S.W.*
1868	Wardlaw, Col. Robert, C.B. *United Service Club, S.W.*
1864	Warner, E., Esq. 49, *Grosvenor-place, S.W.*
1859	Warre, Arthur B., Esq. 109, *Onslow-square, S.W.*
1869	Warre, Col. H. J., C.B. *United Service Club, S.W.*
1869	2250Warren, Charles, Esq. 17, *Hanover-street, Peckham, S.E.*
1862	Warren, Capt. Richard Pelham. *Worting-house, Basingstoke.*
1867	Washbourn, B., Esq., M.D., &c. *Eastgate-house, Gloucester.*
1867	Waterhouse, George Marsden, Esq. *Care of Messrs. Morrison and Co., Philpot-lane, E.C.*
1852	Watkins, John, Esq., F.R.C.S., F.S.A. 2, *Falcon-square, Aldersgate-street, E.C.*
1862	Watney, John, Esq. 16, *London-street, Fenchurch-street, E.C.*
1859	Watson, James, Esq. 24, *Endsleigh-street, W.C.*
1860	Watson, James, Esq., Barrister-at-Law. 13, *Circus, Bath.*
1861	Watson, John Harrison, Esq. 28, *Queensborough-terrace, Kensington-gardens, W.*
1868	Watson, Robert, Esq. 32, *Inverness-road, Bayswater, W.*
1867	2260Watson, Robert Spence, Esq. *Moss Croft, Gateshead-on-Tyne.*
1870	Watson, Thos., Esq., Portuguese Vice-Consul, Capetown. *Care of J. R. Thomson and Co., St. Peter's-chambers, E.C.*
1868	Watson, Wm. Bryce, Esq. 5, *Lime-street-square, E.C.; and 29, Duke-street, St. James's, S.W.*
1853	Watts, J. King, Esq. *St. Ives, Huntingdonshire.*
1857	*Waugh, Maj.-General Sir Andrew Scott, Bengal Engineers, F.R.S., late Surveyor-General and Superintendent Great Trig. Survey. *Athenæum Club, S.W.; and 7, Petersham-terrace, Queen's-gate-gardens, South Kensington, W.*
1868	Webb, Edward B., Esq., C.E., &c. 34, *Great George-street, S.W.*
1858	*Webb, Capt. Sydney. *Oriental Club, Hanover-square, W.; and 24, Manchester-square, W.*

Year of Election.	
1862	*Webb, William Frederick, Esq. *Army and Navy Club, S.W.*
1836	*Webber-Smith, Maj.-General James, 95th Regiment. 23, *Devonshire-terrace, Hyde-park, W.*
1865	Webster, Alphonsus, Esq. 44, *Mecklenburg-square, W.C.*
1864	2270 Webster, E., Esq. *North-lodge, Ealing, W.*
1858	Webster, George, Esq., M.D. *Dulwich, S.E.*
1866	Webster, George, Esq. 40, *Finsbury-circus, E.C.*
1860	Weguelin, Thomas Matthias, Esq., M.P. *Peninsular and Oriental Steam Navigation Co., Moorgate-street, E.C.*
1851	Weller, Edward, Esq. 34, *Red-lion-square, W.C.*
1853	*Wellington, Arthur Richard, Duke of, Major-General, D.C.L. *Apsley-house, W.; and Strathfieldsaye, Hampshire.*
1870	*Wells, Arthur, Esq. *Nottingham.*
1864	Wells, Sir Mordaunt, late Chief Puisne Judge, Bengal. 107, *Victoria-st., S.W.*
1862	Wells, William, Esq. 22, *Bruton-street, W.; and Redleaf, Penshurst, Kent.*
1863	Welman, Chas., Esq. *Norton-manor, Taunton.*
1868	2280 Wentworth, William Charles, Esq. *Combe-lodge, Pangbourne, Berks.*
1857	West, Lieut.-Colonel J. Temple.
1870	West, Raymond, Esq., Bomb. Civ. Serv. 12, *Lower Fitzwilliam-street, Dublin.*
1861	West, Rev. W. De Lancy, D.D., Head Master, Grammar School. *Brentwood, Essex.*
1863	*Westlake, John, Esq. 16, *Oxford-square, W.*
1853	Westmacott, Arthur, Esq. *Athenæum Club, S.W.*
1852	Weston, Alex. Anderdon, Esq., M.A. 18, *Rutland-gate, Hyde-park, S.W.*
1862	Westwood, John, Esq. 8 and 9, *Queen-street-place, Southwark-bridge, E.C.*
1830	*Weyland, John, Esq., F.R.S. *Woodrising-hall, Norfolk.*
1866	Wharncliffe, Lord. 15, *Curzon-street, W.*
1861	2290 Wharton, Rev. J. C. *Willesden-vicarage, N.W.*
1858	Wheatley, G. W., Esq. 150, *Leadenhall-street, E.C.*
1859	Wheelwright, William, Esq. *Gloucester-lodge, Regent's-park, N.W.*
1869	Whichelow, James Sherer, Esq. *Zero-house, Hammersmith.*
1853	*Whinfield, Edward Wrey, Esq., B.A. *South Elkington-vicarage, Louth.*
1839	*Whishaw, James, Esq., F.S.A. *Oriental Club, W.*
1867	Whitaker, Thomas Stephen, Esq. *Everthorpe-hall, East Yorkshire; and Conservative Club, S.W.*
1868	Whitby Rev. Thomas, M.A., &c. *St. John's-terrace, Woodhouse Moor, Leeds.*
1857	White, Arthur D., Esq., M.D. 56, *Chancery-lane, W.C.*
1865	White, Lieut. Arthur Wellesley, R.A. *Kingston, Canada. Care of Capt. T. P. White, R.E., Ordnance Survey, Edinburgh.*
1855	2300 *White, Charles, Esq., J.P. 10, *Lime-st., E.C.; and Barnesfield-house, Dartford, Kent.*
1857	White, Henry, Esq. *The Lodge, Hillingdon Heath, near Uxbridge.*
1862	White, Maj.-Gen. Henry Dalrymple, C.B. 39, *Lowndes-square, S.W.*
1869	White, Robert Owen, Esq. *The Priory, Lewisham, S.E.*

Year of Election.	
1869	Wilson, Capt. Charles William, R.E. 4, *New-street, Spring-gardens, S.W.*
1865	Wilson, E., Esq. *Hayes-place, Bromley, Kent.*
1862	*Wilson, Robert Dobie, Esq. 15, *Green-street, Grosvenor-square, W.*
1869	2340 Wilson, Samuel King, Esq. 3, *Portland-terrace, Regent's-park, N.W.*
1869	Wilson, Rev. T. Given, B.A. *The Parsonage, Halstead, Essex.*
1854	Wilson, Captain Thomas, R.N.
1860	Wilson, Thomas, Esq. 3, *St. George's-villas, Cricketfield-road, Lower Clapton, N.E.*
1866	Wiltshire, Rev. Thomas, M.A., F.G.S., F.L.S. 13, *Granville-park, Lewisham, S.E.*
1868	*Winch, W. Richard, Esq. *Chislehurst, Kent.*
1846	*Winchester, Right Rev. Samuel Wilberforce, Lord Bishop of, F.R.S., F.S.A. 19, *St. James's-square, S.W.*
1862	Wing, Commr. Arthur, R.N. *Care of Messrs. Case and Loudensack.*
1863	Wingate, T. F., Esq. 44, *Albion-street, Hyde-park-square, W.*
1870	Wiseman, James, Esq. 1, *Orme-square, Bayswater, W.*
1870	2350 Wiseman, Lieut. W., R.N. 88, *Belgrave-road, S.W.; and Lagos, W. Africa. Care of Messrs. Case and Loudensack,* 1, *James-street, Adelphi, W.C.*
1864	Wodehouse, J. H., Esq., H.M.'s Commissioner and Consul-General for the Sandwich Islands. *Care of F. B. Alston, Esq., Foreign-office.*
1870	Wodehouse, Sir Philip, K.C.B. *Care of E. R. Wodehouse, Esq.,* 17, *Half-moon-street Piccadilly, W.*
1865	Wolfe, Capt. William Maynard, R.A. *Arts Club, Hanover-square, W.*
1866	*Wolff, Sir Henry Drummond, K.C.M.G. 15, *Rutland-gate, S.W.; and Athenæum Club, S.W.*
1868	Wood, Capt. John. *Oriental Club, W.*
1863	Wood, Hy., Esq. 10, *Cleveland-square, Hyde-park, W.*
1865	Wood, Lieut.-Colonel Wm., R.M. 4, *Hyde-park-terrace, Cumberland-gate, W.*
1868	*Wood, Richard Henry, Esq., F.S.A. *Crumpsall, near Manchester.*
1870	Wood, Capt. T. P. *Holly-bank, Rusthall, Tunbridge Wells.*
1857	2360 Woodhead, Major H. J. Plumridge. 44, *Charing-cross, S.W.*
1867	Woodfield, Mathew, Esq., M.I.C.E. *General Colonial Manager, Cape Copper Mining Co., Namaqualand, Cape of Good Hope.*
1862	Woods, Samuel, Esq. *Mickleham, near Dorking, Surrey.*
1864	Woolcott, Geo., Esq. 78, *Palace-gardens-terrace, Kensington, W.*
1863	*Worms, George, Esq. 17, *Park-crescent, Portland-place, W.*
1845	Worthington, Rev. James, D.D. 27, *John-street, Bedford-row, W.C.*
1856	Worthington, J. Hall, Esq. *Alton-hill, Oxton, near Birkenhead.*
1866	*Worthington, Richard, Esq. 7, *Champion-park, Denmark-hill, S.*
1857	Wortley, Rt. Hon. Jas. Stuart, Q.C. 29, *Berkeley-sq., W.; and Sheen, Surrey, S.*
1866	Wotton, William G., Esq., M.D. 15, *Clement's-inn, W.C.*
1839	2370 *Wyld, James, Esq. *Charing-cross, W.C.*
1863	Wylde, W. H., Esq. *Foreign-office, S.W.*
1867	*Wythes, George Edward, Esq. *Epping, Essex.*

Year of Election	
1869	Yardley, Sir William. *Hadlow-park, Tonbridge, Kent.*
1854	Yeats, John, Esq., LL.D. *Clayton-place, Peckham, S.E.*
1861	York, Most Rev. William Thomson, Archbishop of. *Bishopsthorpe, York.*
1859	Yorke, Lieut.-General Sir Charles, K.C.B. *19, South-st., Grosvenor-squar*
1830	*Yorke, Colonel Philip J., F.R.S. *89, Eaton-place, S.W.*
1857	*Young, Capt. Allen. *Riversdale, Twickenham, S.W.*
1838	*Young, Charles Baring, Esq. *4, Hyde-park-terrace, W.*
1830	2380*Young, George Frederick, Esq. *Limehouse, E.*
1830	*Young, James, Esq.
1858	Young, James, Esq. *Lime-field, West Calder, Midlothian.*
1866	Young, John, Esq., F.S.A. *Vanbrugh-fields, Blackheath, S.E.*
1865	Young, Rev. R. H., B.A. *Royal Naval School, New Cross, S.E.*
1868	Young, William, Esq. *7, Vernon-place, Bloomsbury-square, W.C.*
1857	Yule, Col. Henry, Bengal Engineers. *Messrs. Grindlay & Co., 55, Parlia street, S.W.*
1864	2387 Zwecker, J. B., Esq. *55, Patshall-road, Kentish-town, N.W.*

LIST OF PUBLIC INSTITUTIONS, &c.,

TO WHICH COPIES OF THE 'JOURNAL' AND 'PROCEEDINGS' ARE PRESENTED.

[Those marked with an asterisk * receive the Proceedings only.]

GREAT BRITAIN AND IRELAND.

ADMIRALTY (Hydrographic Office)
AGRICULTURAL SOCIETY (Royal)
ANTIQUARIES, SOCIETY OF
ARCHÆOLOGICAL SOCIETY
ARCHITECTS, INST. OF BRITISH (Royal)
ARTS, SOCIETY OF
ASIATIC SOCIETY (Royal)
ASTRONOMICAL SOCIETY (Royal)
ATHENÆUM CLUB
BRITISH MUSEUM, LIBRARY OF
CAMBRIDGE UNIVERSITY. THE LIBRARY
COLONIAL OFFICE
DUBLIN TRINITY COLLEGE LIBRARY
—— GEOLOGICAL SOCIETY (Trin. Coll.)
EDINBURGH, ROYAL SOCIETY OF
——, THE LIBRARY OF ADVOCATES
——, GEOLOGICAL SOCIETY OF
EDUCATION DEPARTMENT, LIBRARY OF
ENGINEERS, INSTITUTION OF CIVIL
ETHNOLOGICAL SOCIETY
FOREIGN OFFICE, LIBRARY OF
GEOLOGICAL SOCIETY
GEOLOGY, MUSEUM OF PRACTICAL
HER MAJESTY THE QUEEN, LIBRARY OF
HORTICULTURAL SOCIETY (Royal)
HUDSON BAY COMPANY'S LIBRARY
INDIA OFFICE, LIBRARY OF THE
LANCASHIRE AND CHESHIRE, HISTORIC SOCIETY OF
LINNEAN SOCIETY
LITERATURE, ROYAL SOCIETY OF
LIVERPOOL LITERARY AND PHILOSOPHICAL SOCIETY

*LIVERPOOL MERCANTILE MARINE ASSOCIATION
*LONDON LIBRARY, THE
MANCHESTER CHETHAM LIBRARY
—————— FREE LIBRARY
*—————— LITERARY AND PHILOSOPHICAL SOCIETY
METEOROLOGICAL OFFICE
NEWCASTLE-UPON-TYNE LITERARY AND PHILOSOPHICAL INSTITUTION
OXFORD, THE BODLEIAN LIBRARY AT
*———, RADCLIFFE OBSERVATORY
*POST OFFICE LIBRARY AND LITERARY ASSOCIATION
ROYAL ARTILLERY INSTITUTION, WOOLWICH, S.E.
—————— LIBRARY, WOOLWICH, S.E.
ROYAL DUBLIN SOCIETY
ROYAL INSTITUTION
—————— SOCIETY
SALFORD ROYAL MUSEUM AND LIBRARY, PEEL PARK, SALFORD.
STAFF COLLEGE, FARNBOROUGH STATION, HANTS.
STATISTICAL SOCIETY
TRADE, BOARD OF, LIBRARY OF
TRAVELLERS' CLUB
UNITED SERVICE INSTITUTION (Royal)
WAR DEPARTMENT, TOPOGRAPHICAL DEPÔT
ZOOLOGICAL SOCIETY

EUROPE.

AMSTERDAM .. Royal Acad. of Sciences
ATHENS ...... University Library
BRUSSELS .... Royal Acad. of Science
—————— .... Geographical Society
BERLIN .... Academy of Sciences
—————— ... Geographical Society
CHRISTIANIA . University Library
COPENHAGEN . Hydrographic Office
—————— . Royal Danish Ordnance Survey
—————— . Royal Society of Sciences
—————— . —————— of Northern Antiquaries
DIJON ...... Académie des Sciences, Arts et Belles Lettres
DARMSTADT .. Geographical Society
DRESDEN ... Statistical Society
FLORENCE ... Italian Geographical Society
—————— . . Ministry of Public Instruction
—————— .. National Library of

FRANKFORT .. Geographical Society
GENEVA ..... Geographical Society of
—————— ... Society of Natural History
*GOTHA .... Perthes, M. Justus
HAGUE (THE) . Royal Institute for Geography and Ethnology of Netherland India
HALLE AND LEIPZIG } German Oriental Society
JENA ...... University of
LEIPZIG .... Verein von Freunden der Erdkunde zu
LISBON .... Royal Acad. of Sciences
MADRID .... Royal Acad. of Sciences
MILAN ..... Lombardo-Veneto Institute of
MUNICH .... Bibliothèque Centrale Militaire
—————— ... Royal Library
PARIS ..... Institut Impérial
—————— ..... Académie des Sciences

EUROPE—*continued.*

PARIS . . . . . Annales de l'Agriculture et des Régions Tropicales (Madinier. M.)
——— . . . . . Bibliothèque Impériale
——— . . . . . Dépôt de la Guerre
——— . . . . . Dépôt de la Marine
——— . . . . . Ministère de la Marine et des Colonies
——— . . . . . Société Asiatique
——— . . . . . Société d'Ethnographie
——— . . . . . Societé d'Encouragement pour l'Industrie Nationale
——— . . . . . Societé de Géographie
PESTH . . . . . Hungarian Academy of Sciences
*PRAGUE . . . . Bohemian Royal Museum
ROME . . . . . . Accademia dei Lincei
ST. PETERSBURG Imperial Academy of Sciences

ST. PETERSBURG Imperial Geographical Society
STOCKHOLM . . Bureau de la Recherche Géologique de la Suède.
——— . . Royal Academy of Sciences.
STRASBURG . . . Société des Sciences Naturelles
TÜBINGEN . . . University Library
*UTRECHT . . . Royal Dutch Meteorological Institute
VENICE . . . . . Armenian Convent Lib.
VIENNA . . . . Imperial Academy of Sciences
——— . . . . Imperial Geographical Society
——— . . . . Imperial Geological Institute
ZURICH . . . . . Society of Antiquaries
——— . . . . . Society of Naturalists

ASIA.

BOMBAY . . . . Geographical Society
——— . . . . Asiatic Society
CALCUTTA . . . Asiatic Society of Bengal
——— . . . Geolog. Survey of India
CALCUTTA . . . Public Library
DEHRA DHOON . Great Trigonometrical Survey of India, Library of

KURRACHEE . . Gen. Lib. and Museum
MADRAS . . . . Literary and Philosoph. Society
SHANGHAI . . . Royal Asiatic Society (North China Branch)
SINGAPORE . . . Journal of Indian Archipelago

AFRICA.

CAIRO . . . . . Egyptian Society
CAPE TOWN . . The Public Library

AMERICA.

ALBANY . . . . New York State Library
BOSTON . . . . American Society of Arts and Sciences
——— . . . . . Massachusetts State Library
——— . . . . . Public Library
——— . . . . . Society of Nat. History
BRAZIL . . . . Historical and Geographical Institute of
CALIFORNIA . . Academy of Sciences
CHILE . . . . . University of
MEXICO. . . . . Geographical and Statistical Society of
NEW HAVEN . . Yale College Library
*——— . Silliman's Journal
NEW YORK . . Geographical and Statistical Society

PHILADELPHIA, Academy of Natural Sciences
———, American Philosophical Society
———, Franklin Institute
QUEBEC . . . . Library of the Parliament of Canada
TEXAS . . . . . Soule University.
*TORONTO . . . Department of Public Instruction for Upper Canada
———, Canadian Institute of
WASHINGTON . . Congress Library of
——— . . Smithsonian Institution
——— . . National Observatory
WORCESTER . . Antiquarian Society

AUSTRALASIA.

ADELAIDE . . . . . South Australian Institute
MELBOURNE . . . . . Public Library
*——— . . . . Mining Department
*VICTORIA . . . . Royal Society
NEW ZEALAND . . . Library of the House of Representatives
SYDNEY . . . . University Library
TASMANIA . . . . . Royal Society

NAMES OF INDIVIDUALS TO WHOM THE ROYAL PREMIUMS AND OTHER TESTIMONIALS HAVE BEEN AWARDED.

1831.—Mr. RICHARD LANDER, for the discovery of the course of the River Niger or Quorra, and its outlet in the Gulf of Benin.

1832.—Mr. JOHN BISÇOE, for the discovery of the land now named "Enderby Land" and "Graham Land," in the Antarctic Ocean.

1833.—Captain Sir JOHN ROSS, R.N., for discovery in the Arctic Regions of America.

1834.—Sir ALEXANDER BURNES, for the navigation of the River Indus, and a journey by Balkh and Bokhara, across Central Asia.

1835.—Captain Sir GEORGE BACK, R.N., for the discovery of the Great Fish River, and its navigation to the sea on the Arctic Coast of America.

1836.—Captain ROBERT FITZROY, R.N., for the survey of the Shores of Patagonia, Chile, and Peru, in South America.

1837.—Colonel CHESNEY, R.A., for the general conduct of the "Euphrates Expedition" in 1835-6, and for accessions to the geography of Syria, Mesopotamia, and the Delta of Susiana.

1838.—Mr. THOMAS SIMPSON—Founder's Medal—for the discovery and tracing, in 1837 and 1838, of about 300 miles of the Arctic shores of America.
—— Dr. EDWARD RÜPPELL—Patron's Medal—for his travels and researches in Nubia, Kordofán, Arabia, and Abyssinia.

1839.—Col. H. C. RAWLINSON, E.I.C.—Founder's Medal—for his travels and researches in Susiana and Persian Kurdistán, and for the light thrown by him on the comparative geography of Western Asia.
—— Sir R. H. SCHOMBURGK—Patron's Medal—for his travels and researches during the years 1835-9 in the colony of British Guayana, and in the adjacent parts of South America.

1840.—Lieut. RAPER, R.N.—Founder's Medal—for the publication of his work on 'Navigation and Nautical Astronomy.'
—— Lieut. JOHN WOOD, I.N.—Patron's Medal—for his survey of the Indus, and re-discovery of the source of the River Oxus.

1841.—Captain Sir JAMES CLARK ROSS, R.N.—Founder's Medal—for his discoveries in the Antarctic Ocean.
——Rev. Dr. E. ROBINSON, of New York—Patron's Medal—for his work entitled 'Biblical Researches in Palestine.'

1842.—Mr. EDWARD JOHN EYRE—Founder's Medal—for his explorations in Australia.
——Lieut. J. F. A. SYMONDS, R.E.—Patron's Medal—for his survey in Palestine, and levels across the country to the Dead Sea.

1843.—Mr. W. J. HAMILTON—Founder's Medal—for his researches in Asia Minor.
—— Prof. ADOLPH ERMAN—Patron's Medal—for his extensive geographical labours.

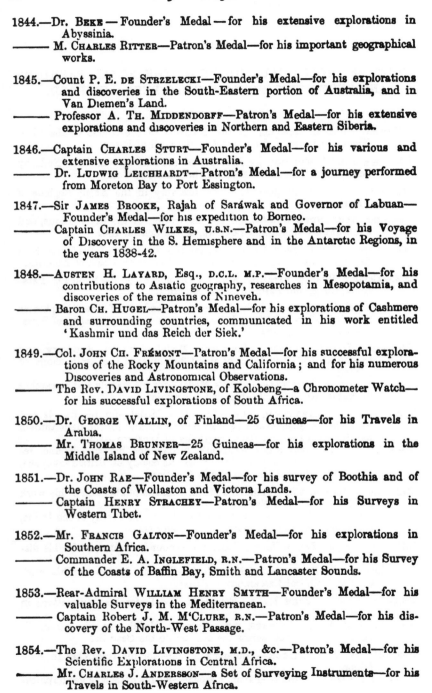

1844.—Dr. BEKE — Founder's Medal — for his extensive explorations in Abyssinia.

——— M. CHARLES RITTER—Patron's Medal—for his important geographical works.

1845.—Count P. E. DE STRZELECKI—Founder's Medal—for his explorations and discoveries in the South-Eastern portion of Australia, and in Van Diemen's Land.

——— Professor A. TH. MIDDENDORFF—Patron's Medal—for his extensive explorations and discoveries in Northern and Eastern Siberia.

1846.—Captain CHARLES STURT—Founder's Medal—for his various and extensive explorations in Australia.

——— Dr. LUDWIG LEICHHARDT—Patron's Medal—for a journey performed from Moreton Bay to Port Essington.

1847.—Sir JAMES BROOKE, Rajah of Saráwak and Governor of Labuan— Founder's Medal—for his expedition to Borneo.

——— Captain CHARLES WILKES, U.S.N.—Patron's Medal—for his Voyage of Discovery in the S. Hemisphere and in the Antarctic Regions, in the years 1838-42.

1848.—AUSTEN H. LAYARD, Esq., D.C.L. M.P.—Founder's Medal—for his contributions to Asiatic geography, researches in Mesopotamia, and discoveries of the remains of Nineveh.

——— Baron CH. HUGEL—Patron's Medal—for his explorations of Cashmere and surrounding countries, communicated in his work entitled 'Kashmir und das Reich der Siek.'

1849.—Col. JOHN CH. FRÉMONT—Patron's Medal—for his successful explorations of the Rocky Mountains and California; and for his numerous Discoveries and Astronomical Observations.

——— The Rev. DAVID LIVINGSTONE, of Kolobeng—a Chronometer Watch— for his successful explorations of South Africa.

1850.—Dr. GEORGE WALLIN, of Finland—25 Guineas—for his Travels in Arabia.

——— Mr. THOMAS BRUNNER—25 Guineas—for his explorations in the Middle Island of New Zealand.

1851.—Dr. JOHN RAE—Founder's Medal—for his survey of Boothia and of the Coasts of Wollaston and Victoria Lands.

——— Captain HENRY STRACHEY—Patron's Medal—for his Surveys in Western Tibet.

1852.—Mr. FRANCIS GALTON—Founder's Medal—for his explorations in Southern Africa.

——— Commander E. A. INGLEFIELD, R.N.—Patron's Medal—for his Survey of the Coasts of Baffin Bay, Smith and Lancaster Sounds.

1853.—Rear-Admiral WILLIAM HENRY SMYTH—Founder's Medal—for his valuable Surveys in the Mediterranean.

——— Captain Robert J. M. M'CLURE, R.N.—Patron's Medal—for his discovery of the North-West Passage.

1854.—The Rev. DAVID LIVINGSTONE, M.D., &c.—Patron's Medal—for his Scientific Explorations in Central Africa.

——— Mr. CHARLES J. ANDERSSON—a Set of Surveying Instruments—for his Travels in South-Western Africa.

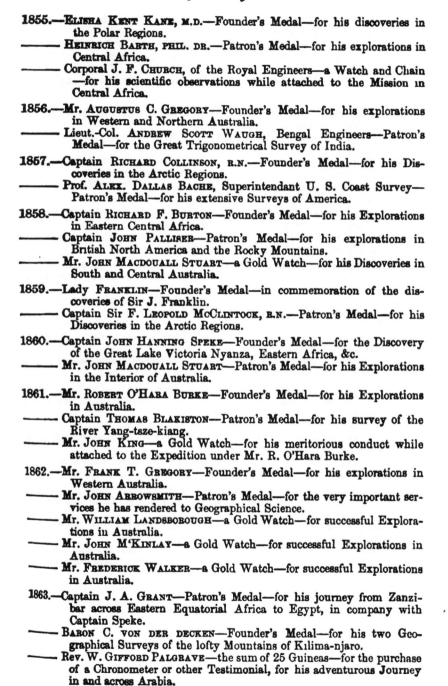

1855.—ELISHA KENT KANE, M.D.—Founder's Medal—for his discoveries in the Polar Regions.

—— HEINRICH BARTH, PHIL. DR.—Patron's Medal—for his explorations in Central Africa.

—— Corporal J. F. CHURCH, of the Royal Engineers—a Watch and Chain —for his scientific observations while attached to the Mission in Central Africa.

1856.—Mr. AUGUSTUS C. GREGORY—Founder's Medal—for his explorations in Western and Northern Australia.

—— Lieut.-Col. ANDREW SCOTT WAUGH, Bengal Engineers—Patron's Medal—for the Great Trigonometrical Survey of India.

1857.—Captain RICHARD COLLINSON, R.N.—Founder's Medal—for his Discoveries in the Arctic Regions.

—— Prof. ALEX. DALLAS BACHE, Superintendant U. S. Coast Survey— Patron's Medal—for his extensive Surveys of America.

1858.—Captain RICHARD F. BURTON—Founder's Medal—for his Explorations in Eastern Central Africa.

—— Captain JOHN PALLISER—Patron's Medal—for his explorations in British North America and the Rocky Mountains.

—— Mr. JOHN MACDOUALL STUART—a Gold Watch—for his Discoveries in South and Central Australia.

1859.—Lady FRANKLIN—Founder's Medal—in commemoration of the discoveries of Sir J. Franklin.

—— Captain Sir F. LEOPOLD MCCLINTOCK, R.N.—Patron's Medal—for his Discoveries in the Arctic Regions.

1860.—Captain JOHN HANNING SPEKE—Founder's Medal—for the Discovery of the Great Lake Victoria Nyanza, Eastern Africa, &c.

—— Mr. JOHN MACDOUALL STUART—Patron's Medal—for his Explorations in the Interior of Australia.

1861.—Mr. ROBERT O'HARA BURKE—Founder's Medal—for his Explorations in Australia.

—— Captain THOMAS BLAKISTON—Patron's Medal—for his survey of the River Yang-tsze-kiang.

—— Mr. JOHN KING—a Gold Watch—for his meritorious conduct while attached to the Expedition under Mr. R. O'Hara Burke.

1862.—Mr. FRANK T. GREGORY—Founder's Medal—for his explorations in Western Australia.

—— Mr. JOHN ARROWSMITH—Patron's Medal—for the very important services he has rendered to Geographical Science.

—— Mr. WILLIAM LANDSBOROUGH—a Gold Watch—for successful Explorations in Australia.

—— Mr. JOHN M'KINLAY—a Gold Watch—for successful Explorations in Australia.

—— Mr. FREDERICK WALKER—a Gold Watch—for successful Explorations in Australia.

1863.—Captain J. A. GRANT—Patron's Medal—for his journey from Zanzibar across Eastern Equatorial Africa to Egypt, in company with Captain Speke.

—— BARON C. VON DER DECKEN—Founder's Medal—for his two Geographical Surveys of the lofty Mountains of Kilima-njaro.

—— Rev. W. GIFFORD PALGRAVE—the sum of 25 Guineas—for the purchase of a Chronometer or other Testimonial, for his adventurous Journey in and across Arabia.

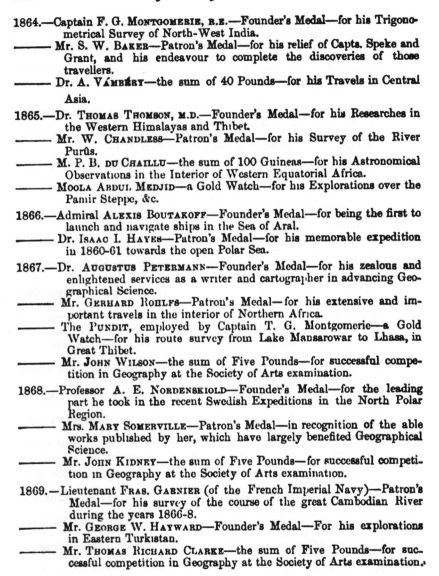

1864.—Captain F. G. MONTGOMERIE, R.E.—Founder's Medal—for his Trigono-
metrical Survey of North-West India.
——— Mr. S. W. BAKER—Patron's Medal—for his relief of Capts. Speke and
Grant, and his endeavour to complete the discoveries of those
travellers.
——— Dr. A. VÁMBÉRY—the sum of 40 Pounds—for his Travels in Central
Asia.

1865.—Dr. THOMAS THOMSON, M.D.—Founder's Medal—for his Researches in
the Western Himalayas and Thibet.
——— Mr. W. CHANDLESS—Patron's Medal—for his Survey of the River
Purûs.
——— M. P. B. DU CHAILLU—the sum of 100 Guineas—for his Astronomical
Observations in the Interior of Western Equatorial Africa.
——— MOOLA ABDUL MEDJID—a Gold Watch—for his Explorations over the
Pamir Steppe, &c.

1866.—Admiral ALEXIS BOUTAKOFF—Founder's Medal—for being the first to
launch and navigate ships in the Sea of Aral.
——— Dr. ISAAC I. HAYES—Patron's Medal—for his memorable expedition
in 1860-61 towards the open Polar Sea.

1867.—Dr. AUGUSTUS PETERMANN—Founder's Medal—for his zealous and
enlightened services as a writer and cartographer in advancing Geo-
graphical Science.
——— Mr. GERHARD ROHLFS—Patron's Medal—for his extensive and im-
portant travels in the interior of Northern Africa.
——— The PUNDIT, employed by Captain T. G. Montgomerie—a Gold
Watch—for his route survey from Lake Mansarowar to Lhasa, in
Great Thibet.
——— Mr. JOHN WILSON—the sum of Five Pounds—for successful compe-
tition in Geography at the Society of Arts examination.

1868.—Professor A. E. NORDENSKIOLD—Founder's Medal—for the leading
part he took in the recent Swedish Expeditions in the North Polar
Region.
——— Mrs. MARY SOMERVILLE—Patron's Medal—in recognition of the able
works published by her, which have largely benefited Geographical
Science.
——— Mr. JOHN KIDNEY—the sum of Five Pounds—for successful competi-
tion in Geography at the Society of Arts examination.

1869.—Lieutenant FRAS. GARNIER (of the French Imperial Navy)—Patron's
Medal—for his survey of the course of the great Cambodian River
during the years 1866-8.
——— Mr. GEORGE W. HAYWARD—Founder's Medal—For his explorations
in Eastern Turkistan.
——— Mr. THOMAS RICHARD CLARKE—the sum of Five Pounds—for successful competition in Geography at the Society of Arts examination.

ACCESSIONS TO THE LIBRARY,

From May 27th, 1869, to May 27th, 1870.

[*When London is the place of publication, the word London is omitted.*]

Titles of Books.			*Donors.*
A???, P.—Colonial Handbook. Victoria. 1869.			
A???, C. B.—Queensland and her Gold Fields. 1867	..		The Author.
Allen's S????.—Geography of India. 1869	..	..	The Author.
A???????, G. F.—Journey to Musadu, West Africa. New York, 1870.			The President.
A?????, W. P.—Railway System, Valley of the Indus. 1869			The Author.
A?????, J. R.—Life of Oliver Cromwell, etc. 1870	..		The Author.
A????, G. F.—South Australia illustrated. 1847	..	..	By Purchase.
... ...New Zealand illustrated. 1847	..	..	By Purchase.
A?????, Phares de la Mer, etc. Paris, 1869 ..		..	Ministre de la Marine.
A?????????? S.—L'Oraison Dominicale en Cent Langues differentes. 1870.			The President.
A???????'s Railway Guide, 1867. New York	..		Cyril Graham, Esq.
A?????; R. A.—Levant, Black Sea, and the Danube. 2 vols. 1868.			By Purchase.
A????; ?? P.—A Fortnight in Egypt at the Opening of the Suez Canal. 1869.			The Author.
A???? Papers relating to the Tea Plant	..	..	.. By Purchase.
A???????, J.—Expedition into Affghanistan. 1842 ..		..	By Purchase.
A??, Two Years in. 1824–1826	..	..	.. By Purchase.
B????? Die, in Wort und Bild geschildert. Vol. I. Leipzig, 1869.			The President.
B?????, Alex. von Humboldt.—			
B???????, J. H. B.—Voyage à la Côte de Guinée, etc. 1853. La Haye.			By Purchase.
B???, J.—An Essay on Emigration. N. D. ..	..	..	The Author.
B????, H. W.—Revised Edition of Mary Somerville's Physical Geography. 1870.			H. W. Bates, Esq.
B???, ???.—Jacob's Flight. 1865. *Map by Dr. Beke*	..		By Purchase.
B?????, F. C.—Travels of Macarius. (Translated by.) 1829.			By Purchase.
B???, W. A. and J. Collinson.—The Denver Pacific Railway. 1870.			W. A. Bell, Esq., M.D.
B?????, H. W.—Political Mission to Affghanistan in 1857			By Purchase.

Titles of Books.	Donors.

BENTHAM, G.—Address Anniversary Meeting Linnean Society, 1869.
The AUTHOR.

BERTOLOTTI, A.—Passeggiate nel Canavese. 1867-70 .. The AUTHOR.

BICKMORE, A. S.—Ainos, or Hairy Men of Yesso. 1868 .. The AUTHOR.

BLANDFORD W. T.—Notes on a Journey in North Abyssinia. 1869.
The AUTHOR.

BOLLAERT, W.—Maya Hieroglyphic Alphabet of Yucatan .. The AUTHOR.

BONER, C.—Transylvania; its Products and its People. 1865 The AUTHOR.

BONWICK, J.—The Last of the Tasmanians. 1869 .. .. The AUTHOR.

Bordeaux.—Notice sur les Vins de Bordeaux. 1834.. .. S. M. DRACH, Esq.

Boston.—Public Library Bulletin. Boston, 1870 .. .. The COUNCIL.

BOUQUET DE LA TRYE.—Note sur les Sondes, etc. Paris, 1869.
DÉPÔT DES CARTES ET PLANS, ETC.

BOWLES, S.—Our New West. Chicago, 1869 .. .. By PURCHASE.

BOYLE, F.—A Ride across a Continent. 2 vols. 1868 .. By PURCHASE.

................The Dyaks of Borneo. 1865 .. .. .. By PURCHASE.

BRACE, C. L.—The New West. California in 1867-68. New York, 1869.
By PURCHASE.

Brazil.—Brasilianische Zustande und Aussichten in 1861 .. J. J. STURZ, Esq.

BAEMOND, A.—Le Guide Toulousan. Toulouse, 1868 .. S. M. DRACH, Esq.

BRIDET, M.—Les Ouragans de l'Hemisphere Austral. Paris, 1869.
DÉPÔT DES CARTES, ETC.

BRINTON, D. G.—Guide-book to Florida. Philadelphia, 1869 By PURCHASE.

BROOKE, T. H.—Island of St. Helena. 1824 .. .. .. By PURCHASE.

BROSSET, M.—Histoire Chronologique. St. Petersburg, 1869 The AUTHOR.

BROWN, E.—Travels through Germany. 1677 .. .. By PURCHASE.

BROWN, R.—Geographical Distribution of the Coniferæ and Gnetaceæ. Edinburgh, 1869 .. .. The AUTHOR.

................Mammalian Fauna of Greenland. Edinburgh, 1869.
The AUTHOR.

BUCHAN, A.—Mean Pressure of the Atmosphere .. .. The PRESIDENT.

BULLOCK.—Across Mexico in 1864-65, 1866.. .. .. The AUTHOR.

BURGESS, J.—Notes of a Visit to Somnath, Girnar, etc. Bombay, 1869.
The AUTHOR.

BYRON.—Voyage of H.M.S. *Blonde* to the Sandwich Islands. 1826.
By PURCHASE.

CALLANDER, J.—Terra Australis Cognita .. .. .. C. ENDERBY, Esq.

CAMPBELL, J.—Thirteen Years amongst the Wild Tribes of Khondistan. 1864.
By PURCHASE.

CASARETTO.—Geo. Discorso del Soc. Economica di Chiavari. 1868.
The AUTHOR.

CASSINI, M. de.—A Voyage to California to observe Transit of Venus. 1778.
By PURCHASE.

CHAMBERS, R.—Ancient Sea Margins. 1848.. .. .. By PURCHASE.

CHAMBERS, WM.—Things as they are in America. London and Edinburgh, 1854.
By PURCHASE.

CHAMBERS, W. and R.—Arctic Regions and Arctic Explorations. Edinburgh.
N. D. .. .. The PUBLISHERS.

| *Titles of Books.* | *Donors.* |

CHAPPE D'AUTEROCHE, an Enquiry into the Merits of his Book. 1772.
Capt. C. ENDERBY.

CHARLEVOIX, P. DE.—Voyage to North America. 1761 ..Capt. C. ENDERBY.

CLARKE, Lieut.-Col.—(See Richards.)

CLAYTON, J. W.—Scenes and Studies. 1870 ..   ..   ..   The AUTHOR.

CLAYTON, R.—Grand Cairo to Mount Sinai. 1753 ..   ..   By PURCHASE.

CLEGHORN, H.—Botany, etc., of Malta and Sicily. Edinburgh, 1870.
The AUTHOR.

COFFIN, C.—Our New Way Round the World.

COLLINSON.—(See Bell.)

COLTON, F.—Route-book through the United States and Canada. New York,
1851 ..   .. C. GRAHAM, Esq.

COOLEY, W. D.—Mémoire sur Tacuy de Barros. Paris, 1869.

Corsica.—Guide du Voyageur en Corse. Ajaccio, 1868   .. S. M. DRACH, Esq.

COX, J. H.—Observations during a Voyage to Teneriffe, &c., and Canton. 1791.
GEO. MORTIMER, Esq.

CROLL, J.—On Ocean Currents, 1870   ..   ..   ..   The AUTHOR.

DAA, L. K.—Om Forholdet mellem det, etc. 1857. Christiania.
The UNIVERSITY.

DANVERS, J.—Indian Railway Reports   ..   .. JULAND DANVERS, Esq.

DAVIS, N.—Ruined Cities. (North Africa.) ..   ..   .. By PURCHASE.

DAVIDSON, G. F.—Trade and Travel in the Far East. 1846   By PURCHASE.

DECKEN, O. C. VON DER.—Reisen in Ost Afrika. Leipzig und Heidelberg. 1869.
By PURCHASE.

DERANNANT, C.—La Situation Economique de L'Espagne. Paris, 1869.

.. .. Panslavisme, etc. 1869.

DELESSE.—Notice sur les Travaux Scientifiques de M. Delesse. Paris, 1869.

DESJARDINS, E.—Rhone et Danube. Paris, 1870 ..   ..   The AUTHOR.

DICKIE, Prof.—On Diatomacea collected in Danish Greenland by R. Brown.
Edinburgh, 1870   ..   ..   R. BROWN, Esq.

DIXON, H.—Life of Robert Blake   ..   .. By PURCHASE.

DODWELL, E.—Classical Tour. 2 vols. 1819   ..   .. By PURCHASE.

DRACH, S. M. Origin of Names of Places in Europe. (MSS.)   The AUTHOR.

DUNCAN, G.—Geography of India. 1868   ..   ..   By PURCHASE.

DAWSON, E. B.—Venezuela. 1868 ..   ..   ..   By PURCHASE.

DAWSON, M. B.—Through Spain to the Sahara. 1868   .. By PURCHASE.

DUVEYRIER, ESKEDDEE-EL—Journey in Interior N. Africa.. By PURCHASE.

ELLIS, W.—Madagascar revisited. 1867   ..   ..   .. By PURCHASE.

ROBINSON, E. P.—(See Jephson.)

DURAND V.—A retrospect of the Affghan War. 1869   .. The AUTHOR.

WILSON, J.—H. R. H. The Duke of Edinburgh in India. Calcutta, 1870.
The AUTHOR.

FINLAY.—On Some Greek Antiquities. Athens, 1869   The AUTHOR.

FINLAYSON.—Mission to Siam, 1826. ..   .. By PURCHASE.

People of the Finnish Race, etc. 2 vols. (in Russian).
W. EGERTON HUBBARD, Esq.

FLACOURT.—Histoire de la Grand Isle Madagascar. Paris, 1658
By PURCHASE.

Titles of Books.	Donors.

FLEMING, G.—Travels in Manchu Tartary. 1863 .. .. By PURCHASE.

FOOT, R. B.—On the Distribution of Stone Implements in Southern India. 1870.
The AUTHOR.

FORD, R.—Murray's Handbook of Spain. 2 vols. 1869 .. By PURCHASE.

FORSYTH, T. DOUGLAS.—Trade Routes between N. India and Central Asia. 1869.
The AUTHOR.

FORTUNE, R.—Yedo and Peking. 1863 .. .. .. By PURCHASE,

FOSTER, J. W.—The Mississippi Valley. Chicago, 1869 .. By PURCHASE.

FOX, C. R.—On a Coin of Glauconnesus. 1869. .. .. The AUTHOR.

FRANCE.—Tableaux de Population, etc. Paris, 1868.

FRESHFIELD, D. W.—Travels in the Central Caucasus. 1869 The AUTHOR.

GAUSSIN, M.—Annuaire des Marées des Côtes de France, pour 1870. Paris, 1869.
MINISTRE DE LA MARINE.

GEIKIE, A.—The Scenery of Scotland. 1865 .. .. .. By PURCHASE.

GIFFARD, E.—The Ionian Islands, Athens, etc. 1837 .. By PURCHASE.

GONNEVILLE, Capt. DE.—Voyage du. Paris, 1869 .. .. By PURCHASE.

GORDON, Lady DUFF.—Letters from Egypt. 1863-65 .. By PURCHASE.

GORMAZ, F. S.—Exploracion del Rio Valdivia. Santiago de Chile, 1869.
Dr. PHILIPPI.

GORRIE, D.—Summers and Winters in the Orkneys. 1868 .. By PURCHASE.

GRABHAM, M. C.—The Climate and Resources of Madeira. 1870.
The AUTHOR.

GRANT, COLESWORTHY.—Trip to Rangoon in 1846. Calcutta, 1853. By PURCHASE.

GRANT, J. A.—Walk across Africa. 1864 .. .. .. The AUTHOR.

GRELOT.—Relation nouvelle d'un Voyage de Constantinople. Paris, 1780.
By PURCHASE.

GRIESBACH, C. L.—Die Erdbeben in den Jahren 1867-68. Wien, 1869.
The AUTHOR.

GRIEVE, J.—History of Kamtschatka, etc. 1864 .. .. By PURCHASE.

GRIFFIN, J.—Active List of Flag Officers, etc., of the Royal Navy. Portsea, 1870.

GRIFFITH, W.—Assam, Burma, etc., Journals of, arranged, by J. McClelland.
Calcutta, 1847.

GRÜBER, W.—Anatomie des Schadelgrundes .. .. .. The AUTHOR.

HALEVY, J.—Excursion chez les Falacha en Abyssine.

HALL, C. F.—Life with the Esquimaux. 2 vols. 1864. .. By PURCHASE.

HALL, E. H.—The Great West. New York, 1867 .. .. J. BATE, Esq.

BALL, M.—The Terraces of Norway of Prof. Kjerulf. 1870 The AUTHOR.

HARRIS.—Occasional Papers of the Boston Nat. Hist. Society. Vol. I. 1869.
The SOCIETY.

HARRISON, J. P.—Inductive Proof of the Moon's Insolation The AUTHOR.

HAYDEN, F. V.—Geological Report of Yellow Stone and Missouri Rivers. New
York, 1869 .. .. U. S. GOVERNMENT.

.................Geological Survey of Colorado and New Mexico. 1870.
SMITHSONIAN INST.

HEADRICK, J.—Island of Arran. 1807. .. .. .. By PURCHASE.

HECTOR, J.—New Zealand Geological Survey. 1868-69 .. The AUTHOR.

HELMERSEN, VON.—Diluvialgebilde Russlands. St. Petersburg, 1869.
The AUTHOR.

Titles of Books.	Donors.

HELPS, A.—Las Casas, Life of. 1868 .. .. .. By PURCHASE.

HEUGLIN, TH. V.—Reise in das Gebiet des weissen Nil, etc. 1862-64. Leipzig und Heidelberg, 1869 .. .. By PURCHASE.

HILL, ED. S.—Lord Howe Island. Sydney, 1869 .. .. Sir D. COOPER.

HIND, J. R.—Path of the Total Phase of the Solar Eclipse. Dec. 21-22, 1870. J. R. HIND, Esq.

HIRSCH. A. et E. PLANTAMOUR.—Nivellement de Précision de la Suisse. Genève et Bâle, 1870 .. .. M. J. M. ZIEGLER.

HOMMAGER, J. C.—Remarks on the Abyssinian Expedition. 1870. The AUTHOR.

HOWARD.—Climate of London. 3 vols. 1833. .. .. The PRESIDENT.

HOWELL, A. P.—Jails in India. Calcutta, 1869. .. .. INDIA OFFICE.

HUTCHINGS, J. H.—Tourist's Guide to the Yosemite Valley. California, 1870. The AUTHOR.

INCOGNITO.—The Earth, etc. 1869. .. .. The AUTHOR.

IRMINGER, C.—Strømninger og Isdrift ved Island. Kjøbenhaven, 1861. The AUTHOR.

JACKSON, J. G.—The Empire of Marocco. 1814. .. .. By PURCHASE.

JAMES, Sir H.—Plan der Stadt und Umgebung von Jerusalem in 1864. Breslau, 1869.

JAUBERT, Comte AUGUER ELOY.—Voyages en Orient. Paris, 1843. By PURCHASE.

JELINEK, C. and C. FRITSCH.—Meteorologie und Erdmagnetismus. Wien, 1868.

JENKINS, H. L.—Burmese Route from Assam to the Hookoong Valley. 1869. The AUTHOR.

JENKINS, R.—State Emigration. 1869. .. .. .. J. BATE, Esq.

JOHNSON, H. C. ROSS.—Long Vacation in the Argentine Alps. 1868. By PURCHASE.

JOHNSTON, K., jun.—Map of Lake Region of E. Africa, with Notes, etc. 1870. The AUTHOR.

India.—Madras Selections. No. 11, 15, 16. .. .. .. INDIA OFFICE.

" " .. Madras Medical College Reports .. .. .. INDIA OFFICE.

" " .. East Indian Colonies, History of .. .. .. By PURCHASE.

" " .. Selections from the Records of the Government of the N. W. Provinces. INDIA OFFICE.

" " .. Annals of Indian Administration in 1867, 68. Edited by G. Smith. Vol. XII. Serampore, 1869

" " .. Selections from the Madras Government Records .. INDIA OFFICE.

" " .. Uncovenanted Civil Service Examinations. 1869. MADRAS PRESIDENCY.

" " .. Scind Railway Reports for 1867-69 .. .. W. P. ANDREW, Esq.

" " . Records for India. Punjab. No. 5. Kangra Tea District. INDIA OFFICE.

" " .. Selections—Home Department. No. 74.

" " .. Trigonometrical, Topographical, and Revenue Survey. 1867-68. Calcutta, 1869 .. .. INDIA OFFICE.

KRUSE, M. DE—Samarkand, 1869 .. .. .. The AUTHOR.

KJERULF'S Terraces of Norway, translated by M. Hall. Christiania, 1866. The PRESIDENT.

KNIGHT, C.—Supplement to English Cyclopædia .. .. A. RAMSAY, Esq.

Titles of Books.	*Donors.*

KOLBEN, P.—Present State of the Cape of Good Hope. 2 vols. 1731.
                                                    By PURCHASE.

KOTSCHY, THEO.—Reise in den Cilicischen Taurus. Gotha, 1858.
                                                    By PURCHASE.

KRUSE, F.—U. J. Leetzens Reisen durch Syrien, etc.   ..   By PURCHASE.

LARTET, E. and H. CHRISTY.—Reliquiæ Aquitanicæ.
                          The Executors of the late Mr. CHRISTY.

LASCARIS, M.—Lo Avvenire del Commercio Italiano. 1870   The AUTHOR.

LATHAM, R. G.—Native Races of Russian Empire. 1854 .. By PURCHASE.

LATHAM, W.—States of the River Plate. 1868   ..   ..   By PURCHASE.

LEIGH, W. H.—Reconnoitring Voyages and Travels. 1836-38. 1839.
                                                    By PURCHASE.

LEMPRIERE, C.—Notes in Mexico 1861-62. 1862 ..   ..   By PURCHASE.

LETRONNE, A. DICUIL.—Recherches Géographiques. Paris, 1814. By PURCHASE.

LINDSAY, W. L.—Lichen Flora of Greenland. 1869   ..   R. BROWN, Esq.

LIVINGSTONE, D. and C.—Expedition to the Zambesi. 1865   By PURCHASE.

LLOYD'S COMMITTEE.—Wrecks and Casualties List for 1868   LLOYD'S.

LOSTSKVITSCH, T.—La Ligne Militaire du Syr-Daria.   St. Petersburg, 1868.
                                                    R. MICHELL, Esq.

LOCKHART, W.—Medical Missionary in China. 1861   .. Rev. J. G. WOOD.

LONG, J.—Travels of an Indian Trader, etc. 1791 ..   ..   By PURCHASE.

LORENZ, L.—Experimentale og theoretiske Undersogelsen, etc. Kjobenhaven,
     1869 ..   ..   The AUTHOR.

LORD, W. B.—Key to Fortune in New Lands. 1869   ..   The AUTHOR.

LUBBOCK, Sir J.—Prehistoric Times. 1869 ..   ..   .. By PURCHASE.

LUGARD, Lieut.-Gen. ED.—Report on the Survey of the Abyssinian Operations.
     1869 ..   ..   WAR OFFICE.

LULKE, C. T.—Additamenta ad Historiam Ophiuridarum. Kjobenhaven, 1869.
                                                    The AUTHOR.

LYELL, Sir C.—Principles of Geology. 10th Edition. 1869   By PURCHASE.

LYON, G. F.—Attempt to reach Repulse Bay in H.M.S. *Griper* in 1824.
                                              R. U. S. INSTITUTION.

MACARIUS. (*See* Belfour, F. C.)

MACGREGOR, J.—The *Rob Roy* on the Jordan. 1869   ..   The AUTHOR.

MACINTOSH, D.—Scenery of England and Wales. 1869   ..   By PURCHASE.

MACKAY, A.—Facts and Dates. 1870 ..   ..   The AUTHOR.

MACKENZIE, C.—Six Years in India. 1854 ..   ..   ..   By PURCHASE.

M'CLINTOCK, Sir L.—Fate of Franklin. New Edition   .. Lady FRANKLIN.

Madrid.—Discurso leido en la Universidad Central. Madrid, 1869.
                                                    The UNIVERSITY.

MAIN, R.—Second Radcliffe Catalogue of Stars. Oxford, 1870   The AUTHOR.

MALFATTI, B.—Scritti Geografici ed Ethnografici. Milano, 1869
                                                    The AUTHOR.

Malta.—12 vols. of Catalogues of the Malta Library   Sir JOHN ROBINSON.

MALTE BRUN.—Voyages et Travaux de M. Le Comte S. d'Escayrac de Lauture.
     Paris, 1869 ..   ..'   The AUTHOR.

MARIS CARNEYRO, ANTONIO DE.—Hydrografia, etc. S. Sebastian, 1675.
                                                    By PURCHASE.

Accessions to the Library

| Titles of Books. | Donors. |

NOVARA.—Reise der Osterreichischen Fregatte *Novara* um die Erde. Wien, 1868.
The AUSTRIAN GOVERNMENT.

OLMSTED, F. L.—The Cotton Kingdom. 2 vols. 1861 .. By PURCHASE.

Omaha.—The Union Pacific Railway to Chicago. 1868 .. C. GRAHAM, Esq.

Orkneys.—Hand-book to the Orkney Islands. 1869 .. By PURCHASE.

ORMSBY, J.—Autumn Rambles in North Africa. 1864 .. By PURCHASE.

Orthographia Samaritanorum. N. D. .. .. .. S. M. DRACH, Esq.

Pacific.—Discoveries in, previous to 1764 .. .. CHAS. ENDERBY, Esq.

... ...Northern Pacific Railroad. Report by Government J. J. STEPHENS.

... ...Washington, 1854 .. C. GRAHAM, Esq.

... ...Union Pacific Railroad. Chicago, 1868 .. .. C. GRAHAM, Esq.

... ... Boston, 1868 .. .. C. GRAHAM, Esq.

... ... New York, 1868 .. · .. C. GRAHAM, Esq.

PALMER, W. J.—Report of Surveys extending the Kansas Pacific Railway to the Pacific Ocean. 1867–68. Philadelphia, 1869 .. .. W. A. BELL, Esq.

PARISH, J.—A French Officer's Voyage to the Island of Mauritius. 1775.
By PURCHASE.

Paz Soldan.—Géographie du Peru. Translated by P. A. Mouqueron. Paris, 1863.

PELLICER, J. DE TOVAR.—Mission al Reyno de Congo. Madrid, 1649.
By PURCHASE.

PENISTON, W. M.—Catalogue of Contributions from the Colony of Natal. 1869.
Dr. MANN.

Persia.—Regni Persici Status, etc. Amsterdam, 1633 .. J. POWER, Esq.

PESCHEL, O.—Neue Probleme der Vergleichenden Erdkunde, etc. Leipzig, 1869.
The AUTHOR.

PETHERICK, Mr. and Mrs.—Travels in Central Africa. 2 vols. 1869.
The AUTHOR.

PEYTON, J. L.—The American Crisis. 2 vols. 1867 .. The AUTHOR.

PEYSSONNEL, M. DE.—Obs. Hist. et Geograph. sur les Peuples qui ont habité les Bords du Danube et du Pont Euxine. Paris, 1765 .. By PURCHASE.

PLANTANOUR, E.—(See Hirsch.)

PLATE, J.—Sermon on XLIst Psalm. Tronheim, 1734 .. S. M. DRACH, Esq.

PODIEBRAD, D. J.—Andenken an die Alterthümer der Prager Josefstadt. Prague. N. D. .. .. S. M. DRACH, Esq.

POWER, W. T.—Three Years' Residence in China. 1853 .. By PURCHASE.

PRATT, N.A.—The Marls of South Carolina. Philadelphia, 1868.
G. GOLDNEY, Esq.

PRESCOTT, W. H.—Conquest of Mexico. 3 vols. 1844. .. By PURCHASE.

PRICE, F. G. H. C.—Hamilton's Sketches of Life and Sport in S. E. Africa, edited by. 1870. .. .. The EDITOR.

QUITO.—Memoria sobre las Oscilaciones de la Brujula, etc. By A. M. D. G. Quito, 1868. .. .. The PRESIDENT.

RAULIN, V.—Description Physique et Naturelle de l'Ile de Crete. 2 vols. and Atlas. Paris, 1869 .. .. By PURCHASE.

RAYNAL, F. E.—Les Naufrages (Iles Aukland). Paris, 1870. The PRESIDENT.

RAVENSTEIN, E. G.—Denominational Statistics of England and Wales. 1870.
The AUTHOR.

Titles of Books.	*Donors.*

RAVENSTEIN, E. G.—Reisehandbuch für London, England, und Scotland. Hildburghausen, 1870 .. .. The AUTHOR.

RECLUS, E.—La Terre. 2nd edit. Paris, 1870. .. .. The AUTHOR.

RENNIE, D. F.—Peking and the Pekingese. 2 vols. 1865. By PURCHASE.

RESEARCH.—The True Theory of the Earth. *N. D.* .. .. The AUTHOR.

RICKARD, P. J.—Mineral and other Resources of the Argentine Republic in 1859. The AUTHOR.

RICHARDS, Capt., and Lieut.-Col. CLARKE.—Report on the Suez Canal. February, 1870 .. .. Capt. RICHARDS.

RICHARDSON, R.—Travels along the Mediterranean. 2 vols. 1822. By PURCHASE.

RICHTER, R.—Das Thüringsche Schiefergebirge. 1869. .. The PRESIDENT.

ROCHON, Abbé.—Voyage to Madagascar .. .. .. C. ENDERBY, Esq.

ROCKFORD.—Rock Island and St. Louis Railroad. New York, 1868. C. GRAHAM, Esq.

ROHLFS, G.—Land und Volk in Afrika. Bremen, 1870 .. The AUTHOR.

ROORDA, T.—De Wagangverhalen van Pálá, etc. Gravenhage, 1869.

ROUSSILLON, le Duc du.—Origines, Migrations, etc. 1867. The AUTHOR.

RUSKIN, J.—The Future of England. 1869. .. .. .. The AUTHOR.

Royal Society Catalogue of Scientific Papers.

SAULCY.—Gharghis Mahomed, etc. Jericho, 1867. .. The AUTHOR.

SAMPSON, Low, & Co.—Book Lists. 1866-70. .. .. The PUBLISHERS.

SANDS, B. F.—Total Solar Eclipse of Aug. 7, 1869. Washington, 1869. SMITHSONIAN INST.

SAUNDERS, T.—Sketch of the Mountains and River-basins of India. 1870. The AUTHOR.

SCHLAGINTWEIT, H. von.—Neue Daten uber den Todestag von Adolph v. Schlagintweit. München, 1869 .. .. The AUTHOR.

SCHRANK, C. F. P. v. MARTIUS.—Ein Lebenbild. Berlin. 1869. J. L. STURZ, Esq.

Schools Enquiry Commission Report of 1868 .. .. By PURCHASE.

Scinde.—Railway Company's Progress. 1867-60. The SECRETARY SCINDE RAILWAY.

SCOTT, Col.—Travels in Morocco and Algiers. 1842. .. By PURCHASE.

SCULLY, W.—Brazil: its Provinces and Chief Cities.

SEETZEN, U. J.—Reisen durch Syrien, etc.; herausgegeben von Prof. Kruse. By PURCHASE.

SEMMES, R.—Adventures Afloat. 2 vols. 1869 .. .. By PURCHASE.

SEMPER, G.—Die Philippinen und ihre Bewohner. Wurtzburg, 1869. H. W. BATES.

SHELVOCKE, G.—Voyage round the World. 1757 .. .. CHAS. ENDERBY.

SIEBOLD, P. F. von.—Manners and Customs of the Japanese. 1841. By PURCHASE.

.. .. Nippon, archiv zur beschreivung von Japan. 6 vols. Leyden, 1869 .. .. By PURCHASE.

SINCLAIR, J.—Correspondence of. 2 vols. 1831. .. ..S. M. DRACH, Esq.

SLEIGH, B.—Travels in Canada, by J. G. Kohl (translated). 1861. By PURCHASE.

SKINNER, J. E. H.—Roughing it in Crete. 1868. .. .. By PURCHASE.

Titles of Books.	Donors.

SMITH, D.—A True Key to the Assyrian History, etc. .. The Author.

SMITH, G.—Annals of Indian Administration in 1857-68. Serampore, 1869.
  INDIA OFFICE.

SMYTH, C. P.—A Poor Man's Photography at the Great Pyramid, 1869. 1870.
  The Author.

SNOW's Pathfinder Railway Guide. New York, 1867. CYRIL GRAHAM, Esq.

Solar Heat: Papers on, by various Authors.

SOLOYNS, F. B. Les Hindoûs. Paris, 1808. .. C. H. BRACEBRIDGE, Esq.

Spa.—Guide aux Eaux et aux Jeux. Spa, 1865 .. .. S. M. DRACH, Esq.

SOMERVILLE, MARY. Physical Geography. New edit. Edited by H. W. Bates.
  1870 .. .. The EDITOR.

Spain: a Winter Tour in Spain, 1868 .. .. .. By PURCHASE.

SPRUCE, R.—Palmæ Amazonicæ. 1869 .. .. .. The Author.

SQUIER, E. G.—The Primeval Monuments of Peru .. .. The Author.

STEEN, A.—Om Œndringen af Integrator af irrationale.

STEINSCHNEIDER, M.—Memoirs Al Farabi. St. Petersburg, 1869.

STEPHENS, J. J.—Northern Pacific Railroad. 1854. Washington.

St. Genevieve Glass-Sand of S. Missouri, Report on. London, 1869.
  C. GRAHAM, Esq.

ST. JOHN.—Village Life in Egypt. 2 vols. 1852 .. .. By PURCHASE.

STOEHLIN, v. J.—An Account of the new Northern Archipelago discovered in
  the Seas of Kamschatka. London, 1774. .. CHAS. ENDERBY, Esq.

STOKES, G. S.—Address of President of the British Association, 1869.
  The Author.

STOCKWELL, G. S.—The Republic of Liberia .. .. .. The Author.

STURS, J. J.—Die Deutsche Auswanderung, etc. Berlin. 1868. The Author.

.. .. Der Nord und Ostsee Kanal durch Holstein. Berlin, 1864.
  The Author.

.. .. Notizen uber den Minenbetreib in Bolivien. Matto Grosso, etc.
  Berlin, 1868 .. .. The Author.

.. .. Neue Beitrage uber Brasilien und die La Plata landern.
  Berlin, 1865 .. .. The Author.

Sweden Meteorologiska Jakdagelser, etc. 3 vols. fol. Stockholm, 1868
  ROYAL ACADEMY.

Syria: Rambles in the Syrian Deserts, etc. 1864. .. JOHN MURRAY, Esq.

TAINE, H.—Voyage aux Pyrenées. Paris, 1860 .. .. By PURCHASE.

TAINOR, E. C.—Geographical Sketch of the Island of Hainan. Canton, 1869.
  The Author.

TACUY DE BARROS.—(See Cooley.)

TCHIHATCHEF, P. DE.—Preface de la Nouvelle Edition de 'L'Asie Centrale' de
  Humboldt .. .. The Author.

Thames Gold Fields, Province of Auckland. New Zealand, 1869.

THOMPSON, G.—The War in Paraguay. 1869 .. .. The Author.

THOMSON, J.—The Antiquities of Cambodia. 1867 .. .. By PURCHASE.

THOMSEN, J.—Thermochemiske Undersogelser, etc. Kjobenhaven.

THOREL, C.—Notes Medicales du Mekong et de Cochinchine. 1870.
  The Author.

Tide Tables for the British and Irish Ports. 1870.
  The HYDROGRAPHER TO THE ADMIRALTY.

Titles of Books.	*Donors.*

TINDALE, J. W.—The Island of Sardinia, etc. 1849 .. By PURCHASE.

Treaties.—Collection of Marine Treaties with Great Britain from the Year 1546 to 1763. 1779 .. .. CHAS. ENDERBY, Esq.

TRENCH, Capt. F.—The Russo-Indian Question. 1869 .. The AUTHOR.

TOZER, H. F.—The Highlands of Turkey. 2 vols. 1869 .. The AUTHOR.

VARNHAGEN, F. A.—La Verdadera Guanahani de Colon. Santiago, 1864. The AUTHOR.

.. .. Carta de C. Colon. Marzo. 1493. Vienna, 1869. The AUTHOR.

.. .. Sull' Importanza d'un Manoscritto inedito della biblioteca Imp. di Vienna, etc. Vienna, 1869 .. .. .. The AUTHOR.

.. .. Das Wahre Guanahani des Columbus von F. A. de Varnhagen übersetzung von * *. Wien, 1869.. .. .. The AUTHOR.

VENEGA, W. T.—El Dorado. 1867 .. .. By PURCHASE.

VERDAM, D.—Wiesbaden, etc. Wiesbaden .. .. .. S. M. DRACH, Esq.

VERNEY, Sir H.—Route from Atlantic to Pacific.

Victoria.—Mineral Statistics of, for 1868, Melbourne. SURVEY OFFICE, MELBOURNE.

VOLKER, P.—Samarkand-par M. de Khanikof traduit. Paris. 1869. The AUTHOR.

VOSSIUS, J.—De Nili, etc. Hague, 1666 .. .. .. By PURCHASE.

Voyages made by the Portuguese and the Spaniards in the Fifteenth and Sixteenth Centuries .. .. CHAS. ENDERBY, Esq.

WADDINGTON, A.—Overland Railroad through British North America. 1869. The AUTHOR.

WAGNER, M.—Reisen im tropischen Amerika. Stuttgart. 1870. By PURCHASE.

WALKER, Col.—Great Trigonometrical Survey of India—Report for 1868-69. Dehra Doon .. .. INDIA OFFICE.

WARRE, H.—North America and Oregon Territory.

Washington Report of the Commissioners of Agriculture for 1867. Washington, 1868 .. .. U. S. CONGRESS.

Washington Astronomical Observations. 1866.

WEIGEL, T. O.—Catalogue, livres anciens et rares. Leipzig, 1869. The PUBLISHER.

WESTGARTH, W.—The Colony of Victoria. 1864 .. .. By PURCHASE.

WHITE.—Travels in the Interior of Africa. 1781 to 1797. Dublin, 1801.

WHITNEY, J. D.—Geological Survey of California. California, 1869.

WILD, H.—Repertorium für Meteorologie. St. Petersburg, 1869. The AUTHOR.

WILKINSON, G. B.—South Australia. 1848 .. .. .. By PURCHASE.

WILLIAMS & NORGATE.—Book Circular. 1869 .. .. The PUBLISHERS.

WILSON, C. W. W.—Plan der Stade, etc., Jerusalem. (See James, Sir H.)

WILSON, D.—Prehistoric Man. Cambridge, 1862 .. .. By PURCHASE.

WILSON, M.—Imperial Gazetteer of England and Wales. A. FULLARTON & Co.

YOSEMITE.—Description of the Yosemite Valley. California, 1869. By PURCHASE.

YOUNG, F.—Transplantation the True System of Emigration. 1869. J. BATE, Esq.

Titles of Books.	Donors.

ZIEGLER, I. M.—Ueber das Verhaltniss der Topographie zur Geologie. 1869.
The AUTHOR.

ZEUSCHNER.—Silurformation im Südlichen Polen. Warschaw, 1869.
The PRESIDENT.

ZUCPERT, J. F.—Die Naturgeschichte und Bergwerks verfassung der Ober
Harzes. Berlin, 1762   ..   .. S. M. DRACH, Esq.

PERIODICALS.

Academy	.. The PUBLISHERS.
Alpine Journal	.. The PUBLISHERS.
American Naturalist.	
Artizan	.. The EDITORS.
Assurance Magazine	The INSTITUTE OF ACTUARIES.
Athenæum	.. The EDITORS.
Ausland (Das). Augsburg	.. PURCHASED.
Bibliothèque Universelle et Revue Suisse ..	.. .. The PUBLISHERS.
Bookseller..	.. .. The PUBLISHERS.
Canadian Journal of Industry, Science, and Art.	The CANADIAN INSTITUTE.
Church Missionary Intelligencer. To date ..	.. .. The PUBLISHERS.
Colonial Intelligencer	.. .. The PUBLISHERS.
Food Journal	.. .. The PUBLISHERS.
Journal de l'Agriculture des Pays chauds ..	CHAMBER OF AGRICULTURE.
Malte-Brun's Nouvelles Annales des Voyages ..	.. .. The EDITOR.
Mercantile Marine Magazine. To date ..	.. .. The EDITOR.
Nature. To date	.. The PUBLISHERS.
Nautical Magazine	.. .. PURCHASED.
Newton's London Journal of Arts and Sciences	.. .. The AUTHORS.
Notes and Queries on China and Japan ..	.. .. PURCHASED.
Petermann's Mittheilungen. Gotha	.. M. JUSTUS PERTHES.
Photographic Journal.	
Publishers' Circular	.. PURCHASED.
Quarterly Review	.. The PUBLISHERS.
Revue des Cours Scientifiques	.. The PUBLISHERS.
Scientific Opinion. To date	.. The PUBLISHER.
Trübner's Literary Record	.. PURCHASED.
Wesleyan Missionary Notices	.. The PUBLISHER.

TRANSACTIONS OF SOCIETIES, &c.

EUROPE.

GREAT BRITAIN AND IRELAND—

Titles.	Donors.
Anthropological Review. To date. 8vo.	The ANTHROPOLOGICAL SOCIETY.
Archæologia: Proceedings and Transactions of the Archæological Society.	The SOCIETY.
Horological Journal	HOROLOGICAL INSTITUTION.
Journal of the Photographic Society	The SOCIETY.
Journal of the Proceedings of the Linnæan Society. To date.	The SOCIETY.
Journal of the Society of Arts	The SOCIETY.
Journal of the Royal Agricultural Society of England.	The SOCIETY.
Journal of the Royal United Service Institution	The INSTITUTION.
Journal of the Statistical Society of London. To date.	The SOCIETY.
Journal of the East India Association	The COUNCIL.
Journal of the Royal Geological Society of Ireland	The SOCIETY.
Manchester Free Library Report.	
Memoirs of the Royal Astronomical Society	The SOCIETY.
Memoirs and Philosophical Society of Manchester	The SOCIETY.
Philosophical Society of Glasgow	The SOCIETY.
Philosophical Transactions of the Royal Society of London.	The SOCIETY.
Proceedings of the Aborigines Protection Society	The SOCIETY.
Proceedings, &c., of Institution of Civil Engineers.	The INSTITUTION.
Proceedings of the Geological and Polytechnic Society of the West Riding, Leeds	The SOCIETY.
Proceedings of the Literary and Philosophic Society of Liverpool.	The SOCIETY.
Proceedings of the Royal Artillery Institution	The INSTITUTION.
Proceedings of the Royal Horticultural Society. To date. 8vo.	The SOCIETY.
Proceedings of the Royal Institution	The INSTITUTION.
Proceedings of the Royal Society	The SOCIETY.
Proceedings of the Royal Society of Edinburgh	The SOCIETY.
Proceedings of the Society of Antiquaries	The SOCIETY.
Proceedings and Transactions of the Royal Dublin Society.	The SOCIETY.
Proceedings and Transactions of the Royal Irish Academy.	The ACADEMY.
Proceedings and Transactions of the Zoological Society.	The SOCIETY.
Quarterly Journal of the Geological Society	The SOCIETY.
Radcliffe Observatory	The RADCLIFFE TRUSTEES.
Report of the Committee of Council on Education	The COMMITTEE.
Report of the British Association for the Advancement of Science.	The ASSOCIATION.

Titles.	Donors.

Sessional Papers of the Royal Institute of British Architects.

The INSTITUTE.

Transactions of the Ethnological Society .. .. The SOCIETY.

Transactions of the Historic Society of Lancashire and Cheshire.

The SOCIETY.

Transactions of the Woolhope Naturalists' Club .. The SOCIETY.

Transactions of the Royal Society of Literature .. The SOCIETY.

Transactions of the Edinburgh Geological Society .. The SOCIETY.

War Office Library Catalogue Index .. .. .. WAR OFFICE.

FRANCE—

Académie des Sciences de Dijon .. .. .. The SOCIETY.

Annales Hydrographiques for 1869 .. DÉPARTEMENT DE LA MARINE.

Bulletin de la Société d'Encouragement pour l'Industrie Nationale.

The SOCIETY.

Bulletin de la Société de Géographie .. .. .. The SOCIETY.

Comptes Rendus de l'Académie des Sciences .. .. The ACADEMY.

Journal Asiatique de la Société Asiatique, Paris. The SOCIETY.

Rapport Annuel fait à la Société d'Ethnographie .. The SOCIETY.

Revue Maritime et Coloniale.

MINISTÈRE DE LA MARINE ET DES COLONIES.

Memoires de la Societe des Sciences Naturelles, Strasburg.

GERMANY—

Abhandlungen und Sitzungsberichte der K. B. Akad. der Wissenschaften
zu München. Munich. .. .. Dr. W. KOCH.

Abhandlungen für die Kunde des Morgenlandes Leipzig. The SOCIETY.

Abhandlungen der Naturwissenschaften zu Bremen. The ACADEMY.

Akademie der Wissenschaften zu München. Munich. The ACADEMY.

Beiträge zur Statistik der Stadt Frankfurt am Main von der statistischen
Abtheilung des Vereins für Geographie und Statistik. Frankfurt.

The SOCIETY.

Jahresbericht der Schlesischen Gesellschaft. Breslau. The SOCIETY.

Mittheilungen aus dem Osterland. Altenburg .. The SOCIETY.

Monatsberichte der K. Akad. der Wissenschaften zu Berlin. The ACADEMY.

Notizblatt des Vereins für Erdkunde. Darmstadt. The SOCIETY.

Physikalisch-Œkonomische Gesellschaft zu Königsberg. The SOCIETY.

Schriften der Universität zu Kiel .. .. .. The UNIVERSITY.

Zeitschrift der Deutschen Morgenländischen Gesellschaft, Leipzig.

The SOCIETY.

Zeitschrift der Gesellschaft für Erdkunde zu Berlin .. The SOCIETY.

BELGIUM—

Annales Météorologiques. Bruxelles.

Annuaire de l'Académie Royale.. .. .. .. The ACADEMY.

Bulletin de l'Académie Royale des Sciences, etc., Bruxelles.

Titles.	Donors.

AUSTRIA—

Jahrbücher für Meteorologie und Erdmagnetismus .. CARL JELENEK.

Jahrbuch der Kaiserlich-Königlichen geologischen Reichsanstalt, Wien.

Mittheilungen der Kaiserlich-Königlichen geographischen Gesellschaft. Wien .. .. F. FOETTERLE.

Mittheilungen der Anthropologischen Gesellschaft. Wien. The SOCIETY.

Sitzungsberichte der Kais. Akademie der Wissenschaften. Wien. The ACADEMY.

SWITZERLAND—

Bibliothèque Universelle et Revue Suisse .. .. The LIBRARY.

GENEVA—

Mémoires de la Société de Physique de Genève .. The SOCIETY.

Le Globe: Journal Géographique de Genève .. .. The PUBLISHER.

ZÜRICH—

Die Naturforschende Gesellschaft zu Zürich .. The SOCIETY.

HOLLAND—

Mittheilungen der Antiquarisch. Gesellsch. Zurich.

Jaarboek van de Koninklijke Akademie van Wetenschappen. Amsterdam. The ACADEMY.

Bijdragen tot de Taal-land en Volkenkunde Nederlandsch Indie

Verslagen en Mededeelingen der Koninklijke Akademie van Wetenschappen. Amsterdam .. .. The ACADEMY.

Nederlandsch Meteorologisch Jaarboek .. .. OBSERVATORY.

DENMARK—

Forhandlingen og dets Medlemmers. Copenhagen .. The SOCIETY.

Mémoires de la Société Royale des Antiquaires du Nord. Copenhagen. The SOCIETY.

Oversigt det Kongelige Danske Videnskabernes Selskabs. Copenhagen. The SOCIETY.

ITALY—

Atti della Academia Pontific. dei Nuovi Lincei .. The ACADEMY.

Atti del Reale Instituto Lombardo. Milan .. .. The INSTITUTE.

Atti della Societa Economica di Chiavari .. .. The SOCIETY.

Bolletino della Societa Geographica Italiana. Firenze The SOCIETY.

Memorie del Reale Instituto. Milan .. .. .. The INSTITUTE.

SICILY—

Atti della Academia di Sciensi e Letteri di Palermo .. The ACADEMY.

Bulletin Météorologique.

Meteorological Reports .. .. The OBSERVATORY.

SWEDEN AND NORWAY—

Christiania meteorologiske jagllagelser .. CHRISTIANIA OBSERVATORY.

Statistics and Reports, &c. .. .. .. CHRISTIANIA UNIVERSITY.

Titles.	Donors.
Acta Universitatis Lundensis; etc.	The UNIVERSITY.
Kongliga Svenska Vetenskaps-Academiens Handlingar. To date. 4to. Stockholm	The ACADEMY.

PORTUGAL—

Boletin e Annaes do Conselho Ultramarino.
The ROYAL ACAD. OF SCIENCES, LISBON.

Historia e Memorias da Academia Real das Sciencias de Lisboa, Classe de Sciencias Moraes, Politicas e Bellas Lettras .. The ACADEMY.

Quadro Elementar das Relações Politicas, &c., de Portugal.
ACAD. OF SCIENCES, LISBON.

RUSSIA—

Annales de l'Observatoire central de Russie .. The OBSERVATORY.

Comptes Rendues de la Société Impériale Géographique de Russie.
The SOCIETY.

Jahresbericht d. Nicolai-Hauptsternwarte, St. Petersburg.

Mélanges Physiques et Chimiques de l'Acad. Impériale de St.-Pétersbourg.

Mémoires de l'Acad. des Sciences de St.-Pétersbourg.

Proceedings and Transactions of the Russian Geographical Society.

SPAIN—

Annuario del Real Observatorio de Madrid .. The OBSERVATORY.

Resumen de las Actas de la Real Academia de Ciencias Exactas.

Almenach Nautico. Cadiz .. .. Dr. DAA Y VELA.

Observations de Marina .. .. .. The OBSERVATORY.

ASIA.

INDIA—

Bombay Meteorological Reports..	INDIA OFFICE.
Ceylon Branch of Royal Asiatic Society	The SOCIETY.
China Branch of Royal Asiatic Society. Shanghai ..	The SOCIETY.
Journal of the East India Association	The ASSOCIATION.
Journal of the Royal Asiatic Society of Bengal ..	The SOCIETY.
Journal of the Royal Asiatic Society of Madras ..	The SOCIETY.
Madras Journal of Literature and Science.	
Reports of Bombay Presidency	INDIA OFFICE.
Selections from the Records of the Government of India.	INDIA OFFICE.

CHINA—

North China Branch of Royal Asiatic Society .. The SOCIETY.

AFRICA.

Proceedings of the Meteorological Society of Mauritius. The SOCIETY.

Revue Africaine. To date. LA SOCIÉTÉ HISTORIQUE ALGÉRIENNE.

Tableaux de la Situation des Établissements Français dans l'Algerie, 1865-66.

L'Academie d'Hippone. Bone. .. .. .. The ACADEMY.

AMERICA.

Titles.	*Donors.*
Anales del Museo Publico de Buenos Aires	H. BURMEISTER.
Anales de la Universidad de Chile. Santiago	The UNIVERSITY.
Annual Report of the Trustees of the Museum of Comparative Zoology.	The TRUSTEES.
Boletin de la Sociedad de Ciencias fisicas y Naturales de Caracas. 1868. Nos. 1-3	The SOCIETY.
Boston Society of Natural History	The SOCIETY.
Journal of the Franklin Institute	The INSTITUTE.
Memoirs of Peabody Academy of Science. Salem ..	The ACADEMY.
Proceedings of the American Academy	The ACADEMY.
Proceedings of the American Association for the Advancement of Science. Cambridge, U. S.	The SOCIETY.
Proceedings of the American Geographical and Statistical Society of New York	The SOCIETY.
Proceedings of the Academy of Natural Sciences of Philadelphia.	The ACADEMY.
Proceedings of the American Philosophical Society ..	The SOCIETY.
Proceedings of the Essex Institute	The INSTITUTE.
Reports of U.S. Sanitary Commission. Washington.	
Smithsonian Contributions to Knowledge	The INSTITUTE.
Smithsonian Miscellaneous Collections..	The INSTITUTE.
Smithsonian Report. 8vo.	The INSTITUTE.
The Canadian Naturalist and Geologist, with Proceedings of Natural History Society of Montreal	The SOCIETY.
Transactions of the Connecticut Academy of Arts and Sciences.	The ACADEMY.

MEXICO—

Boletin de la Sociedad Mexicana	The SOCIETY.

AUSTRALIA.

Sydney Meteorological Reports	G. W. SMALLEY.
Transactions of the Royal Society of Victoria	The SOCIETY.
Transactions of the New Zealand Institute	The SOCIETY.
Transactions of the Royal Society of N. S. Wales ..	The SOCIETY.
Victoria Mining Reports	SECRETARY OF MINES.

ACCESSIONS TO THE MAP-ROOM.

From May 24th, 1869, to May 23rd, 1870.

ATLASES.

Maps, Charts, &c.	Donors.
Imperial Atlas of England and Wales. Sheets 1, 16, 16B, and Index. By A. Fullarton and Co. London, 1869	The AUTHORS.
Galliæ, tabulæ geographicæ. Par Gerardum Mercatorem. Duysburg, 1585. 51 maps, with letterpress	PURCHASED.
Die beiden ältesten General Karten von Amerika ausgeführt in den Jahren 1527 und 1529 auf befehl Kaiser Karl's V. im besitz der Grossherzoglichen Bibliothek zu Weimar erlautert von J. G. Kuhl. Weimar, 1860. 2 maps, with letterpress	Sir R. I. MURCHISON.
Charts showing the Surface Temperatures of the South Atlantic Ocean in each Month of the Year. Compiled from Board of Trade Registers, and the Charts published by the Royal Meteorological Institute of the Netherlands. Issued under the authority of the Committee of the Meteorological Office. London, 1869. 12 maps, with letterpress.	The METEOROLOGICAL OFFICE.

THE WORLD.

Political and Physical Chart of the World on Gall's Projection. Scale 1 inch = $4\frac{1}{4}°$. By J. Bartholomew, Esq. 1870.

The World. On Gall's Projection. Scale 1 inch = 29° (equatorial). 10 copies. 1870.

Chart of the World. On Gall's Projection, showing Winds and Storms. W. and A. K. Johnston. Edinburgh, 1870. Scale 1 inch = 45° (equatorial). 9 copies.

Chart of the Stars. On Gall's Projection. Scale 1 inch = 10°. .. Mr. GALL.

Geographische verbreitung der Hirsche über die Erde, Genealogie der Verbreitung Von Dr G. Jaeger & Dr E. Bessels. Geographische verbreitung der Hirsche über die Erde. Gegenwärtige Verbreitungsbezirke. Scale 1 inch = 22°. Von Dr G. Jaeger & Dr E. Bessels. By A. Petermann. Gotha, 1870.

Geographische Verbreitung der Schmetterlinge über die Erde. Von Gabriel Koch. Scale 1 inch = 30°. Von A. Petermann. Gotha, 1870.

Karte zur Übersicht von A. von Humboldt's Reisen in der Alten und Neuen Welt 1799-1829. Zusammengestellt von A. Petermann. Scale 1 inch = $7\frac{1}{4}°$. Gotha, 1869 .. .. A. PETERMANN, Esq.

THE POLES.

Karte der Arktischen und Antarktischen Regionen zur Übersicht der Entdeckungsgeschichte. Scale 1 inch = 560 miles (geo.). Von A. Petermann. Gotha, 1868 .. .. The AUTHOR.

Maps, Charts, &c. *Donors.*

Nord Polarkarte zur Ubersicht einiger geschichtlichen Momente und der jetzigen
Hauptplätze der Grossfischerein (Walfishfang und Robbenschlag) Von A.
Petermann. Gotha, 1869. Scale 1 inch = 550 miles (geo.).
The AUTHOR.

Karte des Europäischen Nordmeeres zur Ubersicht der Geschichte und des jetzigen
Standes der Grossfischerein (Walfischfang und Robbenschlag). Scale
1 inch = 140 miles (geo.). Von A. Petermann. Gotha, 1869.
The AUTHOR.

Karta öfver Hafvet emellan Spetsbergen och Grönland Utvisande Angfartyget-
Sofias Kurser under den Svenska Polar Expeditionen 1868. Äfvensom
drifisens läge under olika tider af äretlodningar. m. m. Scale 1 inch =
25 miles (geo.). 2 copies. .. Professor J. E. VON NORDENSKIOLD.

Ubersicht des Kurses und der Tiefsee Messungen der Schwedischen Expedition
unter Nordenskiold und v. Otter. 20 Juli—19 Oktober, 1868. Scale 1 inch
= 75 miles (geo.). By A. Petermann. Gotha, 1870. 2 copies.

Die Meeres-Temperatur des Grönlandischen Meeres nach den Beobachtungen der
ersten Deutschen Nordpolar Expedition 1868. Von W. von Freeden.
Scale 1 inch = 140 miles (geo.). By A. Petermann. Gotha, 1869.

Fahrt des Cap' E. H. Johannesen aus Tromsö im Karischen Meere im Sommer.
1869. Scale 1 inch = 54 miles (geo.). By A. Petermann. Gotha, 1870.
A. PETERMANN, Esq.

Süd-Polar Karte. Scale 1 inch = 540 miles (geo.) By A. Petermann. Gotha,
1870 .. .. The AUTHOR.

EUROPE.

Barometer Scale —

Ordnance Maps—1-inch scale (Kingdom)—

England and Wales—

Sheets 106, s.w.; 107, s.w. and s.e.; 108, s.w.; and 109, n.w.

Scotland—

Sheet 25 (Hills).

Ireland—

Sheets 17, 40, 41, and 75 (Hills).

Ordnance Maps—6-inch scale (Counties)—

England and Wales.

Co. Cumberland. Title.
Co. Devon. Title.
Co. Durham. Title.
Co. Middlesex. Sheets 20 and 25.
Co. Northumberland. Title.
Co. Westmoreland. Title.

Scotland—

Co. Aberdeen. Sheets 41, 49, 50, 64, 65, 67, 72, 73, 74, 76, 77, 78, 79,
83, 84, 85, 87, 88, 89, 90, 91, 93, 96, 97, 98, 99, 100, 102, 103, 104,
105, 106, 107, 108, 110, and 111.
Co. Argyll. Sheets 162, 163, 172, 173, 182, 183, 184, 193, 194, 195, 203,
204, 214, 215, 216, 225, 226, 227, 228, 237, 238, 241, 243, 244, 247,
248, 249, 250, 251, 252, 253, 254, 255, 256, 257, 258, 259, 260, 261,
262, 263, 264, 265, and 266.
Co. Ayr. Title.

Maps, Charts, &c.   *Donors.*

Co. Forfar. Title.
Cos. Perth and Clackmannan. Titles and Indexes.
Co. Selkirk. Title.
Co. Stirling. Index.

ORDNANCE MAPS—25-inch scale (Parishes)—

England and Wales—
  Berkshire, 15 sheets; Buckingham, 25 sheets; Hampshire, 258 sheets;
    Isle of Man, 169 sheets; Kent, 251 sheets; Surrey, 161 sheets.

Scotland—
  Aberdeen, 264 sheets; Banff, 48 sheets.

Ireland—
  Co. Dublin, 101 sheets.

ORDNANCE MAPS—5 and 10-feet scales (Towns)—

England and Wales—
  Douglas, 28 sheets; Petersfield, 5 sheets.

Scotland—
  Elgin, 14 sheets; Forres, 8 sheets; Inverness, 42 sheets; Peterhead, 17
    sheets.

Ireland—
  Dundalk, 25 sheets; Pembroke Township, 12 sheets.
                                        Total, 1546 sheets.
                    The ORDNANCE SURVEY OFFICE, SOUTHAMPTON,
                            through Sir H. James, R.E., Director.

London. A new Map of Metropolitan Railways, Tramways, and Miscel-
  laneous Improvements deposited at the Private Bill Office, Nov. 30th,
  1869, for Session 1870. Scale 3 inches = 1 mile. By E. Stanford, Esq.
  London, 1870 ..   .. The AUTHOR.

How to Travel in and around London by Railway. Published by
  M. Vigers. London. 5 maps, for the months of June, July, August,
  and November, 1869, and January, 1870 ..   .. The AUTHOR.

FRANCE—

Seven Photographs of Relief Models. By M. Bardin. Paris, 1868.
  1. Photographie du Plan Relief du Mont Blanc, à gradins. Scale
     2 sheets.
  2. Photograph du Plan Relief Topographique des Alpes Dauphinoises.
     (3 sheets.) Scale
  3. Photographie du Plan Relief, à gradins, des Alpes Dauphinoises.
     3 sheets. Scale
  4. Photographie du Plan Relief Topographique des Alpes Dau-
     phinoises. Scale (2 copies.)
  5. Photographie du Plan Relief, à gradins, des Alpes Dauphinoises.
     Scale
  6. Photographie du Plan Relief Topographique des Puys d'Auvergne.
     Scale
  7. Photographie du Plan Relief, à gradins, des Puys d'Auvergne.
     Scale
  With descriptive letterpress.
                    Madame A. BARDIN, through Mr. R. H. Budden.

Maps, Charts, &c. *Donors.*

GERMANY—

Die Freihafen-Gebiete von Hamburg und Bremen. Nach den Bestimmungen von 1868. Scale 1 inch = 12½ miles (geo.) By A. Petermann. Gotha, 1869 .. .. The AUTHOR.

NORWAY—

Norwegian Government Maps.—

1. Topografisk Kart over KongerigetNorge. Scale ⟶. Sheets 14*b*, 14*d*, 10*a*, and 10*d*.
2. Generalkarte over det Sydlige Norge. Sheets 1 and 2. Scale ⟶.
3. Kart over Lister og Mandals Amt. Scale ⟶. 1862. .
4. Kart over Sondre Bergenhus Amt. Scale ⟶. (Sydlige Blad.) 1867.
5. Kart over Stavanger Amt. Scale ⟶. 1866. On 2 sheets.
NORWEGIAN GOVERNMENT.

SPAIN AND PORTUGAL—

Spain and Portugal. Scale 1 inch = 30 miles (geo.). By Robert Wilkinson. London, 1804.. .. .. PURCHASED.

SWEDEN—

Topografiska Corpsens Karta ofver Sverige. On 102 sheets. Scale 1 inch = 1½ miles (geo.). Four sheets, viz.:—

I. V. 36 Borås.
II. O. 39 Huseby.
II. O. 40 Carlshamn.
IV. O. 32 Westerås.

Two copies of each map .. Major-General J. A. HAZELIUS, Chief of the Royal Topographical Corps of Sweden.

SWITZERLAND—

Karte des Kantons Glarus. Scale 1⅓ inch = 1 mile (geo.). On 2 sheets. By J. M. Ziegler. Winterthur, 1869 .. .. The AUTHOR.

Meteorological maps. 12 in number :—

1. Entwurf einer Karte über die Geschichte der Flüsse und See'n in der Schweiz.
2*a*. Mittlere tägliche Barometerstande und Temperaturen. December and January, 1868-9.
2*b*. do. do. October and November, 1868.
2*c*. do. do. August and September, 1868.
2*d*. do. do. June and July, 1868.
2*e*. do. do. April and May, 1868.
2*f*. do. do. February and March, 1868.
2*g*. do. do. December and January, 1867-8.
3. Niederschläge. September and October, 1868.
4. Meteorologische Stationen der Schweiz.
5*a*. Gewitter im Mai. 1868.
5*b*. Gewitter im Juli. 1868. Winterthur, 1869.

TURKEY—

Die Ausdehnung der Slaven in der Türkei und den angrenzenden Gebieten. Nach den neuesten Untersuchungen. Scale 1 inch = 50 miles (geo.). Von A. Petermann. Gotha, 1869 .. .. .. The AUTHOR.

Map of the Vilayet of the Danube. Scale 1 inch = 15 miles (geo.). By Sir Robert Dalyell, Bart. 1869 .. .. E. C. OULET.

ASIA.

AFGHANISTAN—

Map of Afghanistan compiled in the Office of the Surveyor-General of India. Scale 1 inch = 7 miles (geo.). Calcutta, 1840. On 15 sheets with some duplicate sheets ..   ..   PURCHASED.

BURMAH—

Geographical Sketch of the Burmese Empire. Compiled in the Surveyor-General's Office. Calcutta, 1864. Scale 1 inch = 14 miles (geo.).
                                                       PURCHASED.

CHINESE EMPIRE—

Karte vom Ostlichen China und Korea, zur Übersicht der Chinesischen Dialekte nach Edkins und der Reisen von Oxenham and Markham, 1868–69 Scale 1 inch = 102 miles (geo.). Von A. Petermann. Gotha, 1869 ..   ..   A. PETERMANN, Esq.

Map of the Head-Waters of the Kin-Char-Kiang, Lan-Tsan-Kiang, Now-Kiang, and Great River of Thibet. Laid down from Chinese Maps by T. T. Cooper. A tracing..   ..   ..   T. T. COOPER, Esq.

The Tsien-Tang River, from Hangchow-Foo to Kinchow-Foo. Scale 1 inch = 2 miles (geo.). 2 sheets (*MS.*). By Mr. Elias. November, 1867 ..   ..   The AUTHOR.

Plan and Views of the River Yang-tsze-Kiang from Ichang (Hoopeh) to Kwei-Foo (Sze-chuen). Scale 1 inch = 4 miles (geo.). By Francis Ingram Palmer, R.N. (A photograph.) 1869. R. SWINHOE, Esq.

Chart of the New Course of the Yellow River from the Bar to the Old Bed. Scale 1 inch = 2 miles (geo.). 5 sheets (*MS.*). By Mr. Elias. October, 1868 ..   ..   The AUTHOR.

INDIA—

Map of the Hyderabad Collectorate. 4 sheets in 25 parts. Quarter-Master General's Office. Bombay, 1851. Scale 1 inch = 2 miles.

Map of the Shikarpoor Collectorate. On 2 sheets. Scale 1 inch = 4 miles (stat.). Surveyor-General's Office, Calcutta, 1864.

The Khyrpoor Territory of H. H. Meer Ali Moorad. On 2 sheets. Scale 1 inch = 4 miles (geo.). Surveyor-General's Office, Calcutta, 1862.

Map showing proposed Route of the Indus Valley Railway. Scale 1 inch = 17 miles (geo.). By John Brunton, C.E. London, 1863.
                                                   INDIA OFFICE,
                                          through Sir Bartle Frere.

Map of the Punjaub and protected Sikh States, including the British Provinces to the South. Scale 1 inch = 21 miles (geo.). Compiled in the Office of the Surveyor-General of India. Calcutta, 1846.

Geographical Plan of the north-east Frontier of Bengal, with part of Assam. Scale 1 inch = 7 miles (geo.). Surveyor-General's Office, Calcutta, 1864.

A Map showing part of Cachar and Munnipoor, also the direction of one of the Routes leading from Doodpullie towards Munnipoor. Scale 1 inch = 3 miles (geo.). (*MS.*) By P. Matthews. 1825.

Map showing the Eastern Boundary of the Province of Chittagong and the relative situation to Arracan. Scale 1 inch = 10 miles (geo.). With letterpress. By A. Fitzpatrick. Calcutta, 1824.

Map of Malwa and adjoining Countries constructed by order of Major-General Sir John Malcolm from the Routes of his division and the

Maps, Charts, &c.	*Donors.*

Surveys of Officers under his command. By Lieutenant Robert Gibbings. Calcutta, 1845. Scale 1 inch = 7 miles (geo.).

A Sketch of the conquered Provinces of Martaban, Tavoy, and Mergui. By Captain P. W. Grant. Scale 1 inch = 7 miles (geo).

Geological Map of the Southern Portion of Zillah Shahabad. Scale 1 inch = 2 miles (geo.). By Lieutenant W. S. Sherwill. 1846.

District Maps of the Revenue Survey of India. 18 maps; viz.:—

Agra. Captains Wroughton and Fordyce. 1837-9.
Allahabad. Captains Lawrence and Stephens. 1838.
Allygurh. Captain R. Wroughton. 1837.
Bandah. Lieutenants Abbott and Stephen. 1840-1.
Boolundshubur. Captain W. Brown. 1828-40.
Cawnpoor. Lieutenant S Abbott. 1840.
Dehra Doon. Captain W. Brown. 1838-9.
Etawah. Captain R. Wroughton. 1838-9.
Furruckabad. Captains Lawrence and Wroughton. 1838-9.
Futtehpoor. Lieutenant H. Stephen. 1839.
Humeerpoor. Lieutenant H. Stephen. 1839-40.
Jaloun. Lieutenant S. Abbott. 1841-2.
Mirzapoor. Captain R. Wroughton. 1840-1.
Muthra. Captain R. Wroughton. 1835.
Mynpooree. Captain R. Wroughton. 1837-9.
Pillheebheet. Captain B. Browne. 1838-40.
Pooree. Lieutenant H. Thuillier. 1837-41.
Ramgurh and Sohagpoor. } Captain R. Wroughton. 1842.

Map of the Route from Commilla, via Silhet, towards Cachar. Scale 1 inch = 6¼ miles (geo.). (*MS.*) By P. Mathews. 1845.

River Ganges, from Allahabad to Mohungunjee. On 6 sheets. By T. Prinsep. 1825.

River Bhag'ruttee. By T. Prinsep. 1828.

River Hooghly, from Calcutta to Nuddya. By Captain T. Prinsep. 1828.

River Jellinghee. By J. S. May. 1828.

River Matabhunga. By J. S. May. 1827.

River Ganges, from Mohungunj to Rajapoor. By J. S. May. Scale 1 inch = 2 miles.

Plan of Prome. Scale 1 inch = 200 yards. (*MS.*) 1825.

Plan of the Siege of Rangoon. Scale 1 inch = ½ mile. (*MS.*) By Major J. Jackson.

Country round Rangoon. Scale 1 inch = 6¼ miles. Surveyor-General's Office. Calcutta, 1825. .. . . PURCHASED.

Russia—

Karte der neuesten Russischen Forschungen im Thian-Schan-System besonders der Reise zwischen Issyk-Kul und Kaschgar Von Seewerzow und E. V. Osten Sacken. 1867. Scale 1 inch = 20⅔ miles (geo.). Von A. Petermann. Gotha, 1869. .. A. PETERMANN, Esq.

Russian Map of the Province of Khokan, with part of the Kirghiz Steppe (in Russian character). Scale 1 inch = 28 miles (geo.). 1868. On 3 sheets .. .. Lady STRANGFORD.

Thuringia—

.... und dem Sophienstift in Weimar gewidnet von C. .... Weimar, 186. Scale 1 inch = 4½ miles (geo.).
                                        Sir R. I. MURCHISON, President.

h 2

Maps, Charts, &c.   *Donors.*

Plan der Stadt und Umgebung von Jerusalem nach den Englischen
Aufnahmen in den Jahren, 1864 u. 65, durch Capitain W. Wilson,
R.E., und seiner Brigade unter der Direction von Colonel Sir Henri
James, R.E. F.R.S., &c., Director of the Ordnance Survey, reducirt
von $\frac{1}{10000}$ auf $\frac{1}{25000}$ und herausgegeben von Wurster, Randegger and
Co. Winterthur.

The same Coloured Geologically. By Dr. Oscar Fraas.
                                        Prof. J. M. Ziegler.

The Ordnance Survey of Jerusalem. Five Plans, viz.:—

  1. Plan of Jerusalem. Hill shaded. Scale $\frac{1}{10560}$ or 6 inches = 1 mile.
  2. Plan of Jerusalem, with Contours. Scale $\frac{1}{2500}$ or 25·344 inches =
     1 mile.
  3. Plan of the Haram, with Contours. Scale $\frac{1}{500}$ or 10·56 feet =
     1 mile.
  4. Plan of the Church of the Holy Sepulchre. Scale $\frac{1}{200}$ or 1 inch
     = 17 feet.
  5. Plan of Kubbat as Sakhra (Dome of the Rock) and Sections;
     Plan of the Citadel (Al Kala); House of Caiaphas; David's
     Tomb; Greek Church in the Convent of the Cross; Armenian
     Chapel in the Convent of St. James; Church of the Ascension;
     Church of the Tomb of the Virgin; Church of St. Anne; and
     Church of the Flagellation. On various scales.
  By Captain Charles W. Wilson, R.E., under the direction of Colonel
     Sir H. James, R.E., Director of the Ordnance Survey. South-
     ampton, 1865.
                    Ordnance Survey Office, Southampton,
                         through Col. Sir H. James, R.E., Director.

AFRICA.

General—

  Carte d'Afrique. Sur les Observations Astronomiques de Membres de
  l'Académie Royale des Sciences et sur les Mémoires les plus recentes.
  Par Mr. L'Abbé Clouet. Scale 1 inch = 4°. Rouen, 1780.
                    Charles Thomas, Esq., Lima, Peru.

North-Eastern—

  Zoo-Geographische Karte des Nilgebietes und der uferländer des Rothen
  Meeres. Von M. Th. v. Heuglin. Scale 1 inch = 140 miles (geo.). A.
  Petermann. Gotha, 1869. ..   ..   ..   A. Petermann, Esq.

  Isthmus of Suez and Lower Egypt. Scale 1 inch = 12½ miles (geo.). By
  James Wyld. London, 1869. (5 copies.)

  Map and Section of the Suez Maritime Canal. Scale 1 inch = 4⅞ miles
  (geo.). By James Wyld. London, 1869. (11 copies.)
                                                  The Author.

  Map of the Suez Canal, Lower Egypt, and Palestine. Scale 1 inch =
  10 miles (geo.). By Keith Johnston, Jun. London, 1869.
                                                  The Author.

  Plan of the Suez Canal
  Plan de Port Said.
  Plan d'Ismäilia.
  Plan de Suez.
     By D. A. Lange. Paris, 1869. ..   ..   ..   The Author.

Eastern—

  Neue-karte des Landes zwischen Suäkin und Berber nach den Aufnahmen
  von Heuglin und Schweinfurth. Gez. und entw. von Dr. G. Schweinfurth. Scale 1 inch = 13¾ miles (geo.). Dr. A. Petermann. Gotha, 1869.
                                                  A. Petermann, Esq.

Maps, Charts, &c. *Donors.*

Ethiopia. Carte No. 2: Tigray. Carte No. 3: Simen et Zimbila. Scale 1 inch = 6½ miles (geo.). Par Antoine d'Abbadie. Paris, 1868.
W. D. COOLEY, Esq.

Umgegend von Axum und Adoa in Tigre. Trigonometrisch aufgenommen von Wilh. Schimper. With sections. Berlin, 1869.
Dr. H. KIEPERT.

202. Map of the Country between Mombas and Lake Victoria Nyanza. Scale 1 inch = 23 miles (geo.). By Rev. T. Wakefield. 1869.
The AUTHOR.

CENTRAL.—

Originalkarte der Portugiesischen Reisen in Inner Afrika seit 1798. Nach den Portugiesischen Quellenwerken zusammengestellt. Nebst Skizze von Livingstone's Reise 1866-9. Scale 1 inch = 60 miles (geo.). Von A. Petermann. Gotha, 1870 .. .. .. The AUTHOR.

SOUTH.—

Chronologische Uebersicht von Livingstone's Reisen in Süd-Afrika, 1840-1869. Scale 1 inch = 170 miles (geo.). Von A. Petermann. Gotha, 1870 .. .. A. PETERMANN, Esq.

Sketch-Map showing the Journey of Discovery of the Bembe, Limpopo, or Oori River and adjacent Country, as explored by St. Vincent W. Erskine, of the Surveyor-General's Office, Natal. 1869. Scale 1 inch = 20 miles (geo.) .. .. ST. VINCENT W. ERSKINE.

Originalkarte von C. Mauch's Reisen im Innern von Süd-Afrika zwischen Potchefstroom und Zambesi 1865-9. Nebst Uebersicht aller anderen Forschungen. Scale 1 inch = 27½ miles (geo.). Von A. Petermann. Gotha, 1870 .. .. A. PETERMANN, Esq.

Skeleton Sketch-Map of Mapoota River and neighbouring Country. Scale 1 inch = 17½ miles (geo.). By Captain Cockrane, H.M.S. *Petrel.* 1869.
ST. VINCENT W. ERSKINE, Esq.

AMERICA.

NORTH.—

General—

Map of North America, showing proposed Railway. Scale 1 inch = 200 miles (geo.). By J. H. Colton and Co. New York, 1854.
Mr. CYRIL GRAHAM.

British—

203. Map of British Columbia. Scale 1 inch = 8½ miles (geo.). By Alfred Waddington, Esq. .. .. The AUTHOR.

UNITED STATES—

Nord-West Amerika mit dem vom Russland an die Verein Staaten cedirten Territorium Alaska. Mit Benutzung der neuesten Amerikanischen Aufnahmen besonders der unter W. H. Dall im Coast Survey Office bearbeiteten Karte. Von A. Petermann. Gotha, 1869. Scale 1 inch = 120 miles (geo.) .. .. A. PETERMANN, Esq.

Sketches accompanying the Annual Report of the Superintendent of the United States Coast Survey. 1851.

Maps and Views to accompany the Message from the President of the United States to the Two Houses of Congress, at the commencement of the First Session of the Thirty-third Congress. Washington, 1853.

Ditto ditto, at the commencement of the Second Session of the Thirty-third Congress. Washington, 1854.

Ditto ditto, at the commencement of the Third Session of the Thirty-fourth Congress. Washington, 1856.

Maps, Charts, &c.   *Donors.*

Maps and Views, at the commencement of the Third Session of the Thirty-seventh Congress. Washington, 1862.

Map of the Public Land, States, and Territories, constructed from the Public Surveys and other Official Sources in the General Land Office. Drawn by J. H. Hawes. Philadelphia, 1865. Scale 1 inch = 53 miles (geo.).

Maps to accompany Israel D. Andrew's Report to the Secretary of the Treasury. New York, 1853.

Railroad Map of the United States. Scale 1 inch = 26¼ miles (geo.). By J. T. Lloyd. New York, 1861.

Plans of the Cities of Washington, Louisville, Jeffersonville, New Orleans, and Cincinnati. 1838.

Eighteen Railway Maps and Prospectuses   ..   Mr. CYRIL GRAHAM.

Map of the State of Virginia, from actual Surveys made by order of the Executive, 1828-1839. Scale 1 inch = 9 miles (geo.). J. T. Lloyd, New York, 1862 .. .. G. T. ARCHER, Esq.

CENTRAL—

Originalkarte von Costarica von A. von Frantzius, enthaltend die Resultate der neuesten Aufnahmen und Beobachtungen von Valentini, L. Daser, F. Kurtze, K. v. Seebach, Raf Alvarado, A. Oerstedt, T. A. Hull, u. a. Scale 1 inch = 13¼ miles (geo.). Von A. Petermann. Gotha, 1869.
A. PETERMANN, Esq.

Mapa de Guatemala la Nueva. Levantado par Herman Au, Licenciado en Mayo de 1868.
Professor J. M. ZIEGLER.

SOUTH—

Karte von Chile in 2 blättern nach der Landesaufnahme in ~~maps~~. Reduceit auf ~~1500000~~ or 1 inch = 20 (geo.). Von A. Petermann. Gotha, 1870.
A. PETERMANN, Esq.

AUSTRALIA.

EASTERN—

Map of the Colony of Queensland. Compiled by Parrot and Teage, from the latest Government and other reliable Surveys. Scale 1 inch = 36 miles (geo.). Melbourne, 1869 .. .. .. Dr. MÜLLER.

CENTRAL—

Übersicht der neuesten Reisen und Aufnahmen Seegebeit von Central Australien. Scale 1 inch = 13¼ miles (geo.). Von A. Petermann. Gotha, 1870 .. .. A. PETERMANN, Esq.

WESTERN—

John Forrest's Reise im Innern von West Australien, April—Aug. 1869. Scale 1 inch = 25 miles (geo.). Von A. Petermann. Gotha, 1869.
A. PETERMANN, Esq.

NEW ZEALAND.

Reconnaissance MS. of the Interior of the Province of Canterbury, New Zealand. Scale 1 inch = 4 miles (geo.) By Julius Haast, PH.DR., F.R.S. Christchurch, 1869 .. .. The AUTHOR.

OCEANS.

ATLANTIC—

Chart of Temperatures on the Surface of the Atlantic Ocean between Shetland and Greenland (by the Thermometer of Reaumur). Scale 1 inch = 73 miles (geo.). By Admiral C. Irminger. 1869. MS.
The AUTHOR.

CHARTS.

Charts, &c. *Donors.*

Section 8
No. 415 Port of Matanzas (Cuba).
483A
483B } Trinidad Island and Gulf of Paria.
2097 Port Spain and Bocas de Dragos (Trinidad).

Section 9.
No. 537 Ceara Bay (Brazil), South America.
547 Harbours and Anchorages in Magellan Strait.
1336 East Entrance of Magellan Strait, from Cape Virgins to the First Narrows.
1337 Magellan Strait, from the First Narrows to Sandy Point.
2544 Rio de la Plata (South America, East Coast).

Section 10.
No. 85 English Narrows (South America, West Coast).
555 Goletas Channel (Vancouver Island).
566 Coman, or Leteu Inlet (South America, West Coast).
574 Coquimbo Bay and Port Herradura (Chilé).
1923A and 1923 { Cape Caution to Port Simpson, including Hecate Strait and part of Queen Charlotte Islands (North America, West Coast).
2431 { Port Simpson to Cross Sound, with Koloschensk Archipelago (Russian America).

Section 11.
No. 397 River Volta (Africa, West Coast).
641 Port Elizabeth (Algoa Bay).
1843 Buffalo River (Africa, South-East Coast).
2086 Waterloo Bay to Bashee River (Africa, South Coast).

Section 12.
No. 40 Karáchi Harbour (India, West Coast).
2413 Rhio Strait (China Sea).
2483 Indian Ocean and Western part of Pacific.
2597 Banka Strait (China Sea).

Section 13.
No. 135 Sagitsu-no-Ura (Japan).
139 { Northern Entrance of Hirado-no-Seto, with the Harbour of Yebuko-no-Minato (Japan)
140 { From Mats'sima to Atsuai-no-o-sima, including Hirado Island and Spex Strait (Japan).
141 Yobuko Harbour (Japan).
915 Anchorages in the Moluccas.
943 Philippine Islands and adjacent Seas.
1116a b c d e { The Upper Yang-tsze-Kiang, from Yoh-Chau-Fu tu Kwei-Chau-Fu (on 5 Sheets).
1257 Ping-Yang Inlet and Ta-Tong River (Korea).
1258 Approaches to Sévul (Korea, West Coast).
1261 Saigon River to Phan-Rang Bay (Cochin-China).
1395 Ting-Hae Harbour (Chusan Island).
2409 West Coast of Formosa and Pescadores Channel.
2432 { Tumen Ula River to Strelok Bay, including Peter the Great Bay (Russian Tartary).
2577 St. Bernardino Strait and parts adjacent (Philippine Islands).
2578 Eastern part of the Sulu or Minaoro Sea (Philippine Islands).

Section 14.
No. 18 Port Darwin (Australia, North Coast).
1020 Beecroft Head to Port Jackson (Australia, East Coast).
1029 Danger Point to Cape Moreton (Queensland).
1068 Moreton Bay to Sandy Cape do.
1070 Port Stephens (Australia, East Coast).

Charts, &c.	Donors.

1792 Port Adelaide (Australia, South Coast).
2152 { Macdonnell Sound and Port Wakefield (St. Vincent Gulf, South Australia).
2166 Broken Bay (Australia, East Coast).
2176 Jervis and Bateman Bays (Australia, East Coast).
2375 Torres Strait (Sheet 1).
2422 ——————— (Sheet 2).

Section 15.

No. 896 Távaroa, or Uturoa Harbour (Society Islands).
977 Harbours in Ualan Island (Caroline Islands).
978 Ualan Island (Caroline Islands), South Pacific.
979 Christmas, Rierson, and Humphrey Islands, do.
1382 Otaheite and Eimeo (Society Islands).
1386 Rapa-Nui, or Easter Island (South Pacific).
1640 Marquesas Islands, do. do. Total, 82 sheets.

The HYDROGRAPHIC OFFICE, ADMIRALTY,
through Admiral G. H. Richards, R.N., Hydrographer.

Chart of the North Atlantic Ocean, showing Deep-sea Soundings, and the
Tracks of the Telegraph Cables laid between Europe and America by
the Telegraph Construction and Maintenance Company in 1865, 1866,
and 1869. With Sections of the Bed of the Ocean, and Journals of
each Expedition. Compiled by F. le B. Bedwell. August, 1869. Scale
1 inch = 90 miles (geo.). 5 copies.

Captain SHERARD OSBORN, R.N.

NORWEGIAN GOVERNMENT—

1. Den Norske Kyst fra Faeder til Udsire udgivet af den Geografiske
   Opmaaling. Scale ₁/₅₀₀₀₀₀. (On 2 sheets. 1865.)
2. Karte over den Norske Kyst fra Ekersund til Stavanger og Hvidingso
   Fyr. Scale ₁/₁₀₀₀₀₀. 1860.
3. Karte over den Norske Kyst fra Hvidingso til Espevaer. Scale
   ₁/₁₀₀₀₀₀. 1865.
4. Kart over den Norske Kyst fra Espevaer til Korsfjord. Scale ₁/₁₀₀₀₀₀.
   1865.
5. Kart over den Norske Kyst fra Korsfjord til Helliso. Scale ₁/₁₀₀₀₀₀.
   1866.
6. Indsaling til Kristiansand, &c., og Salor, &c. ₁/₅₀₀₀₀. (On 2 sheets.)
7. Specialkart over den Norske Kyst fra Jaederens Rer til Tanangerhaug. Scale ₁/₅₀₀₀₀. 1862.
8. Specialkart over den Norske Kyst fra Tanangerhaug til Skudesnaes.
   Scale ₁/₅₀₀₀₀. 1864.
9. Specialkart over den Norske Kyst fra Skudesnaes til Rambeskaarfjeld. Scale ₁/₅₀₀₀₀. 1864.
10. Specialkart over den Norske Kyst fra Rambeskaarfjeld til Ryvardens
    Fyr. Scale ₁/₅₀₀₀₀. 1865.
11. Specialkart over den Norske Kyst fra Ryvardens Fyr til Hisken.
    Scale ₁/₅₀₀₀₀. 1865.
12. Specialkart over Hardangerfjord. (On 2 sheets.) Scale ₁/₅₀₀₀₀. 1865.
13. Flakekart over den Indre del af Vestfjorden i Lofoten udgivet af den
    geografiske Opmaaling. Scale 1 inch = 1₁/₁₀ mile (geo.). On 4
    sheets. Kristiania, 1869. (2 copies.)
    Profiler til flakekartet over Vestfjorden.
14. Kart over Havbankerne Langs. Den Norske Kyst fra Stadt til Harö.
    Scale ₁/₅₀₀₀₀. On 4 sheets. Kristiania, 1870.
    With 2 books of Pilotage. The NORWEGIAN GOVERNMENT.

NON-OFFICIAL—

Chart of the Coast of Arracan from Chittagong to Sandoway, from the
Observations of the Captain and Officers of the H.E.I. Company's

Charts, &c.   *Donors.*

Flotillas under Commodore Hayes on the Arracan Expedition. (On 2 sheets.)   Collated and compiled by Captain John Crawford. Calcutta, 1825.   Scale 1 inch = 5½ miles.

Coast of Arracan from Chittagong to Cheduba Island. (*MS.*) Scale 1 inch = 13 miles (geo.).   ..   ..   PURCHASED.

MISCELLANEOUS.

Seventy-eight Photographs taken in Abyssinia, from Zoulla to Magdala, during the British Campaign, by the Photographers of the 10th Company Royal Engineers.   1868-9.
                                SECRETARY OF STATE FOR WAR.

Five Drawings of South-Eastern Africa, viz. :—
  1. "The Meeting of the Waters," Confluence of the Lipalule and Limpopo Rivers.
  2. The Mouth of the Bembe or Limpopo River.
  3. Man of the Knob-nosed Tribe.
  4. Woman of the Knob-nosed Tribe.
  5. Chief's Wife (Inkosigazi) of the Tribe of Umzeila.
     By St. Vincent W. Erskine   ..   ..   ..   The AUTHOR.

Six Photographs, illustrating the Earthquake at Arica (Peru), August 13th, 1868.   Lieutenant S. C. HOLLAND, R.N., H.M.S. *Malacca*.

Formulæ of Navigation, Nautical Astronomy, &c.   A series of 14 cards, arranged by Charles F. Shadwell, Esq., C.B., F.R.S., Rear-Admiral. London, 1869. ..   ..   The AUTHOR.

To the late Mr. I. DUNCAN, Vice-Consul at Whydah, in 1849—

    Telescope.
    Two Compasses.
    Aneroid Barometer.

Dr. P. C. SUTHERLAND, M.D., F.R.G.S., at Natal—

    Brass Sextant (14-inch), with Silver Arc, by Troughton and Simms.
    Strong-framed Artificial Horizon, by Troughton and Simms.
    Two Barometers (Mountain) with Improved Iron Cistern, by Newman.

The late Dr. E. I. IRVING, M.D., F.R.G.S., at Abeokuta—

    Pocket Chronometer, by Barraud and Lund.
    Barometer (Mountain), by Troughton and Simms.

Dr. D. LIVINGSTONE, M.D., F.R.G.S., Zambesi, Eastern Africa—

    Taylor's Hypsometrical Apparatus, No. 1, with Sling Case, by Casella.
    Standard Thermometers, 0 to 212, in Brass Cases,   .
        „   „ in Maroon Cases,   :
    Artificial Horizon, with Sling Cases,   :
    Prismatic Azimuth Compass, silver ring, with leather Sling Case, „
    Rain Gauge.

Dr. D. WALKER, M.D., F.R.G.S., Russian America, Dec. 6, 1862—

    Sextant, 4-in. radius, by Cary.
    Artificial Horizon, circular, by Cary.
    Azimuth Compass, by Elliot.

The late Mons. JULES GÉRARD, Upper Guinea, towards Timbuktu, Feb. 4, 1863—

    Sextant, 3-inch radius, by T. Jones.
    Aneroid, white metal, by Spencer, Browning, and Co.
    Artificial Horizon, spirit-level, by Elliot.
    Boiling-water Apparatus, and three Thermometers in brass tubes.
    Azimuth Compass, by Barnier.
    Two small Pocket Compasses.
    Protractor, brass, 3-in. radius.
                        (The above in Leather Case.)
    Measuring Tape, 50 feet.
    Thermometer, on metal, in Morocco Case.
    Protractor, horn, circular.

H. WHITELY, Esq., in South Peru, March 28, 1867—

    Pocket Aneroid, No. 99, graduated to 15 inches, by Cary.
    Hypsometrical Apparatus, and 3 Boiling-point Thermometers, by Casella.

Rev. F. W. HOLLAND, Sinai, June 25, 1867—

    Prismatic Compass and Stand, by Cary.
    Pocket Aneroid, graduated to 15 inches.
    Hypsometrical Apparatus, and 3 Thermometers, B.P.
    Two Thermometers, divided to 230° for hot springs.
    Three Alpine minimum Thermometers.

PRESENTATION

OF THE

ROYAL AWARDS.

(At the Anniversary Meeting, May 23rd, 1870.)

ROYAL MEDALS.

THE Founder's Medal was awarded to Mr. George J. W. Hayward, the Society's Envoy to Central Asia, for the Map of his Journey across the Kuen Lun into Eastern Turkistan, and for the perseverance with which he is endeavouring to carry out his object of reaching the Pamir Steppe. The Patron's, or Victoria Medal, to Lieutenant Francis Garnier, of the French Navy, second in command of the French Exploring Expedition from Cambodia to the Yang-tsze-Kiang, for the part he took in the extensive Surveys executed by the Commission, for his Journey to Tali-fu, and for the ability with which, after the death of his chief, Captain de Lagrée, he brought the Expedition in safety to Hankow.

In presenting the Medal to Major-General Sir H. C. Rawlinson, on behalf of Mr. Hayward, the PRESIDENT spoke as follows :—

"The Founder's Medal for the year 1870 has been awarded to Mr. G. J. W. Hayward, late of H.M. 72nd Regiment, for the valuable services he has already rendered to Science in improving our acquaintance with the Geography of Central Asia ; and also in acknowledgment of his zeal and energy in entering at the present time on another perilous expedition for the same purpose. Mr. Hayward having proposed, in 1868, to proceed as a private traveller on an exploring journey into Central Asia, if the Royal Geographical Society would provide him with instruments and contribute to the expenses of the expedition, such assistance was readily afforded ;

and the Society has every reason to be satisfied with the results of the Journey, which was thus undertaken under their auspices and with their encouragement. The countries to which Mr. Hayward's attention was particularly directed were the plains of Eastern Turkistan on the one side, and the contiguous Pamir Plateau on the other; the hydrography of the Upper Oxus, of which our knowledge is very imperfect, being considered an object of especial interest. It was recommended to endeavour to penetrate from the Cabul River by the Valley of Chitral to the head streams of the Oxus, and from thence to pass over the Pamir Steppes to the cities of Yarkand and Kashgar; but if this route, which has never yet, it is believed, been followed by a European traveller, proved impracticable, he was authorised to pursue the easier line by Cashmir and Little Tibet. Finding, accordingly, on his arrival at Peshawar, that the tribes to the north-west were in arms, and that the mountain-passes were entirely closed, he proceeded direct to Leh, the capital of Ladak, and from thence took the high road to Yarkand. Here his geographical researches and discoveries commenced. He tracked the upper courses both of the Karakash (or Khoten) and Yarkand rivers, and rectified important errors in the official maps; and he pointed out the immense importance of securing the Yangi-Davan Pass, beyond the Yarkand River, against the inroads of the Kúnjút robbers, as the only road across the Kuen Lun which was practicable to laden horses or mules. Subsequently Mr. Hayward pursued his journey to the cities of Yarkand and Kashgar taking observations for latitude at almost every stage, and, by careful measurements of distance and a continuous series of angles, determining with very considerable accuracy the longitudes also of all the principal stations. The map of Eastern Turkistan, which Mr. Hayward forwarded to the Geographical Society on his return to India, is a most creditable and valuable document. It has already received high commendations both from the Government and the Survey authorities in India, and as it will be published in the Society's 'Journal,' together with the elaborate Memoir with which it was accompanied, it will soon be available for general reference.

The Council of the Royal Geographical Society, as a scientific body, can take cognizance officially only of geographical services; and it is expressly on this ground that they have awarded the Founder's Medal for the present year to Mr. Hayward; but they cannot lose sight of the fact that the travels of Messrs. Hayward and Shaw have been at least as valuable in a public as in a scientific point of view. They have removed causes of distrust and alarm which gave rise to disquietude in India; they have opened out a new field to British trade and enterprise; they have laid the foundation of what may prove in the sequel to be a valuable political alliance.

" With regard to Mr. Hayward's present position, nothing positive can be announced. All that is known is this, that being stimulated rather than disheartened by his failure to reach the Pamir Plateau from Turkistan (the Kashgar authorities having placed an absolute interdict on his proposal to return to India by Badakhshan and Chitral) he had no sooner recruited his strength by a few months' rest on the Indian frontier, than he resolved to make another attempt to carry out his original design. Assisted, accordingly, by a further supply of funds from the Geographical Society, and having obtained the good wishes and support of his Highness the Maharajah of Cashmir, he started towards the close of the year for the valley of Gilgit, which is now held by his Highness's troops. He intended to winter in Gilgit or some of the adjoining valleys, and to endeavour in the early spring, or as soon as the passes were open, to push his way across the mountains to Badakhshan and the Upper Oxus. From thence the road would be open to Pamir, and he hoped, after thoroughly examining the hydrography of the Upper Oxus, to cross into the Russian territory of Samarkand, where, at the instance of the President of this Society, instructions have been sent from St. Petersburg to receive him with kindness and hospitality, and facilitate his return to England. In the course of the next few weeks, it is probable that something definite will be learnt as to his present and prospective movements."

Sir HENRY C. RAWLINSON, having received the Medal, spoke as follows :—

"I feel an especial pride and satisfaction in receiving on behalf of Mr. Hayward this day, at your hands, the Patron's Gold Medal of the Royal Geographical Society. I feel a pride because it was my good fortune to introduce Mr. Hayward in the first instance to the notice of the Society, and I feel a satisfaction because I know that Mr. Hayward has fairly earned the distinction which has been conferred upon him, and because I also foresee that his successful example will stimulate many other travellers to similar exertions in the cause of science. Perhaps it may not be out of place if, in a very few words, I briefly state how Mr. Hayward has come to gain the Medal of the Society, and in how far I was instrumental in sending this promising explorer on his travels. For the last thirty years I have taken a great interest in the geography of Central Asia, and have striven to encourage and promote discovery in those regions. Personally I was unacquainted with Mr. Hayward until a very few years ago, when, on his return from India, he waited on me one day at the India Office, and stated that having retired from the army, and being desirous of active employment, he proposed to undertake any exploratory expedition that I could suggest. He added that he had some experience in such travels, that he was a fair surveyor and draughtsman, and that he was ready, in fact, to proceed on any expedition that I could recommend. I at once suggested to him that the cities of Eastern Turkistan and the Pamir Steppes were regions of great interest with which we were comparatively unacquainted. They were of interest, I told him, not only geographically but commercially and politically. He readily fell in with the suggestion, and offered to proceed by the next mail to India, provided, as he was not in affluent circumstances, the Geographical Society would contribute something to the expenses of the expedition. That contribution was at once accorded, thanks to the liberality of the Council, and he left by the next mail for India. You have already explained, Sir, how on his arrival in the Punjaub, he found the passes into Tartary by the Chitral and Bajore valleys to the west, which a European had never threaded, to be impassable, owing to disturbances among the mountain tribes, and was thus compelled to abandon his first project; but, nothing daunted by this failure, he soon struck out another line further east, and in due course, in company with Mr. Shaw, he reached the cities of Yarkand and Kashgar, which had never before been visited by an Englishman. I will notice one great disadvantage under which he laboured—a disadvantage which I think it is infinitely to his credit that he was able to overcome. He travelled as a mere private gentleman; he was not officially recognised by the Government; he had no profession, no occupation. Now, a private traveller, although that character is perfectly understood in Europe and in Western Asia, is quite unintelligible to the suspicious inhabitants of Central Asia. They regard every one who is not an avowed servant of the Government, or a merchant, or a doctor, as necessarily a spy; they cannot appreciate the desire we have to obtain new geographical information; and, therefore, I think it does greatly redound to Mr. Hayward's credit, and testifies to his tact, temper, and diplomatic skill, that he was able to disarm suspicion, and not only to reach the cities of Kashgar and Yarkand, but to return in safety, and bring back such ample and correct materials with regard to the physical features of all the country between the British dependencies on the one side and the Russian dependencies on the other. I can add nothing, Sir, to what you have already stated as to Mr. Hayward's present position and prospects; but of one thing I am assured, and that is that the same indomitable will, the same fertility of expedient, the same disregard of dangers and hardships, the same iron constitution and great bodily activity, which carried him successfully through the snowy passes of the Karakorum and Kuen Lun, will stand him in good stead in his present still more hazardous undertaking; and that if any Englishman can reach the Pamir Steppe, and settle the geography of that mysterious region, the site of the famous Mount Méru of the Hindoos, and the primeval paradise of the Aryan nations, Mr. Hayward is the man. Sir, with such a hope, I gladly accept this Medal on Mr. Hayward's behalf; I accept it as the reward of daring and enterprise, combined with skill, accomplishments, and intelligence; and knowing as I do Mr. Hayward's ardent and impressible nature, I feel assured that he will receive the Medal as an ample return for his labours in the past, and as a happy augury of his success in the future."

The PRESIDENT then addressed Lieutenant Francis Garnier, of the French Navy, the recipient of the Victoria Medal :—

" The Patron's or Victoria Gold Medal of the Royal Geographical Society is presented to you, Sir, as the accomplished and intrepid traveller who accompanied, as second in command, the late Captain de Lagrée on the great expedition of exploration from the French territory in Cochin China, along the Mekong River, and through the heart of China, to the Yang-tsze-Kiang. In the course of this expedition, from Cratieh in Cambodia to Shanghai, 5392 miles were travelled over, and of these, 3625 miles, chiefly of country almost unknown to us, were surveyed with care, and the positions fixed by astronomical observations.

" In carrying out this important and truly scientific mission, your commander succumbed to the fatigues and privations of the harassing march between the head-waters of the Mekong and Tong-chuan, in the centre of Yunan. Through his illness the progress of the undertaking was for a time arrested, for one of the chief objects—a visit to Tali-fu, the head-quarters of the formidable Mahomedan insurrection against the Chinese authorities—seemed little likely to be realised. But you, Sir, nobly volunteered to undertake this hazardous journey, and, your commander having consented, you made a rapid march to the rebel stronghold, satisfactorily fixed its geographical position, and, escaping a threatened attack by the jealous inhabitants of the place, returned in safety to the capital of Yunan, where, alas! you found your chief had died in your absence. Disinterring his remains for conveyance to your native country, you crossed to the nearest port on the Yang-tsze, and, embarking in a native boat, you brought the remainder of your party in safety to the mouth of the river.

" In my Address of last year, I spoke, M. Garnier, of the most remarkable explorations of yourself and your associates, as having developed not only the true physical geography of vast tracts hitherto undescribed, but also in having contributed much fresh knowledge respecting the philology, antiquities, zoology, botany, and geology of these regions. I then also said, that you and your associates had traversed a greater amount of new country than, according to my belief, had been accomplished for many years by any travellers in Asia, and I confidently anticipated that our Council would at this Anniversary award you our highest honour. In short, as France has the fullest right to be proud of these doings of her gallant naval officers, so on my part I can assure you, M. Garnier, that every English traveller and geographer rejoices in seeing you honoured with the Medal which bears the likeness of our beloved Queen Victoria. It gratifies me to learn that the great work descriptive of your remarkable explorations is about to be published, under the auspices of the Imperial Government; and I shall be delighted to learn that your enlightened Emperor should reward you by promotion to a higher rank in the French Navy."

M. GARNIER replied as follows :—

"Messieurs,—Je regrétte vivement de ne pas connaître assez la langue anglaise pour adresser dans cette langue mes remerciements à la Société de Géographie d'Angleterre. Je ne fais sans doute que

recueillir l'héritage scientifique du noble officier dont j'étais le
second, et qui, après avoir mené à bonne fin un long et périlleux
voyage, a malheureusement succombé au port.   Permettez-moi donc
de rapporter à la mémoire du Commandant de Lagrée tous vos
glorieux suffrages.

"Ne dois-je pas rappeler aussi que c'est à l'initiative anglaise
qu'ont été dues les premières tentatives faites pour pénétrer de l'Inde
en Chine, et qu'il y a plus de trente ans le lieutenant MacLeod
reconnaissait un point du Mekong situé presque aux frontières de ce
dernier empire.   Sur quelque partie du globe que l'on se trouve,
au seuil de toutes les contrées inconnues, ne sont-ce pas presque
toujours des voyageurs anglais qui s'avancent, qui s'exposent pour
étendre le cercle des connaissances géographiques ?

"Cette conquête scientifique du globe est la seule qui doive
exciter aujourd'hui l'émulation des peuples.   Le monde appartient
à qui l'étudie et le connaît le mieux ; et comme Français, je ne puis
m'empêcher d'envier à l'Angleterre et de souhaiter à mon pays,
cette ardeur de découvertes, ce besoin d'expansion qui fait flotter
le pavillon britannique sur tous les rivages, et a fait de son commerce le premier commerce du monde.   La noble récompense que
décerne aujourd'hui à un Français la Société de Géographie de
Londres, prouve que votre pays, Messieurs, sera le premier à
applaudir et à encourager les efforts qui auront pour mobile le
progrès des sciences et l'appel à la civilisation des régions restées
jusqu'à présent en dehors du mouvement général.   C'est là le plus
grand des devoirs qui incombent aux nations civilisées ; c'est dans
son accomplissement qu'il y a le plus de gloire à recueillir, et
cette gloire la Société de Géographie d'Angleterre se l'est acquise
entre toutes, par la féconde et puissante impulsion qu'elle a su
imprimer aux recherches géographiques."

The President then presented the medals awarded to the
successful competitors in the geographical examination of the year,
held, at the invitation of the Society, at the chief Public Schools.
The names of the medallists were :—In Physical Geography, Gold
Medal, Mr. George Grey Butler, of Liverpool College ; Bronze
Medal, Mr. Martin Stewart, of Rossall School.   In Political Geography, Gold Medal, Mr. George William Gent, of Rossall School ;
Bronze Medal, Mr. James Henry Collins, of Liverpool College.   The
young men having been introduced by Mr. Galton, the PRESIDENT
thus spoke :—

" In presenting these Medals, I may remind the Meeting that this act on the part of the Society, which was decided on by our Council at the suggestion of Mr. Francis Galton, is now brought into operation for the second time. The working of this system is due to two Members of the Council, Mr. Francis Galton and Mr. George Brodrick, and the awards are made by two eminent men of science and letters, Mr. Alfred R. Wallace and the Very Rev. Dr. Howson, Dean of Chester. It is needless for me to indicate that every young reader of classical history must infinitely better understand his subject if the proper boundaries and political geography of ancient kingdoms are brought to his mind's eye in maps. Again, an acquaintance with physical geography is an essential part of the instruction of every well-educated person. It is to these two classes of our subject that we assign medals of different values, according to the Report of the Examiners, two of bronze and two of gold. On this point I have to observe that those who have gained medals this year were all competitors for such distinctions last year, as will be seen by the Report of the Examination when published. On this occasion, as at the last Anniversary, the chief honours have been won by educational establishments in Lancashire, viz., the Liverpool College and Rossall School.

" It is much to be regretted that the leading public schools, Eton, Harrow, and Rugby, have not as yet competed for these juvenile honours; but I feel confident that they will ere long prepare youths who will pass with credit through the ordeal of our Examiners.

" On this occasion the Gold Medal for Physical Geography has been won by Mr. George Grey Butler, to whom I now deliver the Prize, adding these observations on my own part. That it is a hopeful sign of the reform in modern education to find that a representative of the name of Butler, a family which has gained so many successes at the Universities in purely classical studies, which has contributed two Head Masters to Harrow and two to other great public schools, should have competed for and won our Gold Medal for Physical Geography. I congratulate you, Mr. Butler, on being the worthy recipient of this distinction. Lastly, when I turn to the subject of Political Geography, I am much gratified to present to you, Mr. George William Gent, the Gold Medal for this important branch of knowledge, for I am happy to remind the

Assembly that you gained the Bronze Medal last year for Physical Geography."

The annual prize of five pounds for proficiency in geography, granted to the Society of Arts for the examination held under their direction, was afterwards handed by the President to Mr. Critchett, on behalf of the successful candidate, Mr. Thomas Richard Clarke.

ADDRESS

TO

THE ROYAL GEOGRAPHICAL SOCIETY.

Delivered at the Anniversary Meeting on the 23rd May, 1870.

By Sir Roderick Impey Murchison, Bart., k.c.b., PRESIDENT.

GENTLEMEN,

I address you once more in this Theatre of the Royal Institution, which, by the kind consideration of its President and managers, has been placed at our disposal. In expressing my thanks for the use of it, I am glad to say that our meetings have been frequented by many members of that distinguished body, and that thus a mutual good feeling has been established, in which I rejoice, as it was in this building that I acquired my earliest scientific knowledge, as taught by Davy, Brand, and Faraday.

I have so fully explained our position, as to the acquisition of a separate local habitation and hall of our own, in the opening of my last Address, that I have only now to add that, by the authority of the Council, I have made a strong written appeal to the Prime Minister, to grant to us apartments similar to those given to six other Societies, but as yet have received no reply.

The numbers of our Society have steadily increased; they amount now to 2263 Fellows, exclusive of 74 Honorary Members, and our warmest thanks are due, as in former years, to our Assistant-Secretary and Editor, Mr. H. W. Bates, for having laid before us the annual volume of our 'Journal' again so much earlier than it used formerly to appear. With regard to the contents of this volume, I shall have to make some remarks in the course of this Address.

OBITUARY.

Captain CHARLES STURT.—I commence the melancholy record of our losses, by a notice of one of the most distinguished explorers and geographers of our age, as prepared by my friend Mr. George MacLeay, simply adding, on my own part, that I heartily applaud every expression in this just tribute.

Of the many hardy and energetic men, to whose bravery and intelligence we owe our knowledge of the interior of Australia, Charles Sturt is perhaps the most eminent. To him we are indebted for the discovery of the great western water-system of that vast island, between the 25th and 35th parallels of latitude, and 138° and 148° of longitude; a discovery which not only speedily led to the occupation of enormous tracts of valuable pasture country in New South Wales, but very shortly resulted in the settlement of the magnificent gold-producing colony of Victoria and its not much less successful neighbour the colony of South Australia. To him we are also indebted for the solution of the great geographical problem, the true character of the Eastern Interior of Australia, which, until he undertook his third expedition in 1844, was, by the colonists at large, as well as by many geographers, believed to be the receptacle of all the western waters, and to consist of one huge inland sea. And, further, to Sturt's instructive example we owe the series of distinguished explorers, such as Eyre, McDougall Stuart, and others, who have since so worthily and successfully trodden in his footsteps.

Charles Sturt, the eldest son of Thomas Napier Lennox Sturt, of the Bengal Civil Service, and grandson of Humphry Sturt, of More-Critchill, Dorsetshire, was born in India in 1796. After receiving his education at Harrow, he obtained a commission in the 39th Regiment, and served with it in America, France, and Ireland.

In 1827 he accompanied the 39th to New South Wales, and very shortly after his arrival in Sydney, though holding a high staff appointment, he volunteered to lead an expedition of discovery into the interior. At that time one of those droughts, which periodically afflict Australia, was at its very worst; and in the ignorance which prevailed as to the nature of the back country, of which nothing was known beyond Oxley's investigation, the deepest anxiety was felt with respect to the pent-up and struggling colony. The then Governor, General Darling, was but too glad to avail himself of the proffered services of the young soldier, and, accordingly, a well appointed party, under Sturt's command, was soon prepared and sent off. Following Oxley's track down the Macquarie, Sturt was more successful than that officer, owing to the very dry season, in turning the marshes in which the river becomes lost; and shortly after, having struck upon the Castlereagh, he followed the course of that stream down to its junction with the noble river—until then unheard of—to which he gave the name of the "Darling." This river, though draining an immense extent of country, was at this time, in consequence of the extraordinary drought which prevailed, very low, and, owing to strong local brine-springs, its water was found to be utterly unfit for use. Sturt was thus compelled for the time to relinquish all further investigation and to return to head-quarters. Conceiving, from the course the Darling was taking at the two different points at which he had touched upon it, that it would eventually be found to unite with the waters of the Lachlan, and the fine never-failing mountain-stream the Morumbidgee, and so form too large a body of water to be absorbed in swamps, he obtained leave in the following year to pursue the course of this last most promising river, and thus to test the accuracy of his theory. His party, on this occasion, consisted of a friend, two soldiers, and eight convicts, specially selected for the service. Running down the Morumbidgee for some weeks, the party came on the junction of the Lachlan, which was found to have re-united its waters beyond Oxley's supposed inland-sea; and in a few days after, having taken to a boat, which had been carried in frame, the party, now reduced to eight, were launched on the wide bosom of that magnificent river, to which Sturt gave the name of the "Murray." Ten days subsequently, after encountering some difficulty from the navigation, and very great peril from the aborigines, who were found all along the banks in great numbers, and who had never seen or heard of a white man before, they came on the mouth of the Darling, which had maintained a pretty direct course from the spot where Sturt had left it some 400 miles to the north-east,—a fact most satisfactorily verifying his prediction. Sixteen days later, after much toil, tugging at the oars from morning till night, the party came upon the great lacustrine expanse—half fresh, half salt—named by Sturt "Lake Alexandrina," the surplus waters of which find vent through narrow channels into Encounter Bay. Not being able to launch their boat through the surf which they found rolling into the bay, there was nothing left for the party but to work their way back up the very streams which they had found it laborious enough to descend, and this they had to do, on a very straitened allowance of food; their supplies, indeed, altogether failed them for some days before they reached

CXXXVI   Sir Roderick I. Murchison's *Address*.

the teams which had been sent from Sydney to meet them. The
sufferings of the party on their return were very great. Sturt never
afterwards had good health, his eye-sight, in particular, becoming
very seriously affected. But a great success had been achieved,
which to the present day is spoken of in the colony with very great
and natural pride.

After some years' employment in the public service in South
Australia, the settlement of which followed very closely upon his
discovery, and of which colony he was regarded as the "Father,"
Sturt, in 1844, volunteered another expedition, and undertook to
penetrate the very centre of the "Island-Continent." This expedition was unfortunate enough. Its failure, however, was in no
way to be attributed to any deterioration in the qualities of its
leader. The season was one of severe drought, and, by a strange
fatality, always hitting upon the most barren strips of country for
his route, he again and again found himself in a hopeless desert,
which it was utterly impossible to get through. A line taken a
degree or so to the east in a far more favourable season, however,
enabled Burke and Wills—though at the expense of their lives—and
also his own Lieutenant, McDougall Stuart, to pass through the very
centre of the land, and so to reach the Gulf of Carpentaria. This
great achievement, the grand object of his ambition, was thus
vexatiously lost to him, who would have been deemed by all the
most worthy of the honour: but no man more warmly expressed
his appreciation of the labours and deserts of those who subsequently
succeeded in this wonderful feat than Sturt. Even as it was, his
discovery of the Barcoo, or Cooper's Creek, led in no small degree
to the success of those who, in more genial and suitable seasons,
followed in his path. Being overtaken by the great heats of
summer, in the neighbourhood of this last-named watercourse, and
knowing that they should not be able to find any other water for
hundreds of miles on their route homewards, Sturt's party excavated
a cell under the ground, in which they had to pass six most
miserable summer months; being thus compelled, in order to
mitigate the frightful heat, to adopt a course analogous to that
made use of in winter by Arctic voyagers to escape the effects of
extreme cold.

Sturt has published narratives of these several expeditions, remarkable for succinctness, modesty, and general intelligence.

Calm and collected, this brave man never failed to inspire perfect
confidence in his followers, while he secured their love and respect

sy and consideration, and the cheerful happy
ra met the difficulties and privations which
expeditions. Like all brave men, Sturt was
compassionate almost as a woman. Though
treme provocation, he never permitted the
ed otherwise than with most humane for-
have boasted, had he been a man to have
lat not one single drop of blood had been
) expeditions. Owing to the hardships and
gone, his constitution (which naturally was
ast expedition, completely broke down, while
d; a state of things which, of course, neces-
rom public life. Some years ago he returned
land, to live on the liberal pension awarded
olony, the people of which reciprocated his
rays delighted to do him honour.
ling, he lived here among us in complete
ing notice, and certainly never seeking dis-
Yet surely such a man, when others without
living honours from the State, ought to have
early period for public recognition! For a
) was quite neglected. It was reserved for
eman, Earl Granville—when the Order of
'eorge was remodelled—at length to show
his deserts; and he received notice from his
that he was to be included in the list, then
of the Knights Commanders of that Order.
Gazette' appeared, Sturt had breathed his

e this kind, gentle, modest, and brave man
rney. No traveller, so bound, could have
)pier or more promising auspices.
UNWIN ADDINGTON.—By the decease of this
!0th year, the Crown and country have lost
lost able and conscientious diplomatist and
gan his career in the Foreign Office in 1807,
e find him serving as an Attaché to Lord
ie Court of Naples and Sicily. Having been
ns at Stockholm, Switzerland, Denmark, and
g negotiated on two occasions between Spain
'as promoted to be the British Minister at

Frankfort, and afterwards held the same office at Madrid. He was twice employed as a negotiator between this country and the United States. Eventually, after all these services, he became Under-Secretary of State for Foreign Affairs in 1842, and continued in that important office for 12 years, when, on retiring in 1854, he was created a Privy Councillor.

In his varied journeys and missions Mr. Addington witnessed many remarkable scenes. Thus, when attached to Sir Edmund Thornton, at Berlin, in 1813, he was present at the capture of Leipsic by the Allied Forces, and, on entering that city with the suite of Prince Blucher, he saw the meeting of the Allied Sovereigns in the great square of that city. Subsequently he was attached to the head-quarters of General Bernadotte, then adopted as Crown Prince of Sweden.

In his last mission to Madrid he served from the autumn of 1830, during the eventful period which witnessed the abolition of the Salic Law in Spain and the succession of Queen Isabella to the throne, on the death of her father, Ferdinand VII.

As Under-Secretary of State his services were thoroughly and warmly appreciated by Lords Aberdeen, Palmerston, and indeed by every one connected with the Foreign Office, and by no one more so than by his distinguished successor the Right Hon. Edmund Hammond, to whom I am indebted for some of the above details.

For the last 10 years of his life Mr. Addington was a very constant attendant at our meetings, serving with great efficiency as one of our Council; for he was by study, as well as keen observation, an accomplished geographer.

Among the traits of character which won for him the attachment of his friends, I may here mention that, in preparing my obituary sketch of Lord Palmerston in 1866,* I was indebted to Mr. Addington for a just delineation of the leading official attributes and habits of that lamented statesman.

As a proof of his manly sincerity and loyalty, I may state that, when my valued friend, our Associate Ex-Governor Eyre, was prosecuted through what I considered to be a misdirected and unjust movement of certain persons, Mr. Addington sent to myself, as the Chairman of the Eyre Defence Fund, a sum of 50*l.*, and added personally, when bewailing the fate of the Governor of Jamaica, "If my uncle, the late Lord Sidmouth, had been Minister at

---

* *See* 'Journal of the Royal Geographical Society,' vol. xxxvi.

the time, he would at once have sent to Governor Eyre some mark of honour from the Crown, for having saved a British Colony from insurrection and ruin."

Lord Broughton, G.C.B. This highly accomplished nobleman, who died in his 84th year, on the 3rd June last, lived a most eventful life. He was also, as will be explained, one of the founders of the Royal Geographical Society. Educated at Westminster School, and afterwards at the University of Cambridge, he was there associated with Byron, and other young men destined to rise to great distinction. In 1809 and 1810 he was the companion of Lord Byron in his travels through Albania and other parts of Turkey, as well as Greece. He published his well-known work, entitled 'A Journey through Albania and other Provinces of Turkey in Europe and Asia,' before he had entered upon public life, in 1813; but he improved and brought it out again in his maturer age, and when he had attained the dignity of the peerage (1855). Few works of travel have obtained a more lasting reputation; inasmuch as it is justly and equally prized by the scholar, the antiquary, and the geographer.

As the eldest son of Sir Benjamin Hobhouse, Bart., he succeeded to the baronetcy in 1812, and was the colleague of Sir Francis Burdett in the representation of Westminster from 1820 to 1833. It was during that period that I formed an acquaintance with him, which ripened into friendship. Thus it was that, in the years 1821-2, I followed the chase in Leicestershire with him and Sir Francis, just before I embarked on a scientific career. Beginning to reside in London in 1823, I became a member of the Raleigh Club of real travellers, of which Sir John Hobhouse was a member and a pretty constant attendant; and it was at that club (since converted, on my suggestion, into the Geographical Club), in 1828, that the origin of the Royal Geographical Society was broached, and in 1829 made its real start. At that time Sir John Barrow was the President of the Raleigh Club; and he, with several others, all of them except myself being now dead, held meetings at the Admiralty, and there drew out the Resolutions on which the Society was afterwards established, at a public meeting in 1830, under the Presidency of the Earl of Ripon. I have always regretted that this preamble, which I now offer, was not inserted in the first volume of our Journal.

The persons, then, who really founded the Society were Sir John Barrow; Sir John Cam Hobhouse; Robert Brown, the Prince of botanists; the Honourable Mountstuart Elphinstone; Mr. Bartle

Frere; and myself. And of these no one was more active than Sir John Hobhouse.

In a long and successful public career, as Sir John Hobhouse, he filled successively the offices of Secretary for Ireland, Chief Commissioner of Woods and Forests, and President of the Board of Control. In this last and most important station, his administration was marked by great vigour, during several of those crises which affect at intervals our Indian empire. For his long services he first received the Grand Cross of the Bath, and subsequently was advanced to the Peerage in 1851.

During the latter portion of his life, Lord Broughton resided much in Wiltshire, and was there, as in earlier days in Leicestershire, a keen fox-hunter. Even when he had passed his seventy-fourth year, I have seen him ride with a loose rein down the steep slopes of the downs, near Tedworth House, his last country residence, where many of his old friends enjoyed, as in Berkeley Square, his true hospitality and ready wit, enlivened by the presence of his charming daughters, the Hon. Mrs. Dudley Carleton and the Hon. Mrs. Strange Jocelyn.

Admiral of the Fleet Sir WILLIAM BOWLES, K.C.B.—This well-known officer was born in 1780, and entered the Navy on board the *Theseus*, 74, in September, 1796; and after serving in six different ships of war, on many stations, with much credit to himself, he obtained the rank of Lieutenant as early as 1803, and subsequently that of Commander in 1806.

Being appointed to the *Zebra* (bomb), stationed in the North Sea, he saw some service, and was frequently engaged with the Danish batteries and flotilla. Having been shortly promoted to the envied rank of Post-Captain, followed by a series of commands, Captain Bowles, then in the *Medusa* frigate, co-operated with the Spanish forces, under General Porlier, and contributed to the destruction of nearly all the batteries between San Sebastian and Santander; and in the following July, " particularly distinguished himself by his zeal, ability, and activity, as second in command of the Naval Brigade, in a successful engagement with a strong detachment of the enemy's troops near Santona." In 1811 this indefatigable officer, ever on the alert where work was to be done, being again in the Baltic, in the *Aquilon*, 32 guns, completely destroyed " seven large merchant-ships, in the face of 1500 French soldiers."

He was then employed for several years on the South American station, latterly as Commodore in the *Amphion* and *Creole* frigates;

and from his great attention to the interests of British commerce, received, on one occasion, " a complimentary address, and, subsequently, a piece of plate, from the mercantile representatives of Buenos Ayres."

Besides other commands, in 1822 Captain Bowles was appointed Comptroller-General of the Coast-guard, and was universally esteemed by that force for his courteous demeanour, impartiality, and strict regard to discipline. He retained it, with much advantage to the service, till November, 1841, when he attained flag rank. In May, 1843, he was selected for the purpose of conducting a particular service, and hoisted his flag on board the *Tyne*, 26, at Queenstown, but in a short time shifted it to the *Caledonia*, 120 guns, where he remained till May, 1844. He became Admiral in 1857, and afterwards was Commander-in-Chief at Portsmouth.

The well-known administrative qualities of Admiral Bowles gained for him, on two occasions, the position of a Lord of the Admiralty, and in after years he was often selected to preside over difficult and delicate enquiries requiring equal discrimination and judgment. He married, 9th August, 1820, the Honourable Frances Temple, sister of the late Lord Palmerston, but became a widower in 1838. He represented Launceston in Parliament, was created a K.C.B., and ultimately was raised to the highest rank in his profession, Admiral of the Fleet.

Sir William Bowles was for thirty-seven years a Fellow of this Society, and took the deepest interest in promoting its objects, either by aiding in the organisation of one of the Land Arctic Expeditions, or in numerous other ways of a substantial description.

Naturally benevolent, his name was greatly respected by the many institutions over which he presided, especially those of the "Sailors' Home," in Wells Street, and the "Seamen's Hospital" (the *Dreadnought*). In short, wherever the claims of the distressed mariner were advocated, he was ever ready with his pen and his purse to assist. In a brief memoir of this nature it is unnecessary to enumerate the many charitable societies to which he contributed; but it would be unpardonable to omit the Royal Naval Female School, for whose welfare he manifested an unwearied earnestness, as well as that useful establishment, the Royal Naval School at New Cross, which may be aptly termed a nursery for sailors. For upwards of twenty-one years he was the vigilant President of its Council; and, in addition to other bounties, he generously gave 1000*l.* to the fund for its chapel.

Still in the enjoyment of health, and an almost unimpaired memory, an accidental fall caused his death, at the ripe age of eighty-nine; and it may be truly said that few men have left a more estimable name than the good Admiral Bowles.

I owe this truthful sketch to my valued friend, Admiral Sir George Back.

M. ADRIEN BERBRUGGER, one of our Honorary Corresponding Members, died at Algiers on the 2nd of July last, in his 68th year. He was known chiefly for his great special knowledge of the Archæology of Northern Africa, where he resided during the greater part of his life, and where he wrote his 'Algérie historique, pittoresque, et monumentale,' his 'Grande Kabylie sous les Romains,' and other works. He was President of the Historical Society of Algiers, and had been a member of all the various scientific commissions appointed for special investigation in the French colony during the past thirty years.

Mr. J. W. S. WYLLIE, a gentleman who had gained distinction in the public service of India, and whose future career seemed full of promise, died on the 17th of March last, at the early age of 35 years. He was the son of General Sir William Wyllie, and was born in India in 1835. After completing his education in England, first at Cheltenham, and afterwards at Trinity College, Oxford, he returned to India, and was one of the first men appointed to the Civil Service of our great Eastern Empire by public competition. He served throughout the Mutiny in the Bombay Presidency, but was transferred in 1860 to the Presidency of Bengal, where he acted successively as private secretary to the Commissioner of Oude, Under-Secretary to the Government of India, and Secretary for Foreign Affairs. He entertained decided views regarding the Foreign Policy of our Indian Government, and, since he quitted the service, advocated the principle of non-interference in the affairs of states and tribes beyond our frontiers, with great force and eloquence, in various articles contributed to the leading reviews, to one of which I had occasion to refer in my Address of 1868.* Mr. Wyllie returned finally to England in 1868, and, in the General Election of December of that year, was returned to Parliament as Member for Hereford. He was enrolled in the same year as Fellow of this Society, and, shortly before his untimely decease, had taken part in the discussion at one of our evening Meetings.

---

* 'Journal of the Royal Geographical Society,' vol. 38, p. clxxvi.

THE EARL OF DERBY, K.G.—This great statesman and brilliant orator, who died on the 23rd October, 1869, had been a Fellow of our Society since 1833. Educated at Eton and Oxford, he entered Parliament at the age of 21, and thenceforward pursued that most remarkable career by which he has been distinguished. Previous to his official life he was a zealous traveller in India and the United States, and so far we claim him as a Geographer.

It would be presumptuous on my part to attempt to sketch even the outlines of the political life of Lord Derby; but I may state that, whilst he represented one of the most ancient of our noble families, he was one of the most distinguished classical scholars of our age. As such he was most appropriately elected Chancellor of the University of Oxford; and I shall ever consider it one of the greatest honours I have received in life that, upon his installation in that Office, he was pleased to select me as one of those persons worthy of being admitted to the Degree of Doctor of Civil Law in that ancient seat of learning.

The career of this illustrious man has been dwelt upon in all the public journals; and, in anticipation of a full Memoir of his life, I cannot better sum up those salient features of his character, which won for him so high a place in the regard and estimation of his countrymen, than by quoting the following paragraph from the 'Times' of the 25th of last October, which concludes a very striking and animated sketch of his life:—

"We have spoken of Lord Derby chiefly as a statesman. But, after all, it is the man—ever brilliant and impulsive—that has most won the admiration of his countrymen. He was a splendid specimen of an Englishman, and whether he was engaged in furious debate with demagogues, or in lowly conversation on religion with little children, or in parley with jockeys, while training 'Toxophilite,' or rendering *Homer* into English verse, or in stately Latin discourses as the Chancellor of his University, or in joyous talk in a drawing-room among ladies whom he delighted to chaff, or in caring for the needs of Lancashire operatives, there was a force and a fire about him that acted like a spell. Of all his public acts none did him more honour and none made a deeper impression on the minds of his countrymen than that to which we have just alluded—his conduct on the occasion of the cotton famine in Lancashire. No man in the kingdom sympathised more truly than he with the distress of the poor Lancashire spinners, and, perhaps, no man did so much as he for their relief. It was not simply that he gave them a princely donation; he worked hard for them in the committee which was established in their aid; he was, indeed, the life and soul of the committee, and for months at that bitter time he went about doing good by precept and example, so that myriads in Lancashire now bless his name. He will long live in memory as one of the most remarkable, and indeed irresistible, men of our time—a man privately beloved and publicly admired, who showed extraordinary cleverness in many ways, and was the greatest orator of his day."

The Marquis of WESTMINSTER, K.G.—This good, accomplished, and benevolent nobleman joined the Society in 1844, under one of my former presidencies. Educated at Westminster School, and afterwards at the University of Oxford, he entered the House of Commons as Lord Belgrave, and sat as member for Chester for twelve or thirteen years.

In 1845 he succeeded to the princely estates and titles of his father, the first Marquis of Westminster, and during his subsequent life he made good use of his vast wealth by giving largely and munificently to public hospitals and charities, besides laying out vast sums in the erection of churches.

Those who knew Lord Westminster well, could not but be struck with the simplicity and ingenuousness of his character, and his constant desire to do all justice to those with whom he was in any way connected. He was, besides, a liberal patron of the Fine Arts; whilst, like a true English nobleman, he supported the breed of our race-horses, which, by his predecessor, had been so much improved.

RALPH WILLIAM GREY.—By the decease of this most amiable and accomplished man, I have lost one of my most esteemed friends. Serving many years as a Member of Parliament, and having been successively secretary of Lord Palmerston and of Earl Russell, his conduct was ever such as to gain for him the esteem and, I may also say, the love of all who had any communication with him. He was for some time a member of our Council, and always took a warm interest in all our proceedings. At the time of his decease, on the 1st October last, he occupied the post of Commissioner of H.M. Board of Customs.

JOHN HOGG.—This gentleman, who died on the 16th September last, was a zealous antiquary and historical geographer, who served on our Council in former years, and was, in the years 1849 and 1850, one of the Secretaries of this Society. His published memoirs in those departmnets of our science which he cultivated, were very numerous; among them I may notice 'Gebel Hauran, its adjacent

Districts and the Eastern Desert of Syria, with Remarks on their Geography and Geology,' published in 1860; 'On some old Maps of Africa, in which the Central Equatorial Lakes are laid down nearly in their true positions' (1864); 'The Geography and Geology of the Peninsula of Mount Sinai and the adjacent Countries' (1850); and 'Remarks on Mount Serbal' (1849). Mr. Hogg was a Fellow of the Royal and Linnæan Societies.

Dr. PETER MARK ROGET, F.R.S.—This venerable philosopher, who died on the 13th September last, in the 91st year of his age, had long occupied a distinguished place among the men of science of our country, and was one of my oldest scientific friends.

His chief contributions to science were physiological, and were communicated in a series of memoirs to the Royal Society, of which body he was the Secretary during many years, in association with numerous Presidents, from Sir Humphry Davy downwards. Courteous and affable in manners, he was an excellent man of business, and on more than one occasion presided over the Physiologists at the meetings of the British Association for the Advancement of Science. His reputation, indeed, stood so high, that when the Earl of Bridgwater bequeathed 10,000*l.* to be given to those authors who should best demonstrate the glory of God in the works of creation, Dr. Roget was selected by the President of the Royal Society to write that 'Bridgwater Treatise on Animal and Vegetable Physiology' which was so well received by the public.

The last work of Dr. Roget's with which I am acquainted—the completion, indeed, of his laborious studies on this subject during fifty years—was entitled a 'Thesaurus of English Words and Phrases,' and in it we trace the same fulness, perspicuity, and closeness of research which are apparent in all his productions.

ARTHUR KETT BARCLAY. — This benevolent gentleman, who, since the decease of his excellent father, so long M.P. for Southwark, has been at the head of the great Southwark Brewery, was from his youth an ardent pursuer of various branches of Natural Science.

He cultivated for many years the science of Astronomy with success, and established a very effective observatory at his country seat of Bury Hill, near Dorking.

By his death I have lost a friend of forty-three years' standing, and who through life was respected and beloved by the large circle of those who had the privilege to know him.

SAMUEL S. HULL.—This gentleman, who died in his 72nd year,

spent the earlier period of his life in Prince Edward's Island, where his father possessed a large tract of land. After a tour through the United States and the Canadas, he published a thoughtful and useful book, entitled 'The Emigrant's Introduction.' He subsequently commenced a series of travels through the Old World, and his journeys through Greece, Syria, and Egypt having been published, his travels through Russia and Siberia, ending with a voyage round the world, justly attracted very considerable notice.

Though unacquainted with Mr. Hill myself, I learn from those who knew him well that his manners were gentle and winning; whilst his writings convey to the reader an impression of the perfect truthfulness and the guileless simplicity of his character.

Mr. CORNELIUS GRINNELL.—Cornelius Grinnell first came to this country in the year 1856, and whilst he was received with the cordiality which was due to the son of the eminent New York merchant who contributed in so princely a manner to the American expeditions in search of Sir John Franklin, his own kindness of manner and generous disposition rendered him a general favourite with all who came in contact with him.

Besides the aid and assistance rendered by Mr. Henry Grinnell in the search for our missing countrymen, we are indebted to him, in a great measure, for the equipment of those expeditions under Kane, and Hayes, and Hall which have added so much to our geographical knowledge of the Arctic seas; but, in addition to these services in the aid of science, there breathes throughout his correspondence a constant desire to promote goodwill between the two countries. Thus in March, 1855, he writes:—

"I have a letter from Sir F. Beaufort, in which he, in the most honourable manner, states that the Americans have the right to the name Grinnell Land: not that I care an iota about it myself, but this little circumstance has more weight than one would suppose; the result will be to create in the minds of many a kindly feeling towards your country."

Again, in 1855, on the departure of Hartstene's expedition to relieve Dr. Kane:—

"I believe there has been nothing left undone, on the part of the British officers, to give every possible information that could be of service. If nothing else resulted from it, it will create a good feeling between the two countries."

And in 1856, on the restoration of the *Resolute*:—

"You are, no doubt, aware that a resolution has passed both

Houses of Congress, without a dissentient voice, to restore to your Government the barque *Resolute.* I think your Government will receive her in the same kind spirit that she is tendered in, and that the act itself will have the effect to increase the friendship of the two countries."

Cornelius Grinnell was present when her Majesty the Queen received the *Resolute* from the American officers.

These short extracts will, I feel sure, induce the Fellows of the Royal Geographical Society to join with me in sympathy with the father on the loss of a son who, during his residence among us, so ably personated the feelings of his parent towards this country. We have the melancholy consolation that his untimely end was occasioned in the act of doing a kindness to a friend.

Colonel GEORGE GAWLER, K.H.—This meritorious public servant, who died on the 7th May, 1869, was very favourably known to geographers by the lively interest he took in promoting researches in South Australia, from Adelaide, during the period he acted as Governor of that colony. On his return to this country he took a deep interest in our proceedings, and during the last twelve years was a frequent attendant at our evening meetings and a very instructive speaker whenever Australian discovery was the topic. He always produced the impression that he was a sincere and truthful observer, and several of his observations respecting the probable condition of the interior of Australia have been proved to be correct by recent discoveries. Colonel Gawler was born in 1796 and served during the eventful years from 1811 to 1814 in the Peninsular war, where he led one party at the storming of Badajos.

Mr. JAMES MACQUEEN.—As I was closing these obituary notices, I received the news of the death of that distinguished veteran geographer, my old and respected friend, Mr. James Macqueen, who died on the 14th inst., at the very advanced age of ninety-two. He was born in the year 1778, at Crawford, in Lanarkshire, and used to relate that his attention was first drawn to African geography, in the study of which he was chiefly occupied during his maturer years, by the perusal of 'Mungo Park's Travels.' During the time he was resident in Grenada, in the West Indies, as manager of a sugar-plantation, whilst reading the exciting narrative aloud to a friend one night, he noticed that a negro boy in the room stood listening very attentively, especially to those passages in which the Joliba was mentioned. The boy being afterwards asked why he showed such interest, said that he knew all about the Joliba, and that he was a Mandingo, born in the country of the Upper Niger. The information obtained through this intelligent boy was afterwards of great use to Mr. Macqueen, when he was engaged in bringing together all that was known about the geography of the Niger, a subject on which he became a leading authority. He was the first, I believe, who demonstrated, before the discovery was actually made, that the Niger emptied itself into the Bight of Benin. Subsequently he published, through Mr. Arrowsmith, the first map, approaching to correctness, of the interior of Africa. He was a trenchant and vigorous writer, and a keen critic; but his literary productions were chiefly confined to articles in newspapers and periodicals. Some of his geographical memoirs were read before our own Society, and published in the ' Journal.' He was known also as a political and historical writer, and was, in the early part of the present century, the proprietor and editor of the 'Glasgow Herald.' As a man of action, he distinguished himself in the projection and organization of two of the most useful and prosperous chartered companies, the " Colonial Bank " and the " Royal Mail Steam-Packet Company." In making the preliminary arrangements for the latter, he visited the various countries embraced in the intended operations, and, on his retirement, received the most flattering testimonials from the Company. His memory and interest in geography and public questions were preserved, in scarcely diminished freshness, almost to the hour of his death, and his last moments were passed in great peacefulness. In him the Society has lost one of its most attached members.

The other Fellows who have departed this life, but who have not taken an active part in geographical inquiries, are Colonel W. Anderson, c.b.; Mr. F. D. P. Astley; Mr. Hugh G. C. Beavan; Captain A. Blakeley; Captain Harby Barber; Mr. H. Blanshard; General Sir Wm. M. G. Colebrooke, k.h., c.b.; Mr. Alfred Davis; Major J. W. Espinasse; Mr. A. Findlay (one of the few remaining members of the Society who joined in the year of its foundation, 1830); Mr. G. F. Harris, m.a.; Mr. John M. Hockly; Mr. R. Jardine; Mr. Wm. John Law; Mr. D. Meinertzhagen; Dr. Charles James Meller; General Alex. F. Mackintosh, k.h.; Mr. G. T. Miller; Dr. David Macloughlin; the Bishop of Manchester (Rev. Dr. Lee); Mr. Frederick North, m.p.; Mr. C. O'Callaghan; Mr. Samuel Perkes; Mr. Thomas Rawlings; Mr. Arthur Roberts;

Captain Wm. Strutt; Lord Sudeley; Mr. Theodosius Uzielli; Captain G. Whitby; and Mr. Champion Wetton.

Lastly, I am proud to record that our Society was honoured by the Fellowship of the late Mr. George Peabody, the good and meritorious philanthropist of the United States, to whom our country is so deeply indebted.

ADMIRALTY SURVEYS.[*]

The hydrographical surveys under the Admiralty have made their usual satisfactory progress during the past year, and, in connexion with them, the exploration of the deep sea, which was commenced in H.M.S. *Lightning*, in the summer of 1868, has been most successfully followed up in the *Porcupine* during 1869. This vessel, ordinarily employed under Staff-Commander E. K. Calver in the survey of the coasts of the United Kingdom, was placed by the Admiralty at the disposal of the Council of the Royal Society for this special and most interesting research, and being fully equipped, and supplied with the necessary instruments and scientific apparatus, she left Woolwich on the 17th of May, and continued to explore the deep-sea bed, from the northern part of the Bay of Biscay, round the west coasts of Ireland and Scotland to the Faroe Isles, until the end of September. During this period important discoveries were made in various branches of physical science, sounding and dredging operations were successfully carried out to the extraordinary depth of 2345 fathoms, or nearly three miles, and very valuable observations on deep-sea temperatures made. The expedition was divided into three separate cruises, and the scientific operations were presided over, respectively, by Dr. Carpenter, Professor Wyville Thomson, and Mr. Gwynn Jeffreys. An account of the results will be found in the 'Proceedings of the Royal Society,' as well as in a lecture delivered at the Royal Institution by Dr. Carpenter, and printed in its 'Proceedings.'

Home Coasts.—In consequence of the many calls for re-surveys of certain portions of the coasts of the United Kingdom, owing to considerable changes which are taking place, especially on the western shores of England between Anglesea and the Solway Firth, a second vessel, the *Lightning*, has been equipped to meet these demands, and will immediately commence her operations under Staff-Com-

---

[*] Furnished by Capt. G. H. Richards, R.N., Hydrographer.

mander John Richards, late in charge of the Channel Island Survey, which, as was anticipated in our last report, has now been completed. During the last summer the off-shore soundings from these islands were obtained and carried as far west as the longitude of 3° 20′; a series of tidal observations were also made throughout the whole extent of the group, and diagrams placed on the chart, by which the precise direction and strength of the stream can be seen at a glance for each hour,—a matter of considerable importance in a region so exceptionally dangerous to strangers.

The charts of the whole group are now published in a complete state, on a scale of 4 inches to the mile, with suitable sailing directions.

Portsmouth.—Staff-Commander Hall, with a steam launch and small party, has been occupied in making a large-scale survey of the whole of the harbour,—a work much required. Accurate tidal observations have been made, and levellings carried through to Langston Harbour, in order to ascertain the probable effect of the tidal scour on Portsmouth Harbour and its bar, when the gun-boat channel connecting the two shall have been completed and opened.

A re-survey of Portsmouth Bar will next be made, with the view of ascertaining whether any change has taken place in the depth of water since it was last deepened by dredging.

Mediterranean.—Captain Nares and the officers of the *Newport* have completed the survey of the coast of Tunis and its off-lying banks, from Cape Carthage, to Tabarca Island, about a hundred miles to the westward, up to which point the south coast of the Mediterranean had been surveyed by the French. The *Newport* has also surveyed the island of Pantellaria in the Malta Channel, and re-surveyed the port of Alexandria; she passed several times through the Suez Canal, at its opening, and subsequently with the Hydrographer of the Admiralty and Director of Engineering Works, who were sent to report on that great work, when soundings and sections were taken throughout the length of the canal, and a survey made of Port Said and its approach.

Strait of Magellan.—Since the last report, the *Nassau*, Captain Mayne, C.B., has returned from this survey. The result of her last season's work has been the examination of 255 miles of the channels leading from the straits into the Gulf of Pênas, and the survey of twenty anchorages or havens, most of which were previously little known; ships of any size may now pass from the Atlantic to the Pacific by this route in safety, with no lack of convenient stopping-

aphic Department will shortly be in a posi-
new series of charts, from Cape Virgin in
f of Pênas in the Pacific, on scales which
m easy and free from risk.

the *Nassau* was employed in searching for
doubtful dangers which still disfigure our
id whose origin in many cases it is difficult,
te.

.—The *Sylvia*, employed on these coasts, has
in surveying the intricate portions of the
pan, through which so great a trade now
mail and passenger vessels, as well as the
ries.

rn shore of the Gulf of Yeddo has also been

survey a considerable portion of the Upper
en explored and mapped by Lieutenant
p of the *Sylvia*.

a this river previously explored was the
Tung-Ting Lake, about 120 miles above
l upwards of 700 miles from the sea. The
have now provided us with maps, which
ry the Admiralty as far as the city of
arly 1000 miles from the sea, and where
e smallest class of Native boats, may be said

ander Brooker, who had ably conducted this
was compelled to resign from ill health and
avigating-Lieutenant Maxwell remained in
since been re-commissioned in China for a
and Commander H. C. St. John has been
Survey.

nander Reed and the officers of the *Rifleman*
ear made an excellent survey of Balabac
lorneo and the Island of Palawan, from the
r Sulu Sea. During this survey upwards of
ndings have been obtained, in the examina-
fs and dangers which lie in and above the

been found defective was disposed of in
g officers returned to England at the close

of last year. H.M.S. *Nassau*, under Commander Chimmo, is about
to leave England in further prosecution of this survey, and its
extension into the Sulu Sea and eastern passages, of which almost
nothing is known, except that at present it is a most dangerous
though necessary highway for sailing-ships.

Newfoundland.—Staff-Commander Kerr, with one assistant, has, in
a hired vessel, during the past year completed 300 miles of the
eastern shores of this colony, in that dangerous locality north and
west of Togo Island.

During the early part of the season, while the ice was closely
packed on the shore north of Cape Freels, the party were employed
in surveying portions of Bona Vista Bay, until driven out of it by
the pack, which ultimately drove into the bay on the 12th of June,
and filled its arms up with ice 10 feet thick.

During the laying of the French Atlantic Cable, the surveyors
were enabled to render valuable assistance to the *Great Eastern* and
her fleet, among the banks in the vicinity of St. Pierre, and in
laying the shore end of the cable from that island.

West Indies.—Staff-Commander Parsons with two assistants, in a
small hired sailing-vessel, has made a complete survey of the Island
of Barbadoes, including a plan of Carlisle Bay, the principal
anchorage, on a scale of 20 inches to the mile.

The Survey has been lately removed to the Colony of British
Guayana, which, combining as it does an extensive coast-line with
outlying shallow banks and the mouths of important rivers, is a
work of considerable magnitude and difficulty to be undertaken with
such narrow means.

British Columbia.—The surveyors in this colony, under Staff-
Commander Pender, have been usefully employed in examining the
exposed western shores of the off-lying islands northward of Vancouver Island, and in sounding the outer and rocky entrance to
Queen Charlotte Sound.

One hundred and twenty miles of coast, from Cape Calvert to
the south-east point of Banks, Island have been surveyed, with the
various passages leading to the main inner channel. It is considered that by the close of the present year sufficient will be done
to meet all the requirements of navigation and commerce for a very
long period, and it is intended that the Survey shall be withdrawn.

Cape of Good Hope.—The survey of the west coast of this
colony has progressed very favourably during the past year, under
Navigating-Lieutenant Archdeacon: the shore from Table Bay

ve, a distance of 130 miles, has been
torough survey made of Saldanha Bay,—
alt.
langers, which extend in some instances
coast between Saldanha and St. Helena
of which have been very doubtful, have
' at all times to seamen approaching
be set at rest by the publication of a
r has also been made of False Bay, re-
some hitherto unknown dangers,—a cir-
importance in this much frequented

e now working northward towards the
suffered much inconvenience from the
f the water, and the almost entire absence
nity of the coast, where the country is

rveying party in this colony have been
ion of Nepean Bay and the Southern
also in completing the survey of Back-
t island and Cape Jervis.
that we have to record the deaths of
lately in command of the Survey, and
assistant, which occurred in July last,
in five days of each other, from illness
are in the execution of their duties. In
se officers the Naval service has lost two
vants.
ig carried on by Navigating-Lieutenant

ast season the survey of the Colony of
on principally in an easterly direction
coast is now completed from a few miles
rt Albert, a town about 80 miles north-
ry. A survey has also been made of
Cape Otway.
reinment steamer, in which the survey is
a employed also in assisting the laying
etween Victoria and the north coast of

Coast Survey of New South Wales is now

complete. The work of last year has been entirely confined to deep-sea sounding, the limit of the hundred-fathom line having been determined from off Point Danger, the northern boundary of the colony, to Cape Howe, its southern extreme, a distance of 600 miles; thus enabling the navigator, by the use of the lead, to determine his position with accuracy,—an advantage not to be over-estimated on approaching a coast where easterly gales and thick weather are by no means infrequent. Although the surveying party have been withdrawn from the coasts of New South Wales, Navigating-Lieutenant Gowland, lately in charge of it, has been kept, at the request of the Government, to examine and survey the rivers and inner waters of the colony, the expense of which they have determined to defray from colonial resources.

Queensland.—The survey of the Coast of Queensland, under Navigating-Lieutenant Bedwell, has progressed very favourably during the past year. The outer coast of Great Sandy Island from Indian Head, northward round the dangerous Breaksea Spit, and the western shore of Hervey Bay, amounting in all to 100 miles of coast-line, have been closely examined, and thickly and carefully sounded; and perhaps on no part of the Australian continent has a survey been so much needed or been more skilfully executed. The work is carried on by two officers in a small colonial sailing-vessel.

West Coast of Africa.—The very imperfect and fragmentary surveys which existed of the entrances of some of the rivers on this coast, frequented by ships employed in the oil-trade, had become so detrimental to the interests of commerce, that last year the Admiralty attached a surveying officer to the senior officer's ship on that station, in order that he might take advantage of any opportunities which might offer, during the visits of our cruisers, to rectify the erroneous charts; and Navigating-Lieutenant Langdon, who was selected for this duty, has already performed very good service in the examination of the mouths of the Binon and Brass rivers, the Bonny, New and Old Calabar, and the Cameroon rivers, the corrected surveys of which will shortly be published.

Mr. Langdon is at present engaged in correcting the survey of the Sherbro River.

Summary.—During the preceding year seventy-one new charts have been engraved and published, and upwards of 1200 original plates have been added to or corrected, while 139,000 charts have been printed for the use of the Navy and the public. Sailing Directions have been published for the West Coast of England and for the Channel Islands, as well as various hydrographical notices, and the usual annual works, such as Tide Tables, &c. Light Lists have been issued.

In concluding the present notice, it will not be considered out of place to record that two names well known to the nautical world in connexion with hydrographical labours have lately disappeared from the rolls of the Hydrographical Department, in the retirement of Commanders Edward Dunsterville and John Burdwood; the name of the former associated for nearly thirty years with all matters relating to charts, and the latter for a scarcely less lengthened period with the annual tide-tables and other useful compilations.

It is due to these old and valued public servants to record, and it is believed it may be done with strict truth, that in the management of their respective important departments there has never been a default throughout their lengthened term of office, and to replace them will not be an easy task.

NEW PUBLICATIONS.—*Journal of the Society*, Vol. 39.—In noticing some of the chief Geographical works published during the year, I may justly commence with the volume of our own 'Journal,' which contains the more important Memoirs presented to the Society, and is properly classed among the chief contributions to the Geographical literature of each year. The number of papers published in the volume is seventeen, of which twelve are accompanied by maps. Among those to which attention may more particularly be called are the following :—' Notes on Manchuria,' by the Rev. Alexander Williamson, illustrated by a map, in which routes are inserted from a sketch furnished by the author, who travelled from the Gulf of Liau-tung to Sansing, the most northerly city in this direction of the Chinese empire; 'From Metemma to Dumot,' in Western Abyssinia,' by Dr. H. Blanc, in which is conveyed much new information regarding this region, and especially the configuration of Lake Dembea, as depicted on the accompanying map; 'Journey in the Caucasus, and Ascent of Kasbek and Elbruz,' by Mr. Douglas W. Freshfield; 'On the Basin of the Colorado and Great Basin of North America,' by Dr. W. A. Bell; 'Account of the Swedish North Polar Expedition of 1868,' by Professor A. E. Nordenskiöld and Captain Fr. von Otter, accompanied by a map, in which the bays of the northern part of the islands are laid down according to tracings supplied by the authors; 'Report of the Trans-

Himalayan Explorations during 1867,' by Captain T. G. Montgomerie, containing the visit of the Pundits to the gold-mines of Western Tibet; 'Narrative of a Journey through the Afar Country,' by M. Werner Munzinger, the zealous and able agent of this country, previous to and during the Abyssinian war—a most valuable contribution to the geography and ethnology of a part of Eastern Africa of which scarely anything was previously known, the accompanying map being drawn from the author's own sketch; 'Journey of Exploration to the Mouth of the Limpopo,' by St. Vincent Erskine; and lastly, 'Notes on the Map of the Peninsula of Sinai,' by the Rev. F. W. Holland.

Petermann's 'Geographische Mittheilungen.' — The principal Geographical publication of the continent of Europe, as I have had occasion in previous years to remark, is the 'Geographische Mittheilungen,' edited by our Medallist and Honorary Corresponding Member, Dr. Petermann, and published by Justus Perthes, of Gotha. The large number of maps, so attractive for their fulness of detail and the amount of new information they impart, is a well-known feature of this important and truly scientific serial. During the past year I remark, in the first place, highly-finished maps of the English surveys made in Abyssinia during the war, with corresponding text, in which the march of our army, and the new information gleaned concerning Abyssinian geography, are given in a clear and attractive manner. In the fifth part for 1869 there is also an account of the most recent scientific expeditions of the Russians in Central Asia, and a map of the Thian-Shan system between Issyk-Kul and Kashgar, both of which ought to be consulted by those interested in the geography of Turkistan. A sketch-map of the great French Expedition from Cambodia to the Yang-tsze-Kiang is also given in the same part, in anticipation of the French official map not yet published. Other Memoirs of value are the following :—'Scientific Results of the first German Arctic Expedition,' by W. V. Freeden (Part VI.); 'Eduard Mohr's Astronomical and Geognostic Expedition in South Africa' (Parts VII. and VIII.); 'Latest Travels and Explorations in China: Baron von Richthofen's Geological Investigations since September, 1868' (Part IX.); 'The Telegraph Expedition on the Yukon in Alaska,' with map (Part X.); 'New Guinea: a German Appeal from the Antipodes,' with map of New Guinea (Part XI.), a communication worthy of attention by those who take an interest in our settlements in tropical Australia, and in the prospect of a German colonization of New Guinea and the neighbouring islands; Mauch's 'Travels in the Interior of South Africa,' with map; 'Sketch of the Physical Geography of the Sutlej Valley,' by Dr. F. Stoliczka (Part I., 1870), and others, which the limited time at my disposal precludes me from enumerating.

Keith Johnston's last Works.—What Petermann is to the Continent of Europe, our associate Keith Johnston is to the British Isles and the Colonies. His last works, to which I have alluded in a former Address, demonstrate the results of pertinacious and exhaustive labours, which bring out in salient relief, in clear tables and beautiful maps, all the latest geographical acquisitions.

I shall allude elsewhere to the treatise of his son on the discoveries of Livingstone, and I hope that at our next Anniversary Meeting the father of this family of geographers, of whom Scotland is so proud, will be placed on the same footing as Arrowsmith of England and Petermann of Germany, by being assigned one of our Royal Geographical Medals.

Kohl's 'History of the Discovery of Maine.'—I must not here omit to notice a work of great value, on the history of geographical discovery, which has been issued during the past year under the auspices of the Maine Historical Society. Under the general designation of a 'History of the Discovery of Maine,' this elaborate work is, in truth, a history of the discovery of the East Coast of North America, from the time of the Northmen in 990, to the Charter of Gilbert in 1578. Its author is Mr. J. G. Kohl, of Bremen, whose name is already well known to us by his numerous books of travels—works, not antiquarian only, but based upon his personal observations in America and most of the principal countries of Europe.

In this new volume Mr. Kohl has given, in a compact and lucid manner, the results of a most laborious investigation of the scattered and often obscure documents which survive from the early times of which he treats. The work is illustrated by extracts from no less than three-and-twenty maps, the latest of which is Mercator's of 1569. It may well be imagined that when a volume, embodying such documents as these, has the abstruse subjects of which it treats dealt with by a man of extensive reading, untiring industry, and remarkable critical sagacity, such as Dr. Kohl, I am drawing your attention to a work of no ordinary importance. It is, in truth, a handbook to the history of Western discovery; and it is much to be regretted that its circulation should be limited to the members of a private Society in America. I may observe that it was at the suggestion of one of our Secretaries, that Mr. Kohl was invited to undertake this responsible and laborious task, and Mr. Major is justly proud of so successful and honourable a result.

Marcoy's 'Voyage à travers l'Amérique du Sud.'—Although not coming within the definition of scientific geography, illustrated books of travel are deserving of some notice in a summary like the present, as tending greatly to diffuse a knowledge of distant regions and a taste for geography among the great body of the public. In our early days the copiously illustrated quarto books of voyages, which were then the usual form of publication, were the delight of young readers imbued with the spirit of adventure; but the production of this class of works seems of late years to have been abandoned by our English publishers. In France such books continue to appear, and with a profusion of beautiful engravings and a luxury of type and paper which excite our astonishment, more particularly as they appear intended for, and succeed in obtaining, a wide circulation. One of these works, published by Messrs. Hachette & Co., is the 'Voyage à travers l'Amérique du Sud,' by M. Paul Marcoy. It contains a narrative of travel and adventure across the continent of South America at its broadest part, commencing with Islay on the Pacific Coast, and passing by Arequipa and Cuzco to the head-waters of the Ucayali, and so on to the River Amazons, and down that great stream to the Atlantic. The illustrations, apparently from drawings by the traveller, are to the number of many hundreds, most beautifully engraved and printed, and the landscape views more particularly convey a vivid idea of the wonderful and varied scenery through which the author passed. A work of this nature, in two large quarto volumes, and evidently intended for popular reading, could scarcely be undertaken by an English publisher, although one would think that such books, as conveying much knowledge of distant regions, by the pictorial illustrations alone, would be well received by the British public. Another work of the same class, and by the same publishers, the 'Japon illustré,' by M. Humbert, has already attracted some attention in England, and deservedly so. The author was the Swiss Minister in Japan, who made good use of his exceptional opportunities in studying the singular country and people amongst whom he lived. Most of the engravings, which thickly stud the two handsome volumes, appear to have been copied from photographs, and are most satisfactory for their evident fidelity. The text, too, forms pleasant and instructive reading, and stamps M. Humbert as a thoughtful observer and pleasing writer.

Millingen "Wild Life among the Koords.'—An interesting volume, with this title, has recently appeared from the pen of Major F. Millingen, whose earlier work on Turkey was noticed in one of my former Addresses. Together with some curious and entertaining descriptions of wild life in the remoter districts of Koordistan and Armenia, this work gives us geographical notices of many little-known parts of this region; such as the valley of the Ennis and its junction with the Upper Euphrates, Lake Nazik; the navigation, harbours, &c., of Lake Van; Lake Ertjek, with its poisonous waters; and the tract of territory lying being Lake Van and the Persian frontiers, forming the watershed between the Caspian and the Persian Gulf. A map accompanies the work, in which these various new features are delineated from information furnished by the author.

Italian Geographical Society.—Our distinguished Foreign Associate, the Commander Cristofero Negri, who worthily presides over the geographers of Italy, has, in his recent instructive Address, dilated with much eloquence on the progress of geography, and on the recent discoveries in many of those distant regions in which we take the deepest interest. Following our example, he laments in his obituary list the death of Count Lavradio, so long the Portuguese Minister at our Court. Although this highly-cultivated and much-respected man was not in our Society, he belonged to the affiliated body, the Hakluyt Society, and took a warm interest in eliciting every portion of knowledge relating to the earliest discoveries of the Portuguese in Africa and the Indies. Our Secretary, Mr. Major, in his memorable work, the 'Life of Prince Henry of Portugal,' of which I spoke to you in a former Address, has, indeed, done full justice to Count Lavradio, and I now add my tribute to the memory of this learned man.

The geographical knowledge of Count Lavradio was so extensive, his heart so thoroughly devoted to the advancement of the cause, and he was so justly proud of being the descendant of Francisco d'Almeida, the first Viceroy of India, that he well deserves due praise from the hands of a Geographical President. The maps which, by his exertions, were extricated from the archives in Portugal, threw great light on mediæval geography; and among them are those maps of Africa which were constructed at the period when the Pope allotted so very large a part of that vast country to his faithful Portuguese.

Signor Negri's comments on the recent discoveries in Africa are very attractive, and I rejoice in knowing that the Italian Society has now reached the large total of upwards of 1000 members.

The Canal of Suez.—As the opening out a navigable communication between the Mediterranean and the Red Sea is unquestionably the greatest work of our age, let us offer our warmest congratulations to M. de Lesseps for having conceived and completed a project which was at first thought impossible by many; but in which—much to their honour—his countrymen, the French, have throughout been his vigorous supporters. Still, without the hearty concurrence of the Khedive of Egypt and his munificent aid, this very difficult operation could never have been realised. This water-communication, or Bosphorus, which has insulated Africa, has been well styled by Cristoforo Negri, the President of the Italian Geographical Society, the "Straits of Lesseps," just as the Straits of Cook, Magellan, and Behring bear the name of those who first navigated in those waters.

For the honour of our Society, it is right to record that the ruler of Egypt specially invited your President to attend the great ceremony of the opening of the Canal, and nothing more mortified me than being obliged, from the state of my health at that time, to decline the proposed distinction. Anxious, however, that our body should be well represented, I induced my friend, Lord Houghton, one of the Trustees and a permanent member of our Council, to represent the Society on this memorable occasion, and the manner in which his Lordship has executed this duty met with our entire approval.

The elaborate Report on the Canal, by Captain Richards, the hydrographer, and Colonel Clarke, R.E., recently published by the Admiralty, will be reprinted in the third number of our 'Proceedings.'[*]

CENTRAL ASIA. — In this year, as in the last, the chief advances in geographical knowledge have been made in Central Asia, and especially in those parts of the great mountain back-bone of the Old World which lie to the north-west of our Indian empire, and

---

[*] In reference to this subject, I may state that our Associate, Captain H. Spratt, so distinguished by his former communications on this branch of Mediterranean hydrography, which I have noticed in former Addresses, informs me that he adheres to his views respecting the inevitable direction of the silt as carried eastward from the mouth of the Nile by the steady marine currents, and hence he believes in the eventual silting up of Port Said.

in the vast territory so recently opened up to us, which is now designated as "Eastern Turkistan," in contradistinction to Western, or what really is at present "Russian Turkistan." When I addressed you at the last Anniversary, I could only speak of Mr. Shaw as having successfully penetrated by Yarkand to Kashgar with his cargo of tea from Kangra, and of his having been well received by the great chief Yakoob Kushbegi, who has since been recognised under the much grander title of Ataligh Ghazi, or "Leader of the Faithful." A residence of several months in Eastern Turkistan enabled Mr. Shaw to establish friendly relations with that powerful ruler, who, as we now know, has sent a special Envoy into British India with a letter for the Queen, and another for the Viceroy of India. The latter, in a letter to myself, has expressed his gratification at the prospect of establishing friendly intercourse with this new nation, as leading to an interchange of the products of Eastern Turkistan with those of the British empire.

So long as China held that fine region in thraldom, which was the case during a whole century, down to 1864, the native Mussulman population were never more than partially subjugated; so that, as soon as a brave and sagacious leader appeared in the person of Yakoob Kushbegi, the Chinese yoke was easily thrown off, and a country, which previously was a continual hotbed of insurrection, and subject to every sort of anarchy, has now, we learn, become a well-regulated and orderly state, under the stern, yet just, rule of this one leader.

There is something quite refreshing and encouraging in the fact that the Envoy of this great ruler will, in his progress to Calcutta, have witnessed a great Durbar of Indian Princes assembled under the presidency of our Queen's son, the Duke of Edinburgh, and that, after having seen some portions of our army and of our marine, of the latter of which the *Galatea* frigate will be a favourable specimen, he will return to his native Turkistan, impressed with a deep sense of the value of an intimate alliance with a magnificent empire possessing such colossal resources.

The accidental meeting, at Shadula on the Himalayan frontier of Eastern Turkistan, of Mr. Hayward, the Envoy from our Society, and Mr. Shaw, which at first sight was naturally viewed with suspicion by the Yarkandi people, has in the end proved very advantageous to us as leading to a great addition to our knowledge. For so soon as the Ataligh Ghazi had satisfied himself that our countrymen were respectively engaged in very different occupations—the one seeking to open out a trading intercourse, the other endeavouring to delineate the features of a region quite unknown to Europeans—he acted with great kindness, and has since shown his good feeling by the transmission, already alluded to, of a special Envoy to the Governor-General of India.

As Mr. Shaw is now among us, and has already communicated, at an evening meeting, an animated sketch of his travels, we may feel assured that the work he is preparing on the subject of his journey will attract, in the most lively manner, the British public.

The mission which the Council confided to Mr. Hayward has been already attended with highly important results. For although we know not as yet whether he has succeeded in entering the great lofty plateau of Pamir, which was the main object of his travels— for it is possible he may have been by native tumults deflected from that purpose, or by the impracticability of traversing the mountainous tracks east of Gilgit, which lie to the west of the territories of Cashmere and the British outposts—yet he has already well employed his time in taking a route which led him to Yarkand, and in course of which he fell in with Mr. Shaw. Whilst waiting in a state of surveillance at Shadula on the frontier, he contrived to escape the vigilance of his guards, and crossed the mountain ranges near the sources of the Yarkand River to survey the country. It was during this rapid excursion that he was enabled to make very great additions to our geographical knowledge. He demonstrated, for the first time, the true course of the Yarkand River, as well as that of the Karakash, ascended to the sources of the former, on the northern slopes of the Karakorum, and obtained information of a better pass, the Yangi, over the Kuen Lun, than the one at present used by traders. He was, moreover, enabled to sketch the outlines of this remarkable mountain-region, with its glaciers and fertile valleys, and also to lay down, for the first time, a number of positions of latitude, longitude, and altitudes, which were hitherto entirely undetermined.

We cannot too much admire the zeal, talent, and singular courage displayed by Mr. Hayward in carrying out these researches, in a country in which, had the inhabitants discovered the only small scientific instrument he possessed, they might at once have killed him.

Nothing daunted by his first failure to penetrate to the Pamir Steppe, he is now endeavouring, since his return from the Yarkand and Kashgar journey, to traverse the country occupied by those warring and savage tribes who hold the passes which lie to the south of that great plateau. If he should succeed in this traverse, he apprehends that he will have comparatively little difficulty in exploring the Pamir Land, its nomad Kirghis inhabitants not being savage or warlike. In case, however, that he should find it impossible to run the gauntlet once more by repassing the hostile tribes lying between Pamir Land and British India, he expresses a hope that we should endeavour to obtain the sanction of the Russian Government, that in that case he might be permitted to return to England by passing through Russian or Western Turkistan.

Acting in the name of the Society, and by the authority of the Council, I have had the satisfaction of learning that, in virtue of the appeal which I made to the Geographical Society of St. Petersburg, the Imperial Government has sent the requisite order to the Governor-General of Turkistan to offer to Mr. Hayward all aid and assistance, and a free passage through these territories to Europe.

If, then, I couple this gratifying fact with the very successful result mission of our Associate, Mr. Douglas Forsyth, to the Court of St. Petersburg, in order fully to lay before the Emperor and his Ministers the exact state of affairs in regard to the great region which, on the north-west, lies between British India and the Thian Chan Mountains, which have hitherto been the Russian boundary, I see in these circumstances cause for rejoicing that there is every prospect of a harmony of views between the Russians and ourselves regarding this great region. I rejoiced when I learnt from Mr. Forsyth himself, that both the Emperor and his enlightened Minister, the Prince Gortschakoff, are willing to maintain the boundary of the Thian Chan, and to undertake not to advance the Russian forces into Eastern Turkistan. As I have long suggested that, for the benefit of Britain and Russia, the large Musulman territory of Eastern Turkistan—now completely independent of China—should be allowed to lie as a neutral region, which may prove thus a source of lucrative trade both for Russia and England, I am the more rejoiced at the present aspect of affairs than in any preceding year. And now that such intermediate country is in a well-ordered condition, thanks to the unflinching power of the Ataligh Ghazi, we may look to a durable arrangement and good understanding on this northern frontier of British India.

If successful trade should be established between British India and Eastern Turkistan, we must ever recollect that the first step taken in it was the work of our able Associate, Mr. Douglas Forsyth, who, by the transmission of a single horse-load of Indian tea, propitiated the great chief, who is now our ally; and we must all feel much indebted to the present Viceroy of India, the Earl of Mayo, for the warm interest he has taken in sustaining and supporting the enterprises which have led to so desirable a result. It is indeed my pleasing duty to inform the Society that the Viceroy has charged Mr. Forsyth with a special mission to the Ataligh Ghazi, in which he is to be accompanied by Mr. Shaw, who for this purpose has been recently recalled from this country by a telegram from the Viceroy. The friendly intercourse between British India and Eastern Turkistan will thus be permanently settled.

In considering the value of the intercourse between British India and Eastern Turkistan which has recently been brought about, I must not omit to do justice to Sir Henry Rawlinson, who on previous occasions has drawn our attention vividly to the important results which must follow from explorations of our North-Western frontiers. I refer you to our 'Proceedings' for the able delineation, in which, quoting the letters of our accomplished and zealous Associate, Mr. Douglas Forsyth, he places the whole subject before us in a masterly style. Sir Henry Rawlinson's speech, delivered at our first meeting of the past session, is so pregnant with knowledge, and so clear in describing the advances made by our envoy, Hayward, and the other explorers, that I commend you to peruse the report of it in our 'Proceedings,' followed, as it is, by the last speech ever made to us by our ever-to-be-lamented Associate, Lord Strangford, as a compendium of nearly all that can be said upon this broad subject geographically, commercially, and politically.

I have already referred to the fact that the Chinese had held Eastern Turkistan in subjection for about a hundred years; and, indeed, their latest conquest of that country dates only from about the middle of the last century. But it may not be without interest, as an illustration of the great antiquity of the Chinese power, and the vitality that it possessed through a great series of ages, to observe that this was by no means the first time that the regions in question had formed a part of the empire. I learn from Colonel Yule that Chinese scholars date the spread of its influence in that direction from the second century before our era; and in the first century after the birth of Christ the Chinese power extended across the Bolor even to the shores of the Caspian! In the following ages it was subject to great fluctuations; but under the great Thang dynasty, in the seventh century, the whole of the country east of the Bolor was under Chinese authority; and even west of the mountains, provinces extending to the frontiers of Persia were claimed as subject, and organized, at least on paper, with all the elaboration of the Chinese system. The conquests of Chingghiz and his successors again brought the states of Turkistan under the same supremacy with China. When they fell, the indigenous dynasty which succeeded them in China held little beyond the limits of China Proper; and it was not till the existing Manchu dynasty was in the height of its power that Eastern Turkistan for a third or fourth time, and, probably, for the last time, became united to China. Such a long series of vicissitudes almost reminds one of geological and ante-historical successions and oscillations.

Russians in Central Asia.—Whilst our own countrymen have thus been largely adding to our acquaintance with Eastern Turkistan, the Russians have extended geographical knowledge throughout Western Turkistan, a large portion of which has been all but annexed to the Russian empire, the chiefs of the principal Khanats, still called independent, being to a great extent subordinate.

We learn from the communication of Baron Osten Sacken to the Imperial Geographical Society, that among the most recent of these surveys are those made by Baron Kaulbars in the central part of the Thian Chan chain, on the upper course of the River Narym, and extending to the edge of the country of Eastern Turkistan, *i.e.* from the borders of the Khanat of Kokan to the mountain Khan Tengri, near the western extremity of the Lake Issyk-kul. In ascending the affluents of the Narym, Baron Kaulbars determined that its principal source was a glacier in the mountains of Ak Schirah, on the same meridian as the east end of Lake Issyk-kul. He also explored the grand snowy chains of Sery Yassy and Kok-schal, extending south-westwards to the Valley of Aksai.

The topographical surveys in the district of Zerafshan, under the direction of M. Scobélew, extend from Urmittan for 80 versts up the Zerafshan, in the valley of which river is situated the renowned city of Samarkand. This survey, and the measurement of the elevations of the so-called *Starved Steppe,* between Tschinaz and Dizakh, show that that arid tract was formerly enriched by the waters of the Zerafshan, through a grand canal of irrigation of Tartar origin, which may be considered one of the greatest hydraulic works of that formerly energetic people.

Besides other surveys, one of which has extended in the direction of the caravan route from Bokhara to Kasilinsk, the Russian topographers have prepared a map of the whole of Russian Turkistan, on a very large scale. When this document reaches our mapmakers, it will doubtless give quite a new geographical face to large portions of Central Asia.

Remains of extensive former brick constructions, found in the great Lake Issyk-kul, were brought to light through the exertions of General Kolpakovsky, and they have excited much curiosity. One of these masses of brick presented on its surface the form of a human figure, and weighed near 500 lbs. Already in 1857 M. Semenof had called attention to some of these ancient ruins, the existence of the places of which they are the remains is recorded in the annals of Chinese history.[*]

The existence of a city on the north end of Lake Issyk-kul 200 years before Christ, and also the remains of an Armenian monastery having been alluded to by Humboldt, I have no doubt that my friend M. Pierre de Tchihatchef will, in his proposed new edition of the 'Asie Centrale' of that illustrious man, develope still more all the topographical and antiquarian knowledge which has been elicited by the Russians, who thus have brought to light many ethnological data, which, through the long continuance of barbarous Turcoman rule, has remained so long unnoticed.

In terminating these observations on Central Asia, I must again express the gratification I have experienced in witnessing the highly praiseworthy efforts of the Russian geographers to lay open to the world of science the true physical features of the vast region of Western Turkistan, of which they are now, to a great extent, the rulers. In former years, I have alluded to the labours of Semenoff, Struve, and others, and very recently we have received an excellent translation, by Mr. Delmar Morgan, of a very remarkable memoir by Baron Osten Sacken, describing the mountainous region between Turkistan and the Russian boundary near Kashgar, the result of an exploratory expedition by General Poltoratsky, when accompanied by Baron Osten Sacken himself. The clearness and spirit of this memoir are such, that a sketch-map might almost be constructed from the author's word-painting; whilst the description of the flora of this highly-diversified country, and its analogy to the flora of the Himalayan mountains, will be highly appreciated by all botanists.

---

[*] See the 'Mittheilungen' of Petermann, 1858, p. 360.

Such a work, independently of other obligations conferred on us (and especially by his kind intervention, which procured the promise of the Russian Government that our envoy, Mr. Hayward, should be well received if he penetrated into Western Turkistan), influenced our Council in unanimously electing Baron Osten Sacken, the Secretary of the Imperial Geographical Society, as one of our Honorary Corresponding Members, of whom we may well be proud.

The day, indeed, has now arrived, and to my great delight, when the Russian Imperial Government on the north, and the British Government on the south, are rivals in thoroughly exploring and determining their respective frontiers, leaving between each dominion wild tracts, which will probably be for ever independent, but whose chiefs will well know how to respect their powerful neighbours.

These geographical operations are also, I doubt not, the forerunners to the establishment of good commercial intercourse, and are, I venture to think, the surest pledges of peace.

In the discussion which followed the reading of the memoir of Baron Osten Sacken, I was most happy to find that my eminent friend, Sir Henry Rawlinson, completely coincided with the views on this point which I have long entertained. It was also a source of true pleasure to me that, at the same meeting, the Chancellor of the Russian Embassy in London, M. Bartholomei, was a witness of the sincere expressions of gratification we all experienced in seeing the cordial and unreserved communication which now happily exists between the geographers of both countries. The earnest and graceful manner in which the Russian diplomatist addressed us was, I am happy to say, duly appreciated by the assembly.

WESTERN ASIA.—In dwelling upon the advances in geographical knowledge which have been made in Central Asia, we must notice in a marked manner the journey of Mr. Consul Taylor to the sources of the Euphrates, as communicated in a letter to Mr. P. K. Lynch, and published in our 'Proceedings.'* The line of exploration taken by our enterprising and learned Associate, Mr. Taylor, was intermediate to the routes taken by previous travellers, for he proceeded from the north of Lake Van, between Diadin and Beegir Kalah. By following this line, Mr. Taylor

---

* See 'Proceedings,' vol. xiii., p. 243.

ascertained that this region, so replete in ancient times with igneous action, is still the seat of an active volcano and many hot sulphureous springs and geysers, besides valleys which have been well filled with basalt, and subsequently deepened into abrupt gorges with precipitous sides.

AUSTRALIA.—Of the recent explorations in Australia it is right to observe that the expedition of Mr. Forrest in the interior of Western Australia, where he penetrated to E. long. 122° 45', though productive of no great geographical results, was zealously conducted, in the hope of tracing some account of the bodies of certain white men who had been heard of, and which were supposed to be the remains of Leichhardt and his associates.   No clue, alas! was found to identify this report, but a large additional area of salt lagoons and pebbly and sandy beds was added to the Western Colony.   The feature which comes out strongly in these tracts is, that whenever granite rocks appear, water is in much greater abundance than in the sandy tracts.

As to the great mass of land forming the northern part of Australia, which, as geographers, we have termed North Australia, in distinction to South-west and East Australia, the new data that have come to our knowledge are of comparatively slight importance. For although considerable tracts on the northern sea-board opposite to Melville Island are found to be well grassed, and will ultimately, perhaps, be capable of occupation, the efforts which have been made by the inhabitants of South Australia to annex and settle in them have not been fortunate.

These subjects have recently been well illustrated by our Associate, Sir Charles Nicholson; but the chief merit of his communication, as recently given to us, consisted in the clear comparative sketch of the rise and progress of the several great Australian colonies, and the remarkable explorers they have produced.

To myself this general sketch was very refreshing, inasmuch as nearly all the adventures he traced have been dwelt on, in more or less detail, at the meetings of the Royal Geographical Society since our foundation in the year 1830.   From that time we have seen little Port Phillip, then a mere dependency of New South Wales, rise into the grand and wealthy colony of Victoria; Port Adelaide become South Australia; the Moreton Bay, or northern settlement of New South Wales, expand into the grand and intertropical colony of Queensland.   This general view is the more valued as coming from one of our Fellows who occupied for many years the high post of Speaker of the House of Representatives at Sydney. But we have also to thank Sir Charles Nicholson, not for the first time, for having incited geographers to explore and do some real work in that vast region of New Guinea which lies between our northernmost Australian settlements and the rich islands of the Malay Archipelago.

Judging from the little we as yet know of the southern portion of this vast equatorial *terra incognita*, that is, the country on the south-west, it is inhabited by ferocious and savage natives, and the climate appears unfavourable to Europeans. My lamented friend, the late Mr. John Crawfurd, so well versed in the Eastern Archipelago, lost, indeed, no opportunity of recording his decided objection to an attempt at colonization in any part of New Guinea. But, after all, when we consider the high probability of a rising commercial intercourse between Cape York and other parts of North Australia, particularly those in and around the Gulf of Carpentaria, with the British Indian settlements, it must be admitted that the future interests of Britain would be greatly damaged if any other Power were to possess itself of the south-eastern shores of New Guinea, or make any settlement whatever in our own territories of North Australia.

By recent intelligence from Australia, it appears that the Papuans are not so irreconcileably hostile to Europeans as previous accounts would lead us to believe. Mr. Chester, the Police Magistrate at Somerset, our new settlement at Cape York, reports that Captain Delargy, of the trading schooner *Active*, engaged in the *béche-de-mer* fishery, whilst in search of a missing boat in the month of August last, was induced to try the hospitality of the natives of the south-eastern shore of New Guinea. He had a large and well-armed party with him, and was met on the beach by about 100 warriors armed with bows and arrows, who ranged themselves in order of battle; but on his making peace demonstrations, the Papuans laid aside their bows and vied with each other in showing hospitality to the strangers. They prepared a sumptuous feast of pigs, yams, taro, and a kind of jungle-fowl, and sent a portion on board the boats for those who remained in them. The chiefs, after the feast, accompanied Delargy through their village, and the most friendly relations were established. In communicating this very interesting information to me, Sir Charles Nicholson suggests that our own Admiralty might be recommended, after this proof of friendliness on the part of the

Natives, to employ the vessel of war on the Cape York station in an
attempt to improve our geographical knowledge of this wonderful
island, and cultivate amicable relations with its spirited inhabitants.

SOUTH AMERICA.—Since my last Address, the account of the
exploration of the River Juruá, by Mr. Chandless, to which I then
alluded, has appeared in the 'Journal' of the Society, accompanied
by a map, drawn by that enterprising and painstaking explorer,
with his usual completeness of topographical detail. This memoir,
and the map appended to it, was sent by him from Manaos, on
the Rio Negro, from which place he afterwards sailed to explore
another great tributary of the Amazons, the River Madeira. It
was his intention, on this journey, to explore the large westerly
affluent of the Madeira, the Beni, up to its sources in the Andes of
Southern Peru, and thus set at rest the vexed question of the course
of the Madre de Dios; but his attempt to penetrate into this
difficult region was not rewarded with his usual success. The
country on the banks of the Mamoré and Beni rivers was almost
impassable, owing to the hostility of a tribe of wild Indians, who
attacked the canoe of one of Mr. Chandless's travelling companions,
and killed its owner with several of the crew. We now learn, for
the first time, that an expedition sent by the Bolivian Government
in 1846 to explore the Beni, and consisting of thirty-two well-
armed men, besides canoe-men, was driven back by the wild
Indians. Besides this formidable obstacle, it was found next to
impossible to hire civilized Indians for the journey. Nothing
daunted, however, our traveller entered the Beni with his canoe
and small party of seven men, and ascended the little-known stream
as far as a rapid, 14 miles from the mouth, which he was unable,
with his weak party, to pass, and re-descended the Madeira to the
main Amazons. A curious feature in the physical geography of
the interior of South America is brought to light by the researches
of Mr. Chandless, and those of the Peruvian and Brazilian Boundary
Commission, in which Senhor Paz Soldan was engaged. This is, that
all the chief southern tributaries of the Amazons, between the
Madeira and Ucayali, flow nearly parallel to the main stream, and
have exceedingly tortuous courses; showing that the western
interior of the South American continent consists of a vast nearly
level plain, sloping gradually from west to east, and with very little
slope from the south, towards the centre of drainage.
In other parts of South America explorations and surveys are

being carried on, by the various States, with more or less activity. The Government of Brazil, as we are informed (through Mr. Chandless) by Senhor Pereira de Andrada, Secretary to the Brazilian Legation in London, has appointed an Imperial Commission to draw up a general map of Brazil, in which the costly surveys lately carried out along the great rivers of the empire will be utilized; and the Commission has offered to the Society copies of the maps and official reports on which their great work will be founded. According to the information which Mr. Chandless has received from the same quarter, an intrepid missionary of Bolivia, Padre F. Samuel Mancini, made an exploration of the River Madre de Dios, in the years 1868 and 1869, from the farthest point reached by Lieutenant Gibbon to the junction of the river with the Beni, and has proved that the course of the river is to the south of that of the Purus, of which it had formerly been considered the head-waters.

Further north, a scientific expedition, under the auspices of the Smithsonian Institution of the United States, has explored, with good results, a large portion of tropical South America. An account of one part of the expedition has recently been published by Professor Orton, who, with his companions, descended the eastern range of the Andes from Quito, and made his way, through the dense forests of the Napo and its tributaries, to the head of canoe-navigation on these rivers. The most important result of this journey appears to be a careful barometric measurement of heights, from the Andes, down the valley of the Amazons, to the Atlantic.

The Government of Chili, which has always honourably distinguished itself by the promotion of scientific investigation and the publication of the results in the completest manner, is now preparing a map of its central provinces on a scale of 1 : 250,000, embracing the most populous portions of the country, from the River Copiapo, in 27° 20′, to Angol, in 37° 48′ s. lat. Besides the official surveys, however, much useful geographical work is being accomplished in Chili by independent scientific explorers—amongst whom our Honorary Corresponding Member, Dr. R. A. Philippi, is one of the most active—who are gradually clearing up the doubts which have long hung over the position of mountain ranges, passes, and the courses of streams in the less-known parts of the Republic.

At the southern extremity of America our own Naval Surveyors have been well employed during the past three years in completing the examination of those intricate and difficult passages of the Straits of Magellan which occupied King and Fitzroy years ago in the voyage which has been made classical by the pen of Mr. Charles Darwin, who sailed in the *Beagle*, one of the vessels, as Naturalist. The commander of the Expedition which has recently returned from the Straits—Captain R. C. Mayne—gave an interesting account of the survey in a Paper read before the Geographical Section of the British Association at Exeter, and described more particularly the narrow passages leading northward from the western end of the Strait, which was carefully surveyed with a view of rendering safe an interior route towards Valparaiso, free from the heavy seas of the open Pacific. The work of the survey which Captain Mayne commanded, in the *Nassau*, commenced in December, 1866, and ended last May.

ARCTIC AND ANTARCTIC RESEARCHES.—The last year has been altogether unproductive of any explorations of the Arctic or Antarctic regions. The spirited expedition of Mr. Lamont, undertaken at great cost, and which proceeded in the summer of 1869 to the coasts of Spitzbergen, entirely failed to penetrate to the eastern side of the islands, from the unusual severity of the season and the enormous increase of sea-ice.

In regard, however, to Antarctic researches, we have been reanimated by a well-reasoned memoir by Captain Hamilton, R.K. He discussed the superior advantages which the use of steam-vessels in the wide Antarctic Ocean would give us, as compared with their utility in the Arctic Seas, and gave us a very able analysis of a work by a Mr. Morrell, of New York, which may now be said to have been thoroughly discussed for the first time; for, though a copy of this rare work existed in our library, no one had published any account of the curious information which it contained regarding high southern latitudes. When in command of a small schooner, Mr. Morrell described himself as having traversed the Antarctic Ocean to a greater extent than any other navigator, in the most rapid manner, and this before the voyages of Commodore Wilkes and Sir James Ross! Comparing the accounts of the successive Antarctic researches with each other, Captain Davis, an Antarctic explorer himself, was of opinion that great scepticism must prevail as to the authenticity of this work. It would appear, indeed, that our revered authority, the late Admiral Sir F. Beaufort, rejected Morrell's story as spurious; and, in truth, the vast spaces traversed with such rapidity, and the absence of all allusion to Wilkes's Land and Sabrina Island, which the voyager is stated to have approached, induce me to consider the work to be an ingenious Robinson Crusoe tale, fortified by some striking geographical data.

The author's exaggerated description of birds and vegetation in some of the parts visited (Auckland Islands, &c.)—productions which only exist in tropical regions—seems to demonstrate the unsoundness of the narrative regarding many parts said to have been visited.

A few more words on Physical Geography as dependent on Geology.— In the Address of last year I endeavoured, as a geologist, to define the great extent to which the present outlines of land had been determined by internal elevatory forces at various periods, by which the earth's crust was not only broken, upheaved, and depressed, but was, consequently, subjected to enormous denudations. Referring you to my former disquisition, I revert for a moment to this topic to make a few additional observations on a memoir which is now published in our ' Journal.' Accounting for the formation of fjords, cañons, and benches in North America, the author, Mr. Robert Brown, has faithfully and well described these openings which he has seen in the crust of the earth; but I take leave to express my disbelief in his explanation of the manner in which he refers them to agencies like those which now prevail. Seeing the existing fjords of North America occupied by great icebergs which have descended from glaciers, and also seeing the sides of the precipitous flanks of these fjords striated and polished by ice-action, he rushes to the conclusion that these enormously deep and broad cavities have been excavated entirely by the action of ice. This, however, is a hypothesis which rests on no sort of evidence. To disprove it, I ask, where in any icy tract is there the evidence that any glacier has by its advance excavated a single foot of solid rock? In their advance, glaciers striate and polish, but never excavate rocks.

Again, in explaining the origin of those remarkable cañons in the limestone mountains of North-West America, in which rivers flow for great distances, he infers that such cavities have been entirely worn out by the waters which flow through them, and which were formerly of vastly greater dimensions.

Now, in both these cases I think the writer errs. The plain and unmistakeable geological, and, therefore, geographical fact, is, that wherever the earth's crust was broken up from beneath, it necessarily underwent great transverse cracks, which opened into fissures and caverns; and these openings, made at different times, were then left to be operated upon in subsequent ages by all the waters which fell upon the surface, or by rivers above and below that surface, to be by them abraded and fashioned.

The true origin, however, of all such great transverse fjords or cañons, or, in short, of all abrupt fissures in hard rocks into which bays of the sea enter, or in which rivers flow, was never produced by such sea or river, but must be referred to original breaks in the crust, of which the waters have taken advantage, and have found the most natural issue.

Again, in illustrating this subject, Mr. Brown refers the origin of the great "benches" or banks of *débris* at various altitudes of North America to a letting off of waters from higher levels: on the other hand, I consider them to be distinct proofs of a subterranean upheaving of the land by which former lakes were desiccated, leaving their shores in the form of ledges or shingle-benches.

The author uses a phrase which, after all, implies an admission of my own view, when he writes: "These breaks may have been (indeed no doubt were) assisted by the volcanic disturbances which at a comparatively late period seem to have riven all the country in that region, and volcanoes in the mountains, through which these rivers flow, were the active agents of disruption."[*]

The author further seems to me to demolish the theory of modern causes by showing that the channel of the Golden Gate at San Francisco has a maximum depth of 50 fathoms, which great chasm he shows is in the line of the axis of the elevation of the main chain.

How, then, with the plain evidences of the origin of such vast fissures by pure geological subterranean agency, is it possible to refer them to the superficial action of ice and rivers, which, geologically speaking, are modern agents, and have only modified the old breaks and cavities of geological times?[†]

---

[*] See 'Proceedings of the Royal Geographical Society,' vol. xiii., No. 3, p. 148; and 'Journal of the Royal Geographical Society,' vol. xxxix., p. 125.

[†] Since the above was written, I have found that a paper on this subject by Mr. J. W. Tayler, a gentleman who has spent the greater part of the last 18 years in Greenland, has been communicated to the Society, and will be published in the 2nd Part of the 'Proceedings' for the present session. In this paper Mr. Tayler combats the views of Mr. Brown, and declares, as the result of his examination of the fjords themselves, that glaciers, instead of excavating fjords, are continually filling them up. He adds that some of the largest glaciers, as that north of Frederickshaab, do not exist in fjords at all. As a conclusive argument, he gives a diagram of a fjord south of Aksut, having two arms, which could not possibly have been cut by a glacier.

AFRICA.—*Great Salt Desert at the Eastern Foot of the Abyssinian Alps.*—In former allusions to the structure of that grand eastern edge of the Abyssinian highlands along which the British army advanced, no sufficient notice has been taken by myself of a very remarkable journey made by Mr. Werner Munzinger in exploring the route which leads from Hanfila on the Red Sea to the Abyssinian highlands. A brief account of this adventurous trip was given in our 'Proceedings,' and the narrative *in extenso* is now published in our 'Journal.' The lower country passed over appears as if it had been raised up from the Red Sea itself, for it consists of coral-reefs, sandy and shelly deposits, enlivened only with a few palm-trees, and containing in its central part a vast basin of salt which lies below the level of the sea. This is the country of the Afars, who occupy a triangular tract the apex of which is Annesley Bay. Volcanic rocks abound in it, and rise into mountains at its southern end.

All the streams which descend from the Abyssinian Alps to the east are absorbed in the low sandy region, the evaporation from which, under the great heat which prevails, accounts for the desiccation.

Mr. Clements Markham has borne ample testimony to the admirable manner with which Mr. Munzinger executed the arduous duties assigned to him, whether as an explorer penetrating far into the interior of Abyssinia, or in accompanying Colonel Grant on his mission to the chief of Tigrè at Adowa, or again in reconnoitring to within sight of Theodore's army at Dalanta, far ahead of the advanced posts of the British army.

In making these references to Mr. Munzinger, who is a distinguished Swiss naturalist, I am glad to find that the Queen has rewarded his services by conferring upon him a Companionship of the Order of the Bath.

DR. LIVINGSTONE.—Throughout the past year we have been kept in a state of anxious suspense respecting the position of our great traveller, Livingstone; and I grieve to close this Address without being able to offer some encouraging sentences on the prospect of speedily welcoming him home. At the same time, there is no cause for despondency as to his life and safety. We know that he has been for some time at Ujiji, on the Lake Tanganyika, whence he wrote home on the 30th May last, though unable to make any movement for want of carriers and supplies. These were, indeed,

forwarded to him by Dr. Kirk from Zanzibar, when alas! an out-
break of cholera stopped and paralyzed the relieving party.
Recent intelligence, however, has reached the Foreign Office to the
effect that the pestilence had subsided to so great an extent, that
we may presume the communication between the coast and Ujiji
has before now been re-opened.

The work which still lies before Livingstone has been often
adverted to, and it is hoped that he will live to advance to the
north end of the Tanganyika, and there ascertain if its waters
flow into the Albert Nyanza of Baker. If the junction should be
proved, we may indulge the thought that, informed as Livingstone
must now be of the actual carrying out of the great project of Sir
Samuel Baker, he may endeavour to meet his great contemporary.
The progress of the great Egyptian expedition of Baker having been
delayed in its outset, we know that it only left Khartoum to ascend
the White Nile in February. After reaching Gondokoro, as was
expected to be the case, in the first days of March, some time must
necessarily elapse in establishing a factory above the upper rapids,
and beyond the tributary Asua, where the steam-vessels are to be
put together before they are launched on the Nile water, on which
they are to pass to the great Lake Albert Nyanza. As soon, however, as a steamer is on that lake, we may be assured that Baker,
with his well-known energy and promptitude, will lose not a
moment in the endeavour to reach its southern end, in the expectation of there giving hand and help to Livingstone. Let us therefore
cherish this cheering hope, which would indeed be the most happy
consummation our hearts can desire.

The British public will be much better informed than they have
been on this subject when they examine a recent small work by
Mr. Keith Johnston, jun. In this pamphlet the author has given a
succinct history of all the explorations in South Africa, and has also
put together from the best authorities (Petermann and others) a map
which shows clearly to what extent the rivers which flow from the
southern highlands, on the south and s.s.w. of Lake Tanganyika,
are for the most part independent of that lake, and may prove to be
tributaries of the Congo. On the other hand, the streams which
enter the Lake Tanganyika through the Lake Liemba of Livingstone, are probably the ultimate sources of the Nile itself, while
the Kasai and other streams which feed the Lakes Bangweolo and
Moero may be found to issue in the Congo.

If this last hypothesis should prove to be true, the waters

which Livingstone has been the first to explore will be found to be the sources both of the Nile and the Congo. As respects the Nile, however, my sagacious friend must feel that, until he proves that some of these waters of the Tanganyika flow into the Albert Nyanza, the problem in regard to the Nile remains unsolved.[*]

In the mean time the Nile hypothesis of Mr. Findlay and others (that the Lake Tanganyika will be found to unite with the Albert Nyanza) is, according to the now estimated relative altitudes of these southern waters, the most probable. God grant that the illustrious Livingstone may demonstrate this to be the case, and that we shall soon see him at home as the discoverer of the ultimate sources of both the Nile and the Congo.

On this important and exciting subject it is gratifying to state that our Medallist, Dr. Petermann, has laid down, on a general map of South Africa in the last number of his 'Mittheilungen,' that which he terms a chronological sketch of all Livingstone's wonderful and arduous travels from 1841 to 1869. In respect to the tributaries of the Congo, the map of Petermann differs hypothetically from that of Mr. Keith Johnston, jun., inasmuch as he indicates that the waters of the Bangweolo, Moero, and Ulenge lakes probably point to north and by east; and, if this should prove to be the case, they also will fall into the great Albert Nyanza of Baker.

In concluding the consideration of this absorbing topic, I rejoice to be enabled to state, that in consequence of my representing to Lord Clarendon the isolated position of Livingstone at Ujiji, where he was without carriers or supplies, whilst he was, comparatively, near his ultimatum, the north end of the Lake Tanganyika, Her Majesty's Government have kindly afforded the means whereby the great traveller may be effectively relieved before he returns to his admiring country.

CONCLUSION.—At the last anniversary I was placed in this chair for the usual term of two years, and, in thanking my associates for this repetition of their never-failing kindness, I informed them that, if at the end of the first of the two years I should be incapacitated by infirmity, I flattered myself they would allow me then to retire, with thanks for my long continued devotion to their cause. I also

---

[*] As an ardent young geographer, Mr. Keith Johnston, jun., lays it down too broadly on the title-page of his clever work, that the sources of the Upper Nile basin are settled. Granting that this is not only the hopeful, but also the probable solution of the question, the ultimate proof, as stated above, is still required, and on that proof being obtained the return of Livingstone depends.

said that I accepted the office in the ardent hope that my dear friend Livingstone might soon return to us, so that I might have the joy of presiding at the national festival which would then unquestionably take place in his honour.

Although the first of my two years of office has passed without this happy realization of my hopes, I trust that before our next annual meeting the great traveller will have determined the grand problem of the ultimate southern sources of both the Nile and the Congo; and if I live to witness this completion of my heartiest aspiration, I will then take leave of you in the fulness of my heart, and with my warmest thanks to you, my friends, the Fellows of this Society, who have so long and so kindly supported me.

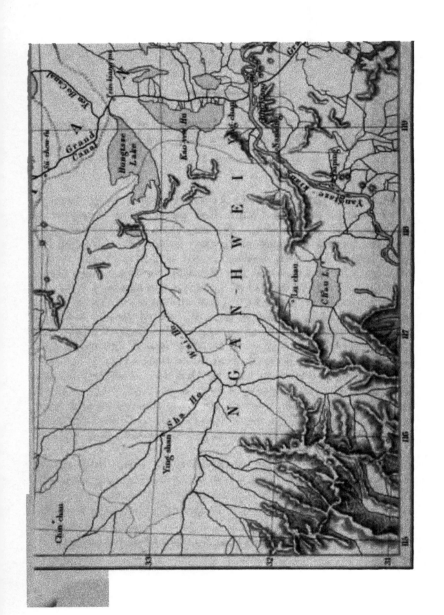

PAPERS READ

BEFORE THE

ROYAL GEOGRAPHICAL SOCIETY

DURING THE SESSION 1869-70.

[FORMING VOL. XL. OF THE SOCIETY'S JOURNAL.
PUBLISHED MAY 5TH, 1871.]

I.—*Notes of a Journey to the New Course of the Yellow River, in 1868.* By NEY ELIAS, Esq., F.R.G.S.

Read, November 22, 1869.

I.

IT is well known that the lower Yellow River, flowing through the great eastern plain of China, has many times changed its course during the historic era. No less than nine such changes are recorded by the Chinese as having taken place during the last 2500 years, the first dating about 602 B.C.;[*] the ninth 1851-3 A.D.; the positions of the mouths resulting from them having ranged over an extent of coast-line, comprised between some five degrees of latitude. Thus the most northern is recorded to have been in about lat. 39°, which would coincide approximately with the present mouth of the Peiho; whilst the most southern is represented to be that which existed before the last change, which is marked on all modern maps in lat 34°. There is reason, however, to believe that this southern mouth was the outlet at some periods of only a portion of the Yellow Waters, another portion finding its way simultaneously still further southward, viz., through the Hungtsze Lake into the Yangtsze, which, as will be shown below, is the case at the present day.

The causes of the earlier changes, their dates and other details I will not go into here, but will limit myself to a few remarks on the last diversion and the resultant course.

This diversion was first brought to the notice of foreigners in

---

[*] See 'Geological Researches in China, &c.,' by Raphael Pumpelly.

China, by Dr. Macgowan, in the 'North China Herald,' of 3rd
January, 1857, and was ascribed hypothetically to various
causes, all of which have been shown by the late examination
to have had little or no connection with it.

The date of the occurrence was for some time a matter of
uncertainty, some authorities placing it in 1851, others in 1852
and 1853, and even later. But on a short journey which I
made in 1867 to Tsin-kiang-pu, and the neighbouring portion
of the old bed of the Yellow River, I was enabled, after questioning numbers of different persons, living in the vicinity, at
different times and under different circumstances, to ascertain
with, I think, some certainty, that the change was gradually
accomplished and extended over the years 1851 to 1853. This
information was corroborated on my last journey, and might be
summed up somewhat as follows. During the summer flood of
1851, the first rupture took place in the north bank near Lan-
Yang-hein, in Honan, and a portion of the water flowed through
the breach on to the plain; the flood of 1852 extended the
breach, and further diminished the supply on the lower river,
and that of 1853 enlarged it to such an extent as to allow the
whole body of water to flow over the lowlands to the northward
and eastward, until it found a channel in the Tatsing River,
which conducted it to the sea in the Gulf of Petcheli. Thus,
not until after the flood of 1853 can the new course be said to
have wholly established itself, and the old one to have become
entirely dry.

This new course having become a subject of interest to a
portion of the foreign community of Shanghai, it was decided in
the early part of last year to send an exploring party to examine
and lay it down, all information upon it received up to that
time having been of a vague and unsatisfactory nature.

In accordance with this decision, I had the honour of being
requested to undertake the task, and my plan for carrying it
out being approved of, a small party was formed for the purpose,
consisting of Mr. H. G. Hollingworth, two Chinese, and myself.

The plan of the journey was to go to Chin-kiang by steamer;
from there to proceed up the Grand Canal until reaching the
Yellow River; to follow the river down to the neighbourhood of
its mouth in the Gulf of Petcheli, and up again as far as the
point where it diverges from its old course (which was reported
to be near I-fung-hein in Honan), returning by the river and
Grand Canal to Chin-kiang. This plan having afterwards been
found practicable, it was carried out.

As most of the rivers in China are known to be in flood
during the summer months, and consequently in an unfavourable
state for exploring operations, and as those in the northern

districts are generally frozen over by about the middle of
December, the autumn was considered the most advantageous
season for the journey, and consequently the party started from
Shanghai on the 24th September.

The Grand Canal between Chin-kiang and Tsin-kiang-pu, or
in other words, between the Yangtsze and the old bed of the
Yellow River, has been visited and described so frequently of
late years by foreigners, that it is almost unnecessary to touch
upon it here. Suffice it to say that it is everywhere in good
repair, and the adjacent country well-irrigated and apparently
in a thriving state both as regards cultivation, and to judge by
the aspect of the towns on and near its banks, as regards trade
also.

After crossing the Old Yellow River, however, a part of the
canal somewhat less known is reached, and the flourishing condition of the country is no longer noticeable; on the contrary,
for a distance of about 150 miles, though the canal itself is in
tolerably good working order, the country in its vicinity has an
arid, sterile appearance, and is but thinly populated. There are
few towns or villages, and what there are seem neither populous
nor busy, though they are not in ruins, and bear but few traces
of the rebellion—general poverty being the prevailing features.
Although the country bordering on this portion of the canal is
a part of the district lately infested by the Nienfei, yet these
would appear to be less the cause of this general poverty than
one of the effects of it, the source of both evils being more probably the want of irrigation which has existed since the Yellow
River has flowed to the north of the Shantung ranges. The
canal which at one time was so deep that in many places the
level of the water was above that of the adjacent country, is
now everywhere considerably below it, rendering irrigation at
even a short distance from its banks, without mechanical appli-
ances, almost an impossibility—even the dry bed of the Loma
Lake is scarcely cultivated, on account of its elevation above the
level of the canal, though it is only separated from it in some
parts by a bank of a few yards in width. It is true that this
lake appears never to have been more than a shallow flood
lagoon, nevertheless, it was some feet below the general level of
the country, and was connected with the canal by means of
water-courses and sluice-gates, and if this is difficult to irrigate,
how much more so must be the country above and beyond it?

This 150 miles being passed over, the Wai Shan (sometimes
called Yü Shan) Lake is reached at a small village, called
Hanchuang-cha. This is the most southern of a chain of lakes
or rather lagoons, which stretch from far to the south of Hanchuang-cha (I believe from near Sü-chan-fu on the old Yellow

River) to within a few miles of Tsi-ning-chow, and which constitute the only important feeder of the Grand Canal to the southward. In the summer they merge one into the other, and form a continuous sheet of water though very shallow in parts; in winter, when the water is low, these shallow parts are mere morasses, which divide the sheet into three or four separate lagoons.

In former days the canal ran, in some places, by the side of these lagoons, and in others through portions of them, but being everywhere embanked on both sides, it was only dependent upon them for its supply of water, the canal itself forming an unobstructed means of communication through the year. Of late years, however, this section of the canal has been allowed to go to ruin, and those portions only are used which run through the morasses existing in the dry season, the lagoons themselves forming elsewhere the only channel for navigation,

Near the northern limit of these lagoons stands the city of Tsi-ning-chow, the first place of any importance on the canal north of Tsin-kiang-pu; it is said to be a place of considerable trade in ordinary times, but for the greater part of last year it was made one of the principal camps of Li Futai's soldiery, and is consequently not in a very flourishing condition at present. It has an inner and an outer wall, the former apparently new, and on the plain outside the city are numbers of stockades of different dimensions dotted about in every direction. The number of junks and boats seen at this place was very large. but all were in the hands of the soldiers, and apparently carried on no trade.

Still proceeding northward, a distance from Tsi-ning-chow of about 25 miles, the summit level of the canal is reached near a small town, called Nan Wang. It is here that the River Wen falls into the canal, a portion of its waters flowing to the south, and the rest to the north, precisely as described by Staunton and other writers. The Wen is a small stream scarcely 20 yards broad at the confluence, the canal at the same point being even less than that—its course is from the north-east, and it is said to take its rise amongst some hills which are plainly visible in that direction. The currents of both are very inconsiderable, certainly under one mile an hour.

About 30 miles beyond Nan-wang, we come to the New Yellow River, the canal for that distance being extremely narrow and shallow; a mere ditch in fact, running between embankments large enough to confine a stream of infinitely greater volume.

The banks along nearly the whole of the Grand Canal between the old and new bed of the Yellow River, excepting those portions bordering on or traversing the lagoons, are surmounted by earthen walls crenallated after the fashion of city walls, behind which are stockades at intervals of every few miles. All this work has the appearance of being recently constructed, though in many places it is already being broken up by the country people to make room for cultivation, for they can ill afford to lose that strip of land immediately adjacent to and irrigated by the canal. The villages also make an attempt at fortifications, most of them being surrounded by earthen or mud walls and moats, and, indeed, many solitary farms have some species of defensive works round them, and in most cases a small, square, brick tower within. These towers are rarely met with to the south of the province of Shantung—they are probably the "Water Castles" mentioned by the historian of the Dutch embassy.

A journey of nearly 400 miles on the Grand Canal, such as I have here attempted in a few words to describe it, brought our party, on the 17th October. to the southern bank, or rather "limit," of the New Yellow River, near a small but busy town called Nan-Shan. The river at this point has no defined bed, but flows over a belt of country some 10 to 12 miles in width, having merely the appearance of a flat, level district in a state of inundation; patches of ground, trees, and even villages, cropping up here and there, the Grand Canal traversing it in a general north-westerly direction until it reaches the northernmost channel of the river at Pa-li-miau, some 15 miles from Nan-Shan. Along this 15 miles the canal-banks have been carried away in a number of places by the Yellow River breaking across them. The gaps are sometimes half-a-mile or more wide, and the current rushing through these almost obliterates the course of the canal and renders navigation upon it difficult. - For dreariness and desolation no scene can exceed that which the Yellow River here presents; everything, natural and artificial, is at the mercy of the muddy dun-coloured waters as they sweep on their course towards the sea—a flood not likely to subside, and a doubly mischievous one from the fact of its ever moving onward with a swift current.

The Grand Canal is now dry, from the Yellow River northwards, as far as Lin-Tsin-chow, or, in other words, it ends where it meets the river, that portion north of the summit level being merely a tributary of the Yellow River. During the two months of the year, however, when the river is in flood and at its highest level, enough water, it is said, flows into the dry bed of the canal to form a navigable stream as far as Lin-Tsin, where it connects with the Wai-ho. Thus, for some ten months of each year, there is no water-communication towards the north beyond the Yellow River.

Near the southern "limit" of the river is a channel, running in a general north-north-easterly direction, down which junks of a considerable size were seen to be sailing. Being informed, however, that a more important one existed near the northern "limit," it was decided to cross at once and commence the exploration by this latter channel, leaving the southern one until a better opportunity should offer. Having crossed, accordingly, to Pa-li-miau (a small village 8 li from Chang-tsin), that place was made the first station, and the necessary observations for fixing its geographical position being obtained, the journey down the river was commenced from there on the 20th October, a date, by the way, so far advanced in the season as to render the greatest expedition necessary in order to complete the journey before the closing of the river by ice.

After sailing down the northern channel for about 19 statute miles, a point is reached where it is joined by the southern one, and consequently also by all the water which higher up floods the country lying between the two. This point is called Yü-Shan, and the deep, narrow, clean-cut river-bed that receives the converging waters and leads them to the sea is that which, fifteen years ago, formed the course of the Tatsing. Although still narrow—some 250 yards hereabouts—there are everywhere many indications of the river having been less than this before the advent of the Yellow waters, and, to judge by the velocity of the current and other circumstances, it has most probably become deeper also during the same period.

Proceeding down stream, we pass through an open, well-cultivated country, with every here and there low dome-shaped hills, sometimes detached, sometimes in groups, and backed up by a range some 400 to 600 feet high, running about east and west, which abuts on the river at Yü-Shan, but diverges from it gradually, the course of the stream being in general a north-easterly one. The small hills near the river are of a limestone formation, the strata perfectly horizontal. The main range was not visited, but the hills composing it being similar in shape are probably also of the same formation. The stone is quarried, but to a very limited extent, and villages within a few hundred yards of a quarry are built almost entirely of mud and chopped-up straw.

Fifty-six miles from Yü-Shan, by the windings of the river, bring us now to the town of Tsi-ho-hien, a small, newly-walled, unbusinesslike-looking place, which, except from the circumstance of its being the site of a serious obstruction in the river, would hardly call for a word of notice. This obstruction consists of the ruins of a stone bridge of some seven arches, which at one time spanned the Tat-sing, but which now would reach only about three-quarters of the distance across the river. There is a space between one extremity of it and the left bank of about 100 yards, which is used by boats as the only navigable channel. The deepest portion of this 100 yards is close under the left bank, where, at the time our party passed down (Oct. 21st), there was a depth of 5 feet, and no stones to be felt with the lead. The outer portion of the 100-yards channel, however, would probably not be practicable even at 3 feet. The bridge evidently stands in deep water, 6 fathoms having been found immediately above it, and 5 a few hundred feet below it. The right bank is the steep one, and the left—near which is the channel—the shelving one, and, naturally, the shallow side of the reach. Its being now nothing but a wreck is, of course, due to the additional force and volume of water in the river for the last fifteen years, which it has been unable to withstand.

It is evident that the ruins of this bridge might be removed, and, if no other obstruction existed, the river rendered navigable as far as Yü-Shan, or within 19 miles of the Grand Canal. Unfortunately, however, about 3 miles below this one there occurs another, though a less formidable, obstruction, in the shape of a shoal extending right across the river. In this case, too, the deepest side of the reach is the right; and here, on the 21st October, only 11 feet was found, the bottom rising gradually towards the left bank. On the 6th November, when this spot was passed a second time, there was but 5 feet of water in mid-stream, and, allowing for the fall since the 21st October, we should have only about 6 or 7 feet in the deep passage near the right bank. The length of the shoal would be about 200 to 300 yards, and is the only place above the bar where less than 2 fathoms was found in the deep channel.

The next point of interest we arrive at is Lokau, the port of Tsi-nan-Foo, a long, straggling unwalled town, on the right bank. Tsi-nan itself stands 12 li from the river, and not far from the foot of the main range of hills, which hereabouts average probably from 800 to 1200 feet, and form a rather picturesque background to the low, thickly-wooded plain upon which the city is built, and which extends for many miles on both sides of the river, giving to the country its characteristic feature of flat lowland. This plain is essentially alluvial, yet there rise from it in this neighbourhood several small wedge-shaped, jagged hills, or rather masses of rocks, in some cases heaped up into fantastic shapes, and the fragments near the bases worn into rounded boulders by the action of water. Their height is inconsiderable, but, being perfectly isolated—sometimes several miles of plain intervening between two of them, or between one and the main range—stamps them at a glance as the direct result of igneous action.

The trade of Tsi-nan-Foo is said to be of great importance, but, as a large proportion of it is carried on by means of cart-roads, a traveller on the river has but little opportunity of forming an opinion of its magnitude. The number of boats seen at Lokau was not large, and many of them appeared to be only passing through towards the Grand Canal. The only article of commerce noticed in any quantity was salt, which had come up the river from Tië-mên-quan. Coal is met with as an article of trade, both here and at other places on the Yellow River, and is used for cooking and other purposes to, I believe, a considerable extent; it is of a rather bituminous nature, and is sold at the rate of 1200 cash per picul in Tsi-nan-Foo. The principal mines are said to be at Tsan-fan, a place in the hills 90 li to the eastward, where the coal is sold at a very much lower price than at the city.

We pass on now through a thickly-wooded, well-cultivated country for about 150 miles, flat but dry, and the soil very light and friable. The river's banks are steep, and indicate a rise in summer of 8 to 14 feet, according to the distance from the sea. The re-entering angles are everywhere much eaten into by the current, and large masses of soil are continually falling away. In many places, the grain of this year having been sown up to within a short distance of the water, portions of the fields supporting the crop already sprouting have been undermined, and fallen into the stream below—thus showing that the undermining process is a very rapid one, probably more rapid this year than the experience of the inhabitants who sowed the grain led them to anticipate. The graves near the river, or rather the coffins from them, have generally been removed to some distance back, and often to the opposite shore; the exhuming and removal being sometimes attended with great ceremony. Near the course of the river are extensive vegetable gardens, growing carrots, onions, celery, turnips, Shantung cabbages, brinjalls, capsicums, &c., &c.; also regular plantations of fruit trees, such as the pear, the date, and others of less importance, and in some places many square miles of land are occupied by the plant whose root is the ground-nut. The date and the ground-nut are two of the staple products of the district. At the time our party passed down the former had already been gathered and dried, but it was harvesting-time for the latter, and the inhabitants of nearly all the villages were at work in the fields digging them up and sifting them. A third staple product is cotton, which, though of an inferior description, is rather largely cultivated; and lastly, we have the Shantung silk. The mulberry near the Yellow River appears principally, if not entirely, to be cultivated on the left bank, and for a distance of scarcely a hundred miles; the trees are standards, and have much the appearance of those grown in England, being larger and older-looking than those in the more southern provinces of China. They are planted in lines at regular intervals, some of the plantations covering a large area of ground. Some specimens of the silk were obtained at a village in the heart of the district, also some eggs and cocoons. The best silk is yellow, and very long-reeled, much resembling the Sze-chuen, and is, I believe, often sold at the treaty ports as the products of that province. It is produced by worms fed indoors, as in the southern provinces, the wild, or out-door-fed worm not being cultivated in this district, but amongst the hills to the southward, in the neighbour-hoods of Mêng-Yin and I-Shui, as I am informed, where the food is not the mulberry leaf but that of a species of oak.

The principal towns situated on the river, and within this 150 miles of garden-like country, are Tsi-Yang, Tsi-Tung, Pu-Tai, and Li-tsin, all " hiens." The first is of no importance whatever; the second is a large, busy, and apparently thriving place, and would probably rank next to the capital in the matter of trade, though it certainly surpasses Lokau in every attribute of a port; the third is small, and apparently but a poor place of trade, and were it not the site of the principal custom-house on the river, it would be scarcely more noticeable than Tsi-Yang. The fourth town, Li-Tsin, appears also to be of no great importance as regards trade, though there is a circumstance connected with it which renders it remarkable—viz. the inroad made by the river into the city. Situated on the concave bank of a sharp bend in the river, the swift current, after eating its way through the foreshore, at length reached the south-west angle of the city wall, which it has carried away, together with some 300 feet of the wall on each side of the angle, up to the present time, and is still at work cutting deeper into the breach—no efforts being made by the natives to arrest its progress. Some of the ruins of the wall and outlying buildings are now visible above water near the middle of the river, but they would form no obstruction to navigation, 7 and 8 fathoms having been found within a few yards of them and towards the right bank.

A few miles below Li-Tsin the country begins to change its character; the well-wooded and well-cultivated district above described giving place to boundless tracts of mud and marsh, but poorly cultivated and thinly inhabited, and the whole aspect one of a bleak, swampy, treeless waste, scarcely fit for man to dwell in. Nevertheless, on the river's banks are villages at short distances from one another down to within about 20 miles of the sea, which causes the traveller on the water to form an exaggerated opinion of the population of the district, though he is easily undeceived by walking a short distance away from the river, when it becomes apparent that on the whole the population is sparse. In fact, the only fairly habitable region is that belt of land immediately skirting the river, and from which the water of the annual flood drains itself off naturally, whilst on the tracts lying farther back it is either absorbed by the soil or remains on the surface in the shape of marshes and ponds, rendering habitation without artificial drainage almost an impossibility except on a most limited scale.

At the limit of the habitable region—viz., about 20 miles above the sea by the windings of the river—stands the village of Tië-mên-quan, the port of the Yellow River, and though only a village, composed like others in the neighbourhood of mudbuilt houses, it has every appearance of being a most important place. It is not a centre of trade, but consists chiefly of hongs to which traders from the different towns in the neighbourhood come to transact their business, and during the winter months, when the river is closed by ice, it is said to be nearly deserted. Now, although it is called a port, Tië-mên-quan is only used as such by small Pei-ho and Yellow River junks. Larger vessels, such as those from Ningpo, Shanghai, Swataw, &c., never come within 20 or 25 miles of it, but remain at an anchorage outside the bar called Tai-ping-Wan, where they discharge their cargoes into Yellow River boats, receiving their homeward freight by the same means—thus, for these junks, Tië-mên-quan can scarcely be considered a port. The direct trade which exists between places high up the river and Tien-tsin, Chefoo. and other ports on the Gulf, is carried on by boats of a lighter draft and of a different construction to the seagoing junks of the southern provinces, but well·suited, of course, to the rivers and shallow seas on which they are employed. The voyage to Taku is said to occupy these boats about two days with a fair wind; that to Chefoo, about four days. In both cases the journey is performed by coasting round the Gulf, and as the water for some distance from the shore is very shallow the sea never rolls heavily, and it is always possible to anchor in the event of a foul wind.

The principal trade of Tië-mên-quan appears to be with Tientsin, which is the nearest open port, though junks bound to and from all parts of the Gulf are to be found there. The exports are chiefly salt, cotton, dates, &c.; the imports, paper, timber, sea-weed, beans, sugar, and a few English cotton goods and lead. More than three-parts of the whole trade, however, would seem to be in salt, which being produced in the neighbourhood, is not only exported but sent up the river in large

quantities.   The salt manufactories, if such they can be called,
for they consist generally of nothing but a few ponds and a mud
hovel or two, are dotted about here and there over the waste
marshy lands lands before alluded to, and though constituting
on the whole a considerable industry, yet it appears to be one
that supports but a small proportion of the population, and
brings but a limited area of land under subjection.   The country
having only recently been left dry by the sea, the soil still contains a certain quantity of salt, and by digging to a depth of
about 2 or 3 feet the salt water is obtained from which the
brine is produced by evaporation.   A manufactory, or saltern,
consists of a series of shallow ponds connected one with the other
by means of narrow ditches.   The water is first collected in the
outside pond, and after being allowed to evaporate for a few
days, is conducted through a ditch into a second, where it
remains for a few days more and evaporates still further; this
process being continued till in the fourth or fifth pond the salt
is seen lying at the bottom in crystallised layers as white as,
and very much of the appearance of, snow, when it is scooped
out and stacked, and covered with a thatchwork of reeds and
mud..   The water in the first or outside pond, is but slightly
salt, but the saltness increases with the amount of evaporation
until arriving at the last or inner pond, when it becomes so
intense as hardly to admit of putting the tongue to it.   It also
acquires a peculiar bitter taste as the evaporation proceeds, and
becomes beautifully clear towards the end of the process.   In
such a concentrated solution of salt as is the fluid in the inner
ponds of these salterns, it would be thought hardly possible that
any animal could exist; yet in the clear water above the crystallised layers, thousands of small transparent shrimps may be seen
darting and gliding rapidly about in every direction, and resting
sometimes on the salt itself.   I believe this little animal to be
the Brine-shrimp, or *Cancer salinus* (Linn.), well known as the
inhabitant of saltpans in England, where the workmen believe
that it is of use in clearing the brine of impurities and cultivate
it accordingly.   Whether this belief obtains on the coast of
Shantung I was unable to ascertain, but possibly it does, for
certainly no attempt is made to expel the little creatures from
the ponds.

Tiĕ-mên-quan, as before remarked, is on the lower limit of the
habitable and salt-producing region; the country between it
and the sea is an uninhabitable region—an immense mud-flat—
stretching away on both sides of the river as far as the eye can
reach.   In the summer and autumn the greater part is covered
with reeds, the more accessible of which are collected for fuel
by a race of miserable reed-cutters, whilst the rest afford cover
to vast numbers of wild fowl—swans, geese of two kinds, peli-

cans, &c., &c.   In the winter, when the reeds are gone, it must
be a desert of mud, and when the river is in flood it is, of course,
totally submerged.   About 12 miles below Tiĕ-mên-quan and
half a mile from the river's left bank is a little knoll, about
10 feet above the general level, formed of sea-shells and *débris*,
evidently at one time an island, and upon which stands a small
brick joss-house, apparently new, and a few mud-hovels, the
dwellings of reed-cutters.   This place is called Lau-Ye-Miau—
the only habitable spot for many miles in every direction—and
is probably the point reached by the naval surveyors in 1860,
and called by them Miau-Shing-pu.   About 4 miles below this
again we come to the bar, an object that has excited a great deal
of interest amongst residents in China, it having been thought
that the navigability of many hundreds of miles of the Yellow
River hinged upon the depth of water to be found there.   This
view, however, as will be seen immediately, is not a correct one,
worse obstructions existing higher up.   At the date I visited it,
October 27th, the least depth found was about 5 feet near the
middle of the river, the water at the time being, according to
the pilot who accompanied me, about a foot or 18 inches above
low water-mark.   The deepest channel is near the right bank,
though there is one almost as deep near the left, the shallowest
part being in the middle.   In the former I found about 9 feet,
and in the latter about 7 feet, which, at low water springs would
give little over 7 feet and 5 feet.   Several junks, drawing, it
was said, 2½ feet of water, were seen sailing through the left
channel.   The range of the tide would appear on the average
to be about 2 feet, rather more at springs and rather less at
neaps.   Ordinary neap floods, when the river is not in flood,
are said to be perceptible for about 20 or 30 li above the bar,
and springs, when favoured by the wind and a low state of the
river, are sometimes noticeable as high up as Tiĕ-mên-quan,
some 60 li above it.   It is, of course, obvious that a sufficiently
long stay to make personal observations on the tides was
impossible; and my information on this subject is derived from
a number of junk-skippers, pilots, and others, questioned at
different times and under different circumstances, and who,
strange as it may appear in this country, agreed remarkably in
their statements.   It is possible, therefore, that some approxi-
mation to the truth has been arrived at.[*]

---

[*] I may here remark that it was the object of the journey to examine the river
as far down only as the highest point reached by the naval survey in 1860, viz. to
about Lau-Ye-Miau, and I wish it, therefore, to be distinctly understood that I
did not go to the bar with any intention of surveying it, and do not pretend for a
moment to have done so.   My only object in visiting it was to carry my work
down to a known fixed point, a few lines of soundings being taken across the bar
itself merely by way of attempting to verify information previously received on
the subject.

From all I can gather, then, on the subject of the bar, I am inclined to believe that the draught of water of the southern junks is no obstacle to their ascending the river, but that the almost total absence of tides and the narrowness of the channel constitute the principal difficulty; this more especially as the anchorage at Tai-ping-wan is stated to be safe and convenient and the transhipment there of cargo easily performed, whilst to work a large junk up a narrow and tideless river for 20 miles would be a slow and risky process, even though the depth of water would admit of it. In the case of steam-vessels, of course, these objections do not apply, the depth of water being then the only matter for consideration.

II.

After returning to Tiĕ-mên-quan from the bar on the 28th of October, our party had thus far examined only the section of the river included between the Grand Canal and the sea, and there still remained that portion above the Grand Canal to be explored. It was already late in the season and before us was a journey on the Yellow River alone of some 550 miles, upwards of 400 of which was to be performed against a strong current and in a craft scarcely suitable to the navigation. Every effort, therefore, was made to push on as rapidly as possible, and no special halts were made for any purpose, except a short one for longitude observations, until arriving at Pa-li-miau on the Grand Canal (our first station), on the 10th November.

The river here, as before noticed, has no defined bed, but presents the appearance of a belt of country, 10 to 12 miles broad, in a state of flood—trees, ruined villages, and patches of bare mud, being all that is left of a once fertile and prosperous district. We have already seen that this is the aspect of the river for the 19 miles immediately below the Grand Canal, viz., as far as Yū-Shan, and in proceeding up stream we find another 76 miles (more or less according to the season) of a precisely similar character, making in all a section of 95 (stat.) miles scarcely worthy the name of a river—bed there is none, and at some periods of the year scarcely a channel for boats of a moderate size. It is true the natives speak of two channels, and indeed use them—a northern one and a southern one—but both were gone over during the month of November; and when I say that our boat, drawing only 15 inches of water, had often difficulty in finding a passage, little more need be said concerning the practicability of this portion of the Yellow River. During the high-water season junks, drawing, it is said, as much as 3 or 3½ feet, can use the southern channel; but the journey is slow and laborious in the extreme, and whole days are frequently spent in kedging over shoals or through places where the deposit, having found a group of trees or some other object to silt against, has commenced the formation of a mud-bank.

That there can be no great depth in this lagoon-like section of the river is at once apparent when we consider that the same volume of water which lower down is contained between the lower banks of the narrow Tatsing, is here spread out over a belt of country, ranging from 10 to 15 miles in width. Had this belt at any time been the site of a fairly deep river or even a deep-dug canal, the water of the Yellow River, although at first of too great a volume to be contained in the bed of such river or canal, would in time have so enlarged it, by means of its scouring power, as to have rendered it of the necessary capacity. This, indeed, is what took place in the case of the Tatsing, for, as we have already seen, the bed of that stream has become both deeper and broader since the advent of the Yellow River, and now contains the whole of the latter's waters in addition to its own, and only overflows its banks at the height of the flood season. Above the Grand Canal, however, there was no river-bed of sufficient size to form the basis of a course for the Yellow River, and hence the wide-spread shallow flood instead of a defined stream.

There were, it is true, two small canals falling into the grand canal within seven miles of one another; the more southerly of these, the Sun Kiang was a very small one, only 90 li long, it is said. A portion of it was examined, and the banks in some places found to be hardly distinguishable, but everywhere the waters of the Yellow River stretching away like an over-flow on both sides. To judge by the ruins of bridges, houses, and "pylows," the region through which it flows must have been a prosperous one; at present a few mud and reed hovels are the only habitations, and a few patches of wheat sown on the mudbanks left temporarily dry by the yearly recession of the waters the only sign of cultivation.

The second or northerly canal was of much more importance than the Suu-kiang, and, though shallow and narrow, was about 400 li (133 miles) in length. It was, and still is, called the Chun-wang-ho, and led from the old Yellow River to the Grand Canal, near Pa-li-miau. The point of junction with the old Yellow River I have never been able to ascertain with any certainty, but I believe it to have been a short distance to the east of the place now called Lung-mên-kau, or the breach in the old river's bank through which the Yellow waters leave their former bed. It presents on the whole much the same appearance as the Sun-kiang; viz., an embanked watercourse running through an

inundated country. Its artificial banks were at one time at
some little height above the level of the country, but they have
now, in most places, been either carried away by the floods, or
worn through by the current of the Yellow River.

The villages and bridges are mostly in a state of ruin, and
the latter, as they reach now little more than half-way from
bank to bank, are additional evidences of the power of the river
to form for itself a bed, provided only that it finds a sufficiently
durable basis to work upon. Durability in this case, however,
is wanting, and even had the Chun-wang been many times its
original breadth it would still have been useless as a channel
for the Yellow River, the artificial embankments being naturally
unfitted to withstand the scouring process. As it is, the canal
is only traceable here and there for a few miles at a stretch,
and as its course through the belt of country at present occupied by the river was a winding one, the portions now left are,
as it is only natural to suppose, those whose direction was identical, or nearly so, with that taken by the Yellow River.

A distance of 76 miles then, by the southern channel, from
the Grand Canal brings us to a point where we find the Yellow
waters again flowing in a defined channel, which is traceable as
far as the old bed, a distance of about 52 (stat.) miles. At the
low-water season this channel contains the whole Yellow River,
but as the banks in the highest places are no more than about
10 feet above the November level, it can contain during the
flood season only a portion of it, for, though broad (in some
places over a mile), it is everywhere exceedingly shallow, and
its capacity much contracted by shoals and mudbanks. Now,
although when the banks are at 10 feet above the level of the
water, this channel has the appearance of being the permanent
bed of the river, yet, so far from thinking it e manen, I
should hesitate even to call it a "bed" at all; for the banks,
and indeed the country for miles on each side, are composed of
the river's own deposit, which seems rather to have silted to a
certain elevation above the river level than that the water had
cut a bed for itself in the soil to a corresponding extent—or, in
other words, it appears that the river here flows but little, if at all,
below the general level of the country, a fact which is at once
demonstrated by considering, for example, that at a point in
the lagoon-like section just below the lower end of the defined
channel, and where there were no banks apparent, *old* trees
were growing on about the level of the water, and ruins of
houses were standing on patches of mud only just awash, whilst
at a point 20 miles higher up, and within the defined channel,
old trees were also growing on mudbanks about flush with the

waterline, though the river banks were 10 feet high and little more than the roofs of houses were to be seen aboveground.

It is almost superfluous to say that the country thus formed of the river's deposit is a perfect level, and that the soil is very light and mobile, and though the flood of each successive year, by adding more deposit, increases the stability, yet a powerful stream like the Yellow River can, I imagine, hardly be thought to have adopted a permanent course when the nature and height of its banks, the character of the adjoining country, the extent of its annual overflows, &c., are taken into consideration.

Perhaps the most striking proof that the banks and neighbouring country are the creation of the river's deposit rather than that the channel is a natural excavation, is that of the buried or silted up houses, which, besides, is a circumstance of interest in other respects; as, for instance, that it goes to show the power of the Yellow River in changing the configuration of the country with which its waters come in contact, but also the effects produced by it in the economical condition of those portions of the population whose misfortune it is to inhabit regions coming within its influence. Such a region is that through which this section of the river flows, and where we find many entire villages half-buried in deposit and deserted by the greater portion of the inhabitants; those who remain being in a poor and miserable condition. The houses are frequently silted up nearly to the eaves, and have generally been abandoned, but few dug out. As an example of this, I may mention a joss-house within a few yards of a point on the river where the level of the deposit was some 10 feet above that of the water. To enter this joss-house it was necessary to crawl under the eaves, and, when inside, it was evident from alterations that had been made in the doorway, &c., that for some time the inhabitants had attempted to accommodate themselves to the constantly diminishing height of the building, though since the last year or two apparently they had been compelled to abandon it. The deposit on the inside was at precisely the same level as that on the outside, and was said by the villagers in the neighbourhood to be 12 Chinese feet in depth (say 13 feet English), and to have been the work of 15 years, or 15 successive floods of the Yellow River. The heads of some of the larger josses still remain visible above the mud-level, but judging by them the size of the rest of the figure to which they belonged, I should say that the statement of the villagers was rather exaggerated, and that 9 or 10 (Eng.) feet depth of deposit would be nearer the mark; and in this opinion I am borne out by the proportions of the building, the height of the river, and other circumstances.

The houses, it may be remarked, in western Shantung and southern Chihli are built of brick, and are more solid and of altogether superior construction to those in Kiang-nan and eastern Shantung; many have two stories, and in every small village are to be found one or more of those square, castellated, little towers sometimes called "water-castles."

So little used is the Yellow River above the Grand Canal, and the navigation on it so little understood, that the people living near its banks, and even the boatmen themselves, seldom know the distance from one place to another by the river, but always speak of distances by the road, and even then two rarely agree, showing of what little importance the river is regarded as a means of communication, even though no roads worthy of the name exist in its neighbourhood—nothing, in fact, but mere tracks over the mud. In the same way information regarding the direction and distances of towns lying back from the river can never be obtained with any certainty, and though the boundary line between Shantung and Chihli was known approximately, that between Chihli and Honan was a subject of the most vague statements by the people living near the spot.

The absence of towns along the course of the river may to some extent account for this kind of ignorance. The only one on the river is the old or former Fan (hien), of which there is nothing left now but a small village and a ruined pagoda, the present town of Fan, being, it is said, some 30 li to the north-ward. Zung-ming-hien is near the river, but not visible from it; the nearest point is one from which the town is said to be distant 12 li in a south-easterly direction. I believe it to be a place of no importance, and there is certainly no trade carried on there by means of the Yellow River.

A somewhat tedious journey of a fortnight from the Grand Canal brought our party on the 24th November to Luug-mên-kau, the diverging point of the old and new courses of the Yellow River and the upper limit of the exploration.

The breach in the embankment of the old river is about a mile in width, and the present channel runs, as it were, diagonally through it. The two banks at this point are about 3 miles apart; near the northern one there is a depression about ¼ of a mile broad, full of small sand-hills, the only part of the old bed having the appearance of a dried-up watercourse; this was the main, or low-water, channel of the old river, the artificial outer embankments marking only the limit attained during the annual floods. The course of this low-water channel, as indicated by the present river to the west, and the line of sand-hills to the east, of the breach, was not always parallel to the flood banks, but made a winding tortuous line between them, apparently like a natural river, and the point where the breach now is, was one where the current impinged on the north bank. Those parts of the bed of the old river lying between the low-water channel and either bank are at a considerable elevation above the general level of the neighbouring country, and this is particularly apparent at the breach, where the bank is seen in section, the outer slope being some 40 feet in vertical height, whilst the inner would be about 20 or 25 feet—showing an elevation of the bed of 15 or 20 feet *near* the bank, though, as it slopes somewhat towards the low-water channel, the average for the whole breadth, exclusive of that channel, would probably not be more than about 15 feet. Thus, by a mere cursory inspection of the neighbourhood of the breach, the cause of the Yellow River's change of course is at once apparent. The river had so diminished the capacity of its bed (which, by the way, was always an artificial one), by depositing the alluvium with which its waters were charged, that the main pressure during the flood season had come to bear on the upper, or weaker, part of the embankments, and no measures having been taken to strengthen these, or deepen the channel, the great catastrophe happened which, with its consequences, had been predicted by the Abbé Huc some years before*—a catastrophe which has caused not only the devastation by flood of that line of country through which the river now flows, but has also impoverished to such an extent the districts through which it formerly flowed, and which were dependent upon it for irrigation, as to render them almost uninhabitable, and to throw a great portion of the population out of employment.

Lung-mên-kau is a small village built along the north bank of the old river east of the breach. About 90 li w.s.w. of it stands Kai-fung-fu (or Pien-liang-ching, as it is more generally called), the capital of Honan ; about 25 or 30 li to the south-east is Lan-I- (or Lan-Yang) hien, a place that has gone entirely to ruin during the last few years; it consists now of nothing but a few mud-and-reed houses, and is said to have been abandoned by the mandarins. A road leads down the old bed, through Lan-I, towards the south, and boats bringing cargoes down the river sometimes discharge them at Lung-mên-kau for conveyance by waggon to towns in that direction. Some of these boats

---

* Car le lit actuel du Fleuve Jaune, dans les provinces du Honan et du Kiang-Su sur plus de deux cents lieux de long, est plus elevé que la presque totalité de l'immense plaine qui forme sa vallée. Ce lit continuant toujours à s'exhausser par l'enorme quantité de vase que le fleuve charrie, on peut prévoir pour une époque peu reculée une catastrophe épouvantable, et qui portera la mort et le ravage dans les contrées qui avoisinent ce terrible fleuve. See ' Voyage dans la Tartarie, &c.,' tome i., p. 223.

bring small quantities of good anthracite coal from the neighbourhood of Hoai-king-fu; others iron ware, such as cast pots
and pans, wire, &c., from places not far distant, all of which are
sent south or east by road, Lung-mên-kau having no trade
whatever of its own.

After leaving Lung-mên-kau the river was followed down as
far as the Grand Canal, which was reached on the 30th November. It had been intended, as before remarked (see p. 13), to
have examined the channel which leads from Nan-Shan (on the
Grand Canal) to Yü Shan and runs near the southern "limit"
of the river; but the fall of water had been so great since the
first view of this channel was obtained on the 17th of October,
that it had become impracticable for all but the smallest boats;
and seeing that the ice had already begun to form on the shallow
waters of the river, it was thought advisable rather to leave this
channel unexamined than incur the risk of being frozen in and
having to transport the timekeepers overland. However, as its
length can be scarcely 20 miles, and as it is impracticable for
large boats except during the summer, but little would have
been gained by visiting it, more especially as the northern
parallel channel had already been thoroughly examined. At
Nan Shan, therefore, on the 1st December, the exploration came
to an end, and the party returning by way of the Grand Canal
arrived at Chinkiang on the 15th December.

To sum up shortly the capabilities of the Yellow River for
navigation, it would seem that a vessel of sufficiently light
draught to cross the bar would have no difficulty in ascending
the river during all but the lowest point of the season as far as
Tsi-ho, a distance approximately of 210 statute miles from the
bar, and were the ruins of the bridge at that place removed, a
further distance of 56 miles would be rendered navigable,
making in all 266 (stat.) miles from the bar to Yü Shan. To
such craft the shoal 3 miles below Tsi-ho would scarcely be an
obstacle, though, with the river at its lowest and the vessel loaded
to cross the bar at high water, it would probably prove to be impassable. In many places the bends in the river's course are
very sharp, with spits sometimes projecting from the salient
angles, but this would certainly be no obstacle to vessels under
200 feet in length; and considering the navigation on the Pei-ho,
where the curves are still sharper and the river narrower, even
a greater length might be found practicable. All beyond Yü
Shan as far as Lung-mên-kau must be regarded as totally unnavigable, except perhaps the 19 miles between Yü Shan and the
Grand Canal, which could be used during the high-water season.
The difference in the river's level between the highest and lowest

points in the year is something very considerable, but until the fluctuations shall have been observed throughout a whole year, it will be impossible to obtain accurate information on this subject. The yearly rise or fall is sometimes greater than at others, but taking the November level of last year (1868) as a base, and judging of the former by the indications of the banks and other signs, and of the latter from native information, I should think that a yearly range of 20 or 22 feet would not be far from the mark. Both rise and fall take place very irregularly, and it is said that a fall of 3 feet in one day is a common occurrence, especially towards the approach of winter.

The current during the flood season is, of course, far stronger than when the water is low, but on this subject also little can be said in the absence of a whole year's observations. On the 26th of October, at Tie-mên-quan, it ran about 2½ or 3 knots an hour; at Lo-kau, on the 5th November, rather under 4 knots; and near Yü Shan, on the 9th, about the same; whilst on the lagoon-like section above the canal it was generally under 2 knots. Such solitary instances as these, however, form but a poor guide for making a general estimate, and though I believe that at the lowest point of the season it is not much less than in November, yet during the height of the floods it must be nearly double.

As in the first paragraph of this paper mention was made of a southern outlet of the Yellow River existing at the present day, it must be explained here that this was caused directly by a rupture of the south bank of the old river near Yang-kiau, in Honan, about 150 li above Kai-fung-fu, which occurred in July last (1868). Not having visited the spot, my information on this subject is, of course, derived from the natives, who report that the bank is carried away for about 3 li, and that the water of the Yellow River flowing through the breach floods a large tract of country outside the bank, and then finds its way into a small river called the Sha, a tributary of the Wai, which latter flows into the Hungtsze Lake. This breach is called Lin-Lung-mên-kau, the one lower down being sometimes known as Lan-Lung-mên-kau. It is said that the authorities are trying to repair it in time for next summer's flood, about a third of the work being finished in November; and the common belief is, that when the upper one has been closed the lower one will be taken in hand, and the river made to flow along its old course to the sea. This, however, would appear to be impossible as long as the old bed remains at its present level, and to deepen it or to raise the embankments would be equally impossible in the present disorganised and impoverished state of the country. On the 22nd November, while proceeding up the Yellow River, I had a corroboration of the statements of the natives, regarding communication with the southern waters, by meeting five or six Hungtsze boats, whose people said they had come from Ying-chow (or Hing-chow), in Kiang-nan, by way of the Sha. These boats were said to draw about 1½ feet of water, but, as they were dropping down with the current, no information could be obtained concerning the Sha or the communication between the Yellow River and the Hungtsze,—a matter of great interest, as, should a permanent communication be found to exist, the Yellow River will have to be regarded as *in part* nothing more than a tributary of the Yangtsze, for into the Yangtsze the Hungtsze Lake discharges. The amount of water parted with through the new breach I believe to be very inconsiderable.

Shanghai, 20th March, 1869.

Subsequent Visit to the Old Bed of the Yellow River.

In laying before the Society the following short remarks respecting a journey which I have recently made to the *old* Yellow River, I would crave its attention for a few moments to allow of a word or two in explanation of the objects of the journey, and also to be able to advert to some points mentioned in the account of my visit to the new course in 1868, and of which the Society is already in possession.

It may be remembered that, on that occasion, the highest point reached was Lung-mên-kau, then called the *Lower Breach,* or that through which the Yellow waters finally left their old bed,—the commencement, in fact, of the new river. It may also be remembered that, in the course of that journey, information was received from the natives of a breach having shortly before occurred in the south bank of the river, about 150 lí above the one then reached, but which it was impossible to visit and examine at that time, owing to the lateness of the season and to other circumstances, and all that could be done was to describe the course taken by the overflowing waters as derived from native information.

It was greatly to be regretted at the time that so important a matter as a communication between two such rivers as the Yellow River and the Yangtsze could not be investigated at once ; but, as it was then impossible, I resolved to take the earliest opportunity of revisiting the neighbourhood and making a thorough examination, or only a cursory inspection, according as circumstances would permit. Of those two alternatives it was my misfortune, as I thought at starting, to be obliged to accept the latter, my time and resources admitting of no more. In a word, however, there was nothing to examine; for, before I had gone half-way on. my road, I received information that the breach had been closed after several months' labour, and that no outflow from the Yellow River then existed at all, and, as will be seen below, I was shortly afterwards able to confirm this information personally, Thus the immediate object of the journey was to obtain information respecting the communication · between the two great rivers, and so far it was a blank; yet there was to me, though possibly not to geographers generally, a matter of far greater interest to be investigated, although one which I well knew on starting I had not the opportunity of doing justice to. I allude to the examination of the old bed and embankments, with a view to forming an opinion on the subject of turning the Yellow waters, or part of them, back into their former channel—a work the importance of which in its political and economical aspects can scarcely be over-estimated, and one which the very Chinese themselves, in deploring the degraded state of their country, are in the habit of declaring to be one of the first steps necessary for it to recover its former condition of prosperity and order. A satisfactory and final report on this question could, of course, only be drawn up after an accurate survey of the whole of the old bed, and after a year's observations on the annual rise and fall of water above the breach, and of the tides at the old mouth ; but this it was impossible for me to undertake, and I have thus had to be content with collecting what crude materials for forming a mere opinion on the subject the opportunities of a rapid journey afforded me. Moreover, there was no middle choice: a rough route-sketch of the bed, or even a traverse-survey and imperfect observations on the yearly range of water, would have been utterly valueless when applied to the solution of such a problem.

But a few words will suffice to describe the journey and places visited on the way, as these have but little interest in themselves. My route lay up the Grand Canal from Chin-kiang by boat to Tsing-kiang-pu, and from there, as nearly as could be managed, directly along the old bed of the river by road to the neighbourhood of Sü-chow-foo. Almost immediately after leaving the vicinity of the Grand Canal and the Hungtsze Lake, and until approaching the latter city, the country passed over is flat, poor, and sandy, and though cultivated, perhaps, to the utmost—for the population is comparatively large, yet it bears on every side an air of want and poverty directly traceable to the absence of irrigation. Near Sü-chow-foo occur a group of low limestone hills and a small creek or two, which somewhat change the scene, though the condition of the country thereabouts is but little, if at all, superior to that between it and the Hungtsze. The city and suburbs form a large but ill-built town, and, considering that it is almost entirely cut off from water-communication, it was rather surprising to me to find a thriving and business-like place. It will not, of course, compare with the larger trading towns in the south of the province, but, taking its situation into account, it must be reckoned a place of importance. It was here that I was first given to understand that the southern or upper breach in the Yellow River had been closed, and this information was further substantiated, a day or two later at Tang-shan-hien, by my falling in with a person who had been actually engaged upon the works. It appeared that they had been undertaken after the subsidence of last summer's (1869) flood, and had been finished about the middle of February, 1870; that the Sha, into which the Yellow River water had been flowing, was now in its normal condition; and that the through navigation which had sprung up was, of course, entirely stopped. The information was undoubted, but, having started with the object of investigating the matter, I determined to satisfy myself by visiting the spot.

On leaving Tang-shan, I was unable to continue along the line of the old bed of the river, but had to proceed by the main road through Kuei-te-foo, Sü-chow, &c., postponing the further inspection of the river-bed until the return journey.

The aspect of the country continues in a great measure the same as that previously passed through—flat, sandy, and sterile. The products are almost exclusively wheat and poppy, both, I believe, of a very inferior description. The cultivation of the latter, however, is said to be greatly on the increase; and as each year larger tracts of the poor, sandy soil have to be devoted to it, so less and less remains available for the food-producing crops,—an evil which the authorities seek to check by means of squeezes and otherwise, but in which, to judge by native report, they are singularly unsuccessful. Opium manufactured from this poppy is but little esteemed by the native smokers, and though sold at a very low cost is scarcely ever used by even the poorest classes without some admixture of the foreign imported drug.

Neither Kuei-te-foo nor Sü-chow are places of any importance, for, though comparatively populous, their isolated inland position precludes their inhabitants from carrying on any but a mere local trade. In both places—and, indeed, in all towns of this district—the energies and wealth of the people appear to have been chiefly spent upon the city walls and oth^{er} defensive works, having for their object the exclusion of Nien-fei, or bands of starving country people, and of Imperial braves, and where this is the case it is unreasonable to look for a large commerce or thriving populations.

A few days' journey from Sü-chow brought me to the town of Chu-sien-chên, on the left bank of the Sha, or rather the Kiu-lu, as it is here called, formerly the chief mart of the province, but now a half-deserted and dilapidated country town. The Kiu-lu is a small stream, in summer probably about 40 yards broad and 8 or 10 feet deep at this point, and in the winter almost entirely dry. For the year and a half during which the breach in the Yellow River bank was open the inhabitants testify to large Hungtsze Lake boats being navigated through into the Yellow River, and during the whole of the winter of 1868-69 the through communication was complete. At the time of my visit, however, it had been restored to its normal state, and on the 6th April I had the satisfaction of riding across its dry bed and thus finally confirming the information already received regarding the closing of the *upper breach.*

From Chu-sien-chên I proceeded to the capital of the province, Kai-fung-foo, a larger and more important place than any yet passed through; but as it has been visited several times of late years by foreigners proceeding from Peking or Hankow, and as notices of these visits have been made public, I will not go into any further description of it here. The city is situated about 30 li south of the Yellow River, and the country surrounding it may be almost termed a desert: so much so that it is nearly entirely destitute of those busy suburbs which are generally to be found outside the walls of large cities in China, and where, as a rule, the chief part of the trade is carried on. Within the walls the aspect is entirely northern: broad, unmade roads, a foot or more deep in mud or sand, through which bullock-waggons and mule-carts are dragged with the greatest difficulty and noise; low mud-built or mud-plastered houses and hongs, and even yamêns; and, above all, the quaint signboards of the Mahomedans conspicuously paraded in every direction. There is direct trade with Tientsin and Hankow, the former of by far the greater importance, about nine-tenths of the distance being accomplished by water, viz., up the Grand Canal and Wei-ho as far as Wei-kue-foo (We-kyun) in Honan, and about 180 li to the northward of Kai-fung. It is by this route that foreign imports reach the city, and, though an up-stream voyage, yet it is the easiest and least costly mode of performing the journey;

also the barrier taxation on the way is considerably lighter than by the other route—so light, indeed, as to be almost inappreciable. A large proportion of the inhabitants are Mahomedans, including many of the most influential citizens; they are mostly avowed rebels, and make no concealment of their sympathy with the Mahomedan cause in the western provinces; indeed, at the time of my visit the greatest satisfaction was being openly expressed at the defeat of the Imperial troops near Tung-quan, in Shensi, and close to the border of Honan, the news of which had only just been received. Any further decided advance towards the east, I was frequently assured, would lead to the rising of all the Mussulman communities in those parts of the northern provinces where a junction with the rebel army would be practicable; and though this army is, in all probability, but roughly organised and poorly armed, yet it possesses an element which renders it most formidable, namely, the religious fanaticism of the Moslem—an element in warfare which its half-hearted Chinese and Manchu enemies are but little accustomed to deal with, and by nature but ill able to withstand. To this also must be added the prestige of five years' successful fighting; for I take it that this uprising is but a continuation of' that movement which finally drove the Chinese from Khoten and Yarkand and lost them the whole *cordon* of garrisons and fortresses situated along the great line of road between Kashgar and Hami, and with these their last hold on the countries of Central Asia.

On leaving Kai-fung-foo, one day's march in a north-easterly direction brought me again to the Yellow River, at a small village on the south bank nearly opposite to Lung-mên-kau, or the point of divergence between the old and new beds; and from here the old bed was followed, as nearly as was practicable, down to Tang-shan, where I last left it, and from there again down to my starting-point at Tsing-kiang-pu, on the Grand Canal.

In the report of my former visit to the breach of Lung-mên-kau, I described shortly how the channel and free-sides within the embankments were at a higher level than the plain outside; and how that it was evident the deposits of successive floods on the free-sides had so diminished the capacity of the so-called bed that the inner side of the embankments had come to measure but a few * feet in height, and were totally inadequate to

---

* I mentioned 20 or 25 feet as the approximate height of the inner slope whilst the outer one was about 40 feet. On my late journey, however, I found the disproportion to be generally much greater—say 40 to 15 or even more in some places. Only near impinging points it is much less, and the neighbourhood of Lung-mên-kau, seen in 1865, is one of these points.

pen up the immense volume of water brought down by each summer's flood. The few miles then seen formed a true specimen of the whole of the old bed as far down as the Grand Canal, with the exception of a short section near Sü-chow-foo, where it runs through the group of hills before-mentioned, and where its general characteristics are somewhat different. The main, or low-water, channel is a winding line, generally of easy curves, impinging sometimes on one embankment and sometimes on the other, the channel being, of course, deeper at the impinging points, and the difference of level between the inside and outside of the embankments nearly disappearing at these points. The slope of the free-sides, as far as one can judge by the eye, is *from* the embankments *towards* the main channel; and it is this slope, I am inclined to believe, that has cost the Chinese enormous trouble and expense for hundreds of years, and has eventually landed them in the lamentable state of things now existing, for it is *from* the channel *towards* the embankments that the free-sides should slope, thus forming a deeper and more rigidly scoured channel for the main river on the one hand, and a natural reservoir of lagoons and ponds upon the free-sides on the other. It is evident that the advantages of this arrangement would not be confined to the river itself, but would extend to the country outside the embankments, inasmuch as the reservoirs on the free-sides might be made to supply water for artificial irrigation for a great portion of, if not throughout the whole of, the year, without in any way impairing the main channel by drawing off even the smallest quantity of its water. That the main channel should be kept intact is necessary for its own preservation; for, although it might be of temporary advantage to navigation to draw off sufficient water from it to reduce the current to a more convenient velocity while not reducing the volume of the stream, yet I venture to think that to do so would be a most fatal mistake, and one which would entail enormous expense in the end for dredging and embanking. It is true that increased facilities to navigation might more than counterbalance this expense; but the question is whether, in the peculiar state in which the country now finds itself, a channel with as swift a current as can be made to flow through it would not be the greater boon; for, the swifter the current, the greater the scouring power, and, consequently also, the more permanent the channel; whilst on the other hand, the slower the current, the greater the depositing power and the greater the danger to its permanent efficiency; and, moreover, in this exceptional case of a poor and over-populated country, a difficult navigation and one requiring a large amount of manual power, though of itself expensive, might finally prove of more real value than one attended with greater ease and economy.

Some movement in the matter of turning, or at all events regulating the Yellow River, has already taken place among the Chinese both at Peking and in the provinces, and it is stated that the plan which finds the most favour is that proposed by the Kai-fung authorities, who recommend a division of the river into two, or in other words, to cause a certain proportion of the water to drain off down the old bed, after deepening the main channel therein, whilst allowing the breach to remain open and the new river to continue flowing.

The disadvantages of this plan are very great.

In the first place, one of the principal objects sought is to redeem from its present inundated condition the country through which the new river flows between the breach and the Tatsing, and it is conceived by those who advocate this plan that, by dividing the river, a stream of small and manageable volume will result in each case, whilst each line of country will be provided with a navigable channel. This scheme, on the face of it, is plausible enough; but there is a fallacy underlying it which, by calling to mind the well-known experiments of Gennete at Leyden, can be easily demonstrated. This celebrated engineer showed, amongst other things, *that to divide or "derive from" a river, was to diminish its velocity, but not its volume;* and the correctness of his precepts has since been frequently proved, especially in Holland and North Italy. His law applies directly to the case of the Yellow River; and if further it be admitted, as above, that a diminished velocity implies an increase of deposit, we have here the double disadvantage of finding that the belt of country at present inundated would remain of the same extent, whilst the prospect of obtaining a permanent and efficient channel would be more remote than ever. In the second place, it would be obviously a waste of resources to build embankments for the 147 miles of new river between the breach and the Tatsing, when those already existing along the course of the old river would also have to be kept in repair; and lastly, the regulation of that one diverging point would in itself prove to be an engineering problem of great magnitude—a circumstance which neither the Chinese nor foreign supporters of the plan for dividing the river appear to have sufficiently weighed.

A second plan that has been proposed, and one of considerable reason, is to utilise the new course exclusively, to consolidate its bed, and to avail of past experience in making it a permanency.

The advantages claimed for this plan are somewhat as follows :—

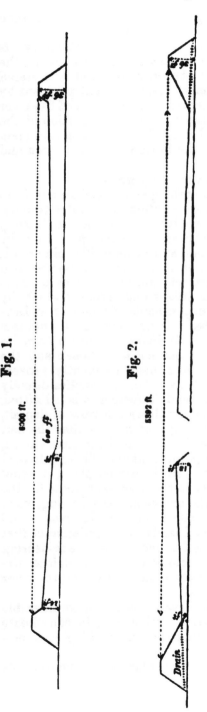

Fig. 1.

6500 ft.

Fig. 2.

5592 ft.

Drain.

Fig. 1 is an example of a section of the old bed at a point estimated at about 6500 feet broad, and where the embankments were taken to be 36 feet in vertical height above the plain and 18 feet above the free-side of the river-bed. The dotted depression represents the low-water or main channel.

Fig. 2 represents the same point with the slope of the free-sides reversed, and both them and the main channel excavated, as proposed in the third plan mentioned in the accompanying paper. The former are shown as excavated to within 6 feet of the level of the plain (thus providing for a fall for irrigation-drains leading through the embankments); and the latter is left open, it being impossible to recommend any exact depth until after an accurate survey.

The Vertical Scale throughout is .. .. 144 feet to 1 inch.
The horizontal ditto for Embankments .. .. .. 144 " " " "
Ditto ditto Bed within Embankments .. .. 1290 " " " "

Firstly, that it would be only necessary to raise and maintain some 147 miles of embankments, viz., from the breach to the Tatsing at Yü-shan, whence the river flows in a natural bed to the sea, and that the alluvium already deposited along the line would supply a great portion of the material necessary for the work.

Secondly, that the Tatsing being already navigable to near Tsi-ho, and easily rendered so as far as Yü-shan, that a more useful river would be gained, and one that at some future time might be availed of by steamers for purposes of foreign trade.

This second consideration would, in a civilised country, be one of great weight; but, as I have attempted to show above, an *easy* navigation is in this country a matter of no moment—sufficient water to float a junk, drawing 4 feet or so, throughout the year being all that is required, and as the introduction of steamers is an event not to be looked for, it would be useless to commence costly works in order to provide for them. A far greater benefit than *easy* navigation is derived by the bulk of the people in any district from the irrigation afforded by the rivers that traverse it; and in this case the district between the breach and Tatsing was a well watered and fertile one before the advent of the Yellow River, and had navigable canals running through it, whereas that district through which the old bed runs is naturally a poor and sterile one. Should the Yellow River be diverted from its present course, the navigable canals which existed there will certainly be found to have suffered severely from the current and deposits of the river, but the all important matter of irrigation would remain on the whole unchanged.

As to the first advantage claimed for this plan, I conceive that it falls to the ground with the second, though even if taken on its own merits it would meet with the same fate; for the embankments along the old bed as far as I have seen them are still in good repair, and though some 400 miles in length, they would have to be maintained only, and no new ones raised unless possibly for a short distance near the mouth, the only part which I have not visited—but even there, to judge by native report, it would not be necessary—and again whatever material were required for repairing could be most conveniently and advantageously excavated from the free-sides or the main channel.

There remains then but the one plan of turning the entire river back into its old course, and a little consideration will show that this offers the greatest and most numerous advantages from every practical point of view.

Firstly, a channel sufficiently navigable for the wants of the Chinese and their commerce would be obtained.

Secondly, irrigation would be supplied to the extensive district at present in a state of poverty and disorganisation for the want of it.

Thirdly, another large district, as large in square mileage as the counties of Kent, Surrey, and Middlesex together, would be delivered from an inundation which has caused the greatest distress and ruin, and which when without the Yellow River is still well watered and traversed by navigable channels.

Fourthly, that by restoring the balance of water-supply in these two districts or lines of country, the condition of a large population would be greatly improved and the labour of thousands, who at present have no employment and can barely earn an existence, would be in demand, not only for the works while in progress, but also for the increased agriculture and industry which would follow on their completion. By this means too the Nien-fei* would be most effectually rooted out and their power utilised.

This last plan, thus roughly sketched and shortly summed up, is, I venture to think, the only one likely to produce results of enough importance to induce the Chinese to make the effort necessary for carrying out a great national work, and one which would only repay them in an indirect manner. Unlike enterprises of a novel character introduced by foreigners, such as railways and telegraphs, which no amount of reasoning or "friendly pressure" will ever induce them to adopt, a work such as regulating the Yellow River, which in one form or another they have been used to for hundreds of years, and the necessity for which at present is acknowledged on all sides, appeals to them in the strongest manner, and would require, I

---

* The Nien-fei insurgents, in spite of much that has been said and written to the contrary, have I believe, never had, either in the last or any previous uprising, any political element in their movement until inspired by a series of successes in the neighbourhood of the capital, as in 1868. They have always been the inhabitants of the north of Kiangsu, Anhnei and Honan, and of southern Shantung, districts through which the Yellow River formerly flowed and which have been left dry and impoverished since the change of course; they have been composed chiefly of the agricultural classes, who, not finding a means of livelihood in their ordinary calling, have, to save themselves from starving, banded together for the purpose of attacking villages or towns where the means of subsistence might be obtained by force. These bands have been most miserably armed and organised, and I have been crediblyinformed by an Englishman serving with the Imperial troops operating against them, that in the most successful band—that which overran eastern Shantung in 1868—there was scarcely one in a thousand who possessed a firearm of any description. On my late journey I passed through the heart of the so-called " Nien-fei country," but as the only characteristic of both people and country was extreme poverty, it was evident that no such formidably organised movement as the " Nien-fei rebellion " has been represented, could have emanated from such a region. Want of irrigation in a land where frequently no rain falls for several months at a stretch, and the consequent lack of employment for the inhabitants, is alone, in my opinion, sufficient to account for a rising of half-starved agriculturists in such a country as China, and it is in this way that it is looked upon by those Chinese who have the most intimate knowledge of the subject.

imagine, but little urging at Peking to put into train, provided only the right strings were pulled in those provinces having the greatest interest at stake.

In the above sketch I have taken care to leave out all mention of the important matter of cost, thinking it less misleading to do so than to attempt to give vague estimates, founded as they must necessarily be upon the most slender native information obtained on rapid journeys. Should any *boná fide* movement be made towards the commencement of works, a certain amount of information under this head could be readily gained even without going into exact estimates, which last could only be done after an accurate survey.

That English engineers be employed and the whole works be placed under the superintendence of Englishmen for organisation, would seem to be an absolute necessity if the result is to be successful; for over and above all the ordinary corrupt influences at work amongst Chinese tending to the non-success of any enterprise they undertake, and also to the enhancement of its cost, it is well known that no more productive " pagoda tree " ever existed in China than the Yellow River, and the commencement of fresh works upon it under Chinese superintendence would be a signal to the whole tribe of mandarins in the empire to regulate things so that as many as possible might have an opportunity of " taking a pull." No less then for the efficiency of the works, than for saving expense to the country, should the whole enterprise be planned and carried out by Englishmen in the pay of either the central government or the provincial authorities.

NOTE 1.—The accompanying chart * of the Yellow River has been constructed on an ordinary traverse survey, checked at several points by observations for latitude and longitude by chronometer; every course being roughly protracted, and the soundings and every other particular sketched in on the spot, as well as noted in their respective columns in the log-book. Since returning to Shanghai, the whole has been accurately re-protracted and corrected for the observations.

It is divided into two sections, for two reasons; firstly, because the characteristic features of that portion flowing in the bed of the Tatsing, and those of that portion at present without a permanent bed, differ so essentially; and secondly, because the soundings on the lower section having been taken a month earlier than those on the upper, I have thought it better to give them as first recorded rather than attempt to reduce them for

* Mr. Elias's original charts (from which the accompanying map is engraved) are deposited in the Map-Room of the Society.—[ED.]

the fall of water which had occurred during the month, and the amount of which I had no means of ascertaining exactly. To have shown the whole under one date would have had some advantages, but it could only have been done at the risk of inaccuracy.

Latitudes are from stars north and south of the zenith, except in the case of the single latitude 35° 25' on the second section, which is from an ex-meridian altitude of the sun.

Longitude. I carried two chronometers not with a view of combining their results, but in order to have two to choose from after having satisfied myself of the value of their respective performances; and this I was enabled to do by rating before leaving Shanghai, after returning to it, and by finding a temporary rate several times, whilst on the journey from the difference of longitude shown at certain points passed two or more times at certain intervals. In this way one of the instruments (by Barraud) was found to have gone with such an extremely small and steady rate that I have not only adopted its results in preference to those of the other (by Hutton), but have based the absolute longitude of the chart on them; for although I determined one point on the river by observations of the moon for absolute longitude, I have not utilised these, thinking it unlikely that a position would be more decisively fixed by such means than one deduced from a chronometer whose performance appears to have been so satisfactory. Should a correction at that one position become necessary, the same will apply to all the longitudes throughout.

Variation was marked at three points, but as at each one I only had the opportunity of taking one set of sights, I am not inclined to look upon these as possessing any great accuracy. I have marked them all, however, as they were taken rather than omit any—thinking it the fairest way.

Breadth of the River.—On account of the narrowness of the river on the lower section, I found it impossible to show the soundings, sandbanks, &c., with the breadth laid down to scale, and I have therefore been obliged to draw it on double its natural breadth, at the same time giving the number of yards at intervals along the bank. These last were determined in some instances by sextant measurement, in some by rifle shot, and in others by estimation.

Current.—For the measurement of the velocity of the current a Walker's patent harpoon log was taken, but although tried on many occasions and under different circumstances, it gave no result—a five-knot current had no effect upon it, and it must therefore be looked upon as unsuited to such work.

Height of hills, &c.—Some of those marked on the first section

are from rough measurements with an aneroid barometer, and the others are estimated by comparison with those measured. A boiling-point apparatus which Mr. Hollingworth carried proved useless in determining the elevation of the river above the sea level; but considering that the whole course of the river is through an alluvial plain, whose slope must be very gradual, this elevation, even at the uppermost point visited, cannot but be very slight.

II.—*Journey from Leh to Yarkand and Kashgar, and Exploration of the Sources of the Yarkand River.* By G. W. HAYWARD.

Read, December 13, 1869.

1. JOURNEY FROM LEH TO SHADULA, AND EXPLORATION OF THE KARAKASH RIVER.

October and November, 1868.

I ARRIVED at Leh, the capital of Ladak, on the 21st September, 1868, having left Murree, in the Punjab, on the 26th August. The distance is 390 miles, and I made double marches every day, being anxious to get off for Turkistan before the snow fell and made the passes difficult.

From Leh to Yarkand there are three routes open to the traveller to choose. The first is the Zamistânee, or winter route, which from Leh crosses the Digur Lá Pass and ascends the valley of the Shayok River to near the Karakoram Range; the second, the Tabistânee, or summer route from Leh, crosses the Kardong Pass, 17,574 feet above the sea, and the Shayok River at Suttee, from where, ascending the Nubra valley, it crosses the Karawal Pass, and then the difficult Pass of Sasser, 17,972 feet above the sea, joining the former route at Moorghoo. From here they continue together across the Karakoram Pass, 18,317 feet above sea-level, to Aktâgh, where they separate; the Zamistânee route conducting down the valley of the Yarkand River and across the Yangi Pass to Kugiar, Karghalik, and Yarkand.

The Tabistânee route from Aktâgh crosses the Aktâgh Range by the Sooget Pass, 18,237 feet above the sea, and, following the course of the Sooget stream, joins the valley of the Karakash River 4 miles above Shadula. The Kilián range of mountains has then to be crossed by either the Kullik, Kilián, or Sanju passes, which are all very difficult and impracticable for laden horses, whose loads have to be carried on yaks (tame buffaloes), and all of these routes join at Karghalik, 36 miles from Yarkand.

The third route from Leh is *viá* Chang Chenmo and the Chang Lang Pass, 18,839 feet above the sea, and across the series of high plains lying between Chang Chenmo and the Kuen Luen Range, below which it enters the valley of the Karakash River, and, conducting down that valley, joins the Tabistânee route at Shadula.

With reference to these routes, it may be stated that the Tabistânee route is the one now most frequently traversed by the merchants, although the passes on the Ladak side of the Karakoram are very severe. Still grass and fuel are everywhere obtainable on this road, with the exception of some 50 miles from below the Karakoram to Sooget. The Chang Chenmo route possesses the advantage of having only one pass to cross after reaching the Chang Chenmo valley, and this, the Chang Lang (18,839 feet, as above stated), is a very easy pass, the ascent being only on the south side. Beyond this pass the road traverses some elevated plains, called Lingzi Thung, where no grass and but little fuel and water is to be met with for seven to eight days' journey to the Karakash River, below the Kuen Luen Range. Down the valley of this river grass and fuel are everywhere obtainable. As far as the road is concerned, this route is excellent; but the fact of there being little fuel and grass at such high elevations renders it a severe undertaking for the caravans of the merchants, and there seems at present but little chance of its becoming the main trade-route. The Yarkand merchants appear to be somewhat loth to strike out a new line of traffic in this direction, since, notwithstanding a reward has been offered in Ladak to any trader who will proceed by this route, they all seem to prefer the old road *viá* the Karakoram Pass.

The distance from Leh to Yarkand, by the Zamistânee route, is 530 miles; by the Tabistânee, some 480 miles; while by the Chang Chenmo route it is 507 miles; Shadula being distant from Leh by this latter route 316 miles, from where it is 191 miles to Yarkand.

The great desideratum to insure an increasing traffic with Central Asia is the opening out of a shorter and easier trade-route, leading direct from the north-west provinces of India to Yarkand. A good road, avoiding both Kashmir and Ladak, would offer greater facilities to the Yarkand traders for reaching India direct, and have the desired effect of insuring an easier transit, as well as doing away with the difficulties, both political and geographical, which attach to the old Karakoram route.

The desirability of such an event was so evident that the ascertaining if such a route existed was one of the main objects kept in view by the present expedition.

I remained a week in Leh, making the necessary arrangements to proceed to Yarkand. Ponies and horses to carry the baggage and supplies for the road had to be purchased. By far the best animals for the purpose are the Turkistan ponies, as the Ladak ones are small, and hardly fit for the severe work attending so long a journey. The Yarkand ponies, however, can carry from 200 lbs. to 250 lbs. across most of the passes met with on the way, and there is generally a large demand for them in Leh. Their price varies according to the demand and their capability of carrying weight. A good pony should be bought in Leh for 80 to 120 rupees (8l. to 12l.), while the best ones, used for riding purposes, fetch higher prices, many of them selling readily for 200 to 300 rupees (20l. to 30l.) Several Afghan horsedealers make a yearly trip into Turkistan, and return with horses and mules for sale. Some of these mules are exceedingly fine; and were any inducement offered to dealers to import a larger quantity for the Palampore fair, which is largely frequented by merchants from Yarkand and all parts of Northern India, a certain mart would be established for mules suitable for transport-trains and mountain-batteries.

Here, at Leh, I engaged two men to look after the horses, one of whom had been to Yarkand before and was competent to act as a guide on the road. Warm clothing, lined with sheepskin, had to be obtained for self and servants, as the cold to be met with would be intense, also " pubboos," or Ladak snowshoes, and supplies for the journey, carried on hired yaks.

Intending to proceed by the Chang Chenmo route, we got off from Leh on the evening of the 29th September, and made a short march to Tiksee, a village situated in the valley of the Indus. At 20 miles from Leh the road leaves the Indus valley and turns to the north, up a ravine, to the village of Sakti, where yaks can be obtained for crossing the pass into Tanksee, There are two passes across this range of the Himalaya, known as the Kilás Range, one of which leads across the Durgoh Pass to the village of Durgoh, some miles below Tanksee, while the other, the Chang Lá, is more to the eastward, the road across which leads direct into Tanksee. The Chang Lá Pass is 18,368 feet above the sea, and, though not impracticable for laden horses, it is severe enough to render its passage preferable with the aid of yaks, which can always be obtained at the villages below the pass.

. The descent on the north side is at first steep, and the road conducts down a ravine to Seefrah, a Bhoot encampment in the valley. Tanksee, 12 miles distant, is reached early the next day. This village, which is 49 miles from Leh, is the last place in the Maharajah of Kashmir's territory where any supplies can be obtained. With the exception of a few stone huts near Chang Chenmo no habitation is met with, nor can any supplies be obtained until reaching Turkistan, 400 miles away. A delay of two days here was, therefore, unavoidable, in order to make the final arrangements for the long journey before us. Four Bhoots, or Ladak villagers, were engaged to accompany me to the borders of Turkistan, and their yaks laden with grain for the horses.

Chang Chenmo is a district lying about 50 miles north-east from Tanksee, from which it is reached in three or four days' journey.

Chang Thang, with Rudok, lie more to the eastward. The greater part of the Chang Thang district contains salt-mines, from which the whole of Ladak and part of Tibet are supplied with salt, while a large traffic is also carried on with Kashmir. The salt is brought down from the mountains on sheep, which are extensively used throughout Ladak and Tibet for carrying light loads. I met a flock of several hundred coming down the Chang Lá Pass, all laden with salt, placed in small bags across the back, the average weight which one sheep will carry being about 30 lbs. The wool of these sheep is considered to be excellent, and is in great demand at Leh for transportation to Kashmir; besides which, the valuable wool of the shawl-goat, abounding in Chang Thang, is the main article of traffic sent to Kashmir.

Leaving Tanksee on the 5th October, we proceeded towards Chang Chenmo, marching that day to Lukong, a place consisting of a few stone huts, situated at the head of the Pangong Lake. Already thus early we had warnings of the inclemency to be expected from the lateness of the season, for a snow-storm came on towards evening, during which we wandered from the track, and, not at once regaining it, did not reach Lukong until late at night. Between here and Chang Chenmo another high pass, the Masimik, has to be crossed. It is nearly 18,500 feet above the sea, and is generally covered with snow. Although at such an elevation, it is a very easy pass, but laden horses suffer somewhat when crossing it from the rarefaction of the atmosphere. The road to the pass is gradually on the ascent for 6 to 7 miles, and the mountains on either side the valley were this day quite covered with snow. Crossing the pass, we encamped that night at Rindee, 2 miles below, at an elevation of 17,200 feet. The cold was severe, the thermometer at 7 A.M. the next day marking 3$\frac{1}{4}$° Fahrenheit.

At 14 miles below the pass the Chang Chenmo Valley is struck at Pumsul. The whole country is very barren, and the mountain ranges quite bare, save of snow and glacier. In a few places in the ravines and valleys a little grass alone is met with. Wood, too, is very scarce, the only fuel obtainable being " boorsee," a short thick shrub, with ligneous roots. This is found on the lower slopes of the mountains, at some distance from the main valley, otherwise it would be impossible to exist for any length of time in these elevated regions. All the streams were dried up at this time of the year, with the exception of the main river, which was a mere brook. When, however, the snows are melting during the summer months this stream attains to a considerable size. It receives the waters from the ranges on either side the Kugrang Valley, at the head of which it has its source, and, flowing with a course nearly due west through the Chang Chenmo Valley, joins the Shayok River near Lauxakeuti.

Chang Chenmo is now well known, being visited every year by at least half-a-dozen officers on long leave to Kashmir. The game to be found is the wild yak, ovis ammon, burrell, Tibet antelope, and *wild horses*. The wild yak is met with, more or less, all along the high table-land of Tibet up to the borders of Turkistan. Eastward, they range the high country near the sources of the Indus and Sutlej rivers, and are there, with the ovis ammon, found in much greater numbers than in Chang Chenmo. They live at the higher elevations, being generally seen right up in the snow, and only descend to feed on the scant grass found in the ravines and valleys. Water does not appear to be a necessity to them, as they eat snow. It is not known whether the wild yak is found on the Pamir Steppe, although the Kirghiz who frequent that range possess large herds of the tame species. I have never seen any traces of them further west than in the valleys near the head-waters of the Yarkand River, and the 78th meridian of longitude may be fixed as the limit of the range of the Tibet species in this direction.

Having marched from Pumsul, we ascended the Kugrang Valley, with the intention of crossing the range at its head and following the stream rising there, which is represented on our maps as the Yarkand River, down to Aktâgh. The pass at the head of the valley was found to be a very practicable one, but no feasible route into the valley of the supposed Yarkand River was discernible. We, therefore, returned down the Kugrang Valley, losing a horse from cold and inflammation on the way, and camped at Gogra for a few days previous to going on up the Chang Lang Valley on our way to Turkistan. More supplies were got up from Tankaee, and farewell letters written to England, as all communication between civilisation and the wilds of Central Asia was about to be severed.

We left Gogra on the 25th October, making a march to some hot springs in the Chang Lang Valley. These springs are at an elevation of more than 16,000 feet above sea-level, and gush out from orifices in the summit of these rocks, situate in the bed of the stream which flows through the valley. The whole ground is white with incrusted saltpetre, while a fantastic pile of earth indicates the position of an old spring now extinct.

The mountains round the valley are chiefly volcanic, and many rocks have been formed by the continuous action and accumulation of springs—the spar from these accumulations presenting features of various and beautiful texture. The prevailing formation of the mountain-ranges of the Kugrang Valley is basalt and greenstone, with schists, while immense landslips of shale and *débris* fill the ravines and transverse-valleys. The features of the Chang Lang Valley are strata of argillaceous shale, more or less thick, reminding one of thick roofing-slates, as seen in English quarries, while layers of schists, much laminated, are seen low down, where the ground has been worn away by the action of the water. These strata are vertical or oblique, and in many places reversed or turned over. Beds of conglomerate, alternating with layers of finer gravel, fill up the valley between the stream and the foot of the mountains, while. volcanic rocks are scattered about in the forms of boulders and irregular fragments, chiefly composed of granite, on the higher ridges, in the landslips and ravines.

I endeavoured to ascertain the temperature of the warmest spring here at Hot Springs; but a thermometer would not steady itself on account of the force of the stream, but it indicated a heat of upwards of 140° Fahr.

From Hot Springs to the Chang Lang Pass the road is up the bed of the stream, which, frozen over, had to be crossed several times. The ascent of the pass lies up a ravine, filled with loose stone and *débris*, and is very gradual and easy to within 500 feet of the summit. This pass, which is at an elevation of 18,839 feet above the sea, is generally known as the Chang Chenmo one, and is said to be the easiest of all the passes leading across the Karakoram and Hindu Kush ranges. It is quite practicable for laden horses and camels, and would offer no great impediment to the passage of artillery; indeed the ground is so favourable, that a little labour expended on the construction of a road up the Chang Chenmo Valley to the pass would render it practicable for two-wheeled carts and conveyances. Geographically, the pass is remarkable as being across the main range of the Karakoram, forming the watershed between the Indus and the Turkistan rivers, and constituting the natural boundary of the Maharajah of Kashmir's dominions to the north.

In travelling to the pass it might be preferable not to halt at Hot Springs, but to march from Gogra to near the pass. There is a good encamping-ground in the valley, some 5 miles above Hot Springs, where grass and fuel are met with, and from here the pass can be crossed on the following day. From here to the Karakash River below the Kuen Luen Range is some 120 miles, and the route lies across the series of high plains to the north of Chang Chenmo, where little or no grass is obtainable on the way, whilst water is very scarce, especially late in the year. A little "boorsee" is here and there met with; but it is very small and hardly available as fuel for cooking purposes. There is no regular road as yet, and the mere track of a few merchants and travellers who have ever gone this route is easily missed. The road from the pass is level and good down the open valley between the low hills to Nischu, where we camped without finding grass or fuel, or even water, so late in the year. The cold was intense, the thermometer at 7 A.M. marking 11° below zero. I found it most difficult to keep anything liquid without being lost; everything froze at once and burst the bottles. Trying to paint in water-colours was out of the question; water, brushes, and colours all froze together, and the enamel on the tin paint-boxes cracked from the intense cold. The country beyond the pass until the Lingzi Thung Plains are reached consists of low hills and broken ridges of a sand and clay formation. It is evidently covered with snow during the winter, since the surface of the ground shows signs of the action of running water from the melting of the snows.

Some 16 miles from the pass is the descent to the Lingzi Thung Plains, which are nearly 17,000 feet above sea-level, and extend for 40 to 50 miles from north-west to south-east. Their breadth is some 25 miles, being bounded on the south by the Karakoram Chain, and on the north by a somewhat irregular and lower range, called the Lak Tsung Mountains. They are covered with snow during the winter, and in the summer many lakes and pools of water must be formed by its melting. At this time of year, however, not a drop of water was to be found, all the pools having dried up or infiltrated into the sand. The plains are of a gravel and clay formation, with tracts of sand in the slight depressions of the general surface, and are covered with small pebbles and stones, mostly of angular form, and composed of limestone, flint, &c. Near the foot of the range bordering it on the south, where the slopes of the mountains subside into the plains, the clay and alluvium formation prevails, and appears to be of considerable depth. The country here is broken up into water-courses and nullahs, which, when rendered soft and slippery by the melting of the snow, much impede the passage of travellers.

The wind blowing across these elevated plains was intensely cold; and directly after leaving the low hills the full force of it was felt—my servants complained most bitterly, and seemed to be quite incapable of doing anything. The weather was generally fine, with a clear sky, during the months of October and November; but the wind which came on to blow daily from noon until sunset was most intensely cold. The only way to cross these inhospitable regions in any comfort is to bring wood and water from the Chang Lang Valley; and this we failed to do, as the Bhoots, with the usual obtuseness of Ladak villagers, never mentioned the total absence of these requisites until after we had crossed the pass.

About half way across the plains we passed a dried-up pool, by which were the carcases of a horse and yak; and further off, wrapt up as if asleep, the dead body of a poor Yarkandi. I afterwards ascertained that this poor fellow was one of the party who had accompanied Mahomed Nuzzur, the Yarkand envoy, back to Turkistan, and remaining behind from illness, had missed the road, and thus died of cold and starvation. I wished to bury the body, but none of the men would approach it; and besides we had no implements.

We lost another horse this day, and again no fuel or grass was met with. For water we had to melt snow, which was found in the nullahs and hollows. T e e is a remarkable round peak in the Lak Tsung Range, whicb is a good fixed point to march by in crossing the plains, the road going down a broad sandy valley through the range to the east of this peak.

Beyond the Lak Tsung Mountains is a second series of plains, with low ranges running through them, extending up to the spurs of the Kuen Luen Range. They are very similar to the Lingzi Thung, but some 1000 feet lower. Late in the evening of the day we entered them; we arrived at Thaldat, where is a frozen lake and spring. The water here was very brackish, but the animals drank it eagerly, being the first they had had for four days. There was no grass, however, at Thaldat; but the day we left the place some was discovered in a ravine lying west of it, about a mile away.

As I had failed to find a pass from the head of the Kugrang Valley into the valley supposed to contain the head source of the Yarkand River, I determined to attempt a route across the mountains from Thaldat, though, from the probable absence of grass and water, it was a somewhat hazardous undertaking for our animals so late in the year. At Lome, 50 to 60 miles distance direct north, I knew we should strike the valley of the Karakash River below the Kuen Luen Range, and the route which I intended to explore might lead us anywhere. When I gave orders to strike camp and prepare to march, the Bhoots and my own servants were anything but pleased at going off to explore a new route. I had this morning ascended the ridge lying west of Thaldat, and obtained a good view of the country around. Looking north was seen the sunny range of the Kuen Luen, with its highest peaks glistening in the morning sunlight, while eastward stretched the wide expanse of desert, known as the Aksai Chin. In many places the appearance of a mirage indicated the position of a former lake, the water of which had now evaporated, leaving an extensive saline incrustation, while a large lake was distinctly visible to the south-east. Beyond this again some high sunny peaks occurred; but whether situated in the main chain of the Kuen Luen, or in a secondary spur of that range, could not be determined with accuracy. The impression at the time favoured the supposition that the main chain of the Kuen Luen terminates as such somewhat abruptly to the eastward, and at about the 82nd meridian radiates in lower spurs running down into the high table-land of the Aksai Chin, or White Desert.

A high range, in which are peaks of upwards of 20,000 feet above sea-level, bounded the view at the distance of 80 miles to the south-east. This range—either the continuation of the main Karakoram Chain, or a spur from it—was visible, stretching from the head of Chang Chenmo, and trending with a direction of E.N.E. towards the spurs of the Kuen Luen to the eastward.

Looking to the west, it was evident that a journey of 25 to 30 miles in that direction would strike the head-waters of the supposed Yarkand River, if an easy pass could be found across the range forming its east watershed. A valley running westward appeared to offer the best line of route, and getting into this we went up to its head, and crossing a low ridge, descended into a wide sandy valley, flanked by irregular detached ridges. We encamped here for two days, in order to give the animals a rest, as fortunately there was a little grass and fuel obtainable, and I went off alone to explore the country ahead. The features of the mountains about here are irregular, with broken ranges of red clay and sand formation, while the valleys and ravines are filled with sand and conglomerate. No water was to be seen in any of the valleys or ravines, excepting in one or two places where a deeper depression in the valley had accumulated a little water, which was one mass of ice. It was gratifying to find a

very easy pass across the range, beyond which should be the
valley of the Yarkand River; and all the animals were safely
got over across the watershed into a branch valley late on the
evening of the 4th November. The pass was found to be 17,859
feet above the sea by the temperature of boiling water, and is a
mere ascent of a few hundred feet from the valley below, with
an equally easy descent on the north side. It is hardly worthy
the name of a pass in the general acceptation of the term; still
no less is it across a watershed into the head of one of the
Turkistan rivers. I then discovered that the direct road to
come from Chang Chenmo to this pass would have been direct
from the Chang Lang Pass, skirting the Lingzi Thung Plains,
and that a route across from these direct was shorter and easier
than the one which we had followed from Thaldat. Our going
to Thaldat was, however, a necessity, in order to obtain water
for our yaks and horses. Should one have no cause to make
the detour, the shortest road is gained by marching due north
across the corner of Lingzi Thung Plains, which brings one into
a wide open valley leading up to the Kizil Pass. The pass was
so named from a prominent hill of red earth, situated to the left
of our route above the pass. To the right is a high conspicuous
peak (Δ 5, *vide* Maps), 20,992 feet above sea-level, of saddle-
back shape, very irregular and broken, and surmounted by snow
and glacier. This peak is a most prominent object from the
Lingzi Thung, and in marching across from there at once serves
as a conspicuous landmark, and indicates the direction of the
line of route. Its south side presents the features of a perpendicular escarpment, and is one mass of granite, while the slopes
and ravines at the foot of the precipice are covered with large
masses of granite-rock, which have fallen from the summit of
the mountain.

At 10 miles below the Kizil Pass we struck the junction of a
large valley coming in from the south-west, in lat. 35° 16′ 25″ N.,
and camped here, calling the place Kizil-jilga. This was evidently the upper waters of the Karakash River, now nearly
frozen over. At the time I imagined this stream to be the
main branch of the Yarkand River, which it should have been
were our present maps correct; but eventually, by following
this river down to Shadula, it proved to be the real Karakash,
which, instead of rising in the Kuen Luen Mountains, has its
source where the Yarkand River is represented as rising, in the
valley lying west of the range bordering the Lingzi Thung
Plains in that direction, which range forms its east watershed.
There was plenty of grass and fuel at this place, Kizil-jilga, and
though a cold and desolate place at this time of year, and 16,200
feet above sea-level, it was still infinitely preferable to a camp

on the Lingzi Thung Plains. The valley of the Karakash River above Kizil-jilga is flanked by snowy ranges, that to the west being the main chain of the Karakoram, which here forms the watershed between the Shayok and Karakash Rivers. The valley we had descended from the Kizil Pass is one of the main ones running into the upper part of the river, and as grass and fuel are here obtainable, it is evident that a route direct from Chang Chenmo to Kizil-jilga would be far preferable to one *viâ* Thaldat to the lower Karakash Valley.

The next day we made a long march down the main valley, which runs north-west, and is wide and open, and the road excellent. Again the wind came on to blow, and surveying was certainly accomplished under difficulties. When on some high ridge of mountain, after taking the bearings of the different peaks around, it was often difficult enough to write down the observations legibly in one's field-book. Notwithstanding the extreme inclemency of the weather, I enjoyed the exploration thoroughly, for all the country was totally unexplored; and it was interesting in the extreme, since, at the time, I did not know what river it was that we were following; and, furthermore, the road was so good, and quite practicable for laden horses and camels, that it was probable I was then traversing what in future would become the main trade-route between India and Eastern Turkistan.

Having marched some 17 miles, we encamped at Khush Maidan, in a wide part of the valley, where there was plenty of good wood for fuel as well as grass for the yaks and horses. Here we lost a yak, which was unable to travel further on account of sore feet, therefore the Bhoots killed him, and were soon busily engaged in gorging pounds of almost raw flesh. Although the yaks will not eat grain, they do not appear to suffer so much as the horses from the privations of the road, for so long as their hoofs do not crack they go well. Having suffered on the march, however, from tender feet, they are useless for the onward journey, and may be killed at once.

The morning we left here I crossed the river, and ascended the range on the opposite side of the valley, wishing to obtain a good view of the country around. The curious features of this range were particularly observable towards the summit. The hill for miles is in layers of laminated schist and slate, some of which are as thin as paper. These layers, projecting vertically or obliquely, and being much broken, cause the ground to assume a most curious appearance. From the station I reached at the summit of the range, nearly 19,500 feet above sea-level, a magnificent view was obtained of the peaks and ranges around. To the west and south the snowy chain of the Karakoram bounded the view, while north was the Kuen Luen, between which range and the range I had reached extended an interminable mass of mountains, sloping gradually to the eastward to the plains of the Aksai Chin.

Immediately below Kush Maidan the Karakash River increases in volume, being apparently fed by some internal springs in the valley. Some distance above this encampment, towards Kiziljilga, the water had entirely disappeared, leaving the bed of the river quite dry. This singular occurrence could only be explained by its having been infiltrated into the sand and gravel, extensively developed in the widest part of the valley. The stream below Kush Maidan was still partly frozen over on its surface, and we found crossing it to be somewhat of a difficulty. A thick sheet of ice having formed on either side, necessitated a straight drop from the edge of this into the centre of the stream. The Turkistan ponies, however, are capital beasts and very seldom come to grief, whereas the yaks often take it into their heads to lie down in the middle of the river, much to the detriment of one's baggage and provisions.

At eight miles below here a large valley effects a junction from the westward, and immediately beyond the river winds round to the north, and steep spurs from the ranges on either side running down into the valley, form a narrow gorge for three miles. The road was here somewhat indifferent; and, being over the rocky ground close to the water's edge, must in the summer, when the volume of water is great, be difficult compared to what it is above.

Arriving at some hot springs in this defile we encamped behind them. These springs are at an elevation of 15,482 feet above sea-level, and, as at Chang Lang, the ground on the river's bank is white with a calcareous incrustation. Above the river are two small pools of water situated at the foot of the mountain, the temperature of which marked 92° Fahrenheit, while on a lower level, immediately above the water's edge, small mounds of earth and tufa show a calcareous deposit. The heated water issuing from their sides and base indicated a temperature of about 130° Fahr.

Six miles beyond the hot springs the river suddenly turns to the north-east, and from this bend resembled a frozen lake for three miles, of about half a mile in width. The journey was here over the ice, since the steeper sides of the mountains, and the rocky ground, rendered a road along the bank more difficult than one over the frozen river.

The river at this point diverging to the north-east was at first unaccountable, since, if it were the Yarkand River, its course from here should have been north-west, yet it was soon evident that this could not be the Yarkand River, but the real Kara-kash. It was now optional to follow the river along its down-ward course, or attempt a·route across the Karatâgh Range into the basin of the Yarkand River to the westward, and join the regular road from across the Karakoram Pass at Aktâgh.

The latter course would be desirable as proving the feasibility of a trade-route in that direction or otherwise, while the former offered the greater inducement of exploring the course of the Karakash down to Shadula. It seemed certain that a road conducting up the ravine joining the main valley at this bend, or one ascending the wide valley noticed just above the hot springs, would lead across the range bounding the Karakash here on the north—and named the Karatâgh—and join the Karakoram route near Aktâgh, which place lay at a distance of 35 to 40 miles in a direct line from this point. Judging from the configuration of the country, the pass across the Kara-tâgh would probably be found to be a very easy one, and assuredly not more difficult than the famed Karakoram Pass, which, notwithstanding its notoriety, is a very easy one, although at the high elevation of 18,317 feet above the sea. The inter-est attaching to the course of the Karakash, however, pre-vailed; and I determined to follow the river downwards to Shadula.

Having marched until dusk we encamped in a ravine on the right bank of the river, at the foot of a moraine, which has car-ried immense quantities of rock and *débris* into the valley from below a glacier. The bed and slopes of this ravine, which extends for upwards of a mile and a half from the foot of the glacier to the river's bank, are covered with large boulders and fragments of granite rock, piled up in masses one above the other. The whole of the country passed during this day's march was wild and rugged in the extreme. Deep ravines between precipitous heights were seen from where the Karakash, forcing its way between abrupt spurs of the ranges on either side, rushed on over its rock-bed to the bend, where, assuming an easterly course, it emerges into the more open valley, and was now held arrested in its frozen expanse. From a lime and slate formation near the hot springs the mountains lower down the valley change to strata of grey and yellow sandstone, while rocks of grey and dark granite with fragments of felspar lie interspersed upon the beds of conglomerate, which fill the valley and extend from the foot of the mountain to the water's edge. These lighter granites seemed to be of an inferior, coarse texture, and much worn by the action of water.

A mile below our camp, at Zinchin, immense moraines have fallen from the high ranges and blocked up the valley, causing the river to form the lake alluded to.   The river has worn its way through these, and for some distance flows on through narrow gorges much confined.   The scenery was still very rugged and beautiful.   High mountains, surmounted by snow and glacier, towered above the valley on either end, their sides terminating abruptly in steep heights and precipices, while every ravine running into the main valley is filled with moraines of *débris* and granite boulders.

The river from here winds round more to the eastward, and it was now certain that it would emerge near the Kuen Luen Range, the snowy peaks of which were already in sight far down the valley.   Granite was still the prevailing formation of the mountains, and at the foot of a precipice of granite terminating a spur from the Karatâgh our camp was fixed for the night.

The next day, the 11th November, we made a march of 17 miles further down the valley, which widens as the elevation decreases and the mountains are less steep and precipitous. The breadth of the valley had here increased to upwards of a mile, and the river flowed in several streams over its more open bed.   Platforms of conglomerate and sand occur on either side the stream, sloping gradually to the foot of the higher mountains.   A snow peak of 19,615 feet above sea-level overlooks the valley, where we encamped that night at Mulgoon—the Turki name of a description of wood which is met with in abundance from here downwards.   The valley here was found to be 14,458 feet above the sea, by the temperature of boiling water.   The cold, too, was not nearly so great, indeed, with a huge fire blazing, it was quite the contrary.   Near here some fresh springs issuing from the ground add to the volume of water in the river; and the temperature of these was sensibly above that of the stream, for many fish were seen in their shallower parts.

Some 12 miles below Mulgoon the river suddenly turns to the north-west and runs through the valley of Sarikee to Shadula. The name Sarikee is applied to the valley of Karakash from here downwards, which is evidently the Sarka of Moorcroft and the Chinese itineraries.

We were now under the Kuen Luen Range, some high peaks in which rose immediately to the north-east; and coming in at this bend is a valley from the south-east, down which the road from Thaldat conducts, which route we should have followed had we not diverged from there.   I had now proved the river we had been following to be the real Karakash, and thus to have its rise, not in the Kuen Luen Range, but in the main chain of the Karakoram.   The valley effecting a junction here from the south-east has hitherto been represented as containing the main branch of the river; and the error has apparently arisen from Mr. Johnson not having seen the point of junction of the real stream when he crossed this valley on his way to Khotan in 1865. Mr. Johnson, it is known, went into the valley of the upper Karakash, but never so far down the river as to be able to see its upper course for any distance. Had he done so, any observations for altitudes would have shown that this could not be the same stream as that which passes Aktâgh, on account of the differences in the elevation of the several places. Any one not following the river downwards would probably make a similar mistake, for the configuration of the country, as seen from a distance, would lead one to suppose that the river continued the general direction of its upper course in the same line as far as Aktâgh.

From this point the Karakash runs with a general curve bearing w.n.w. to Shadula, some 75 miles distant; and skirting the southern base of the Kuen Luen which rises in a high, rugged range to the north, some of the higher peaks attaining to an altitude of 21,000 and 22,000 feet above the sea.

We had come a great round following the course of this river, and were nearly out of provisions, but expected to meet some of the Kirghiz with their flocks and herds lower down the valley. Our horses and yaks were quite exhausted from the severe work and privations, and we marched but slowly. Grass and fuel are met with everywhere in abundance, and game is plentiful all down the valley. Wild horses and a few yak frequent the best pasture-grounds, while ducks, hares, and partridges are to be had in plenty. Before we had reached halfway down to Shadula, we fortunately met a couple of Kirghiz who had come up the valley to look for game. The sound of my firing at some wild duck had attracted their attention, and they galloped up on their rough-looking ponies. In answer to our inquiries I learnt that they belonged to an encampment 20 miles further down the valley, where, they said, we could obtain some supplies.

Near lhere, at Ak-koom, a wide valley known to the Kirghiz as Karajilga, joins from the eastward, the peaks at its head being in sight, situated in the main chain of the Kuen Luen. The Karakash valley is here upwards of a mile and a half in breadth, and is bounded on the north by the steep rocky heights of the spurs from the Kuen Luen. The mountains behind rise in snow-capped peaks to the height of 20,500 feet above sea level, with their ravines and ridges filled or crowned with moraines and glaciers. The spurs of the Aktâgh range to the southward are more even and less abrupt, while their slopes are covered with accumulations of drifted sand.   The lower stratum of this range is sand and argillaceous rock.

The road continues down the right bank of the river, and is level and good, excepting where the mouths of the ravines running into the main valley have to be crossed.   Large beds of conglomerate occur all down the valley, wherever these openings have given passage to the detritus brought down from the higher mountains.

Continuing our journey, we reached a camping-ground of the Kirghiz, called Gulbashem, on the 18th November.   The Kirghiz had moved up a branch valley in the Aktâgh range, in order to find a better pasture ground for their flocks of sheep and goats, and herds of yak; but on hearing of our arrival the head man of the encampment came down to Gulbashem with supplies of flour, milk, and butter.   By means of my interpreter we carried on a conversation; and I obtained some information from him respecting the names of the different valleys and encamping places.   A very easy pass was said to lead across the Aktâgh range, from the head of the valley where the Kirghiz were encamped to Muliksha, on the Karakoram route; and another pass, difficult for laden animals, but still practicable, crossed the Kuen Luen range near the junction of the Karajilga valley, above Ak-koom, from where a road conducts down the valley of the Khotan River to Ilchi, the capital of that province.

In the ravines above Gulbashem are situated the jade quarries, which were formerly extensively worked when the Chinese were in possession of Eastern Turkistan.   They are now quite neglected, and have been so since the expulsion of the Chinese. The jade is called "sang-i-kash," and was manufactured into cups, snuff-boxes, vases, and other curiosities, and transported to Pekin, where it commanded a high price.

This quarry is probably the same as that mentioned by Mir Uzzet Ullah, Moorcroft's explorer, as being situated about half way between the Karakoram Pass and Yarkand.   The distance hardly coincides, for the Karakoram pass lies some 60 miles due south from here, whereas Yarkand is upwards of 200 miles away.   There are other jade quarries situated lower down the Karakash valley towards Khotan, which may also be identified with the "Causanghi Cascio," or Stone Mountain, mentioned by the mediæval traveller, Benedict Göes, as being famed for its supply of jade.

The valley of the Karakash, at Gulbashem, is 12,645 feet above sea-level.   Ten miles further down is another encamping place of the Kirghiz, called Balakchee, below which the Sooget Valley effects a junction from the south.   Down this valley the

Tabistânee route conducts from across the Sooget Pass and the Karakoram. Some two miles below this junction the Karakash River turns to the north, and piercing the main chain of the Kuen Luen, again assumes an easterly course; until nearing the meridian of Khotan, when it diverges to the northward, and enters the plains of Turkistan.

Shadula, in lat. 36° 21′ 11″ N. and long. 78° 18′ E., and 11,745 feet above sea level, is situated on the left bank of the river, immediately below this bend, at the junction of the Kirghiz Pass Valley from the westward. It consists of a stone fort and several ruined huts, originally built by the Ladak wazeer of the Maharajah of Kashmir. The fort was occupied by a detachment of Kashmir troops from 1863 to 1866, during the disturbances caused by the rebellion against the Chinese in Eastern Turkistan; but on the invasion of Khotan by the Kush Begie in the autumn of the latter year, it was evacuated by the Kashmir troops, who retired across the Karakoram. The Kush Begie sent a detachment of troops to occupy the fort, and it has since remained in the hands of the Yarkand ruler.

The Maharajah of Kashmir, it is believed, considered his territory to extend up to the Kilian range, north of Shadula, doubtless from the fact of having had a fort built there; but the last habitation now met with in his territory is at the head of the Nubra valley, in Ladak. The boundary line is given on the latest map of Turkistan as extending up to Kathaitum, in the Kilian Valley; but not only this valley, but the valleys of the Yarkand and Karakash rivers are frequented by the Kirghiz, who all pay tribute to the ruler of Turkistan.

The natural boundary of Eastern Turkistan to the south is the main chain of the Karakoram; and the line extending along the east of this range, from the Muztâgh to the Karakoram, and from the Karakoram to the Chang Chenmo passes, may be definitely fixed in its geographical and political bearing as constituting the limit of the Maharajah of Kashmir's dominions to the north.

Yarkand, January 29, 1869.

2. Exploration of the Yarkand River,

December, 1868, *January,* 1869.

Before narrating our onward journey to Yarkand, it is necessary to briefly mention the aspect of affairs as I found them on my arrival on the borders of Turkistan.

We reached Shadula on the 20th November, and found the fort occupied by a Panja-bashi (sub-officer) and some dozen soldiers of the Yarkand ruler.

As I had come openly as an Englishman, the news that I was on my way to Yarkand had reached there many days before; and the time that had elapsed in following the Karakash river down to Shadula had given the guard ample opportunity of making arrangements to allow me to proceed or stop me here, according to their orders.

Arrived at Shadula, I found that Mr. Shaw, who had travelled up from Kangra with a large caravan of tea and other goods, had reached here by the direct Chang Chenmo route a few days earlier. The guard would not allow us to communicate in any way; and it was at once evident that they were immensely suspicious at the almost simultaneous arrival of two Englishmen. Unfortunately, Mr. Shaw and myself had been in ignorance of each other's intention and movements, and were therefore unable to combine our plans and act in concert. After some conversation with the Panja-bashi, by means of an interpreter, I began to perceive how matters stood, which may be thus explained. A Moghul at Yarkand, who had lately arrived from Ladak, had spread there a report that fifty Englishmen were coming, and that he had seen them himself. Consequently the greatest amount of suspicion prevailed in Yarkand, whence messengers were daily despatched to the king at the camp, beyond Kashgar, where, it was reported, he was holding the Russians in check on the northern frontier of Turkistan. Some Punjabi merchants arriving a few days later had greatly relieved the fears of the suspicious Yarkandies, by assuring them that the report about fifty Englishmen coming was entirely false; and the Moghul who had caused the alarm was at once imprisoned, and would probably be executed.

Still their distrust, so easily aroused, was not to be at once allayed, and an extra guard was immediately despatched to Shadula with strict orders to stop any one there coming from Ladak.

On my expressing a wish to the Panja-bashi to have a letter sent off to the king asking permission to proceed, he ordered a mounted sipahi to be in readiness; but as none of the men could write, and of course English was unknown in Turkistan, a difficulty presented itself. This was at length got over and arranged, by my writing a letter in English to the king, and giving it to my interpreter to take, accompanied by the sipahi. A horse was also provided for my man, who had strict instructions as to what he had to say—that I had travelled a distance of 8000 miles, occupying six months; and now, having arrived on the borders of Turkistan, sent forward asking permission to enter his country and have the honour of an interview. This was the substance of my letter to the king. The Panja-bashi strongly urged our returning to Ladak without delay, and offered to furnish supplies for the journey, expressing an opinion that Mr. Shaw and myself would probably be kept here until some decisive action with the Russians took place. If the Russians were repulsed, we should be allowed to proceed under strict surveillance as far as Yarkand, and sent back thence. Should the Russians be victorious, we should probably not be allowed to proceed at all. I trusted this might not be the case; but, any how, there did not seem to be the faintest chance of any exploration being allowed, or any departure from the direct road to Yarkand. Were permission to enter the country refused, it would be impossible to force one's way; and, having now been seen, equally impossible to penetrate by this frontier in disguise. Still, if even allowed to go on, there was the strong chance that an Oriental despot accustomed to daily bloodshed, perhaps smarting under a defeat, might order the intruders' heads to be cut off, since he had done so before to men whom he took to be spies of the English. The Moghul officers and soldiers here at Shadula were also in terror of their lives from the king, who was likely to visit on them his displeasure at an Englishman's arrival. This was how matters stood on my despatching a letter to Mahomed Yakoob Beg, the "Atalik Ghazee" and ruler of Eastern Turkistan.

An answer to my application could not be expected to arrive within twenty days; and during the next few days I considered what other plans lay open to me to endeavour to carry out should permission to enter Turkistan be refused. To return to Ladak across the mountains in December would be sufficiently unpleasant; but as the Zojji Lá Pass into Kashmir would be closed by the snow, there would remain the only alternative of wintering in Ladak, and in the spring endeavouring to penetrate to Turkistan and the Pamir Steppe by some other route. The idea of passing a winter in Ladak doing nothing was not to be entertained, and to be turned back now, after having travelled 800 miles from our own frontier, would be most unpleasant. The great suspicions of the Yarkandies seemed absurd and needless, although the advance of the Russians from the north had no doubt alarmed them; while the simultaneous arrival of two Englishmen on the southern frontier, with the avowed intention of penetrating to Yarkand, where no Englishman had ever been before, added to the doubts and alarms excited at the prospect of their being brought into contact with Europeans.

Having discovered the source of the Karakash to be where all our maps make the head-waters of the larger river, the Yarkand one, to have their rise, it was most desirable to ascertain the

real course of the Yarkand River, as being the chief river of
Eastern Turkistan. I was, therefore, most anxious to undertake
this expedition, knowing the time could not be better employed
while awaiting the return of my messenger from Yarkand. The
difficulty in accomplishing it lay in the close surveillance of
the guard of Turki sipahis, which rendered any attempt at getting away on an exploring expedition unlikely to be successful;
and if the sipahis suspected my object, they would be sure to
accompany me, in which event using surveying instruments
openly would be out of the question, and any exploration further than two or three days' journey also impossible. There
was a chance that I might be able to get away for the day, for
the purpose of shooting, without being accompanied; and this
seemed to be the only way of shaking off the guard. The men
with me at this time, besides my own servants, were the Bhoots,
who had accompanied me from Ladak. They were awaiting
my interpreter's return from Yarkand, when, if I was allowed to
proceed, they would be dismissed to their homes, or else accompany me back to Ladak should I have to return. They regarded
the Turki sipahis in no very friendly light, and were, therefore,
not likely to disclose my plans which were carried out successfully.

Leaving my own servants in charge of camp, and taking
three of the Bhoots with a week's supply of provisions, we
started from Shadula at the first streak of daylight on the
morning of the 26th November, without the guard being aware
of our departure. Marching up the valley leading to the
Kirghiz Pass, beyond which lies the valley of the Yarkand
River, we encamped that night at Kulshish Kun, a famous place
for wild yak, but this day found without any large game upon
it. From Shadula the road runs up the right side of the
stream, and is gradually on the ascent to Kulshish Kun, which
is 13,965 feet above sea-level, or some 1800 feet higher than
Shadula, from which it is 15 miles distant. To the north of the
valley the Western Kuen Luen Range rises into lofty peaks,
while, to the south, it is bounded by a long spur from the
Aktâgh Range, across which lies the Sooget Valley. In order
the better to distinguish the geographical features of the great
Kuen Luen chain of mountains, it has been divided into eastern
and western ranges, from where the Karakash River pierces
the chain on the meridian of Shadula. Any remarks on the
Kuen Luen, therefore, will be understood to apply to that division of the range, as it bears relatively eastward or westward
from Shadula.

Above Kulshish Kun the valley widens and opens out, being
bordered by thick beds of conglomerate, furrowed and inter-

sected by watercourses, on the slopes of which large boulders of granite rock are strewn, while in the ravines above are several glaciers, which have worn their way downwards until nearly level with the base of the steep heights above them. The road continues over the more even ground on this side the valley, and skirts the long spurs of the Aktâgh Range on the left, after it has crossed the stream.

Approaching the pass the valley bifurcates, the northern branch containing the main source of the stream, which rises under a large glacier lodged at the head of the ravine, between two high snowy peaks in the Western Kuen Luen.

The Kirghiz Pass, 17,093 feet above the sea, lies at the head of the western ravine, up which the road winds, with a gentle ascent, to the summit. The pass commands an extensive view of the country far and near, and I was able to fix the bearings of some of the highest peaks in the Eastern Kuen Luen, lying 90 miles away, which had already been mapped in, and thus ascertain the value of my survey up to this point, as these peaks are visible from the southward on entering the Lingzi Thung plains, at a distance of upwards of 100 miles.

The Karakoram and Muztâgh mountains, with the range of the Western Kuen Luen, were in sight to the westward, and one was at once struck with the very wild and rugged scenery in this direction. Amongst the interminable mass of precipitous ridges, deep defiles, and rocky ravines, it was difficult to distinguish the exact course of the Yarkand River, but its general direction could be easily determined as flowing through the long longitudinal valley between the two main ranges. Not a tree, bush, or shrub, met the eye anywhere. It was solely a magnificent panorama of snowy peaks and glaciers, as the last rays of the setting sun tinged their loftiest summits with a ray of golden light.

It was dusk as we commenced the descent down the lateral ravine leading from the pass. The road, a mere track, winds down the steep side of the ridge to the head of the ravine, the bed of which is blocked up with *débris* and rocky boulders, while the stream in it was entirely frozen over. We were now in the basin of the Yarkand River, since the Kirghiz Pass leads across a depression in the Aktâgh Range, immediately below its point of junction with the main chain of the Kuen Luen. Marching up to 9 o'clock by moonlight, we got down to near the valley of the Yarkand River. The spot chosen for our camp for the night lay in the gorge of a lateral defile, where running water was found, while a few stunted bushes, which fringed the stream, were soon appropriated and kindled into a cheerful blaze. We had descended nearly 3000 feet from the summit of

the Kirghiz Pass, since our camp lay at an elevation of 14,225 feet
above sea-level.   Starting early the next morning, we continued
down the narrow ravine to its junction with the valley of the
Yarkand River, which was struck at a distance of 33 miles
west of Shadula.   An observation obtained one mile lower down
the valley showed the latitude to be 36° 22" 7' N.   The river
here comes down from the south, winding between precipitous
spurs of the Karakoram and Aktâgh ranges, the valley being
here much confined, and varying from 300 to 500 yards in
width.

Immediately overlooking the river, on its right bank, stand
the ruins of several stone huts, which were formerly inhabited
by some of the Kugiar villagers, when a copper-mine, which lies
to the west of the valley, was being worked.   Our road here
joined the Zamistânee route, which conducts down the valley of
this river from the Karakóram Pass viâ Kufelong.   Three miles
below the junction of the Kirghiz Pass Ravine is a pasture-
ground, called Kirghiz Jangal, where are several springs of fresh
water.   The valley widens at this point, and is upwards of a
mile and a half in width, while the slopes of the mountains are
still abrupt, and are covered with accumulations of sand, and a
detritus of shale and shingle.

The Kirghiz of Sarikol and Pamir Khurd formerly frequented
this pasture ground, and used to commit many depredations on
the caravans of the merchants trading between Leh and Yarkand.   They have now quite abandoned their plundering excursions, and for many years have not even frequented the upper
valleys of the river.   Their place has been taken by the Kunjooti robbers of Hunza and Nagar, who, crossing by the
Shingshâl Pass from Hunza, ascend the river, and are in the
habit of lying in wait for the caravans as they traverse this
desolate valley.   Though eager for rapine and plunder, they do
not relish fighting, and a well-appointed caravan has little to
fear from these robbers, who make it their especial aim to surprise any small party unawares.

Their last great raid took place three years ago, when a band
of 120 armed men attacked a large caravan, chiefly composed
of Kashmir merchants, returning to Ladak with many horse-
loads of merchandise near Koolunooldee, when the whole of their
property was seized, and the merchants themselves carried off
to Kunjoot, from where they were sold as slaves into Badakhshân.   The Kunjooties committed so many depredations during
the disturbances in Turkistan, that Mahomed Yakoob (Kush
Begie) has ordered this route, which is by far the best one, to be
closed, until he shall have had time to send an expedition to
punish them, the penalty being that any merchant who shall

travel this road without express permission shall lose his head.

Below Kirghiz Jangal, where the elevation of the valley is 13,684 feet above the sea, the river turns to the westward, and continues with a general course in this direction towards Sarikol. The valley from here downwards is full of low jungle, grass, and herbage, which become more profuse as the elevation decreases, while saltpetre and rock-salt occur in many places. Deep, long ravines between the high spurs running down from the Karakoram Chain come in from the south, while the shorter valleys of the Western Kuen Luen to the north narrow as they reach the crest of the range, and are closed in by rocky heights and glaciers.

At 14 miles below Kirghiz Jangal is a camping-ground, called Koolunooldee, where the road leaves the valley of the river and ascends a confined, somewhat difficult, defile leading to the Yangi Pass. It is practicable for laden horses and camels throughout, and there can be no doubt that this is by far the easiest and most direct route from across the Karakoram into Eastern Turkistan. Kugiar is reached in five days' journey from Koolunooldee, and Yarkand in from seven to eight days. East of the Yangi Pass, on the northern slope of the Western Kuen Luen, rises the Tiznáf River, which joins the Yarkand River to the east of where it is crossed on the road between Karghalik and Posgâm, and is one of its principal tributaries. The Yangi Pass leads across a remarkable depression in the Western Kuen Luen Range, and is about 16,500 feet above sea-level in elevation.

I did not reach the summit of the pass, since it was out of my line of exploration, but, when returning up the river, I ascended the Western Kuen Luen, and, attaining to a station on the range, at an elevation of nearly 19,000 feet above the sea, had a full view of the pass below me. A long spur, running down from near the pass, bounds the ravine, up which lies the road, to the westward, the prevailing features of this ridge being red earth, sand, and shingle. The west side of the ravine itself rises in successive terraces and platforms of conglomerate one above the other, presenting a steep scarp to the eastward.

Continuing down the main valley we left Koolunooldee behind us, and, walking up to dusk, reached to near where the Muztâgh Pass stream joins the river.

The Yarkand River from here bears somewhat more to the south, and skirts the precipitous and rocky spurs running down into the valley from a group of high snowy peaks in the Western Kuen Luen. The highest peak in this group was found to attain to an elevation of 22,374 feet above sea-level. It may here be mentioned that the heights of inaccessible peaks were calculated from the angles of altitude found with sextant and artificial horizon at two stations fixed by triangulation, the peak also being fixed by triangulation, and the heights of the stations known from observations of the temperature of boiling water. They have no pretension to being very accurate, but are fairly approximate, and may be considered to be within 300 or 400 feet of true altitude.

Arriving at dusk at the junction of a large stream coming in from the south, we prepared to halt for an hour. This stream, of considerable size, is one of the largest of the upper branches of the river flowing from the northern slope of the Karakoram Range. Its banks are very precipitous, and the continued action of the water on the beds of pebbly conglomerate, which fill the exit of the valley, has abraded their sides, until a series of caverns have been formed, extending far under the bank. Immediately beyond the mouth of this valley we came upon the fresh tracks of camels and horses, which indicated Kugiar men being about, or perhaps Kunjooties. It was necessary, therefore, to proceed with caution, since if seen down here by even the harmless Kugiares, the report would spread like wildfire that another Englishman had turned up in these valleys, and cause the suspicious Yarkandies to believe that the original rumour of fifty Englishmen coming from Ladak was, after all, correct; and, if Kunjooti robbers, to be carried off by them and sold into slavery, would most effectually put a stop to further exploration. Halting beyond this we lighted a fire, taking care to choose a favourable spot from where it could not be seen by any one, if about; and as soon as a full moon rose above the mountains and was shedding her silvery light far down the valley, we went on again down the left bank of the river for about 9 miles, until stopped at a place where the stream runs deep and strong under a high bank to the left of the valley. We wasted an hour trying to invent something on which to cross; but the long poles cut from the jungle close by, with which we endeavoured to form a temporary bridge, were washed away at once. Going back for a mile we climbed the steep slope of the hill above the river, consisting of loose sand and shingle—the ground that gives way and lets one down about as fast as one progresses upwards. At length, descending again to the bank immediately above the river, we were arrested again a short distance further down, where a stream comes in from the south. This stream has carried into the main valley immense quantities of earth and *débris*, and now flows down out of sight between precipitous and over-topping banks, as if split by an earthquake. Steep heights enclose the valley on either hand, while above to the north rise the lofty snow-capped group of peaks in the Western Kuen Luen. The valley, some 2000 yards in width, was here found by observation of the temperature of boiling-water to be 12,130 feet above sea-level. This was the furthest point down the Yarkand River which was reached. There was every probability that the guard of Turki sipahis would follow us from Shadula, and arriving at the junction of the Kirghiz Pass Valley before we could return, thus cut me off from going up the river to its source; consequently I determined to march back up the valley during the night. Retracing our steps we reached the spot where we had lighted a fire the evening before, and, as soon as day broke, started back again up the valley. Crossing to the north side, I left the men with me at the foot of the mountain and commenced the ascent of a steep spur of the Kuen Luen. It was evident that a station on this range would command an extensive view, and what appeared to me the most accessible point was fixed upon for the attempt. The steep slope of the mountain, covered with loose shingle and sand, was most unfavourable for climbing, and very different from the Kashmir Mountains, which, although steep, afford firm footing on the grass and rocks. When the crest of the ridge, however, was gained, the difficulty decreased, and though the higher slope was steeper, the ascent was more rapidly and easily accomplished. After five hours' hard climbing I reached the summit of the mountain in time to fix the latitude of the range by the sun's meridian altitude.

The magnificent view which this station commanded was an ample reward for the toil of the ascent. Far away to the south and south-west stretched the high peaks and glaciers of the Karakoram and Muztâgh Range, some of whose loftiest summits attain to the height of from 25,000 to 28,000 feet above the sea. One peak, situated to the east of the Muztâgh Pass, reaches the stupendous elevation of 28,278 feet above sea-level, and is one of the highest mountains in the world. Beyond where the river sweeps out west, the snowy peaks above the Kunjoot country were in sight towards Sarikol. East and west extended the whole chain of the Kuen Luen and the Kilian Mountains, the last range to be crossed before the steppes and plains of Turkistan are reached, while immediately below lay the confined ravine up which the road ascends to the Yangi Pass, now full in sight beneath me. The extent of view of the main Karakoram or Muztâgh Chain comprised a length of 200 miles, stretching from near the Karakoram Pass to the head of the Tashkurgân territory north of Hunza and Nagar.

The valleys that traverse the mountains between the crest of the chain and the longitudinal valley of the Yarkand River appear to narrow into ravines towards the head of the range, and are filled with glaciers; and the whole surface of the ground to the north of the chain is probably more elevated in its average altitude than the mountain system, embracing the southern slopes of the range, in the watershed of the Indus.

The cold at this elevated station, nearly 19,000 feet above the sea, so late in the year was very severe, the thermometer sinking to 5° Fahr. in the shade, notwithstanding it was mid-day, and a bright sun was shining. I had reached many higher altitudes, but never any commanding so extensive a view of such a stupendous mass of mountains; and it was with a feeling of regret that we turned to leave a spot from where the peaks and glaciers could be so well seen, stretching far away on every side in their solemn grandeur.

Descending into the ravine beneath, I went on down its rocky bed, and at 4 miles below again struck the valley of the Yarkand River, and being joined by the men who had awaited my return, ascended the valley of our camp near Kirghiz Jangal, from where we had started the morning of the day before, having walked incessantly since that time a distance of more than 55 miles.

On the next day, the 1st December, we went 16 miles up the river, thus getting above the junction of the Kirghiz Pass Valley, and found that no sipahis had as yet followed us from Shadula. I sent off one of the men by this route to Shadula, with orders to my servants there to send provisions for us to a camping-ground, called Aktâgh, some 50 miles further up the river, as the supply with which we had originally started was nearly consumed. This place, Aktâgh, is the third stage from the Karakoram Pass on the Turkistan side, where the Shadula route separates, and the Kugiar or Zamistânee one conducts down to Kufelong and thence down the valley of the Yarkand River. We were now on this road, and never doubted but that we could reach Aktâgh in three or four days, at the latest. Keeping on up the valley, we encamped that night at an elevation of 13,882 feet above sea-level. From here, ascending the river, the road is up the right bank, skirting the steep spurs of the Aktâgh Range; it then crosses to the left bank and goes over the spur of a hill, round which the river winds. Continuing up the left bank the road is good; the valley again widens and the slopes of the mountains are more gentle and less precipitous. Keeping on up the valley and mapping the whole way, on the morning of the 4th December we arrived at Kufelong, where the Karakoram Pass stream, passing Aktâgh, joins the main river. At this place, Kufelong, which is in lat. 36° 4′ 48″ N., long. 77° 57′ E., and the valley here 14,340 feet above sea-level, the main river comes down from the south-west, and the Karakoram Pass stream, much smaller and now entirely frozen over, joins from the south-east. The latter stream is represented on some of our maps as the head of the Tiznâf River, and on others as the Yarkand River, whereas the real main stream of the Yarkand River is not down on maps at all. From Kufelong I followed the main stream up to its source, but at the time was not aware that Aktâgh lay up the valley to the south-east, on account of the error on the maps, imagining it to be on the main stream so represented.

Thirty miles a-head up the main valley the snow-covered spurs of the Karakoram were in sight, and the foot of these was reached on the evening of the day after we had left Kufelong—the valley up to here being wide and open, and at an elevation of some 15,000 feet above sea-level. The mountains on either side are broken up into ridges, with rugged and irregular crests, showing strata of clay, alternating with sand. The clay-formation prevails, and is of a highly ferruginous appearance. The bed of the valley is composed of deposits of conglomerate and gravel, above which, in places, occurs an alluvium of fine clay, covered with pebbles of every shape and size of different compositions, such as flint, limestone, slate, &c. Being open and level, the river runs in several streams over its bed. A hot spring, running out of the base of a cliff on the right side of the river, is passed at an elevation of 14,900 feet above sea-level. On the evening of our second day's journey from Kufelong we encamped in a wide part of the valley, opposite to the entrance of a deep narrow ravine effecting a junction from the south-west. At the head of this ravine a pass leads across the Karakoram Range into the Nubra Valley in Ladak, and to Chorbut, in Baltistan. It is apparently at a very high elevation, probably not less than 19,000 feet above the sea, and is closed for nine months in the year by the snow. It is impracticable for anything but foot-travellers, and perhaps for yaks; and although not in use for many years, was formerly traversed by the Balties, carrying their own loads of merchandise into Yarkand. This pass appears also to have been used by the Kalmâk Tartars in their successful invasions of Ladak and Tibet towards the close of the 17th century.

From this point to the summit of the pass the distance is from 25 to 30 miles, the road ascending gradually up the ravine, flanked by the snow-capped spurs of the Karakoram.

The main valley here turns and the river comes down from the south-east. Skirting these high ranges our road lay up the open valley through the wildest and most desolate country, where nothing but snowy peaks and glaciers and the barren slopes of the mountains met the eye. Not a blade of grass was to be seen, and it was with difficulty sufficient "boorsee" could be collected wherewith to light a fire. The valley again turns to the south, and we were now evidently near the source of the river, since it was rapidly decreasing in size and nearly entirely frozen over as we ascended.

· On the afternoon of the 8th December I reached the source of the Yarkand River. This is in an elevated plateau or basin, surrounded by high snowy peaks, with the ravines at their base filled with glaciers. The centre of this plateau forms a depression of about 2½ by 1¼ miles in area, which must contain a lake when the snows melt and drain into the basin, in which the little water now accumulated was a solid mass of ice. The outlet is to the west, in which direction the stream issuing from the basin runs through a narrow ravine for 2 miles to the head of the open valley, where, joined by two other streams from the high range lying west, they form the head-waters of the Yarkand River, commencing here and flowing with a course of nearly 1300 miles into the great Gobi desert of Central Asia. I found the source of the river to be in lat. 35° 37′ 34″ N., and by its distance and bearing from the Karakoram Pass to be in long. 77° 50′ E., while the mean of three observations of the temperature of boiling water gave an elevation of 16,656 feet above the sea.

The cold in this inclement region, in the depth of winter, was most intense; the thermometer, at 8 o'clock the following morning, showed the mercury to have sunk to a level with the bulb, or some 18° below zero.

By exploring the country eastward, I ascertained that I had reached to near the summit of the main range of the Karakoram, and west of the Karakoram Pass. The Karakoram Chain here loses the great altitude to which it attains in that portion of the range lying between the Muztâgh Pass and the source of the Yarkand River; and from here eastward, to beyond the Karakoram Pass, is much broken; presenting features assimilating to the crest of an irregular and detached range bordering a high table-land; while higher summits occur in the more elevated spur which, branching from the chain near the head of the Yarkand River, forms the watershed between the Shayok and its tributary, the Nubra River. The main range continues eastward beyond the Karakoram Pass to where a remarkable double peak occurs in the chain; and at this point throws out a somewhat irregular spur, named the Karatâgh, towards the Kuen Luen, which forms the eastern crest of the high central plateau of Aktâgh. At this double peak the Karakoram range, after running with a general direction of E.S.E. from the "Pusht-i-Khar," a distance of 320 miles, suddenly turns to the south, and, again rising into a lofty chain of snowy peaks considerably above 21,000 feet above sea-level, forms the watershed between the Shayok and Karakash rivers, until, in the parallel of 34° 43′ N., it trends again to the eastward, and runs along the head of Chang Chenmo; and here constitutes the southern crest of the elevated table-land known as the Lingzi Thung plains and the Aksai Chin.

After exploring the country at the head of the Yarkand River, it only remained for us to make the best of our way back to Kufelong; and, as Aktâgh lay up the branch valley joining there, we had thus missed the man sent off to Shadula to bring supplies for us. The only yak with us had succumbed from hard work and the want of grass half way up the valley, and the Bhoots had killed him for food, as we were quite out of provisions. The weather had been threatening snow for the last few days, and an immediate return was imperative. Already heavy clouds were breaking up amongst the high peaks of the Karakoram, obscuring their summits, while the sun set angrily and threw a lurid light through the higher masses of thick cloud, as we returned to camp on the evening of the 9th December. It commenced snowing as we started at dusk and retraced our steps down the valley, marching up to midnight through the falling snow. The next day it again snowed heavily, as we kept on down the valley from noon till dark, when we camped under a rock, which afforded shelter for the night. At length the 55 miles back to Kufelong were traversed, though the last part of the way but slowly, as one of the men with me became ill. This was not surprising, as, where no wood was obtainable for fuel, the Bhoots, nothing discouraged, set to and devoured the yak-flesh, like wolves.

On the evening of the 18th December we reached Kufelong again at the moment a long string of laden horses appeared coming in down the valley from Aktâgh. They were on the opposite side of the river, and five men who were with the caravan immediately commenced shouting to us in the most vociferous manner across the ice, at the same time waving sticks and making other hostile demonstrations. At length, after a good deal of beckoning and persuasion, they were induced to venture across the ice to where we were, when the cause of these proceedings was soon explained, as I learnt from one of the men who could talk a little Hindustani, that they had mistaken me and my men for Kunjooti robbers. Mutually satisfied, we proceeded to their camp for the night, where they entertained us most hospitably, and endeavoured in every way to atone for the privations endured during our late trip up the Yarkand River.

They were a party of Kugiar merchants with some thirty horse-loads of merchandize, come from Ladak, and now on their way to Yarkand; and, regardless of Kunjooties and captivity, were travelling down the Zamistânee route with the express per-mission of the Governor of Yarkand, when we fortunately met them here at Kufelong.

On the following morning, after bidding them farewell, we started for Aktâgh and Shadula, having received from our hos-pitable entertainers three days' provisions for the road. They promised to renew our acquaintance, should I proceed to Yarkand, and were delighted at receiving a flask of powder and a couple of hunting-knives, as tokens of our friendship.

From Kufelong to Aktâgh, the road is excellent; and runs up the right bank of the Karakoram Pass stream, now en-tirely frozen over, which had to be crossed twice in the defile at Kufelong. Aktâgh—or "White Mountain" in Turki—as the name implies, is a peculiar hill of white sandstone, standing out prominently in the open valley at the foot of the Aktâgh range. From here, eastward, the country slopes up into a high plateau, traversed by low ridges of hills, until it reaches the foot of the Karatâgh range, forming its eastern crest. This central plateau of Aktâgh is geographically remarkable as con-stituting the table-land intervening between the main chains of the Kuen Luen and Karakoram, and the Yarkand and Karakash rivers. It is bounded by the Aktâgh range on the north, the Karakoram on the south, and the Karatâgh on the east; and has an average elevation of 15,800 feet above sea-level, while the whole surface of the table-land slopes down gradually to the north-west to the valley of the Yarkand River. The Karatâgh running out from the Karakoram, and Aktâgh range branching from the western Kuen Luen, converging towards the east of this table-land, determine its triangular configuration, round the apex of which the Karakash River winds. The extent of this plateau, including the ranges which bound it up to the Karakash Valley on the north, and to the valley of the Yarkand River at Kufelong, embraces an area of 3400 square miles. At Aktâgh itself, which is an elevation of 15,252 feet above the sea, several springs of water add to the volume of this branch of the Yarkand River. The stream from here down to Kufelong was one mass of ice, which caused the breadth of the river to appear greater than when not frozen over, as the accumulations of layer upon layer of ice had expanded the river beyond its usual size, when the temperature of the air does not admit of its being entirely frozen over. I saw the stream afterwards, in the month of June, and instead of the valley being filled up with a sheet of ice, as when seen in December, the stream flowing through it was a mere brook; while above Aktâgh, towards the Karakoram Pass, that branch of the stream was not of sufficient volume even to reach Aktâgh. There is a very little grass, but no fuel whatever here; consequently travellers and merchants, going across the Karakoram towards Ladak, have to carry wood with them from Kufelong for this part of the journey, since no fuel is obtainable until reaching the valley of the Shayok River, on the Ladak side of the Karakoram Pass. As before stated, the road here bifurcates,—the Zamistânee route going down the valley of the Yarkand River, and being the road we followed up from Koolunooldee,—while the Tabistânee route crosses the Aktâgh range by the Sooget Pass, and conducts down that valley to Shadula. From here I traversed this latter route to Yarkand throughout.

Coming into Aktâgh we met two of the Turki sipahis, who had been sent out to search for us from Shadula. From what I could understand, I gathered from them that permission for me to proceed to Yarkand had arrived, and that no slight disturbance had been caused by my sudden disappearance; that the Panja-bashi, in despair, had sent out all the sipahis to search for us in different directions, who had never reached to within 50 miles of where we were, being themselves obliged to return after consuming the little provisions they were able to carry.

We remained the night at Aktâgh, with the thermometer below zero, and no fuel wherewith to light a fire to be had; and started the following morning at daylight, intending to go straight into Shadula, 40 miles distant. The road is level and good as far as Chibra, a camping-place at the foot of the Sooget Pass. No grass or fuel is obtainable here, the place being at an elevation of 16,812 feet above sea-level. A stone hut in ruins stands on the right bank of the stream, outside of which, in the ravine close by, lay the skeletons of half-a-dozen horses. The whole way from the Karakoram Pass is covered with the whitened bones and remains of horses which have perished in traversing this severe route. The great commercial enterprise of the Turkistani traders is fully evinced by their efforts to carry their merchandize for hundreds of miles over what would be thought in Europe such impracticable mountains. Their losses in horses must be considerable, for a caravan during its journey from Yarkand to Leh, and back to Turkistan, generally loses a third, at least, of its horses.

The want of grass, and often of water, near the Karakoram Pass, but more especially the difficult passage of the Sasser and Kilian passes, are the chief causes of the losses thus suffered.

Camels are much better adapted to endure the privations of

such a journey; and should a feasible route for such animals be opened out, there can be no doubt but they will be extensively used.

From Chibra to the summit of the Sooget Pass, which is at an elevation of 18,237 feet above sea-level, the ascent is very gradual and long, the road leading up the open ravine which runs up to the pass.

The view from the summit is limited, since the spurs of the Aktâgh range close the view to the south, and do not admit of a sight of the more distant Karakoram. The Western Kuen Luen is, however, partly in sight to the north, and the Sooget Valley lies below. The pass is across the Aktâgh range, dividing the basins of the Yarkand and Karakash rivers; and, as above remarked, forming the northern crest of the table-land of Aktâgh.

Descending the pass, the road winds down to the head of the Sooget Valley, which is here, as far as it runs north-west, some 12 miles in length, and enclosed by snow-capped ranges. The valley then narrows, and turns to the north-east, and the road is rapidly on the descent over rough ground, until it enters the valley of the Karakash River, four miles above Shadula.

It was dusk soon after we had crossed the pass, but we kept marching on down the valley, one of the sipahis leading who knew the way. The road is very indifferent, and certainly not the sort of ground one would voluntarily choose to march over during a dark night. For many. miles our way lay over the lower slopes of the mountains, and then down the centre of the ravine blocked up with rocks and granite boulders. Thus stumbling on over the rocks and stones, and being consoled by the sipahi to my inquiry, where wood and water were to be found, with the answer, " Su yoak, utum yoak "—in Turki " No wood or water for miles yet "—we kept marching on up to past midnight. At length, coming to wood and water, we halted for the remainder oï the night near the Sooget camping-ground, 13,905 feet above sea-level, and 33 miles from Aktâgh. The following morning I sent off one of the men into Shadula, 7 or 8 miles distant, to bring out a horse to meet me; and, following soon afterwards, met one of my own servants with the horse and a most welcome breakfast. Fording the Karakash River twice, Shadula was soon in sight, a dreary and desolate place at any time, but it appeared almost charming just then.

As I rode up to the fort the Panja-bashi and sipahis were waiting to receive me, and seemed to be in utter astonishment at my sudden re-appearance. They had quite concluded that we were lost amongst the mountains, or had gone back to Ladak, for the sipahi who had come on in the morning had fortunately arrived just in time to prevent their starting without me for Yarkand. They had everything ready for the march, horses and yaks were loaded, when he came in, so thoroughly convinced were they that we should never return; and in dreadful fear lest the King should visit on them his displeasure for their remissness in allowing me to get away on an exploring expedition unaccompanied. They were delighted, therefore, at my re-appearance; and equally pleased was I at the prospect of seeing Turkistan, although the fact could not be ignored that, hitherto, it had proved to be to others "the country from whose bourne no traveller returns."

We had been absent from Shadula just 20 days, and, during that time, had traversed more than 300 miles of mountainous country.

The result of the expedition was very satisfactory from having determined the geographical features and relative bearing of the Karakoram and Kuen Luen chains of mountains, as well as the true course of the Yarkand River.

Hitherto, our maps have represented the Karakoram and Kuen Luen to be one and the same great chain, whereas a distinct watershed and the Yarkand and Karakash rivers intervene between the two ranges. On the other hand, the Tiznâf River has been defined as rising in the Karakoram Pass, and flowing through the Kuen Luen range to its junction with the Yarkand River, which stream has been represented to have its source at the head of the Sarikol territory, near the source of Wood's Oxus. This is entirely wrong. The Tiznâf is but a tributary of the Yarkand River, and rises on the northern slope of the Kuen Luen, to the east of where the Yangi Pass crosses that range.

The main breadth of the Yarkand River has its source to the west of the Karakoram Pass; and at Kufelong, 55 miles below its source, receives its tributary rising in the Karakoram Pass; thence flowing with a course north-west to Kirghiz Jangal, it makes a bend, running west towards Sarikol between the main ranges of the Karakoram and Western Kuen Luen. From the respective slopes of these two ranges it receives innumerable tributaries, and the whole drainage of the Sarikol district, conveyed to it by the Toong, Tashkurgân, and Charling rivers; and then sweeping round with a gradual course to the north-east, flows through the confined gorges of the Kurchum Hills, and debouching into the plains of Eastern Turkistan pursues its course past Yarkand.

It would be satisfactory were a definite geographical name assigned to the great watershed dividing the basin of the Indus from the Turkistan rivers, and which is comprised in the great chain known, in the different portions of its length, as the

Karakoram, Muztâgh, and more anciently Belortâgh, and Pololo. To the inhabitants of Eastern Turkistan the whole chain is known as the Muztâgh range, which in Turki means the "Glacier Mountain," or range; the word Karakoram being merely applied to the pass of that name. That the name Karakoram should be given to the range indefinitely is desirable for the sake of distinction, and the Karakoram range, as specified in this report, will be understood to refer to the whole mountain system included in the chain stretching from the "Pusht-i-Khar" to the head of Chang Chenmo.

3. FROM SHADULA TO YARKAND.

Having reached Shadula on the 17th December, on my return from exploring the Yarkand River, we immediately prepared to start for Yarkand. The Panja-bashi had received orders to make arrangements for the journey, so that everything was in readiness, and a mounted sipahi was despatched on ahead to have the yaks of the Kirghiz awaiting us at a Kirghiz encampment down the valley, for the passage of the Sanju Pass. The Bhoots with me were dismissed and received provisions for their journey back to Ladak. I was disappointed in not being able to send news or letters by them; since it was not advisable to risk any correspondence being found on them by the Ludki sipahis. Nor was I now able to survey openly as before, for in the eyes of a Central Asiatic exploration and surveying is spying, the utmost caution was, therefore, necessary; and henceforward all observations had to be made *sub rosâ*.

As we rode away from Shadula, every one was in high spirits at the prospect of leaving these inclement mountains, the sipahis testifying their joy by firing at a mark as they passed it at a gallop. I had given a pistol to the Panja-bashi, and we each followed with 5 barrels from a revolver. It was amusing to witness the delight and wonder of the Turkies at inspecting a revolver. They could not understand how a small weapon could shoot so many times in rapid succession, and they were never tired of looking at European firearms and expressing their desire to possess such weapons.

Our road from Shadula lay down the left bank of the Karakash River, which here runs with a northerly course piercing the main chain of the Kuen Luen. The mountains on either side the valley are, consequently, very high and precipitous, and many glaciers and moraines occur at the heads of the steep ravines. Much schist appears in the higher strata; and long slopes formed by a detritus of shingle and drift in the transverse valleys. Between the river and the foot of the mountains the valley is filled with thick platforms of coarse indurated conglomerate, which occasionally vary in height and the angle of their slope, the whole being strewn with angular fragments of rock, and rounded boulders, and pebbles.

Passing Ulbuk, a camping ground of the Kirghiz, we continued down the valley to Tograssu, 10 miles below Shadula, where a wide valley containing a large stream of water effects a junction from the north-west. A road conducts up this valley to the Kullik Pass, which can be crossed on the third day's journey from Tograssu. The pass, a difficult one, is little used by the traders, who traverse the Kilian or Sanju passes in preference. When the rivers are melting the stream in this valley attains to so considerable a size as often to debar the passage of the caravans for many hours. It was now, however, easily fordable on horseback. Two miles below Tograssu, where the river again turns to the east, are the ruins of a stone fort, situated on the summit of a low mound in the centre of the valley, and known as Ali Nuzar Kurgân ; and here the Kilian route leaves the Karakash valley, and ascends a ravine in a north-westerly direction to the Kilian Pass. As we were to cross this range by the pass leading into Sanju, we continued down the valley to Mazar Badshah Abubekr, 19 miles below Shadula, where the road to the Sanju Pass leaves the Karakash Valley, and ascends up a very confined ravine leading to the pass. An observation for latitude obtained here showed Mazar Abubekr to be in lat. 36° 33' 14" N.; and the height of the valley at the water's edge was found by the temperature of boiling water to be 11,095 feet above sea-level. The Karakash from here flows eastward through narrow gorges, formed by the steep slopes of the mountains, which do not admit of a road from here down the river to Khotan. This place, being a camping ground of the Kirghiz, takes its name from a Kirghiz chief, named Abubekr, who formerly held sway over the whole of the mountain country down to the borders of Khotan, and used to levy black mail on the caravans of the merchants passing through his territory. His grave is to be seen here at the place which bears his name, surrounded by a rude wall of loose stones, and surmounted by pieces of red, yellow, and white cloth, and yaks' tails.

Whilst the Panja-bashi and the sipahis were regaling themselves on a mess of stewed horse flesh, a most popular article of consumption throughout Turkistan, and I was engaged in taking an observation for latitude from amongst the thick jungle which here fringes the river, the loads were transferred by the Kirghiz to the backs of the yaks awaiting our arrival, and our saddles changed to the best ones to ride. To be seen riding one of the animals in England would no doubt create a sensation, but here in the mountains of Central Asia they are the regular beasts of burden of the Kirghiz, and are invaluable, since they can safely cross a mountain pass which is impracticable for laden horses. Our horses from here consequently were unladen.

Leaving the valley of the Karakash River, we proceeded up the narrow ravine leading to the Sanju Pass, the stream in which was quite frozen over, while our journey lay between rocky precipices towering above the narrow defile. As no wood for fuel is obtainable near the pass, it was necessary to load one of the yaks with wood gathered in the lower part of the valley; and grass for our horses was also carried by the Kirghiz who accompanied us. We encamped that night 2 miles below the pass at an elevation of 14,474 feet above the sea. The last part of the way was over some difficult ground, where the ravine is much contracted, and the road lay over the frozen surface of the stream. Our camp for the night was formed under some overhanging rocks in the defile well sheltered from the wind; and Roza Khoja, the Panja-bashi, at once commenced dispensing Turki hospitality by spreading out a "dastarkhan" of bread, dried fruits, and cakes, as we sat by a blazing fire. Already the Turkies had impressed me with a favourable opinion of their good intentions towards their visitor; and from their frank and courteous, yet independent hearing, I was inclined to regard them in a most friendly light. We went on again up the pass at daylight, the last part of the ascent being very steep and over rocky ground, but the yaks we were riding carried us well right up to the summit, which is 16,612 feet above the sea. From the summit of this, the last pass into Eastern Turkistan, the country on the north side lies far below. Looking back are seen the snowy peaks of the Kuen Luen beyond the Karakash River, and the Sooget Hills beyond Shadula. I was disappointed in my expectations of being able to see the plains of Turkistan in the distance, since a haze overhung the lower country, and light clouds drifting over the intervening mountains obscured the view. Down the north side of the pass the descent is very steep, and many accidents occur from horses slipping on the ice, which lies during the winter on this side the summit.

On my interpreter going to Yarkand, bearing my letter to Mahomed Yakoob, the King, the horse of the sipahi, who had accompanied him, had slipped on the ice when crossing this pass, and falling down the precipitous ravine had been killed. We therefore waited on the pass until the horses came up, in order to see them safely over; I had already lost two horses during the journey from Ladak, and could not afford more accidents. I much doubt if one could ride a horse over the

Sanju Pass, on account of the steep ascent and rarefied atmosphere.

When the merchants cross this pass with their caravans, they are obliged to obtain yaks from the Kirghiz to carry their goods over; and thus often experience serious delay in procuring them at once. The Kilian Pass is quite as, if not more difficult, while the Kullik Pass is even worse. They are all simply impracticable for laden horses and camels, and for any animals except yaks; and there can be no doubt that the true road into Eastern Turkistan is that conducting down the valley of the Yarkand River, and across the Yangi Pass to Kugiar, Karghalik, and Yarkand.

Mounting the yaks below the pass we again rode on down the valley, and striking the head of the Sanju River, continued down it to a Kirghiz encampment, at 14 miles from the pass. We had been descending rapidly the whole way, as this place is at an elevation of 9123 feet above sea-level, the lowest altitude which I had reached during nearly four months' wandering, having for that time lived at elevations varying from 13,000 to 17,000 feet above sea-level. We were now evidently nearing the plain country, since the mountains here now slope rapidly to the north. The range of the Kilian Mountains, which we had crossed by the Sanju Pass, is a spur from the Western Kuen Luen, and bounds the Karakash Valley on the north, after that river has pierced the main chain, at the same time throwing out transverse spurs running down into the high table-land of Tartary, dividing the minor rivers of Oglok, Oshokwas, Kilian, Sanju, and Arpalak. These streams all run to the northward, the Oglok and Oshokwas joining the Tiznâf River; and the Sanju, Kilian, and Arpalak streams, after fertilising the country which they traverse, being finally lost in the sandy wastes of the Takla Makan Desert.

Our approach to the Kirghiz encampment was greeted by a loud barking of dogs and the shouts of the young Kirghiz, who came forward to lead away our shaggy steeds as we dismounted. I was then conducted to an empty tent, and seating myself by the fire, tea and cakes of Indian corn were placed before me, while a sheep was brought and killed as a token of Kirghiz hospitality. The firearms with me attracted the attention of the Kirghiz above everything; and the leading men of the encampment were delighted at being shown how to fire off a revolver. Powder seemed to be in most request with them, as they possess matchlocks and are keen hunters. Their chief sport is in hunting the ibex found in the Kilian and Kuen Luen mountains, which they pursue with the aid of their dogs. This species of ibex seems to differ from the Kashmir kind, and is the same as the black ibex of Baltistan, judging from the description of it given by the Kirghiz. I was unable to obtain a skin of this species, which is found on the lower slopes of the Kuen Luen, but 1 saw some ibex horns in several places; and these differed somewhat from the horns of the Kashmir and Ladak species in being thinner, and not having the knots so well defined. As this is a peculiarity of the Skardo ibex, it is probable that the Kuen Luen and Baltistan ibex are the same species.

There are some 20 tents of Kirghiz in the Sanju Valley, under a head-man, or chief, Aksakal. The valley from the Kullik Pass is frequented by some 25 tents, whereas the valley of the Tiznâf River from the Yangi Pass downwards contains about 120 tents of a tribe of Kirghiz, called Phakpook. Besides flocks of sheep and goats and herds of yaks, these latter possess large numbers of fine dromedaries, or the two-humped Bactrian camel, which might be used extensively for traffic between Yarkand and the north-west provinces of India, on the route up the Yarkand River to Aktâgh, and thence to Chang Chenmo direct. The Kilian Valley is frequented by a totally different race to the Kirghiz, being Wakhanees from Wakhan in Badakhshan. There are about 40 tents of these people who came over *viâ* Pamir Khurd from the Wakhan some forty years ago, and they and their descendants have remained here since. They at different times frequent this valley, and some of the lower valleys of the Yarkand River and the Sarikol district. Speaking the dialect of Wakhan besides Persian, and being of the Sheeah sect, they will not associate or intermarry with the Kirghiz, who alone speak Turki, and are of the Sunnee sect of Mahomedans.

The Kirghiz are the Bedouin Arabs of Central Asia, and like those children of the desert possess no fixed habitation, but move about amongst the valleys of the mountains with their flocks and herds. Their tents are made of felt, in shape circular, supported on cross pieces of wood constituting a framework against which rush matting is fixed in the interior of the tent. Felt carpets cover the floor, in the centre of which is the fireplace. A circular opening in the roof of the tent which can be covered over and closed at will, gives escape to the smoke. The tents are very comfortable and impervious to rain, and will last a dozen years. Each family possesses one, and when a young couple marry, a separate tent is provided for them, and they henceforward constitute a separate family in the encampment. An average sized tent is about 16 feet in diameter, and when moving to a new encamping-ground all at once assist in striking the tent, which is then packed on three yaks, and the Kirghiz move off to "fresh fields and pastures new." They are an excessively hardy race, as men, women, and children alike brave the rigours of winter and the heat of summer amongst their native mountains. Their food is the produce of their flocks and herds, which are driven to pastures and tended by the men and boys during the day time; and on the approach of evening return to their folds in the encamping-ground. Here the women milk the yaks and goats, make butter and curd cakes, and are employed during the day in carrying water, or weaving the warm material which the wool of their flocks affords them, into articles of wearing apparel. In appearance they are seldom attractive, and are short and robust. The men are low in stature, and generally of spare, wiry frame, with high cheek bones, a low and slanting forehead, and a broad flat nose. Their complexion is a yellowish brown, with a ruddy tinge, and they are mostly devoid of beard, with very little hair on the face, while their features unmistakeably exhibit the true Mongolian Tartar type.

The following day we reached the small village of Kibris, the first habitation met with on entering Turkistan from the Sanju Pass. It is 19 miles distant from the encamping-ground of the Kirghiz, and 36 from the pass. Our road this day lay down the Sanju stream, which had to be crossed and recrossed from 20 to 30 times. It was in part frozen over, and in many places the passage was difficult. In the summer this road is often rendered impracticable by the large volume of water in the stream, which is then unfordable. Another route is then used by travellers, which crosses a low pass 11,847 feet above sea-level to the east of the Sanju Valley, and conducts down the Arpalak Valley to Sanju.

On the 21st December we arrived at Sanju, a district containing some 3000 houses, comprised in several villages situated on each side of the stream in the Sanju Valley. Ilchi, the capital of Khotan, lies east from here at the distance of some 66 miles, or three days' journey.

The day we entered here was " Du Shamba," or Monday, on which day the bazaar, or market, is held. Each town and village in Turkistan has its fixed market-day once a week, and the Sanju one being on a Monday, is called the "Du Shamba" bazaar. The place was, therefore, more astir than usual, and we passed many villagers riding in with their country produce. They all wore the costume peculiar to the agricultural classes throughout Turkistan, consisting of a round cap, lined with sheep or lambs' wool; a loose " choga," a description of loose coat confined by a roll of cloth at the waist, and lined with wool or sheepskin; and felt stockings, with boots of untanned leather. Their costume is nearly all of a grey or drab colour; but on the occasion of some festivity they perhaps don a more gaudy coat, and wear a turban of white or coloured material.

Some distance from the principal village we were met by some thirty mounted sipahis, sent forward by the Hakeem Beg, or chief official, to escort us in. The dress of the sipahis is somewhat similar to that of the villagers, with the exception of being much gaudier, most of their long chogas being of silk of the most striking and absurd colours. They wear the pointed cap and turban similar to the Afghans, but wind the turban, which is generally of white muslin, in a different way.

For arms they possess a matchlock and tulwar (sabre). The former are long, rather rude-looking, weapons, of inferior manufacture, with hardly any stock, consequently ill-balanced, all the weight being at the muzzle. Fitted to the barrel is a cross-piece for a rest to fire from when dismounted, and in general appearance the weapon approaches closely to the native matchlock of the north-west frontier of India, save that the stock is more curved. Their accoutrements consist of two leather pouches, and a wooden powder-flask. A coil of thin rope is carried, attached to the waistbelt, with which, lighted, they fire off their matchlocks, and notwithstanding it is often a tedious process, as perhaps the rope will not light, the powder ignite, or there is a hitch somewhere, they manage to make very fair practice. Like the Afghans and frontier tribes, they would be found to be no despicable enemies in mountain warfare, in which their strength would lie, since in the plain country, however brave they might prove, they could not contend successfully against European arms. They are excellent horsemen, and at once strike one as good irregular cavalry. Their horses are good, though rough and small, while for horse-accoutrements they carry a small wooden Andijâni saddle, with a thick felt saddle-cloth, embroidered with silk, and a plain bit with leather reins. These saddles, made in Khokând, are of good manufacture, and are very durable and lasting. I have seen a horse fall down a ravine and killed, when the saddle on his back was not in the least injured. In front the saddle rises to a small peak, which is used for slinging the matchlock and sabre upon when dismounted, or on the march; and this custom, as also the shape of the saddle, would indicate a similarity to the old Scythian manner of slinging the mace and battle-axe.

The Panjsads (officers commanding 500 men), Yuzbashies (commanding 100 men), Panja-bashies, Jemadars, and people of higher rank, are better dressed and mounted, many of them riding good Turkoman and Andijâni horses. Their weapons are the same, but better finished, whilst their accoutrements are superior, being mounted in gold and silver and studded with

precious stones. Their horses carry silver-mounted bridles, with
silk and embroidered saddle-cloths. Their dress is the same
universal choga of silk, lined with fur or lambswool, or two or
three perhaps, one worn over the other, which are all of the
most striking colours, large stripes and cross patterns. A pointed
velvet and gold embroidered cap adorns the head, round which
a snow-white turban is twisted in many folds. They wear long
leather riding-boots, which are called "ittook," and all carry
the short Kalmâk whip. Their sabres are similar to the Turkish,
having a curved blade and cross hilt, often inlaid with gold and
silver, and stones. It would be difficult to imagine a more picturesque sight than that displayed by a band of these wild
Tartars, as they gallop in their silk array over the plains of
Central Asia, or engage in their national game of "oglok."
True descendants of those conquering hordes which burst like a
torrent upon Europe in the thirteenth century, and overran Asia
Minor, Persia, and India in the next century under Tamerlane,
their restless love of adventure and wandering render them good
irregular troops, ever ready for a hostile combat, or a raid of
rapine and plunder; and in the hands of a stern commander
and sagacious tactician, like their own victorious ruler, Mahomed
Yakoob, the "Atalik Ghâzee,"* or defender of the (Moslem)
faith, they are likely to prove themselves formidable foes.

It was most refreshing to see houses and cultivated land, with
all the appendages of rural life, after so many months' wandering
amongst the mountains; and though I saw Sanju at the most
unfavourable time of year, it was still striking as a picturesque
landscape. The villages are very different to those met with in
India, and instead of being crowded and huddled together, as is
seen there, two or three houses together form a small homestead,
surrounded by enclosures for cattle and horses, while outside are
rows of fruit-trees, thus giving to a village a truly agricultural
appearance. Cattle, horses, sheep—the broad-tailed sheep of
Central Asia—the shawl-goat, and numbers of fowls and pigeons,
fill up the pleasant picture. The houses, though built of unburnt
brick, are good, comfortable, and clean. Entering the outer
door of one of them the visitor finds himself in an open enclosure,
or court-yard, surrounded by a verandah, on each side of which
are the rooms, while a third, or open space, leads to the enclosures
where horses and cattle are kept. The rooms are generally small,
with recesses in the walls, shelves and small cupboards, while
the walls are perhaps coloured white or yellow. The roof is of

---

* Atalik is an old Turki title, like "father chief," the *Khanbaba* of the
Afghans, and presumably the original of the Hun Attila; and Ghazi is a title
(literally " ravager ") assumed only by those engaged in war with infidels.

wooden rafters, fixed close together and carved in many places, and the floors covered with mundas, or felt cloths, and Khotan carpets, for the production of which that city is celebrated. A fireplace occupies one side, facing which are the windows, generally small and covered over with glazed paper, which is the substitute for glass, as that article is unknown in Turkistan. The carpets and mundas are also used by the family for sleeping on, since a bed is a luxury almost unknown, or at least unappreciated in Eastern Turkistan, the few that were ever seen being rude manufactures of hard planks! Altogether a good house is picturesque and comfortable; and although no display of costliness appears, an air of comfort and homeliness pervades the place. The housework is all performed by the women, assisted by a Mahrum-bashi, who carries water and wood, and is answerable for the minor offices of the household. In the principal towns and cities many of the chief people possess several slaves. These, both male and female, have either been captured in war, or bought openly in the bazaars when brought for that purpose from Badakhshân and Chitrâl, in which latter country an open traffic in slaves is carried on, parents bartering their own children in exchange for merchandise with the Badakhshi traders. During the rule of the Chinese this disgraceful traffic was carried on to a great extent, secretly encouraged and supported by the Chinese officials in Yarkand, where the fame of the beauty and fair complexion of the Chitrâl women had preceded them, and caused them to be eagerly sought after in the slave-market. The Badakhshi merchants found the traffic to be profitable, for a beautiful girl purchased for 12 tillas (6*l.*) in Chitrâl, would sell for four times that amount in the slave-market in Yarkand. Besides Chitralies, the people kidnapped from Kunjoot, Gilgit, and Kafiristan were brought to Bokhara and Yarkand for sale. Since the expulsion of the Chinese from Eastern Turkistan, this inhuman traffic has entirely ceased in Yarkand, thanks to the prompt and vigorous measures taken by the Atalik Ghâzee to put a stop to it, who, immediately he had firmly established his rule, abolished the slave-market in Yarkand, and prohibited further importation from Chitrâl and Kunjoot. It is believed that the open purchase of slaves has ceased even in Chitrâl, since the merchants have now no mart in which to carry on their nefarious practice—Bokhara, as well as Yarkand, being now closed to them. In the country, however, amongst the agricultural classes, no slaves are met with. The house-proprietor's wife and daughters manage the household and do the cooking. The country women also go about with their faces uncovered, which is never the case in the towns and cities, where a female, though allowed perfect liberty, is forbidden to be seen out of doors without her veil, and, if caught transgressing this decree, is severely punished by the "Kazi," or head judge. To enforce the due observance of this order, the Kazi may be seen daily parading the streets of Yarkand and Kashgar, preceded by six or eight men armed with long leather whips, whose duty it is to seize and severely chastise any man neglecting to say his prayers at the hours appointed by the Kôran, and any female seen out of doors without her veil.

The costume of the women of Eastern Turkistan is in accordance with all Oriental ideas, and perhaps somewhat assimilates to the dress of their sisters of Constantinople and Turkey in Europe. The country women wear a loose cholah, somewhat like that worn by the men, but shorter and more adapted to their stature. This is of light material for summer, or of warmer material, lined with wool, suitable to the rigour of an Asiatic winter. A round silk cap in summer, or a cloth one, lined with fur or lambswool, in the winter, covers the head, under which a white veil is worn, which is allowed to flow behind. They wear long leather boots of different colours, green and red prevailing. A town beauty, however, is much more gorgeously dressed. She is arrayed in one or two silk cholabs, sure to be of gaudy colours, lined and trimmed with fur and lambswool, while her head-dress consists of a circular silk cap, richly embroidered, for summer wear; and for winter, of a high black lambswool cap, turned up and trimmed with fur, or beaver-skin. Under this a white veil flows behind, when in doors, or is worn over the face when without. Her feet are encased in long red leather boots, embroidered with silk tassels, which are quite as long as riding-boots. They wear no ornaments, like most Eastern women, which is certainly an advantage; for in Ladak and Kashmir, for instance, a female is often so bedecked with rings and ear-rings, or a heavy weight attached to the back of the head, that she appears to walk with difficulty. They have not the fine eyes and gait of the Cabul or Chitrâl women, but are certainly handsome, with round pleasant faces, rather low in stature, and robust, while their complexion is of the healthiest. Not content, however, with the charms which nature has given them, they plait long masses of horsehair in their own hair and wear it in two long thick tails flowing down the back; consequently horsetails are at a considerable premium in Turkistan.

The hospitality of the Turkies is far-famed, and I experienced it everywhere to the utmost. Every day on the journey into Yarkand I received a " dastar khan," or present to a guest, consisting of a sheep, fowls, eggs, bread, butter, dried fruits of every description, &c., sent by the Hakeem Beg, or chief official of the towns and villages, where we remained the night. The hos-

pitality everywhere tendered was simply as uubounded as the
" dastar khaua" were profuse. Everything was at once supplied
in far greater abundance than sufficed for myself and servants.
On dismounting in the court-yard of the house prepared for me
at Sanju, I was conducted into a room where tea and cakes were
served, and the usual preliminaries of receiving a guest gone
through. It is imperative, on entering a house, to invoke a
blessing on its inhabitants, in order to secure their good-will;
and to express good wishes for their health and prosperity, by
the usual salutations of courtesy; and on seating oneself in the
room it is usual with visitors—at least with those of the Mahomedan religion—to accompany such expressions by stroking
the beard and uttering the formula, prescribed by the Koran, of
"Allah ho Akbár." The Turkies never sit cross-legged,* like
the natives of India, but assume a kneeling position, with the
feet behind. This custom should be known to a visitor, or in
ignorance of their manners he may probably sit down in the
fashion adopted by Indian Rajahs and Nawâbs, and thus unin-
tentionally offer a deadly insult to his host.

Soon after I had arrived at Sanju the Hakeem Beg came to
an interview. He is an Uzbeg, and a good specimen of the
country. By means of my interpreter we conversed together;
and shortly after the usual conversational etiquette had been
gone through, he expressed a wish to see the firearms and
weapons which I had with me. Himself a soldier, and one who
had seen service, and now suffering from the effects of a severe
wound, received at the capture of Khotan in 1866, he said he
was delighted to see and handle weapons of European manufacture; adding, that although Allah had given to an Uzbeg
his horse and sabre, yet he had made no provision for furnishing
him with a rifle, wherewith to fight his foes, thereby implying
the inferiority of the Turki matchlocks compared with a rifle
of European manufacture. Before leaving he made a request
for some medicine, which I gave him, as he was suffering from
bad rheumatism, and he sent the next morning to say that,
having taken the medicine the night before, he had experienced
much relief from the effect of it. I had brought a stock of
medicines, and was besieged with applications for them wherever
I went. A doctor would have plenty of work to do in Turkistan.
Having written instructions for dispensing the medicines, I
endeavoured to give them accordingly, as the complainant stated
his wants; and so never-failing was the virtue of physic considered to be, that the patient invariably declared that the remedy
had been effectual.

: * This is the rule throughout the East, with the exception of India.

In only one instance did the patient seem dissatisfied with the result. An old man, afterwards at Kashgar, who was suffering from fever, applied to me for some medicine; I accordingly dosed him with some quinine, and he was well again in four days, but he most ungenerously and ungratefully retorted that the "Feringhee Kafir" had given him some awfully strong physic—I believe he used the word poison—which had nearly killed him. The chief complaint seemed to be goître, from which the native Moghul population suffer more or less, while the Uzbeg invaders from Andiján and strangers do not incur it. It prevails in both towns and villages, though in a greater degree in the populous places, and its cause may be traced to the bad water accumulated in the tanks, which the inhabitants use. The story of its first appearance in Turkistan is as follows:—Several years ago a holy man possessed a wonderfully sagacious camel, which was in the habit of going into Yarkand daily for provisions, and returning unaccompanied by any one. The Posgâm people, having a spite against the owner, one day took counsel together to seize the camel as it was returning; and not only did they waylay the unfortunate beast, but killed and ate him on the spot. On hearing of this the holy man prayed that the perpetrators of the deed might become known by some sign in their throats, immediately upon which all the Posgâm people suffered from terrific goître. This is as the story is related, and with about as much truth in it as other Eastern ones.

Crossing the river at Sanju, the Yarkand road ascends the range of sand-hills bounding the valley on the north, and traverses a sandy steppe to the village of Langar, from where it goes forward across the dry bed of a river, coming down from the Kilian Mountains. Beyond this it again traverses an open plain, and descends a low ridge to the village of Koshtok, pleasantly situated on the right bank of the Kilian River, which village lies some 12 miles to south-west from here. The Kilian stream flows from here, with a north-east course, through the sandy steppe to Guma, a large place, containing some 6000 houses, situated on the direct road from Yarkand to Khotan. The sandy steppe which we crossed this day slopes down very gently to the plain country, which is here a broad belt of sterile soil stretching away to the eastward, and known as the Taklá Makán Desert, beyond which again lies the "Dusht-i-Tâtar;" towards the Great Gobi Desert of Central Asia. The whole of this vast tract of country is a sandy or stony plain, with tracts of clayey soil, impregnated with a saline incrustation, rendering the little water here and there obtainable brackish and unwholesome. The surface is by no means a dead level, but in places is raised into ridges and hillocks of sand, drifted by the action of the wind.

As we journeyed on towards Koshtok, the sipahis of the escort amused themselves by chasing each other on horseback over the sandy plains, or firing at a mark as they passed it in full career, while the Punja-bashi was relating to me the history attached by the Turkies to this Taklá Makán Desert. "Many hundred years ago," he said, "all this wilderness was a flourishing province, possessing 160 towns and cities, now entirely overwhelmed with sand and buried. During their populous and thriving state, a great Moulvie, or priest of the Moghuls, visited them, and requested the ruler of the province to persuade the inhabitants of this territory to embrace the Mahomedan religion. After some demur, this the ruler agreed to, on the condition that the priest should furnish him with as much gold as would suffice to convert the doors of his palace into that precious metal. On hearing this demand, the priest invoked his deity to such effect, that on the instant the ruler's palace and everything in it was turned into solid gold and silver. Upon this the ruler laughed, and turned the priest away, saying, there was no necessity for him to force his subjects to become Mahomedans, since he had now obtained his desire. Angrily the great Moulvie turned to depart, but had not yet disappeared from view when black clouds overshadowed the glittering palace, and a tremendous sand-storm came on which buried the palace, the cities, and all the inhabitants beneath its overwhelming fury. There are said to be fabulous quantities of treasure buried in this desert, and many people, attracted by the hope of gain, have endeavoured to reach the buried cities, but though many have departed into the Taklá Makán, no one, so the Turkies say, has ever returned to relate his experience. Their camels have been lost in the sandy wastes, and the owners in search of an *ignis fatuus* have perished with them. Not only do wild camels frequent the desert, but even human creatures are said to have been seen, that are covered with hair, wear no apparel, and live on what they can pick up in the jungle." The Punja-bashi was quite indignant at my suggesting monkeys; and declared he knew people who had seen them frequently, but that these creatures always ran away laughing on being approached.

The Karakash River skirts the edge of this vast tract of uninhabited country, and the gold found in that river is said to be washed up from the buried treasures in the Taklá Makán Desert. So strangely do superstition and the love of gain actuate the mind of a Central Asiatic.

To return to our journey from Sanju. I had obtained an

observation for latitude about a mile from the principal village, at the foot of the range of sand-hills, bounding the valley on the north; which gave a value of 37° 12' 30" N., and by observation of the temperature of boiling water, an elevation of 6420 feet above sea-level. The utmost caution was now necessary in using scientific instruments, when amongst such a suspicious race as the inhabitants of Turkistan. To display them openly would have been highly dangerous to the success of the enterprise, and to be detected doing anything in secret would have been as sure to lead to some unpleasant complication, if not absolute danger to life. As opportunity, however, offered, I succeeded in taking observations whenever the escort left me alone, or by riding away from the road for a few minutes; nevertheless from this time until I left the plain country, and once more reached the mountains, on the return journey, it was impossible to use the scientific instruments openly as before; and all observations had to be made with the utmost caution and circumspection. It was advisable even not to be seen writing too much, for the suspicion of an Asiatic once aroused, it is hard to allay, and correspondence in his eyes is conspiracy. Let him once imagine he has cause to look upon any one as endeavouring to compass an object, upon which he could put the construction of harbouring the remotest intention of anything apparently hostile to his rulers, and consequently to himself, and he will ever afterwards prove, at least, a faithless friend, if not an actual foe.

From Koshtok, the Yarkand road skirts the sandy steppes; and passing through the small village of Oitogrok, situated in a valley, watered by a branch stream from the Kilian River, ascends another range of sand-hills, and traverses the steppe to where it descends to the larger village of Borah, pleasantly situated in an open valley, through which runs a stream coming down past Oshokwas, a large place of some 1300 houses, lying south from here. The Kilian route joins the Sanju one at Borah and from here they are the same to Yarkand throughout. From Borah the road traverses another sandy steppe; and then descends to the true plain country of Eastern Turkistan, or more strictly speaking the high table-land of Central Asia, which has here an elevation of about 4500 feet above sea-level. From the foot of the low hills our way lay across a barren tract of plain country, the soil of which is gravelly, and covered with loose stones, until passing the widely scattered village of Beshiruk, the more fertile country is reached.

On the 25th December, Christmas-day, I had hoped to have reached Yarkand, but we did not enter the capital of the

Moghuls until two days later. On Christmas-day we arrived
at Karghalik, situated 79 miles from Sanju, and 36 miles from
Yarkand. This is a large town and district, comprising some
20,000 houses, and possessing a large bazaar and several cara-
vanseries; and is a place of considerable importance from being
situated at the junction of all the roads, debouching across the
Karakoram Range into Turkistan, from Kashmir, Ladak, and
India, as well as the Khotan road through Guma. Outside the
town, at the junction of the Khotan road, stands a new earth
fort, with four towers, and a dry ditch, which was originally
built by the Tungâni in 1865, when they feared Habibula Khan,
the ruler of Khotan, invading Yarkand. The fort lies to the
west of the road, and is now unoccupied. Karghalik is a thriving
place of trade, since all the traffic coming from Khotan and the
district eastward passes through the town, which is watered by
a canal cut from the Tiznâf River.

I was conducted to a most comfortable serai, and immediately
afterwards the chief official of Karghalik, a fine-looking old
man, by name Ibrahim Beg, came to an interview. The
"dastar khan," which he sent, was most profuse; and exhibited
the most unbounded hospitality. It comprised two sheep, a
dozen fowls, several dozens of eggs, large dishes of grapes, pears,
apples, pomegranates, raisins, almonds, melons, several pounds
of dried apricots, tea, sugar, sweetmeats, basins of stewed fruits,
cream, milk, bread, cakes, &c., in abundance. In fact, it was
enough to feast thirty or forty people, and, although there is a
saying in Turkistan, that whoever has once tasted Turki hospitality is so charmed therewith, that he never wishes to leave
the country afterwards, which means that he is not allowed to;
still one could not but confess that however treacherous the
Atalik Ghazee might be, he certainly had no intention of killing
his guest by starvation.

The following morning, after receiving Ibrahim Beg's pro-
found salâms, and being the observed of all observers, as we
rode through the bazaar of Karghalik, we proceeded on towards
Yarkand. From here the distances are marked by "tashes,"
being a rude sort of sign-post, with a flat board nailed to it,
on which the number of tashes are written in Persian. The
"tash," * so named from the pile of stones at the foot of each
post, "tash" meaning stone in the Turki; corresponds, I believe,
to the farsang of Bokhara and Western Turkistan; and is said

---

* *Tash* may be understood as a "mile-*stone*;" *Aghatch*, in Azerbijan, used in
the same sense, is a "mile-*post*." These distances are really the space traversed
by a horse walking in an hour, and usually, in Turkey and Persia, range from
3 to 4 miles; but in Khorassan, where the horses amble, the farsang reaches
5 miles, and Hayward's estimate, therefore, is not excessive.—[NOTE BY SIR H. C.
RAWLINSON.]

by the Turkies to measure 12,000 paces. The farsang has been variously stated to be 3½ to 4½ miles; and I calculated the "tash" of Eastern Turkistan to be from 4½ to 4¾ miles. As a general standard for computing distances roughly they are reliable. Thus from Yarkand to Khotan is 37 tash, or about 176 miles; Yarkand to Kashgar, 26¾ tash, or about 125 miles; but observations for latitude and bearing of points on the route are the more reliable data to depend upon for obtaining correct positions.

The whole country from Karghalik is profusely irrigated by the Yarkand and Tiznâf rivers, and is well cultivated and thickly populated. Large villages are seen on every side, embosomed in fruit trees of every description, while the road itself is flanked by mulberry and poplar trees. Rice, wheat, barley, Indian corn, carrots, turnips, clover, &c., are grown in great abundance, while cotton is largely cultivated. Flocks of sheep and goats are everywhere seen, and the quantities of fowls and pigeons are very great. I noticed very few ducks and geese, but quantities of wild fowl in the streams and rivers. The sheep are all the broad-tailed species, and one specimen was seen which was quite a curiosity. This is a species of sheep with four horns, one pair curving backward like an ibex's horns, and the other pair forward over the ears. The cattle appeared to be small and indifferent, and in colour mostly black and red. Eleven miles beyond Karghalik we crossed the Tiznâf River in lat. 37° 51' 35" N.

The stream at this time of year was of inconsiderable size, partly from the volume of water carried away by the canals and dykes to irrigate the cultivated lands; and partly from its diminution caused by the little snow now melting on the southern slopes of the Western Kuen Luen, from which it receives the drainage.

Continuing our journey the road passes the villages of Khojerik, Alamakun, Boghorlok, and Meklah, immediately beyond which is " Yak Shamba " Bazaar, a large market, and, as the name implies, crowded by the country people on Sundays. Beyond this is the town of Posgâm, at a distance of 21 miles from Karghalik. It is a large place, and with the immediate suburbs comprises some 16,000 houses, with a long bazaar and a large caravanserai. The town is watered by the Beshkun Canal cut from the Yarkand River, a wooden bridge crossing this canal in the centre of the main street leading through the bazaar.

A considerable amount of traffic appears to be carried on. As we rode through the main street it was crowded with people hurrying through the bazaar, while articles of merchandise were being carried in every direction, laden on horses, camels, and donkeys, which latter animal abounds in Turkistan, and is made use of for conveying everything transportable.

The main street or bazaar is covered over with a rude roof of matting, which affords a shelter from the sun. On each side of the way the shops are placed, consisting of mere booths ranged in front of the houses, and generally mixed up with no particular regard to the distribution of wares. Butchers and bakers, silk and cap vendors, vegetable and fruit sellers, all ply their several vocations together amidst the din and hubbub peculiar to an Oriental mart. After passing through part of the bazaar, the road runs up the right bank of the canal to the caravanserai situated on some slightly elevated ground. The Serai itself is a large open enclosure, flanked by rows of trees, and surrounded by long sheds for stabling horses, while the east side of the enclosure is occupied by buildings containing several comfortable rooms for travellers.

The plain country extending from Karghalik to Yarkand seems to slope very gently to the banks of the Yarkand River. Observations of the temperature of boiling water showed the elevation of the town of Karghalik to be 4570 feet above the sea; that of Posgâm 4355 feet; and the bed of the Yarkand river near Posgâm, 4180 feet.

On the 2'th December, I entered Yarkand, the capital of Eastern Turkistan, so long deemed unapproachable and impracticable to Europeans.*

No European in later times had succeeded in penetrating to Yarkand and returning. The only one who had made the attempt, besides Mr. Shaw and myself, was the unfortunate Adolph Schlagintweit, who in 1857, after undergoing many dangers in Yarkand, succeeded in reaching Kashgar, where he was foully murdered by a scoundrelly robber named Wullee Khan, who had temporarily seized the supreme power; and who, strange to say, met his death at the hands of the Atalik Ghazee, the very man we were going to see. The high honour of being the first British subject to have reached Yarkand belongs to Mr. Shaw, who had preceded me from Shadula, whilst I was exploring the Yarkand River, and had arrived at the capital some ten days before.

After leaving Posgâm, the road continues in a northerly direction, and at four miles from that town crosses the Yarkand River. The bed of the river is here not less than a mile in width from bank to bank, and three branches of the stream have to be crossed. At this time of the year they are not deep,

---

* Serjeant Ephraimoff, a Russian captive who escaped from the Uzbegs, passed through Yarkand at the close of the last century, and published his travels on his return to Europe.—[H. C. R.].

and are fordable on horseback; but in the summer this is a very large and rapid river. Taking into consideration the whole of the drainage which it receives from the northern slopes of the Karakoram Range, the Western Kuen Luen, and the Sarikol district, and estimating the number of canals and branches carrying away water to irrigate the extensive tracts of cultivated land, the volume of water brought down from the mountains by this river during the spring and summer months must be very great. From its source to this point, where it is crossed on the road to Yarkand, the river has a course of nearly 420 miles, which gives a mean fall of about 30 feet per mile. The Yarkandies themselves say that six other large rivers unite with it below Aksu before it enters the Lob Nor. During the summer it is, where we forded it this day, quite impassable, and is then crossed in boats at Aigachee several miles lower down, which then becomes the direct road from Karghalik to Yarkand.

Our road this day lay through the most fertile country, pleasantly situated villages surrounded by cultivated land appearing on either hand. The road was crowded with people and animals, strings of camels and donkeys carrying bales of silk and goods from Khotan, country produce going in to market, with men and women riding ponies, the latter astride the saddle like the men.

Passing Otunchee, a widely-scattered village to the east of the road, and crossing the Yulchak Canal conveying water to the capital, the road continues on across the open to Yarkand. Meeting the Wuzeer and *suite* sent forward to escort me in, a great salutation took place. We both dismounted and shook hands; and after the usual mutual inquiries after one's health, and his expressing much concern in hoping I had experienced no difficulties *en route* from the cold and the mountains, and had received everything required since being their guest, we again mounted and rode on together, being now followed by a retinue of some 40 mounted men.

The Wuzeer was a good-looking man, of about 35 years of age, but with the unquiet, suspicious eyes of the Asiatic— gorgeously arrayed in silk and embroidery—with jewelled accoutrements and sabre, and riding a fine grey Andijâni horse. A short distance further, and the city of Yarkand was in sight. The day was very gloomy, since it had snowed slightly in the morning, and a thick haze overhung the horizon. Consequently the scene did not appear to the best advantage, for it simply seemed to be a confused mass of mud houses surrounded by a high fortified earth wall. A few mosques and higher edifices reared their summits above the general level of the walls and houses, whilst gardens and trees, now quite devoid of foliage, were interspersed amongst the mass of buildings.

The city itself lies in the form of a parallelogram, being some two miles in extent from north to south, and 1½ mile from east to west; the walls thus embracing a circumference of nearly 7 miles. They are from 40 to 45 feet in height, of great thickness, with bastions at each corner, and intermediate flanking defences, and run nearly parallel with the four points of the compass. The city contains some 40,000 houses, and not less than 120,000 inhabitants. It is entered by five gates; from the entrance of the one in the west wall the main street runs nearly due east to the Aksu gate in the east wall. This street is very narrow, being not more than 12 feet in many places. There are 160 mosques, many schools, and 12 caravanserais, which are always crowded with merchants from every country in Asia.

Both the city and fort are supplied with water from several tanks, into which it is conveyed by canals cut from the river. These are frozen in the winter, and the supply is then stopped, but the tanks contain sufficient water for the consumption of the inhabitants until the regular supply is renewed in the spring.

The fort lies at the distance of about 500 yards to the west of the city, with which it is connected by a bazaar. The walls are 40 feet in height, 12 feet in breadth at the summit, and of great thickness at the base, are entirely of earth, and also run parallel with the four points of the compass. The fort is about square, each side measuring from 650 to 700 yards. It has a bastion and tower at each corner, with eight intermediate flanking defences in the several walls, the parapet of which is loopholed all round. The dry ditch, or moat, surrounding the walls does not follow the configuration of the flanking defences, but extends in a straight line, at the distance of some 25 feet from the base of the main wall, while a lower intermediate wall runs along the inner crest of the ditch. The moat itself is some 25 feet deep, 30 feet in width at the summit, and about 18 feet at the bottom. There are three gates—the east gate facing the city; the Khotan gate on the south side; and the Kashgar gate some 80 yards from the south-west corner of the fort, facing the west, and immediately behind the place of residence occupied by the chief authorities. These two latter gates are closed, the Khotan one being also barricaded with unburnt bricks. The only gate through which ingress and egress is permitted is the east one, and this is closed every evening at sunset, and opened daily at daybreak.

In the south-west corner of the fort is situated the " Urdoo," or place of residence of the chief authorities, surrounded by a wall of 30 feet in height, the entrance to which is on the east side, facing the main road which leads up to it through the bazaar of the fort. Occupying the north-west corner of the fort is the inner fort formerly the residence of the Chinese Governor and his officials. This is surrounded by a lower wall and dry ditch, while the main wall is 35 feet in height, fortified and loopholed. The walls of this inner fort are in ruins in many places, and the old palace of the Amban presents the same dilapidated appearance. The north-east corner of the fort is one mass of ruined walls and houses. No guns are anywhere mounted on the walls, for which, however, there are embrasures, as, when the Chinese held Yarkand, many were in position.

Clearing the south side of the city wall, we came in sight of this fort, or old Chinese quarter, which, in appearance, presents similar features to the city. The road crosses the drawbridge over the moat, and turning, leads through the east gate into the interior of the fort. As we rode up the main street, or bazaar, the place was crowded with people—sipahis leading their horses out to exercise, merchants passing to and fro from the city, women closely veiled, walking, or riding on horseback, while a lively traffic appeared to be carried on in the shops on either side the way.

Near the centre of the street we passed several guns drawn up in regular order on the south side of the road. They consisted of five long swivels, two small mortars, and five apparently 4-pounders, all mounted on carriages, with their ammunition waggons drawn up in rear, and ready for instant use. The gunners on guard pacing in front of them are immediately recognised as Hindustânees, nearly the whole of the Atalik Ghazee's artillery being served by natives of India. I afterwards conversed with several of these men, and heard related their antecedents and adventures. Many had come round from Peshawar to Cabul and Bokhara, and thence to Khokand and Kashgar, serving the different rulers of those countries, and then changing their allegiance, as fate or fortune ruled for or against them. Several of them had come over to the Atalik when he captured Khotan in 1866 from Habibula Khan, whom they had accompanied from India on his return from pilgrimage to Mecca; and a few, no doubt, were escaped mutineers of 1857. Mr. Johnson reported, on his return from Khotan in 1865, that he thought he had recognised the Nana Sahib of 1857 in one of Habibula's artillerymen; but there is not the slightest ground for such a supposition. The individual pointed out by Mr. Johnson took service with Kush Begie in the following year. This man I afterwards saw at Kashgar, but further than his being a native of India, and a military adventurer, there is not the slightest evidence to connect him with the Nana.

Dismounting immediately beyond the guns, I was conducted up a long open passage to the door of the courtyard of the house prepared for my reception, or confinement, as it may be termed, since during a stay of two months in Yarkand, I never went outside of the garden attached to the house I occupied, excepting when proceeding to interviews with the Governor, and on one occasion when I rode round the fort. I entered the house, and found it to consist of two rooms, small, but very comfortable, and the floors covered with excellent Khotan carpets. Shortly afterwards the "dastar-khan" of Mahomed Yanus Beg, Dad Khwah, the Shâghâwal, or governor of Yarkand, was brought in by the Mahrum bashees sent from the palace. It was very profuse, and I returned my best thanks, and sent to request the honour of an interview, which was accorded. Having dined and dressed in appropriate Oriental costume, I started for the "urdoo," or palace, escorted by a person of rank. At the distance of about 150 yards from the entrance to the passage of the house I occupied, the main entrance to the place of residence of the chief authorities is reached. The road to it is a prolongation of the main street of the bazaar, and passing through the gateway, a guardhouse is first noticed. A covered verandah occupies the front of the guardhouse, and extends over the way to the outer wall. Some twenty Turki sipahis were pacing the raised platform under the verandah, or were lounging about in different places; and preciseness and military order were at once apparent, as exhibited by their neat and soldierly bearing, and the display of their arms and accoutrements.

Passing from under the covered entrance, the visitor finds himself in a large open enclosure, comprising a garden and tank of water, flanked by rows of trees. The enclosure is subdivided by an intermediate wall, through which lies a way leading to the Kashgar gate immediately opposite. From this enclosure the inner side of the defences is seen. The main wall is crowned by a parapet, below which a broad way runs all round the fort. Steps at the corners and several gateways lead to the summit of the wall, while higher flights of steps conduct from the walls to the watch-towers at each corner. Facing the embrasures in the flanking defences, or bastions, are situated a row of wooden huts, formerly used as a shelter and cover for their guns by the Chinese. A second gate and guard-house conducts to a paved court of about 50 yards square, surrounded by a verandah, passing across which, an inner court of the same size is reached. The second court is also surrounded by a verandah on three sides, opposite to the entrance to which under the verandah, on the west side, are the rooms of reception. Not the least elegance or display appeared, but the place seemed to be excessively clean and neat. The official who escorted me stopping at the entrance to the inner court, a Yusawal-bashèe, dressed in scarlet silk and embroidery, came forward, and, wand in hand, led the way across the court, and up the steps of the verandah to the door of the reception room. With the exception of two or three Mahrum bashees (pages) the inner court and verandah were quite empty, and a deep silence reigned around. The room, to the entrance of which I was ushered, was a long, plainly decorated apartment, with a bright fire at the further end, in front of which two carpets were spread, covered with scarlet silk cushions. On one of these was seated a little man, plainly yet splendidly dressed in green silk cholah lined with fur, and a high fur and velvet cap. This was the Dad Khwah, Shâghâwal, who rose and came forward as I advanced, receiving me very graciously, and shaking me by both hands. Motioning me to be seated, I assumed a sitting posture on one of the carpets, while he resumed his own, and an interpreter was summoned. This man just entered the doorway, and bowed towards the governor to the very ground, the utmost fear being depicted on his face. By means of this interpreter and my knowledge of Persian we carried on a conversation; and before leaving, after half an hour's conversation, I concluded that the Shâghâwal was a very pleasant, agreeable, and well informed man. He was evidently well read, while his fund of anecdote was inexhaustible, and he appeared to be very keen and eager to acquire information regarding India and Europe in general. Tea, fruit, and sweetmeats were then served, brought in by a file of Mahrum bashees, and shortly afterwards I asked permission to leave. As I rose a "khillut," or silk dress from Khokand, was brought forward by an attendant, and in this I was enveloped. I then took leave, again shaking hands, and was conducted back to the house I occupied by the official who had escorted me. Before leaving, I had presented the governor with some firearms, ammunition, &c., and shortly afterwards a second "dastar-khan" from him arrived, and I was informed that provisions for myself, servants, and horses would be supplied regularly every day.

By the 1st January, a few days afterwards, it was evident that I should be well treated, and was in no immediate danger; but although not officially informed that I was not permitted to go about, the presence of a guard, or escort, outside the house was a sufficient hint, and I determined to wait a few days to see what would come to pass. My servants were allowed to proceed to the bazaar in the fort to purchase anything required, but not until after they had been nearly a month in Yarkand were they permitted to go outside of the fort into the city.

On asking to go about on horseback accompanied by an escort, I was told that it was not the "custom of the country"—the "Andijâni rusmee," the "more Usbeco"—to be allowed to do so until an interview with the king, who was at Kashgar, had taken place. The confinement was excessively irksome after such an active life amongst the mountains, but it was in vain to urge the plea of exercise being needful. The Yuzbashee and sipahis quoted the orders of the Shâghâwal, and he the higher decree of the Atalik Ghazee, at the same time endeavouring to atone for their apparent want of courtesy to their guests by dispensing the most open-handed hospitality. Mr. Shaw was at this time under the same surveillance as myself within 100 yards of the house I occupied; but during the five months we were in Yarkand and Kashgar together, we were not permitted to see each other, nor could we correspond openly; not until after we had left Yarkand on the return journey, and the imaginary danger of any conspiracy we might hatch was removed from the Turki mind, were we once allowed to meet. As Mr. Shaw had brought a caravan of goods, and had entered the country to trade, the Yarkandies were somewhat less suspicious of his motives, although the strict surveillance precluded all possibility of trading. There could be no doubt that they were very suspicious of my motive in entering the country; and as Yarkand is full of spies, many of whom are bitter enemies, false representations placed on the cause of our presence were not wanting. As an Asiatic alone travels from political, commercial, or religious motives, and can conceive no others, they were at the greatest loss to account for my coming amongst them on fair grounds, since I was excluded from all these; and it was impossible for me to give them a reason for the true cause of my presence amongst them, for exploration in their eyes is spying.

Opportunities, however, were not wanting of being able to look about a little during a two months' stay; and there were many secret friends who were willing to volunteer information respecting the country and the true state of affairs. Foremost amongst these were several Hindoo merchants who had arrived from the Punjab, and were living in one of the caravanserais appropriated to Hindoos in the city. The Hindoos themselves when in Turkistan are subjected to different rules to the Mahomedans, in not being permitted to wear turbans, and being obliged to conform to the order which renders it imperative on them to appear in dark-coloured robes, with ropes twisted round the waist. It is the duty of the Kazi Kalan to see to the due observance of this order by strangers, as well as at the same time to punish any Mahomedan found in the public streets without his turban. The order respecting the dress of the Hindoos appears to have somewhat relaxed of late, and there is every probability, should merchants of Hindoo caste continue to frequent the cities of Turkistan for purposes of trade, that this custom will become a dead letter. It is very desirable the order should be abolished, for the number of Hindoo merchants who yearly visited Yarkand is very small; and any custom they have to conform to which would be considered derogatory, at least, in their own country, tends to lessen their inclination to renew their visit. Still, the profits must be very great which can induce them to come so far, and endure all the hardships inseparable from such a journey.

Amongst the men from whom I obtained news was an Afghan, named Kureem Khan. Through the medium of my men servants, he sent me information respecting affairs as they daily occurred in Yarkand and Kashgar, and was of much use. Being unprejudiced, and regarding the Turkies in no very friendly light, he was not likely to exaggerate matters on their behalf; and the information he gave proved afterwards to have been generally correct. It was hopeless to expect to obtain any accurate news from the Yarkandis themselves; for if they volunteered any statement, it was generally in such a confused mass of contradiction as to render it almost impossible to get at the truth of anything, even after discarding the indiscriminate exaggerations.

Kureem Khan first stated, regarding himself, that he had left Cabul during the late disturbances there, and had come by way of the Punjab and Kashmir to Yarkand. with the intention of taking service under Kush Begie. He had hitherto been deterred from doing so by the representations of many of his fellow-countrymen in the employ of the Atalik, who had warned him, that should he enter his service, he would never be allowed to return to his own country. Alarmed at this, he had requested permission to return, but this had been refused; and he was now awaiting the Atalik's arrival from Kookhar to make his application in person, or enter his service involuntarily, should permission to leave the country be refused. He eventually proceeded to Kashgar, and there took service under Kush Begie. After volunteering some information about Khokand and Samarcand, Kureem Khan stated that the alarm and suspicion created in Yarkand by the report that fifty Englishmen were coming from Ladak was immense, and only somewhat relieved on the after discovery that the rumour was without foundation.   Still, so suspicious were the authorities that strict orders had been sent by the Atalik Ghazee, from Kashgar, to the effect that his English guests were to be treated well, and receive everything they required, but were not to be allowed to go about, for fear of carrying back any information respecting the condition of the country and the true state of affairs.   Kureem Khan, also, gave an interesting account of Badakshan and Chitral.   He had travelled through those countries himself, and averred there was a much shorter and easier route to India direct, *viâ* the Chitral Valley, than any of the longer routes by Ladak.   He stated his opinion, however, that it was impossible for an Englishman to traverse Chitral, as such, in safety ; but that, by assuming the disguise of an Afghan or Turki merchant, it might be accomplished from this side, since the Chitralis alone regard with suspicion travellers coming from the direction of the Indian frontier ; and not those arriving from the Bokhara and Turkistan side.   With a view to eventually returning to Peshawar by Chitral, it was my intention to make an application to the Atalik Ghazee, should my interview with him prove favourable, to be allowed to visit the Pamir Steppe from Kashgar, for if I found him open and in good faith, he might materially assist my views by affording me his assistance and protection as far as his territory extended.   Still, the strict surveillance under which one was placed foreboded but little chance of being able to obtain such a concession.   On the contrary, the very suggestion of such a step might cause him to become more suspicious, and make him feel sure that I had entered his country for a purpose hostile to himself.   Should, however, nothing of the sort be allowed, and I be permitted to leave the country by the route by which I had come, there would be no alternative but to retrace one's steps, and endeavour to penetrate to the Pamir by the way of Hunza and Nagar, or of Gilgit.

During my stay in Yarkand I succeeded in obtaining eleven observations for the latitude of that city, the mean value of which gave a resulting position of 38° 21' 16" N., and long. 77° 28' E., while several observations of the temperature of boiling-water showed an elevation of 3830 feet above the sea. These results all closely coincide with the values obtained by Major Montgomerie's unfortunate explorer, Mahomet Hameed, who died in Ladak on his return from Yarkand, under somewhat suspicious circumstances.   The position of Yarkand as deduced by Major Montgomerie, from the papers of Mahomet Hameed, was given as in lat. 38° 19' 46" N., long. 77° 30', and an elevation of 4000 feet above sea-level.

4. From Yarkand to Kashgar.

March, 1869.

During my stay in Yarkand I had several interviews with the Dad Khwah, and at length got off from Kashgar on the 24th February. An escort accompanied me under the command of Mahomed Azeem Beg, an Uzbeg who had followed the fortunes of Kush Begie since he had left Khokand. I found this man very communicative, and he never tired of relating their late campaigns, and extolling the military prowess and bravery of his leader and ruler, the Atalik Ghazee. Passing along the northward of the Torr, the storms of war and siege, which the ill-fated Chinese underwent, have left their traces in the marks of bullets and cannon-balls with which the wall is perforated.

From here the Kashgar road bears away west, passing the villages of Karakoom and Bigil to where, at four miles from the city, it crosses the Urpi Canal by a wooden bridge. The road is deep in dust, and the traveller is covered with it as it is kicked up by the horses. The road to Sarikol, and thence to Uakhan and Badakhshan, lies up the left branch of the Urpi Canal. It is regularly traversed by Badakhshi merchants, residing in Yarkand, who yearly take their caravans of goods across the Pamir Steppe to Badakhshan. Tash Kurgân (or Stone Fort), the capital of the Sarikol district, lies in a w.s.w. direction from Yarkand, at about 175 miles' distance; while the total distance to Fyzabad, the chief town in Badakhshan, is some 460 miles. A journey of from 7 to 8 days to Tashkurgan, and of 18 days to Badakhshan, is considered very rapid travelling; but the caravans of the traders seldom accomplished the whole distance under the period of one month. The road traverses a plain country for nearly 70 miles from Yarkand; and then crosses a low range into the Sarikol district; and, ascending the valley of the Charling River, crosses the Chichilik Pass, leading across a high spur of the main Pamir range into the Tashkurgan Valley. From Tashkurgan it crosses the pass at the head of the Sarikol territory, and conducts through Pamir Khurd into the valley of the Oxus. The road is practicable for laden horses throughout, and for camels as far as the foot of the Chichilik Pass from the Turkistan side; and from Badakhshan up to the head of Pamir Khurd from the westward. A second route conducts from Yarkand to Tashkurgan more to the southward, but it lies through a very mountainous country, and, as there high passes have to be crossed, it is consequently but little frequented, except by Kirghiz and foot travellers. This road bifurcates near Tashkurgan, one branch conducting down the valley to that place, while the other crosses the Karachunker Pass, and joins the main road to Badakhshan in Pamir Khurd.

Continuing our route towards Kashgar, the road skirts some marshy ground, lying 12 to 15 miles to the west of Yarkand, which is surrounded by low sandy steppes covered with reeds and a coarse kind of grass. The marsh gives rise to several streams, which flow with a north-easterly course across the road; and appear to continue into the wide expanse of desert lying in that direction. A low forest fringes the edge of the steppe, while the country more immediate to the streams is frequented by wild pigs, a few deer, and quantities of wild-fowl and pheasant are to be found in the more retired portion of the marshy ground. The Urpi Canal, which is crossed within an hour's journey from the city, is a branch from the Yarkand River, and fertilises all the country lying north and north-east of the city towards Aksu.

At dusk this day we arrived at Kokrubat, a village containing about 200 houses, the first stage from Yarkand, from which it is 22 miles distant; and the following day rode 27 miles further to Kizil, a large village of 500 houses. From Kokrubat the road skirts the Dusht-i-Hameed, a large barren tract of stony plain extending up to the Kiziltâgh range on the west. This plain presents the usual features of desert country, consisting of a hard gravelly and clayey soil, furrowed and intersected by watercourses, and covered with loose stones. A few stunted bushes and a little coarse grass are here and there seen, otherwise it is quite barren. Halfway between Kokrubat and Kizil, a halting-place, called Ak-Langar, is reached. It lies to the east of the road, and consists of a few houses in an open enclosure with the usual musjid, or mosque. The water here is derived from two wells in the enclosure, and is palpably brackish. Iron-ore abounds in the lower slopes of the Kiziltâgh

traverses a low ridge, and descends to the banks of the Sargrak, or Yanghissar River, coming down from the Kizil Yart range of the Pamir steppe, which is in sight to the south-west. On the left bank of the river the town of Yanghissar is situated. It comprises some 11,000 houses, and appears to be a considerable place of traffic. The market-day is Friday, on which day we arrived here; and the main street was so crowded that it was with difficulty we could ride through it to the caravanserai. The streets, like those of other towns and cities in Central Asia, are narrow and confined, while the shops consist of mere booths, upon which the shopkeepers display their goods for sale, which appear mixed up in the greatest confusion. Next to the shop of a vendor of silks and caps will be seen the stall of a butcher, reeking with horseflesh, which is the most popular article of consumption throughout Eastern Turkistan, many more horses than cattle being killed for food. At the distance of 600 yards from the town, towards Kashgar, lies the fort of Yanghissar. This was the first place which the Atalik Ghazee captured on his invasion from Khokand; and as we rode past it to the caravanserai, Mahomed Azeem, the Yuzbashee, related the different phases of the siege. The Chinese made a better defence of this fort than of any which remained in their hands during the rebellion against their rule in 1863 to 1865; and the Uzbegs themselves acknowledge that they here lost many men before effecting its capture. Nor does the treachery, which was a leading feature in the capture of the cities of Yarkand and Khotan, appear to have here run its usual course; since the Chinese defended the fort gallantly to the last, until starvation and hunger compelled them to succumb. Nearly all perished in the siege, and the few survivors were constrained to embrace Mahomedanism. The fort is much smaller than those of Yarkand and Kashgar, being only some 230 to 250 yards in extent. Each wall possesses three flanking defences, intermediate between the bastion and tower at the several corners. The moat is some 40 feet in width at the summit, and 36 feet in depth, which is also the height of the main wall. We remained five days in Yanghissar, living in a most comfortable serai, which the Atalik has lately had built for his own especial use, since he is in the constant habit of visiting Yanghissar from Kashgar. This was by far the most picturesque place which was seen in Turkistan; the great cause of its attraction being the magnificent view of the lofty Kizil Yart range of the Pamir, which is full in sight lying south-west and west. Contrary to the usual supposition that the eastern crest of the Pamir slopes down very gradually into the high plateau of Eastern Turkistan, or the high plain country of Central Asia, the range forming its eastern crest rises into a chain of lofty peaks of 20,000 to 21,000 ft. above sea-level, the spurs from which run down most abruptly into the high table-land below. The range thus presenting a steep face towards the plains of Eastern Turkistan, the slope of the water-shed will be found to be very gentle and sloping to the west-ward, while the waters issuing from the lake-system of the Pamir must, of necessity, drain into the basin of the Oxus. The Kizil Yart range is crossed by high passes leading on to the true Pamir, and it is exceedingly unlikely that any of the Pamir lakes drain to the eastward into the Kashgar River and its tributaries. A high peak in this range, known by the name of Taghalma, lies at the distance of 63 miles w.s.w. from Yang-hissar. This Taghalma Peak is the most conspicuous of any in the range, as seen from the eastward, and its approximate height was estimated by observations to be 21,279 feet above the sea. At the foot of this peak the Sargrak or Yanghissar River takes its rise, and another stream—the Hosun River—has its source to the north of the intermediate range, and flowing eastward joins the Khanarik River, which, also rising in the Kizil Yart range, effects a junction with the Kashgar River to the east of that city.

The town of Yanghissar stands on the left bank of the Sargrak River at the foot of the low sandy steppes lying between it and the Kizil Yart Range. These low hills appear in gentle undulations or broken up into long and narrow ridges. The soil is clayey and sandy, and but little cultivated, save in the immediate neighbourhood of the town and its adjacent villages. Broken up into ridges, as seen along the banks of the Sargrak River, this clayey soil appears in their horizontal strata covered over with drifted sand. The steep and rugged slopes of the Kizil Yart Range, sheering from precipitous scarps towards the crest of the range, meet the gentle slope of the table-land and close in the plain country to the west.

The town of Yanghissar was found by observation of the sun's meridian altitude to be in lat. 38° 52′ 3·4″ N., and by triangulation, and from its distance and bearing from Yarkand, the meridian of 76° 18′ E. has been assigned for its longitude. It is situated nearly south of Kashgar, from where it is 43 miles distant, and north-west of Yarkand, to which city the road distance is nearly 83 miles. The elevation of 4256 feet above sea-level was estimated from careful observation of the temperature of boiling water; and this elevation may be fixed as the mean height above the sea of the high plateau of Central Asia near the mountain ranges which bound it on the west.[*]

[*] Boiling water alone is a very fallacious test, so much depending on the normal state of the thermometer, and a single degree of temperature giving 550 feet of elevation; so that, unless the thermometer is finely graduated, differences of 200′

The successive gradations in the elevation of the principal towns of Eastern Turkistan have been based on data obtained from actual observations of the boiling-point of water: and these calculations should be of especial interest as tending to remove all doubt regarding the actual elevation of the plateau or table-land of Central Asia.

Continuing our journey towards Kashgar we left the serai at Yanghissar on the morning of the 4th March in the midst of a heavy snow-storm, which continued until we had accomplished nearly the half of our day's march. The road passing the Fort runs with a direction bearing E.N.E. to the village of Koomlok, surrounded by low domes of drifted sand, and situated six miles from Yanghissar. From here it bears to the north, and then again inclines to the north-west, passing a wide expanse of marshy ground occupied by small pools and streams which are enclosed by low sandy hillocks covered with reeds and coarse grass. The country beyond this is more fertile, and is irrigated by the Hosun River, which is crossed by a wooden bridge before reaching the village of Yupchan. This village is at an elevation of 4055 feet above sea level; and the land around it is well cultivated, being fertilised by the water conveyed to it by dykes cut from the Hosun River. The village comprises some 700 houses, and is considered to be the fourth regular stage from Yarkand, from where it is 104 miles distant.

Shortly beyond Yupchan we crossed the Khanarik River, coming down from the westward, and rising, it is believed, in the Kizil Yart Range of the Pamir. This must be a stream of considerable size when the volume of the water is increased by the melting of the snows which lie on the range. The width of the bed of the river is here not less than 700 yards from bank to bank. It must not be supposed, however, that the actual stream occupies a space approaching to such a breadth. On the contrary the banks of the river are very low, and somewhat resemble in appearance the banks of one of the Punjah rivers, in being fringed with strips of grassy land covered with low tamarisk jungle. The two separate streams flowing through the bed of the river were crossed by rude wooden bridges, while the actual breadth of the streams did not at this time of the year here exceed 90 feet. This Khanarik River in its course and direction most closely represents the Yaman Yar River of our maps, which is defined as having its source in the Karakul (Lake), the dragon lake of the Chinese mythology. Any

---

or 300 feet cannot be detected. Thermometrical, when compared with barometrical observations, are of course of value. All the recent confusion about the Upper Nile basin has arisen from so-called boiling-water observations.—[H. C. R.].

reliable information, therefore, which could be obtained with reference to this stream was of particular interest, as tending to decide the vexed question of the direction of the drainage from the Karakul—or Black Lake. Putting aside the genuine, or fabricated, account of the ascent of this river to the Karakul Lake, all evidence goes far to prove that no stream issuing from the Karakul is the source of any of the Eastern Turkistan rivers. But not until some explorer can really succeed in reaching the Karakul, will this problem be definitely solved. The people with whom I was mostly brought into contact during my stay in Turkistan, were naturally the ruling powers, the Uzbegs of Andijan; who, no doubt, gave their opinion, not from an actual visit to the Pamir, but merely from hearsay. The Kirghiz who frequent the Pamir, and know every lake and valley in the mountains, were but seldom met with, and then under the circumstances of such strict surveillance as to render any attempt at a lengthy conversation out of the question. I was, however, fortunate in meeting the son of an influential Kirghiz chief when at Kashgar; and I obtained information from him with reference to the Karakul which, until further facts can be proved by an actual visit to the lake, is worthy of careful consideration. I will, however, defer relating what was stated with reference to the Karakul until the account of our journey to Kashgar is concluded.

Having crossed the Khanarik River the road continues past the large village of Tasgam, through fertile and well-irrigated country, to near Kashgar. The Fort is the first conspicuous object which meets the eye; and situated prominently in the open country on the right bank of the Kashgar River, its high walls and poplar trees attract the attention when yet at some distance. Two streams have to be forded before reaching it, one a canal from the Khanarik River, and the second a broad shallow watercourse supplied with water from the Kashgar River. Immediately beyond this the road passes several serais, and skirting the east wall of the Fort turns the N.E. corner and runs along the northward to the entrance gate facing the city. The walls are somewhat longer on the N. and S. sides, and measure about 600 yards in length. They are 40 feet in height, and surrounded by a lower wall and dry ditch, which moat is 25 feet deep and nearly 40 feet broad at the summit — this measurement making the ditch of the same depth as the Yarkand one, but broader at the summit. The main gate is situated in the centre of the north wall facing the city, while the east and south sides of the Fort are also pierced with a gateway in the centre of their length, which have flanking defences but are both closed. The Fort possesses six flanking defences in the north and south walls, and the usual bastion and tower at the corners, while the east and west sides have but four. Between the east gate and the N.E. corner, a distance of upwards of 250 yards, there is no intermediate defence; and this is the weak point of the Fort, as no flanking fire can command the space outside except such as could be delivered from the N.E. tower. The walls are entirely of earth, and loopholed all round, with embrasures also for guns. Passing through the north gate the main street runs through the centre of the Fort due N. and S., while branch streets diverge on either side between the houses. In the S.E. corner is situated a large mosque, the tower and upper part of which commands the ground outside. In the centre of the west wall a Chinese pagoda, now converted into a guard-house, rears its roof above the level of the main wall, and commands the exterior ground in that direction.

The place of residence of the chief authorities consists of a large enclosure surrounded by high walls, embracing three separate courts, in the inner one being the palace, or "urdoo," occupied by the king. The entire Fort appears to be in better order and more available for defence, were it not for its one weak point, than the Fort of Yarkand. The moat, as before stated, is much broader at the summit than the one of Yarkand, which being deep and narrow could be more easily bridged by temporary constructions. At the same time this moat being broader and shallower could be more easily crossed by an attacking force. Although the Chinese defended this Fort against the Kirghiz and Üzbegs in 1864 and 1865 for eighteen months, and only surrendered when compelled to do so by hunger and starvation, it is improbable that a European force provided with a siege train, or even attempting its capture by an assault, would experience much difficulty in effecting an entrance when defended solely by Asiatics.

Four miles off, lying north, across the Kizil Daria, or Kashgar River, is situated the city of Kashgar, now distinguished by the inhabitants from the "Yangishahr," or new city, which name is applied to the Fort, by its general appellation of the old city. The derivation of the name Kashgar is involved in some obscurity, nor does it appear to follow the usual formation terminating the names of the towns of Turkistan in the affix of كَنْر, or قَنْر, as represented in Samarcand, Tashkend, Khokand, and Yarkand. The word is generally written by Europeans as Kashgar, but the name as written and pronounced by the inhabitants is كاشقَر Kâshkar. The city is surrounded by a high fortified earth wall, in the south side of which 17 flanking defences were counted. It is entered by five gates, and having rapidly increased in size and prosperity since the expulsion of the Chinese, now contains some 28,000 houses and a population of 60 to 70,000 souls. Although, perhaps, little known to European statesmen, it must eventually play an important part in Asiatic politics, since the power in possession of Kashgar holds the key of Eastern Turkistan from the North. Wonderfully well and centrally situated, it is a place of the utmost importance both in a political and military point of view. Here all the roads from the Khanates of Central Asia converge, and in the hands of any European power it would be a place of immense commerce. As it now is, however, external trade is in utter stagnation. From the East and China few caravans ever arrive; and the large traffic in tea, which formerly flowed through Central Asia to Bokhara and Western Turkistan, has now entirely ceased. On the other hand, owing to the hostility existing between the Atalik Ghazee and his former master Khodayar Khan, the Khan of Khokand, and the near advance of the Russians, the road to Khokand, *via* the Terek Pass, is all but closed from that side. Recognising the great importance of Kashgar as a base for developing the tea trade with China, the Russians entered into a treaty with the Government of Pekin in 1861, as is well known, by which it was agreed they should be granted sites of land in Kashgar for building purposes, and allowed to establish a manufactory and warehouses under Consular authority.* The insurrection of the Tungânies against the Chinese, and the overthrow of the Tungânies in their turn by the present ruler, prevented this treaty from being carried into effect. Now, however, overtures have already been made to the Atalik Ghazee to carry out the terms of this treaty of 1861; but as he has decidedly refused to entertain such views, and holds that he is not liable for the fulfilment of any treaty entered into with the Chinese, we must, in the natural course of events, expect to see the whole of this splendid country up to the Karakoram absorbed as a Russian province; for it cannot be doubted that Russia, if she cannot obtain her end by diplomacy, will coerce him by force of arms; and will not allow an Eastern despot, whom she is pleased to consider an usurper,† to debar her from pursuing the career

---

* In the Treaty of Pekin, concluded by Russia in imitation of the English and French treaties with China, three places are named for the establishment of Consuls,—Kashgar, Kulja, and Chuguchak, the two last being in Zungaria.

† Russia declines categorically to recognise the Atalik Ghazee. She considers Kashgar and Yarkand still belonging to China, and may perhaps, on an invitation from Pekin, reconquer them some day for the Chinese Emperor. In the mean time Khodayar Khan, of Khokand, will probably be pushed forward to try the Atalik's metal; but Russia will not commit herself to any direct interference.

opened to her in this zone of Central Asia, a career which, in its aggressive phase, will be immortalised by future historians as a parallel to that of the conquest of our Indian empire.

The day we reached Kashgar I proceeded to a caravanserai, lying between the fort and old city, and situated on the right bank of the river; and the following morning went to an interview with Mahomed Yakoob Beg, the Atalik Ghazee and ruler of Eastern Turkistan. Passing through the north gate into the Fort, a body of Tungáni soldiers, armed with long lances, were first noticed, drawn up on each side of the way, while a guard of Turki sipahis, in scarlet uniform and high sheepskin caps, were grouped around some few pieces of artillery in position, near the main entrance. It was evident that Kush Begie had ordered an extra gathering of his followers, in some sort of review order, with a view to exhibiting a military display. Dismounting at the entrance of a large court-yard, I was conducted by the Yuzbashee across this enclosure to the gate of an inner court, where a Yusawal-bashee, dressed in the costume and chain-armour of the Egyptian Mamelukes, came forward to say, that if I would sit down for a few minutes, the Atalik would be prepared to see me. I accordingly waited until he returned and ushered me across the second court, which, with the first, was filled with men all dressed in silk and armed. Nothing could be more picturesque than the gaudy display, showing the outward glitter of Oriental pomp and splendour, in the courts where but lately all the horrors of siege and starvation had been endured by the ill-fated Chinese. Their Moslem conquerors had, however, effaced all traces of the tragedy, and if cruel and merciless in their religious fanaticism to their foes, their frank and manly courtesy, and warlike bearing, contrasting most strikingly with the degenerate and effeminate Chinese, win the good will as well as excite the admiration of the stranger. Arrayed in every variety of coloured costume, with bright arms and studded accoutrements, they sat, or stood in rows under the verandahs as I passed to my interview with the King. Having reached the entrance of the innermost court, I found it to be quite empty, save of a piece of ordnance, in . position, with muzzle pointed towards the entrance-gate. At the further end of this court, sitting under the verandah in front of his apartment, was the Atalik Ghazee himself, and here, as at Yarkand, no display or decoration appeared in the plain and unadorned buildings of his palace. As if scorning any costliness but that of military display, everything about him is in keeping with his simple and soldier-like habits. Never so happy as when living the hard life of the soldier in camp, or assisting with his own hands to erect forts on his threatened frontier, it is not too much to predict, that were Asia alone in the hands of its native rulers, he would prove the Zenghis Khan, or Tamerlane of his age. But, with more sagacity and foresight than those conquerors, he admits the inevitable contact of the strong European races, and bends himself to the overpowering force of circumstances.

The Yusawal-bashee who escorted me retiring, I advanced alone, bowed, and then shaking hands sat down opposite to the Atalik. He was dressed very plainly in a fur-lined silk choga, with snow white turban, and in the total absence of any ornaments or decorations, presented a striking contrast to the bedecked and be-jewelled rajahs of Hindustan. I was at once favourably impressed by his appearance, which did not belie the deeds of a man who in two years has won a kingdom twice the size of Great Britain. He is about forty-five years of age, in stature short and robust, with the strongly marked features peculiar to the Uzbegs of Andijâu. His broad, massive, and deeply seamed forehead, together with the keen and acute eye of the Asiatic, mark the intelligence and sagacity of the ruler, while the closely knit brows, and firm mouth, with its somewhat thick sensuous lips, stamp him as a man of indomitable will, who has fought with unflinching courage; and never sparing his own person, has, in the hour of success, been alike stern and pitiless in his hatred to his foes. Although an adept in dissimulation and deceit, the prevailing expression of his face was one of concern and anxiety, as if oppressed with constant care, in maintaining the high position to which he has attained. His manner was, however, most courteous, and even jovial at times. If report speaks true, his bed can hardly be one of roses, as it is said that the danger from some secret assassin's hand is so great that he never remains for more than one hour in the same apartment during the night. The few presents which I had bought for the Atalik were delivered, and a man was summoned to interpret, who remained standing at some short distance off, on the ground below the verandah. The conversation was at first the usual Oriental etiquette; and shortly afterwards the Atalik Ghazee expressed a hope that the English would in future visit his country, as hitherto they had been prevented from entering Central Asia by the Bokhara tragedy, when Colonel Stoddart and Captain Connolly were murdered by the Ameer of Bokhara in 1842. He then proceeded to say that another European—meaning Schlagintweit —had also been killed in this very place, Kashgar, by a robber named Wullee Khan, who relying on his spiritual influence as one of the seven Khojas, overran the northern provinces of Eastern Turkistan with a wild rabble of unscrupulous followers in 1857 and 1858, executing and murdering the most innocent people, for the mere purpose of shedding blood. The Atalik, however, never mentioned that he himself had involuntarily avenged the murder of Schlagintweit, and this he might have averred, for he cut Wullee Khan's throat two years ago. After a short conversation I took leave, and was conducted to the house of the Yusawal-bashee, in which quarters were assigned to me during my stay in Kashgar. It is almost needless to say that the same strict surveillance was exercised here as at Yarkand, and I was not permitted to go about. This was more especially the case when staying in the larger towns, but when on the march, and moving from place to place, the surveillance of the escort somewhat relaxed, and greater liberty was enjoyed.

I remained in Kashgar for upwards of a month, from the 5th of March to the 13th April; and during this time took observations as opportunity offered. The resulting position obtained for the fort was in lat. 39° 19′ 37″, N. (uncorrected) and by its distance and bearing from Yarkand, it was found to be in long. 76° 10′, E., while the elevation of 4165 feet above sea-level was determined from observations of the boiling-point of water. The position of the city of Kashgar, lying directly north from the fort across the river, was estimated to be in lat. 39° 23′ 9″, N., and in the same meridian of 76° 10′, E.

Meteorological observations were conducted from the month of October, 1868, to June, 1869; and the data thus obtained have been added to this paper; in the hope that they may prove of interest in throwing light upon the climatic features of the country.

I found that any representation of my intentions to endeavour to return to India by way of the Pamir Steppe and Chitral were ineffectual to obtain the desired object of being permitted to make the attempt from the side of Eastern Turkistan. The Atalik Ghazee would not hear of such a step for one moment; and it was evident that no such expedition would be allowed, and the only alternative remained to endeavour to leave the country as soon as possible, and make the attempt to reach the Pamir by another route.

From the roof of my house, on a clear evening, I could see the snow-covered peaks of the Kizil Yart Range of the Pamir in the far distance, some 60 miles away, beyond which lay the true Pamir, the "Bâm-i-duneeah," or Roof of the World, as it is called, the very name of which makes the mouths of geologists and geographers to water; while beyond that again lay Badakhshan—the ancient Bactria—and Trans-Oxiana, and the disappointment felt at being debarred from visiting all this unexplored ground was enhanced by its very proximity, after having succeeded in penetrating thus far.

It was during my stay in the fort that I obtained information regarding the Karakul. A Kirghiz chief, by name Togrok Kholoh, and his son, Mâhmur Khan, were detained in Kashgar, by order of the Atalik, on account of some dispute with reference to territory, on the northern frontier. The latter was living in the serai, attached to the house of the Yusawal-bashee, which I occupied; and as several opportunities occurred of conversing, I made it an especial object to cultivate his acquaintance, knowing that a local Kirghiz should be good authority on the subject of the Pamir. He stated with reference to the Kizil Daria, or Kashgar River, that this stream takes its rise in a small lake situated in the angle, where the Thian Shan Range (known to the local Kirghiz as the Artush Range) intersects the transverse chain of the Pamir. At the distance of 170 miles to the west of Kashgar, the stream rising in the Terek Pass, effects a junction with the main branch. The largest lake of the Pamir is the Karakul, and no stream, he averred, issuing from its eastern flank, joins the Kashgar or Khanarik rivers. On the contrary, he affirmed that the sole outlet of the Karakul is to the west, and that the affluent flowing out of its western side, has a course to the south and west through the hills of Karatigeen. Although any information acquired from native hearsay is unsatisfactory, and very often unreliable, still in the absence of proved facts from the result of further research and investigation, it is at least worthy of being recorded. It is confidently believed that on further exploration the greater slope of the watershed of the Pamir will be found to be to the westward; and that its lake-system drains solely into the Oxus. The Karakul itself must embrace a considerable circumference, since it was stated by different authorities to be a lake of fourteen, twelve, and ten days' journey round.[*] According to the general definition of a day's journey, this calculation would make the lake to be of vast extent; but even deducting considerably from this measurement, and allowing for native exaggeration, a sufficient margin is left to justify a conclusion that the lake is of no mean size; and that when its topography shall have been determined by some explorer, it will be found to embrace an area of at least 600 square miles.

The incessant turmoils and disturbances, with their accompanying scenes of violence and bloodshed, which have swept over the soil of Eastern Turkistan during the last five years,

---

[*] Abdul Mejid, who camped on the Karakul, says in his report it is four days' journey in extent, that is, probably, from west to east; so that it may well be ten days' journey in circumference. —[H. C. R.]

causing a change of dynasty in its rulers and their religion, may be made the subject of a brief notice, as being closely connected with the career of its present sovereign, the Atalik Ghazee.

The causes which led to the rebellion, commenced in 1863 against the Chinese, may be traced to many sources. Foremost amongst these was the incorporation of a large body of Tungânies, as a military force, with the regular Chinese soldiery. Rigid Mahomedans, and with some love of nationality, their fealty to the rule of foreigners was always untrustworthy; and since they had become a disciplined force their power was so greatly increased as to constitute an element of constant danger to the State. The Hakeems, or Uangs, holding authority over the Chinese, and ruling many of the outstanding provinces, together with the native Moghul population of the country, had become irritated and disgusted at a partial administration of justice, and the unscrupulous manner in which an exorbitant revenue was extorted. The Chinese soldiery in the large towns had fallen into a lax state of discipline, and enfeebled and demoralised by a constant use of opium and noxious drugs, they offered an easy prey to their more vigorous and simple neighbours, and the warlike Uzbegs of Andiján. When the standard of revolt was raised by the Tungânies in the eastern provinces of Urumchee and Karashahr, the Chinese at once shut themselves up in their forts, and preferring to endure the privations of a siege, in no instance endeavoured to retrieve their falling fortunes by attacking the rebellious forces. Cut off from all communication with China, and without hope of relief, or reinforcement from the Government of Pekin, they were not only expelled, but exterminated during the struggle, for with the exception of a few who preferred to save their lives by embracing Mahomedanism, they perished to a man.

The Tungânies first seized Karashahr, and being joined by a large body of Kucharies advanced on Aksu and Yarkand. Khotan had already been wrested from Chinese rule, as Habibula Khan, instigating a revolt, had attacked the Chinese who were in the city, most of whom were killed, while a few escaped to Yarkand, and others embraced Mahomedanism. Besieged in the Fort of Yarkand, the Ambân, or Chinese Governor, who was the chief authority throughout Eastern Turkistan, held out in the vain hope of receiving succour from Pekin. Every one of the inhabitants of the city had sided with the Tungâni and Kuchar allies; and the fort was closely invested. A sally against such overwhelming numbers would have been useless, nor did the Chinese once attempt it, or move outside the fort. Their enemies were content to wait, knowing that relief was hopeless, and they must eventually succumb. There was no occasion to precipitate their ruin, since hunger and starvation were soon doing their deadly work. The siege had continued for forty days, when terms of capitulation were offered to the besieged by the Kuchar chief, Jamal-ud-deen Khoja, the one extreme demand being that they should all embrace Mahomedanism. On hearing this excessive demand, the Ambân, who had previously formed his plan of action, assembled his officials and courtiers under the pretence of consulting with them as to the terms of the capitulation. The consultation took place in an upper room of the palace, under which had been stored a large amount of gunpowder, while by orders of the Ambân a train had been laid connecting it through the floor, immediately from under his chair of state. Assembled here the old Ambân laid before his officers the terms of the capitulation, and requested them to state what amount of ransom each was prepared to give. The last act in this tragedy was described by a Mahrumbashee, formerly in the service of the Ambân, with a glow of feeling quite Oriental; and the scene that follows is the subject for an artist.

The courtiers, in ignorance of their approaching doom, are wrangling and disputing amongst themselves as to the amount of ransom which each shall give, at the same time the sons of the Ambân are handing round tea and sweetmeats to them as guests.

Regarding the scene before him with the calmness of a stoic, sits the grey-bearded old Ambân in his chair of state, quietly smoking a long pipe, while beside him kneel his weeping daughters, all conscious of their coming fate. Suddenly the clamours of the disputing courtiers are hushed, as the report of cannon is heard, and the deep-muttered sounds of "Allah-ho-Akbar!" burst upon the ear. It is the enemy shouting their war-cry, as all impatient of delay they rush to storm the gates. The old Ambân has taken his resolution of dying sooner than fall into the hands of the cruel foe; and the courtiers, now aware of his dread intention, start up in the wildest despair, but all too late! With one word of farewell to his trembling children, the old Ambân slowly reverses his long pipe, and allowing the ashes from the bowl to fall upon the fatal train, in this act of self-immolation perishes with all around him; and thus ends the rule of a dynasty which had held sway for upwards of a century.

The Tungânics being now in possession of the chief city of the country, at once turned their attention towards Kashgar, where the Chinese garrison and population were bein besieged by Kush Begie, who with a following of some 500 men had fled from the Russians now advancing on Khokand; and had blockaded the Chinese in the fort. A large force of Kirghiz from near Samarcand, under their leader Sedeek Beg, had also descended from the Pamir, and were attacking the Chinese when Kush Begie arrived from Khokand.

Buzurg Khan, one of the seven Khojas of Andiján, accompanied the force with Kush Begie, and was nominally, from his spiritual influence, the principal leader of the expedition, having been invited by the Kashgaries to their city to resume his spiritual sway over them. Kush Begie, however, intriguing with his followers, soon won them over to his own designs, and seizing the supreme power imprisoned Buzurg Khan; and from his energy and name as a soldier at once obtained a complete ascendancy over the Kashgaries, as well as the Kirghiz.

His former career had been marked by a conspicuousness which had caused him to be looked on as one of the few able and energetic men who had striven to arrest the waning power of Khodayar Khan the Ameer of Khokand, now threatened by the advance of the Russians up the line of the Syr Daria, or Jaxartes. The son of a villager near Namagân, Mahomed Yakoob had risen by ability and perseverance to the rank of Kush Begie under Khodayar Khan; and in the summer of 1852 was the Governor of Ak Musjid, now Fort Perofski on the Syr Daria. In that year he repulsed the attack of the Russian army, nor was the town taken until the following year, and not then until after "a most heroic defence," as was stated in the Russian despatch.

During the European war with Russia, in 1854 and 1855, little advance was made up the Jaxartes after the capture of Ak Musjid; but in the next few years the Russians continued to encroach upon Khokand, and during the contests which ensued, Kush Begie was ever the foremost in opposing them.

After receiving five wounds from Russian bullets, he relinquished his command under Khodayar Khan, and accompanied Buzurg Khan to Kashgar, in order to try his fortune in Eastern Turkistan. Not only the Chinese, but mostly the Tunganies, have had cause to acknowledge his ability as a military commander.

When the fort of Kashgar was invested, it is said that there were 32,000 Chinese within the walls and adjoining enclosures, but for humanity's sake it must be hoped that this number is greatly exaggerated. The horrors which they endured from starvation have not, however, been over-estimated. Not only were they compelled to eat horses, and every animal within the walls, but when these were consumed, they killed their own wives and children for food in their extremity of hunger.

Hundreds of famished wretches died every day, and of the whole number who perished in the course of the siege it is impossible to form any accurate computation. While they were yet besieged, and before they had surrendered, the Tunganies, flushed with success at their late capture of Yarkand, and the downfall of Aksu, which had likewise fallen into their hands, advanced against Kashgar with a large force, variously estimated at from 28,000 to 40,000 men.

Kush Begie, with all the followers he could collect, moved out from before Kashgar to oppose them. They met on the banks of the Khanarik River, near Yupchan, and although it is unlikely the Tungâni were in anything like the numbers represented, yet the rival forces must have been out of all proportion. On this occasion Kush Begie surpassed anything which he had before achieved. They fought for eight hours with varying fortunes, during which he had two horses killed under him, and received two severe bullet-wounds. Victory was for a long time doubtful, notwithstanding the repeated and vigorous charges made by his men. Almost despairing of success, they wished to induce him to withdraw, but he declared his intention to win the battle or to perish in it.

At length the Tungânies were routed by the furious charges of the Uzbegs, and victory declared for Kush Begie. His wounds were unknown to his followers until after the day was won, when, in escorting him to his camp, he fainted from loss of blood.

Following up his success, he attacked and captured the fort of Yanghissar, where the Chinese made a stubborn resistance, and fought better than their fellows in Yarkand and Kashgar. The Chinese, too, in Kashgar, had now been nearly exterminated after enduring an eight months' siege, and the survivors, embracing Mahomedanism, threw open the gates of the fort, and surrendered.

Concerting measures to attack Yarkand, he proceeded there with only 500 followers. The small force with which he was accompanied had no chance of success against the Tungânies in the fort; and entering the city, he nearly paid the penalty of his rash attempt to win the capital of Eastern Turkistan by a *coup-de-main.* The inhabitants were in favour of the Tungânies, who held the fort, and creating an *émeute* while Kush Begie and his men were in the city, the gates of the town were closed upon them, and they were assailed by the Tungânies from the walls. Two hundred of them were killed or wounded, and he himself received a slight wound. All hope of escape seemed gone, when he rode his horse up the steps leading to the summit of the fortified city wall, which is here some 12 feet in width.   From the summit of the wall they leapt their horses into the open ground outside, and riding up an accessible part of the moat, escaped.   Kush Begie was the last man to leave the walls.

Shortly afterwards he a second time attacked Yarkand, and again unsuccessfully.   A third attempt, however, proved successful.   The measures he now took to effect its capture were better matured, and with a force of 6,000 to 7,000 men, two thousand of whom were Kipchaks who had fled from Khokand, and a reinforcement of Badakhshi troops sent by Jahandar Shah, the Meer of Badakhshan, he marched to Chinibâgh, situated two miles south of Yarkand, and besieged the Tungânies within the forts.   The Tungânies who had sallied out during Kush Begie's second attack, and being caught in an ambuscade had suffered some losses, were not to be induced to venture outside of the fort.   Treachery, however, was at work within the walls.   One of their chief men, Mahomed Niyaz Beg, was in secret correspondence with Kush Begie, with whom he arranged that on an appointed day he would send the Tungânics out against him ; and it was agreed that Kush Begic should feign flight in order to delude them.   In the mean time Kush Begie had made it his constant endeavour to win over to his side the population of the city ; and by promises and numerous presents he had succeeded in attaching to his cause most of the influential people in Yarkand.   According to the agreement with Niyaz Beg, the Tungânies sallied out of the fort and attacked Kush Begie and his force in their camp at Chinibâgh.   Some guns which Dad Khwah, Nubbee Buxsh, the commander of his artillery, had loaded with broken pieces of iron, swords, and guns—having no cannon-balls—did great execution among the assailants.   As previously arranged, Kush Begie then feigned flight with all his followers, and the Tungânies were soon busily engaged in plundering his camp.   During this the Uzbegs returned at full gallop, and surrounded them.   Taken by surprise, it was scarcely possible to offer much resistance ; and those who did were trampled down and cut to pieces, while the remainder, who fled for refuge to the fort, found the gates closed upon them by the traitor Niyaz Beg.   They surrendered at once, and Kush Begie then entered the fort, where he was joined by Niyaz Beg and those who had been induced to come over to him.   The Kuchar allies of the Tungânies were well treated, and their leader, Jamal-ad-deen Khoja, professing allegiance to the new ruler, after receiving numerous presents, were dismissed to their homes.

The year 1866 opened under different auspices and much altered circumstances for the successful invaders.   Now in full

possession of the provinces of Yarkand and Kashgar, Mahomed
Yakoob, no longer the Kush Begie, assumed the title of the
Atalik Ghazee, or Defender of the Faith, while adventurers
from every part of the country flocked to seek service under
his standard. The fertile city and province of Khotan now
attracted his attention, and he secretly concerted measures for
attacking its new ruler, Habibula Khan, who, since the expulsion of the Chinese, had enjoyed a short duration of power.
It was late in the autumn of the previous year that Mr. Johnson visited Khotan, at the invitation of Habibula Khan; and
thus established the claim to the undeniable merit of being
the first European to have penetrated into Eastern Turkistan
since the sad fate of Adolph Schlagintweit. The Tungánies
were at this time in possession of Yarkand, which rendered
any attempt of Mr. Johnson to visit that city out of the question. Habibula Khan was in fear of his turbulent neighbours,
having already repulsed an attack of the Tungánies in an action
fought in August, 1865, near the Karakash River, and was thus
anxious to cultivate the friendship of the English, in the hope
that he might be able to ward off the coming storm. Nor had
he long to wait. Kush Begie, now the Atalik Ghazee, sent to
invite Habibula's son to his court, and by many presents and
representations sent to Habibula, through his son, succeeded in
disarming all suspicion regarding his ultimate designs. Under
the now friendly protestations towards the Khotan ruler, an
old man of nearly eighty years of age, he left Yarkand with a
small force, and proceeded to Khotan, giving out that he had
come to seek a blessing at the hands of Habibula, who was
venerated as a Hadji and spiritual leader. The Khotan ruler
had a large force under his command, and was prepared to
resist any hostile demonstration, but the deceit practised by
the Atalik Ghazee completely misled him as to his real intentions. Having encamped near Khotan, Habibula's son was
despatched by his father to Kush Begie's camp, with a request
that the Atalik Ghazee would explain the cause of his being
accompanied by an armed force, when he had professed to be
proceeding solely on a friendly visit. This was met by the
answer from Kush Begie that he was about to march to attack
Aksu, and had come to entreat Habibula's blessing for the
success of his expedition. All further suspicion was disarmed,
by Kush Begie swearing on the Koran that he meant him no
injury. The son went into Khotan, and Habibula was so
firmly convinced that Kush Begie entertained none but friendly
views that he at once proceeded to his camp, accompanied by
a small escort. The deceit was maintained to the last, until
Habibula rose with the intention of returning to the city, when

at a sign from their leader, the Atalik guard immediately surrounded him and his attendants, and made them prisoners. Before the news could reach the city, Kush Begie's troops marched against it, and the Khotanies believing in their friendly intentions, and unable to act without their leaders, offered but little resistance. Khotan was captured ; and the tragedy completed by the late ruler, his son, a nephew, and his wuzeer being carried off to Yarkand, and there secretly executed. Their graves may be seen behind the Tungâni Ziarat in the fort, where the Khotan sipahis, in memory of their old leader, proceed every morning to scatter flowers upon his grave.

From this time the Atalik Ghazee continued to enjoy an uninterrupted career of conquest. The downfall of Aksu followed the capture of Khotan, and shortly afterwards Kuchar was attacked. The Khucharies had thrown off the allegiance which they had declared for Kush Begie, when he had permitted them to return to their homes from Yarkand, and were now in open revolt under their leader Jamal-ud-deen Khoja. They made a gallant defence of the town, and it was not until the Uzbegs had suffered a loss of 163 men that its capture was effected. Jamal-ud-deen Khoja was seized, and met the fate which had been dealt out to so many others. Usk Turfan and Bai Sairam were brought under the Atalik's sway, and he advanced as far as Ili, where the Kalmak population agreed to pay him tribute, without an actual occupation of the province. He then returned to Kashgar, where he has since remained, after pursuing his successes uninterruptedly for a period of two years. His latest acquisition is the district of Sarikol, which was brought under his rule so lately as during the autumn of 1868. Recognising the danger to be expected from the near advance of the Russians on the Naryn River immediately to the north of Kashgar, he has busily employed himself in collecting a large force at Kashgar ; and by erecting forts commanding the roads debouching from the passes across his threatened frontier, has prepared to meet the coming storm to the best of his power. Meanwhile the country under his rule presents a striking contrast to its condition when under the sway of the Chinese. The cities are increasing in wealth and prosperity, and a large internal traffic is carried on, while a most favourable opportunity is now occurring for developing its external commerce. Merchants are turning their views towards India, for the supply of such general articles of consumption as tea and sugar. Owing to the greater facilities of the road communication with Khokand, goods and articles of Russian manufacture find their way in far greater quantities than British merchandize into the bazaars of Yarkand and Kashgar, but with the present obstacles to

commerce removed, and the opening out of a new and easier
trade-route, a thriving and increasing traffic will be carried on
with India alone. The agricultural industry of the country is
progressing; and a prosperous future would appear to be in
store for Eastern Turkistan, but for the standing menace of a
Russian invasion from the north. The Atalik Ghazee has quite
consolidated his power; and the only enemies he has to fear
are his powerful neighbours on the Naryn. He is certainly
prepared to resist an invasion, and his courage cannot be doubted
for one moment. If the villany and deceit which he has practised during his career stand in strange contrast with his forti-
tude and unflinching bravery, he has also fought well, for he
has been twelve times wounded. And as an Asiatic will never
hesitate to stoop to treachery and deceit, it would be hopeless
to look for any trait of generosity and magnanimity displayed
by him during his rapid rise to power. He is now proving
himself to be an able and energetic statesman, and a fit ruler
of the somewhat turbulent subjects whom he has to govern.
Should the peace and quietness now prevailing throughout his
dominions continue, he is likely to enjoy a lengthy time of
power; but already the storm is gathering, which threatens to
hurl him from his high position. If Russia advances to the
south of the Thian Shan Chain, the Atalik Ghazee must be the
first enemy with whom she will have to contend. Eastern Turkistan possesses a splendid mountain frontier, especially to the
north, the roads of which are impracticable for guns, and even
difficult for mounted men. The Russians were long delayed
in the Caucasus, and our own Affghanistan disasters are not yet
forgotten; and as savages have before now defied disciplined
armies in the mountains, it is even possible that he may succeed in beating his enemies back from the passes. Once in the
plains, however, the Russians must carry all before them, for
the discipline that makes men act with coolness and collected-
ness amidst scenes of greatest danger, is wanting here, and in
the plains of Turkistan he could not oppose them successfully
for one hour.

I left Kashgar on the return journey, as the sun rose on the
morning of the 13th April. It was one of those perfectly clear
days, so characteristic of the climate of Eastern Turkistan; and
in the grand display of the mountain-masses around offered an
ample compensation for the long detention and delay which
had been experienced.

Lying north immediately beyond the Kashgar River appears
a low, undulating ridge of ground, from which the transverse
slopes run down very evenly and gently into the level plain
beyond the river. Beyond this again an irregular rocky-range

occurs, presenting a steep face to the south, an opening in which admits the exit of a stream flowing with a south-easterly course to its junction with the Kashgar River. A road conducts up the valley to the village of Tajend, beyond which is situated the fort of Aksai, commanding the route debouching across the snowy range to the north, by the Pass of Tailâb. To the north-east, in the far distance, appear the slopes of the Artush Range, branching from the Great Thian Shan Chain of Central Asia, while conterminous with the horizon to the north, this Snowy Range stretches with an even crest at nearly 70 miles' distance from Kashgar. The direction of the range is from w.s.w. to e.n.e., while the spurs slope evenly and with a regular alternation to the south and east. The Artush Valley is seen throughout a considerable portion of its length to where it deflects to the northward. The stream rising in the pass at its head has, at first, a course to the south-east, and then to the southward; and again flowing eastward after leaving the lower hills, forms one of the tributaries of the Kashgar River. But very few peaks in the Snowy Range appear to attain to a greater height than 18,000 or 19,000 feet above the level of the sea; and the crest of the chain, as before mentioned, presents no alternate lofty summits and deep depressions so remarkable in the chains of the Kuen Luen and Karakoram. The appearance of the range, as seen from the southward, is somewhat desolate, since no forests occur to break the interminable view of the bare slopes of the mountains, with their snow-crowned summits. Although forests are found on the northern slopes in the basin of the Naryn, yet no trees are visible from the south, or at any rate no timber of sufficient height to be seen at the distance of Kashgar. It is not known with any degree of certainty to what altitude the passes across the range attain; but if the mean elevation of 15,800 feet is assigned to them, this measurement is, in all probability, sufficiently accurate for an approximate calculation.

Looking south and west from this point of observation is seen the whole Kizil Yart Range, forming the eastern crest of the Pamir, surmounted by snow-capped peaks and glaciers. It would be impossible for any scene in nature to surpass the vast grandeur of these mountains, as seen towering up like a gigantic wall, with the well-defined outline of their lofty summits cutting the clear azure of the sky. The lines of Pope at once occur to the observer with striking appropriateness:—

" Eternal snows the growing mass supply,
  Till the bright mountains prop the incumbent sky ;
  As Atlas fixed each hoary pile appears
  The gather'd winter of a thousand years."

It was a scene that could not fail to be indelibly impressed upon the memory; and the more so from the circumstances under which it was beheld. The Russians, our friendly rivals in the noble science of geography, had already reached to the crest of the range now in sight to the north; and here, in the very heart of Central Asia, it was gratifying to know that, at length, through the medium of British enterprise, had been determined the much-vexed question of the position of Kashgar.

From here again was noticed the very abrupt and rugged declivities of the lofty Pamir Range, while, trending westward to its junction with the Artush, the Thian Shan Chain was visible at the head of the open valley, through which flows the Kashgar River. The point of junction of the two chains could not be seen at such a distance, but some lofty isolated peaks were discernible towards the Terek Pass, as the rays of the morning sun lit up their snowy crests.

Continuing our journey, we arrived at Yanghissar the following day, and a further delay of thirteen days here occurred, since the Atalik Ghazee, who had preceded me from Kashgar, did not grant me a farewell interview till the evening of the 25th April. I had here an opportunity of seeing the interior of the fort, as, instead of living in the serai as before, a house in the fort belonging to the Yusawal-bashee, in command of the garrison, was assigned to me during my stay. A few observations of the peaks and principal points in the range of the Pamir were here taken to check the bearings already fixed from Kashgar and intermediately; and as the sun, when on the meridian, was now out of the range of the sextant, I could make no further observations for latitude at mid-day. The weather had now changed from that of a severe winter to the more genial climate of spring.

Only on two occasions was any rainfall noticed, and then in but very slight showers towards the end of March in Kashgar. No rain fell at Yarkand, although it snowed slightly on several days in the month of January. The snow was very fine-grained, and was melted within a few hours, with the exception of where it had drifted and lay in shady places and hollows. The greatest fall of snow upon the mountains would appear to have taken place during the first week in March, while thunder-storms with rain were prevalent towards the Sarikol Hills throughout May. The predominant winds were gentle breezes from the west and south-west, and a few heavy dust-storms, accompanied by a strong south-east gale, occurred towards the end of April and the beginning of May.

The thermometer at Yarkand rose from a temperature of 23° Fahr. at noon in the commencement of January to 71° and

72° Fahr. at the end of May. As the mercury probably indicates a temperature of 82° or 85° during the months of July and August, which is undoubtedly the hottest time of the year, Eastern Turkistan thus experiences alternate periods of great heat and excessive cold. And as in countries where ranges of mountains intercept the course of the prevalent winds, being enclosed on the north, west, and south by lofty chains of mountains, a peculiarly dry climate is here met with.

We remained another month in Yarkand, since all the passes on the southern frontier were reported by the Kirghiz to be impracticable up to the end of May. The welcome news at length arriving that the Sanju Pass was practicable for laden yaks, we bid farewell to the Dad Khwâh—the courteous and hospitable Governor of Yarkand—and started on the return journey to Ladak on the 30th May. I had still hoped to have been able to return to India by way of the Pamir and Chitral, but on my again representing my desire to make the attempt, the governor would not bear of my running the risk attending such a route. Although the Karakul lay within 150 miles of the furthest point to which I had penetrated, it was imperative to return by way of Ladak, and travel over again some 1600 miles before one could hope to reach it by another route.

The country now presented the most blooming and fertile appearance. The trees were all in full foliage, and the gardens and fruit trees fragrant with blossom. Encamped in the picturesque gardens of Karghalik, under the spreading walnut trees, Mr. Shaw and myself were now allowed to meet, and were soon discussing together the various events of our sojourn in the land of the Moghuls, and our rides through the pleasant plains and steppes of Turkistan.

We finally left Sanju on the 12th June by a different route to that by which we had entered. The swollen state of the Sanju River did not admit of a road up that valley to the pass, and a divergence was necessary in order to strike the stream nearer to the pass. Leaving the village of Sanju we crossed the low sand-hills, and ascending the valley of the Arpalak stream encamped at Kizil Aghil for the night. The following day a journey of 14 miles further up the stream brought us to the junction of a branch valley leading to the Chuchu Pass across the transverse spur from the Kilian Range dividing the Sanju and Arpalak valleys. The pass is at an elevation of 11,847 feet as calculated from hypsometric measurement, and the ascent immediately before reaching the crest of the ridge is steep for a short distance. Mazar, an encamping-ground in the Arpalak Valley, where the road turns off to the pass, was found to be at an elevation of 8617 feet, while the distance from that

The Sooget Pass, which had been crossed on the 24th June, was covered with a fall of new snow to the depth of six inches, but when this pass, which is at an elevation of 18,237 feet above sea-level, was traversed on the 15th December, it was observed to be quite free from snow on the summit, while but a slight bed of snow had accumulated at the head of the open ravine on the north side. I am inclined to think that the mean elevation of 18,400 to 18,600 feet should be assigned for the snow line, or limit of perpetual snow on the northern slope of the Karakoram chain. The snow-line on the northern declivity of the Kuen Luen may be estimated to extend to a somewhat higher level, since observations made at different stations on the range show the height of the snow line to be 18,800 to nearly 19,000 feet above sea-level. Many glaciers may be observed at a lower elevation, which have gradually moved down into the head of the ravines. The descent from the Karatagh Pass to the eastward is as gradual and gentle as the approach from the valley of the Yarkand River. Having now crossed the crest of the plateau of Aktagh, we had descended into the basin of the Karakash River; and continuing down the open valley, struck that river the following morning immediately above the hot springs. The valley we had descended from the pass was the one originally noticed in November when exploring the course of the Karakash, as likely to afford an easy route to Aktagh, which proved to be the case, and the much desired connecting link was thus established. The ascent from the Karakash River to the pass which lies at a distance of 18 miles from this point, is a very gradual one of 2400 feet, which gives the exceedingly easy gradient of 133 feet per mile. The descent from the pass to Wahâbjilga above Aktagh, on the regular Karakoram route, presents the same gentle gradient, while the natural advantages offered by the open character of the country are such as to render this line of communication perfectly practicable for two-wheeled carts and c

nned our journey up the Karakash Valley, and
as of Kizil found ourselves once more on the
Lingzi Thung. Knowing the position of
Pass we marched across the open plains
mped for the night at an elevation of
level of the sea. Towards morning it
l continued uninterruptedly for the next
o height of summer, the cold experienced
d was severe—the thermometer at 9 P.M.
re of 26° Fahr. and sinking to 11° Fahr.
ould have been impossible to distinguish

passing Aktagh and Muliksba, we had crossed the high table-land of Aktagh, and ascended the wide open valley of the table-land up which lies the road to the pass leading across the main range of the Karakoram. A most favourable opportunity now presented itself of exploring the small strip of country lying between here and the upper valley of the Kara-kash. The discovery of this connecting link in the new route traversed during the preceding November was not only desirable, but the exploration was one of much personal interest to myself, since it would afford the means of testing the value of the survey hitherto executed, and prove an admirable check on all the previous observations which I had conducted. The valley at Châdartâsh was found to be at an elevation of 16,190 feet above the sea, by the temperature of boiling water; and as it was known that the valley of the Karakash River at the point where I expected to strike it at a distance of 30 to 40 miles to the south-east, lay at an elevation of about 15,600 feet above sea-level, the probability that an easy pass would be found to lead across the intervening range was very evident.

During the course of a long day's exploration, what appeared to be the most practicable route was fixed upon; and returning to camp at Châdartâsh long after dark, I perceived that misfortunes had already begun, as my best horse was lying dead in the moonlight before my tent. Moving at daylight the next morning we struck across the low hills to the eastward, and skirting the short slopes of the series of broken ridges which the northern face of the Karakoram here presents, encamped for the night at an elevation of 16,905 feet above the sea. A little of the lavender-like plant called "boorsee" was here found, but at no higher elevation than about 17,000 feet above the sea, which altitude would appear to be the utmost limit to which any vegetation attains on the high lands lying between the chains of the Karakoram and Kuen Luen.

Proceeding eastward, our road lay up an open ravine leading with a very gentle ascent to the summit of a depression in the ridge, here branching from the Karakoram range and named the Karatagh. This pass was found to be at an elevation of 17,953 feet above the sea from data obtained by hypsometric measurement. A very little snow was noticed on the slopes of the summit of the ridge immediately above the pass, while the open face of the pass itself was quite free from it on this day, the 28th June. Many of the conical-shaped crests of the lateral slopes of the Karakoram, lying at an elevation of 18,500 feet, were also observed to be free from snow, save where a few patches had accumulated on their shady sides. Below the pass, to the eastward, the open ravine was quite free from snow.

difficult passes of Sasser and Kardong on the Ladak side of
the Karakoram, as well as the Karakoram Pass, but possesses
the great desideratum of affording grass and fuel on that portion
of the route where it is most essential.   It is desirable to draw
especial attention to this line of communication, since what is
capable of being converted into an easy trade-route, may be
made equally available for military purposes. In all discussions
regarding the Central Asian question, the feasibility of this
route has been ignored.   In a late excellent work which
discusses the question in all its bearings, it is said that " this
latter route may be entirely dismissed from consideration, as
being impracticable for several reasons, the chief of which
would be the sterility of the country through which an army
would have to pass."   Admitting the difficulties which a hostile
force advancing across these mountains would have to overcome,
yet the undertaking is very far from being impracticable, and
history affords for example the successful invasions of Ladak
and Tibet by the Kalmuks towards the close of the 17th
century.   It is equally true that the invaders were eventually
overwhelmed in their attempt to return through these very
mountains, but the analogy cannot be maintained between a
host of wild Tartars and what would be a disciplined European
force, equipped with every material and appliance of the art of
war.   An army attempting a passage across the mountains
from Eastern Turkistan to India would have no great impediment to encounter until they had entered the deeper defiles of
the lower Himalayas.   The portion of the line intervening
between the crest of the Karakoram range and the plains of
Turkistan is quite practicable, and, as in all human probability,
it is here that the Russian and Indian empires will first come
into contact, and the frontiers run conterminous, this fact is
deserving of especial consideration.

It may be interesting to briefly notice the results of the geographical research which has been conducted by the present
expedition.

The theory advanced by the explorers, the Schlagintweits and
Johnson would indicate a system in which the Karakoram and
Kuen Luen ranges are the northern and southern crests of the
same great chain, but the mountain system assigned by Humboldt to Central Asia, which divides them into great mountain
chains, " coinciding with parallels of latitude," is strictly the true
one.   Whether regarding the Karakoram as a separate chain,
or as a prolongation of the Himalaya to the northward, it forms
a distinct watershed between the Indus and the river-systems
of Tartary or Eastern Turkistan, while the Kuen Luen consti-

tutes a parallel chain bounding the high table-land of Tibet to the north. To the west of this elevated plateau or table-land the extensive tracts of level plain, which are its characteristic features, are no longer met with, for here they break into detached ranges, and the general level of the country sinks into the basin of the Turkistan rivers.

With the relative bearing of these two main chains, the more immediate points of interest which have been discovered are the true courses of the Yarkand and Karakash Rivers. The former river, rising on the northern slope of the Karakoram, flows through the longitudinal valley between the two chains, while the Karakash, rising in the same range more to the eastward, pursues a parallel course until it pierces the main chain of the Kuen Luen.

Before discussing the positions of the chief towns of Eastern Turkistan, I may be allowed to quote an extract from the Report of the Great Trigonometrical Survey Department of India, of 1867-68, by Colonel Walker, which illustrates the compilation of their new map of Turkistan, and bears closely on the subject:—

"(87.) The map of Turkistan, with the adjacent portions of the British and Russian territories, which was stated in my last report to be under compilation, has now been completed, and published by the photo-zincographic process. A great deal of valuable information has been incorporated into this map from the Punjab 'Report on the Trade and Resources of the Countries on the North-West Boundary of British India,' as to the routes from Afghanistan, *viâ* Kokan and Kashgar, and *viâ* the Oxus River, the Pamir Steppe, and the Sarikol or Tashkurgan district to Yarkand; from Peshawur, through Swat, Panjkora, and Chitral, to the Pamir Steppe and the sources of the Oxus, and from Leh and Iskardo into Tashkurgan and Yarkand.

"In the regions beyond the British frontier, which no Europeans could safely enter unless backed by a strong army, there are many hill-peaks whose positions and heights have been determined with accuracy by the operations of this survey in previous years. These points furnish the basis on which the geographical details, as obtained from oral information, or travellers' itineraries, or explorations by native surveyors, have been fitted; where they are numerous, the map is probably fairly accurate; where they are scanty, it is necessarily less reliable. The regions of which least is known are those lying between the Oxus and the southern frontier of Kokan; nothing is known of the configuration of the Pamir Steppe, and very little of the positions of places on it.

"(88.) The determination of the much-questioned positions

of the chief towns of Altyshahar, or Little Bokhara, is approaching solution. The position of Ilchi, the capital of Khotan, may be considered to have been definitely fixed by Mr. Johnson, while that of Yarkand has probably been very approximately fixed by Captain Montgomerie's explorer, Mahomed-i-Hamid. Adopting these positions, and collecting all the evidence available in this office as to its distance and bearing from Yarkand and from Ilchi, Kashgar would appear to be in lat. 39° 25', and long. 75° 25'. This value of the latitude agrees with what has generally been adopted hitherto, but the longitude is 1½° east of the position adopted by Klaproth, Humboldt, and Ritter, and no less than 3° 35' (nearly 200 miles) east of the value adopted by the Messrs. Schlagintweit. On the other hand, a new and entirely independent value of the position of Kashgar has been recently obtained in the summer of 1867, by the Russian General Poltoratsky, in the course of a reconnaissance of the regions to the south of Lake Issikkul and the Naryn River, down to the border of the plains of Altyshahar, the resulting position of Kashgar was lat. 39° 35'; and long. 76° 22', or still more to the east than the value adopted in this office. I am indebted to Baron Osten Sacken, Secretary to the Imperial Geographical Society of Russia, who accompanied General Poltoratsky's expedition, for the above information, as well as for several of the latest and most correct maps of the regions on the south of the Russian frontier. They have been of great assistance in compiling the new map of Turkistan."

In order to facilitate the interesting discussion of the positions of Yarkand, Kashgar, and Khotan, I subjoin a table of several values for longitude of those cities adopted by different authorities, many of which have been taken from Col. H. Yule's work, entitled 'Cathay, and the Way Thither':—

Authorities.	Yarkand.	Kashgar.	Khotan (Ilchi).
	° '	° '	° '
Chinese Tables	76 3	73 48	80 21
Veniukhof	76 10	73 58	..
Kiepert	74 56	72 53	79 12
Colonel Walker	76 24	73 58	79 13
Schlagintweit	73 58	71 50	78 20
Montgomerie	77 30	75 25	79 26
General Poltoratsky (67)	..	76 22	..
Present Expedition (69)	77 28	76 10	80 5

The values adopted by Major Montgomerie are the results of the observations of his explorer, Mahomed-i-Hamid, for the meridian of Yarkand, and of Mr. Johnson for that of Khotan. Those assigned to Col. Walker are the values adopted in the map of Central Asia. The position of Yarkand may be considered to have been definitely determined, since the value obtained by Mahomed-i-Hamid, viz. lat. 38° 19′ 46″ N., long. 77° 30′ E., and that by the present expedition, lat. 38° 21′ 16″ N., long. 77° 28′ E., very nearly coincide. The difference in the latitudes may be accounted for by presuming the observations to have been made from different points, as, for instance, from the north and south walls of the city, which would afford this variation.

With reference to the position of Khotan, the latest data which have been obtained are the observations of Mr. Johnson, whose value is 79° 26′ E. for the longitude of that place.

Taking the longitude of 77° 28′ to represent the position of Yarkand, this value of Khotan is only 1° 58′ to the east of the meridian of Yarkand. Now the road distance from Yarkand through Karghalik and Guma to Khotan is about 175 miles, and allowing a margin for the difference of latitude between the two cities, viz. about 1° 13′ 16″ (Yarkand being in 38° 21′ 16″, and Khotan in 37° 8′ of north latitude), this would appear to be insufficient. The value of 80° 21′ E., as given in the 'Jesuit Register,' is probably more accurate, though deviating too much in the opposite direction.

Although Khotan has not been visited by myself, the results of all observations taken of the country lying between Sanju and Khotan tend to show its position as more to the eastward than the meridian adopted by Mr. Johnson. A meridian of 79° 26′ would approach too close to Sanju—would, in fact, nearly cut the head of the Arpalak Valley, which has been visited and surveyed. The road distance, also, from Sanju to Khotan is not less than 60 miles. Judging, then, from the distance between Yarkand and Khotan, and Sanju and Khotan, as well as the course of the Karakash River, the longitude of Khotan would appear to be from 79° 55′ to 80° 5′ E., and the latter meridian has been approximately adopted in the present map.

This gives a difference of 2° 37′ to the east of the meridian of Yarkand, and is believed to be correct within small limits. Still it must be remembered that this result is merely based on calculation, and not from an actual visit to the place; whereas Mr. Johnson was in Khotan for several days, and, it is believed, took astronomical observations in order to ascertain its position. It is much to be regretted that the general inaccuracy of his map should tend to throw doubts upon the correctness of the position assigned by him to Khotan.

The great geographical puzzle, however, appears to have been in identifying the true position of Kashgar.

As given by the Jesuits, Veniukhof, Kiepert, and the Schlagintweits, it has never been placed to the east of the meridian of 74°. Our latest map has it fixed by Major Montgomerie as in long. 75° 25' E. Even this is not far enough eastward. The Russian General Poltoratsky has recently obtained a value of lat. 39° 25', long. 76° 22'. General Poltoratsky did not actually reach Kashgar, or probably any point within 80 miles of that city, but his value places it still more to the east of the position adopted by our latest map. The position of Kashgar, as based on data obtained from observations by the present expedition, would appear to be in lat. 39° 23' 9" N., long. 76° 10' E.

This position is 4° 20' (or 240 miles) to the east of that adopted by the Schlagintweits. Those explorers, however, can hardly maintain any claim to be considered authorities on the subject, for it can be shown that the brothers Herman and Robert never reached any point within 191 miles of Yarkand and 317 miles of Kashgar. Their treatment of the observations of so good a traveller and careful observer as Wood is unjust and their alterations unreliable; but, not content with dislocating the map of Turkistan, they summarily place Wood's Sir-i-kul Lake 2° 22', or some 135 miles, to the west of its true position. There is no direct evidence of the longitude of Sir-i-kul as fixed by Wood being correct, independent of Wood's observations connecting it with Badakhshan.

Yarkand, in long. 77° 28', and Sir-i-kul Lake, in long. 73° 50', as given by Wood, show a difference of 3° 38', or just 200 miles between the two meridians. As Yarkand lies about 41' N. of the latitude of lake Sir-i-kul, and the road distance, including curvatures, is from 240 to 250 miles from Yarkand to Sir-i-kul, *viâ* Tashkurgan and the Sarikol district, this should prove the correctness of Wood's valuation.

With reference to my own value of the position of Kashgar, I would observe, that it may be considered to be as fairly accurate as could be expected to be obtained, under circumstances attending an expedition into a country which has hitherto been so jealously closed to Europeans.

What has now become the great question for decision in the geography of Central Asia is, the exact configuration of the Pamir Steppe, and the identity of the main source of the River Oxus. So much of the configuration of the Pamir Steppe has been determined, as would go far to prove it to be of greater breadth than is conjectured, and that none of the streams issuing from its lake-system drain through the range forming its eastern crest into the rivers of Eastern Turkistan. That the stream flowing out of the Karakul to the west is the main source of the River Oxus, there can be little doubt. I am indebted to Colonel Yule's work for the following interesting information on the subject of the sources· of the Oxus. The extracts from Edrisi, as quoted by Colonel Yule,˙are as follows :—

"The Jihun takes its rise in the country of Wakhan, on the frontier of Badakhshan, and there it bears the name of Khariâh. It receives five considerable tributaries, which come from the countries of Khutl and Waksh. The Khariâh receives the waters of a river called Aksura, or Mank, those of Than or Balian, of Farghan, of Anjâra (or Andijâra), of *Wakshab*, with a great number of affluents coming from the mountains of Botun (it also receives) ; other rivers, such as those of Sâghaniân and Kawâdian, which all join in the province of the latter name and discharge into the Jihun." "The Waksbah takes its rise in the country of the Turks, after arriving in the country of Waksh it loses itself under a high mountain, where it may be crossed as over a bridge. The length of its subterranean course is not known; finally, however, it issues from the mountain, runs along the frontier of the country of Balkh, and reaches Tarmedh. The river having passed to Tarmedh follows on to Kilif, to Zane, to Amol, and finally discharges its waters into the lake of Khwarizm (the Aral)."

The stream mentioned as Sâghaniân may be identified with the tributary of the Oxus, flowing through the modern province of Shagnan ; but where does the main stream mentioned as Waksbah obtain its source ? Edrisi says,—" it takes its rise in the country of the Turks," or the mountains of Pamir. And as the main source of the Oxus is represented by this Wakshab, all evidence would show it to have its rise in the Lake Karakul. If, then, the affluent flowing from the Karakul to the west and south-west drains through the hills of Karatigeen and Kolab, it must be the main source of the Oxus. But where does this stream, the Wakshab of Edrisi, effect a junction with the Panja Daria of Wood ? Must it not be in that bend of the river which Wood did not see, or near the Hagrat Imam ?

It is interesting to note the difficulty which occurs in endeavouring to make all the parts of a map agree with information acquired from hearsay, if that information happens to be at all ample. It must have been this confused mass of evidence, collected at various times from native reports, which has made the errors in the extant maps of Central Asia so possible. And yet it is strange that we have really no correct map of the interior of the oldest continent of the world. The causes of the absence of this desideratum, now appreciated in the science of geography, may be at once traced to that antagonism of race and religion, which has hitherto been the deadly barrier to the acquisition of such knowledge. It is the fanatical tribes and bigoted Mahomedans inhabiting Central Asia, which alone offer a bar to its successful exploration, for no country possesses a finer climate, grander scenery, or places of more attractive interest. Although during the extended sway of the Moghul Empire, mediæval travellers from Europe were enabled to traverse the land in safety, yet a fatality would appear to have attended those who have more lately followed. But as civilization advances, so the hope may be entertained, that the entire physical geography of Central Asia will soon be as well known as that of Europe.

To the explorer is here offered a field of surpassing interest, whether regarded from its extent of unknown country, or the history which attaches to it, both past and future. With such an incentive to enterprise, he may well toil on; and, though in such a life many privations and hardships will have to be endured, they should, at least, teach him to sacrifice every consideration of personal ease in pursuit of the objects of what, he may rest sure, is a noble ambition; an ambition which excites to arduous enterprise and scientific labour, in so fair a field as that embracing the grand mountains and wilds of Central Asia.

REMARKS ON EASTERN TURKISTAN.

Eastern Turkistan, the country comprised in the elevated table-land of Central Asia, extending from the mountain-ranges which bound it on the north, south, and west, to the borders of the great desert of Gobi, is more generally known to Europeans by the name of Little Bokhara, Chinese Turkistan, and Chinese Tartary. Turkistan, or the country of the Turks, includes the provinces of Bokhara, Samarcand, Tashkend, Khokand, and Andijan, comprised in Western, or Russian Turkistan, whilst the provinces of Kashgar, Yarkand, Khotan, Aksu, &c., constitute the country called Eastern Turkistan.

This interior plateau is defined within the meridian of 74° to 88° E. longitude, while it varies in breadth from north to south from 380 to 520 miles, between the parallels of 35° to 43° of N. latitude, and embraces an area of 360,000 square miles. If we extend its boundaries into the Gobi Desert, and include the extent of uninhabited country lying to the east of its most fertile tracts, a total area of 430,000 square miles will be obtained.

The surface of this extensive table-land presents alternate

features of desert country, and well irrigated and fertile tracts. In its general elevation it attains to a height of about 4000 feet above the level of the sea on the west, and probably decreases to 1200 feet on the confines of the Gobi Desert. The boundaries are natural, and remarkably well defined.

On the north extends the great Thian Shan Chain of Central Asia, which, from its point of intersection with the Altai, in its prolongation westward, forms the natural boundary of Eastern Turkistan to the north. It is known to the inhabitants of the country in that portion of its length lying immediately north of the province of Kashgar, as the Artush or Kokshal Range.

On the south the Kuen Luen and Karakoram chains stretch eastward, from the point of intersection with the transverse range of the Pamir, near the meridian of 74° E., while this transverse range, or Pamir Steppe, now the object of so much geographical interest, forms the natural boundary of Eastern Turkistan on the west. Thus enclosed on three sides by immense chains, comprising independent mountain-systems, an uninterrupted natural frontier line of 2000 miles is presented, affording to the country the greatest elements of security within itself. The routes debouching across these mountain ranges into the plains of Eastern Turkistan, include the roads from Khokand and Andijan, those from the sources of the Naryn River, over the Thian Shan Chain to the north; the two roads from Ladak, Tibet and India, across the southern frontier; and the main road from Badakhshan, in the valley of the Oxus from the westward. They are thus few in number, and are chiefly traversed by merchants and adventurous pilgrims on their way to Mecca.

That portion of the country under more immediate consideration, may be defined as the west zone of Eastern Turkistan, including the fertile provinces of Kashgar, Yarkand, and Khotan. Fertilised by three considerable rivers, which unite on the 80th meridian of E. longitude, these provinces form a distinct zone of the country, which embraces an area of some 193,000 square miles, included in the basins of these rivers, to where they discharge into the Lob Nor, the united stream of Tarim Gol.

The Karakoram chain of mountains extends from the meridian of 74° E., with a general direction from W.N.W. to E.S.E., to near the sources of the River Indus. It intersects the Hindu Kush Chain, at the head of the Gilgit Valley, at a point known by the synonym of the "Pusht-i-Khar," or Ass's Back. Of its prolongation eastward beyond Chang Chenmo, nothing is very definitely known. Whether it joins the lofty group of Kailás Peaks of the Himalayas, overlooking the sacred sources of the Indus and Brahmaputra, or loses its character as a single chain, but merging into the high table-land of Tibet, in a series of radiating spurs, has yet to be determined. The most elevated summits occur in that portion of the chain lying between the Karakoram Pass and the head of Gilgit, where some peaks attain the height of 25,000 or 26,000 feet above the sea. The crest of the range reaches a mean elevation of 20,000 to 21,000 feet above sea-level, and the most lofty summit is found near the Muztagh Pass, where a peak near the 77° meridian of E. longitude, rises to the stupendous height of 28,278 feet above the level of the sea. The chain to the north is here penetrated by long transverse valleys, while the southern face in the watershed of the Indus presents steeper declivities, and is more rugged than the northern slope. Thus the ground to the north is visibly more elevated in its general surface, than to the south of the chain in the basin of the river Indus.

Extending with a general direction from w.n.w. to e.s.e., the crest of the chain continues from the Pusht-i-Khar for a distance of 420 miles to beyond the Karakoram Pass. Here, at where a double peak occurs in the chain, it deflects to the southward, and again rises into loftier summits. From hence eastward it forms the southern crest of the high table-lands which extend to the Kuen Luen, at a mean elevation of 16,500 feet above the sea; and continues eastward from Chang Chenmo to the north of the Pangong Lake and Rudok. By following the crest of the chain up to this point, a length of 650 miles is reached. The snow-line on the northern face of the chain would appear to attain an elevation of 18,600 feet above the sea; and on the southern declivity a mean height of 18,200 to 18,400 feet, may be assigned for the limit of perpetual snow.

The height which the passes attain is very considerable. The two principal passes over the more central portion of the chain are the Muztagh and Karakoram, the latter reaching an elevation of 18,317 feet above the sea. The road across this pass ascends from the southward from the head of the Shayok River, one of the principal tributaries of the Indus, and crossing the pass thence descends into the plateau of Aktagh, and conducts down the valley of the Yarkand River to the capital of Eastern Turkistan. The third pass, that of Chang Lang, or Chang Chenmo, crosses the range more to the south-east, at an elevation of 18,839 feet above the sea, and is remarkably easy. Ascending from the head of the Chang Chenmo Valley, the road across the pass descends gradually into the elevated plains lying between the pass and the chain of the Kuen Luen. The chief difficulty connected with the passage across this range is caused by the distress of laden animals from the rarefaction of the atmosphere at such high elevations, and the general sterility of the surrounding country. No natural obstacles exist on the Chang Chenmo Pass to the formation of a road to admit the passage of light conveyances.

The Kuen Luen forms a system of mountains in a long *narrow* chain which extends on the parallel of 36° to $36\frac{1}{2}$° N. latitude from east to west. It bounds the high table-land of Tibet on the north, and in its western portion runs along the Karakash and Yarkand rivers. The southern declivity of this portion of the range appears broken up into short transverse valleys, in distinct lines. Towards the eastern extremity of the chain the mountains decrease in altitude, and are more sloping. Between the meridians of 77° and 81° E. long., the chain attains greater height, and here the mountains are rugged and pre-cipitous. The summits of the loftiest peaks reach an elevation of 22,000 to 22,500 feet above the level of the sea, while the average height of the crest of the chain is considerably above 20,000 feet.

The Kuen Luen thus constitutes a long and narrow chain, the eastern portion of which is a single ridge of heights and sunny peaks, while the western throws off branches which accompany the main chain in a parallel direction, or run down as transverse ridges into the high table-land of Central Asia. The Kilian Mountains constitute a subordinate range to the north of the main chain, and as a secondary spur commence to the west of the meridian of 78°, and stretch eastward to the meridian of Khotan.

The mountain-system of the Pamir Steppe, the transverse range which bounds Eastern Turkistan on the west, rises into a high elevated plateau of probably not less than 16,000 feet in its average altitude above the level of the sea. From the " Pusht-i-Khar," the point of its junction with the Hindu Kush, it extends to where it intersects the Thian Shan chain to the west of the Terek Pass in a direction bearing N.N.W., and then trending more to the westward on the parallel of $38\frac{1}{2}$° N. lati-tude, and is comprised within the parallels of 36° 40' and 40° 20' of north latitude. The eastern crest of the plateau rises into the lofty range of Kizil Yart, presenting a steep face towards the plains of Turkistan, into which the mountains descend in steep and rugged slopes. The high country lying to the west of the Kizil Yart Range, which is the true Pamir, embraces the lake-system in which the several branches of the Oxus take their rise. The largest of these lakes is the Karakul, which is supposed to be the main feeder of that river. The Pamir throws off several high spurs to the eastward ; the main one being the Chichiklik Range, dividing the province of Yarkand from the Sarikol district.

Eastward of the 79th meridian of longitude the country situated between the Kuen Luen and Karakoram chains forms a high table-land, which is the prolongation of that of Tibet to the westward. The general appearance of this elevated plain is characterised, by Tibetian features—low undulating hills, and broken, irregular ridges, occur to vary the monotony of the general level, while numerous salt-lakes are found in the depressions of the surface, many of which evaporate or infiltrate into the soil at certain periods of the year, and leave an extensive saline incrustation. On such elevated plains vegetable life almost ceases to exist, and it is only on a few more favoured spots that a few blades of grass, or the lavender-like plant called "boorsee," can be found springing to welcome the eye.

The Yarkand River, which fertilises the province of that name, has its source on the northern slope of the Karakoram chain. The main branch of the river issues from the basin of a small plateau in lat. 35° 37' 34" N., long. 77° 50' E., which lies at an elevation of 16,656 feet above the level of the sea. Flowing with a northerly course to Kirghiz Janjal, 90 miles below its source, it here deflects at right angles to its former course, and enters the long longitudinal valley between the Kuen Luen and Karakoram ranges. Traversing this valley with a westerly course towards Sarikol, it winds round gradually to the N.E., and issuing into the plain country of Turkistan, continues in that direction to where it receives its large tributaries, the Kashgar, Karakash, and Aksu rivers. After the junction of these tributaries it flows eastward in a united stream, known as the Tarim Gol, and finally discharges itself into the lake of Lob Nor, on the confines of the Gobi Desert. From its source in the Karakoram to the Lob Nor, it has a total length of 1230 miles, and of some 680 miles to where it receives the Kashgar River, while the area of its basin embraces an extent of about 85,000 square miles. Zimmerman has shown that the elevation of the lake of Lob is probably about 1280 feet above the level of the sea, thus the river has a fall of nearly 15,500 feet during its entire course of 1230 miles, or a greater fall than from the summit of Mont Blanc to the level of the Mediterranean Sea. During the first 150 miles of its upper course it descends at the rate of 34 feet per mile, while the mean fall of the bed of the river in 400 miles is shown to be nearly 30 per mile by hypsometric measurement. In that portion of its course where it traverses the low country as the Tarim Gol, before entering the Lob Nor, it must have a very gentle fall of about 3¾ feet per mile, since throughout its length from below Yarkand to the Lob Nor the total fall would appear to be about 2,900 feet. Thus the river throughout its entire course has a mean fall of

12½ to 13 feet per mile. The river shows little diversity in the country which it traverses, since the general character of the bed is level and open, except where it flows through the gorges of the Kurchum Hills before debouching into the plains of Eastern Turkistan. From the level character of its bed the river more often flows in several channels than in a single united stream, and the valley is seldom contracted to less than a mile in width, save where the stream winds along the spurs of the Aktagh Range, and the steeper declivities of the Western Kuen Luen. From immediately below Yarkand, where the width of its bed increases to upwards of 5000 feet, it is navigable during the months of June, July, and August. Its tributaries are numerous. It receives innumerable streams from the northern slope of the Karakoram chain, the largest of which is the Shingshâl and Toong rivers, and the Tashkurgan and Charling rivers draining the Sarikol district. While on the right bank the Ruskum and Tiznâf rivers with their minor tributaries effect a junction. The larger rivers of Kashgar, Aksu and the Karakash unite with it near the 80th meridian of east longitude. The vegetation found above the banks during the first 200 miles of its upper course is scanty, and is principally confined to low brushwood, with patches of coarse grass. Where the elevation as decreased to 11,000 and 10,000 feet, good pasture grounds fringed with bushes and timber is met with in the valley, and the slopes of the mountains are less bare. Here the Kirghiz encamp, and graze their flocks and herds, and ascend to the heads of the transverse valleys during the summer months. The banks of the river, where it traverses the desert country to the east of Yarkand, consist of sandy tracts covered with a dense forest of underwood, coarse grass, and tares.

The river freezes during the winter throughout its course to below Yarkand, wherever the stretches of water flow evenly and gently. In its upper course, where the fall is as great as 34 feet per mile, a thick mass of ice accumulates on either bank, so intense is the cold; and in many places the edges of the ice approach close enough to admit of a passage, and in others have joined so as to form a bridge. The river is fordable throughout its upper course during the winter months, in those places which are in use as regular fords. The great increase in the volume of water renders these impracticable during the months of June, July, and August.

The principle of irrigation is extensively developed along the banks of the Yarkand River. The water, conveyed by numerous canals and dykes cut from the river, fertilises extensive tracts of cultivated land throughout the province. A few of the prin-

cipal canals are the Beshkum, Urpi, and Yulchak, which irrigate the country around the capital and Posgâm.

The Kizil Daria, or Kashgar River, takes its rise in the angle
formed by the intersection of the prolongation of the Thian
Shan chain with the transverse range of the Pamir. It receives
a tributary from the Terek Pass, and flowing, with a course
nearly due east, along the southern slope of the Thian Shan
chain, unites with the Yarkand River at a distance of about 500
miles from its source. Its tributaries on the right bank are
the Khanarik, Hosun, and Sargrak Rivers, descending from the
Kizil Yart range of the Pamir, and the Aksu, Artush, and
other streams which it receives from its southern slope of the
Thian Shan chain. The province of Kashgar is comprised in
the basin of this river with its tributaries, which embraces an
area of from 55,000 to 58,000 square miles.

The third principal river of the west zone of Eastern Turkistan is the Karakash, which fertilises the more southerly province
of Khotan. Rising in the northern slope of the Karakoram
chain, in about lat. 34° 45′ N., long. 78° 45′ E., and at a probable
elevation of 16,800 feet above the level of the sea, it flows with
first a northerly and then an easterly course to under the Kuen
Luen chain, where, on the parallel of 35° 54′ N., at a distance
of 120 miles below its source, it bends to the north-west, and
flows through the valley of Sarikia, along the southern slope of
the Kuen Luen, for a further distance of 80 miles. Arrived at
Shadula, 200 miles below its source, it turns to the north and
pierces the main chain of the Kuen Luen on the meridian of
78° 18′ E. long., and then, meeting the southern declivities
of the Kilian range of mountains, deflects at right angles to the
eastward, and having thus turned the chain of the Kuen Luen
continues an easterly course to near the 80th meridian of E.
longitude, where, diverging to the north, it emerges into the
plains of Turkistan, and fertilises the fair province of Khotan.
The Karakash continues with a northerly course through the
Taklá Makán and Dusht-i-Tatar, where it unites with the Yark-
and River. From its source to this point of junction it has
a length of about 590 miles, while the basin of the river,
including the province of Khotan, embraces an area of 48,000
to 50,000 square miles. Its principal tributary is the Kakka
or Khotan Daria, and it also receives several smaller streams
from the slopes of the Kuen Luen. The bed of the river has a
fall of 27 feet per mile from its source to Shadula, where it
pierces the Kuen Luen, and flows at the rate of 200 yards per
minute, or nearly 4½ miles per hour, as observed at a point
220 miles below its source. In the general character of the
country which it traverses throughout its upper course it greatly

resembles the Yarkand River, save that the fall of its bed is more gentle, and the general surface of the valley less sloping. Where it skirts the base of the steeper mountains of the Karatagh and the Kilian Mountains, the course of the river is more confined. At as high an elevation as 15,800 to 16,000 feet grass and the lavender-like " boorsee " are found in the valley and lower slopes of the mountains, and below an elevation of 12,000 feet vegetation, with bushes and trees, occur along its downward course. The bed of the river consists chiefly of gravel and conglomerate, while an alluvium and fine sand formation is developed in many parts of its course. Nearly the whole volume of water which it rolls down is utilised for irrigation throughout the province of Khotan. Like the Yarkand River the stream is frozen during the winter months.

In the plain country of Turkistan we meet with several tracts of marshy ground, which contain small ponds or larger accumulations of water, with running streams of greater or less size, while the prevailing features of the surrounding country are hillocks of drifted sand. The margins of these marshes are fringed with high coarse grass and reeds, and are the favourite haunts of large quantities of wild fowl. Between the more fertile tracts and the foot of the mountains wide sandy steppes occur along the southern bases of the Thian Shan chain and on the eastern base of the Pamir.

From its central position Eastern Turkistan enjoys a peculiarly dry climate. The rain-fall at any distance from the mountains must be excessively small, while towards the interior of the plateau, on the confines of the Desert of Gobi, the moisture must be reduced to a minimum. The inhabitants being deprived of periodical rains to fertilise the cultivated tracts, are thus dependent for their well-being and means of subsistence upon the waters of the rivers, brought down from the mountains during the spring and summer months. By numerous branches, canals, and dykes led off from the main rivers, the cultivated ground is irrigated, and as the soil must be naturally productive, for the rudest implements of agriculture are everywhere employed, abundant harvests are generally yielded. It is only in exceptional years when the fall of snow upon the mountains has been but little, and the volume of water brought down by the rivers during the ensuing summer consequently less than usual, that anything like a scanty harvest occurs. Thus the welfare of the country depending upon the fertilising powers of its rivers, the towns and villages of Eastern Turkistan will be found situated at greater or less distances along the courses of these streams. The population of the country is therefore accumulated on certain tracts running parallel with the course of the rivers, whilst between them large barren tracts of uninhabited country are met with. They consist for the most part of bare, open plains, whose level expanse is broken by dunes and low hills of drifted sand, or here and there marshy ground surrounded by sandy steppes. A journey between two distant towns can only be accomplished by crossing some of these uninhabited deserts, where supplies for the road have to be carried. The usual halting-place, or langar, is invariably met with at the end of the day's journey. All the roads in the plain country are practicable for wheeled carriages, and are regularly traversed by two-wheeled carts and conveyances. Of these main roads may be mentioned that from Yarkand to Khotan, Yarkand to Aksu, and Yarkand through Yanghissar to Kashgar. The true beasts of burden in the plain country are the ass and dromedary, while the mountains are traversed by the sure-footed yaks of the Kirghiz.

Besides the Moghul population of the country the inhabitants of the towns and cities consist of Chinese who have become Mahomedans, Tungâni descendants of Chinese mothers, with a sprinkling of Kalmuks, while amongst the mercantile classes a large infusion of Tadjiks from Andijan has taken place. Emigrants from Badakhshan, Afghanistan, and the more adjacent countries have taken up a permanent residence in many of the large towns. The ruling powers and military are chiefly Uzbegs and Kipchaks who have invaded from the north.

The mountain districts are frequented by nomadic Kirghiz who possess no fixed habitation, or cultivate an acre of ground. They possess large flocks of sheep and goats, and herds of yaks and camels, with which they frequent the pasture grounds of the higher mountains during the summer months, and in the winter descend to the rivers of the lower valleys. These Kirghiz periodically visit the nearest towns for the purpose of disposing of some of their live stock, and in exchange purchase grain, flour, and other necessaries sufficient for their wants during the ensuing season.

The chief towns of the country present no diversity in their uniformity of feature and situation. Enclosed by a fortified earth wall the houses appear in regular narrow streets or enclosures, surrounded by small gardens, in which fruit and other trees thrive. No stone is anywhere used for building purposes, as it is not obtainable in the plain country. The houses built of mud and unburnt bricks, are of one storey in height, with flat roofs and small apartments. The shops consist of open booths displayed on either side the narrow streets. On account of the total absence of stone buildings and erec-

tions, inscriptions and antiquities which might throw a light
upon the early history of the country are in no place to be
met with.

The go e nmen of the country presents the usual forms of
Oriental despotism. Rigid Mahomedans, the inhabitants all
conform to the ordinances exhorted by the Koran, and hold
their priests, or moulvies, in much veneration. The hand of
charity is ever ready to be extended to the many mendicants
who, in the dress of dervishes and fakeers, incessantly roam the
land, while the numerous mosques are daily and hourly crowded
with devout Moslems. The Sheik-ul-Islam is the chief head of
the religious community, while the Kazi Kalan dispenses perhaps a somewhat partial administration of justice in law and
civil cases. They are the only persons who have the will or
power to differ from the king, and to them, out of good policy, the
sovereign generally defers, since in all disputes, especially on
points of religion, the popular feeling would be with the Kazi.
The king himself is very particular in observing all ceremonies
connected with the Mahomedan religion.

Each outstanding town and district is under the authority of
a Hakeem Beg, or chief official, who is bound to pay a certain
fixed revenue yearly to the king. The several villages in a
district are under a head man, or Diwan Beg, who is responsible
to the Hakeem Beg of the district for the yearly revenue of his
village. These appointments are given by the king to his
favourites, or to such as are able to pay a bribe for the
somewhat doubtful honour of holding them, since they are
often displaced on the slightest ground of complaint or even
at the mere caprice of the sovereign, to be given to a later
favourite.

The revenue is collected in kind, while a fixed land-tax is
levied. Each family can be called upon to furnish one individual as a soldier to the State in time of war, or in case of
emergency. This institution, like the law of the Medes and
Persians, is one against which there is no appeal.

The products of the country are various and abundant.
Wheat of two descriptions is produced. One crop called
"Khuzgha bagdai," sown in September, the other, "yuzgha
bagdai," which is sown in April. Rice, barley, Indian corn,
four kinds of oil-crop called "zagboon," "zâghee," "zeranghza,"
and "muskar," are yielded in abundance. Clover, root-crops,
as turnips, carrots, onions, are grown extensively. Cotton of a
fair description is produced in large quantities, and manufactured into material for native wear, or exported to Khokand
and Russia.

The fruit crops consist of the pear, apple, apricot, almond, pomegranate, walnut, peach, and melon, while the vine is cultivated—the yield from which is excellent. The mulberry is everywhere seen, and prevails in the province of Khotan. The principal exports from Yarkand to India, Kashmir and Tibet, are felt cloths, silk, churrus or hang, pushmeena wool, gold, silver, and cotton. Yarkand imports from India opium, spices, sugar, tea, linen cloths, Kinkhâb, English broad cloths, muslins, Kashmir shawls, arms, leather, brass utensils, and indigo.

Khotan produces excellent carpets and felt cloths, silk in large quantities, gold, &c.

Through Kashgar the imports from Khokand are silk, Russian prints and calicoes, iron, silk caps, cochineal, indigo, porcelain, Russian knives and padlocks, Russian broad cloths, tobacco, snuff, &c. From Afghanistan and the valley of the Oxus come horses and the Bactrian camel.

The country is rich in minerals.

Gold is found in the northern slopes of the Kuen Luen, and extensive gold washings occur on the banks of the rivers east of Khotan. The country adjacent to Khotan Daria and the Karakash furnishes a moderate supply. Copper and iron ore are found in the Karakoram and Kizil Yart Ranges—while rubies, turquoises, lapis-lazuli, and other precious stones, are met with in the mountain ranges or are imported from Badakhshan.

The animals are the horse, two-humped camel or dromedary, ox, ass, sheep, and goat, from which the valuable shawl wool is extracted. Fur-yielding animals are found in the Pamir Steppe, as the fox, lynx, sable, and hare.

Wild animal life is not very profuse, on account of the scant pasture and herbage in the more uninhabited tracts. The ibex is found in the Kuen Luen chain, and probably also in the Kizil Yart, the stag abounds in the forests towards Aksu, and antelopes roam the plains, while wild camels are met with in the desert tracts of the Taklá Makán, and "Dusht-i-Tatar." The beasts of prey are two species of wolf, and the tiger is met with in the forests towards Aksu.

APPENDIX I.

Main Road from YARKAND to KASHGAR.

This road is regularly traversed by two-wheeled carts and conveyances.

Number of Marches.	Stages or Halting Places.	Estimated Distance in Miles.	REMARKS.
1	Yarkand to Kokrubat ..	22½	Road conducts along the north wall of the Fort, and at 4½ miles from the city crosses the Urpi Canal by a wooden bridge. Passing the village of Kara-Koom and Bigil, it skirts some marshy ground to Kokrubat, a village of 200 houses with a caravanserai.
2	Kizil	27½	Road skirts the "Hamed-i-Dusht," a large barren tract of country, extending up to the Kiziltágh Range on the west. At 14 miles from Kokrubat is a halting-place, called Ak-Langar, where is a musjid and two wells of water. Kizil is a village of 500 houses, with a large tank and caravanserai.
3	Yanghissar ..	32	The road passes the villages of Chamalung, Kheduk, Koshimbash, and Toblok, to Kelpun, an old Chinese "urtang," or police-station, now in ruins. At 2½ miles, before reaching Yanghissar, it crosses the Sargrak River by bridge. Yanghissar is a commercial town of some 11,000 houses, situated 32 miles N.W. of Yarkand, and 43½ south of Kashgar. The Fort lies at a distance of 600 yards to the north of the town.
4	Yupchan ..	22½	Road passes villages of Koomlok and Toglok, and crosses the Hosun River by bridge. Then continues up left bank of river to Yupchan, a village of some 700 houses.
5	Kashgar (Old City.)	21	At 2½ miles from Yupchan cross Khanarik River by bridge; and passing the village of Tasgam, cross a canal from the Khanarik River, and a branch of the Kashgar River, to the Fort of Kashgar, which lies some 3 miles south of the city. Cross "Kizil Daria," or the Kashgar River, midway between the Fort and City. Kashgar contains about 28,000 houses; and from 60,000 to 70,000 inhabitants.
	Total ..	125½	

YARKAND to KHOTAN.

This road is traversed by two-wheeled carts and conveyances.

Number of Marches	Names of Halting Places.	Estimated Distance in Miles.	REMARKS.
	Yarkand to		
	Karghalik ..	36	
	Guma	52	
	Khotan	88	
	Total ..	176	

No. 1.

LEH to YARKAND.

LEH to YARKAND *via* Chang Chenmo and Valley of the Lower Karakash River.

Number of Marches.	Names of Halting Places	Estimated Distance in Miles.	REMARKS.
1	Leh to Chimray.. ..	21	Road good up right bank of the River Indus to opposite Marsalong, where turn up valley to the left for 4 miles, to the village of Chimray. At 7 miles from Leh is a large village called Tiksu, which is also a halting-place.
2	Sakti	6	Up a ravine to the village of Sakti, where yáks can be obtained for crossing the Chang La Pass.
3	Seeprah	11½	Cross Chang La Pass, 18,368 feet above the sea. From Sakti to the summit of the pass is 7 miles. Ascent steep for the last 2 miles. Descent of 4½ miles to Seeprah, which is a Bhoot encampment in the valley. Pass a lake called Tso Lak.
4	Tanksee	10½	A large village and thannah. The last place in the Maharajah of Kashmir's territory where supplies can be obtained.
5	Lukong	19	At 9 miles from Tanksee Pass a small village named Moglib. Lukong is a small village situated north of the Pangong Lake.
6	Chagra	6½	A Bhoot habitation of a few stone huts. No habitation is met with after this.

LEH to YARKAND—*continued.*

Number of Marches.	Names of Halting Places.	Estimated Distance in Miles.	REMARKS.
7	Rimdee	10	Cross Masimik Pass, 18,457 feet above the sea. Ascent to the pass gradual and easy. Steeper descent of 3 miles. The pass is generally covered with snow, and is closed from December to March.
8	Pumsul	15	Road down valley to Pumsul, where the main Chang Chenmo Valley is struck. Fuel in abundance.
9	Gogra	11½	Road for 4 miles up left bank of Chang Chenmo River, then ford river and cross a low spur on the right side of the valley to Gogra in Kugrang Valley.
10	Hot Springs ..	5	Road up Chang Lang Valley, crossing the stream several times. Hot springs on rocks in the bed of the stream, at an elevation of 16,000 feet above sea-level.
11	Camp South of Chang Lang Pass.		Road up Chang Lang Valley. At 4 miles from Hot Springs, the road leaves the main valley, and leads to the east up a branch valley to the Pass. For the journey onward water and fuel should be carried from Chang Lang, as generally no water, and but very little fuel, is to be met with between here and the lower Karakash River. It might be preferable not to halt at No. 10, Hot Springs, but march from No. 9, Gogra, to near the Pass. There is a good halting-place in the valley, 5 miles beyond Hot Springs, where is fuel and a little grass. By camping here the Pass can be crossed the next day.
12	Kala Pahar ..	14½	Cross Chang Lang (or Chang Chenmo) Pass, 18,839 feet above the sea. Road to the pass up a ravine filled with loose stones and debris. The last part of the ascent is steep for half a mile. From the pass road good, down the open valley of the table-land. At ten miles from pass is a camping-place called Nischu, beyond which is Kala Pahar. No fuel or grass obtainable.
13	Camp, Lingzi Thung Plains.	15	Descent gradual to the Lingzi Thung Plains, which are nearly 17,000 feet above sea-level, across which the road is level and good. Camp in sandy bed of wide nullah. The Lingzi Thung is destitute of water from October to March, but water is to be met with during the

LEH to YARKAND—*continued.*

Number of Marches.	Names of Halting Places.	Estimated Distance in Miles.	REMARKS.
	Camp, Lingzi, &c. (*contd.*) ..		spring and summer months. A little "boorzee," which is available for fuel is found here and there on the plains. Direction of route across the plains is N.N.E. There is a remarkable Round Peak in the Lak Tsung Range, which is a good fixed point to march upon in crossing these plains, the road going down a broad sandy valley to the right of this peak. ..
14	Lak-Tsung ..	13	Camp in valley of Lak Tsung Range, where is a little grass.
15	Thaldat or Paldar.	19	A frozen lake and a salt spring at an elevation of 15,886 feet above the sea. Road good. Some grass is obtainable in a ravine lying due west about a mile off.
16	Pats-alung ..	18	Road good across open plain. At 12 miles cross low pass to Patsalung.
17	Camp Soda Plain	14	Road down open valley, passing some salt lakes.
18	Brungsa	13	At Brungsa camp in valley of a branch of the Karakash River. A little grass and fuel.
19	Mandalik ..	12½	Road down valley for 8 miles, where strike main valley of the Karakash River, continuing down right side of valley to Mandalik. Grass and fuel everywhere obtainable on the march from here downwards.
20	Languak.. ..	9½	Road good down right bank of Karakash River. At half-way cross and recross river. Camp under some granite rocks close by the stream.
21	Ak-koom	10	At 5 miles cross river to left bank. Camp in open valley. Fuel in abundance.
22	Mulbash	16½	Road good down right side of valley.
23	Gulbashem ..	14	A Kirghiz encampment, where the Kirghiz bring flocks of sheep and goats, and herds of yaks to graze.
24	Balakchee ..	10½	Another Kirghiz encampment. Some supplies can generally be obtained from the Kirghis.

LEH to YARKAND—*continued.*

Number of Marches.	Names of Halting Places.	Estimated Distance in Miles.	REMARKS.
25	Shadula	8	A Fort in the Turkistan territory, garrisoned by a guard of Turki sepoys. Permission must here be obtained to enter Turkistan. The guard does not interfere with the passage of the native traders. At 4 miles below Balakchee, the Tabistánee route from the Karakoram Pass down the Sooget valley joins.
26	Pilartákash	14¼	Road down Karakash Valley passing Ulbuk and Tograssu. At Tograssu, 10 miles below Shadula, a valley to the N.W. conducts to the Kullik Pass. At Ali Nazur Kurgan, 2 miles below this, a ravine to the N.W. conducts to the Kilian Pass.
27	Diwanjilga South of Sanju Pass	10¼	At Mazar Badshak, 4½ miles below Pilartákash, the road leaves the Valley of the Karakash River, and ascends up a narrow ravine to the Sanju Pass. A Kirghiz encampment is generally at Mazar Badshak, where yaks can be hired from the Kirghiz for crossing the pass. Fuel should be carried from near Mazar Badshak, as none is obtainable near the pass.
28	Kichikyulak	9	Cross the Sanju-Pass, 16,612 feet above the sea. Very steep ascent and descent. The pass is impracticable for laden horses, whose loads have to be carried on yaks. Horses and camels unladen can cross the pass. The pass is closed by snow during the winter, from December to May. Kichikyulak is a camping-place of the Kirghiz, on the north side of the pass.
29	Tám	12¾	Road down Sanju Valley, crossing and recrossing the stream.
30	Kibris	19	Road difficult, the Sanju stream having to be crossed and recrossed some twenty times. Kibris, a village of a few houses, is the first habitation met with on entering Turkistan.
31	Sanju	10	Road down open valley to Sanju, a large place of some 3000 houses, comprised in several villages situated on each side of the river. The market-day is " Du Shamba," or Monday.

Number of Marches.	Names of Halting Places.	Estimated Distance in Miles.	REMARKS.
	Sanju (*contd.*) ..		Ilchi, the capital of Khotan, lies 66 miles east from Sanju. There is another route from Sanju to the pass, which is used when the Sanju River is impassable between Tâm and Kibrâs. This road conducts from Sanju, *viâ* the Arpalak Valley and Chuchu Pass (11,847 feet above the sea), to its junction with the Sanju Valley near Tâm, and is as follows:— 　Sanju to Kizil Aghil 11½ miles. 　　,,　　Mazar　.. 12　,, 　(Cross Chuchu Pass) 　　to Tâm　.. ..　}15　,, 　　　　　Total　.. 38½ miles.
32	Koshtok	25	Road crosses the Sanju River, and ascends a low range of sand-hills, across which it descends to the small village of Langar, 15½ miles from Sanju. From here it crosses the open to Koshtok, a village of 70 houses, watered by a stream from the Kilian Mountains.
33	Oitogrok ..	18½	A small village in a valley. Road level and good.
34	Borah	11½	Ascend low range of sand-hills, and cross a sandy steppe to Borak, a village of 90 houses. The Kilian route joins at Borak, and from here they are the same to Yarkand throughout.
35	Karghalik ..	24½	At 6 miles from Borak the road leaves the low hills, and descends to the plain country, crossing a barren tract of country, to the village of Beshiruk, beyond which, at the distance of 4½ miles, is Karghalik. It is a large town and district of some 20,000 houses, with a large market and several caravanserais, and is watered by a canal from the Tiznâf River. The following roads join at Karghalik :— The Kugiar route, from the Karakoram, *viâ* the valley of the Yarkand River, and the Yangi Pass. The Kullik Pass route. The Khotan Road through Guma. At the junction of the Khotan Road there is a new earth fort.

LEH to YARKAND—*continued.*

Number of Marches.	Names of Halting Places.	Estimated Distance in Miles.	REMARKS.
36	Posgam	21	At 11 miles from Karghalik the Tisnâf River is crossed. Six miles beyond this is " Yak Shamba Bazar,' a large market, crowded by the country people on Sundays. Posgam is a town and district of some 16,000 houses, with a market and caravanserai, and is watered by the Beshkun Canal, cut from the Yarkand River.
37	Yarkand	15	The capital of Eastern Turkistan, containing some 40,000 houses, and about 120,000 inhabitants. At 4 miles from Posgam the Yarkand River is crossed. It is here fordable during the winter months. In the summer it is crossed in boats at Aigâchee, many miles lower down, which is then the direct road from Karghalik and Posgam. Four miles beyond this the Yulchak Canal, carrying water to the city, is crossed by a wooden bridge.
	Leh to Yarkand.		
	Total ..	507½	Leh to Shadula .. .. 316½ miles. Shadula to Sanju .. 75½ ,, Sanju to Yarkand .. 115½ ,, Total .. .. 507½ miles.

No. 2.

LEH to YARKAND.

NEW ROUTE *viâ* Chang Chenmo and the Valley of the Upper Karakash River to Aktâgh.

Number of Marches.	Names of Halting Places.	Estimated Distance in Miles.	REMARKS.
	Leh to No. 12 Camp, Kala Pahar,	143½	Vide Route No. 1.
13	Camp, Lingzi Thung Plains	11	Descend low hills to Lingzi Thung. Road across plain due north.
14	Boorsee	10½	Continue due north. At 8 miles cross a low ridge, leaving the Lingzi Thung Plains. At Boorsee that wood for fuel. No grass.

LEH to YARKAND—*continued.*

Number of Marches.	Names of Halting Places.	Estimated Distance in Miles.	REMARKS.
15	Karasu	10	Road across open plain and up wide valley to Karasu, where is a little water. No grass; no fuel.
16	Kizil Jilga ..	14½	At 2 miles cross Kizil Diwan, 17,789 feet above the sea. Very easy ascent and descent of a few hundred feet. Road down open valley to Kizil Jilga, where the main valley of the Karakash River is struck. Plenty of "boorsee" for horses and for fuel. A little grass close by. Plenty of good grass on the hill side a mile away.
17	Khush Maidan	16	Road excellent down left side of valley. Khush Maidan is a camping in the valley. Plenty of good wood for fuel and grass here. It would be advisable to carry some wood from here for the journey onwards.
18	Shorjilga.. ..	10	At 5 miles below Khush Maidan is Choongtásh, a perpendicular cliff overlooking the river on its right bank. At one mile below this the road leaves the main valley of the Karakash River, and ascends, in a N.W. direction, up a wide valley leading to the Karatágh Pass. A little grass is met with at Shorjilga.
19	Oglok	21	From Shorjilga to the pass is 13 miles. About half-way is a camping-place, where is some "boorsee" for fuel; so it might be preferable to march from Khush Maidan to this place, and not halt at Shorjilga. Ascent to the Karatágh Pass (which is 17,953 feet above the sea) very gradual and easy. The pass is across a depression in the Karatágh Range, which here forms the watershed between the Yarkand and Karakash river-basins. Road across the pass down open valley to Oglok, where is water, grass, and a little fuel.
20	Wahábjilga ..	10	Road down open valley to Wahábjilga, on the Karakoram Pass route.
21	Aktágh	15½	Regular Karakoram Road down open valley to Aktágh. Camp under rocks at foot of Aktágh, or "White Mountain." A little grass; no fuel; water plenty.
	Leh to Aktágh	262	

No. 3.

LEH to YARKAND—*continued.*

From Aktāgh to Yarkand by the Valley of the Yarkand River and the Yangi Pass.

Number of Marches.	Names of Halting Places.	Estimated Distance in Miles.	REMARKS.
1	Aktāgh to Kufelong.. ..	21	This is the regular ˈZamistānee route to Yarkand. Road level and good down open valley along the Karakoram Pass stream, which is passed twice in the defile near Kufelong. At Kufelong the main valley of the Yarkand River, coming down from the S.W., is struck. Camp below junction on right bank. Grass and fuel everywhere obtainable from here on the line of march downwards. In journeying towards Ladak, fuel should be carried from Kufelong. Two routes, not now in use, one from the head of the Nubra Valley, in Ladak, and the other from Chorbut, in Baltistan, conduct down the main valley of the Yarkand River, and join at Kufelong.
2	Bukhurooldee..	17	Road down valley of the Yarkand River, which is crossed once or twice. There are two intermediate camping-places.
3	Kirghiz Jangal	18	A pasture-ground in the open valley. Several springs of good water. Three miles before reaching Kirghiz Jangal are passed some ruined stone huts, on the right bank of the river, from where a road goes off east, leading up a ravine and across the Kirghiz Pass (17,092 feet above the sea) to Shadula, 33 miles distant.
4	Koolunooldee ..	14	Cross to left bank of river immediately below Kirghiz Jangal. Again cross to right bank, 1½ mile further down. Road continues down open valley to Koolunooldee.
5	Camp S. of Yangi Pass.	9½	At Koolunooldee the road leaves the valley of the Yarkand River, and ascends up a narrow ravine to the Yangi Pass. Road somewhat difficult up bed of the ravine. The Kunjoot robbers of Hunza and Nagar, lying in wait on the hill side, overlooking this defile, sometimes attack and plunder the caravans of the traders.

Number of Marches.	Names of Halting Places.	Estimated Distance in Miles.	REMARKS.
6	Toor-Aghil ..	12½	Cross Yangi Pass. Ascent somewhat steep for the last half mile. The pass is practicable for laden horses and camels. Descent gradual to valley of the Tisnáf River, which rises N.E. of the pass.
7	Mazar Badshah	18	A Kirghis encampment. Cross and recross Tisnáf River several times. Grass and fuel obtainable all down the valley.
8	Chiklik	20	Cross and recross river several times. Near Chiklik two streams join from the eastward, called Oglok and Sanooch. The valley of the Tisnáf River is frequented by a tribe of Kirghis called Phakphook. They generally supply travellers with provisions if required.
9	Ak Masjid ..	15	Near Chiklik the road leaves the valley of the Tisnáf River, and crosses a low pass named Toopa Diwan. Ak-Masjid, a Kirghis encampment in the open valley.
10	Kugiar	18	Road through plain country to Kugiar, a village containing 200 houses. No one is allowed to proceed onwards without permission of the Turkistan ruler.
11	Beshtiruk Langar.	19	Road across a barren tract of country, called the "Beshtiruk Dusht." At Langar the Kullik Pass route joins from Oshokwas, a large village of 1200 houses, lying S.E.
12	Karghalik ..	17	Road across "Beshtiruk Dusht" to near Karghalik.
14	Yarkand	36	Vide Route No. 1.

		Miles.
Aktágh to Yarkand	.. ..	235
Leh to Aktágh	.. ..	262
	Leh to Yarkand	497

No. 4.

AKTÁGH to SHADULA, by the Sooget Pass.

Number of Marches.	Names of Halting Places.	Estimated Distance in Miles.	REMARKS.
1	Aktágh to Chibra	12½	Road good up right bank of stream coming down from the Sooget Pass. A ruined stone hut in the ravine at Chibra. No grass or fuel.
2	Sooget	19½	At 4 miles cross the Sooget Pass, 18.237 feet above the sea. Ascent very gradual and easy up the ravine. Steeper descent of one mile to the Sooget Valley, which runs N.W. for 9 miles. The road passes a camping-place called Kotasjilga, then winds round down the valley, and is on the descent to Sooget, where are grass and fuel in abundance.
3	Shadula	7½	Road on the descent down the Sooget Valley to where it joins the valley of the Karakash River, 4 miles above Shadula. Cross and recross river to fort.
	Total ..	39½	

YARKAND to BADAKHSHAN, *viâ* TASHKURGAN and the PAMIR STEPPE.

Tashkurgan, the capital of Sarikol, is reached in from 8 to 9 days' journey from Yarkand.

The probable distance is about 175 miles.

The journey onwards from Tashkurgan to Badakhshan is accomplished in from 16 to 18 days. The whole distance from Yarkand to Badakhshan is probably about 460 miles.

This road traverses a plain country for nearly 70 miles from Yarkand. It then crosses a low range into the Sarikol district, and, ascending the valley of the Charling River, crosses the Chichiklik Pass, leading across a spur of the main Pamir range into the Tashkurgan valley. From Tashkurgan it leads across the pass at the head of the Sarikol territory into Pamir Khurd, and conducts down the valley of the Oxus into Wakhán.

The road is practicable for laden horses throughout, and for laden camels up to the foot of the Chichiklik Pass, from the Turkistan side; and from Badakhshan as far as Pamir Khurd, from the westward.

The caravans of the merchants seldom accomplish the whole journey under one month.

There is also a second route from Yarkand to Tashkurgan, but it lies through a very mountainous country. On this route three high ranges—the Kandar, Arpatallah, and Oogrhiot—have to be crossed, and it is consequently but little frequented.

APPENDIX II.

METEOROLOGICAL OBSERVATIONS FROM OCTOBER, 1868, TO JUNE, 1869.

MONTH OF OCTOBER, 1868, in LADAK and TURKISTAN.

Date.	Thermometer.			Wind.	Weather.	B. P. of Water.	Air.	Elevation in Feet above Sea-level.	Place, &c.
	7 A.M.	12 Noon.	7 P.M.						
1	35	53	40	W.S.W.	Fine	188·0	47	13,274	Sakti, in Ladak, foot of Chang La Pass.
2	30	27	30	W.S.W.	Fine	183·4	34⅓	15,790	Seeprah, 4 miles below pass. Crossed Chang La Pass, 18,368 feet above the sea. Thermometer on pass at noon = 27°.
3	18	53½	45	W.S.W.	Fine	188·8	45	13,128	Tankse.
4	34	54	45	S.W. by W.	Fine	::	::	,,	Ditto.
5	37	49	::	S.E. by S.	Snow	::	::	,,	Ditto.
6	29	44	29	S.W. by S.	Cloudy	185·8	35	14,394	Chagra, foot of Marsimik Pass.
7	12	::	7	S.W.	Fine	180·8	7	17,208	Rimdee, 2 miles below pass. Crossed Marsimik Pass 18,457 feet above the sea. Thermometer on pass 41½°. B. P. of water on pass = 179°·4¼.
8	3½	::	27	S.W.	Fine	185·5	27	14,780	Pamzal
9	14	::	11½	W.S.W.	Fine	184·0	24	15,598	Kiam
10	7	33	19	W. by S.	Fine	::	::	::	Ditto } Chang Chenmo Valley.
11	4	::	14	W. by S.	Fine	::	::	::	Ditto
12	7	34	18	W.S.W.	Fine	::	::	::	Ditto
13	5	::	14	W. by N.	Fine	::	::	::	Ditto

Date	Morning	Mid-day	Max.	Wind	Weather	Boiling point	Temp.	Height (feet)	Remarks
14	Zero	··	30	W.S.W.	Fine	183·6	··	··	Kung Kong, 10 miles from Kinn.
15	—7	··	··	W.S.W.	Fine	··	··	··	Gogra, in Kugrang Valley.
16	Zero	14	··	W.S.W.	Fine	··	··	15,770	Ditto ditto.
17	—4	9½	··	W.S.W.	Cloudy	182·0	17	16,610	2nd camp, Kugrang Valley, in lat. 34° 27' 42" N.
18	—3	7	··	S.W. by W.	Snow	··	··	··	Ditto ditto.
19	0	7	30	W.S.W.	Fine	··	··	··	Ditto ditto.
20	—3	4	31	W.	Fine	··	··	··	Ditto ditto.
21	—1	15	24	S.W. by W.	Cloudy	··	··	··	Ditto ditto.
22	Zero	12½	29	W.S.W.	Snow	182·4	19	16,408	Camp, junction of branch, Kugrang Valley.
23	—2	11	28	S.W. by W.	Fine	··	··	··	Ditto ditto.
24	—6	9½	30	S.W. by W.	Fine	183·6	27	15,770	Camp near Gogra, lat. 34° 22' 38" N.
25	—5¼	10	28	W.S.W.	Fine	183·0	28	16,028	Hot springs, Chang Lang, lat. 34° 24' 24" N.
26	—2	9½	34	S.W. by W.	Fine	··	··	··	Camp near Chang Lang Pass. Crossed Chang Lang Pass, 18,839 feet above the sea. B.P. of water on pass 178°·3'.
27	—9	1	25	W.	Fine	··	··	··	Camp 7 miles beyond pass.
28	—15	Zero	21	W.S.W.	Fine	181·6	1	16,810	1st camp, Lingzi Thung Plains.
29	—11	—4	··	W.S.W.	Fine	181·5	··	16,892	2nd camp ditto.
30	—9	5	··	W.S.W.	Fine	182·4	13	16,342	Lak Tsung.
31	—2	10	··	S.W. by W.	Fine	183·4	22½	15,896	Camp Thildat, lat. 35° 14' 41" N.

NOTE.—Being on the march almost daily the morning observation of the thermometer as recorded will be that of the place where camped the evening of the day before. Thus: the morning observation of the 29th instant, 11° below zero, was taken before starting from 1st Camp, Lingzi Thung Plains; and the morning observation taken at 2nd Camp, Lingzi Thung Plains, the camp of the night is recorded the next day, the 30th instant, being 9° below zero. Mid-day observations on the march were taken by suspending thermometers in shade, whilst observing for latitude.

MONTH of NOVEMBER, 1868, in EASTERN TURKISTAN.

Date.	Thermometer. 8 A.M.	12 Noon.	8 P.M.	Wind.	Weather.	B.P. of Water.	Air.	Height in Feet above sea-level.	Place, &c.
	Zero.								
1	− 1	23	7	S.W.	Cloudy	188·4	22½	15,896	Camp Thaldat, lat. 35° 14' 41" N.
2	− 5	30	2½	S.S.W.	Fine	189·2	30	16,525	Camp Somah Lam.
3	− 5	..	2	S.W.	Fine	,,	..	,,	Ditto.
4	− 6	25	..	S.W. by W.	Snow	..	..	..	Camp near Kizil Pass. Crossed Kizil Pass 17,859 feet above the sea.
5	−11	..	13	W.N.W.	Fine	182·8	18	16,192	Kiziljilga, lat. 35° 16' 33" N.
6	− 3	..	17	W.	Fine	184·0	..	15,370	Khush Maidan, lat. 35° 27' 12" N.
7	− 9	24	15½	S.W.	Cloudy	..	..	..	Ditto ditto.
8	−10½	25½	..	W.S.W.	Fine	184·2	18	15,482	Hot springs, lat. 35° 35' 7" N.
9	− 5	31	16	W.S.W.	Fine	184·4	19	15,364	Zinchin.
10	2	..	18	W.S.W.	Fine	185·2	27	14,957	Sang Kulan, lat. 35° 48' 25" N.
11	5	..	20	S.W. by W.	Fine	186·2	20	14,458	Mulgoon.
12	3½	..	15	W.S.W.	Fine	186·6	..	14,320	Kyang Jangal.
13	2	34	18	S.W.	Cloudy	186·9	23	14,043	Mandalik.
14	5	..	23	S.W.	Snow	187·4	23	13,848	Longmik.
15	3	22	19	S.W.	Fine	188·1	21½	13,480	Ak-Koom.

Valley of the Karakash River.

Day				Wind	Weather			Elevation	Locality
16	6	35	15½	N.W. by W.	Fine	188·9	30	13,070	Langar.
17	9	..	19	W.N.W.	Fine	189·1	19	12,952	Malbash.
18	9¼	..	18	W.N.W.	Fine	189·6	..	12,649	Gulbashem, a Khirghiz encampment.
19	1½	36	18½	W.N.W.	Fine	..	..	..	Bainkcbee ditto.
20	9¾	..	15	N.W. by W.	Fine	191·0	28	11,745	ditto, lat. 36° 21' 11" N.
21	5	33½	16	W. by N.	Fine	"	"	"	Ditto dito.
22	4½	32	16½	W.N.W.	Fine	"	"	"	Ditto dito.
23	3¼	38	19½	W. by N.	Fine	"	"	"	Ditto ditto.
24	5	34	22	W.	Cloudy	"	"	"	Ditto dito.
25	3¼	30	13	W.	Gldy	"	"	"	Ditto dito.
26	3	26¼	19	W. by N.	Fine	187·2	22	13,065	Kulahiahkun.
27	9	..	10	W.N.W.	Fine	186·6	17	14,222	5 miles below Kirghiz Pass. Crossed the Kirghiz Pass 17,092 feet above the sea.
28	Zero 0	35	9¾	W. by N.	Fine	188·5	22	13,280	Camp below Kirghiz Jungal, lat. 35° 24' 18" N.
29	9	..	18	W. by S.	Cloudy	189·4	18	12,590	12 miles below Koolunooldee.
30	Zero	..	17	W.S.W.	Fine	188·5	..	13,280	Camp of 28th instant.

Valley of the Karakash River. (Langar through Ditto entries)

Valley of the Yarkand River. (Camp below Kirghiz Jungal; 12 miles below Koolunooldee)

NOTE.—Up to the 15th November the direction of the winds was chiefly from the south-west. From the 16th November north-west winds were prevalent.

The winds were always calculated from carefully noticing the direction of the upper clouds with a pocket compass.

The figures showing heights above sea-level are all calculated from Casella's Tables. They are subject to correction, for which purpose the boiling-points of water have been recorded.

Field-books and original observations are all preserved.

15	−18¾	..	3¾	N.W.	Fine	187·3	3¾	13,905	Sooget. Crossed the Sooget Pass, 18,287 feet above the sea. B.P. of water on pass = 180°·6°.
16	2	..	13	W.S.W.	Fine	191·0	22	11,745	Shadula. Lat 36° 21' 11" N.
17	5½	..	25	W.S.W.	Fine	192·0	..	11,293	14 miles below Shadula. Valley of the Karakash River.
18	14	..	7	N.W.	Fine	186·0	11	14,474	Camp south of Sanju Pass. Crossed the Sanju Pass 16,612 feet above the sea.
19	3	..	17	W.N.W.	Fine	195·2	23	9,525	Kirghis Encampment above Thm.
20	15½	..	18½	W.N.W.	Fine	198·2	27	7,685	Village of Kibris,
21	16	..	20	N.W.	Fine	200·4	24	6,420	Village of Sanju, lat. 37° 12' 30" N.
22	12	..	21½	N.W. by W.	Cloudy	200·8	21½	6,298	Village of Koshtok, 25 miles from Sanju.
23	13	..	27	W. by N.	Cloudy	201·2	31	6,100	Village of Oitogrok.
24	15½	..	17	N.W. by W.	Fine	202·1	22	5,554	Village of Borah.
25	8½	..	21½	N.W. by W.	Fine	203·7	..	4,570	Town of Kurghalik.
26	17	..	23	W.	Fine	204·1	27	4,355	Town of Posgam. Crossed the Tiznáf River in lat 37° 51' 35" N.
27	19½	..	25	W.N.W.	Cloudy	205·1	..	3,830	Yarkand, the capital of Eastern Turkistan (crossed the Yarkand River) in lat. 38° 21' 16" N., long. 77° 28' E.
28	20	25	22	S.W. by W.	Snow	..	..	..	
29	14½	23½	19	S.W.	Fine	..	..	..	
30	10	23	15½	S.S.W.	Fine	..	..	..	
31	8½	22	15½	S.	Fine	..	..	..	

MONTH OF JANUARY, 1869.

At YARKAND.   Latitude, 38° 21′ 16″ N.; Longitude, 77° 28′ E.
Elevation, 3830 feet above Sea-level.

Date.	Thermometer, Fahr.						Wind.	Weather.
	Sunrise.	9 A.M.	12 Noon.	3 P.M.	6 P.M.	9 P.M.		
1	8½	11½	22½	27	19½	13½	S.S.E.	Fine
2	9	10	23	27	21	14	S.S.E.	Fine
3	5	11	32½	26½	22	19	W.N.W.	Fine
4	14	22	28	32½	25	18	S.W.	Cloudy
5	5½	13	24½	28	24½	19	S.W.	Fine
6	10	18	25	27	24	21	S.W. by S.	Cloudy
7	9	20	27	28	19	18	W. by S.	Cloudy
8	..	19	30	31	24	19	S.W.	Cloudy
9	10½	19	28	29½	24	20	W.	Fine
10	..	19½	28½	29	25½	23	S.W.	Fine
11	..	22	26½	28	25	23	..	Cloudy
12	13	21	27	31	26	22	..	Cloudy
13	12½	20	29½	32	25	21½	W.N.W.	Fine
14	..	16½	27	30	.27	23	W.N.W.	Cloudy
15	..	17½	32	36	31	25½	W. by N.	Fine
16	..	22	34	35	30	24	W.	Fine
17	15½	19	28	34	28	25	W.	Cloudy
18	16	22	28½	31½	29	26	W.S.W.	Cloudy
19	..	21½	29	30	26	25½	W.S.W.	Snow
20	24	28	30½	33	28½	27	S.W.	Snow
21	..	25	27	28	25	23	S.W.	Snow
22	..	25	31	29	25½	22	W.S.W.	Cloudy
23	..	24½	36	32	31	29	S.W.	Cloudy
24	..	30	35	33½	31	29	S.W.	Snow
25	26	29	34½	35	30	26	S.W.	Snow
26	..	21	33	37	31½	29	W.	Fine
27	..	27	36	40	36	29	W.	Fine
28	..	22	36	38	32½	29½	N.W.	Cloudy
29	26	29	37	38	30½	25½	W.N.W.	Cloudy
30	..	25	34	37	33	30	W.	Fine
31	..	28	31	32	30	24	W.N.W.	Cloudy
..	13	21	29	31	27	23	} Averages.	
..	·633	·225	·725	·790	·096	·322		

Monthly mean deduced from 9 o'clock observations = 22°·273.

MONTH OF FEBRUARY, 1869.

At YARKAND (to 23rd Instant). Lat. 38° 21′ 16″ N.; Long. 77° 28′ E.
Elevation 3830 feet.

Date.	Thermometer, Fahr.						Wind.	Weather.
	Sunrise.	9 A.M.	12 Noon.	3 P.M.	6 P.M.	9 P.M.		
	o	o	o	o	o	o		
..	..	24	38	35	30	28	W.	Cloudy
..	..	29	37	35	32	29½	W.	Cloudy
..	..	31	45	40	35½	29	W.	Fine
..	..	27	35	39½	31½	27½	W.S.W.	Fine
..	..	28	36	37	31	25	S.W.	Fine
..	14	25	36	37½	34	27	W.	Fine
..	..	25½	37½	40	35	28	W.	Fine
..	..	27	32½	37	35	31	S.W. by W.	Cloudy
..	..	31	45	47	42	34	S.W.	Fine
..	..	27	43	45	40	31	W. by N.	Fine
..	..	30½	45½	50½	41	38	W.	Fine
..	..	33	47	51	42	37½	W.	Cloudy
..	..	32	46½	50	43	33½	W.	Fine
..	27	31	48	50½	44	38	W.S.W.	Fine
..	25½	..	43½	46	41	34	..	Cloudy
..	..	33	47½	50	44½	37	S.W.	Fine
..	..	30	45	49	44	38	W.S.W.	Fine
..	..	33	45	50	45	35	W.	Fine
..	..	34	50½	50	45	33	W.	Fine
..	..	34	44½	49½	44	39	S.W.	Cloudy
..	..	36	47	50	44	39	S.W.	Cloudy
..	..	34	43½	47	43½	39½	W.S.W.	Cloudy
..	..	37	46½	49½	44½	39½	W.S.W.	Cloudy

Date.	Therm. Fahr.		Wind	Weather.	B.P. of Water.	Air.	Elevation in feet above Sea-level.	
	9 A.M.	9 P.M.						
	o	o			o	o		
24	36	37	W.S.W.	Cloudy	205·2	43	3,728	Village of Kokru
25	34	39	S.W.	Cloudy	204·8	39	3,932	Village of Kizil.
26	35	39	W.	Fine	} 204·2			{ Town of Yang
27	35	39	W.	Fine	}	42	4,256	Lat. 38° 52′ 3·
28	37	38½	W.S.W.	Fine	}			Long. 76° 18″

Averages .. .. .. { 9 A.M. = 31·444°
{ 9 P.M. = 34·375

Monthly Mean .. .. .. 32·909°

MONTH OF MARCH, 1869.

At KASHGAR (Fort of).   Lat. 39° 19' 37·1" N.; Long. 76° 10' E.

Elevation 4165 feet above sea-level.

Date.	Thermometer, Fahr.				Wind.	Weather.	
	9 A.M.	12 Noon.	3 P.M.	9 P.M.			
1	37	48	56	45	W.S.W.	Cloudy	Town of Yanghissa 38° 52' 3"·4 N.
2	42	53½	55	44	W.	Fine	,, ,,
3	40½	42½	41	37	..	Cloudy	,, ,,
4	36	..	..	37½	W.N.W.	Snow	Village of Tupchan, B.P. of water, 206°
5	..	..	..	35	N.W.	Cloudy	Kashgar, 4165 feet a level. B.P. of wate
6	39	..	..	36½	W.	Cloudy	Lat. 39° 19' 37"·1
7	39½	..	52	36	W.	Fine	
8	39	53	55½	41	W.S.W.	Fine	
9	42	53½	56	39½	S.W.	Cloudy	
10	43	54	57	41	W.S.W.	Fine	
11	44	56	59½	42	W.S.W.	Fine	
12	46	57½	59	40	W. by S.	Fine	
13	48	61	60	48	W.	Cloudy	
14	48	55	56	42	W.	Cloudy	
15	49	59	61	45	W.S.W.	Fine	
16	49	61	64	45	W.	Fine	
17	50½	65	68½	49	W.	Fine	
18	51	66½	67½	52	W.	Fine	
19	52	67	64	49	W.	Fine	
20	..	61	63	45	W.	Fine	
21	49	59	56	50	S.W.	Cloudy	
22	52	61	60	48	W.S.W.	Cloudy	
23	43	45	47	41	S.W.	Rain	
24	..	47	49	42	S.W.	Rain	
25	42	49	49	38½	W.	Cloudy	
26	46½	49	52	41	W.	Fine	
27	45	58	63	45	W.	Fine	
28	48	62	65	49½	W.	Fine	
29	51	63½	67	49½	W.	Fine	
30	55	66½	65	..	W.	Cloudy	
31	58½	70½	72½	53½	W.S.W.	Fine	

MONTH OF APRIL, 1869, IN EASTERN TURKISTAN.

Date.	Thermometer, Fahr.			Wind.	Weather.	
	9 A.M.	12 noon.	9 P.M.			
1	59		59	W.S.W.	Cloudy	Kashgar (fort of).
2	61		52½	W.	Fine	,, ,,
3	61		52	W.S.W.	Fine	
4	60½		58	W.	Fine	
5	62½		56	W.	Cloudy	
6	64		57½	W.S.W.	Fine	
7	64½		59	W.S.W.	Cloudy	
8	63		58	S.W.	Fine	
9	55		41	S.W.	Rain	
10	49½		51	W.	Fine	
11	55		51	W.	Cloudy	,, ..
12	62		58	W.S.W.	Fine	,, ,,
13	62½		59	W.	Fine	Village of Yupchan.
14	..		50½	..	Cloudy	
15	62½		56	W.	Fine	
16	63		53	W.	Fine	
17	63		55	W.	Fine	
18	64		53	W.	Cloudy	
19	..		51	W.S.W.	Rain	Fort of Yangbissar.
20	59½		55½	W.S.W.	Cloudy	
21	60½		57½	S.W.	Cloudy	
22	64½		57½	S.W.	Cloudy	
23	64½		60	W.S.W.	Fine	
24	63		59	S.S.E.	High Wind, Dust Storms	
25	67		..	S.	Fine	Village of Toblok.
26	..		67	S.S.W.	Fine	Ak Langar.
27	..		65½	S.	Fine	
28	66		64½	S.W.	Fine	Yarkand.
29	64½		67	S.W.	Fine	
30	66½		69	S.W.	Fine	

Averages .. .. $\begin{cases} 9 \text{ A.M.} = 61\cdot84 \\ 9 \text{ P.M.} = 57\cdot00 \end{cases}$

Monthly return .. .. = 59·42

MONTH OF MAY, 1869.

AT YARKAND. Lat. 38° 21' 16" N.; long. 77° 28' E.

Elevation 3830 feet above sea-level.

Days.	Thermometer, Fahr.			Wind.	Weather.		
	9 A.M.	12 noon.	9 P.M.				
1	68	72	72½	W.S.W.	Fine		
2	68½	71½	71½	S.W.	Fine		
3	69	73	73	S.W.	Fine		
4	70½	72	74	S.W.	Fine		
5	73	75	74	W.S.W.	Fine		
6	72	73	69	S.E.	Dust Storms		
7	68	72	66	W.S.W.	Fine		
8	68½	73	70½	W.S.W.	Fine		
9	68½	71½	69½	S.W. by W.	Fine		
10	67½	72½	69½	W.	Fine		
11	68	73	71½	W.	Fine		
12	68½	74	73	W. by S.	Cloudy		
13	70	73	72	W.	Cloudy		
14	68½	72	69	W.	Cloudy		
15	69	71	70	W. by S.	Fine		
16	70	74	67½	W.	Fine	Slight rain towards Hills.	
17	69	72	66½	W.	Fine		
18	61½	66	69	W.	Fine		
19	68	69½	71½	W.	Fine		
20	69	72	74½	W. by S.	Fine		
21	70½	73½	75	W. by S.	Fine		
22	73	74½	73½	W.	Fine		
23	72	73	72	W.	Cloudy		
24	71½	72½	72	W. by S.	Fine		
25	69½	71	69	W.S.W.	Fine		
26	69	71	70½	S.W.	Fine		
27	70	71½	70	S.W.	Fine		
28	69	70½	71	W.S.W.	Fine		
29	70	71½	72½	S.	Fine		
30	70½	..	..	..	..	Otunchee	Left Yarkand
31	..	..	..	..	..	Posgam	return journ

Averages .. .. { 9 A.M. = 69·333 / 9 P.M. = 71·017

Monthly mean .. .. .. = 70·175

MONTH OF JUNE, 1869, IN EASTERN TURKISTAN.

Date.	Therm. Fahr. 9 A.M.	Therm. Fahr. 9 P.M.	Wind.	Weather.	B.P. of Water.	Air.	Elevation in Feet above Sea-level.	
	°	°			°			
1	..	..	S.S.W.	Fine	..	..	..	Karghalik.
2	..	..	W.S.W.	Fine	..	..	..	,,
3	..	..	W.S.W.	Fine	..	..	..	,, ,,
4	..	..	..	..	..	..	..	,, ,,
5	..	..	..	..	..	..	..	,, ,,
6	..	..	..	..	..	..	..	,, ,,
7	..	..	..	..	..	..	..	,, ,,
8	..	..	..	..	..	..	..	,, ,,
9	..	..	..	..	..	..	..	,, ,,
10	..	..	..	..	..	..	..	,, ,,
11	..	60	S.W.	Rain	..	..	..	Village of Sanju.
12	65	57½	S.S.W.	Fine	199·1	57½	7,234	Kizil Aghil.
13	56½	..	W.S.W.	Fine	196·7	65	8,617	Mazar (Arpalak \
14	..	..	W.S.W.	Cloudy	196·3	63	8,933	Tâm.
15	47	49	W.	Cloudy	..	..	..	Five miles above
16	53	40	S.W.	Fine	191·2	44	11,528	Kichikyulak.
17	38	39	S.W.	Fine	..	..	..	,, ,,
18	37	..	..	Fine	188·2	52	13,482	{Crossed Sanju Diwânjilga.
19	41	..	..	Fine	..	..	11,215	Silartakash.
20	..	..	..	..	..	..	11,745	Shadula.
21	..	38	W. by S.	Rn. & Snow	..	..	..	,, ,,
22	44	44	S.W.	Fine	..	..	..	,, ,,
23	55	41	W.S.W.	Fine	188·0	52	13,598	Sooget Valley.
24	47	26	S.W.	Fine	182·1	26	16,812	{Chibra. Crossed Pass, 18,237 fe
25	26	22	W.S.W.	Fine	183·1	22	16,190	Châdartash.
26	..	..	W.S.W.	Fine	..	..	..	,, ,,
27	39	23	W.S.W.	Fine	182·0	33	16,905	{Oglok. Crossed tagh Pass, 17,9
28	29	..	W.S.W.	Fine	183·7	39	15,929	Shorjilga.
29	39	40	S.W.	Fine	..	..	15,570	Khush Maidan.
30	41	37½	S.W.	Fine	..	..	16,192	Kiziljilga.
July. 1	° ..	° ..	S.W.	Fine	° 181·0	° 47½	17,655	Karasu.
2	..	..	N.E.	Snow	..	..	..	Camp Lingzi Th
3	..	26	N.E.	Snow	180·9	31½	17,527	Kala Pahar.
4	31½	..	N.E.	Snow	..	..	..	
5			Crossed Chang Lang Pass into Chang Chenmo.					{Chang Lang ,, ,, 18,839 feet.

TOWNS and VILLAGES in EASTERN TURKISTAN.

1. *Kibris.*—The first village met with on entering Turkistan from the Sanju Pass, consisting of 6 or 8 houses, inhabited by some of the Sanju people during the summer months.

2. *Sanju.*—A district consisting of 3000 to 4000 houses, comprised in several villages scattered for several miles on each side of the Sanju River, situated in an open valley running nearly east and west, is 6420 feet above sea-level. Head official, a Hakeem Beg.

3. *Koshtok.*—A village of some seventy houses, lying 25 miles north-west of Sanju, on the road to Yarkand, situated on the left bank of the Kilian stream, is 6298 feet above the sea. Head official, a Yuz bashee.

4. *Langar.*—A small village to the left of the Yarkand road, at 15 miles distance from Sanju.

5. KILIAN.—A village of 200 houses, south-west of Koshtok, situated on the left bank of the stream rising in the Kilian Pass. One of the main routes into Eastern Turkistan crosses this pass. This road conducts up a ravine from Ali Nuzur Kurgán, 12 miles below Shadula on the Karakash River, and crossing the Pass, debouches here at Kilian, and joins the Sanju route at the village of Borah.

6. *Ismasulla.*—A village lying between Koshtok and Kiliar, at 7 miles distance from the former. Is situated on the right bank of the Kilian stream at the foot of a low range of sand hills.

7. *Has-an-Boora.*—A small village lying between Sirzum and Oitogrok.

8. *Sirzum.*—A small village 4 miles north-west from Kilian. The road from Kilian to Yarkand passes through these two villages.

9. GUMA.—A district of 5000 to 6000 houses, lying 27 miles north of Sanju and 23 miles north-east of Koshtok. Is on the right bank of the Kilian stream, which runs from Koshtok past it. The main road from Khotan to Yarkand passes through Guma.

10. *Otunsu.*—A village of fifty houses, lying 1 mile south of Kugiar.

11. KUGIAR.—A village of 300 houses, some 40 miles south of Karghalik. The Karakoram Pass road down the valley of the Yarkand River, crosses the Yangi Pass, and debouches into the plain country at Kugiar.

12. *Yulurik.*—A village of 100 houses, lying 5 miles north-east of Kugiar.

13. *Oitogrok.*—A village of twenty houses, 18 miles north-west of Koshtok, towards Yarkand. Is 6100 feet above the sea.

14. *Borah.*—A village containing ninety houses, situate 11 miles north-west of Oitogrok, on the road to Yarkand. Watered by a stream from Oshokwas. Chief official, a Yuz bashee. The Kilian road here joins the Sanju one, and from here they are the same to Yarkand throughout. Borah is 5554 feet above the sea.

15. *Tonkzi.*—A village 5 miles south of Borah.

16. *Oshokwas.*—A large village of 1300 to 1400 houses, lying south of Borah. Is situated at the foot of the low range of hills running north-west from Kilian past it, and is watered by the stream rising in the Kullik Pass. The road to the Kullik Pass ascends the ravine which joins the Karakash Valley, 10 miles below Shadula at Tograssu, and crossing the pass, follows

the course of this stream to Oshokwas, from where it joins the Kugiar route at Beshtiruk Langar.

17. *Beshiruk.*—A large, widely-scattered village of some 900 houses, lying 4 miles south of Karghalik, watered by a considerable stream coming down from Oshokwas. Is 4665 feet above sea-level.

18. KAROHALIK.—One of the largest towns and districts in Eastern Turkistan, consisting of some 20,000 houses, with a large market, bazaar, and caravanserai. Is a place of considerable importance, being situated at the junction of all the roads leading across the mountains into Turkistan from India, Kashmir, and Ladak. The Kujiar route joins here, and the Khotan road through Guma. Outside the town, at the junction of the Khotan road, is a new earth fort. Karghalik is 4570 feet above sea-level, and is watered by a canal cut from the Tisnâf River. Chief official, a Dakeem Beg. Karghalik is 36 miles from Yarkand.

19. *Alamakun.*—A village passed to the right of the Yarkand road, 12 miles from Karghalik. Is situated on the left bank of the Tisnâf River.

20. *Khojerik.*—A village lying 2 miles north-east of Alamakun.

21. *Boghorlok.*—A village of some sixty houses, lying opposite to Khojerik.

22. *Mekla.*—A small village, 14 miles from Karghalik, on the Yarkand road.

23. *Yak Shamba Bazaar.*—On the Yarkand road, 15 miles from Karghalik, and 6 miles from Posgam is a large market crowded on Sundays.

24. POSGAM.—A town and district of 16,000 houses, 15 miles from Yarkand, possessing a large bazaar and caravanserai. There is an old fort in ruins, 1½ miles to the west of the town, which is 4355 feet above sea-level and watered by the Beshkur Canal, cut from the Yarkand River.

25. *Kurum Togrok.*—A village of 700 houses, lying 6 miles from Posgam, on the left bank of the Yarkand River.

26. *Aigâchee.*—A village of 1000 houses on the right bank of the Yarkand River. In the summer the river is here crossed in boats, and this then becomes the direct road from Posgam to Yarkand.

27. *Otunchee.*—A large, widely-scattered place, containing some 5500 houses, situated 8 miles south-east of Yarkand.

28. *Chinibagh.*—Situated 3 miles south of Yarkand. Formerly the residence of Ahmed Wang.

29. *Shamal bagh.*—Situated one mile east of the city of Yarkand.

30. YARKAND.—The capital of Eastern Turkistan, in latitude 38° 21′ 16″ N., longitude 77° 28′ E., and 3830 feet above sea-level. Contains some 40,000 houses and about 120,000 inhabitants. There are 160 mosques, many schools, and some twelve caravanserais, which are always crowded with merchants from every part of Asia. The city is surrounded by a fortified earth wall, varying from 40 to 45 feet in height, and is entered by five gates, which are as follows :—

1. *Altun Dubza.*—On the west side, leading to the fort and Kashgar, &c.

2. *Moskari Dubza.*—On the south side, leading to Karghalik, Khotan, &c.

3. *Balti Dubza.*—On the south-east side, leading to Aigâchee, &c.

4. *Aksu Dubza.*—On the east side, leading to Lai Musjid, Aksu, Oosh Turfân, &c.

5. *Terek Bagh Dubza.*—On the north side.

The city lies in the form of a parallelogram, being some 2 miles from north to south and 1¼ miles from east to west. The main street runs nearly due east and west from the Altun Dubza (Gate) to the Akan Dubza (Gate), and is very narrow, being not more than 12 feet in many places. The city and fort are supplied with water from tanks, into which the water is brought by canals cut from the Yarkand River. These are frozen over in winter, and the supply of water is then stopped; but the tanks contain sufficient for the consumption of the inhabitants until the regular supply is renewed in the spring.

Yarkand imports from India opium, spices, tea, sugar, kinkâb, English cloth, &c. The principal exports to India and Kashmir are silk, churrus, felt cloths, pushmeena, shawl-wool, gold and silver, &c.

THE FORT is situated some 500 yards to the west of the city. The walls run nearly due with the four points of the compass, and are 40 feet in height, 12 feet broad at the summit, and are entirely of earth. The fort is nearly 700 yards square, and has four bastions with towers at the corners, and eight intermediate flanking defences on each side. It is surrounded by a lower wall and dry ditch. This moat is 25 feet deep, 30 feet broad at the summit, and about 18 feet at the bottom. There are three gates.

The East Gate, facing the city.

The Khotan Gate, on the south side.

The Kashgar Gate, some 80 yards from the south-west corner, facing the west; and immediately behind the place of residence of the chief authorities.

These two latter gates are closed, the Khotan one being also barricaded with unburnt bricks.

The only gate through which ingress and egress is permitted is the East Gate. This is closed every night at 8 P.M. in winter and 9 P.M. in summer, and is opened daily at daybreak.

In the south-west corner of the fort is the residence of the chief authorities, surrounded by a wall of about 30 feet in height, the entrance through which is on the east side facing the main road, which leads up to it through the bazaar of the fort.

Occupying the north-west corner of the fort is the inner fort, formerly the residence of the Chinese Governor and officials. This inner fort is surrounded by a fortified earth wall of 35 feet in height, with a lower wall and dry ditch. These walls are in ruins in many places. The north-east corner of the fort is one mass of ruins. No guns are anywhere mounted on the walls, for which, however, there are embrasures; as, when the Chinese held Yarkand, many were in position.

31. *Beshkun*.—Twenty-two miles east of Yarkand, on the left bank of the Yarkand River, contains about 4000 houses.

32. *Beshwook*.—A large village of 3300 houses, on the road to Sarikol and Tashkurgan; six miles distant from Yarkand, and lying west of it.

33. *Hasrat Peer Mazar*.—Tomb of a Syad and ruined village, lying half a mile north of the fort of Yarkand.

34. *Bigil*.—A small village, situated five miles north-west of Yarkand, on the left bank of the Urpi Canal.

35. *Urpi*.—A village of 1000 houses, 4½ miles north of Yarkand, on the right bank of the Urpi River or Canal.

36. *Tuyarchee*.—A village of some 600 houses, lying 9 or 10 miles north-east of Yarkand, on the right bank of the Urpi Canal.

37. *Karchi.*—A village of 100 houses, 13 miles north-east of Yarkand, situated on the left bank of the Urpi Canal.

38. *Tongus.*—A large village of 1500 houses, situated 22 miles north-east of Yarkand, towards Aksu.

39. *Karakoom.*—A village four miles north-west of Yarkand, on the Kashgar road.

40. *Otun Langar.*—A small village, 15 miles from Yarkand, towards Kashgar.

41. *Kokrubat.*—A village of 200 houses, 22 miles north-west of Yarkand; is the first stage or halting-place towards Kashgar.

42. *Ak Langar.*—A halting-place, 14 miles from Kokrubat, on the main road to Kashgar. There is an enclosure with a musjid, and two wells of water.

43. *Abdoola Khan Langar.*—Eight miles beyond Ak Langar; is now un-occupied.

44. *Kizil.*—A village of 500 houses, the second stage from Yarkand, from which it is 50 miles distant; is 3932 feet above sea-level.

45. *Oordoo Badshah Mazar.*—Tomb of a Syad and small village lying six miles north-east of Kizil.

46. *Hazrat Bakeem Mazar.*—Tomb of a Syad and small village situated 13 miles north-east of Kizil.

POSITIONS OF THE TOWNS OF EASTERN TURKISTAN.

	Lat. N.				Long. E.	
	°	′	″		°	′
Yarkand	38	21	16	.. ..	77	28
Yanghissar	38	52	3·4	.. ..	76	18
Kashgar—						
Fort	39	19	37·1	}	76	10
Old City	39	23	9			
Khotan	37	8	0	.. ..	80	5
Shadula	36	21	11	.. ..	78	18
Source of the Yarkand River ..	35	37	34	.. ..	77	50
Kufelong—						
At junction of main stream and Karakoram Pass stream .. }	36	4	48	.. ..	77	57

Chamalung.—Containing 30 houses lying 3 miles west of Kizil.

Khoduk.—A village of 20 houses 5 miles from Kizil.

Koshimbash.—A small village of a few houses to the right of the road 9 miles from Kizil.

Toblok.—A village of 40 houses 14 miles from Kizil, on the main road. Is a halting place between Kizil and Yanghissar.

Kelpun.—A ruined village 3 miles beyond Toblok, formerly a Chinese " urtang " or police station. There are now only a few inhabited houses and a " mazar."

Sooget Bolok.—A halting-place 5 miles south of Yanghissar.

Koomlok.—A village of 30 houses 6 miles from Yanghissar towards Kashgar.

Shorlik.—A large village lying north-west of Yanghissar.

Syad Masar Khoja.—Tomb of a Syad, or Khoja, in an enclosure passed to the right of the road 9 miles from Yanghissar.

Toglok.—A scattered village of 350 houses, 13 miles from Yanghissar, on the Kashgar road.

Yupchan.—A large village of 700 houses 22½ miles from Yanghissar towards Kashgar. Is the 4th stage from Yarkand, from which it is 104½ miles distant. Is situated on the left bank of the Hosun River, from which the land is irrigated by numerous dykes, and is 4055 feet above sea-level.

Tasgam.—A village of 800 houses 6 miles north-west from Yupchan towards Kashgar. Situated on the left bank of the Khanarik River.

KASHGAR (the Fort), the Yangishar, or New City. In lat. 39° 19' 37·1" N., long. 76° 10' E. Elevation 4165 feet above sea-level. Is nearly square, being somewhat longer in its north and south sides, which are about 600 yards in length. The walls are 40 feet in height, surrounded by a lower wall and dry ditch. This moat is 25 feet deep and some 40 feet broad at the summit. The main gate is in the centre of the north wall facing the city. There are 2 other gates, one on the east the other on the south side. They have flanking defences, but are both closed. The Fort has 6 flanking defences on the north and south sides, whereas the east and west ones possess only four. The walls are entirely of earth, and the whole Fort appears to be in better preservation and more available for defence than the Yarkand Fort. It is distant 123 miles from Yarkand.

THE CITY is 3 miles off, lying north between the Fort, and, which is crossed, the "Kizil Daria," or Kashgar River, coming down from the westward between the Kizil Yart Range of the Pamir and the Terek Pass on the road to Khokand. The city is surrounded by a high fortified earth wall, and is entered by 5 gates. It has rapidly increased in size since the expulsion of the Chinese, and now contains some 28,000 houses and from 60 to 70,000 inhabitants.

PASSES ACROSS THE KARAKORAM AND KUEN LUEN RANGES INTO EASTERN TURKISTAN.

Name of Pass.	Height in feet above Sea-level.	B. P. of Water on Pass.	Temperature of Air at time of Observation.	Across what Range.	Where from, where to.	Character of Pass.
Chang La	18,369	..	..	Khas Range of Himalaya	Leh to Tankse	*Somewhat difficult.*—Practicable for laden horses and yaks.
Marsimik	18,467	179·4	41½	Range north of Pangong	Leading into Chang Chenmo	*Easy.*—Practicable for laden horses, mules and yaks: practicable for artillery.
Kiel	17,649	180·4	57		From Lingzi Thung Plains into Karakash Valley	*Very easy.*—Practicable for artillery for laden camels, horses and yaks.
Chang Lung (or Chang Chenmo)	18,639	179·7	43	Main Karakoram Range north of Chang Chenmo	From Chang Chenmo District into Eastern Turkistan	*Easy.*—Practicable for guns and for laden animals.
Karataßh	17,963	180·0	56	Karataßh Range	From valley of Karakash River to valley of Yarkand River to Aktßh	*Very easy.*—Practicable for guns, laden horses, camels and yaks.
Boogat	18,237	180·2	43	Aktßh Range	Aktßh to Shadula in Karakash Valley	*Easy.*—Practicable for laden animals.
Kirghis	17,092	..	..	Aktßh Range	Shadula to Kirghis Jangal in valley of Yarkand River	*Difficult.*—Practicable for laden horses and yaks: impracticable for artillery.
Yangi	Not known exactly—Approximate 16,400	..	..	Western Kuen Luen	Koolanooldee in valley of Yarkand River to valley of Tisaf River	*Easy.*—Practicable for guns, laden horses, camels and yaks.
Sanju	16,613	182·6	42	Spur of Western Kuen Luen or Kilian Range	Valley of Karakash River to Sanju	*Difficult.*—Impracticable for guns, laden camels and horses: only practicable for laden yaks.
Kilian	Not known exactly—Approximate 17,300	..	..	Spur of Western Kuen Luen or Kilian Range	Valley of Karakash River to Kilian	*Difficult.*—Only practicable for laden yaks.
Chachan	11,847	190·9	51	Lateral spur of Kilian Range	Appato Valley to Sanju Valley	*Somewhat difficult.*—Practicable for laden animals.
Karakoram *	18,317	..	..	Main Karakoram Range	Ladak into Eastern Turkistan	*Easy.*—Practicable for laden animals and guns.

* The two passes on the Ladak side of the Karakoram on the road to Leh, viz., the Sasser Pass, 17,972 feet, and the Karlung Pass, 17,576 feet, are difficult, more especially the Sasser Pass. They are impracticable for guns, but practicable for laden horses and yaks.

APPENDIX III.

Re-computation of Mr. Hayward's Observations for Latitude. By S
Commander C. George, r.n., Map-Curator Royal Geographical Society.

Date.	Name of Place.	Object used.	Result of Observations for Latitude. N.		E.
1868. Nov. 26	Shadula	☉ M. Alt.	36 23 24		78
,, 29	Koolunooldee	Ditto.	36 25 4		77
Dec. 4	Kufelong	Ditto.	36 5 55		77
,, 9	Yarkand River, Source of	Ditto.	35 38 39		77
,, 18	Mazar Badshah	Ditto.	36 34 52		78
,, 22	Sanju	Ditto.	37 15 20		78
1869. Jan. 5	Yarkand	Ditto.	38 23 52		
,, 9	Ditto	Ditto.	38 23 30		
,, 15	Ditto	Ditto.	38 23 27		
,, 28	Ditto	Ditto.	38 21 50		
,, 30	Ditto	Ditto.	38 21 13		
Feb. 4	Ditto	Ditto.	38 21 23	= 38 21 43·3	77
,, 6	Ditto	Ditto.	38 20 44	Mean.	
,, 10	Ditto	Ditto.	38 20 51		
,, 11	Ditto	Ditto.	38 20 29		
,, 13	Ditto	Ditto.	38 21 44		
,, 14	Ditto	Ditto.	38 19 48		
Mar. 1	Yanghissar	Ditto.	38 52 15	= 38 52 12·5	76
,, 2	Ditto	Ditto.	38 52 10	Mean.	
,, 10	Kashgar	Ditto.	39 19 43		
,, 11	Ditto	Ditto.	39 19 32	= 39 19 44	76
,, 16	Ditto	Ditto.	39 19 56	Mean.	

-COMPUTATION of Mr. HAYWARD's OBSERVATIONS for HEIGHTS in EASTERN TURKISTAN.* By Staff-Commander C. GEORGE, R.N.

Date.	Name of Place.	Approximate Position. (From Map.)		Boiling Water.	Tempo-rature.	Resulting Height.†
		Latitude.	Longitude.			
		N.	E.			
		° ′	° ′			Feet.
1868.	Leh	34 6	77 15	191·75	61	11,532
	Sakti. At foot of Chang La Pass	34 1	77 58	188·0	47	13,697
	Seeprali. 4 miles below pass ..	34 1	78 2	183·4	34½	16,320
	Tanksee	34 2	78 13	188·8	45	13,228
	Chagra. At foot of Masimik Pass	34 4	78 28	185·8	36	14,933
	Rimdee. 2 miles below pass ..	34 9	78 42	180·8	7	17,656
	Masimik Pass	34 5	78 39	179·4	41½	18,724
	Pumsul. Chang Chenmo Valley	34 17	78 50	185·5	27	15,077
,, 9	Kiam. Chang Chenmo Valley ..	34 18	78 59	184·0	24	15,930
,, 15	Gogra. In Kugrang Valley ..	34 21	78 56	183·6	20 ?	16,136
,, 17	No. 2 Camp. In Kugrang Valley	34 28	78 57	182·0	17	17,040
,, 22	Camp △. Junction of branch in Kugrang Valley	34 21	78 56	182·4	19	16,818
,, 24	Camp △. Near Gogra	34 23	78 57	183·6	27	16,172
25	Hot springs. Chang Lang	34 24	78 58	183·0	28	16,520
	No. 1 Camp. Lingzi Thung Plains	34 47	79 14	181·6	1	17,164
	No. 2 Camp. Ditto Ditto	34 52	79 22	181·5	1	17,220
	Lak Tsung	35 1	79 30	182·4	13	16,747
	Thaldat. Camp	35 15	79 28	183·4	22½	16,229
	Somah Lam. Camp	.. ..	.. ..	182·2	30	16,965
,, 5	Kizil Jilga	35 17	79 1	182·8	18	16,546
,, 6	Kush Maidan	35 27	78 51	184·0	18	15,872
,, 8	Hot springs	35 35	78 50	184·2	18	15,757
,, 9	Zinchin	35 43	78 55	184·4	19	15,647
,, 10	Sang Kalan	35 48	79 3	185·2	27	15,226
,, 11	Mulgoon	35 48	79 14	186·2	20	14,632
,, 12	Kyung Jangal	.. ..	.. ..	186·6	22?	14,412
,, 13	Mandalik	35 55	79 26	186·9	23	14,247
,, 14	Lungnak	35 58	79 20	187·4	23	13,964
,, 15	Ak-Koom	36 6	79 5	188·1	21½	13,565
,, 16	Langai	.. ..	.. ..	188·9	30	13,116
,, 17	Mulbash	36 19	78 50	189·1	19	13,002
,, 18	Gulbashem, a Kirghiz encampment	36 13	78 40	189·6	25	12,733
,, 20	Shadula	36 23	78 18	191·0	28	11,951
,, 26	Kulshishkun	36 25	78 5	187·2	22	14,147
,, 27	Camp △. 5 miles below Kirghiz	36 23	77 46	186·6	17	14,397
,, 28	Ditto. Below Kirghiz Jangal	36 24	77 41	188·5	22	13,844
,, 29	Ditto. 12 miles below Kool-nooldee	36 23	77 10	189·4	18	12,833
,, 30	Ditto. Of the 28th November	36 24	77 41	188·5	20	13,340
cc.	Ditto. 12 miles up valley ..	36 17	77 46	187·0	7	14,103

* The latitudes, longitudes, and heights in Mr. Hayward's memoir are those computed by nself during his journey, and agree with his map, which has been engraved from his own wing, without alteration.
† The boiling-water observations have been reduced to Leh, the starting point, which has en assumed to be 11,532 feet in height. On Mr. Hayward's map it is marked as 11,740 feet. uis difference of 208 feet in some measure accounts for the variation in the heights between : Table and Diary.

COMPUTATION of OBSERVATIONS for HEIGHTS in EASTERN TURKISTAN—*continued*

Date.	Name of Place.	Latitude. N.		Longitude. E.		Boiling Water.	Tempe- rature.	Resulting Height.
		°	′	°	′			Feet.
	Camp A Near Bukhurooldee ..	36	13	77	47	186·8	12	14,228
	Ditto. 6 miles above Kufelong	36	1	77	50	185·8	8	14,765
	Ditto. 12 miles up valley ..	35	53	77	45	185·0	10	15,212
	Ditto. 13 miles above camp 5th	..	..	..	..	183·8	7	15,862
	Ditto. 12 miles from head of } Yarkand River..	35	45	77	51	183·2	4	16,176
	Source of Yarkand River	35	39	77	54	182·2	4	16,730
	Kufelong	36	6	77	58	186·4	7½	14,651
	Aktagh	35	55	78	15	184·6	3	15,402
	Sooget	36	17	78	15	187·3	3½	13,929
	Sooget Pass	36	7	78	16	180·2	3	17,835
16	Shadula	36	21	78	18	191·0	22	11,942
	Camp A. 14 miles below Shadula	36	30	78	20	192·0	20	11,396
	Ditto. South of Sanju Pass ..	36	43	78	34	186·0	11	14,666
	Kirghiz East, above Tâm	..	..	..	..	195·2	23	9657
	Kibris	37	9	78	41	198·2	27	8029
	Sanju	37	15	78	47	200·4	24	6868
	Koshtok	37	24	78	20	200·8	21½	6671
	Oitogrok	37	31	78	5	201·2	31	6403
	Borah	37	36	77	53	202·1	22	5980
	Karghalik	37	55	77	42	203·7	25	5118
	Posgam. River Tisnâf	38	9	77	34	204·1	27	4891
	Yarkand	38	22	77	29	205·1	25	4384
	Kokrubat	38	28	77	5	205·2	43	4146
	Kizil	38	40	76	46	204·8	39	4391
	Yanghissar	38	52	76	18	204·2	42	4690
	Yupchan	39	10	76	18	204·6	36	4444
	Kashgar	39	20	76	11	204·4	38	4536
	Kizil Aghil	37	6	78	52	199·1	57½	7255
	Mazar	36	57	78	52	196·7	65	8615
	Tâm	36	56	78	34	196·3	63	8855
	Kichik-yulak	36	46	78	35	191·2	44	11,852
	Diwan-jilga	36	39	78	31	188·2	52	13,627
	Sooget Valley	36	15	78	14	188·0	52	13,746
	Chibra	36	4	78	20	182·1	26	17,133
	Châdartash	35	43	78	10	183·1	22	16,515
	Oglok	35	39	78	18	182·0	33	17,236
	Shor-jilga	35	34	78	43	183·7	39	16,255
	Karasu	35	12	79	11	181·0	47	17,950
	Kala Pahar	34	39	79	14	180·9	31	17,901

III.—*A Visit to Easter Island, or Rapa Nui, in* 1868. By J. LINTON PALMER, F.R.C.S., Surgeon of H.M.S. *Topaze.*

Read, January 24th, 1870.

THIS little island, which has been rendered celebrated by the gigantic stone images which are so plentiful in it, is also so isolated as to require the special notice that it is in 27° 8′ s. lat., and 109° 24′ w. long., about 2000 miles from the South American coast, and 1000 from Pitcairn Island, or the Gambier Islands. It is mentioned in the voyages of many navigators, who in their notices of it do not always agree. In the account of the voyages of Captain Cook, the names there given to the island we found to be those of districts in it. I have given the native name, which originates from the fact, that many generations ago, a large migration to it took place from the island of Oparo, or Rapa-iti (Small Rapa). This island is about 1900 miles due west of Easter Island, which from its greater size was called Rapa Nui or Great Rapa. In length it is about 12 miles, and in breadth 4 miles, somewhat like a cocked-hat in shape, the base towards the south ; the ends are high and bluff, and there is a tall hill, 1050 feet, an extinct crater in its centre. It is of volcanic origin, and abounds in craters, but these have been extinct for so long that no tradition of their activity remains. As they are of interest, I may mention the position and names of some of these craters.

1. *Terano Kau.*—This is a very large one at the south end of the island; in diameter it is about a mile, and is 600 or 700 feet deep. The bottom, which is flat and 1200 yards across, is a bog, with reeds and sedge, and many pools here and there; these were found to be 20 or 30 feet deep. There is a zigzag path to the bottom of the crater, as a farm-garden has been made by a settler, Captain Bornier. At the south side of the crater is the gap by which the last lava-flow escaped, and the north side is pretty well clothed with Hibiscus, *Broussonetia*, &c.

2. *Terano Hau,* not far from the centre of the island, is very much smaller, and is dry. This is the source of the red tuff which has been quarried to form the head-dresses, or crowns, of the large trachyte images, as the material can be found in mass, here only.

3. *Otu-iti.*—"The little hill," which is at the north-east end of the island, is very similar to Terano Kau, but of smaller size. It stands isolated in a large plain, and furnishes the grey lava (Trachyte) of which all the images are made. The largest images, and the only ones now erect, are at this hill.

Near the Terano Hau is a rounded hill of obsidian; it is

capped with a white earth apparently argillaceous. I was not on it. All the hills are rounded, and the soil on their slopes, and in the intervening valleys, being nothing but decomposed lava, is very fertile. I should say that there are many small blocks of harder lava mixed with the soil, which render walking over the island very tiring, the paths being just broad enough to put one foot in, and necessitating a swinging gait, very irksome to acquire. The whole island is volcanic; I did not see any sedimentary deposits, nor diatom-earth. Roggewin, who visited the island in 1722, but whose narrative is to be received with caution, says that, "the island was full of trees, which were in full fruit;" this has never been corroborated by subsequent visitors. There were holes of large trees, Edwardsia, coco palm, and hibiscus, decaying in some places, when we visited the island, but, though La Pérouse left fruit trees with the inhabitants, we saw no traces of them. From the size of some of the paddles and rapas, large trees must have existed. Just now the only approach to wood is found in the sheltered nooks, bushes of 10 to 12 feet high, of hibiscus, *Edwardsia*, *Broussonetia*, &c. The rate of growth of these is extremely slow.

As to the supply of fresh water on the island, a good deal of misapprehension has existed. In several of the craters there are many deep pools of it; in those of the Terano Kau these are fully 25 feet deep, and I have tasted it pure and fresh from many places, near the shore. At Winipoo, not only is there a subterranean reservoir (to which a tunnel leads from the face of the cliff), but on the very sea beach the natives have made a cistern to catch the water which distils from a little runnel. I did not see that the natives had sunk any wells. On the road from Otuiti are many pools of small size, but the natives warned us not to drink of them. They chew, to appease their thirst while journeying, sugar-cane, which is even now, though uncultivated, abundant, or sweet potato. At meal times they use salt water as seasoning with their vegetables, and this must have led to the belief that they used it alone, from the absence of fresh water. At Otuiti I was told distinctly that there was no water, except that in the pools of the crater.* As to the water of the sulphur spring mentioned in Cook's voyages (and which is close to Terano Kau), we found that though it had a distinctly mineral taste, it was not very unpalatable, and in sufficient quantity to satisfy our pretty large and thirsty party. The rocks in most of the gullies are evidently stream-worn, but

---

* *Ina voi—ina ina:* no water, none at all—a pleasing notice to us when parched with thirst, and at sundown !

now not the smallest brook exists. The soil, as a rule, seems moist enough not to require particular irrigation.

The coast line is bluff, irregular, yet not much indented. The slope of the land is more gradual to the south-east shore, where the cliffs are of varying heights. At the ends of the island, as much as 800 feet, at Angaroa (Cook's Bay) is a sandy beach. This, though an open roadstead, is the best anchorage. The swell and surf round the island frequently prevent any chance of landing. This was the case with H.M.S. *Portland* in 1852—also with Captain Amasa Delano, in 1808. There are not many outlying rocks—very little seaweed was to be seen, although La Pérouse says it was used as food. Its name was then go-e-mon, it is now au ké. There was plenty of flat sponge on the rocks and boulders at the landing place.

We did not take any fish with the line, but at some time there must be some, and large ones too, if one may judge by the size of the hooks, made of stone, with which the natives used to take them. Large flying-fish are not uncommon, and I saw plenty of small fry, close inshore; several nets we obtained have small (¼ inch) meshes. Crayfish, which are taken by the natives diving for them, and crabs, are common and good; shellfish also. I saw no oysters, but there were plenty of univalves, and, in the stone houses at the Terano Kau, there was an abundance of the shells of a small periwinkle (piripi), *Nerita.*

As in the rest of Polynesia, no quadruped has been found peculiar to the island. The rat is in great abundance. Pigs have been landed by some visitors: but were not allowed to breed. Roggewin says hogs were domesticated: there is no name for such a beast in their language, and I did not find any drawing of such in the mural paintings at Terano Kau. Birds were quite as scarce; some sea fowl were seen, but the ordinary domestic fowl was the only other bird; and these were in sufficient number. Small birds altogether absent.

Reptiles. *I was told* some one had seen a lizard, but this was a solitary instance; and in questioning all those who had been wandering over the island I was answered negatively, nor did I see one myself. No snakes exist.

No coleoptera were collected, but I think I saw one or two species. Centipedes exist. I saw no butterflies except one very like the *Cynthia cardui*, and one like the *Sulphur Butterfly* so common in England. Flies were exceedingly annoying to any one in places out of the free current of wind. There were no mosquitoes. Fleas were in myriads even in a grotto at Anakena (La Pérouse Bay). But no collections were made of the fauna of the island, which is meagre enough.

The vegetables which were cultivated were the sugar-cane, like that of Tahiti, very good, and now found self growing in numerous parts. Several kinds of yams. A remarkably good sweet potato, white, and when raw very like the chestnut in taste, and in this state is used to quench the thirst of the natives when travelling. It is very good also when cooked. There are no coconut-palms now growing, but boles of large ones are to be found. A wild gourd is common, 'it was used formerly for water-bottles. The tii-plant is pretty plentiful, but is put to no other use than for wattling of the grass houses, and as javelin-shafts. Of flowering plants we saw but few. The vervain, *Verbena officinalis*, is common everywhere, growing into bushes of as much as 4 feet in height; but it was imported some years since in a French ship, M. Bornier told me. None of the fruit trees, left by La Pérouse, could be found. I saw no tobacco plants. Of ferns there are some very beautiful, of the genus Asplenium, and several new varieties have been sent to Kew Gardens. Sedges and other bog plants grow in great profusion in the craters which are wet, but I regret that I was not able to collect any for an herbarium. The hill sides are covered with a fine grass which serves capitally to fatten animals, if we may judge by the state of some sheep now there.

The look of these people has been commented on by all visitors. Mendanã (1566) says, some were almost white, and had red hair. They were so well shaped and of such stature that they had much the advantage of the Spaniards. La Pérouse contradicts (1722) Roggewin's account as to their enormous height, and in many cases, singular leanness; but speaks favourably of them, and passes a high encomium on the beauty and form of the women, who he says, resemble Europeans in their traits and colour. Cook coincides. The Jesuit priest Eugène (1864) says the same; that they most resemble the Marquesans of all the other Polynesians,—many quite white, he says. We found them, in 1868, although under great disadvantage at the time of our visit, robust enough, and well grown, and they had a more European cast of countenance than the rest of the natives of the islands we visited. Three of the crania from a burying-place at Winipoo were brought home, two of which are in the College of Surgeons, London. The tracings and measurements of the other were sent to Prof. Huxley. In disposition they are friendly, affable, and merry, excessively indolent, very fond of finery and adorning themselves. La Pérouse says they had an amazing fondness for the hats of their visitors; we found our trowsers equally coveted. The men, says Frère Eugène (1864), were in their habits all thieves, and distrusted one another, and as the island abounds in caves and hiding places, these were always in request for shelter of the filchers. They are very patient of hunger, which they will rather suffer than work. Very dexterous in plaiting and carving both wood and stone; chips of obsidian are used for the former material, instead of chisels; they used obsidian flakes for razors and for their javelin heads. They practised circumcision.

The women, says La Pérouse, were fond of coquetry, which there seemed no disposition on the part of the men to restrain, nor were they jealous at it. Captain Amasa Delano, 1808, bears similar testimony.

Their mode of cookery was very simple. The various materials for the repast were wrapped in leaves, and baked in an underground oven filled with heated stones. They did not shed the blood of any animal, but stunned it; or suffocated it in smoke (like the Fuegians).

Cannibalism was practised. Four or six years since some Spaniards were eaten. From some remains, and native testimony, we were led to infer that human sacrifice took place, and burnt-offering was part of their religious worship.

The ground is so fertile, that a few days' work suffices to keep any family in subsistence for the entire year. Hence "they have no idea of agriculture," says Père Eugène. Yet the whole of the island has formerly been under cultivation, and rahui-stones are met with in eve   direction.

In consequence of the strong winds, the paper-mulberry (*Broussonetia*) was cultivated in small enclosures, with stout stone walls of about five feet high. The inner bark of this shrub served to make the mahuté (matué) for the blanket-coat of the men, which they called nuá.

Both sexes wore the maro, as commonly used in Polynesia. The men wore a cincture of woman's hair, as thick as a finger, and finished at each end by a tassel. The covering was a mantle over the shoulders, and fastened at the throat. This nuá was made of paper-mulberry for the men, and of fine grass for the women; and, says Captain Delano (1808), it was fastened round the waist, for them, and so hung nearly to the ground. The mantle was either white or made with brown patterns on it.

Both sexes used pigment for the skin; the men use not only earth of all colours, but also the sap of plants. The women were permitted red pigment only.

Tattooing was practised by the women more elaborately than by the men, and completely. In 1852 it was noticed, in particular, that they had a row of dots over the forehead, close to the hair, which ran down to the lobes of the ears. The women gather their hair into a knob at the crown of the head.

Both sexes wore ear-ornaments. The lobes were pierced, and

tebræ of sharks, were inserted. Roggewin says the priests wore great balls of wood hanging to their ears; some of these were sold to us, they were of the size of a fist, and carved into faces and joined together. They told us these were used at the dances. At their dances the men wore a gorget made of hard wood, lunate in shape, and each end terminated in a head; the concavity was worn uppermost, the profile of the face in the oldest gorgets was very aquiline. Also coronets of feathers, made like a modern hat without the brim; some we saw had the feathers radiating, like a flat diadem. They were usually made of dark metallic-looking hackles of the common fowl. La Pérouse says they much coveted the hats of the French.

In their hands, in place of weapons as used by the Maori, they carried short double-ended paddles, which they named "rapa." This had some symbolic meaning, as it occurs continually in the carvings and paintings, and also in the tattooing on the women's backs (1852). It would seem to be a human trunk, as at one end there is usually a face, and at the other a short phallus. It was not used for rowing.

Their weapons were the patoopatoo, or meré, a short club like that of the Maori; but I did not see any made of bone or stone, only of wood. They did not know the use of the sling. They used a pike for thrusting, and a javelin for casting; they both had heads of obsidian, the shafts made of pourou (hibiscus) and Tū (*Dracæna terminalis*); the javelin was thrown underhand, with the little finger foremost, and no throwstick was used. When an adversary was disabled, he was knocked on the head. They avoided bloodshed, and as the javelin-head was made for cutting more than for piercing, the legs and arms were more aimed at. The spear-shafts we saw were sometimes made of the stems of palm-leaves.

We saw no large war-clubs. The chiefs carried as a baton of office, a long staff as thick as the wrist, a little expanded and flattened at the lower end, and at the upper carved into a head,

They did not offer any fishing lines for sale, and the only hooks we saw were the large ones called "rou," made of stone, and which were of some age, and scarce, about three inches across the head. They are not in use now. The nets were made with small (half-inch) meshes, and from their size used for the small fry.

The houses are low and long, like a canoe upset; we find a good account of them in Cook; Delano saw them, in 1808, fully 200 feet long; in 1852, the *Portland* saw some fully 120 feet long, generally 60 or 70; in 1868, the *Topaze* found them about 30 feet long (and smaller), 12 to 14 feet broad, in height 5½ feet. The big houses were assembly halls, and were raised on low stone walls, on which a thatched roof was placed.

The ordinary house is made of a framework of sticks, on which grass is thatched. It is windowless, no hearth nor fire; an aperture in the side, about 18 or 20 inches square. This is closed by a net to exclude the fowls, and, as the natives pack pretty closely in these, the heat and noisome smell are indescribable.

There were some massive square buildings, built of unmortared stones, some 20 or 30 feet square, and 6 feet high, with little square apertures of a foot in size, here and there, at the ground level. These, we were told, were hen-houses, and fowls were in them; but it seems unlikely they were made originally for this purpose, as some very similar, but with white-washed tops, were used, we were told, for sepulture.

There were three principal feasts, or occasions of rejoicing, during the year :—

In spring (September), there was a great gathering at Mataveri. The people dressed themselves in their best, and remained there for two months. Athletic sports, running races, &c., were the order of the day. In summer (December), the feast of Paina took place. It is specially noted that each brought his' own provisions. The ceremony ended by the erection of a column of boughs; this was the Paina. In the winter (June, July), the large houses were built, and the people met for dancing, and held choral meetings, chanting songs, in which the same couplet was often repeated. These meetings were called Arcauti.

Their monarchy was elective; after the death of the sovereign, all the High Chiefs met together near the Terano Kau, and the candidates, with the view to prove their capability, descended the cliff there, swam to the islets, and, having got sea-fowls' eggs, returned with them. The successor was chosen by superior dexterity. The son of the last king, Roto-pito, was alive four years since. M. Bornier, the French settler, told us that on one occasion, being storm-bound in his boat on the islet, his crew swam to the island for food, with which they returned; so that this narration may not be fable.

The earlier voyagers thought that idols were worshipped by these people. Roggewin gives the names of their gods as Tau-pi-co and Dago; that fires were lighted before the idols at sun-rise, and that the priests who ministered were shorn. But we found that the Moai or Platform images were not worshipped, and that the people believed in one God—a spirit—sexless, whom they called Make-Make, the Creator—and that mankind, his children, but not by reproduction, were made by him from the earth; not by plastic agency, but by growth, like plants, &c. They repudiated the idea of a female deity. The Jesuit Father Eugène, 1863-65, noticed, in his letter to the Superior of his Order, that although they had "household gods" suspended to the roof of their dwellings, they did not worship them. The priests uttered the wishes of the god, oracularly; also his re-quirement of human sacrifice, and subsistence—by which they lived.

The taboo and rakui were here in full force, as in the other islands. By taboo, I mean that prohibition as regards man; by rahui, as regards property and crops. The symbol of the rahui was a cairn of three or four stones, piled on one another; the upper one very frequently white-washed. If a man planted ground, he immediately dotted the place with these cairns.

We did not find out whether there was any belief in a future state, yet it seems probable. After death the corpse was wrapt up in a bale of sedge and grass, and laid on the papakoo, or cemetery-terrace, the head pointing seawards. There was also another way by wrapping the corpse in tappa (native cloth), and lowering it into the cleft of a rock, or some inaccessible place. Some were seen by our people in such a position at Anakena, La Pérouse Bay. There were also burying-places inland. The small image Hoa Hava was the genius, so to say, of a cemetery at Mataveri. "Plenty, plenty dead here," said the guides; but we saw no platform, so that the corpses must have been buried. Yet so great is their aversion to promiscuous interment, as in Christian burial, that just before our visit a woman (whose child died shortly after birth, and had been so interred), rose in the night, and, after digging up the corpse, carried it two or three leagues to the papakoo of her tribe. Since the Peruvian raid, all the survivors have been massed together at Angaroa. That burnt sacrifices were offered, we found, by there being pillars here and there, on which were marks of fire, and in some in-stances charred bones near them. We were told these were of Heaka—*victimes* (French translation).

The papakoo, or cemetery, is a terrace or platform, generally near the sea, made of rolled sea stones, and faced seawards by a strong wall of large irregularly square stones, fitted together without cement. The ends of the terrace were whitened; they were usually about 100 yards long. One or two had no facing wall, being probably unfinished. There were a few inland, but I have no notes of them, except that on the flank of the Terano Kau, near Winipoo, there was a moated enclosure, on one side of which was a raised terrace overgrown with grass. This, we were told, was a papakoo. Near it a small trunk image—Libi Hoahava; where there was a small image we were led to infer a papakoo had existed.

Some square tombs, but for what class of individuals I could not learn, have been adverted to.

No images were placed on the papakoo terrace in the same way as the structures now to be described. These are to be seen on nearly every headland, as a rule pretty close to the sea, and being built on sloping ground, the sea-front is always the taller. They vary much in size. I will describe a pretty perfect one, which I have called the Fifteen-image Flatform.

Seawards, just where the ground becomes broken as it nears the cliffs, is built a very stout wall. Its height is much obscured by fallen rubbish, broken images which have toppled over, and rank vegetable growth, reeds, &c.; but it seems to have been about seven or eight yards high. The stones of which it is made are large and irregular, both in size and shape, though more or less four-sided. Some are fully six feet in length. They are fitted together very exactly, without any cement. This wall is built flat and level at the top, about 30 feet broad, by 100 paces long, squared at each end, and parallel to the shore in its long direction. This constituted, in fact, the platform, on which were the slabs which served as pedestals for the images.

Landwards it seemed to be not much more than a yard high, and on that side also was much ruinated, especially at about the centre. Before it, in the same direction, was a smooth space, or terrace, of the same length as the platform, but at least four times as broad, and this terminated in front by a low façade, or step, built of stone, and about as high as that of the platform seemed to be from the same point of view. The terrace sloped gently to this step, and the sides were built square and raised above the adjoining ground, so as to join the ends of the platform. The image platform was strewn with bones in all directions. They were old and weatherworn, but bore no marks of fire on them. The images had been thrown down in all directions, and were all more or less mutilated. The debris prevented my seeing if there was any crypt under the image pedestals, or in the platform, as at Winipoo, and the openings must have been at the ends of the platform, or at its sea-front, I think, if any existed.

At the south-west end of the island, at the sea edge of the Terano Kau Crater, are a number, say eighty or more, of houses of great age, now unused, mostly in good preservation, which are built in irregular lines as the ground permits, their doors facing the sea.   Each house is oblong-oval, built of layers of irregular flat pieces of stone, the walls about 5½ feet high.   The doors are in the side, as in the present grass huts, and of about the same size.   The walls are very thick, 5 feet at least, which makes the entrance quite a passage.   On entering, the walls are found to be lined with upright slabs, say 4 feet high, but not so broad.   Above these, small thin slabs are ranged like tiles, overlapping and so gradually arching till the roof-opening is able to be bridged over by long thin slabs of some 5½ or 5 feet, which are not more than 6 inches in thickness and 2 feet in width. The inner dimensions of the "hall" are about 16 paces long by 5 paces wide, and the roof is fully 6½ feet high inside, under the centre slabs.   The passage leading to it is paved with slabs, under which is a kind of crypt, or blind drain, which extends to the distance of about 6 feet outside, where also it is covered with flat slabs and is of the same dimensions as the passage.   It is carefully built of stone, squared and dressed; it ends abruptly and squarely.

In these drains, I was informed, the dead men heated were kept till required for the feasts.   Outside the hall, and at right angles to it, are smaller chambers which do not communicate with it, and each of which has a separate door from the outside. We were told that these were generally the women's apartments. The upright slabs which lined the hall, and those of the roof, were painted, in red, black, and white, with all kinds of devices and figures, some like the geometric figures of the Mexicans, some birds, rapas, faces, Eronié (a curious mythic animal like a monkey with a bird's head); M'hanus, or double-headed penguins. Symbolic figures of Phallic nature (Hiki-Näu), rude tracings of horses, sheep, and ships with rigging were found in a few.   These were very new, and misled some to the idea all were equally recent, and the houses also, which we were told was not the case.   There was no appearance of pavement in the hall, and in many of them enormous quantities of a univalve—a maritime Neritina—which had been used for food.   It was in one of these houses the statue Hoa-haka-nana-Ia was found.   It was the only one there, we are told.

Near these houses are some remains apparently of very great age—the sculptured stones on the brink of the sea cliffs at the

Terano Kau. They are at the place where the last lava-flow issued, and quite overlook the sea, which is directly under them. The blocks are of various sizes, carved *in situ* into rude tortoise-form, or have odd faces shaped on them. The nowaiu bushes and grass much obscure them, and had my visit been at any other time than mid-day, I should have sketched a good many, but I was very pressed for time. I could not learn their significance. These are very numerous, even to hundreds. I began to count them, but found them to be so plentiful as to make it lost time. They are almost always on platforms, but now all have been thrown down; except in the crater at Otuiti, and outside it, where they are in the earth only, and in groups, not in rows, and here even very many are prostrate. They are made of but one material, a grey, compact, trachytic lava, found at Otuiti, where there is a distinct *slide* for them to be removed by, and where there are still imperfect ones to be found. They are trunks, terminating at the hips—the arms close to the side, the hands sculptured in very low relief on the haunches. They are flatter than the natural body. The longest I measured was 37 feet; the usual size, 15 or 18; the small ones, as Hoa Hava, 5 or 4½ feet. These were more boulder-shaped. The head is very flat; the top of the forehead cut off level so as to allow a crown (hau) to be put on. This was not done till the image was on its pedestal on the platform. In the giant images at Otuiti, outside the crater, the head seemed to project before the line of the trunk, which we did not notice in the others. The face and neck of these measured full 20 feet to the collar-bone. They were in the best preservation. Those inside the crater were of large size, but weatherworn, apparently the oldest in the island, and also many were prostrate. They differed a little in profile from those in the other parts of the island. The face is square, massive, and sternly disdainful in expression; the aspect always upward. The peculiar feature is the extreme shortness of the upper lip, or the upthrust of the lower one, which would produce the same appearance. This gesture is sometimes. seen now among the natives. The eye-sockets are deep, close under the brows, and, as far as we could make out, eye-balls of obsidian were inserted in them; but we were not fortunate enough to find any. The nose broad, nostrils expanded, the profile varying somewhat in different images. The ears were always sculptured with very long pendant lobes.

The beautifully-perfect one Hoa-haka-nana-Ia (each image has its own name), now in the British Museum, was found in the stone house called Tau-ra-re-n'ga, at the Terano Kau. It is elaborately traced over the back and head with rapas and birds, two of which much resemble the apteryx. It was coloured red and white when found, but the pigment was washed off in its transit to the *Topaze*. Its height is 8 feet, weight 4 tons. It was buried waist deep in the ground, and had no crown. Its face, like those of the rest, turned from the sea. It was the only one under cover, although it was reported that there were some in a cave on the sea-shore. This arose from the misconception of some mural paintings found there. The house in which it was found was a small circular one (20 feet across) into which two small dark chambers opened.

The crowns were always made of the same red, vesicular tuff found in the Terano Hau, down the outside slope of which as many as thirty were waiting for removal to their several platforms. The largest I measured was 10½ feet in diameter, but they varied very much in size, at Anakena to only 2 feet across. In shape they were short, truncated cones, or nearly cylindrical. Some of the very large images have such small tops to the head that it would seem difficult to fit them with a crown.

The principal track of the images from Otuiti is by the Coast Road, on either side of which they are found, face downwards. On the Mid Path of the island I found but two or three. Many were found also from Anakena; but there was a great part of the island untraversed. All accounts go to the same point, that it is on the coast these images are most abundant.

The implement used for carving these statues was a long boulder-pebble from the shore, like a rolling-pin or huge incisor. The chisel edge was produced by chipping it, and rubbing it down afterwards on obsidian. We saw but one. This was presented to Commodore Powell, and is now in the British Museum. It was called Tingi-tingi. It was noticed that on many of the statues little projections were left; these were portions harder than the chisels.

The number of images on the platforms is very variable, and also their size is by no means uniform. They always faced landwards. At the fifteen-image platform five of them are quite dwarfs in comparison with the rest.

In La Pérouse's voyages it is said that the image platforms were used as Morais, and Cook says that they were the sleeping-places (*i.e.* tombs) of the chiefs. We found that the word Morai was never used in reference to any papakoo or cemetery of the tribe. Each image, and some of the stone houses also, had their proper name. Beechey surmises that these are relics of a past age, as in some now desolate islands he visited he saw similar terraces and images. In Maldon Island I was informed by a visitor that under the guano similar platforms existed, without images. In the Marquesas the images were made of wood, and there is no doubt, from signs on Easter Island, that sufficient wood existed to have made wooden images long since the fabrication of these trachyte statues, which as material would hardly have been chosen for ease or rapidity in working. And besides, only one chisel has been found; nor could any others be procured.

At a little distance from this terrace, and about the centre point, was a short pillar or cylinder of red tuff (vesicular lava), standing in an area paved with large, smooth, sea-worn stones. It stood on a low slab, which was of the same material, and which served as a pedestal. It was about 6 feet high, and as much in diameter; the top was flat, and cut away on each side, so as to make a step or shelf. On it I found two skulls, very much perished, which, from the dentition, I judged to be those of youths of twelve or fourteen years old. The faces of the skulls were directed towards the platforms. At Winipoo there is one similar. The upper part is paved with smooth sea-stones of the size of a dinner plate. The measurement of the pillar, which is oval, is 7 feet by 5, and it was 4½ feet high. It stood also in a paved area.

Cremation Stone.—Again, in a direct line from this, landwards, at about 80 or 100 yards from the platform, is one of the low-slanting saddle-topped pillars used for cremation (burnt sacrifice). It is also of red tuff, but was not more than 4½ or 5 feet high. The finest I saw was at Winipoo, of which I append a description. In a paved area, similar to that of the last-described pillar, is a pillar of red tuff, 3½ feet squared and 8½ or 8¾ feet high. The top projects *forwards*, and ends in two horns, with deep saddle-shaped notch between them; each horn had a face traced on it, in low relief—face surmounted with a crown (hau), but that to the north-west had crumbled away (from the action of fire?). The projecting part is terminated at the breast, and lower down a round projecting navel is marked. Just above, where the pillar joins the area, the fingers are sculptured, in low relief, flat, and clasping the hips, as in the images.

We were told heaka (victims) were burnt here, and at the foot of one of these pillars at Winipoo we found many burnt bones. The pillars were in number at least one for every image-platform.

With respect to the former of these two pillars. In the most excellent description by M. de Bovis, Lieutenant de Vaisseau and Surveyor of Poumotie group, published in the Government *Annuaire de Taiti*, p. 292, he says: " Il y avait sur le parvis une sorte de parvis dallé en pierres plates devant l'autel (before the Morai), une enorme pierre plate, un peu plus élevée que les autres; le prince s'y plaçait tout nu pendant la consecration." It was here the maro-ura (red maro or breech-cloth) was put on the new prince by the priest, as a symbol of royalty, in the sight of all people.   It was at the great Morai of Opoa this was done in grandest pomp.   If, as has been surmised, and it seems warranted, that these flat forms were over the chiefs' tombs or family vault, the images being, as on the Bustum of the Romans, an effigy of the departed, this stone may be the place of hereditary succession, and the Cremation stone the place where slaves or prisoners of war were burnt at the death of the chief, to attend on him in the spirit-world.

M. de Bovis says the missionaries carried away the sacred stone of Opoa to another place, in order that the kings might be consecrated, without idol-worship ; so to say so great was the idea of the natives of its value.

Lares or Household Gods, ' *Domestic Idols* ' *of Père Eugène.*— These are generally male figures, about a foot in length, made of solid dark wood (Toro miro or *Edwardsia*), a little bowed forwards, and suggesting the idea that they represent flayed carcases.   The profile, differing from the images, is strongly aquiline—the mouth grinning, the ears with long lobes, and eyeballs of obsidian are inserted.   There is a small tuft on the chin ; the arms by the side, the hands on the thighs, but not clasping them.   These figures are very well carved.

The female figures are much ruder in execution, flatter and larger—a small tuft on the chin also; the attitude that of a pancake Venus de' Medici.

Besides these, there were a quantity of very odd figures carved, representing lizards, sharks, fowls, nondescripts.   Some of these are in the possession of the Rev. Mr. Dearden.

On the heads of the male images are carved in very low relief the most peculiar figures, evidently mythic—double-headed birds, fishes, monkeys, lizards ; some figures too in which no likeness to anything can be traced.   These are on the male figures.   I saw but one female figure thus adorned.

These lares were not worshipped, and though the present people still carve them, we could not find that they were aware of the significance of the mythic emblems which they copy.

Traditions.—We could learn very little of their antecedent history, and but little of their traditions.   It is a current belief that many generations since a large migration hitherwards took place from Oparo or Rapaiti, the leader of the swarm being Tu-ku-i-u, who, after arrival, abode for some time near Otuiti, where he caused the images to be made.   That subsequently he went to reside at the Terano Kau, in the stone houses.   That the images followed him by night, walking of their own accord. and that that accounts for the places where they are found face downwards about the island (see parallel destruction of giants,

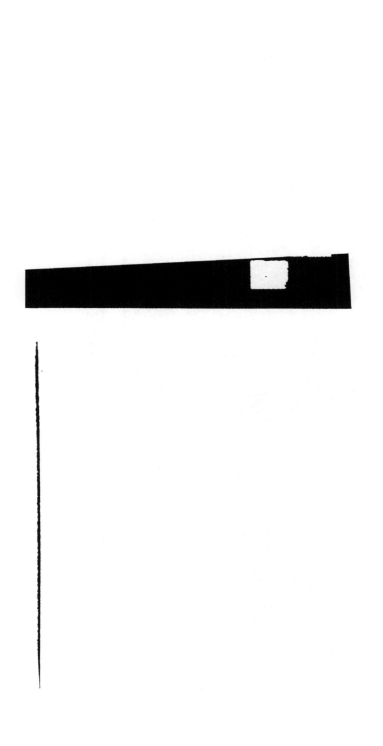

MAP TO ILLUSTRATE SIR H. BARTLE. E. FRERE'S NOTES ON THE RUNN OF CUTCH &c.

Published for the Journal of the Royal Geographical Society by J. Murray, Albemarle Street, London, 1871

at sunrise in the Eddaia Myths). That at his death he vanished from earth in the shape of a butterfly (Puru puru), and this insect is shouted at now by small children, as Tu-ku-i-u, Tu-ku-i-u. There is no hint at his reappearance. The distance due west nearly from Easter Island to Oparo, Rapa or Rapaiti, is about 1900 miles. The last successor of Tu-ku-i-u was named Ro-to-pi-to, and his son died about 1864.

IV.—*Notes on the Runn of Cutch and neighbouring Region.* By Sir H. BARTLE E. FRERE, K.C.B.

Read, Feb. 14, 1870.

THE tract of country to which these remarks refer forms a considerable portion of the great basin of the Indus, which is bounded on the north by the barrier of the Himalayas, west by Suleiman, and Hubb ranges, south by the sea and the hills of Cutch and Kattywar, and east by the Aravulli mountains and their offshoots. But the portion to which I wish now more particularly to draw attention, can hardly be said to form an integral part of the basin, inasmuch as it is in no part watered by the Indus or its tributaries.

It may be more clearly defined as the tract which intervenes between the basin proper of the Indus on the west, and on the east the basin of the Ganges, and the plains of Rajpootana, which are watered by streams from the Aravulli mountains. The length of this tract, measured in a slightly curved line from the hills of Cutch to the northern borders of the Thurr, or what is called the Sandy Desert, is nearly 600 miles. The breadth between the permanently watered and fertile plains which bound it east and west, varies from about 100 to sometimes more than 150 miles. Its limits are well defined. To the south the hills of Cutch bound the flat plain of the Runn. The Runn itself is, in general, clearly distinguished from the plains east and west of it by its lower level, and by the total absence of vegetation.

North of the Runn the east and west limits of the Thurr are in general equally sharply defined as low sand hills which rise abruptly from the level plains. To the north the transition is less abrupt. From the southern slopes of the sub-Himalayan ranges, between the Jumna and the Sutlej, the country sinks very gently to the south-west. The fertile plains gradually become more and more sandy, till south of the direct road from Delhi to the Sutlej *viâ* Sirsa, the country assumes the aspect of a constant succession of sandhills, which continue with very few and brief intervals to the border of the Runn of Cutch. The divisions noted in the most popular manuals of geography within this vast area are political and ethnological rather than physical.

Thus, starting from the north to where the last slopes of the sub-Himalayan ranges sink into the plains, popularly known as the province of Sirhind, the independent Sikh states of Puttiala, Jeend and Nabba, and the British Cis-Sutlej districts, occupy a vast, fertile and nearly uniformly level tract of nearly 17,000 square miles, extending between the Jumna and the Sutlej. This province has been called the Belgium of India, for in it, or its immediate neighbourhood, are the battle-fields on which the fate of India has been repeatedly decided. At a point where the Jumna near Kurnal is about 112 miles from the Sutlej near Loodhiana, the general level of this plain is about 1000 feet above the sea. The summit-level of a line measured between the two rivers is about 90 miles west of the Jumna, where the plain is about 67 feet above the level of the river. The ground then slopes somewhat more rapidly towards the Sutlej, the general cold weather level of which is said to be about 2 feet below the Jumna. The streams which descend from the Sub-Himalayas into this plain are so nearly on a level with each other, and with the surface of the country, that most of them are connected with the others by natural or artificial channels, which fill during the inundation season, and a canal cut by Tiroz Togluk, the Tartar king of Delhi, from the Jumna, carried its waters to Hansi and Hissar, a distance of 150 miles, throwing off branches back to the Jumna near Delhi, and also westward, to the Sutlej. South of Sirhind comes the district known as Hurreeana, estimated at between 3000 and 4000 square miles. The mountain torrents from the Himalayas do not reach so far, even in the heavy rains. But there are tolerably regular showers in the later summer and autumn, and where rain falls the sandy soil yields abundant crops. The wells are much deeper than further north, being seldom less than 100 to 120 feet deep.

Bhutteeana may be regarded as the western half of Hurreeana. It is sandy, much more sparsely inhabited; but the ruins of the numerous towns and villages show that this was not always the case; and there is a tradition that it was once traversed by the Gugzer, a large river of which the dry channel can still be traced, and the sites of large lakes which were filled within the period of authentic history are pointed out. The same character of country continues throughout Bickaneer, gradually becoming more sterile as the traveller proceeds south, and the patches of regular cultivation more infrequent.

The state of Bhawulpore is of comparatively recent origin, formed by the conquests of the Mahomedan tribes of Daood-poohas. It is sharply divided into a strip of level cultivated land on the left bank of the Sutlej and Indus, and a desert tract composed of sand-hills.

The northern part of Jeysulmere presents the same general characteristics. To the south it becomes more rocky, but throughout its whole area, estimated at more than 12,000 square miles, there is no running stream. The rain-water collects in temporary pools or lakes, locally known as "Sars," the water in which is generally more or less salt. One of these (at Kurrod) is said to be 18 miles long. The wells are here as much as 300 feet deep, and the inhabitants depend much on the water retained in masonry tanks.

South and west of Jeysulmere come the desert districts of Sind divided between his Highness Ali Morad of Kyrpoor, and the British districts of Thurr and Parkur, which extend to the borders of Cutch. But none of these provinces or states, or their boundaries, have any reference to any natural or physical features. Indeed throughout the whole of the tract, usually marked on our maps as the "Great Indian Desert," there is an absence of those physical features which are necessary to ordinary geographical classification. There are neither mountain ranges nor river systems. Here and there a few rocks protrude from the surface, and about Jeysulmere, and at some points on the edge of the Desert, as in Nugger Parkur, the rocks may be dignified as hills. But with these very rare and partial exceptions the "Desert" may be traversed in any direction from end to end, without the traveller finding a stone or a pebble which has not been imported. So with regard to rivers. A few streams descend from the Himalayas, and flow south to the desert portions of Hurreeana. There all trace of them is lost in the sand-hills, and for about 500 miles to the hills of Cutch nothing like a running stream, nor any trace of one, is to be found. But notwithstanding this total absence of rivers or mountain ranges, this vast area hardly resembles anything usually classed as a "plain." Very little of the surface is really level for more than a few hundred yards. It is usually ridged into sand-hills, or rather long billows of sand running in directions generally parallel to each other for considerable distances. Nor does the term "Desert" at all aptly describe the tract, for it hardly conveys the notion of a district which is everywhere inhabited, though sometimes very sparsely, which, in parts, supports a considerable fixed population, and vast herds of cattle. All these peculiarities are taken into account by the natives of the tract, and their neighbours, in their classification of the various divisions of what we have been used to call the Great Indian Desert (according to the character of the surface) into "Thurr," "Put," and "Runn."

The portion covered with sand-hills, which we have just been describing, is popularly and officially known on the spot, and in all adjoining provinces, as the "Thurr." The name has been already applied to it, in some of our best and most recent maps and geographical treatises, so that perhaps it may be allowed a place in regular geographical nomenclature as a variety of plain, as remarkable and uniform in its characteristics as "Savannah," "Prairie," or "Pampas."

Not less characteristic is the "Put," or hard level plain, from which the sand-hills rise in many parts of the eastern and western borders of the "Thurr," with the same abruptness as the sand-hills or cliffs of the sea-shore rise from the ocean. The surface of the Put is smooth and shining, and, except where the rain-water lodges, generally free from vegetation. Large tracts are frequently barren from saline, or alkaline efflorescence, but where this is not the case the Put forms a very fertile soil, whenever fresh water, either from showers or from inundation channels drawn from the Iudus, has access to it.

In these respects the Put forms a contrast to the "Runn"—the vast salt plain which intervenes between the Thurr and Cutch. The outline of the Runn is irregular, but its greatest clear length from the ancient channel of the Indus near Luckput, on the west to Soegaon on the east, is about 150 miles. But a sort of arm extending in an irregular south-easterly direction is connected, without any perceptible rise in the level, with a narrow strip of similar formation which fringes the Gulf of Cambay, down to a point nearly another 150 miles distant from what is commonly marked in the maps as the eastern boundary of the Runn. Its general breadth, exclusive of the island-like detached portions of the higher land, which occasionally rise from its surface, is from 40 to 50, or perhaps 60, miles in the wider parts.

The Runn is usually marked on our best and most recent maps as a "salt plain," which gives a much better idea of the character of its surface than some of the terms applied to it in modern maps, such as "morass," "salt-marsh," or "swamp," or "arm of the sea." But it is a salt plain of a very peculiar character, and only for part of the year. The surface is apparently, for all practical purposes, a dead level. Towards the centre of the Runn there is a slight rise, not exceeding a foot or two above the level of the margin of the plain. This rise is of course quite imperceptible to the naked eye—it is only apparent to the surveyor's level, or to the traveller who, in crossing the Runn when covered with water, finds a sensible difference in the depth of water through which his camel wades.

When the surface is dry, so imperceptible is the slope, that a shower of rain falling on the hard, polished surface, neither sinks in nor runs off, but lies, like a vast slop, on the plain, and may sometimes be seen moving along before the wind, till it gradually dries up by evaporation. Not only is there no visible change in the level, but there is a total absence of any sign of animal or vegetable life which could break the uniformity of the surface. There are no trees, no tufts of grass; and the bones of a .dead camel are visible for miles, whether seen in their actual form and size, or drawn up into the likeness of towers, rocks, and houses by mirage.

The general surface is hard and polished. It consists of fine sand and clay, with sufficient salt in it to attract any moisture which the air may possess, and to keep the surface damp when all around is arid. Hence, though sometimes covered with a saline efflorescence, the surface itself never pulverises, even in the hottest weather, and is usually so hard that a horse's hoof hardly dents it in passing. But during the hottest weather the "Runn" is generally under water. When the s.w. monsoon winds begin to blow steadily in May they bank up the waters of the ocean off the coast of Cutch, so as to occasion a considerable general rise of the sea-level in the Gulf of Cutch, and in the Luckput estuary, both of which terminate by a gradual shallowing of the shore, where it joins the eastern and western extremities of the Runn. The whole surface of the Runn is so little raised above the ordinary sea-level, that when this occurs, the first high tide causes the sea to overflow the Runn, across which the waters are blown by the steady s.w. breezes, till the whole is some feet deep under water, the depth of which is generally augmented, about the same time, by the contributions of rain-water, brought down by the Loonee, and a few other smaller streams which discharge into the Runn. This does not interrupt the transit of those who have occasion to cross the Runn. The bottom feels like smooth sand under the water, and is more pleasant to the feet of the traveller and his animals than when dry. With a good guide—and no one attempts to pass without one—parties of travellers wade steadily through miles of water from one to three feet deep, and generally deeper at the edges of the Runn. The night is usually preferred for crossing, to avoid heat and the glare, which during the daytime often renders it very difficult for even an experienced guide to keep a direct course. The stars, too, are generally visible at night; whereas the haze and glare are often so great during the daytime in the hot weather, as to render it difficult to tell whereabouts in the heavens the sun is; and sad stories are told of travellers, who, attempting to cross in the daytime, with the sun almost perpendicularly overhead, have got confused and wandered in circles, till they sank exhausted and died.

Near one of the most frequented tracts a sort of island, with a high rocky hill in the centre, rises from the Runn. The top of this hill used to be the residence of an old fakeer or devotee, who, as an act of charity to travellers, kept a fire constantly burning on the hill-top as a beacon, while a draught of fresh water and a light for the hooka was always to be got in a hut by the wayside, at the foot of the hill. But, as elsewhere in the plain country of Sind, and here more conspicuously, owing to the absence of any prominent natural features or marked tracts, the best guides seem to depend entirely on a kind of instinct—they will generally indicate the exact bearing of a distant point which is not in sight quite as accurately as a common compass would give it to one who knew the true bearing. They affect no mysterious knowledge, but are generally quite unable to give any reason for their conclusion, which seems the result of an instinct—like that of dogs and horses, and other animals—unerring, but not founded on any process of reasoning, which others can trace or follow.

The maps show a few projections, like islands or peninsulas, which rise from the general surface of the Runn. These are in two or three instances rocky outliers of the Cutch Hill Ranges; but generally they are level, and though raised above the surface of the Runn, are composed of the same soil, but free from salt, and hence after rain become covered with a fine crop of grass.

The peculiar conditions of the atmosphere on the Runn and its neighbourhood occasion phenomena, not often observable elsewhere. Besides the dazzling haze already referred to, which at times prevents distinct vision, except at very short distances, the mirage is generally visible by day in one or other of its endless varieties; that of an expanse of water, or a long range of precipitous cliffs, are among the most common. Almost any object which rises, however slightly, above the general surface, is usually made visible by some distorted image, due to mirage, long before it is really above the horizon, and frequently assumes a great variety of fantastic forms before the traveller comes up to it. Thus a mere molehill of earth, thrown up to form a fireplace, or a heap of camel bones, may appear before they are really in sight as a clump of trees, or drawn up into a white minaret or spire; after a while they change to a mass of rocks, a castellated building, or a town with clusters of white houses, often rising till a clear space is seen below the object, which seems suspended in the air; suddenly it vanishes, and the only substantial part of the illusion is seen in its true shape and position on the ground as a heap of earth, a tuft of grass, or a few bleached bones.

The mirage is naturally the subject of many local traditions and legends. The origin commonly attributed to it by the country people is, that it is the spectre of the ancient possessions of a pious king who once reigned here, when the country was all fertile. He had succeeded so completely in restoring a golden reign of virtue, that his capital, purified from all that was unclean or offensive, was in gradual process of elevation to Heaven, when a donkey, regarded by the Hindoos as a most unclean animal, which had been forgotten in an outhouse, betrayed his presence by braying. The elevation of the city then stopped, and it has ever since wandered uneasily between earth and Heaven over the district to which it once belonged.

Another phenomenon, not unfrequently observable among the sandhills bordering the Runn, is a spectral illusion, similar to that known in Germany as the "Spectre of the Brocken." After rain, or in the cold weather, dense white fog often covers the surface of the earth in the early morning. At such a time, if the spectator, on the top of a sandhill, turns his back to the rising sun, he will often see a gigantic indistinct image of himself, and of any other objects within a few yards of him, reflected in the centre of a vast luminous mirror-like space surrounded by a faint halo, in the clouds of white mist in front of him. The images move as he moves, and melt away as the sun rises or the mist or the spectator alter their position. When first seen it is generally some time before the spectator realises the fact that it is his own image which he sees, and the effect is often peculiarly weirdlike.

Just after and before rain the clearness of the atmosphere is often as remarkable as its haziness at other times, and extra-ordinary effects are seen in the endless varieties of rose-coloured and ruddy lights, such as are depicted in Mr. Holman Hunt's picture of the "Scape Goat," or such as are occasionally seen among the snows of the higher Alps.

One of the most remarkable features of the Runn and its neighbourhood is the profusion of salt in every form and in every place. A white saline efflorescence covers the ground for many square miles together, and any water which is found by digging on or near the Runn is generally intensely salt. Wherever a depression admits of water lodging it is saline, and is usually fringed with a thick incrustation of salt, and near the margin of the Runn, especially to the north-east, as at Mokye, near Nuggur Parkur, are salt-lakes, which annually fill during the rains, and as they dry up leave salt crystallised into masses, so clear and hard as to bear the rough transport on bullocks without being pulverised, like the soft salts in common use.

In some parts on or near the Runn, especially on the salt-plains of the old Delta of the Indus west of Luckput, many square miles are found covered with a solid cake of hard transparent ice-like salt, from a couple of inches to in some cases 2 feet in thickness. Various theories have been put forward to account for these thick sheets of solid salt, on a perfectly level surface of dry sand and clay. The most probable and most consistent with observed facts appears to be, that it is formed by the gradual evaporation of the intensely salt water which is always present in the subsoil, and which oozes to the surface by capillary attraction or under pressure, from rain in the upper country, and from high tides in the creeks which intersect the plains where the sheets of salt are found.

It is obvious that the Runn, entirely destitute of any supply of drinkable fresh water, and alternating in condition at different seasons between a hard, perfectly level, uniform dry plain and a shallow inland sea filled to a depth of a foot or two with mixed sea and rain water, is ill fitted for the support of any form of animal or vegetable life, and it usually appears absolutely destitute of both. Nothing living is to be seen on its surface save perhaps an occasional crow or other bird, which has wandered from the shores of the Runn, nor anything like a tree, shrub, or even a tuft of grass.

There is, however, in one part a curious exception to the general absence of animal life. On the eastern and southern borders of the Runn are still to be found a few herds of wild asses, apparently the same animal which is found in the salt deserts and their neighbourhood, in most parts of Western and Central Asia. They appear to feed during the night in the fields and pasture-grounds bordering the Runn, and resort to the Runn itself during the day for the sake of the safety which is afforded by the total absence of all cover. They are extremely wary and so swift that it is said to be as difficult to ride down a wild donkey as a wolf or black buck. The young are often captured by driving the herd during the rains into muddy ground, where the foals, unable to extricate themselves, are secured. They are easily tamed, and may be taught a variety of mischievous tricks; but I have never heard of success attending the most persevering attempts to break them in to carry burdens or to go in harness.

We will now return to the Thurr, the region of sand-hills which bounds the Runn to the north, and which, though inhabited and far from unproductive, forms by its peculiarities of surface a most serious obstacle to communication between the countries lying east and west of it. The general surface has been already described, as an interminable succession of billows of sand, lying in parallel ridges, like the rollers of the Atlantic. A few attain a maximum elevation of 500 feet above the sea, or about 400 measured from the bottom of the intervening valley to the crest of the ridge; but the usual height is less than half this, and in many parts the hills do not exceed 80 or 100 feet. The intervening valleys are narrow, and rarely contain any extent of level ground; and it is only in very few localities, such as Miltee, Islamcote, and Deepla, that a small plain of a mile or two in diameter affords space for a few fields, and gives an indication of the character of the original surface of the country, before it was thrown up into the sandy billows which now characterise the whole region.

In some parts, chiefly on the western border of the Thurr, north of Oomercote, are found a series of remarkable depressions, locally known as "dunds" or lakes, where the valleys between the sand-hills are of unusual depth, and are so far below the general surface that they are filled with water. This is the general character of all the valleys between the sand-hills, on the west edge of the Thurr, for from 40 to 80 miles north of Oomercote.

The dunds here form a series of long, narrow lakes, generally lying parallel to each other, in a direction from south-west to north-east, and often connected, so that some of them form sheets of water of many square miles in extent. They are often of great depth—some are popularly said to be unfathomable, and I was assured of one having been measured 70 feet deep. Many of them have no visible source of replenishment except from rain-water, and these are usually intensely brackish. A few of those which are furthest to the east afford deposits of natron, which is collected and exported. These saline dunds are easily distinguished from the fresh-water lakes by the absence of all animal life. The fresh-waters swarm with wild fowl during the cold weather, and during the hot months attract the few birds and wild animals which are to be found in the neighbourhood.

Most of the dunds near the edge of the Thurr are now or were formerly open to the ancient channel of the Indus, known as the Eastern Narra, which runs from a little to the east of Roree, on the Indus, nearly due south, till it comes out on the plains to the north-east of Hyderabad, whence it skirts the eastern edge of the Thurr till it is lost in the border of the Runn. At intervals of a few years, when a very high inundation has covered the low-lying districts north of Roree, the flood-waters used to

FIG. 1.

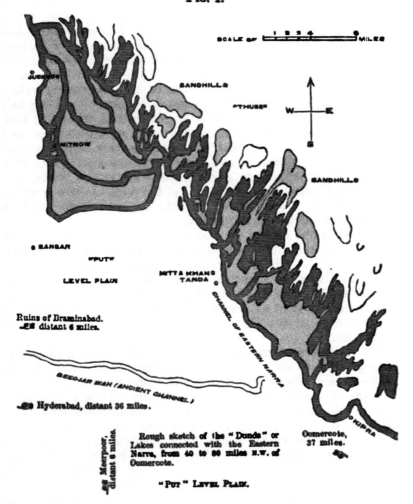

SCALE OF ___ MILES

Rough sketch of the "Dunds" or
Lakes connected with the Eastern
Narra, from 40 to 80 miles N.W. of
Oomercote.

"PUT" LEVEL PLAIN.

descend by this main channel, and sometimes reached as far
south as the road from Hyderabad to Oomercote. But in the
upper part of their course the floods had to pass the entrance to
some of these dunds, into which they poured sometimes for weeks
together till the dunds communicating with the Narra Channel
were all filled, after which the flood-waters passed onwards to
spread themselves over the open plain north-west of Oomercote.
When works were undertaken a few years ago with a view to
render these floods annual, one of the first things done was to
close up the mouths of these channels so as to prevent this
profitless drain of water in the desert hollows. I have not heard

what effect this has had on the dunds, which were formerly thus supplied.

Besides the Eastern Narra, there are other ancient channels of the Indus and Sutlej, or its tributaries, of which clear traces may be found almost continuously, from the frontiers of Sirhind to the Roree hills, generally running in a direction parallel to the western boundary of the desert, and from a couple of miles to 20 miles from its present margin.

Possibly the researches of future geographers and archæologists may determine whether any of these ancient river beds can explain the traditions of the lost "Sariswati"—the sacred river which in the heroic ages issued from the sub-Himalayan mountains between the Sutlej and the Jumna, and flowed to the sea either by the Gulf of Cutch or Cambay, but not far, as local traditions affirm, from the holy shrine of Dwarka.

Some of the dunds present unmistakeable traces of having been craters of eruptions whence the sand has been thrown out by a subterranean explosion, which has deposited it on the margin of the crater in the shape of a sand-hill, locally known as a "bhit." Nor are these the only visible traces of the volcanic action, to the frequent occurrence of which observation, history, and local tradition bear testimony.

Earthquakes.—The neighbourhood of "Cutch," including the Runn and the districts immediately adjoining it, is a region of constantly recurring earthquakes. The inhabitants of the Thurr bordering on the Runn assured me that the shocks were more observable during the warm months, and that not a hot season passed without several shocks being noted. But no earthquake is remembered in recent days exceeding in severity that which, beginning on the 16th of June, 1819, lasted, with a continual succession of perceptible shocks, till the 23rd of November in the same year.

It is recorded * that till August no day passed without a shock, and 100 in all were noted. The first severe shocks were distinctly felt, and identified as far as Calcutta and Pondicherry in India, with a difference in time of seven or eight minutes. An earthquake was felt at the same time in Arabia, and the period was one of great volcanic movements in Southern Europe. The direction of the shocks in Cutch was from north-east to south-west.

Readers of Lyell are familiar with the most marked of the phenomena attending this earthquake, in the subsidence of a large portion of the Western Runn, including the small town

---

* A very full account of the earthquake, by Captain MacMurdo, and others, is in vol. iii. of the 'Transactions of the Literary Society of Bombay,' p. 90, and other authorities quoted by Lyell.

of Sindree, which sank several feet under the surface of the water, in the Luckput Inlet, formerly one of the main mouths of the Indus.   At the same time there was a considerable elevation of a neighbouring part of the Runn, and in one place a mound 10 or 12 miles in length and of considerable height was thrown up.   The mound, which is still visible, and known as the Allah Mound, or " Embankment of God," is of tolerably uniform height and thickness, with a long slope to the north and a more abrupt face to the south, and bears other indications of having been formed by a crack or fissure of the surface (at right angles to the direction of the undulation of the ground), one lip of the longitudinal fissure overlapping the other so as to throw up the edge and form the ridge above described.

Besides the phenomena noted by MacMurdo, Burnes, and the other writers cited by Lyell, every old man in the country has some anecdote of his own relating to this great convulsion, I will only here mention a few which I do not remember seeing noted before.

It was remarked that in the districts south of Hyderabad all the canals drawn from the Fullalee River, an ancient natural channel of the Indus, stopped running when the earthquake occurred, and did not flow again for about three days.   This indicates a general upheaving of the surface in the lower part of the course of the canal, which must have lasted for a day or two, and was followed by a return to its old level, as compared with the point where it branched off from the Indus.

Many spots are pointed out which, during the rockings of the earlier shocks, ejected, some, mud, others sand and water. These generally proved, when examined about 10 years ago, to be crater-like depressions, whence the earth and sand, which previously filled the space, had been ejected by a sudden explosion, and generally deposited on the mound to leeward or northeast of the crater.

In some places we were told of small islets of raised ground, which formerly existed on the Runn, and which had been swallowed up, and we were taken to the spot, near Vingur, where such an islet used to stand out from the surface of the Runn. But in no such case did any hollow or chasm remain visible. The islet seemed to have melted down to the general level, and the description given by eye-witnesses of what they saw indicated the action one would expect, from continued agitation of a mass of wet sand surrounded by water.

A curious proof of the frequency of earthquake tremor on the alluvial lands of Sind was mentioned to me by the late General John Jacob.   He was extremely fond of astronomical observations, and used to amuse himself by taking the height of the sun at noon, daily, wherever he happened to be. He told me that he remarked that hardly a week passed without his detecting on the surface of the artificial horizon, when observing in the low lands in Sind, a slight tremor for which he was puzzled to account, till he came to the conclusion that it could be due to no other cause than a slight tremor of the earth, which was imperceptible to the senses, but of which the surface of the mercury gave sensible evidence.

Tradition in Sind almost everywhere points to an unusual frequency of earthquakes, especially throughout the Thurr and its neighbourhood, and many of the facts indicate a progressive upheaval of the surface, as a process still in constant operation. Thus in many parts of the Thurr, but especially in the northern borders and their neighbourhood, there is a constant complaint of the decrease of water in old wells, which is so general as to indicate some widely operative cause, such as a general rise in the surface.

I am told that this complaint is by no means confined to the Desert, but that it is frequent in the district north of Delhi, and in many parts of the north-eastern Rajpootana and the Gangetre Doab.

From the frontiers of Sirhind to Jeysulmere, the Thurr contains frequent traces of ancient watercourses which once flowed where no stream now flows even in the heaviest rains; and ruins of ancient towns and villages, which have for ages been utterly deserted, prove that the country in former days was more populous, and contained more water and cultivation than at present.

Further south similar indications are very frequent, though owing to the want of permanent buildings in the Thurr itself, they are more frequently met with in the plain country east and west of it. Thus the plain between Hyderabad and Oomercote presents everywhere indications of having been thickly populated as far back as the Bhuddist period, and well irrigated by canals from the Indus, which no longer carry water in consequence of a slight change in the relative levels of the Indus and the plain to be irrigated.

The change is not so great but that it may often be met by deepening the old canals. The probable date of at least one great alteration of level is pretty clearly ascertained by the fate of the large city of Brahminabad, of which very extensive ruins still exist, and regarding the former history and destruction of which many facts are ascertainable.

The ruins of Brahminabad are situated about 40 miles north-east from Hyderabad, on the banks of what was evidently a very large branch, if not the main channel, of the Indus, but which is now perfectly dry. The walls of the city, well-built of bricks, are still tolerably perfect, and include an area of 7 or 8 miles in circumference.   The streets and houses are clearly traceable, and where the ruins are excavated often afford most interesting evidence of the wealth and civilisation of the inhabitants, and of the sudden manner in which the city was overwhelmed.   Great quantities of small copper coins, and various objects in metal, earthenware, ivory, tortoiseshell, and stone have been dug up; many human skeletons, some crushed in the act of crouching in corners, and rows of skeletons of cattle in their sheds, testify to the nature and suddenness of the catastrophe.

The date of the earthquake is ascertained by independent historical evidence as having occurred about 700 or 800 years ago, and though the local traditions are obscured by many picturesque legendary additions, there seems no reason to doubt the general assertion that, from that day to this, the waters of the Indus have never visited the ancient channel, which is still a conspicuous feature near the city, nor the plains in its neighbourhood, until brought ba   by the engineering works of the present possessors of Sind, which have drawn a supply of Indus water from a higher level.*

On the other or eastern side of the Desert similar evidence of destruction by earthquake of a great city is to be found in the neighbourhood of Balmeer.   The ruins have never, as far as I am aware, been visited by any traveller who had leisure to examine them thoroughly; but judging from the descriptions given by casual observers, and from the beautiful specimens of stone carving procured from the ruins by Major George Tyrwhitt, which were for some time in the South Kensington Museum, and are now, I believe, in the new Architectural Museum in Dean Street, Westminster, some of the temples must have been of rare architectural beauty.

I am not aware whether the exact date of the earthquake which destroyed the city has been ascertained, but it apparently corresponds closely with the period of the destruction of Brahminabad on the west side of the Thurr.

Very curious evidence of the gradual elevation of the land, or rather of the constant retrocession of the sea, is afforded by the traditions of the commercial community of Verawow, a small town in the district of Nugger-Parkur, at the south-east corner of the Thurr, formerly called " Pallee Nugger," or the ancient city.   Verawow is the residence of a small Rajpoot chief, whose family has ruled in the neighbourhood for many ages.   It is

---

* Vide an account of Brahminabad, published by Mr. A. F. Bellasis of the Bombay Civil Service.

evidently a place of considerable antiquity, and besides the carved stone tombs of the chief's ancestors, contains some beautiful little white marble Jain temples, which appear from inscriptions to be from 500 to 600 years old. Some of them are said to have been built by ancestors of the present trading community at Verawow, whence in after ages a colony of traders established themselves at Mandavee, in Cutch, which has since become a great emporium of trade with Western India, Persia, Arabia, and Africa. Their account of themselves, which they state is supported by documentary evidence reaching back for many centuries, is, that they were originally settled as a trading community at a spot in the north-eastern angle of the Runn, not far from Barkasir, whence they removed to Verawow, then called Pallee Nugger, more than 800 years ago; that at that time sea-going ships came with ease to the immediate vicinity of the present town, and they still show the stone posts to which the ships were moored, when the present edge of the sand-hills was washed at every tide by the waters of the sea. They add, that in consequence of the progressive shoaling of the water a great portion of their community migrated 300 or 400 years ago to Mandavee, in Cutch, as a more convenient spot for sea-borne commerce, and that since that time the water near Verawow has gone on shoaling, till now it is several generations since any sea-borne ships have been near their ancient port. The truth of the main facts of this tradition seems to be beyond a doubt.

I need not here recapitulate the evidence so ably brought together by Sir Charles Lyell on the subject of the extensive and constantly-recurring changes effected by earthquakes in this particular region, and in very recent times, as well as in remote ages. I would only remark that an extension of observations would everywhere afford the geological observer evidence of great changes in the level of the sea-shore at a comparatively recent period. Travelling westward he would find, on the coasts of Sind and Mekran, especially in the regions marked by the mud volcanoes, near Hinglay, raised sea-beaches of such recent elevation that the shells remaining on them, unfossilized, have not yet lost all trace of their original colours, while if he travelled down the coast to the east of Cutch he would find similar evidence, both of elevation and depression, on the Kattywar coast, and some clearly within the historical period, as in the traces of the ancient city of Wullabah, described by Colonel Sykes, which has evidently been submerged in the ocean and again elevated, since it was a great capital not more than 120^0 years ago. Equally conclusive evidence is found further south, in the Island of Bombay, which it has been satis-

animals living on it, before it finally assumed its present shape.

The question naturally suggests itself—What connection have these convulsions with the peculiar formation of the earth's surface in the neighbourhood of Cutch? How far do they assist us to an explanation of the mode in which either the Runn or the Thurr have been formed?

1st. As regards the Runn. It is obvious that the mode of formation of such an extensive level surface must have differed considerably from any process which we commonly see in action when the bed of the sea is raised by deposits of silt or sand, borne either by ocean currents or brought down to the coast by rivers. However still may be the waters in which the deposit takes place, the material is never uniformly spread over the ocean bed, so as to form a perfectly smooth surface. It is true, that as the coast of a river delta is raised to nearly the ordinary high-water level, a certain uniformity of surface is always observable, owing to the fact that no deposit can take place above the highest water-mark, and as soon as the deposit has nearly reached that level, there is a constant tendency to fill up depressions, as they are the only spots where fresh deposits can take place. Still, a delta always retains traces of the original mode of its formation, in the remains of the channels through which the water used to pass into and off it. But nothing of the kind is observable on the Runn, which is throughout its whole vast surface almost absolutely level, and free from even the smallest traces of water channels. It seems to me, that the constant recurrence of surface agitation from earthquakes, especially during the time when the surface is annually covered with a couple of feet of water, supplies exactly the kind of cause which would account for the uniformity of level.

We have evidence that under the action of an earthquake mounds of such sandy soil as that of the Runn melted down, as it were, into the water which then covered the Runn, and that in place of the mound there is now the usual, firm, smooth level of the rest of the Runn. There seems no reason why the same sort of process, frequently repeated, should not obliterate all traces of creeks and water-courses, and reduce the Runn to the uniform surface which we now find. The Runn is, in fact, a great basin, protected from the direct action of the sea by the rocky ridges of Cutch. and enclosed on the other three sides by higher and firmer lands, and the whole basin is subject to frequent earthquake-agitation, especially at the season when the lowest part is covered with water. The effect is, on a large scale, what would be produced by the gentle agitation of a tub half-filled with sand and water—the sand is shaken down into a nearly uniform, but slightly convex, surface, the water at the edges being a little depeer than in the centre. To a similar cause may, I think, be ascribed, beyond all doubt, both the gradual elevation of the Runn into "Put," and the corrugation of the surface of the elevated Put into the sand-hills of the Thurr. As before observed, the Put differs from the Runn only by its greater elevation, and comparative freedom from salt, and any cause which elevated the mainland in the neighbourhood of Cutch so gradually as not to cause violent disruption or cracking of the surface, would in no long course of ages convert the Runn into Put.

2nd. But the mode in which the sand-hills of the Thurr have been formed appears a much more difficult question. They are usually ascribed to the action of the wind, but I think erroneously so, for the following reasons. They differ entirely in shape from any true wind-formed sand-dunes, such as are to be seen on any sandy part of any sea-coast, and in every part of the Thurr itself. The wind-formed sand dune usually shows unmistakeable traces of one or other of three modes of formation.

(a) A sand-hill formed under the lee of some fixed obstruction, such as a rock or bush, or an older and firmer sand-hill. The characteristic form of this sand-hill is always tapering away from the fixed obstruction, which forms the nucleus, and over which is found the highest point of the sand-hill, the tail or tapering portion of which points away from, or down, the prevailing wind.

FIG. 2.

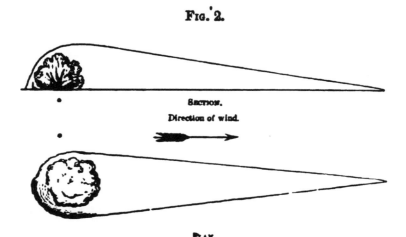

SECTION.

Direction of wind.

PLAN.

* Bush or other obstruction under the lee of which the formation of the sand-hill commences.

If the obstruction which forms the nucleus of the sand-hill be a line of obstacles, such as a hedge, or a ridge of sand originally cast up by a line of breakers, the new sand-hills formed by the wind will then appear a serrated series of tongues of sand, at right angles to the line of obstacles.   But the normal form of the original sand-hill is always to be traced in a tongue, or tongues, of sand tailing away from the obstruction which gave

FIG. 3.

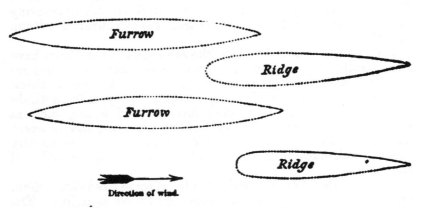

Direction of wind.

rise to the first deposit of sand under its lee, and always pointing to leeward, down the prevailing wind.

(*b*) Sand-hills formed mainly after a kind of furrowing process

FIG. 4.

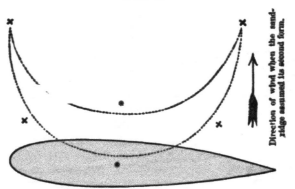

Direction of the wind when the original sand-ridge was formed.

The continuous lines show the shape and direction of the sand-ridge as originally formed. The dotted lines show its shape after the wind changed and blew at right angles to its former direction. At X the particles of sand travel continuously before the wind. At • they move over the surface of the sand-ridge, and having reached the summit, fall down at • to leeward of the sand-ridge, and do not move again till exposed by the forward movement of the whole mass. The particles at X are consequently travelling much faster than at • and •.

by a prevailing wind. The wind sweeps a flat surface of sand, and commences a furrow in any part softer than the rest, prolonging the furrow in the direction of the wind, and depositing the excavated sand in long ridges tapering in the same direction, at a little distance further down the wind. Here, as in the former case, the thickest and highest part of each sand-hill will be found to point to windward, while the tail tapers off to leeward.

(c) The third form is, when from any cause, such as a change of wind, a long ridge of loose sand is at right angles to a steady wind. The sand ridge then quickly assumes a crescent shape; the two ends are most speedily acted on, and move faster than the middle portion; the particles of sand at the ends never get effectual shelter from the wind, and are consequently moved along without intermission, while the particles in the middle of the ridge, when blown from the top, fall down under shelter to the bottom on the leeward side, so that while the particles at the two extremities of the ridge are travelling continuously, the particles in the middle travel intermittently, by falling from the top to the bottom, under the lee of the ridge, and there remaining till, by the travelling of the whole mass, they become exposed on its windward side, and being moved to the top of the mass by the wind, again fall into shelter to leeward. The accompanying diagrams may make my meaning clearer.

To one or other of these forms, namely, the long tongue of sand formed under shelter, the tongues of excavation, and the crescent form, caused by changes in the direction of the wind, may be referred all forms observable on sea-coasts, due allowance being made for disturbing causes, such as accidental hardenings of the surface and diversions from irregular obstacles; and all these forms are observable, in a minor and subordinate scale, in the little sandy hillocks, obviously due to the action of the wind, which are seen on the Thurr. But none of these forms are to be traced in, nor are they capable of being reconciled with. the features of the great sand-ridges of the Thurr. Instead of pointing or tapering off down the wind, the Thurr sand-ridges, in the southern portion where the wind is steadiest and most continuous, always run more or less in a direction E. and W., nearly at right angles to the direction of the prevailing winds, which are generally more or less northerly or southerly. Instead of the tongue-shaped form invariably present in blown ridges, the Thurr billows are long, parallel ridges of nearly uniform height, sometimes for several miles together. The highest portion, instead of being, as in blown ridges, at one end, and that to windward, is usually in the middle of the ridge, often several miles from either extremity.

Lastly, the size of the sand-ridges in the Thurr seems to me
quite inconsistent with any theory of their formation by the
agency of the wind.   I do not know what may be the greatest
measured height of sand-hills clearly formed by the wind on
any exposed sandy coast like those of Northern Europe, but
I believe they never approach to the height of between 400
and 500 feet, which is a not uncommon elevation of the Thurr
sand-ridges; and that, be it observed, not in single sand-
hillocks, but in continuous ranges of parallel ridges, each of
which maintains, for a long distance, nearly uniform elevation.
But all the phenomena of the Thurr sand-ridges are consistent
with the theory of their formation being due to undulation of
a surface like that of the "Runn" or "Put," furrowing the previously smooth surface into billow-like ridges whenever the
undulation caused a crack at right angles to the direction in
which the earthquake wave was proceeding.   We have, indeed,
a well-observed and indubitable recent example of the formation of one such ridge by an earthquake undulation in the
Allahbund, formed on the Runn by the great earthquake of
1819, and so well described from MacMurdo and Burnes by
Sir Charles Lyell.   The Allahbund, in fact, is in all respects
a perfect, outlying specimen of a Thurr sand-billow of moderate
height; and if the process which formed it were repeated, so as
to form a sufficient number of similar parallel ridges connecting it with the Thurr, from which it is now a few miles
distant, the character of the ridges would be in no respect distinguishable from those of the main portion of the Thurr.
Abundant evidence that this is no fanciful theory may be found
in the structure of the rocky ridges which wrinkle the plains
W. of the Indus.   These plains frequently consist of a surface-
bed of calcareous sandstone or conglomerate, full of marine
shells, and often apparently very little changed, except in elevation from the position it occupied when at the bottom of the
ocean.   The substratum consists of marls and clays; the surface,
sandstone, being of various depth, but often a mere shell only
a few feet thick.   This crust frequently lies in large plains

FIG. 5.

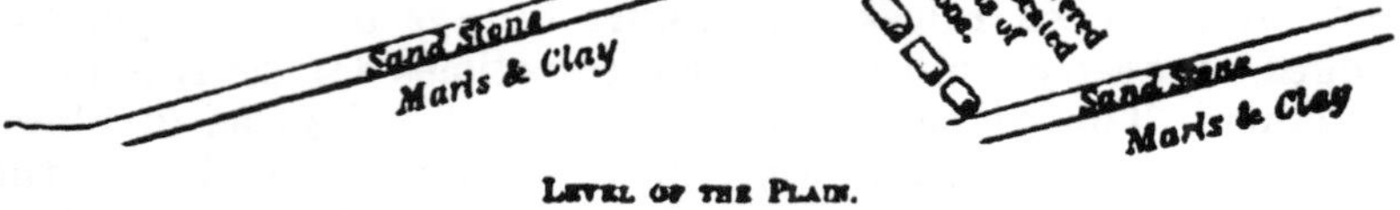

LEVEL OF THE PLAIN.

Transverse section of rocky ridges on the plains west of the Indus.

many miles in extent, and the ridges which traverse it are very generally similar in shape and size to the sand-ridges of the Thurr, or to the Allahbund; while from the presence of the sandstone crust they retain even stronger traces of the mode of their formation. A section at right angles to these ridges generally presents the appearance shown in the annexed diagram, namely, a long slope of almost undisturbed sandstone, rising to the crest of the ridge, and then a steep scarp down to the level of the plain, where another slope commences rising, while the scarp itself bears the fragments of the broken stony crust, often so little dislocated, that they appear almost capable of being restored to their original position, like the pieces of a jointed map.

But I am trespassing rather on the ground of the geologist. I must say a few words on the productions, other than mineral, of the singular region which I have attempted to describe.

Meteorology.—All the productions of the Thurr are greatly modified by its meteorology, which differs, in many important respects, very widely from that of any other part of India or its neighbourhood. There are no mountains either in the district itself or within 150 miles of it, no streams, no forests, and as a consequence there is no regular rain. It cannot be said to be a rainless district, for occasionally very heavy showers fall, and sometimes a heavy downpour continues for many days together; but the rainfall appears to depend more on electrical than geographical conditions, and little has yet been ascertained regarding the laws which govern the fall of rain there. August and September, or the winter months, appear to be the most common seasons for showers, but sometimes three or four years, and, according to tradition, occasionally a much longer period, passes without anything like a general rainfall. Some of the consequences of this scarcity of rain have been already referred to, and this is not the place to enter on any detailed description of its effects, beyond observing that the extreme dry heat acts almost as a prohibition to travelling during the summer months, not only from the effects of the direct rays of the sun during the day, but from the action of the heat or dryness, in some manner not hitherto, I believe, understood or explained, which deprives the air of its power of supporting animal life, and causes death by a process vaguely classed as "sunstroke," though it frequently occurs when the direct action of the sun is excluded, and even when the sun itself is below the horizon. I have unfortunately lost the notes giving dates and other particulars of occurrences of this sort; but the following instances will give the kind of facts to which I refer as of not unfrequent occurrence in this region. A regiment of European dragoons on a march to Ferozepoor, near the northern boundary of the Thurr, during one of the Punjaub campaigns, lost eleven men from what was called " heat apoplexy," seven of them after sunset, most of the men dying in a few minutes after seizure, without previous warning, and with a suddenness which reminded my informant rather of death by poisoning from prussic acid than from anything like ordinary apoplexy or sunstroke. About seventeen years ago a party of about 200 native travellers from Roree to Jeysulmere in the month of August lost eighty of their number in a similar manner, many of them dying at night, and some while asleep. The medical records of the force stationed at Jacobabad, on the Sind frontier, abound with facts illustrative of this cause of mortality, and I have no doubt that Dr. Forbes Watson, and others who have investigated the subject scientifically, could add much to our very imperfect knowledge of the mode in which death occurs from such causes in these rainless countries. I will only mention one more instance, as an illustration of the obstacles thus caused to travelling in the hot weather.

About sixteen years ago the late General Jacob had occasion to move with two squadrons of the Sind Horse across the desert on the north-west frontier of Sind. The weather had just changed from extreme cold to great heat, the change being accompanied by extraordinary electrical disturbance, but the heat did not prevent his marching during the daytime, and was not felt as in any extraordinary degree inconvenient by the men of the detachment, or by their five European officers; but of the horses 106 dropped, without in any case a previous symptom of weakness, and more than 90 of them never moved from the spot where they fell. But only one man suffered, a native orderly, who dropped dead while talking to his officer in the shelter of a tent after the march was over, and without any previous complaint of feeling at all unwell. No hot season passes without the police reporting many deaths apparently due to the same cause, which, though not by any means unknown in other parts of India, nowhere else forms such a barrier to intercommunication. I may remark that irrigation modifies, in some way not well understood, the conditions which lead to death from such causes, and that even fields of growing millet appear to have a perceptible effect in attracting light showers of rain from clouds, which pass over the uncultivated plain without parting with any visible drops of their moisture. It may seem a vain-hope that irrigation should ever influence any appreciable portion of the desolate tract we are considering, but during the past twenty years considerable inroads have been made by irrigation on tracts which had been for many previous ages waterless, and the marks of former human habitation, where there is nothing now to sustain human life, encourage a hope that peace and engineering science may yet do much to render these deserts more habitable.

As might be supposed, the flora of this region differs considerably from that of neighbouring provinces which are blessed with a regular periodical fall of rain. It approximates, in fact, much more nearly to the Flora of Arabia than of India. It might at first seem surprising that any vegetation should be found in a country where there is sometimes no perceptible rain for two or three consecutive years, but some compensation is afforded by the softness of the sandy soil, which allows the roots of plants to penetrate to distances far exceeding what is usual with the same plants in the lightest soils of moister countries. and probably to find sustenance at depths almost protected from the surface heat. Hence the vegetation, such as it is, is less really withered after a season or two of absolute drought than is often the case on the plains of India, after a month or two of ordinary hot weather; and cattle, sheep, and goats may be seen browsing and thriving on what is apparently withered grass, in places where the herdsman has seen no rain for a couple of years. Most of the grasses have prodigiously long succulent roots like our English couch grass, and the cattle in grazing seem to live almost as much on the roots, which are easily drawn from the sandy soil, as on the withered stalks and leaves. A good shower of rain has a magical effect on the whole surface of the desert. All sinks into the deep soft soil, and under the action of a powerful sun, in a very few days, all plants whose roots and seeds are able to stand a couple of years' baking, burst forth in a carpet of luxuriant vegetation: 3 or 4 inches of rain seem to be ample, and 5 or 6, or about a fourth of the average fall in the driest part of England, not only covers the desert with a carpet of deep rich grass, but furnishes the wells which are fed from the sand hills as by vast sandy sponges, with a supply of water which in a few of the best wells is said not to be entirely exhausted even by ten years of continuous drought. The news of a good fall of rain in the Thurr attracts from the neighbouring countries of Sind and Rajpootana thousands of cattle whose owners generally belong to the Thurr, and move their cattle about from province to province as they find pasturage and water. They are not entirely dependant on grazing, for in a good season great quantities of grass are cut and stacked as hay. The people generally refer to this as an ancient branch of industry, which had almost been forgotten during the troubled generations which preceded the British rule. Cattle lifters, disappointed in driving off herds were apt to fire

2. Coolies, who appear to be of later origin than the Bheels, but still anterior to the earliest Hindoo immigrants.

Jŭtts (apparently the pastoral brethren of the agricultural Jâts), who are said to be of Scythian origin, and hold themselves quite distinct from all the Hindoo and Mahometan races among whom they live, and who are hardly ever known to forsake their ancestral occupation as breeders of cattle. Whether from this inability to learn a new trade, or on any better ground, their name has become a synonym for a blockhead in almost every dialect of the provinces they inhabit.

Hindoos, of every tribe and caste, from the priestly Brahmins and warrior Rajpoots down to the lowest of the mixed castes, whose servile position marks the severity with which Hindoo custom visits any disregard for laws framed to insure the absolute purity of the race.

Of later immigrations, subsequent to the Mahometan invasion, numerous representatives are found in tribes of Beeloches and small communities of Affghans, Brahuis, and occasional families of Turcoman, Kurdis, and Arab origin, who appear to have been dropped by some of the waves of invasion from the West.

As one result of the remoteness and desert character of this region, all who come to it appear able to maintain their independence to a degree unknown in more favoured provinces, and to live separate without being fused into the mass of the population around them.

The various Hindoos and the Aryan races form generally the bulk of the population. Many of the tribes claim identity, both in name and origin, with Mahometan tribes among the Sindee inhabitants who cultivate the rich lands on the Indus. Their general tradition is, that when the Mahometans conquered Sind, they offered to the Hindoo tribes who were in possession of the land the alternative of holding their lands as Mahometan converts, or of exile beyond the Great Desert; and that the present Mahometan Sindee zemindars are the descendants of those who accepted the former alternative, while their Rajpoot and Hindoo namesakes in the Thurr, are the offspring of those who gave up their lands and their country to retain their faith.

As a result of their isolation, the Hindoos of the Thurr have retained many peculiar ancient customs, which have been modified in other provinces. Sometimes the difference consists in maintaining restrictions which have elsewhere been relaxed. Thus, the ancient Hindoo prejudice against anything feathered is shewn in a kind of proscription of fowls, which are bred only by Bheels and other low-caste races. One Christmastide we were anxious to have a plum-pudding, and inquired for eggs;

best huntsmen near Oomercote. They had with them a bitch of the common breed of Sindee pariah-dogs; she acted as watch or scout to the troop, and was, the Bheels said, so sagacious and wary that, till she was surprised and killed, they never had a chance of destroying the wolves.

Wild hog, so numerous in the surrounding provinces, require too much water to be found in the Thurr; and of all the tribes of Indian deer and antelopes one only, the common gazelle—which is said to subsist without any moisture beyond what is furnished by the dew and the herbage it feeds on—inhabits the Thurr.

The most characteristic quadruped is a very pretty field-rat, with large eyes and a bushy tail, and a habit of sitting upright like a jerboa. It is a sociable animal and lives in burrows, which often undermine sand-hills to an extent to make riding over them difficult. Their numbers seem to be subject to periodical increase to an enormous extent, on which occasions colonies migrate to a considerable distance, doing great damage to any vegetation which may come in their way.

Snakes are very numerous and destructive, and the sandy soil, retaining every trace of their movements, probably increases their apparent numbers; but I know of no species peculiar to the desert.

Domestic cattle, of every sort, are excellent of their kind. Horses are generally imported from Cutch or Rajpootana; but the horned cattle, sheep, and camels of the desert are all celebrated throughout Northern India, and form the great source of wealth to the inhabitants. Of late years, since peace has been restored, the people of the desert, previously known for many ages as plunderers of their neighbours' cattle, have become the great cattle-breeders for all the markets of the surrounding provinces; and a man was pointed out to me as distinguished among the cattle-farmers in being the first man who, in modern times, had brought to the Oomercote bazaar a bill of exchange from a banker at Ahmedabad, in payment for cattle honestly bred and sold at a cattle-market 300 miles from their original pasturage. These cattle now often find their way by the Ahmedabad Railway to Bombay.

Ethnology.—As might be expected from the security afforded by its desolation, the Thurr contains specimens of every race which now or in former days has inhabited the regions round about. Space admits only of a bare examination of a few of the more prominent :—

1. The wild Bheels, who claim to be Autocthones, and a drop of whose blood is essential to ratify every very solemn ceremony of the Rajpoot dynasties.

2. Coolies, who appear to be of later origin th
but still anterior to the earliest Hindoo immigran
   Jåtts (apparently the pastoral brethren of th
Jåts), who are said to be of Scythian origin, a
selves quite distinct from all the Hindoo and M
one whom they live, and who are hardly eve

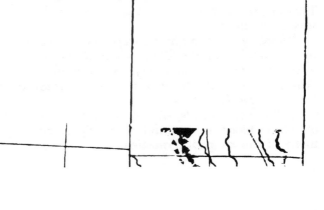

5

Yellow River

but it appeared that there were no fowls in the country, and one local official, after careful inquiry, reported that we were 50 miles by estimation from the nearest egg, which, it was supposed, might be found in the possession of a tribe of Bheels at that distance from us. Sometimes, on the other hand, the preservation of ancient customs proves the comparatively recent introduction of some of the very general and inveterate Hindoo restrictions existing in every other part of modern India. Thus the Desert Rajpoots, of the Soda tribe, till lately, made little objection, when in the field, to eat from the same dish and of the same food as men of other races. I have myself been asked to partake with a Rajpoot of his hunting luncheon,—a thing which could never possibly happen in any other part of India.

It is only one tribe of Desert Rajpoots, the Sodas, who have maintained what was clearly once the universal Rajpoot custom of bringing up all their female children and abstaining from infanticide. Hence for ages past they have had almost a monopoly of the supply of wives to the best families in Rajpootana and Cutch, including most of the royal houses. The demand is so great that large sums are paid to Sodas, to induce them to give their daughters in marriage to chiefs of far higher rank and greater wealth than themselves: 1000*l*. is no uncommon sum to be so paid, and the good fortune of a Soda is reckoned by the number of his daughters, and not, as among other Hindoos, by the number of sons. Soda wives are celebrated not only for their beauty but for their intelligence. The Soda chiefs often pay a round of visits to their sons-in-law in Rajpootana, but are said never to settle there, but always return to their native sand-hills.

V.—*Notes on a Journey through Shantung.* By J. MARKHAM.

Read, February 28, 1870.

UNTIL very recently the province of Shantung, in the north of China, has been a closed book to the civilised world, but now, owing to the travels and researches of the Rev. Alexander Williamson, this rich and most interesting country is better known. As, however, I considered that there was yet a vast amount of information to be gained I made a tour of the province in the early part of 1869, visiting the principal cities, seaports, harbours, and silk growing districts, and had the honour of reporting thereon to Her Majesty's Government. There were, however, many subjects of interest not embodied in my official Report, and which this Society may deem worthy of

perusal. It is with much pleasure, therefore, that I furnish this
Paper, which I trust may not be considered devoid of interest.

Leaving my post at Chefoo—the seaport opened to foreign
trade in lieu of Tungchow, the city named in the treaties, as
that port was found impracticable owing to the gradual filling
up of the harbour—on the 24th of February, 1869, I took a
south-east course, passing through the Tsihya valley, a district
in which the ailanthus silk (*Saturnia Cynthia*) is largely cultivated. This district is mountainous, but the valleys are most
fertile, producing cereals of nearly every description, and also
some cotton. The hills 'are studded with small plantations of
the stunted oak, on the leaves of which the wild silkworm
feeds. The district is watered by a stream of considerable size,
but unnavigable; it runs east and west, and empties itself into
the Fooshan Ho, another river whose mouth is some six miles
west of Chefoo. The first city of any importance I passed
through was that of Lai Yang, lat. 37° long. 120° 55'; it is a
walled city with a population of some 50,000, including the
suburbs, which are large, extending in a long continuous street
from the north and south gates. The walls of Lai Yang appear
to be of very great antiquity, but are in excellent repair, a
rampart about five feet wide surmounts them. The city, but
more especially the suburbs, has a thriving appearance, and the
people give the idea of a well to do community. All the shops
of importance are in the suburbs, as are also the looms where
the ailanthus silk is manufactured into the strong fabric termed
Pongee. In the vicinity of Lai Yang a large quantity of wax
is manufactured from the excrement of a small insect which
feeds on the leaves of the La Shoo tree, and from which tree
the insect takes its name. In the streams and gorges of this
district, after heavy rains, is found gold in considerable quantities, which finds a ready market in Lai Yang. The Chinese
are allowed to wash for gold, but are restricted from digging, or
mining in any way; from the quantity obtained from these
washings it is presumable that the precious metal abounds in
the province; I have myself seen large nuggets which were
picked up in the gorges and ravines of the Tsihya district.
The main streets of Lai Yang are spanned by large paelous, or
monumental arches, some of great age, one or two bearing the
date of the Emperor Tae Ting of the Yüan dynasty (A.D. 1324).
They are formed by an upright pillar of stone on either side of
the street, of some 15 feet in height, resting on a pedestal
some 4 feet high; a horizontal slab connects these pillars, and
the whole is surmounted by a finely carved piece of stone arch-
work. These paelous are generally erected to commemorate
the good works of some deceased citizen.

Proceeding on from Lai Yang my course was S.S.E. through an extensive and highly cultivated plain which trends away to the south-west and north-west through Shantung and Chihli. After travelling some 90 li * we arrived at the seaport of Kin Kia; this is an unimportant town with a population of about 3000. It is situated to the south of a small range of bills; the harbour runs north-east and south-west, and is a safe anchorage for small craft. Kin Kia was once an important port, but the opening of Chefoo has drawn off nearly all the trade. From Kin Kia my route lay due south through the same fertile plain, and over the Laou Shan Hills, a range of mountains which stretch away to the eastward. These hills are the resort of numerous pilgrims, and are said to be rich in precious stones, such as rubies, amethysts, &c. Buddhist and Laouist temples are scattered everywhere over the range, and the priests attached to them are said to derive a large income from the sale of these stones, and the medicinal herbs and roots which likewise abound. Amongst other plants I heard of rhubarb. Proceeding on I came to the walled city of Tsi-mi, a Hsien of the second class, with a population of about 18,000, including the suburbs. The city presents a very poor appearance, though there are numerous and very ancient paelous across the main street; one of these paelous was erected in the reign of the Emperor Shih Tsu, of the Yüan dynasty (A.D. 1280). The district of which Tsi-mi is the Hsien, is 160 li north and south, and 100 li east and west, and contains a population of about 150,000, including the city. The chief products are pigs, fruit, bean-cake, pulse, oil, animal wax, and willow baskets. The fruit trees of this district are very varied, the principal being apples, pears, dates, and walnuts. Tsi-mi is 60 li inland, its seaport being New-kow. I visited this place and found it to be a small unwalled town with some 2500 inhabitants. I noticed warehouses of large dimensions, but I saw very few signs of trade. The harbour is a safe one for small craft, and runs north and south. ·Leaving Tsi-mi I travelled south-west, and on the 3rd of March arrived at the city of Kyau-chow, lat. 36° 16', long. 120° 10', crossing the Ta-Ko-Ho by a flat stone bridge about 20 li from the city. The Ta-Ko-Ho is a rapid stream running N.N.W. for about 50 li, when it joins the Kyau-Ho near the Pi-mo Lake, thence the river runs north and empties itself into the Gulf of Pechihli near the city of Lai-chow-foo. During the reign of the Emperor Kiang Hi a canal was attempted in order to take the grain junks to Peking and obviate the necessity of rounding the Shantung promontory, it was to have

---

* About 3 Li to the English mile.

gone through the province taking the course of the Ta-Ko-Ho for some distance, but insuperable difficulties prevented its completion. The city of Kyau-chow was formerly the most important centre of trade in the east of Shantung; it received its supplies from the south by the sea, and distributed its goods all over the country, but now, owing to the filling up of the harbour, from which it is distant at least 18 miles, and probably also owing to the opening of Chefoo, the trade has greatly fallen off. Ta-pu-tur is now the port of the city; it is six miles distant from it, and is eight miles up a creek, which is nearly dry at low water. Between the mouth of the creek and the anchorage extends a mud bank for nearly four miles, which is also dry at low water. Ta-pu-tur is very unhealthy; it is situated in a marshy plain full of lagoons, and fever and ague are very prevalent. The walls of Kyau-chow are not more than two miles in circumference and about 30 feet high; very old, but in excellent repair. There are three gates, south, east, and west, protected by bastions. The suburbs are extensive, and also surrounded by a wall, much slighter than the city wall and more recently constructed. The city itself is full of large houses, the residences of wealthy gentry and literati. The district of which Kyau-chow is the chow is 180 li north and south and 95 li east and west, containing a population of about 200,000. The country around Kyau-chow, as far as the eye can reach, is a dead flat; the products much the same as in the vicinity of Tsi-mi. From Kyau-chow I travelled west and by north through an undulating and highly cultivated plain watered by the rivers Wen and Wei, which have their source in the Yeh-shan mountains about 300 li to the west of my route. The plain is studded with small towns and large villages, and innumerable orchards of pear, apple, and plum trees; the walnut, chestnut, and persimmon trees also abound, while the tall and stately silver poplar, towering above the others, tends much to enhance the beauty of the landscape.

Kau-Mi, the next walled city I passed through, is prettily situated in a sort of wooded basin, the plain rising above it on all sides. The walls are in good repair; it differs little from other Chinese cities, and with the suburbs contains some 10,000 souls. Tobacco is largely cultivated in this vicinity and brought to Kau-mi, where it is packed in bales and distributed throughout the country.

On the 6th of March I reached the large and important city of Wei-hsien, my route being over what had once been a carefully constructed Imperial road, but which is now in utter decay, and nearly impassable in wet weather; it leads through an undulating and highly cultivated plain, studded with numerous villages and prettily wooded. Some 20 li s.s.w. of Wei-hsien are extensive coal-fields; they lie in the plain with hills to the south and west about six miles off. A great number of pits have been opened, but only ten are now being worked. These pits or shafts are from 15 to 30 feet deep; the miners work on until the water rises over the seam, when the pit is abandoned for another, their only means of getting the water out being by large skin buckets, holding three gallons each, which are hauled up by a huge windlass; hence it is probable that the finest quality of coal is missed. The whole district is honey-combed with old workings. The Wei-hsien coal is principally anthracite, burns bright and clear, with scarcely any ash, and throws out a great heat. The price of the coal at the pit's mouth is about 60 cash a picul, or about 4d. for 130 lbs. The cost of conveying this coal to the city is from 125 to 200 cash a picul, or say from 8d. to 10d. for 130 lbs. The means of transport being carts, drawn by from 3 to 6 mules, wheel-barrows, and donkeys. The carts carry from 6 to 20 piculs, wheelbarrows from 2 to 10 piculs; these are propelled by one man pushing, another in front, in a sort of shafts, or one or more donkeys or oxen, in front of all attached by traces; sometimes with a fair wind a huge square sail is hoisted. For the man behind the work is terrific, and seemed to me to be the very acme of human labour. These wheelbarrow men seldom reach the age of 40 I was told. They are confined to a certain class, and commence the work very early.

The city of Wei-hsien, lat. 36° 40', long. 119° 16', is a first-class Hsien, enclosed by high substantial walls, 8 miles in circumference, and surrounded by a deep moat. The ramparts are 50 feet in height, and 12 broad on the top, with guard-houses all round at every 30 yards; the gates are protected by bastions, and the north gate doubly protected by a strong square tower, with two tiers of ports for guns. A portion of the eastern suburbs is enclosed by walls, and forms a separate city of itself; it is separated from Wei-hsien by the River Peh Tang. Wei-hsien is one of the large emporium cities of the province, and the entrepôt for most of the merchandize landed at Chefoo. It has extensive iron foundries, and great quantities of hardware are made and distributed throughout the province. Although there is iron in the vicinity no mine has been opened, and all the metal used is imported from abroad. The population of Wei-hsien, including the suburbs, is quite 100,000, while the district of which it is the chief city contains another 100,000. To the north-west of Wei-hsien, and on the sea-coast, distant about 95 li, is the sea-port of Kia Ying. The harbour is shallow, and exposed to the north; but I heard that numerous junks are laden with coal and other products at this place.

Tsing-chow-foo, the next city of importance which I visited, is only 140 li west of Wei-hsien, and is the centre of a large silk-producing district, and of a large local trade. It differs little from other Chinese cities which appear to be nearly of the same age. Tsing-chow-foo, besides some fine Buddhist and Taouist temples, has one handsome edifice erected in memory of Confucius, and a splendid Mahomedan mosque, to which is attached a school, where children are taught Arabic. There are over 12,000 Mahomedan Chinese in this city, the entire population being 70,000. Tsing-chow-foo was formerly the capital of Shantung, and the old Tartar city still exists, about 1½ li to the north-west. The roads in the vicinity of the city have been formed with the greatest care, and everything indicates former grandeur; but, as in every portion of the province, all has been allowed to fall into decay.

The plain in which Tsing-chow-foo is situated is most fertile, and is reckoned the healthiest in the province, the people being famous for their longevity. At the back of the city rises a fine range of hills, called the Tae Shan, which trend away to the southward. Over 1000 families are employed in the manufacture of silk fabrics, some of which are of very superior quality to the ordinary products of Chinese looms. I made inquiries as to the obstacles which prevented silks being brought in larger quantities to Chefoo, and what was the reason for the high prices charged at that port, and I was told that all sales had to be made through a "middle man," who is responsible to the local authorities, to whom he had to report all transactions. When an attempt is made to buy silk for Chefoo, the authorities double the squeeze, rendering it impossible to lay down the article at a reasonable rate. This system is common throughout the province. On several occasions the country people and merchants at large cities expressed to me their desire to trade directly with foreigners if they were allowed, but they said that it was hopeless to expect it.

About 25 li west of Tsing-chow-foo are the tombs of the kings or chiefs of the ancient tribes who inhabited the eastern portion of Shan-tung, the same tribes that the great Yu instructed in the arts of tillage and pasturage in the year 2205 B.C. These tribes continued down to within comparatively modern times. The tombs are large mounds of earth, enclosed by what originally had been a high mud wall. Tablets have been erected and renewed by the different emperors; one of them dates as far back as the Emperor Nan's reign, B.C. 314, some of the Tsin dynasty, A.D. 265/419, and many of the Yang and Ming dynasties, A.D. 618/1628. It is a source of regret to me that I did not obtain sketches of these tombs, or of other objects of antiquarian interest I met with, but I trust, in the course of a short time, to get photographs of all, which I shall have much pleasure in presenting to the Society.

I next entered the Laou-foo Valley, passing the walled city of Che-chuen at the entrance. It is important from its vicinity to the coal-mines of Poshan. Here we found thousands of tons of coal stored, and, as I proceeded down the valley, the road was impeded by carts, wheelbarrows, mules, camels, &c., all laden with coal and coke, and I was frequently delayed for some time by the road being blocked up with traffic. The Laou-foo Valley is the chief of the great coal-producing districts of Shan-tung. It is in long. 117° 56′ E., beginning about lat. 36° 50′, and extending south to lat. 36° 30′. The valley runs north and south, and the hills on both sides are perforated with coal-pits. Several varieties of coal are produced. Forty li down the valley are the cities of Poshan and Yen-shih-hsien, only separated from each other by the Laou-fu-ho, a stream which rises in the court-yard of a temple some 10 li distant, where it jets up in a fountain. The mines being situated in the side of the hills, are much easier worked than at Wei-hsien. The coal is said to lie in thick beds, and I saw some very large blocks extracted. The pits are worked in a similar manner to those described before, except that the water is drained off at the foot of the hill into a well, whence it is drawn up in skin buckets. A moderate Government tax is levied, but the squeezes of the local mandarins are said to be very exorbitant. A Shanse merchant who was working these mines told me that, were he allowed, and not charged the local squeezes, he could lay coal down in Chefoo at four dollars fifty cents per ton, and make a good profit. At Poshan are extensive potteries; glass is also largely manufactured at Yen-shih, and forms a considerable article of commerce. The whole of the Laou-foo Valley is rich in minerals; iron in large masses occurs, but is not worked. Some years ago a silver-mine was opened, but soon closed by the authorities, and its working is now prohibited. Saltpetre is likewise produced. Leaving the valley, I once more took the high road, and travelled towards the capital of the province Tsi-nan-foo, which I reached on the 15th of March. It is in lat. 36° 50′, long. 117°, a large and important city, enclosed by high walls in excellent repair. The suburbs are large around each of the gates, another wall surrounds them, and the whole is again encircled by a mud wall or weitage, making the circumference of the defences at least 85 li. The city contains many springs of line water, and in the north part a lake, called the

Ta-ming-hu, in which are artificial islands containing tea-houses,
which are the resort of the citizens in the summer months.
The springs outside the west gate are an interesting feature
connected with the city. They gush up about two feet high,
and fill the moat which surrounds the city with the purest of
water. A large temple, in which are held annual fairs, has been
built near the springs. Fine fish abound in the moat, so that
Tsi-nan-foo could well stand a protracted siege. The streets
are broad and well paved, shops handsome and well stocked.
The city lies at the foot of a range of mountains, and a plain
slopes away to the north, towards the New Yellow River. Besides several fine temples, the city contains the palace of the
viceroy, and the Yamên of the Fan-tae (treasurer), Neitae
(judge), and the chief examiner, Hu-tae, besides several other
fine official residences; all these are situated in extensive and
thickly-wooded grounds. Here is also a new Roman Catholic
cathedral, erected by the Bishop of Shan-tung, this city being
the head-quarters of the Roman Catholics of the province,
numbering, I was told, about 12,000. A large and handsome
Mahomedan mosque is also built in the western suburbs, and
another in the north part of the city. I was told that Tsi-nan-
foo has over 20,000 Mahomedans. The range at the back of
the city contains both coal and iron, and I visited a hill, about
10 li to the east, which was one mass of iron-ore, so magnetic as
to affect my compass held some way off. The north-east part
of the city is very low and damp, and is considered extremely
unhealthy; hence, during the summer months, numbers of the
inhabitants resort to the temples, which are scattered all over
the mountains at the back of the city, and are most prettily
situated. The plain between the city and the Ta-tsing-ho
(whose course has now been taken by the Yellow River), a
distance of 16 li, is inundated twice a year. The year preceding that in which I visited it, the water came up to the city
walls. I went to the old port of Tsi-nan-foo, called Lu-kow,
16 li from the city, on the Ta-tsing-ho. It bears traces of
having once been a fine and important place, but is now in
decay, owing partly to the visitation of rebels, and partly to
the floods. I noticed the ruins of large warehouses and fine
dwelling-houses. The new course of the Yellow River, formerly
the Ta-tsing-ho, at Lu-kow, appeared to be about 360 yards
broad from bank to bank, but the breadth of water at this
time was not more than 200 yards. The stream appeared very
rapid. I saw no junks of any size upon it. Lu-kow has been
given up as a port, and Tsi-tung, 100 li to the east, substituted
for it. Some distance above Lu-kow is an old stone bridge,
which formerly spanned the Ta-tsing-ho, but which is now,

owing to the increased body of water, in mid-channel. From Tsi-nan-foo I took the imperial high road from Nanking to Peking, and travelled on it south, along the foot of the Tai-shan range, to the city of Tai-ngan-foo, distant 160 li from Tsi-nan-foo. I noticed that both hemp and tobacco were largely cultivated. The scenery here is very beautiful. I passed through a lovely valley, with the Tai-shan on the left.

Tai-ngan-foo is a walled city situated at the foot of the Tai Shan, the sacred mountain of China, and the highest of the range bearing that name which stretches between Tsi-nan-foo and this city. In the north part of the city is a magnificent temple, dedicated to the mountain, which occupies the greater part of the north of the city. This building is situated in a fine park of 25 acres. Some of the trees, composed principally of yews, cedars, and cypress, are of very great age, having been planted by emperors of the Sung, Yüan, and Ming dynasties— 960, 1628 A.D. The main temple is a large hall, 120 feet long by 50 broad. It contains, facing the entrance, a huge statue representing the Emperor Shun sitting enthroned in a massive chair. Shun is said to have dedicated the Tai Shan to the God of Heaven, and sacrificed thereon a burnt offering to the Supreme Ruler, during his first tour through the empire, when acting as viceroy for Yaou, in the 76th year of that emperor's reign; this would be B.C. 2281. Shun succeeded Yaou in the year 2255 B.C. The walls of this magnificent temple are covered with a panoramic painting, really well executed, representing an imperial procession, white elephants, camels, and other animals, fabulous and real, are depicted. The painting commences on the east side wall, and continues round along the north, or back wall, finishing on the west wall. The entrance to the hall is on the south side, it is composed of a succession of gates along the whole length, which can be opened and shut at pleasure. A terrace with white marble balustrades, the length of the temple, and some 100 feet broad, with a flight of marble steps, forms the approach. Beyond this is a courtyard, in which were assembled, during the time I was there, at least 70,000 persons, pilgrims to the sacred mountain from all parts of the empire. The front verandah of the temple has eight fine marble pillars, carved with dragons, 10 feet in height, and 2 in diameter. The roof of the temple is tiled with the green and yellow imperial porcelain tile, the ceiling beams and supporters of roof inside are most beautifully carved, and are of cedar-wood. From the temple I made the ascent of the mountain by a road 12 miles in length, consisting of a succession of flights of steps. It commences outside the north gate, and leads up a gorge, at first with a gentle ascent, but it gets gradually greater, until at last it approaches the vertical, and becomes most laborious. The road for the first 2000 feet is lined with handsome cedar and yew trees, but beyond this altitude, for the next 3000 feet, these are replaced by the common flat-topped fir. Numerous small temples are erected on either side all the way, and tablets put up by various emperors, viceroys, and other high personages, occur frequently. Some of them are of very ancient date, the characters being nearly obliterated by time and the number of rubbings taken from them. There are also many inscriptions by emperors of the Chaou and Han dynasties, B.C. 1112—76, and of the Tsin and succeeding dynasties, A.D. 265—1821, the latest being that of the Emperor Tao Kwan, 1820. This emperor also erected a tablet near the summit of the mountain. The scenery along the whole ascent is grand beyond description The mountain is bold and rugged, with numerous ravines and huge boulders, all visible from the road, which is sadly in need of repair. I was nearly eight hours making the ascent, and only stopped once for ten minutes from the time I commenced until I reached the summit. At an altitude of 4500 feet is the large Taouist temple, the Laou Mo Meiou, where women who are barren come to offer sacrifice. The name of the temple signifies " Holy Mother." The grand gate is only opened once each year by the viceroy in person, or by a high official deputed by him, who takes out the offerings deposited there, which sometimes amount to large sums. On either side of this temple are smaller ones; the main building is tiled with glazed yellow porcelain tiles, the two side temples with bronze tiles, and a small temple in the centre of the court which contains the figure of the " Queen of Heaven" most elaborately carved and gilt, is tiled with brass tiles. The main entrance is a covered gateway, and the tiles are of bronze and brass. In the court-yard, and facing the Queen of Heaven's temple, are two handsome tablets of solid bronze, 14 feet in height, 3½ in breadth, and 6 inches thick, most beautifully carved with imperial dragons. They were erected by the Emperor Kiang Leong, A.D. 1736. In front of them are two handsome bronze vases, also elaborately carved. At the back of the temple are ancient inscriptions in the rock, one said to be by the Emperor K'ai Yao, of the Tang dynasty, A.D. 681; others by emperors of the Sung dynasty, while one is dated as far back as the Emperor Tang, sometimes called "Ohing, the Completer," B.C. 1766. The summit called the Yu-hwang-shang-ti is 500 feet above the temple, making the altitude of the peak 5000 feet above the temple at the foot of the hill. Since Shun dedicated this Tai Shan to the true God, it has been held as the sacred worshipping-place of the emperors of each dynasty. All the temples of the Tai Shan are Taouist, and the priests the most dirty, degraded-looking creatures imaginable, dressed in a dirty robe of coarse yellow cloth. The view from the summit of the Tai Shan is most grand; to the north-east and north-west you look down upon range after range of mountains, and to the south-east and south-west the plain in which the city of Tai Ngan is situated is mapped out; in the distance are other cities, and away to the south-east the river Ta-wan Ho is visible, winding its way amidst groves of fine timber. The descent of the hill was made much more rapidly, but we were importuned by beggars all the way down; these unfortunates have their homes in holes and caves all along the road, and solicit alms of the pilgrims who visit the sacred mount. Many great men have attempted to ascend and failed, amongst them Confucius, who only reached half-way; a temple marks his halting-place, and is conspicuous from the others as having nothing approaching to an idol in it. At the foot of the hill is a Taouist nunnery, and also a Taouist monastery; in the latter I was shown the skeleton of a Taouist priest, who ruled over the temple in Kiang Loong's reign, and was famed for his piety; he died in the year 1744, and his remains were embalmed, or rather dried; the skeleton, with the skin on it like parchment, is in magnificent prese- vation, it is seated on a kind of throne in a vault, and dressed in the yellow silk robes of a Taouist abbot. Outside the west gate of the city is a cast-iron Pagoda in the midst of the ruins of a temple. I was told this Pagoda was erected in honour of the Empress Min, wife of the Emperor Seang 5th of the Hea dynasty, B.C. 2146, by a succeeding Emperor Shao Kang, B.C. 2079. It is a curious old structure, 40 feet in height, and apparently one solid piece.

Leaving Tai-ngan, I proceeded towards Kio-fu, the city of Confucius; the country I passed through being fine, undulating and very·beautiful, and full of historical interest. My route was due south, and, some 50 li from Tai-ngan, I crossed the Ta-wan Ho by a fine stone bridge. It is a broad stream, but not navigable. The fruit-trees were very numerous, large orchards of date-trees, besides apple and pear, we saw every-where. We found numerous fine bridges over the dried beds of what had once been considerable streams. Some 20 li outside Kio-fu, we came to the site of Confucius' lecture-hall; nothing now remains but the ruins of what once had been a fine wall, enclosing some five acres of ground, full of splendid yew and cypress trees. The foundation of the hall is still standing, and a stone urn marks the place. There is the trunk of an old tree still standing, on which the fabulous Sphinx is said to have alighted in Confucius' time, and foretold the greatness of the Sage.

Some distance further and towards the city, from which it is distant about 1½ miles to the north, is the burial-place of the Confucins family. The approach is through a fine avenue of cypress and yew trees, planted by Emperors of different dynasties, the last by some of the Ming dynasty. Along this avenue are some very handsome arched bridges, spanning imaginary streams, also elaborately carved Paelous placed at intervals along it. Halfway to the great gates of the cemetery are two pavilions, containing tablets erected by the Emperor Wan-Hi, of the Ming dynasty, A.D. 1573. The cemetery is enclosed in high walls, and consists of some 45 to 50 acres of magnificently wooded grounds. Through the entrance-gates the road turns to the left, and passes between rows of stone lions, elephants, &c., larger than life size, up to a hall where the descendants of the Sage worship and sacrifice once every year. Near this hall, and to the right, is the trunk of an ancient tree, said to have been planted by Tze-Kung, one of Confucius' disciples; and close to it is a handsome pavilion, erected by the Emperor Kiang Loong. Further on, we came to the tomb of Confucius' grandson, Tsi-Sze, the author of the 'Chung Yung,' who is said to have been the preceptor of Mencius. He bears the title "Philosopher Tsi-Sze, Transmitter of the Sage." Advancing on a little further, between two colossal figures in stone, holding scales, we came to the tomb of Confucius. It consists of a high mound of earth covered with brushwood; in front is a stone urn and table, and a huge tablet engraved with the seal characters, giving the names and titles of the Sage. This tablet is 25 feet high, by 6 feet broad. To the west of the tomb is a building, erected on the spot where Tze-Kung is said to have mourned for six years, in a hut of reeds, for his departed master. This disciple is greatly respected by the Chinese, and many temples have been built to his memory. On either side of the tomb are two other mounds, beneath which lie the remains of Confucius' son and mother. These four mounds are the only tombs within the sacred precincts, consisting of an enclosure of about four acres; but all around and within the 50 acres, to the west, are the tombs of the representatives of the Sage, and to the east those of the descendants of less importance. It was very easy to trace on the different tablets the generations from the Sage. All the tombs of the representatives are distinguished by higher mounds, with figures in stone of men and animals forming the approach.

On the 23rd of March I reached the city of Confucius, Kiofu-hsien. This city is chiefly inhabited by the descendants of the great Sage, eight out of ten families bearing his surname; the magistrate's office is hereditary in the family. The city is walled, and differs in nowise from other Chinese towns, except that, besides the usual four gates, it has a second south gate, which is only opened to an Imperial visitor. This gate is in front of the temple of Confucius, and leads directly to it, which, together with the Ducal palace of the Sage's descendants, occupies the greater portion of the north and west of the city. Both edifices are situated in magnificently wooded grounds, those of the temple covering some 35 acres. The temple is in the west, and the chief part of it stands on the spot where Confucius lived. The plan of the temple is somewhat similar to other buildings of this class in China, but on a far grander and more superb scale, and I have never seen anything to compare with it in any part of China. On arriving at the inn, I sent my card to the Representative of the family, intimating a desire to see him; but I confess I little expected that honour, considering the treatment I had received throughout my journey at the hands of the Mandarins; to my surprise and gratification, however, I received a reply that the Duke would see me with pleasure, and I therefore proceeded to the palace, where, on entering the large gates, I was met by a high official and proceeded with him down the avenue, through several courtyards lined with handsomely-dressed retainers, to the gate of the inner palace, where the Duke, with several members of his family, awaited me. After the customary greetings, the Duke ushered me into the reception-room, but, not pausing, conducted me into his private study, where he invited me to be seated. This study was a small room, the walls lined with books on shelves. Here many relics of the Sage were pointed out to me, such as bronze urns, tripods, censers, and ancient manuscripts. I was particularly impressed with the Duke's manner, which was pleasing and gentlemanlike in the extreme. He is about 22 years of age, slightly deformed, and not more than 4 feet 8 inches in height. His countenance, however, is most pleasing and intelligent. His title is Kung-Yeh, equivalent to that of Duke in England. He receives a large pension from the Government, and ranks immediately after the princes of the blood. A Viceroy, on coming into his presence, has to make the nine kow tows, or bows, to the ground. His manner was entirely free from reserve, and he seemed most desirous for information, as were also his immediate attendants, who were all connected with him by blood in some degree. On leaving, the Duke accompanied me as far as the outer gate, and expressed his gratification at having made the acquaintance of foreigners, none having visited him before. On my return to the inn, shortly after the interview, I found that a high officer had been sent by the Duke to inform me that the gates of the temple would be opened for me; this was a mark of great favour, as the day being the anniversary of the death of one of the representatives of Confucius, the temple was closed. We accordingly proceeded to this splendid edifice, accompanied by several members of the Duke's family. The grounds are very spacious and well wooded, and enclosed by high walls. They contain numerous temples, pavilions, and tablets of every date. The main temple is in an oblong enclosure, and is 12 halls deep, each hall having a square to itself, shut off by massive gates; these squares are full of magnificent tall, old cypress-trees, and the sides of the avenue are crowded with tablets in honour of the Sage; every dynasty is represented, therefore many of these tablets are of vast interest and importance.

On the left of the entrance stands a cypress, or rather the trunk of one, said to have been planted by Confucius himself, and certainly its gnarled trunk testifies great age. Close to this is the place where Confucius taught, marked by a large pavilion wherein, on a marble tablet, is engraved a poem in praise of the Sage, composed by the Emperor Kiang Loong (A.D. 1736). The great hall lies third from the entrance; it is two stories in height, 160 feet long, and 88 feet broad. The upper verandah is supported by 34 pillars of white marble, 25 feet high, and 3 feet in diameter, each one solid block, those in front white, and most elaborately carved with the traditional dragons chasing the fly, and those at the side alternate black and white veined marble. The tiles of the roof are of yellow and green porcelain, the eaves beautifully carved and painted, as is all the woodwork. Within this building is a statue of Confucius, in a sitting posture, about 12 feet in height. It represents a strong, thick-set man, with a fine full face and large head. He is attired in yellow silk, handsomely embroidered, and has a square college cap on his head, with strings of beads falling in front and behind to a level with the neck. The seat is a throne raised some 6 feet from the ground, and surrounded by yellow satin curtains magnificently embroidered in blue and gold. The statue is in the attitude of contemplation, the eyes looking upwards, the hands hold a scroll, a slip of bamboo, which in those days was used for paper. On a tablet over the statue is the inscription, "The most holy prescient Sage Confucius,—his spirit's resting-place," while from the ceiling are suspended other tablets to his honour, all in extravagant praise. In front of the statue is a high table containing relics of the Sage, and presents made by different Emperors to the family—amongst them a bronze censer, bearing on the lid the date of the Shang dynasty, B.C. 1700; some magnificent enamels, such as are not seen in the present day; also a rosewood table of very solid make, which I was told was used by the Sage himself. On close inspection signs of its great age were apparent, but it is in excellent preservation; likewise a clay dish, said to be of the Emperor Yaou's time, B.C. 2300; and two bronze elephants, dating from the Chow dynasty, B.C. 1122-235.

In the second hall from the entrance are four marble tablets, erected by Kianghi A.D. 1622, with characters signifying "the teacher of 10,000 ages." In this hall is a marble slab with an engraving of Confucius, said to have been taken during his lifetime, and to be an excellent likeness; also, two other engravings, but of more modern date. The engraving first alluded to is nearly obliterated by age and from the number of rubbings taken from it; the other two represent the sage at different periods of his life, and are perfectly distinct. I obtained rubbings of all three. Here are also 120 marble slabs let into the wall all round the hall. Each slab has an engraving representing some scene, and the whole forms an illustrated life of Confucius, with explanations at the side. These were most interesting; for, apart from their great antiquity, they gave an idea of the houses, carriages, dress, and furniture of that period. I obtained rubbings of all these, but I regret to say that I left them in China. The other balls are erected in honour of Confucius' father, mother, wife, son, grandson, and some of his favourite disciples. Each contains a tablet setting forth the names and titles of the individual to whom it is dedicated. Confucius' father was a man of note in the empire; he governed the cities of Yenchow-foo and Tsou-hsien. To the east of the temple is a huge slab of black marble, some 25 feet in height, on which is engraved the genealogical tree of the family down to the present generation. Near it is a well from which the Sage drank. The grounds of the Confucian Temple are full of objects of the utmost interest to the antiquarian. Tablets of every age are erected throughout. The temple has been renovated within the last six years, and is now resplendent in paint and gilding. The ceilings are really grand, the blue and gilt dragons which adorn them being masterpieces of carving. The balustrades of verandahs to the temple, and the steps leading up to it, are of pure white marble, and exquisitely carved.

The grounds of the Ducal Palace adjoin those of the temple, and within them is the house in the walls of which were found the classics, hidden there for fear of that destroyer of learned men and literature, the Emperor Tsin, B.C. 212. As an instance of the respect and veneration in which the great Sage Confucius is held by the Chinese of all classes, I may mention that when the rebels occupied Shan-tung, and were devastating the country around Kiu-fu, they approached the city, and on being asked if they would destroy the temple of the great Sage they replied that all they wanted was to kill the unjust mandarins, and on being informed that Kiu-fu was governed by mandarins of the Confucian family they at once departed, doing no damage whatever even to the cemetery, although thousands of the country people had taken refuge within the sacred precincts. They entered the grounds certainly, and, it is said, murdered numbers of the refugees, but they carefully abstained from damaging the tombs therein.

I next visited the temple erected in honour of Yên-hwuy or Tze-yuen, a favourite disciple of Confucius, situated just outside the north gate; the grounds are very extensive and handsomely wooded. The temple is very similar to the Confucian, only on a much smaller scale. In it and about the grounds are some very ancient tablets, erected by imperial visitors. In the great hall is a statue of Yên-hwuy seated, like his master, on a throne with handsomely-embroidered silk or satin curtains. It is surmounted by a tablet with the inscription—

" The perfect man who attained to Holiness equal to the Holy Man."

On a table in front of the throne are some very handsome old bronzes. Near the temple is an exceedingly fine silver-pine tree, over 200 feet in height and 15 in circumference. It appeared of enormous age, but in full life and vigour. I inquired its age, and was told that it was 4000 years old, and had been planted by Yên himself! Close to it is a pavilion with an ancient tablet marking the site of Yên's dwelling-place.

Not far from this temple is another, in honour of " Duke of Chow," the great ideal statesman whom Confucius constantly held up for imitation. The building, like the others, stands in a finely-wooded park, with tablets of almost every dynasty; one as far back as the Yang A.D. 618, another of the Yüan A.D. 1280. One, a very handsome marble tablet, was erected by Kianghi, A.D. 1662. The statue of Duke Chow depicts the figure of a large-boned, strong man, with a bluff, good-natured countenance, well-developed forehead, hanging cheeks, and high nose. Chow is said to have been miraculously conceived (*vide* Dr. Legg's 'Shoo King'). He is represented as the son of Shanti (God). After his birth his mother, fearing something inauspicious from his wonderful conception, exposed him to be trodden to death in various ways, but he was always preserved, and at last taken home and nursed, when he grew up a wonder to all. But Chow had no such idea of his origin. In one of his speeches he calls himself the son of King Wan, the brother of King Woo, and the uncle of King Chin. His brother Woo was the first emperor of the Chow dynasty, B.C. 1122, and was greatly helped by his brother Chow's wonderful talents. To Duke Chow is ascribed, by tradition, the invention of the "south-pointing chariot," or mariner's compass.

To the east of Kiu-fu is the tomb of the Emperor Shaou-haou, B.C. 2597. I did not visit this tomb myself, and am indebted to the Rev. Mr. Williamson for the following particulars. The grave is beneath a pyramid built of large blocks of granite compactly placed together; on the top is a small house with upturned eaves in the present Chinese fashion, and covered with porcelain bricks. An old tree grows out of the pyramid and gives the whole thing a most venerable appearance. The pyramid is not at all to be compared to the Egyptian ones for size, but is of the same shape, and instantly reminded Mr. Williamson of them. This Emperor Shaou-haou is said to have been the son of Hwang-ti, the first historical emperor of China, who reigned 2697 years B.C. His mother was called New-tsee, and previous to the birth of her son is said to have seen a star come floating down the stream to the Isle of Wha. When he ascended the throne there was an auspicious omen of Phœnixes. The only other thing recorded to have taken place in this reign is rather startling. Chinese writers say that anciently the people attended to the discharge of their duties one to another, and left the invocation of spirits, and the calling of them to earth, to officers who were appointed for that purpose. In this way things went on with great regularity; but in the time of Shaou-haou the people intruded into the functions of the regulators of the spirits, and tried themselves to bring the spirits down from above. The spirits, no longer kept in check, made their appearance irregularly and disastrously; all was confusion and calamity. Then Chuen (Chuen Hu), who succeeded Shaou-haou, took the matter in hand, and appointed Chung, the Minister of the South, to superintend heavenly things, and Li, the Minister of Fire, to superintend earthly things. In this way both spirits and people were brought to their former regular courses, and there was no unhallowed intercourse the one with the other (see Legg's 'Shoo King'). This recalls to one's mind the passage in Gen. vi. 4, about the sons of God having intercourse with the daughters of men, and a reference to the date on the margin of the Bible will show B.C. 2448.

To the east of Kiu-fu is a range of hills called the Ni Shan, in which is a cave where Confucius is said to have been born. His mother is supposed to have gone to the hill to pray for a son, and although we may discard the fables which surround his birth, there is no reason to doubt that he was really brought forth in the cave pointed out, which is named "The Hollow

Mulberry-tree Cave." There is a small temple standing near it, erected in honour of his mother, Ching-tsai.

Forty li south of Kiu-fu I reached Tsiu-hsien, the city of Mencius, and on the way to it I visited the tombs of that philosopher and his mother, to do which I had to make a *détour* of some 60 li. The tomb of the mother consists of a high mound, having in front the usual stone urn, table, and a tablet stating who and what she was; that she had not lived on the spot, which was only her resting-place. This tablet was erected by Wang-le, an emperor of the Ming dynasty, A.D. 1573. The great charm about this tomb is the magnificent park in which it is situated; some of the trees—oak, yew, cypress, and cedars —are of enormous age and very handsome. I was disappointed with the appearance of the tomb, for it is said that Mencius gave her such an extravagant funeral that his disciples blamed him for the expense which he incurred, and I naturally expected to see something about the tomb which would bear evidence of it. Some 15 li further on, and towards the city, is the cemetery of the Mencius family, in which is the tomb of the philosopher himself, as of those of his descendants. It is enclosed by a low wall, and approached by an avenue, a mile in length and 40 feet wide, of the most magnificent old yew-trees; along it are erected tablets by various emperors. The tomb of Mencius is nearly in the centre of the cemetery, which is splendidly wooded; it is in the usual shape, that of a mound of earth some 25 feet in height. There is a sacrifice-house facing it, of no architectural beauty. On one side of the tomb is a huge tablet, erected by the Emperor Yüen Tung, of the Yüan dynasty, in the tenth year (A.D. 1333) and sixth month of his reign; and another, on the other side, of a much more ancient date. The characters on it were so obliterated by age that we could make nothing of them.

Tsiu-hsien is a small walled city, very prettily situated at the foot of the Yih-shan, a range of hills full of historical interest. One high peak, called E-shan, is famous for its natural curiosities: here is a large stone drum, there a rock in the shape of a bell, another of octagonal shape, and so on.

The temple to Mencius is outside the south gate, in an extensive park, which is enclosed by high walls. The approach to the temple is up a fine old avenue of cypress-trees, on the left of which are tablets of the Han, Lung, and Yüan dynasties, A.D. 947-960 and 1280; on the right, pavilions erected by Kiang Loong and Kiang-hi. The temple is in the usual form, built on a raised terrace; the verandahs are supported by marble pillars, but inferior to those at Confucius's temple. Within the Great Hall is a statue of Mencius, and another of

Yoh-chin-tae, his favourite disciple. In another hall are tablets in honour of Mencius's father, son, and wife. In the court-yard is a curious old well, said to have been opened by a thunder-bolt; it is looked upon as a very great curiosity. Near it is the trunk of an ironwood-tree, said to be of enormous age; it is looked upon with great veneration by the Chinese; a balustrade of white marble surrounds it. In this yard also stands a huge tablet, erected by Kiang-hi in honour of the philosopher. It is a block of marble, 20 feet high and 6 feet broad, by 20 inches thick; it stands on the back of a gigantic marble tortoise, 12 feet long, 6 broad, and 4 high. Both tablets are beauti-fully cut.

The statue of Mencius represents a stout, burly man, with a full, round face, thin lips, and flat nose : it conveyed the idea of a thoughtful, resolute man. His likeness is engraved on a marble slab, and taken when he was very old ; the same fea-tures are there pourtrayed. Behind the main temple is one to Mencius's father, in which is no statue, only a tablet, with the inscription—

"The Spirit's resting-place."

I was not able to see the present representative of the family, for he was mourning for his father, who had died only ten days previously. The family live in a mansion opposite the temple. The rank held by Mencius's representative is much lower than that of the head of the Confucius family, but, like him, he receives a pension from the Government commensurate with the rank. The temple was built by the Emperor Shin-tsung, A.D. 1083.

On the east side of the south gate of the city is a tablet in honour of Mencius's mother, with an engraving on the marble illustrating the famous legend of her cutting through the web she was weaving to point out to her son the risk of neglect of study, and how one bad action might destroy a good life. This engraving is on the back of the tablet; on the front is an in-scription stating that the spot is the site of the dwelling of Mencius and his mother.

The city of Tsiu-hsien is a very poor one; there appeared little or no trade doing, although it is on the Imperial highway between Nanking and Peking. It is remarkable that neither at Kiu-fu or Tsiu-hsien did we see one Buddhist or Taouist temple.

Leaving Tsiu-hsien, I took an easterly course back through the province, my route at first lying through an extensive valley, with the Yu-shan range to the north and the Chiang-shan to the south. I heard that coal was plentiful in the last-named mountains, but no mines have been opened. Iron, likewise, frequently occurs. The country here somewhat resembles that of the Loau-foo valley, in the Poshan district.

The first city of any size I came to was Sze-shui-hsien, a small walled town, with very extensive and thickly-populated suburbs. It is built on the banks of the Sze-Ho, a very historical river. It is spoken of in the 'Tribute of Yu,' B.C. 2205, said to be famous for its "sounding-stones," to be used in payment of taxes. The river is not navigable, owing to sand-banks; it is now a feeder to the Grand Canal. Outside the eastern suburb of Sze-shui is a fine temple to Tsi-loo, a favourite disciple of Confucius. The Sage predicted that this man would die in battle, which prophecy was fulfilled: he was killed while leading an Imperial army against rebels. The Temple is sadly out of repair.

A large quantity of silk is cultivated in this vicinity, both from the mulberry and oak-fed worms. The next city I came to—Mong-yin-hsien, lat 35° 55', long. 118° 5', at the foot of the Pou-yi-shan hills—is a large salt depôt.

I crossed two rivers, the Ta-nan-ho and Woo-ho, striking my outward route near the city of Wei-hsien, travelling north-west, I reached the pretty little walled city of Lai-chow, on the sea-coast, and from thence to Wang-hsien, the great entrepôt and distributing centre for the foreign merchandise landed at Che-foo.

The products of Shan-tung are various. Besides coal, iron, and gold, it contains silver, lead, and other minerals, while quarries of fine marble and granite also occur. Limestone predominates, but slate is very common all over the province. Clays of different sorts suitable for making porcelain and other kinds of pottery are abundant. Silk is very largely cultivated (much more so than is generally supposed), and fabrics therefrom, of very superior quality, are manufactured in many of the cities. Hemp, tobacco, pulse, fruits, and nearly every description of cereals, are extensively grown.

There are three modes of travelling in Shan-tung, namely, on horseback, in carts, or by shansees or mule-litters. With the natives the carts are preferred; but the roads are so exceedingly bad that I should always give preference to the mule-litter in wet weather and horseback in fine. The inns along the main roads are fair, but the traveller must always carry his own bedding, which he spreads on the kang or birch-bed place, which is warmed from beneath in cold weather. The rooms are generally common rooms, for the Chinese traveller has no great objection to share his bed-place with a stranger; I found no difficulty, however, in securing privacy at night, except in some of the small inns off the main road, where once or twice I had to put up with other travellers sleeping on the same kang. The food, as a general rule, is good—that is to say, along the main roads; but when off them I had much difficulty in obtaining anything. Travelling throughout Shan-tung is on the whole cheap, and far from unpleasant. My experience teaches me that a foreigner, so long as he behaves himself, can travel through most of the provinces of China with perfect safety, so far as the people are concerned; I feel perfectly satisfied that they will never of their own accord molest him in any way so long as he conducts himself properly towards them. The only danger a foreigner has to apprehend is from the mandarins, who are so inimical to us that they frequently set on the people to commit acts they would not dream of otherwise. During the whole of my journey, extending over six weeks, I met with the greatest possible civility and kindness from the middle and lower classes, and with the utmost rudeness and contempt from the mandarins, although I was armed with an official passport and a special letter to the governors of some of the principal cities. The exception was the manner in which the representative of Confucius received me; but it must be remembered that he is not one of the ruling mandarins, only an independent noble. He is, however, the head of what is called the Literati—a class whose prejudices the foreigner is always supposed to offend, and who, when anything arises of a disagreeable nature, is the class that apparently takes the lead against us. I maintain that this antagonism is not their own, but that they are made to incur the odium for the mandarins, who are our real enemies. His reception of me I think fully justifies my statement.

It is lamentable to see the utter decay and ruin into which all works of a public nature have been allowed to fall. The roads and bridges all bear evidence of having been works of a magnificent kind. The roads in the vicinity of large cities show traces of a very superior description of pavement, but now are quite impassable for any vehicle except the rough cart of the country. Fine bridges over rivers or streams have likewise fallen in and are replaced by rafts of the Kau-liang stalk (a description of millet), covered with larch, which forms a most dangerous causeway.

The main roads are, on an average, 30 feet broad, and might, with little expense, be made perfect, as the material for repairing them is at hand, and labour as cheap as possible; but such is the apathy of the local authorities throughout China in regard to anything of a public character, that things are allowed to go on from bad to worse, and the consequence is that the fine roads of Shan-tung, once so celebrated, are things of the past. It is most surprising that something is not done for the traffic I encountered everywhere throughout the province, which was very great.

There is no water-communication in Shan-tung—no rivers worthy the name except the Ta-tsing-Ho, now the new Yellow River, which empties into the Gulf of Pichili. The Wei-ho, the Siau-Tsing-ho, and one or two other streams, can be navigated by small flat-bottom boats for about 20 to 40 li from their mouth, but not beyond that distance. The new Yellow River has taken the course of the Ta-tsing-Ho, and is crossed by the Grand Canal nearly half way between the towns of Tong-Chang and Tong-Ping, about 100 li from the borders of the provinces of Shan-tung and Pichili. This river is navigated by Chinese vessels, but its navigation is attended with very great danger, owing to the shifting nature of the bottom. New sand-banks are continually forming, and twice every year the river overflows its banks, and the country for miles round is inundated.

Never was a country better adapted for railroads than Shan-tung, and no province in China needs them more. The present high roads would be an admirable guide for a line, for they avoid every natural difficulty and pass through the most populous districts.

VI.—*On Greenland Fiords and Glaciers.* By J. W. TAYLER.

Read, March 14, 1870.

IN the 'Proceedings' issued July, 1869, which I received in October, I see in a paper by R. Brown, Esq., that he has arrived at the conclusion that glaciers have "hollowed" the fiords of the North. By hollowing, I take it for granted he means causing fiords to be where none were before. Glacier the cause —fiord the effect. This extraordinary conclusion seems to have passed unquestioned, except by Mr. Whymper.

I have spent the greater part of the last eighteen years in that home of glaciers Greenland, exploring the fiords, but have never seen anything to lead to such a conclusion. I maintain that the reverse is the case, that instead of glaciers excavating fiords, they are continually filling them up. It is true that boulders and débris borne along by the ice, scratch, polish, and grind the rocks to a considerable extent; but though strong as a transporting agent, ice alone has but little excavating power —it is like the soft wheel of the lapidary, the hard matter it carries with it does the polishing. I hope to show that the power of ice in excavating has been much overrated.

I have described fiords in Greenland in a former paper. Fiords in general are familiar things to many. I will merely remind my readers that those of Greenland are walled in by rocks averaging 1000 feet in height, their length varies from 10 to 100 miles, breadth 1 to 8 miles, depth of water from a few feet to 200 or more fathoms. The rocks on each side of these fiords are marked by ice action at intervals, but more so near the glaciers.

The deep fiords have for the most part glaciers launching icebergs; the shallow ones have not. Some of the largest glaciers are really not in the fiords; witness that one north of Fredcricks-haab, 15 miles broad, which has not made a fiord, and does not launch bergs—and for this reason—it has brought down a lot of loose material to a reefy coast, and formed a beach at its base; and this great ice power which we are asked to believe has excavated fiords in granitic rocks, 100 miles long and 3000 or 4000 feet in depth, is overcome by loose débris and sand. Why does it not cut its way through these—by far the easier task?

There are numerous fiords in Greenland, nearly filled up by loose material brought into them by the glaciers; the first fiord south of the great glacier I before alluded to is only navigable in boats at high water. No icebergs now come out of this fiord: the glacier is (as a power), if I may use the term, extinct, it has choked itself up, it is mastered by soft mud.

The inland ice from some cause not yet explained (but probably the weight of the interior and higher ice pressing on the lower), moves slowly towards the coast, more like pitch on a roof exposed to the sun, than like a solid body forced forwards, and the glacier finds its way into the deep fiords, simply because they afford an easy outlet. The ice brings with it below loose rocks and stones rounded into boulders, and much sand and mud produced by these in their passage over the rocks beneath. On the surface, angular fragments fallen upon the ice from the sides of mountains and fiord sides of the glacier. Almost all the lower transported material is pushed into the fiord, the mud floats away, most of the boulders and sand remain, and the first iceberg launched in the fiord commences the slow but certain checking of the glacier; for as before shown it has not power to remove its own loose material. The glacier blocked up, the edge of the ice retires inland by melting, and a stream of water brings down the sand and mud left on the land, making the fiord still shallower. The inland ice now seeks another outlet, and then a deep fiord perhaps previously clear of ice becomes encumbered in its turn with icebergs. The ruins of Scandinavian villages may be seen in fiords now almost inaccessible for icebergs, and at the heads of fiords now unnavigable in boats from deposits from the glaciers.

The sides of many fiords are of soft material—sandstone, with coal, blacklead, &c. Why were these not ground away? And then the shape of some fiords is incompatible with the theory of ice cutting, for we could not cut in contrary directions. Take for example the fiord Fig. 1., it is the second south of Arksut.

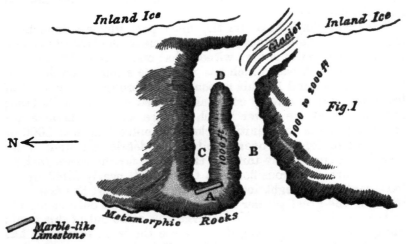

The barrier at A is so high that the existence of the arm C was not suspected until turning the point D.

At the barrier A there is a large vein of crystalline limestone some 20 feet broad, not in the slightest degree marked by ice. How could such a fiord be cut by a glacier? even if we grant it the power to cut the arm B—it will be hard to explain the arm C on the same theory. I maintain that the fiords were in existence prior to their invasion by glaciers. As to their origin, I think the geologist in Greenland will see in the immense number of erupted dykes, and upheaval distortion and fracture of the older stratified rocks, a cause more adequate to the effect than ice "hollowing."

J. W. TAYLER.

Alberton, Prince Edward Island,
    November 4, 1869.

VII.—*Exploring Expedition in Search of the Remains of the late Dr. Leichhardt and Party, undertaken by order of the Government of Western Australia.* By JOHN FORREST, Government Surveyor.

Read, March 28, 1870.

April 15th, 1869.

IN pursuance of instructions received from the Surveyor-General of Western Australia, the exploring party under my command consisted of the following persons, viz.,—Mr. George Monger, as second in command; Mr. Malcolm Hamersley, as third in command; probation prisoner David Morgan, as shoeing smith; and two natives (Tommy Windich and Jimmy Mungaro). The latter native gave Mr. Monger the information respecting the murder of white men to the eastward. We left Newcastle on Monday, 19th, with a 3-horse cart and teamster, and 13 horses, making a total of 16 horses.

* * * * *

May 10th.—Started this morning in company with Tommy Windich and a native boy (one of the nine who joined us at Mount Churchman) to examine the locality called Warne. Steering N. 42° E. magnitude for 7 miles, we came to a grassy flat about half a mile wide, with a stream bed trending south, running through it. The natives state it to be dry in summer, but at present there is abundance of water, and in wet seasons the flat must be almost all under water. After following the flat about 7 miles, we returned to camp about 5 miles, and bivouacked.

12th.—Steered this morning about N. 38 E. (mag.) for 8 miles; we camped on a shallow lake of fresh water—our bivouac of the 10th. Here we met a party of twenty-five natives (friends of my native Jimmy and the nine who joined us at Mount Churchman), who had a grand corroboree in honour of the expedition. They stated that at "Bouincabbajilimar" there were the remains of a number of horses, but no men's bones or guns, and pointed in the direction of Poison Rock, where Mr. Austin lost nine horses. Being now satisfied that the natives alluded to the remains of Mr. Austin's horses, I resolved to steer to the eastward, towards a spot called by my native Jimmy " Ncondie," where he states he heard the remains of white men were.

14th.—One of our horses having strayed, we did not start till 10·40 A.M., when we steered in about a south-east direction for 8 miles, and camped on an elevated grassy spot called Curdilyering, with water in granite rocks. At 7 miles passed a small grassy spot called "Mingan," with water in the granite rocks, probably permanent; the thickets were a little less dense than usual, but without any grass, except at the spots mentioned. By meridian altitudes of Mars and Regulus, Curdilyering is in s. lat. 29° 30′ 30″, and in long. about 118° 30′ E.

15th.—Steering north-east for 4 miles, and N.N.E. for 7 miles, over sandy soil, with thickets of acacia and cypress, we bivouacked on an elevated grassy spot, with water in granite rocks, called "Earroo."

16th (Sunday).—Rested at Earroo; horses enjoying good feed. By meridian altitudes of Regulus and Mars camp at Earroo was in s. lat. 29° 23′ 3″; and in long. 118° 35′ E. Read Divine service; weather cloudy; barometer 29.

17th.—Started at 7·50 A.M., and steered N. 60° E. for about 5 miles; thence about N. 50° E. for 8 miles; thence about N. 85° E. for 5 miles, to a small grassy spot called "Groobenyer," with water in granite rocks. Sandy soil, thickets of cypress, acacia, &c., most of the way. Found camp to be in s. latitude 29° 12′ 43″ by meridian altitudes of Regulus and Aquilæ (Altair). Barometer 21·70.

18th.—Steering N. 70° E. for 2½ miles we saw a low hill called "Yeeramudder," bearing N. 62° 30′ E. mag., distant about 17 miles, for which we steered, and camped to the north of it on a fine patch of grass, with a little rain water on some granite rocks. At 11 miles crossed a branch of a dry salt lake, which appears to run far to the eastward.

19th.—Steering about N. 85° E. mag. for 14 miles, we attempted to cross the lake we had been leaving a little to the southward, making for a spot which we supposed to be the opposite s ore, but on arriving at which I found to be an island, and as we had great difficulty in reaching, having to carry all the loads the last 200 yards, our horses saving themselves with difficulty, and being late, I resolved to leave the loads and take the horses to another island, where there was a little feed, on reaching which we bivouacked without water, all very tired.

20th.—On examining this immense lake I found that it was impossible to get the horses and loads across it. I was therefore compelled to retrace my steps to where we first entered it, which the horses did with great difficulty, without their loads. Was very fortunate in finding water and feed about 3 miles N.N.W. to which we took the horses and bivouacked, leaving all the loads on the island, which we shall have to carry at least half-way viz., three-quarters of a mile, being too boggy for the horses.

21st.—Went over to the lake in company with Messrs. Monger, Hamersley, and Tommy Windich, with four horses, and succeeded in getting all the loads to the mainland, having to carry them about three-quarters of a mile up to our knees in mud, from which point the lake became a little firmer, and the horses carried them out. I cannot speak too highly of the manner in which my companion assisted me on this trying occasion. Having been obliged to work barefooted in the mud, the soles of Mr. Hamersley's feet were completely worn through, and he was hardly able to walk for a fortnight. Seeing a native fire several miles to the southward, I intend sending Tommy Windich and Jimmy in search of them to-morrow, in order that I might question them respecting the reported death of white men to the eastward.

22nd.—Went over to the lake with all the horses and brought all the loads, &c., to camp. Started Tommy and Jimmy in search of the natives, whose fire we saw yesterday. After returning to camp, overhauled all the pack bags, and dried and repacked them, ready for a fresh start on Monday morning. Also washed all the mud off the horses, who appear to be doing well and fast recovering from the effects of the bogging. Tommy and Jimmy returned this evening, having seen some natives after dark, but were unable to get near them.

23rd (Sunday).—Went with Tommy Windich and Jimmy, on foot, to follow up the tracks of the natives seen yesterday. After following the tracks for five miles across the lake, and seeing no chance of overtaking them, as they appeared to be making off at a great rate, and being twelve hours in advance of us, we returned to camp, which we reached at 2 P.M., having walked about 15 miles. Read Divine service. This spot, which I named "Retreat Rock," I found to be in s. lat, 29° 3' 51" by meridian altitudes of Regulus and Mars, and in about long. 119° 16' E.

24th.—Some of the horses having strayed, we did not get a start till 10·40 A.M., when we steered in about an E.N.E. direction for 16 miles, and camped on a piece of rising ground, with very little water. From this bivouac a very remarkable peaked hill called "Wooling," which I named Mount Elvire, bore N. 162° 15' E. mag., distant about 20 miles, and two conspicuous hills close together, called "Yeadie" and "Bulgar," bore N. 105° E. mag. Dense thickets, acacia, cypress, &c., sandy soil with spinifex, most of the way.

25th.—Steering for "Yeadie" and "Bulgar" for 5 miles, came to some water in granite rocks, which we gave our thirsty horses. Leaving the party to follow, I went with Jimmy in advance to look for water, which we found in a rough stream bed, and brought the party to it. This afternoon went with Jimmy to the summit of "Yeadie," and took a round of angles. The local attraction was so great on this hill that the prismatic compass was useless; luckily I had my pocket sextant with me, by which I obtained the included angles. From the summit of "Yeadie" the view was very extensive. The great lake which we had already followed for 40 miles, ran as far as the eye could reach to the east and south, studded with numerous islands. Low ranges and hills in every direction. This immense lake I named Lake Barlee, after the Honourable the Colonial Secretary. By meridian altitudes of Mars and Regulus, camp was in s. lat. 28° 58' 50", and in long. about 119° 39' E. "Yeadie" bearing N. 172° E. mag., distant about 2 miles.

26th.—Steering in about a north direction for 9 miles, we turned to the eastward, rounding a branch of Lake Barlee towards some loose granite rocks, where we encamped, but could not find any water. Sent Jimmy over to another rock 1 mile to the southward, where he found a fine permanent water-hole, to which we took the horses after dark; distance travelled to-day, about 18 miles; Tommy shot a fine emu to-day, which was a great treat to us all.

27th.—Shifted the party over to the water found last night, 1 mile distant, and camped. Found camp to be in s. lat. 28° 53', and in long. about 119° 50' E. Marked a small tree with the letter F. close to the water-hole.

28th.—Some of the horses having strayed, we did not start till 9·30 A.M., when I went in advance of the party in company with Jimmy to look for water. After following Lake Barlee for 9 miles it turned to the southward. After scouring the country in every direction for water without success, we found the tracks of the party (who had passed on), and following them over plains of spinifex and stunted gums, found them encamped with plenty of water, which they had luckily found just at sundown. Distance travelled, 18 miles; about true east. By meridian altitudes of α Bootes (Arcturus), this bivouac is in s. lat. 28° 53' 34", and long. about 120° 9' E.

29th.—Started this morning in company with Tommy and Jimmy to explore the country eastward, leaving the party to remove the horses' shoes, &c. Travelling in an easterly direction for 8 miles, over sandy soil, spinifex, &c., we reached the summit of a high hill, supposed by Jimmy to be "Noondie," which I named Mount Alexander, from which we saw another range about 11 miles distant, bearing N. 82° 15' E. mag., to which we proceeded, and found water in some granite rocks. None of these hills, however, agreed with the description given by Jimmy; and all our expectations that he would be able to show us the spot where the remains of the white men were, were at an end. Returning to camp, 7 miles, we bivouacked on a grassy flat, without water or food.

30th (Sunday).—Started at dawn, with our saddles and rugs on our backs, in search of our horses, and after travelling 1½ mile on their tracks, found them at a small waterhole passed by us yesterday. Saddled up and reached camp at 11 o'clock, and found all well. The dogs caught an emu yesterday morning, off which we made a first-rate breakfast, not having had anything to eat since yesterday morning.

31st.—Started this morning in company with Mr. Monger and Jimmy in search of natives, leaving Mr. Hamersley in charge, with instructions to proceed eastward about 22 miles, to where I found water on the 29th. After starting the party, we steered in a S.S.E. direction towards a high range of bills, which I named Mount Bevon, about 12 miles distant, to the westward of which we found a fine water-hole in some granite rocks, where we rested an hour to allow our horses to feed. Continuing in about the same direction for 5 miles, we ascended a rough range, to have a view of the country, from which we descried a large fire to the westward, 7 miles towards which we proceeded, in hopes of finding natives. When we were within ¼ mile we could hear the hallooing and shouting; and it was very evident that there was a great muster (I should say at least 100) of natives corro-berrying, making a dreadful noise, their dogs joining in chorus. Having stripped Jimmy, I told him to go and speak to them, which he started to do in very good spirits, but soon beckoned us to follow, and asked us to keep close behind him, as the natives were what he called like "sheep flock." He appeared very nervous, trembling from head to foot. After reassuring him, we tied up our horses and advanced through the thicket towards them. After getting in sight of them, Jimmy com-menced cooeying, and was answered by the natives; after which he advanced and showed himself. As soon as they saw him, the bloodthirsty villains rushed at him, and threw three dowaks, which he luckily dodged; when fortunately one of the natives recognised Jimmy (having seen him at Mount Elvire when he (Jimmy) was a little boy), and called to the others not to harm him. Seeing Jimmy running towards the horses, Mr. Monger and I thought it was time to retire, when we saw the mistake we had made in leaving our horses, as the thicket was so dense. We had a difficulty in finding them quickly. On reaching the horses, Mr. Monger found he had dropped his revolver. Had not Jimmy been recognised, I feel sure we should have had bloodshed, and might probably have lost our lives. Mounting our horses, we advanced towards them, and had a short talk with one of them who came to speak to Jimmy, with a guard of eight natives, with spears shipped and dowaks in readiness, should we prove to be hostile; and although I assured them we were friends, and asked them to put down their spears, they took no notice of what I said. This native told us not to sleep here, but to go away and not return, or the natives would kill and eat us, after which he turned away, as if he did not wish to have any more to say to us, and it being now dark, we took his advice and retreated towards where we had dinner, 5 miles. Camped in a thicket without water, and tied up our horses, keeping watch all night.

June 1st.—At daybreak saddled up our tired and hungry horses, and proceeded to where we had dinner yesterday, and after giving our horses two hours' grazing and having had breakfast, started back towards the natives' camp, as I wished to question them respecting the reported death of white men in this neighbourhood. When we approached the natives' bivouac, we saw where they had been following up our tracks in every direction, and Jimmy found the place where they had picked up Mr. Monger's revolver. While Jimmy was away looking for the revolver Mr. Monger saw two natives following up our trail, and within 50 yards of us; we both wheeled round and had our guns in readiness, but soon perceived they were the same as were friendly last night, and I called Jimmy to speak to them. At my request, they went and brought us Mr. Monger's revolver, which they stated they had been warming near the fire, but fortunately for them did not go off. On being questioned by Jimmy, they stated that the place, Noondie (where Jimmy stated he heard the remains of white men were) was two days' journey north-west from here; that there were the remains of horses, but no men's, there; and volunteered to show us the spot. Being now 1 P.M., and having to meet the party to-night at a spot about 23 miles distant, we started at once, leaving the natives, who did not wish to start to-day, but who apparently sincerely promised to come to our camp to-morrow. Reached camp at the spot arranged an hour after dark, and found all well.

2nd.—Rested our horses at this spot, which I called the Two-Spring Bivouac (as there were two small springs at the spot). Re-stuffed with grass all the pack-saddles, as some of the horses were getting sore backs. By meridian altitude of sun found camp to be in s. lat. 28° 51' 45", and in long. about 120° 30' E. Was very much annoyed at the natives not putting in an appearance, as they promised.

3rd.—No signs of the natives this morning. I concluded to steer in the direction pointed out by them, and travelling about N. 206° E. mag. for 15 miles, we found water in some granite rocks, with very good feed around them. Cypress and acacia thickets, light red loamy soil, destitute of grass.

4th.—Steering in a w.n.w. direction for 16 miles, the first six of which were studded with granite rocks, with good feed around them, after which through poor sandy country, covered with spinifex, &c., we bivouacked in a thicket without water or feed, and tied up our horses. Saw a natives' fire, but was unable to get near them. Barometer, 28·52. Fine.

5th.—After travelling in a northerly direction for 7 miles, without finding water and without seeing any hill answering the description given by Jimmy, I struck about east for 16 miles, and camped at a fine spring near some granite rocks, with splendid feed around them. This is the first good spring seen since leaving the settled districts. At 8 P.M., barometer 28·44, thermometer, 72°.

6th (Sunday).—Rested at camp, which I called Depôt Springs, and found to be in s. lat. 28° 36' 34", by meridian altitude of sun. Barometer at 8 A.M. 28·38, thermometer 57° ; at 5 P.M., barometer 28·30, thermometer 77°.

7th.—Started this morning in company with Mr. Hamersley and Jimmy to explore the country to the northward, where we had seen a remarkable peaked hill. Travelled in a northerly direction for about 30 miles, the first twenty of which were studded with granite rocks, with fine feed around them. At about 27 miles crossed a salt marsh about 1 mile wide, and continuing 3 miles farther, reached the peaked hill, which was composed of granite, capped with immense blocks, giving it a very remarkable appearance. Bivouacked on north-west side of hill, at a small water-hole.

8th.—This morning, after saddling up, we ascended the conical hill (which I named Mount Holmes), and took a round of bearings and angles from it, after which we struck N. 81° E. (mag.) to a granite range about 8 miles distant, where we found two fine water-holes, and rested an hour ; thence in about a s.s.E. direction for 12 miles. We bivouacked without water on a small patch of feed. The day was very fine, and the rainy appearances of the weather cleared off, much to our grief.

9th.—At daybreak, not hearing the horse bells, and anticipating they had made off in search of water, we put our saddles, rugs, guns, &c., on our backs, and started on their tracks. After following the tracks for 9 miles, we came to a water-hole, and had breakfast, after which we continued and overtook the horses in a grassy flat about 13 miles s.s.E. from our last night's bivouac. The last few miles our loads appeared to become very awkward and heavy. One of the horses had broken his hobbles. Continuing in about the same course for 6 miles, we struck about w.s.w. for 10 miles, and reached camp, where we found all well. At 6 P.M. barometer 28·64 ; cloudy.

10th.—Started again this morning in company with Mr. Monger and Jimmy to explore the country to the eastward, leaving Mr. Hamersley to shift the party to our bivouac of the 2nd instant, about 24 miles south-east from here. After travelling E.N.E. for 6 miles, we came upon a very old native at a fire in the thicket, but Jimmy could not understand what he said, save he thought he said there were a number of armed natives about. He was very frightened, and howled the whole time we stayed. He was apparently quite childish, and hardly able to walk. Continuing our journey, we camped at a small water-hole in some granite rocks, with good feed around them, about 16 miles E.N.E. from Depôt Springs.

11th.—Started at sunrise, and steered about E.N.E. over lightly grassed country; and on our way came upon a middle-aged native and two small children. We were within 20 yards of him before he saw us, when he appeared very frightened, and trembled from head to foot. Jimmy could understand this native a little, and obtained from him that he had never seen or heard anything about white men or horses being killed or died in this vicinity. Did not know any place named Noondie; pointed to water a little way eastward. I got Jimmy to ask him all manner of questions, but all to no purpose, he stating he knew nothing about it. Upon Jimmy asking him if he had ever heard of any horses being eaten, he answered no, but that the natives had just eaten his brother; while stating this the old man tried to cry. I have no doubt parents have a difficulty in saving their children from these inhuman wretches. He stated he had two women at his hut, a little westward. After travelling 10 miles from our last night's bivouac, and not finding any water, we struck N. 204° E. (mag.) for about 20 miles, through scrubby thickets, without feed, and reached our bivouac of the 2nd, where the party will meet us to-morrow. Reached the **water at** the Two-Springs Bivouac half-an-hour after dark.

12th.—Explored the country around camp in search of a better place for food, but could not find any water. Mr. Hamersley and party joined us at 4 P.M., all well. Tommy shot a red kangaroo, which was a great treat, after living so long on salt pork. Barometer 28·60; fine; cold wind from the east all day.

14th.—Started this morning, in company with Morgan and Jimmy, to examine the country to the southward; travelled in a south-westerly direction for about 25 miles, and camped at the spot where we had the encounter with the natives on May 31st. We found they had left, and there was no water on the rocks; luckily our horses had water 6 miles back.

15th.—Saddled up at daybreak, and steered about south-east towards a high range of hills about 10 miles distant, which I named Mount Ida, from the summit of which I took a round of angles with my pocket sextant. On all the hills in this neighbourhood the local attraction is so great that the prismatic compass is useless. Found a fine spring of water on south side of Mount Ida, in an almost inaccessible spot. After giving our horses two hours' rest, we continued our journey N. 154° E. (mag.) for 8 miles, to a granite range, where, after a diligent search, I found two fine water-holes, and bivouacked, with good feed around the rocks.

16th.—Saddled up at sunrise, and steered to some trap ranges N. 124° E., and about 7 miles distant, from which I could see an immense lake running as far as the eye could reach to the eastward and westerly and northerly, most probably joining Lake Barlee. Not being able to proceed farther southward, on account of the lake, I steered in a northerly direction for 20 miles, and finding neither feed nor water, bivouacked in a thicket without it, and tied up our horses.

17th.—At dawn found that my horse Sugar had broken his bridle and made off towards our bivouac of the 15th. Placing my saddle on Jimmy's horse, we followed on his tracks for 6 miles, when we came to a few granite rocks, with a little water on them, from the little rain that had fallen on them during the night, where I left Morgan with the horses and our guns, while Jimmy and I followed on Sugar's tracks, taking only my revolver with us. After travelling on the tracks for 2 miles, we overtook him, and after a little trouble, managed to catch him. On reaching the spot where we had left Morgan, we found him with the three double-barrelled guns on full cock, together with his revolver, in readiness. On my asking him what was the matter, he stated "nothing," but that he was ready to give them what he called "a warm attachment." After having breakfast, we steered N.N.W. for about 20 miles, and reached camp at 5 P.M., and found all well. Rained a little during the day.

18th.—Having now made an exhaustive search in the neighbourhood where Jimmy expected to find the remains of the white men, by travelling over nearly the whole of the country between lat. 28° and 29° 30′ s. and long. 120° and 121° E., I determined to make the most of the little time at my disposal, and carry out my instructions, which were to attempt to proceed as far eastward as possible. Accordingly, after collecting the horses, steered about E.N.E. for 9 miles to a low quartz range, over tolerably grassy country; not very dense. From this range I saw some bare granite rocks, bearing about N. 120° E. (mag.), for which we steered, and luckily, after travelling 6 miles over a plain, which in severe winters must be almost all under water, found a fine pool of water in a clay-pan, and bivouacked. Rained a little during the night.

19th.—The horses having strayed back on our tracks, we did not start till 12 o'clock, when we continued our journey towards the granite range seen yesterday, about 10 miles distant, and camped on west side of it, with plenty of water from the recent rain on the granite rocks, but with very little feed. At 5 miles crossed a dry stream-bed, 18 yards wide; sandy bottom. Thickets most of the way, but not very dense.

20th (*Sunday*).—Rested at camp. Jimmy shot four rock kangaroos to-day. Read Divine service. Took a round of angles from a bare granite hill N. 50° E. (mag). about 1 mile from camp, which I found to be in s. lat. 28° 57' by meridian altitudes of a Bootes (Arcturus) and a Pegasi (Markab), and in longitude about 120° 55' E. Saw a high hill, bearing N. 81° 30' E. (mag.) about 25 miles distant, which I named Mount Leonora, and another bearing N. 67° E. (mag,) about 25 miles distant, which I named Mount George; intend proceeding to Mount Leonora to-morrow. Marked a small tree (ordnance tree of Mr. Austin) with the letter F at our bivouac.

21st.—Steering towards Mount Leonora over some tolerably grassy country, we reached it at sundown, and, not finding any water, camped without it, with very good feed, in s. lat. 28° 53' by meridian altitudes of a Lyræ (Vega) and Aquilæ (Altair), and in longitude about 121° 20' E.

22nd.—After making every search in the vicinity of our bivouac for water, and the country ahead appearing very unpromising, I concluded to return 10 miles back on our tracks, where we found a fine pool of water in a brook, and camped. I intend taking a flying trip in search of water to-morrow.

23rd.—Started this morning, in company with Tommy Windich, to explore the country to the eastward for water, &c. After travellin 3 miles towards Mount Leonora, saw a natives' fire, bearing ngrth-east about 3 miles, to which we proceeded, and surprised a middle-aged native at a fire. Upon seeing us, he ran off shouting, &c., and the remainder of his companions, who were at a little distance, decamped. The horse I was riding (Turpin, an old police horse from Northam) appeared to well understand running down a native, and between us we soon overtook our black friend and brought him to bay. We could not make him understand anything we said, but after looking at us for a moment, and seeing no chance of escape, he dropped his two dowaks and wooden dish, and climbed up a small tree, about 12 feet high. After securing his dowak, I tried every means to tempt him to come down. I fired my revolver twice, and showed him the effect it had on a tree, and it also had the effect of frightening all the natives who were about, who no doubt made off at a great rate. I began to climb up after him, but he pelted me with sticks, and was more like a wild beast than a man. After discovering we did not like to be hit, he became bolder, and ew more sticks at us, and one hitting Tommy, he was nearbshooting him, when I called on him to desist. I then offered him a piece of damper, showing him it was good by eating some myself, and giving some to Tommy, but he would not look at it, and when I threw it close to him, he dashed it from him, like as if it was poison. The only means of getting him down from the tree was force, and, after considering for a moment, I decided to leave him where he was, and accordingly laid down his dowaks and dish, and bade him farewell in as kindly a manner as possible. Continuing our course, passing Mount Leonora, we steered N. 81° 15' E. (mag.) to a table hill, which I ascended and took a second round of angles. This hill I named Mount Malcolm, after my friend and companion, Mr. M. Hamersley. Saw a remarkable peak, bearing N. 65° E. (mag.), distant about 20 miles, towards which we proceeded, and at 6 miles came on a small gully, in which we found a little water, and bivouacked.

24*th.*—Started early this morning and steered E.N.E. for 6 miles to some low stony ranges, lightly grassed, thence N. 61° 30' E. (mag.) to a remarkable peak, which I named Mount Flora, distant about 9 miles from the stony ranges, ascending which, I obtained a round of bearings and angles. Saw a high range, bearing about N. 106° 15' E. (mag.), apparently about 16 miles distant, towards which we travelled till after dark, searching for feed and water on our way, without success, and bivouacked without water, and tied up our horses.

25*th.*—Saddled up at dawn, and proceeded to the range, which bore N. 93° 30' E. (mag.), about 5 miles distant, on reaching which I ascended the highest peak, which I named Mount Margaret, and took a round of angles and bearings. From the summit of Mount Margaret the view was very extensive. A large, dry, salt lake was as far as the eye could reach to the southward, while to the east and north-east there we e low trap ranges, lightly grassed. A high table hill bore N. 73° E. (mag). Being now about 60 miles from camp, and not having had any water since yesterday morning, I decided to return, and steering about west for 8 miles, we struck a brook, trending south-east, in which we found a small quantity of water in a claypan. After resting an hour, in order to make a damper, and give our horses a little of the feed which grew very sparingly on the banks of the brook, we continued our journey towards camp. Passing Mount Flora, we camped about 8 miles farther towards camp, on a small patch of feed, without water, about a mile north of our outward track.

26th.—Started at dawn, and reached our bivouac of the 23rd, where we obtained just sufficient water for ourselves and horses. Continuing, we found a fine pool of rain water in a brook, 1½ mile west of Mount Malcolm, and reached camp an hour after dark, and found all well. On our way Tommy Windich shot a red kangaroo, which we carried to camp.

27th (*Sunday*).—Rested at camp. Read Divine service. Found camp to be in s. lat. 28° 55', by meridian altitude of sun, Aquilæ (Altair) and Lyra, and in longitude about 121° 10' E. Although we had great difficulty in procuring water on our last trip, I was very loathe to return without making another effort, especially as from the appearance of the country east of our farthest, I had every hope of a change. I therefore concluded to shift the party to the water found yesterday near Mount Malcolm and make another attempt to proceed farther east.

28th.—Steering about N. 81° E. (mag.) over lightly grassed country, thinly wooded, for 16 miles, we camped 1½ mile to the west of Mount Malcolm in s. lat. 28° 51' 19", by meridian altitude of Aquilæ (Altair), and in longitude about 121° 27' E.

29th.—Started this morning in company with Tommy Windich, with seven days' provisions, leaving instructions for Mr. Monger to shift the party back to our last camp, where the feed was much better, in lat. 28° 55' s., and long. 121° 10' E. Travelled about east for 30 miles towards Mount Margaret, our farthest point last trip. We camped in a thicket without water on a small patch of feed.

30th.—Saddled up at dawn, and proceeded towards Mount Margaret, obtaining a little water at the spot where we found water on the 29th (our former trip). Continuing, we found a fine pool of water in a brook (Mount Margaret bearing north-east about 2½ miles) and rested an hour. Hardly any feed near the water. Resuming, we passed Mount Margaret and steered towards the table hill seen on our former trip, bearing N. 73° E. (mag.), apparently about 18 miles distant, over a series of dry salt marshes, with sandy country and spinifex intervening. After travelling 8 miles we bivouacked without water on a small patch of feed. With my pocket sextant I found this spot to be in south lat. about 28° 50' and long. about 122° 11' E.

July 1st.—After travelling towards the table hill seen yesterday for 6 miles, crossed a large brook trending south-west, in which we found a small pool of rain-water, and rested an hour to breakfast; resuming for about 6 miles, reached the table hill, which I ascended, and took a round of angles. I have since named this hill Mount Weld, being the farthest hill seen to the eastward by us. Continuing about N. 77° E. (mag.) for 15 miles, through dense thickets, without any grass save spinifex, we hivouacked without water or feed, and tied up our horses. With my pocket sextant I found this 'spot to be in south latitude about 28° 41' by meridian altitude of Bootes (Arcturus), and in longitude about 122° 37' E.

2nd.—Started at dawn, and steered about east, searching on our way for water, which our horses and ourselves were beginning to be much in want of. At 6 miles found a small hole in some rocks, apparently empty, but on sounding with a stick I found it to contain a little water. The mouth of the hole being too small to admit a pannikin, and having used my hat with very little success, I at last thought of my gun-bucket, with which we procured about two quarts of something between mud and water; and after straining through my pocket handkerchief, we pronounced it first-rate. Continuing for 6 miles over clear open sand plains, with spinifex and large white gums, the only large trees and clear country seen since leaving the settled districts, we climbed up a white gum to have a view of the country eastward. Some rough sandstone cliffs bore N. 127° E. (mag.), about 6 miles distant. The country eastward was almost level, with sandstone cliffs here and there, apparently thickly wooded with white gums, &c. Spinifex everywhere, and no prospects of water. More to the north a narrow line of samphire flats appeared, with cypress and stunted gums on its edges, and very barren and desolate—so much so, that for the last 25 miles we have not seen any grass at all save spinifex. After taking a few bearings from the top of the tree (which I marked with the letter F on south side, which is in south latitude about 28° 41', and longitude about 122° 50' E.), I concluded to return to our last watering-place, about 31 miles distant, as we were now above 100 miles from camp, and our horses had been without water or feed since yesterday morning; therefore, keeping a little to the north of our outward track, we travelled till two hours after dark, and camped without water or feed, and tied up our horses.

3rd.—Saddled up at dawn, and steered westerly towards our last watering-place, about 14 miles distant; but after travelling about 7 miles, came to a small pool of water (at the head of the brook in which we found water on the 1st), and rested two hours, to allow our horses to feed, as they had neither eaten nor drank for the last 48 hours. Resuming our journey along the brook (which I named Windich Brook, after my companion, Tommy Windich) for 10 miles, in which we found several pools of water, but destitute of feed, we bivouacked on a patch of feed, without water, about 2 miles east of our bivouac of the 30th June.

4th.—Travelling about w.s.w. for 12 miles, we reached the pool of water found on our outward track on the 30th June, 2¼ miles south-west from Mount Margaret, where we rested an hour. Resuming, we travelled nearly along our outward track for 18 miles, and encamped without water on a small patch of feed. Tommy shot two wurrongs to-day.

5th.—Started at daybreak, and following nearly along our outward track for 25 miles, we reached the water close to Mount Malcolm (where we left the party, they having shifted, as instructed, 17 miles further back), where we rested an hour; but having finished our provisions, we roasted two wurrongs and made a first-rate dinner. Tommy also shot an emu which came to water, which we carried to camp. Reached camp at 6 P.M. and found all well, having been absent seven days (every night of which time we were without water), in which time we travelled over 200 miles.

6th.—Weighed all the rations, and found we had 283 lbs. flour, 31 lbs. bacon, 28 lbs. sugar, and 4 lbs. tea, equal to 32 days' allowance of flour, 10 days' bacon, 19 days' sugar, and 21 days' tea on a full ration. I therefore now concluded to return to Perth as quickly as possible and reduce the allowance of tea and sugar, to last 30 days (bacon we will have to do without), by which time I hope to reach Clarke's homestead, Victoria Plains, as I intend returning by Mount Kenneth, Nanjajetty, Ningham, or Mount Singleton, and thence to Damperwar and Clarke's homestead, thus fixing a few points that will be useful to the Survey Office.

    •   •   •   •   •   •

13th.—Leaving the party in charge of Mr. Monger, with instructions to proceed to Retreat Rock, our bivouac of May 23rd, I started, with Mr. Hamersley and Jimmy, to attempt to cross Lake Barlee, in order to explore the country on its south side, near Mount Elvire, as well as to try to find natives, as Jimmy is acquainted with these tribes. Steering N. 154° E. (mag.) for 7 miles, we came to the lake, and entering it, succeeded in reaching the southern shore after 12 miles of heavy walking, sinking over our boots every step, our horses having great difficulty. When we reached the southern shore it was nearly sundown; we, however, pushed on and reached the range, where we bivouacked on a patch of feed and a little water—Mount Elvire bearing N. 87° E. (mag.), about one mile distant, and Yeadie and Bulgar N. 8° E. (mag.). Rained lightly during the day. Being wet through from the splashing of our horses crossing the lake, and it raining lightly during the night, and not having any covering, our situation was not the most pleasant. Jimmy informed me that there was a fine permanent spring close to Mount Elvire; we did not, however, go to see it.

14*th.*—This morning, after ascending a range to have a view of the country, steered N. 288° E. (mag.), and after travelling 6 miles, came to a branch of Lake Barlee, running far to the southward, which we attempted to cross; but after travelling 1½ mile, which was very boggy, our horses went down to their girths, and we had great difficulty in getting them to return, which, however, we ultimately succeeded in doing, and made another attempt at a place where a series of islands appeared to cross it, and crossed safely over without much difficulty, reaching the opposite shore at sundown, where we bivouacked on a splendid grassy rise, with abundance of water in granite rocks—Mount Elvire bearing N. 108° E. (mag.), and Yeadie and Bulgar N. 45° E. (mag.).

15*th.*—Having finished our rations last night, we started at dawn and steered towards Retreat Rock (where we are to meet the party), and after travelling 5 miles, came to that part of Lake Barlee which we attempted to cross, without success, on May 19th (our outward track); but leading our saddle horses, with difficulty we succeeded in crossing, and reached camp, all very tired, at 12 o'clock, and found all well. The party were encamped one mile north of our former bivouac, at some granite rocks with two fine water-boles.

16*th.*—Considerable delay having occurred in collecting the horses, we did not start till 10 o'clock, when we travelled nearly along our outward track (passing Yeeramudder Hill, from the summit of which Mount Elvire bore N. 111° 30′ E. (mag.) about 35 miles distant) for about 21 miles, and bivouacked at some granite rocks, with a little feed around them, which I found to be in s. lat. 29° 8′ 47″ by meridian altitudes of *a* Bootes (Arcturus) and *a* Pegasi (Markab), and in long. about 118° 59′ E.

17*th.*—Started at 8·45 A.M., and steering about west for 25 miles, through dense thickets without feed, we camped without water on a small patch of poor feed, in s. lat. 29° 7′ 13″ by meridian altitudes of *a* Bootes (Arcturus). Marked a small tree with F., 1869. Being now in a friendly country, I decided to give up keeping *watch*, which we had done regularly for the last two months.

18*th* (*Sunday*).—After starting the party, went, in company with Tommy Windich, to take bearings from a low hill, bearing N. 289° (mag.), distant about 8 miles, after which we struck in the direction in which we expected to find the party, but, for some reason or other, they had not passed by; and I anticipate they must have met with good feed and water and camped, it being Sunday. However this may be, we kept bearing more and more to the southward, in hopes of crossing the track, till after dark, when we reached the Warne Flats and bivouacked. Not expecting to be absent more than a few hours, we had neither rations nor rugs. Luckily Tommy shot a turkey this afternoon, which we roasted in the ashes, and made an excellent meal. The night was bitterly cold, and, not having any rug, I slept with a fire on each side of me, and, considering the circumstances, slept very fairly.

19*th*.—After making a first-rate breakfast off the remainder of the turkey, we started in search of the party, making back towards where we had left them, keeping well to the southward, and, after spending the whole of the day and knocking up our horses, we found the tracks of the party nearly where we left them yesterday morning, and following along them for 9 miles, found where they had bivouacked last night, and it being now two hours after dusk, we camped also, with an opossum between us for supper, which Tommy had luckily caught during the day. The night was again very cold, and hardly anything to eat, which made it still worse.

20*th*.—Started on the tracks at daybreak, and followed them for about 13 miles, when we found the party encamped on the east side of a large bare granite rock, called Meroin, Mount Kenneth bearing N. 24° E. (mag.), about 15 miles distant. From a cliff about one mile west of camp took a splendid round of angles, Mount Kenneth, Mount Singleton, and several other known points being visible. By meridian altitudes of sun, α Bootes (Arcturus), ε Bootes, and α Coronæ Borealis, camp was in s. lat. 29° 10' 49", and long. about 118° 14' E.

21*st*.—At 7 A.M. barometer 29·10; thermometer 35°. Started at 8.15 A.M., and steering about west for 15 miles over country studded here and there with granite rocks, with good feed around them, but in some places noticed rock poison. Camped at a spring called Pullagooroo, bearing N. 189° from a bare granite hill three quarters of a mile distant, from which hill Mount Singleton bore N. 237° E. (mag.). By meridian altitudes of ε Bootes (Arcturus) and ε Bootes, Pullagooroo is in s. lat. 32° 7' 46". Finished our bacon this morning, and for the future will only have damper and tea.

22*nd*.—Steering a little to the north of west, through dense thickets without grass, we bivouacked at a very grassy spot, called Bunnaroo, from which Mount Singleton bore N. 205° E. (mag.). By meridian altitudes of α Bootes (Arcturus), ε Bootes, and α Coronæ Borealis, camp is in s. lat. 28° 58', and long. about 117° 35' E.

23*rd*.—After starting the party, with instructions to proceed straight to Mount Singleton, distant about 32 miles, I went, in company with Jimmy, to the summit of a high trap-range, in order to take a round of angles, and fix Nanjajetty, which was visible. On our way from the range to join the party saw the tracks of two men and two horses, with two natives walking, and soon after found where they had bivouacked about three days ago. Was much surprised at this discovery; suppose it to be some one looking for country. Continuing we found the tracks of the party, and overtook them encamped at a fine permanent spring, Mount Singleton bearing N. 146° E. (mag.), about 3½ miles distant. Reached the party at 7 o'clock. There was a partial eclipse of the moon this evening.

24th.—There being splendid green feed around Mount Singleton, together with our horses being tired, I concluded to give them a day's rest. Went, in company with Mr. Monger and Jimmy, to the summit of Mount Singleton, which took us an hour to ascend; but on reaching it I was well repaid for my trouble by the very extensive view and the many points to which I could take bearings. Far as the eye could reach to the east and south-east was visible Lake Moore, Mount Churchman, &c.; to the north conspicuous high trap-ranges appeared; while to the west, within a radius of 6 miles, hills covered with flowers, &c., gave the country a very pretty appearance. Further to the west a dry salt lake and a few trap-hills appeared. Returned to camp, which we reached at 2 P.M. On our way shot three rock kangaroos.

25th (*Sunday*).—Rested at camp (near Mount Singleton), which I found to be in s. lat. 29° 24′ 33″ by meridian altitude of sun, and long. about 117° 20′ E. Read Divine service.

26th.—Some delay having occurred in collecting the horses, did not start till 9 A.M., when we steered a little to the north of west towards Damperwar. For the first 7 miles over rough trap-hills lightly grassed, when we entered samphire and salt-bush flats for 4 miles, crossing a dry lake at a point where it was only 100 yards wide, and continuing through thickets, we camped at a spot with very little feed and water, in s. lat. 29° 21′ 48″. From this spot Mount Singleton bore N. 113° 20′ E. (mag.), distant about 20 miles. Here we met two natives, whom we had seen at our outward track at the Warne corroboree, who were of course friendly and slept at our camp. They had a great many dulgites and opossums, which they carried in a net bag, made out of the inner bark of the Ordnance tree, which makes a splendid strong cord. They informed us that a native had come from the eastward, with intelligence relating to the encounter we had with the large tribe on May 31st, and adding that we had been all killed, and that all the natives in this vicinity cried very much on hearing the news. This is another specimen of the narrations of natives, with whom a tale never loses anything by being carried.

27*th.*—Steering a little to the north of west for 18 miles, we reached Damperwar Springs, a clear grassy spot of about 300 acres, on west side of low granite hill. The spring was dry, but by digging a few feet we obtained an abundant supply. From the appearance of the country there has been hardly any rain in this neighbourhood for many months. Took a round of angles from a ra -hill about 2 miles distant, Mount Singleton and many other points being visib e. Met a party of friend y natives here. By meridian altitudes of α Bootes, α Coronæ Borealis, and α Lyræ (Vega), Damperwar Spring is in s. lat. 29° 16' 32", and long. 116° 47' E.

28*th.*—Steering in a southerly direction, and following along the western margin of a salt lake, most of the way over samphire flats, &c., with thickets intervening, more dense than usual, we encamped on a small grassy spot with plenty of water in granite rocks, called Murrunggnulgo, situated close to the west side of the lake, which I named Lake Monger.

August 4th.—Reached Newcastle at 11 o'clock.

I now beg to make a few remarks with reference to the main object of the Expedition, viz. the discovery of the remains of the late Dr. Leichhardt and party. In the first place Mr. Fred. Roe was informed by the native Weilbarrin, that two white men and their native companion had been killed by the aborigines, 13 days' journey to the Northward, when he was at a spot called Koolanobbing, which is in s. lat. about 30° 53', and long. about 119° 14' E.

Mr. Austin lost eleven horses at Poison Rocks (nine died and two were left nearly dead), which is in lat. 28° 43' 23" s., and long. about 118° 38' E., or about 150 miles from Koolanobbing, and in the direction pointed to by the natives. I therefore imagine it to be very probable that the whole story originated from the horses lost by Mr. Austin at Poison Rock, as I am convinced the natives will say anything they imagine will please you. Again, the account given to us at Mount Churchman, on May 5th, appeared very straightforward and truthful, and was very similar to that related by Mr. Roe, but on questioning them for a few days, they at last stated there were neither men nor guns, but only horses' remains, and pointed towards Poison Rock. Again, the native who gave all the information to Mr. Monger was one of our party. His tale, as related by Mr. Monger, appeared very straightforward and truthful, viz. that white

FORREST's *Report of the Leichhardt Search Expedition.* 249

men had been killed by the natives twenty years ago: that he
had seen the spot, which was at a spring near a large lake, so
large that it looked like the sea as seen from Rottnest, 11 days'
journey from Ninghan or Mount Singleton, in a fine country,
the white men were rushed upon while making a damper, and
clubbed and speared ; he had often seen an axe which formed
part of the plunder. All this appears very feasible and truthful
in print; but the question is, of what value did I find it? Upon
telling Jimmy what Mr. Monger stated he told him, he said he
never told him that he had seen these things himself, but that
he had heard it from a native who had seen them; thus
contradicting the whole he had formerly stated to Mr. Monger.
Moreover, the *fine* country he described we never saw, as what
a native calls good country is any place where they can get a
drink of water and a wurrong ; and if there is an acre of grassy
land, they describe it as a very extensive grassy country. This I
have found the case time after time. As a specimen of the
untruthfulness of these natives, I may quote, that my native
Jimmy, who was a first-rate fellow in every other respect, stated
to Mr. Monger and myself at York, that there was a large river
similar, he said, to the Avon at York, to the Eastward, knowing
at the time that we would find out he was telling a falsehood.
He even told Mr. Geo. Monger, before leaving Newcastle, to buy
books, in order to catch the fish that were in the river; and
concluded by stating we would have a great difficulty in crossing it, as it ran a great distance North and South. Almost
every evening I questioned and cross-questioned him respecting
this river; still he adhered to what he first stated. You may
imagine how disappointed we all were on reaching the spot to
find it was a small brook running into a salt marsh, with water
in winter, but dry in summer.

   With reference to the country travelled over, I am of opinion
that it is worthless as a pastoral or agricultural district; and as
to minerals, I am not sufficiently conversant with the science to
offer an opinion, save that I should think it was quite worth
while sending a geologist to examine it.

   It now becomes the very pleasing duty for me to record my
entire satisfaction with the manner in which all the members of
the Expedition exerted their best energies, in the performance
of their respective duties. To Mr. Geo. Monger, and Mr. Malcolm Hamersley, I am indebted for their cc-operation and advice
on all occasions. I am also deeply indebted to Mr. Hamersley
for collecting and preserving all the botanical specimens that
came within our reach, as well as for the great care and trouble
taken with the store department, which came under his immediate charge. To probation prisoner David Morgan my best

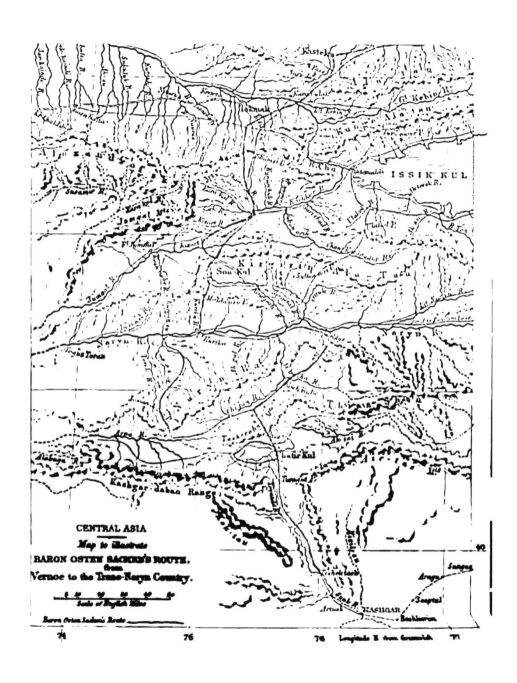

CENTRAL ASIA

Map to illustrate

BARON OSTEN SACKEN'S ROUTE,
from
Vernoe to the Trans-Naryn Country.

Scale of English Miles

Baron Osten Sacken's Route

mountainous district referred to on his way to Kashgar with a trading caravan.

"The work of surveying, which had been actively carried on along the Chinese frontier, was now approaching the Issik-kul. In 1859 the southernmost points in this part of Central Asia fixed by astronomical observation by the late Captain Golubieff were the east and west extremities of the lake. These points are even now the only accurate data upon which to base the topographical surveys south of Issik-kul. In the following year, 1860, M. Veniukoff surveyed and mapped the whole of the Issik-kul country, including the valleys of the Chu and Koshkar, to the west of that lake. He also did good service in publishing the information he had obtained about the country and about Lake Son-kul. Captain Protsenko made further additions to Veniukoff's work in his survey of the country to the river Naryn, a most important feature in which survey was the alpine lake of Son-kul. There was still left an unexplored district, south of the Naryn, along the road to Kashgar. The exploration of this tract of country was the object of Poltoratsky's expedition, in which I was able to join.

"On the 2nd July we left Vernoye by the road to Kastek. As this road has been described more than once, I will confine myself to an historical fact connected with it. Our second night's encampment was situated close to the picket of Uzun-agatch, half-way between Vernoye and the Kastek Pass, at the foot of a hill, on which has been lately erected a simple memorial, viz., a cross covered with tin on a white stone pedestal. This cross has been placed there in memory of the severe engagement which took place on this spot on the 21st October, 1860, between the Russians and Khokandians. It is strange that, with the exception of the notice about it in the 'Invalide Russe," no account has been written of the engagement at Uzun-agatch; and yet it must be reckoned among the most important successes of the Russian arms in Central Asia, and marks the epoch of the solid establishment of our power in the Trans-Ili district. In 1860 we were far from being complete masters of this country. The destruction of the Khokan forts of Tokmak and Pishpek seemed to have only exasperated our adversaries, who collected all their forces for one last desperate attack. The numbers of the Khokandians at the Uzun-agatch engagement are estimated at 40,000. The Russian forces only numbered 1000, including the garrison of Vernoye, who were all in the field, the defence of that fort being left to a few civilians and women. Never before in Central Asia had the Russian population been in such imminent danger from foreign foes, and if victory had not attended our arms on the 21st, a dreadful fate would probably have been in store for Vernoye. The victory of Uzun-agatch, due to General Kolpakoffsky, finally secured to Russia the possession of the Trans-Ili country. Since that time the Khokandians have never again attempted to drive us out of the country.

"On the 6th and 7th July the party crossed the Kastek Pass, which has been admirably described by M. Severtsoff. Along this defile the post-road between Vernoye and the Syr Darian district runs. Here, too, a line of telegraph wires to connect Russia with Turkestan is meditated. As we went through the pass we saw works in progress for making the road practicable for wheel conveyances, an operation attended by great difficulties, owing to the large quantity of boulders which obstruct the way, and the frequent windings of the little river Kastek, whose waters first strike one side then the other, of the narrow defile. The work was being done by soldiers, and five wooden bridges were ready. The newly-planed handrails glistened in the sun, and our Kirghiz horses, startled at so unusual a sight, could hardly be forced across the bridges.

"The view from the summit of the Kastek Pass, the desolate grandeur of which is commented on by Severtsoff, also made a deep impression on us. We could distinguish particularly clearly the long snowy chain of the Alexandroffsky Mountains on the other side of the Chu Valley. These were the first of five ranges which lay across our route.

"The southern slope of the Kastek Pass was much shorter and more abrupt than the northern one. It is worthy of remark here, that in all that portion of the Thian Shan range crossed by our expedition, the northern slopes of the chains of Kastek, Alexandroffsky, Djaman-daban, and Tashrobat, are considerably more gradual than the southern, and therefore the defiles which lead up them are more varied in character, more deeply indented, and more tortuous. The northern slopes, too, have a greater abundance of mountain springs; and owing to better irrigation, or, perhaps, to other causes, are richer in vegetation than the southern slopes. I found pine-trees growing on the northern sides only of the mountains, except in one place, viz., the southern side of the Son-kul plateau, in the defile of Molda-asu, where there is a belt of white pine. This locality, however, is particularly well adapted for arboreal vegetation, being well sheltered from the winds, and affording ample moisture for vegetation. As the defile of Molda-asu widens the trees become scarcer, and near the valley of the Naryn they entirely disappear. The tree-belt commences again on the south side of the last-mentioned valley (*i. e.* on the northern slope of the Naryn range), and even here only at some distance further to the east (near the mouth of the Atbasha), where the belt of white pines resembles a dark green ribbon winding over the slopes of the mountains.

"Opposite the former Khokan fort of Kurtka, on the south side of the Naryn Valley, the height of the mountains diminishes considerably, and there is a total absence of vegetation. Further to the south we saw pine forests at some distance off, on the northern slopes of the Tashrobat range, in the valley of the Atbasha. But I return to the description of our march-route.

"As we entered the valley of the Chu, the chief manap (sultan) of the Saribogishi—Djantai—rode to meet the commander of our detachment. A gold medal, with the ribbon of St. George, and a deep scar on his forehead, were sufficient proofs of Djantai's services to the Russian government, and of his victories over the notorious robber Kenissar Kassimoff, whose exploits from 1840 to 1850 kept the Kirghiz steppes in constant commotion.

"The valley of the Chu presents the same steppe-like, unvarying aspect as the Trans-Ili plains—fields of wheat and millet, cultivated by the Kirghiz-Saribogishi, were flourishing. Along the irrigating trenches white hollyhocks and the blue chicory-plant, common to the eastern portion of the Kirghiz steppe, grew in abundance. These two may be considered as the typical flora of those steppes. The latter further south and west disappears altogether.

"Neither here nor on the southern slopes of the Thian Shan did we meet with any rice-fields, though it is said that in former times the Khokandians cultivated rice in the valley of the Chu, near Tokmak. To the south-west rice-fields are first met with near Tashkend, and to the south not before approaching near Kashgar.

"On the 9th and 10th July our detachment crossed the Alexandroffsky range by the Shamsi defile. Here, for the first time, we met with thick pine-forests. We encamped in a beautiful meadow surrounded by forests of white pine. The dark green of the pine woods contrasted with the bright variegated grass, and reminded us of the fertile mountain scenery of Central Europe. The mountain ash was in full blossom; I saw no other kinds of deciduous trees. Among the shrubs I noticed the berberry, honeysuckle, dogberry, wild rose, &c.

"Of all the mountain passes and defiles we crossed during this expedition the Shamsi is certainly the most striking and picturesque. It can rank with the celebrated pass of the Tête-noire in Switzerland, between Chamouni and Martigny, and even surpasses the latter in grandeur. The beauty of the scenery in the Thian Shan loses considerably by the small number of mountain streams, and almost entire absence of waterfalls, which give so much animation to Swiss scenery. There is, however, one waterfall in the Shamsi of considerable width, and with a succession of descents. It is above the limit of arboreal vegetation, and on the left side of the pass. This was the only waterfall we saw during our seven weeks' excursion in the Thian Shan.

"On the 9th July, at 3 P.M., the thermometer in my yurt (felt tent) marked +17° Reaumur (70° Fahr.); the temperature then gradually diminished, and from 9 P.M. to 11 P.M. remained stationary at +5° R. (43° Fahr.). The following morning, at 5 A.M., the temperature was the same as that of the previous night; at 6 A.M., it was at +7° (48° Fahr.), and then increased rapidly.

"On the 10th July we commenced the last and most difficult part of the ascent of the Shamsi Pass. Before starting I availed myself of the opportunity of collecting specimens of the beautiful alpine flora met with here in profusion. Ranunculi, gentians, potentillæ, corydalis, in every variety of shape and colour—blue, yellow, white, and pink—glittered in the sunlight. I was reminded of the Mayen-wand at the glacier of the Rhone, in Switzerland, a spot familiar to all botanists.* Vegetation became rapidly more scanty as we ascended; the road over the rough shingle became more precipitous, and snow lay in places. We were now at the summit of the pass, which is very marked by the steepness of the ascent and abrupt descent.† The view from the summit is not extensive, as the nearest snowy peaks shut out the horizon. The valleys of the Chu and Koshkar, on the north and south, were, however, easily distinguishable. We remained two hours on the

---

* To give an idea of the botanical wealth of the Shamsi Pass, I will give a list of specimens collected within a radius of a few yards, and carefully arranged and classified by the academician Ruprecht:—*Pulsatilla albana, Ranunculus amœnus, Trollius altaicus, Aquilegia atrata, Papaver nudicaule, Corydalis Gortchakowii, Smelowskia calycina, Taphrospermum altaicum, Physolychnis gonosperma,* nov. sp., *Cerastium trigynum, C. triviale, Hedysarum flavum,* nov. sp., *Hed.,* nov. sp., *Oxytropis amœna, Ox.,* nov. sp., *Potentilla nivea, Alchemilla vulgaris, Schultzia crinita, Erigeron pulchellus, Ligularia heterophylla,* nov. sp., *Doronicum oblongifolium, Leontopodium Alpinum, Carduus nutans, Saussurea sorocephala, Taraxacum Steveni, Galium boreale, G. songaricum, Saxifraga Siberica, Cortusa Matthioli, Gentiana umbel'ata, G. variabilis,* nov. sp., *G. barbata, Myosotis alpestris, Euphrasia officinalis, Veronica biloba, Pedicularis rhinanthoides, P. dolichoriza, Polygonum polymorphum, P. bistorta, Salix marginata, Allium oreophilum, Alopecurus glaucus, Festuca ovina, Poa alpina, P. pratensis, Triticum Schrenkianum.*

† According to Buniakoffsky, the summit of Shamsi Pass is 11,830 feet above sea-level.

pass, in order to allow of the passage of our caravan of 130 camels, with which we experienced considerable difficulty, as they continually stopped and refused to go on, lost their footing on the steep, slippery ground, and at last lay down; no efforts would then make them move, and we were obliged to unload them and carry their packs up to the top of the pass. Profiting by the delay, we climbed the nearest rocky summit, and there found in the shingle a few solitary flowering specimens of the graceful Chorispora exscapa, Leontopodium, Rhodiola, and others.

"In the Shamsi Pass we met a small caravan which had already accomplished a considerable journey that summer. Its first point was Andidjan, whence it had gone to Tashkend, Auliye-ata, and Tokmak, thence across the Shamsi Pass to the Naryn, and was now on its way to Vernoye. Daba (cotton cloth) is the chief article of trade. One cannot help feeling astonished at the boldness and enterprize of these traders, who venture in small numbers into parts of the country where their lives are entirely at the mercy of the cunning Kirghizes. Of course the Kirghizes protect them as much or as little as they consider advisable, in order not quite to discourage them from entering the country. But they do not always do even this. During our expedition we were joined by a Sart, who complained that all his valuable merchandize had been appropriated by a relation of the celebrated Kara-Kirghiz manap (sultan), Umbet Ala. The robber, however, considered it expedient to give the unfortunate Sart a written promise to ay him whenever convenient, to which document he had affixed his seal.

"As already stated, the descent from the Shamsi Pass into the valley of the Koshkar is far more abrupt than the ascent from the north. The valley of the Koshkar is still more dreary and monotonous in aspect than that of the Chu. Crossing the above valley, we commenced ascending the Kizart, a tributary of the Koshkar, and then entered the belt of mountains on the northern slopes of the Son-kul Plateau. We reached the lake of Son-kul on the morning of the 14th July. All the previous night we were forcing our way through defiles and over cliffs in an ever upward direction, till at last, after climbing the last lofty axis of the range, we suddenly saw before us the blue surface of the lake scarcely rippled by the morning breeze. From the eminence on which we were standing, an inconsiderable distance of sloping ground covered with luxuriant grass separated us from the lake. In our rear, *i. e.,* to the north, the snowy peaks and ridges presented the appearance of a troubled sea. On the opposite (southern) side of the lake, the hills separating the Son-kul Plateau from the Naryn Valley, were as one continuous chain, not much elevated above the lake, with peaks here and there speckled with snow. To the north-west of the lake low equidistant ridges of green porphyry radiate from the nearest hills,* terminating at the shores of the lake in picturesque promontories of rock. On the night of the 14th we encamped at the very edge of the lake between two such ridges. The shores and bottom of the lake close to the shore are thickly covered with large pebbles, beyond which is a yellowish clay. I could not judge whether the depth of the lake increased suddenly. The taste of the water, though inferior to that of the rivulets which force their way through the grass to the lake, is fit for use, and our cattle drank it eagerly. There are no large fish in the lake; in the pebbly bed of the lake near the shore a few small fish about an inch long were seen and also crustaceous animals. We only saw one kind of bird on the lake, which the sportsman of our expedition pronounced to be widgeon. The only plant growing in the water was the Myriophyllum. We saw no reeds. ·

" The Son-kul Plateau is above the elevation of arboreal vegetation, it is admirably adapted for Kirghiz encampments. The soil is thickly covered by a short grass (called by the Cossacks kipets), in which are half concealed specimens in miniature of Alpine gentians, Saxifraga, Oxytropis, Leontopodium, Potentilla, &c. The extreme length of the lake is 25 versts (17 miles), extreme breadth 18 (12 miles), the Koidjarti flows out of the south-east corner of the lake, and falls into the Naryn.

" On the 15th July we marched round the north-west end of the lake, and came to a level plain which joins the lake on the west. The pasturage here is excellent, but we found it entirely deserted by the Kirghizes, who had retired to the mountains on hearing of our approach. Son-kul presented a very different appearance when we revisited it on our return in the month of August. Kirghiz encampments were then everywhere scattered along its banks, and presented all the animation of nomad life. The commander of our expedition was received with due honour, fêtes (baigi), with horse races, &c., were celebrated for the occasion. I never before witnessed such prosperity and wealth as we saw in the encampments of the Kirghizes on the Son-kul when we returned from our successful reconnaissance. The numerous clean white yurtas (felt tents) glistened in the sun, and afforded a marked and agreeable contrast to those we were in the habit of seeing on the Kirghiz steppe, where they are almost black, dirty, and smoke-stained.

---

* During the expedition I made a small collection of minerals, which were kindly classified by Nikolsky, Mining Engineer at Tashkend.

" To return to the description of our journey. The detachment continued its march in a southerly direction, and soon the lake was hidden from view by the spurs of hills on its south-west side. We halted on entering the picturesque defile of Molda-asu, whose descent to the Naryn is wonderfully steep. Gigantic orange-coloured rocks of close grained limestone marked the entrance to the defile, in which we first saw bushes of juniper, and as we descended graceful white pine trees. The pine forests grow in one continuous dark green belt along the most inaccessible cliffs of the mountain sides, and render the scenery quite peculiar of its kind. From the nearest heights we obtained a splendid view over the valley of the Naryn and the southern Naryn range, beyond which could be seen the elevated table-land sloping towards the north, and forming part of the water-shed of the Naryn. The distant view to the south was shut out by the snowy peaks, which seemed part of one great range, though in reality portions of a succession of ranges, of which fact we were afterwards convinced. But this great panorama was wanting in brightness—the colouring was monotonous and dreary, owing to the prevailing clay and grey limestone. Verdure could only be seen in the foreground close to the observer.

" We commenced the descent of the Molda-asu pass, rendered difficult for horses owing to the numbers of boulders and shingle strewn on the ground. Vegetation here was plentiful. Besides the white pine and juniper mentioned above there were varieties of leaf-bearing shrubs, wild rose, arbutus, currants, barberry, honeysuckle, dogberry, and mountain ash, the greater number of which were already bearing fruit.

" The first half of the descent and by far the most difficult terminates in a deep gorge, where a small stream from the right unites with the one in the defile, which here takes a sharp bend to the left (*i.e.* to the east) and widens out. The road becomes more even, and the stream of greater volume. Vegetation becomes still more plentiful and diversified. On both sides of the defile are the same orange-coloured walls of hard limestone, to which the dark stems of the white pines cling. In addition to the shrubs already mentioned, the banks of the stream were fringed with willow, birch, Potentilla fruticosa, and finally with poplar.

" We were more than an hour making the descent to the first bend of the defile, and three-quarters of an hour to the second, when we came to another deep valley at the wood of Ike-chat, the scene of Zubareff's memorable skirmish in 1863. Lieutenant Zubareff was sent in 1863 from Vernoye with an escort of 30 Cossacks in charge of a convoy of provisions for the force under

Captain Protsenko, who was then in the Naryn. At the wood of Ike-chat Zubareff was suddenly surrounded by numbers of the Black Kirghizes, who were acting in concert with the Kokandians. Zubareff defended himself resolutely for a whole day, having formed a wall of the bags of dried rusks, under which his baggage animals found shelter, in this manner he withstood 18 attacks. Seeing, however, the overpowering numbers and obstinacy of their opponents, his small force of Cossacks despaired of safety, when most opportunely succour arrived from Captain Protsenko. The approach of the relieving force, who attacked the Kara-Kirghizes in their rear, soon dispersed them.

"On emerging from the defile on the 16th July into the valley of the Naryn, which is 15 to 20 versts (about 12 miles) broad at this point, the vegetation suddenly changed. Poplars, birch, willow, and the different shrubs were no longer visible, and the bare clayey soil was here and there covered with a small heath and tall tufts of a kind of grass vetch (Lasiagrostis). We soon saw the channel of the Naryn conspicuous by the green belt of poplars and willow, which grow on its banks and islands.

"At 11 A.M. we were under the walls of Kurtka, an insignificant Khokan fortress, demolished in 1863 by Captain Protsenko, but since restored, as after that time no Russian forces had appeared on the Naryn. The garrison of the fort took to flight as we approached. It consisted only of 20 men, as we discovered from an interesting report to the Khan of Khokan, forgotten by the commander of the fort in his haste to run away. Kurtka stands on a lofty precipitous cliff, which is being undermined by the abrasion of the river, whose current is here very rapid. Kurtka had more the appearance of a heap of clay and dirt than the habitation of people accustomed to settled life. From a distance, indeed, it is difficult to distinguish it, for the colour of the habitations blends with the yellowish grey tint of the clayey soil. The fort consists of a number of small buildings, crooked lanes, and hedges, &c. The confusion and disorder everywhere apparent proved the hurried nature of the flight; mats, bags, sheepskins, and different domestic utensils were scattered on the ground, and the whole scene reminded me forcibly of a similar desolation, on the taking of the Chinese forts at the mouth of the Peiho by the English and French forces in the summer of 1858.

"On the evening of the same day a ford was discovered across the Naryn, where in its deepest part the water reached the saddle-girths. Here we crossed the following day (17th July). This part of the river is divided into several channels, its total breadth being from 200 to 300 yards.

"Our route on the other side of the Naryn lay through an

unknown country. As far as Kurtka we had an excellent guide in Captain Protsenko's topographical map, but we had now to trust to Kirghiz guides. The beautiful weather we had enjoyed ever since our departure from Vernoye deserted us, and for ten days continuously we were drenched by a small rain. At night there was hail.

"On the 18th we continued our journey down the valley of the Naryn. After making our way through tall cane thickets (Phragmites), we skirted the foot of the hills bordering the valley on the south, where we found willow and the Halimoden-dron argenteum* growing.

"In an hour's time we came to the Terek, a tributary of the Naryn, whose banks were wooded with poplar. This was the last arboreal vegetation we saw till we reached the southern slopes of the Thian Shan, two days' march from Kashgar.

"As we ascended a sloping valley, we were constantly obliged to avoid the large masses of clay which, detaching themselves from the impending heights, came rolling down into the valley. We encamped for the night at the upper end of the valley where it is more open, at the foot of some clayey cupola-shaped eminences totally devoid of vegetation. Far to the south lay stretched before us a gigantic transverse chain of mountains, whose snowy peaks, at first hidden in the clouds, shone out before evening quite clearly. The slopes were covered with grass land. In this chain we could distinguish the entrances to three passes leading to the valleys of the Atbasha and Arpa tributaries of the Naryn, and running parallel to it. We are told they are respectively named, beginning at the easternmost: Airitash, Beibitche, and Chalkudè.

"On the 19th July we came to an open plain, forming a plateau somewhat inclined towards the north, and surrounded on all sides, except on the north, by low spurs, beyond which were the lofty mountains. Our camping-ground was surrounded by a thick growth of phragmites, among which were interspersed clusters of violet-coloured Statice; but when we emerged on to the table-land the vegetation suddenly changed. The Acrop-tilon picris then became the prevailing plant, and in places there grew varieties of wormwood.† At times the vegetation

---

* In the beginning of June I saw this pretty shrub in flower on the Irtish, between Omsk and Semipalatinsk. In the Thian Shan I only found it twice, in the Buam Pass and Naryn Valley.

† This peculiarity in the grouping of the herbaceous plants on the steppes of Central Asia is well described by Semenoff, who says :—" Nowhere do the plants form one continuous sward, but grow in groups at some distance apart from each other ; the intermediate space is very often quite bare of vegetation, and the different kinds of plants are not intermingled equally, but for the most part grow in distinct groups."

entirely ceased, and we rode over the bare clay. The soil indeed was so sterile, that in places it reminded me of an enormous well-trodden exercising ground. The clay is at length mixed with sand and shingle, when vegetation re-commences.

"On this drearily monotonous table land our party had to encamp, for the first time, without water, since their departure from Vernoye. We continued a south-westerly direction, and after six hours' march, towards 3 P.M. we encamped for the night at the entrance to the defile of Djaman-daban.

"The Djaman-daban pass takes two days' march to get through, as does that of Shamsi. The Djaman-daban chain is, therefore, quite as important as the Alexandroffsky. We have no knowledge of the continuation of the chain westwards; towards the east it very soon diminishes in size as it approaches the Naryn.

"The entrance to the Djaman-daban pass is very conspicuous from a distance, owing to the bright red rocks of ferruginous conglomerate, which forms a distinct stratum. This was the wildest and most dreary defile which we crossed during our expedition. The rocks are for the most part of a gloomy grey colour. The ascent at first is very gradual; about the fourth mile up the pass there were bushes of juniper, but white pine does not grow in this defile. On the second day I saw (for the first time in the Thian Shan) a low shrub, known by the name of Camel's tail (*Caragana jubata*). It grows in large clusters, and disappears only when continuous vegetation ceases. On the southern slope of Djaman-daban the Camel's tail again covers all the rocks.

"The Alpine flora was very abundant; among the flowers which attract the eye of the traveller from their size or the brilliancy of their colour I will mention: *Anemone narcissiflora, Ranunculus songaricus, Geranium longipes, Potentilla Lehmanniana,* nov. sp., *Aster flaccidus, Doronicum oblongifolium, Gentiana Kurroo, G. falcata, G. detonsa, Scutellaria alpina,* &c. On the very summit of the pass, where plants hardly find room to grow, owing to the siliceous-sandstone shingle which covers the soil I collected: *Isopyrum grandiflorum, Drapa altaica, Cheiranthus himalayensis, Cerastium lithospermifolium, Rhodiola gelida, Richteria pyrethoides, Jurinea tenuis, Gentiana falcata, Androsace chamæjasme, Veronica Lütkeana,* nov. sp., *Gymnandra decumbens,* nov. sp., *Lamium rhomboideum, Rheum rhizostachium.*

"Some of the above mentioned plants have a particular botanico-geographical interest from their being almost, if not exactly, similar to Himalayan descriptions. The botanists

Royle* and Falconer† were the first to draw attention to the similarity between the flora of Kumaon (borders of Kashmir and Tibet), and those of the Altai. This similarity becomes more apparent as our botanists penetrate into the Thian Shan from the North, while Europeans from the south, that is from the Indian side, have explored the vegetation on the northern slopes of the Himalayas and part of the Karakoram range. Semenoff in his excellent article draws attention to this fact; his rich collection of plants includes no less than ten species,‡ which are common to the Himalayas. Our expedition in 1867 contributed materials for forming a comparison between the Alpine vegetation on the Thian Shan and Himalayan ranges. In this respect two important species were found on the heights of Djaman-daban: *Cheiranthus Himalayensis* and *Lamium rhomboideum*—two very beautiful and original plants first discovered by the celebrated French botanist, Jacquemont, on the southwestern border of the Thibet table land.§ I likewise found a third plant in the Toyanda Valley, described in Jacquemont's travels, and found by him near the village of Chango, viz. the *Artemisia macrocephala*. The *Androsace chamæjasme* obtained on the Djaman-daban pass was also found by the Schlagintweits in the Rotang pass in Cashmere; the *Delphinium vestitum*, often met with in the Thian Shan, belongs also to the Indian flora.

"The altitude of the Djaman-daban pass was not ascertained by barometrical observation, though I conclude that it is not lower than the Tashrobat pass, which is 12,900 feet above sea-level, according to Buniakoffsky. On the top of the pass the cold was intense, and snow fell. Our caravan was so exhausted that we determined on halting upon the slope of the mountain before entering the valley of the Arpa. We pitched our camp on some level ground which terminates in a rocky precipice. From here we had an extensive view over the valley of the Arpa, which is undoubtedly at a considerable elevation. On the opposite side of the valley lay the great chain of Kashgar-daban, on which I counted sixty-three snowy peaks. The snow-covered mountains appeared low to us, and the snow

---

* 'Illustrations of Himalayan Plants,' pp. 34–40.
† 'Travels in Cashmir' (1844), vol. ii. p. 462.
‡ *Anemone Falconeri, A. micrantha, Corydalis Gortchacowii, Oxytropis Kashmirica, Potentilla Salessowii, Sedum coccineum, Carum Indicum, Gentiana Kurroo, Phlomis spectabilis* (found by N. A. Severtsoff on the Chirchik western Thian Shan), and *Rheum speciforme.*
§ The *Cheiranthus Himalayensis* was found by Jacquemont on the Chiunbrunghaut Pass " inter lapides mobiles supra omnes cæteras plantas, alt. 5300 mètres:" *Lamium rhomboideum* on the Santi Pass ; the latter found also by Schlagintweit in Thibet, at an altitude of 17,000 feet.

seemed to continue to the very bed of the valley. We saw through a defile on our right what appeared to be a glacier descending into the valley, very like the Glacier de Bossons in the valley of Chamouni in shape. This eventually proved to be a flight of fancy, or an optical delusion, for the topographer of our party approached close to the very spot, and found no glacier whatever.

"The Kashgar-daban chain visibly decreases from west to east. As we descended the Djaman-daban we could see the great size of the Kashgar-daban chain on the west, with enormous snow-fields between the peaks. Further eastward, though not so far as the Súúk cleft, forming the pass of that name, there seemed much less snow. To the east of Súúk, opposite the rivulet Djamat (where we passed the night of the 23rd July on our way to Tashrobat), the snow again increases in quantity, though I should consider that the Kashgar-daban chain is not so stupendous here as in its western part. One of our companions, Chaldeyeff, Capt. of *Cossacks*, visited the Súúk pass. According to his opinion, this pass is considerably higher than that of Djaman-daban. The Cossacks and Kirghizes who accompanied him suffered much inconvenience from the rarity of the atmosphere, and he himself, for the first time during the whole expedition, experienced difficulty in breathing. However, Buniakoffsky, who visited the Súúk pass in 1868, ascertained the altitude, according to his observations, to be 12,740 feet, *i.e.* about 200 feet lower than Tashrobat.

"Chaldeyeff found on the summit of the pass an elegant *Oxytropis*, which was growing in profusion as a diminutive creeper on the very snow; this species is quite new and very remarkable, and has been named *O. tianshanica*.

"On resuming our march we descended to the valley of the Arpa, turned to the east, *i.e.* up the valley, and crossed the barely perceptible watershed separating the Arpa from another tributary of the Naryn—the Atbasha. On the 24th July we were at the entrance to the Tashrobat defile, which serves as a pass across the snowy chain, extending from the east and terminating abruptly at the head of the Arpa valley. A road frequently used by the caravans going to Kashgar passes through the Tashrobat defile; the caravans cross either by the latter route, or by the Terekti pass, more to the east.[*]

"We crossed the Tashrobat pass in one march, *i.e.* six to seven hours. This pass is very monotonous; the bare grey rocks, from the specimens I collected, consist entirely of siliceous

---

[*] Valikanoff went to Kashgar by the Terekti Pass, and returned by that of Tashrobat.

slate and hard limestone. There are no shrubs of any kind; the last junipers we had seen were at the entrance to the Arpa valley from the Djaman-daban.

"The Tashrobat pass is remarkable from the building to which it owes its name: robat or rovat—caravanserai.' This edifice is built on the mountain side which forms its back wall. The material of which it is rudely constructed is the siliceous slate, found in quantities in the defile; the whole building forms a square 49 paces on all sides; the façade consists of a wall, with an entrance through an arched gateway. The interior of the building contains a circular room (18 feet diameter), on which had been a cupola, now in ruins, and round this centre room a number of small cells.* Altogether the architecture of the building is peculiar, if intended, as it is said, for a halting-place for caravans. The Kirghizes attribute its existence to Abdullah Khan emir of Bokhara, who is known to have constructed simi-lar buildings in other parts of Central Asia.† Valikanoff mentions the Kirghiz myth about the impossibility of counting the number of cells in it. There are no old inscriptions on it. In the principal apartment are traces of fire having been lighted in it, and the walls are scribbled over by travellers accompany-ing the caravans.

"The last part of the Tashrobat pass is very steep. Here I made a valuable collection of beautiful Alpine plants. From the summit of the pass the eastern extremity of the Chatir-kul was visible. The descent to the lake is much shorter than the ascent from the north side. The height of the lake, according to Buniakoffsky's observations, is 11,050 feet, *i.e.* 1850 feet below the Tashrobat pass.

"We suffered a great deal from the cold on the Chatir-kul plateau—large flakes of snow fell at intervals; the following morning we found ice in our cooking utensils. We had no wood fuel, and in damp weather *kiziak* burns badly. We were forced to make our cooking-fire with pieces of old felt tents, &c.

"On the 26th July, in alternate sunshine and snowfall, our party passed round the east side of Chatir-kul, over level ground covered with saline grass, which further to the east joins the Aksai table-land. On our return journey we completed the cir-cuit of the lake: in shape it somewhat resembles the Issik-kul; its length is 21 versts‡ (14 miles), breadth 9¼ versts (6 miles). The snowy mountains are 6 to 10 versts (about 5 miles) distant from the lake on the south side. These mountains are a con-

---

* Here are a few more details about this building: front wall 107½ feet long, entrance 27½ feet in width, height of front wall 12½ feet.

† Grigorieff, 'Cabulistan and Kaffiristan.'

‡ The Issik-kul is 169 versts (113 miles) long.

The Chatir-kul has no outlet; but I think it important, for he sake of future travellers, to mention the following facts, which may be worthy of their investigation :—As we rode over he plateau which joins the lake on its eastern side we crossed a small brook or ditch, about 10 feet wide, which joins the Chatir-kul and has a direction from east to west. There was apparently no current in this brook. Our topographer in surveying went considerably more to the east and found that this brook finishes in a sedgy marsh, about 6 miles from the lake. From this marsh a small stream flows to the eastward, which he was told was the Aksai, but which, on further investigation, proved to have no connection with that river, for he found that it turned to the north and at length lost itself in one of the dry channels which return again to the Chatir-kul. The real Aksai has its source further to the east, about 13 miles (20 versts) from the lake.[*]

"Judging from the grass (*Batrachium* and *Potamogeton* pectinatus) left by the receding waters of the lake in some places on the east shore, at a distance of 20 paces from the water's edge, I should say that the lake rises 1½ foot, and probably a great deal more when the snows melt. At the west end of the lake there were traces of submersion of the shores for a distance of 200 yards from the lake.

"Chatir-kul is on the borders of Russia and Eastern Turkistan; the object of our journey was therefore attained. The commander of our expedition, however, very justly remarked that his survey would be far more valuable if it could be continued to some well known and important point. Kashgar was well suited for our purpose, and, according to the march-routes of the caravans, was only 5 marches' distant from Chatir-kul. We had no intention of going as far as Kashgar itself; but we

---

[*] Here is another circumstance I find recorded in my diary which is difficult to reconcile with the foregoing : As we rode along the plateau to the east of the lake, we suddenly lost sight of the lake as soon as we had gone some distance from it, which proves that the plateau on which we were rising slopes to the east.

thought that if we could cross the mountains in front of us we should see that city, even though at a distance from it, for we believed it to lie in an open plain like Strasburg, when seen from the heights of Schwartzwald, or Berne, from Mount Niesen on the lake of Thun. But we were wrong in our conjectures, for after three forced marches to the south we found ourselves still among the mountains, which, though lower, gave us no view of Kashgar.

"Seventeen versts (11 miles) from the eastern extremity of Chatir-kul is the pass of Turagat, forming the watershed at no great elevation above the lake. The streams running to the south unite with the small river Toyanda, part of the basin of the Kashgar-daria, Eastern Turkistan. A sloping defile, in which is the dry channel of a mountain torrent (100 yards wide), leads to the Turagat Pass; the bed of this dried watercourse is not covered with the shingle generally found in these mountain torrents, but with a red clayey sand. We met with several such watercourses during our excursion in an easterly direction from Turagat, where we saw numbers of small hillocks, between which the aforesaid dry water-courses of red sand wind. The whole district has a peculiar character. Judging from the appearance of the cliffs and sandstone rocks which form the sides of these watercourses, one is led to infer that in spring, when the snows melt, the quantity of water in them is very considerable.

"We rode more than once along the channels of these dried-up streams, whose beds form admirable bridle-paths, and found on the soft damp sand innumerable traces of animal life. M. Skorniakoff, our naturalist and sportsman, showed us tracks of bears, wolves, deer, argali, and hares; of the latter we saw numbers. What particularly struck us was the large number of skulls of argali, with great twisted horns, lying on the ground. This animal is timid by nature, and often falls a victim to birds of prey, who pursue it over the rocks, and by hitting them with their wings endeavour to throw them down. The frightened animal thus is often forced to jump from the rocks, and always falls head downwards. M. Skorniakoff said he was himself a witness of an occurrence of that sort. The body, with the bones, is very soon devoured by beasts and birds of prey, the skull and ponderous horns alone remaining. It sometimes happens that the argali becomes entangled by the horns in a narrow defile, when he falls an easy prey to his enemies.

"Another characteristic of these dry mountain torrents along which we rode for considerable distances is the total absence of all vegetation in them, with the exception of a low broad-leaved rhubarb-plant (*Rheum rhizostachium*), which was abundant in some places and nowhere visible in others. In all probability the seeds of this rhubarb are brought down from the mountains by the water.[*]

"On the 29th July we continued our journey south from the Turagat Pass. At first we descended the valley of the Toyanda Rivulet, bounded by low hills, and flowing almost due south; there are no steep descents and the ground slopes gradually. At our first night's encampment we found bushes of hippophae and a prickly heath. The temperature at night was warmer than the preceding nights. The following day, 30th July, we finally entered the zone of shrubs; the banks of the stream were covered with tamarisk, lycium, &c. Of herbs we found quantities of lagochilus, a pretty white flower belonging to the Labiatæ.

"About midday we came to the junction of the Toyanda with the Súúk; the latter flows from the west and probably has its source in the Súúk Pass in the Kashgar-daban. We this day overtook a small caravan on its way to Kashgar with sheep. The further we went the narrower grew the valley and the loftier its rocky walls; we were tired of this wild scenery. At last the defile took a turn to the east; the slates were the prevailing strata in the rocks, and their gloomy colour added still more to the dreariness of the aspect.

"The night of the 30th was bright and warm, and we slept comfortably under thin canvas tents, as we had left our yurtas (felt tents) at the Turagat Pass.

"The 31st July was our last day's advance. After two hours' ride we saw the first tree, a poplar; it is first met with singly, then in groups mingled with willow. The nature of the scenery reminded me a good deal of parts of Egypt, near Cairo: there was the same yellowish grey colouring, with the striking contrast of the verdure of the trees in the dazzling sunlight; the only difference was that in the Súúk Valley the trees are poplar and willow, while in Egypt they are sycamore and date palms. The Kirghiz burying-places reminded me too of Egyptian buildings.

"A little further on it became apparent we were approaching an inhabited country. There were fields of excellent wheat, and people, who did not seem at all disturbed by our appearance, quietly engaged in agricultural labours. They called themselves Chou-bogishi. Here too was a small caravan, whose proprietor, a Sart, was bringing cotton cloth from Kashgar into Russia; he

---

[*] This variety of rhubarb was found on the heights of the Djaman-daban Pass, *i.e.* nearly at the extreme limit of vegetation.

was much elated at seeing us. According to his tale, caravans are more than ever subjected to the exactions of the Kirghiz chiefs through whose country they may pass, and more particularly to those of the Sari-bogish Umbet-Ala. This Sart had now every prospect of reaching the Naryn safely with our escort.

" At a bend of the valley to the south-east we had a view of a small fort or outpost, situated on some elevated open ground. This was the outpost of Tessik-tash, where the Chinese authorities met Valikhanoff in 1858. The Dungan insurrection, however, put an end to the Chinese government in this country. In the stone quadrangle forming the fort were only a few dirty wretched yurtas, from which the frightened faces of some Kirghiz women and children appeared. Tessik-tash was the ultima-Thule of our expedition.

" According to the Kirghizes, we were 12 versts (8 miles) from Artush, and 30 versts (20 miles) from Kashgar. These distances are only approximate. In the march-route of Humboldt's ' Asie Centrale' (vol. iii. page 370), the distance from Kashgar to Artush is marked 30 versts, and from Artush to the first Chinese outpost 25 more, making 55 versts in all. It is doubtful whether in Humboldt's march-route the first Chinese outpost is understood to refer to Tessik-tash, though the march-route is certainly directed on Chatir-kul and Tashrobat. In any case, it was interesting to know that Kashgar is situated to the east of where we were. This fact determined Poltoratsky, in reading a paper before the Physical section of the Geographical Society, to lay down the longitude of Kashgar 2° farther east than that assigned by the Jesuits—a statement, however, made by him with some reserve, because all surveys south of the Issik-kul were based on the only astronomical point fixed, viz., the western extremity of the Issik-kul.

" Certainly, if we take into consideration Captain Reinthal's map, made at the end of 1868, of the road from Fort Naryn through the Terekti Pass to Kashgar, the position of the latter town would be again west of that assigned to it by our survey. In Walker's map of Central Asia (1867), Kashgar stands on the 75° 25' long. E. of Greenwich, *i. e.* 1½° further east than Klaproth's map, and this is based on Montgomery's calculations (see vol. xxxvi. ' Journal London Geograph. Soc.'). With regard to the astronomical positions fixed by the brothers Schlagintweit, who assigned the positions of Kashgar and all the towns of Eastern Turkistan 2° further to the west than those laid down by the Jesuits (an opposition to which was first made by the Russian Society in 1861 by the late Captain Golubieff, Fellow of the Society), we must exclude them altogether from the cartography of Central Asia.

"We have only to hope that the position of Kashgar may be finally determined as soon as possible by sending an astronomical expedition there."

Note.—Since the above was written, Mr. Hayward has made observations in Kashgar, and found its position, as stated in the present volume, to be in longitude 76° 20' E. of Greenwich.

IX.—*Special Mission up the Yang-tsze-Kiang.*   By R. Swinhoe, H.M. Consul.

Read May 9th, 1870.

I was employed last spring on a special mission of inquiry into the trade of the River Yang-tsze. On the close of the mission I forwarded a series of Reports to the Government through Her Majesty's minister at Peking. A printed copy of these Reports has been communicated by the Foreign Office to this Society, and at the request of our President, Sir Roderick Murchison, I have attempted to reduce and modify the Reports into the following paper.

I was instructed by his Excellency Sir R. Alcock to apply to Vice-Admiral Sir Henry Keppel for naval assistance, in prosecuting the inquiries connected with my mission on towns and marts on the Yang-tsze and in its neighbourhood. The Admiral, in reply to my application, said that he was himself going up the river in her Majesty's ship *Salamis* as far as Hankow, and perhaps beyond, and intended to have a look at the Poyang Lake on his way, and desired me to accompany him. He added that at Hankow he would have the gun-boat *Opossum* in readiness to take me in continuation of my explorations. At this juncture the Chamber of Commerce at Shanghae expressed an interest in the inquiries, and asked the Admiral if he would order a survey of the waters of the Upper Yang-tsze, to ascertain how far up they were navigable for steamers, and at the same time asked me if I would have any objection to be accompanied in my wanderings by two delegates that they desired to send to gather commercial information for them. I had every reason to be in favour of the assistance and companionship that gentlemen of commercial training would afford me, and, with the Admiral's sanction, I signified the same to the Chamber, and they appointed Mr. Alexander Michie, of Shanghae, and Mr. Robert Francis, of Kiu-kiang, to

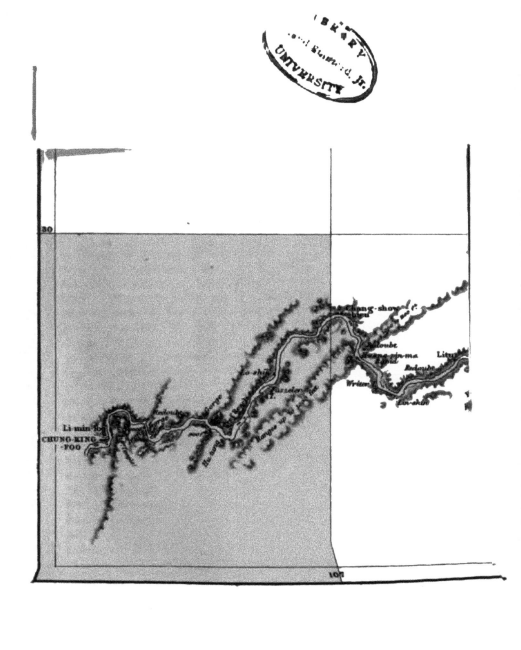

LIBRARY
UNIVERSITY

be my companions in travel. Messrs. Jardine, Mattheson, and Co., of this port, put their little steamer *Faust* at the Admiral's disposal, and the Chamber of Commerce undertook to put her in order, and have her towed by an early steamer of the Shanghae Steam Navigation Company to Kiu-kiang to take the admiral and our party about the shallow waters of the Poyang Lake. The admiral sent directions to two surveying officers to meet us at Hankow, and it was arranged that the *Salamis* should start on the 8th of March, and proceed leisurely to Kiu-kiang, Messrs. Michie and Francis to follow by the first Shanghae Steam Navigation Company's steamer. The *Salamis* was very crowded, and the Shanghae Steam Navigation Company had been so good as to offer free passages to all employed on the expedition.

I embarked on board her Majesty's ship *Salamis* on the 8th March, and she left Shanghae the same morning. We visited on our way Chin-kiang, Nanking, and Woo-hoo, and arrived at Kiu-kiang on the 14th March. At Kiu-kiang we waited the arrival of the next steamer from Shanghae, which was to tow up the little steamer *Faust*. Sir H. Keppel was anxious to try the navigability of the Poyang Lake to Jao-chow-foo, Shuy-hung, and Woo-ching—the three towns which are desired as landing-places. The passenger steamer arrived on the 17th, with the news that the *Faust's* repairs were not completed, and that she would not be in readiness for some days. It was useless for the *Salamis* to attempt the lake without a light draught tender in company, and time pressed, so at 4 p.m. the same day we were on our way to Hankow, where we arrived the following day. Messrs. Michie and Francis, the delegates of the Chamber of Commerce, were before us at Kiu-kiang, and passed on by the passenger packet.

The Shanghae Chamber of Commerce had pointed out the desirability of ascertaining the water route across the Tung-ting Lake up the River Seang to Seang-tan above Chang-sha-foo, the capital of Hoo-nan; and they also wished the River Han examined as far as Fan-ching (Seang-yang-foo) and Lao-ho-kow. These three places are great marts connected in trade with Hankow, and the right of residence and steam approach was much desired for them all. But the trade of Szchuen offered greater attractions, and the greatest desire was manifested for an inquiry into the navigability of the river up to the chief mart, Chung-king-foo. Captain Blakiston had already shown that sea-going steamers could reach Ichang. Foreigners could send their goods throughout the provinces of Hoopeh and Hoonan by means of transit passes issued at Hankow, but these were rejected at the customs barrier at Kwei-foo (the first Prefecture in Szchuen). The Chamber hoped that if we found the rapids insuperable to the class of river steamers at present in China, that a choice for the present should be made between Ichang and Shasze (King-chow-foo) whereat to establish a consular port, which would by overcoming the barrier Kweichow, throw open the Szchuen trade to us, and if the steam should eventually reach Chung-king, would form a port of tranship-ment from the lower river steam craft to those capable of stemming the rapids. The Admiral therefore consented to our exploring the river first, and to doing the rest of the work afterwards as time allowed. The two surveying officers, Messrs. L. S. Dawson and F. J. Palmer, ordered to accompany us, arrived at Hankow, and the Admiral determined to take our party up himself in her Majesty's ship *Salamis* as far as she could go, when he would transfer us to her Majesty's gun-boat *Opossum*, which he had specially detached and sent on in advance for this exploration service.

On the 23rd the *Salamis* started with our party, having the small steamer *Faust* in tow. There is an Admiralty chart of the river as far as Yo-chow-foo, at the mouth of the Tung-ting Lake, and so we had no difficulty on the road to that city, which we reached at 2 P.M. on the 26th. A crowd assembled on the bank below the town, and the surveying officers landing in the midst of it with their instruments to take sights, got rudely jostled and pelted by the mob. The people of Honan Province have long professed a hostile spirit to foreigners. At the Admiral's desire I called on the officer commanding the guard-boat, who was entertaining some friends, and brought him to see the Admiral. He made profuse apologies, and offered to send to the Prefect of the city for a guard to protect us if we wished to go on shore again. The Admiral was very anxious to get away, and all that could be done farther was to address a hurried note to the Prefect, quoting the treaty and complaining of the outrage. We left the city, and anchored at dark off King-ho-kow, the mouth of the Upper Yang-tsze. We travelled on nearly to She-show-hien, where the water shallowed, and the pilot told the Admiral it was unsafe for the *Salamis* to attempt to advance any further. The gun-boat was anchored not far above us round the bend, and the Admiral sent our party in the *Faust* to join her. This was on the 1st of April. On the 3rd the *Opossum* reached the large mart of Shasze, where we stayed till the afternoon of the 5th, prosecuting our inquiries on its trade, &c. On the 9th we anchored off Ichang. Above this the river contracts and runs through a series of gorges with rocky bottom, forming in places rapids and whirlpools. The Chinese pilot engaged at Hankow refused to take the gun-boat any higher, and, of course, her Commander, Lieutenant J. E. Stokes, naturally objected to attempt unknown waters against the pilot's warnings. We knew before that the river was navigable as far as Ichang, though the *Opossum* was the first vessel that had attempted it. But it was the rapids that we wanted the surveyors to examine and give an opinion on. We accordingly sent to engage a native boat to take us as far as Kwei-foo, the first prefectural city in Szchuen, situated just beyond the series of gorges and worst rapids. Passenger boats were at the time somewhat scarce, and there was such small difference between their charge for taking us to Kwei-foo, and that for advancing all the distance to Chung-king, that we thought it as well to engage one to Chung-king, in case we might find it necessary to push on so far. The captain of the boat required some days' preparation, so, not to lose time, Commander Stokes, the delegates, and I made a cruise in the *Faust* through the first or "Ichang gorge" (Hing Kwan Hea), and moored at the foot of a village called Nanto on the left bank, within a few miles of the Woo-e-tan or first rapid. The surveyors remained behind to complete their survey of Ichang. A local pilot that we had engaged was afraid to venture the *Faust* over the rapid, so we returned to Ichang the next day. On our way back we passed several large Szchuen boats, with their large crews of rowers pulling down the stream. The *Faust* steamed past them with ease.

The Szchuen boat was ready by the 15th April, and we at once embarked in her for the up-river voyage. Our party consisted of Lieutenant and Commander Stokes with two of his crew, Messrs. Dawson and Palmer, Messrs. Michie and Francis, and myself. Our boat was a Szchuen passenger or Kwa-tsze boat with a crew of forty-five all told, and had in company a Kwa-tsze or sampan to carry the trackers from one tow-path to another, and to serve at night as a sleeping-place for a good many of them. We started in the afternoon of the 15th. It is needless here to describe the slow and painful mode of tracking, the difficulties the boats encounter at each rapid, or, in fact, any of the details of this kind of travelling. We found our experience an almost exact repetition of what Captain Blakiston underwent, and has so well described in his work on the Yang-tsze. We passed the cities of Kwei-chow and Patung-hien (the latter a city without a wall) in Hoopeh, and Woo-shan-hien in Szchuen, and at length reached Kwei-foo, the barrier city, at which all boats are examined on their way up and down the river. Before arriving at this town the surveyors gave an unfavourable opinion on the rapids; but I thought their experience should extend to Kwei-foo before it should be accepted as decisive. They were very industrious in making a careful sketch survey so far, and took every opportunity of getting observations and fixing points, and Mr. Palmer made some truthful sketches of many of the most interesting spots on this enchanting portion of the river. Mr. Dawson, who was the senior officer of the two, represented that, as their opinion was against the navigability of the river for steamers until a thorough and separate survey of each rapid should be made, which would be a work of time, it was useless continuing their sketch survey any farther, when they might be doing more serviceable work at Ichang and downwards towards Hankow. I agreed with him, and he resolved to return with Mr. Palmer. Lieutenant and Commander Stokes had also accomplished all he wished to do, viz., to form an opinion of the rapids, and he also made up his mind to return with his two men. With the delegates and myself it was different. We had heard the adverse opinion expressed on the navigation above Ichang by the naval officers, and their doubts as to whether steamers could ever make use of the water way between Ichang and Chung-king : but Chung-king was reported to be the great mart of Eastern Szchuen, and it was highly desirable that we should by personal observation confirm this. At Ichang, just as we were starting in the junk for the upper waters, I had received a despatch from Peking, ordering for special reasons the return of the expedition, and the only consideration with me was whether I should be acting against the minister's wishes by progressing further. Her Majesty's minister objected to the gun-boat advancing, for fear the Chinese should misinterpret our intentions, and imagine that we were going to assist the French missionaries in their quarrels about their Christian converts. The gun-boat was safely anchored below Ichang, and the party proceeding comprised only three foreigners with Chinese attendants. We had two visits from M. Vincot, the Roman Catholic missionary resident at Kwei-foo, and he assured us that the disturbances had in no way extended to Chung-king, or to any part of our river route. We sent the passes that Li-hung-chang, the Viceroy at Hankow, on the application of her Britannic Majesty's Consul at Hankow, had supplied Messrs. Mitchie and Francis, to the Prefect, and after taking a copy of them he gave our boat a clearance and made no objection to our proceeding. The Szchuen people were friendly and well-disposed, and we had every hope before us of a successful cruise. I therefore determined to carry on my investigations as far as Chung-king, and thence to return with all speed to Hankow. A month would be required to get to Chung-king and back to the gun-boat, and I begged Lieutenant and Commander Stokes to return as soon as possible to Hankow, and leave us to find our way back by native means. This, however, Lieutenant and Commander Stokes declined to do, as he considered he would be acting against his instructions if he did not carry us safely back to Hankow. It was then agreed that we should make our journey to Chung-king and back to Ichang with as much speed as possible, and that the *Opossum* would remain at Ichang in readiness to carry us down to Hankow. A junk was hired at Kwei-foo to take the naval party back to the gun-boat, and they left us on the 26th of April; and the same day we started in the original boat on our further voyage.

At Kwei-foo a tribute-bearing boat of Lamas anchored just below us. The interpreter was a native of Yunan, and very Oriental looking. He said he was a small mandarin appointed by the Chinese Kinchai at Hlassa to escort a tribute-bearing embassy from the Great Lama to Peking, and as soon as this duty was over he expected to get office in Kwang-tung. The Lama was of a tribe near Si-ning-foo, in Kanshuh, and was one of the forty-eight Lamas that composed this mission. He spoke scarcely any Chinese. The interpreter gave us the names of places they had passed on their journey to Szchuen, through Tibet, and they tallied very fairly with the route on Williams' map of China. He told us that there was one head Lama among the forty-eight, and the rest were all Tootees. They prayed very frequently. That he himself accompanied them the whole way, but a local Wei-yuan was given them from town to town, and their expenses of travel paid by the local authorities of the towns they passed through. They were to receive 300 taels, 100*l.* at Kwei-foo, which would be divided among them according to scale.

Every six years a large tribute of this kind was sent to Peking, and the route followed was the present one—down the river to King-chow, and by canal across to the River Han, and up it to Seang-yang-foo, &c.

There was also a three years' tribute which went by another route, mostly overland through Shensi. The interpreter had met Mr. T. T. Cooper, who had stayed with his family. He had given Mr. Cooper a mule in exchange for a gun. His family were now journeying with him.

At Kwei-foo we also called on General Pao, who assisted in suppressing the Taiping rebellion, near Shanghae. Pao returned from the campaign to this his native town, a wealthy man. His house and grounds are by far the finest within the walls of Kwei, and recalled to mind some portions of the Yuen-ming-yuen, or summer palace. He has not recovered from the wounds he received. He was very glad to see us, and showed us about his premises, talking pleasantly of Gordon and other foreigners he had met, and saying what pleasure it gave him to be visited by Englishmen. He pressed us to come and spend a few days in his house. We were of course too hurried to accept this invitation.

On the 28th of April we reached and passed Yun-yang-hien, and on the 30th arrived at Wan-hien. Here we moored close to three boats of the Prefect of Kia-ting-foo, who was on his way to Peking. This Mandarin's brother, a merchant, called on us, and no little surprised us by telling us that he had a cargo of white (insect) wax, which he was taking down on speculation to Shasze. He added, that as he travelled in the suite of a Mandarin no questions were put, and his goods escaped duty. Thus, I presume, the Mandarins pay expenses of travelling on service.

On the 1st of May we made a start, but before proceeding far the boat struck against a rock, and made a hole, which required the greater part of the day to patch up. On the 4th we got to Chung-chow, and on the 6th passed Fung-too-hien. On the 8th we passed Foochow, on the 9th Chang-show-hien, and on the 12th we were at Chung-king-foo. We had already sent on to the delegates' Chinese assistant (a Hankow man, and one that had visited and knew well the chief trading cities of Szchuen) in a small boat from Foochow to prepare quarters for our reception, and this man met us with chairs at the landing-place, and we were soon installed in a big empty hong in the city. We expected to have a large excited crowd thronging our doors, but we were agreeably disappointed. People did come in some numbers, but there was no rudeness. It was necessary, however, to obtain some Mandarin recognition, and I sent next morning my writer to the Hien (the Taoutae and Prefect were both away at the capital Ching-too-foo, together with the French Bishop of Chung-king, about the Roman Catholic troubles in Yew-yang-chow), with Li Hung-chang's pass, to explain the object of our visit, and to ask for a guard for our door. The Hien sent back a polite message to say that he had heard through the Prefect's office from the Prefect of Kwei-foo, announcing our passing that city with the intention of visiting Chung-king to make commercial inquiries. He took a copy of the pass, and at once sent a guard of five Braves. With the help of these men our house was kept clear of all except those that were able and willing to give the information we sought; and with one or two of them following we were able to visit what part of the city we wished to see without molestation. The people were curious but not ill-behaved. They spoke of us as the "Yang-jin" (foreign man), and we very seldom heard the term "Yang-kwei-tze" (Foreign devil). We received a visit from M. Favard, the Roman Catholic Procureur, the only Frenchman then in the city, and in returning the visit Mons. Faurie, the Bishop of Kwei-chow, was found at their mission with two missionaries, one from Yunnan, and the other from Kwei-chow. The bishop was on his way to Rome to be present at the Œcumenical Council, but intended *viâ* Shanghae to go first to Peking. Many of the respectable native merchants, when they heard the object of our mission, called on us, and we busied ourselves nearly the whole time of our stay with collecting information on all matters connected with trade. The Chamber of Commerce had supplied Messrs. Michie and Francis with a bale of musters of foreign goods, and these we daily exhibited and heard opinions on. We also got specimens of all the various native goods that we thought would interest the Chamber of Commerce. Thus we occupied ourselves most thoroughly until the 19th of May, when, having arranged with the boat we came in to take us back, we again embarked and moved down three miles to a large temple called Ta-fut-sze (or Monastery of the Great Buddha), and prepared to devote the whole of the next day to the study of the country, while our men were settling accounts in the city prior to returning. All boats upward-bound stop at this temple, and the sailors worship and give thanksgiving for safe voyage before a gilt giant idol seated on its shrine, conspicuous in a building with open front below the temple wall, above high-water mark. The hills in this neighbourhood were dotted with hamlets and farms, and thoroughly cultivated. The great crop of the winter—opium— had been gathered, and rice, tobacco, cotton, maize, millet, ground nuts, and runner beans were now springing up in luxuriance. The people met us everywhere with smiles, and talked of us as simply "Yang-jin" or "Yang-tsze." No opprobrious epithets were heard from them, nor did they shout derisively at us.

On the 21st we left for our downward voyage. And passing rapidly over the ground which had been so tedious and laborious to ascend, in spite of the north wind, which blew pretty constantly against us, we reached Kwei-foo at 10 A.M. on the 25th of May.

We were delayed here till 4 P.M., for inspection, and the grant of a pass by the customs' officials.

The water had risen considerably, and most of the towns before perched so high, were now not far removed from the water level. Most of the rapids, too, had changed their character : some had disappeared, and others had arisen where before the stream ran smoothly.

After a few short delays, owing mostly to the trouble the captain of the boat had in controlling our crew, who, the moment the boat touched at a city, were off to do small trading speculations of their own, beyond the reach of the gong, which was beaten to recall them; and in one case, to the carelessness of the steersman in allowing an upward-bound junk to run foul of the boat, and break its bow sweep, we arrived before noon on the 27th at Ichang, and were delighted to see the gun-boat anchored a couple of miles below the city. We had to stop a short time at the Lekin Station to have our Kwei-foo pass inspected: and before long, were once more on board the "Opossum." The Bishop of Kwei-chow and his two attendant missionaries came alongside soon after us, and spent a good part of the afternoon on board.

On the 28th the "Opossum" weighed anchor: and on the night of the 31st May, we were landed at Hankow.

A week before our return to Ichang, Lieutenant and Commander Stokes, on a visit to the city, had his attention drawn to a proclamation posted on its walls, which professed to emanate from certain scholars and people who wished for the extermination or expulsion of the foreigners. He called the attention of the mandarins to it; and he tells me a counter proclamation was issued. He speaks in high terms of the civility he received from the people of Ichang during his stay. Large numbers of them used to crowd off, and showed great interest in the steamer and things foreign.

Mr. MacFarlane, the second master, was, on the gun-boat's arrival at Ichang, suffering from a rheumatic complaint; and a young man at once offered to bring a noted doctor to attend on him. The doctor came, and Mr. MacFarlane was soon cured of his painful ailment. This, and other signs of attention were numerous. Lieutenant and Commander Stokes gave me an original copy of the proclamation, which has been carefully translated by Mr. Oxenham, of the Hankow consulate. The proclamation is well composed; but I think, from its tenor, is either a squib or an official bugbear to scare the foreigner. It had no effect, for we found, as we passed the place, that the people were no worse disposed than before.

The following extract will convey some notion of its style:—

" Now there is a country, by name England, in a corner of the ocean, insignificant yet offensive, and with a people obstinate in their lawlessness, sudden as pigs in their appearance, and as destructive as wolves in their ravages, with a rooted desire to injure our people for their own glorification. The sea monster swallows large pieces at once, but the silkworm nibbles; so these think to add to their territory, and proclaim themselves a great nation. The doctrine they most esteem is that of Jesus, which leaves nought for those who worship at chapels but ruin. In assembly, too, they honour the Lord of Heaven; but those who worship beneath the Cross are all but as demon elfs. Moreover, what they rely on is an inaccessible port, and wickedly do they deceive men. Their fire-wheel vessels go up and down like wind; and no matter whether a place is prosperous or otherwise, nowhere do they not do their best to entice away men to ravage, in accordance with their wishes. They have encroached upon Hankow, and now desire to disturb Ichang, so as to deprive us of the sources of our livelihood, and have designs on our land and property, with various other evils, innumerable as the hairs on the head.

"For place of such reformed manners and prosperity as ours, how, in these enlightened days, are we to grant permission to such wicked demons as these to come? On account of this, we must all be bound over to be unanimous, and with care found a secret society; and, in order to make a clearance of these imps, establish certain rules, as below, for their destruction."

Then follow some childish rules which are not worth wasting time on.

Mr. L. S. Dawson, the senior surveyor, gave me at Hankow the following extract from his letter to his commanding officer, stating his views on the difficulties of navigation above Ichang:—

"The part of the river between Ichang and Kwei-foo was particularly examined, more especially in the vicinity of the rapids, and I regret to have to give it as my opinion that steam navigation cannot be carried on above Ichang. The force of current, want of anchoring ground, intricacy of navigation, and changeable condition of the river's bed, are, I consider, sufficient reasons to preclude the possibility of anything beyond a native junk being able to ascend these rapids. The descent would be, if anything, more difficult, as should a vessel fail to answer her helm at the exact moment, nothing could prevent her being dashed upon the rocks.

"To make a proper survey of these rapids would be, I consider, at any time a matter of much danger, if not of sheer impossibility, as I found on making the attempt in a boat with ten rowers, that she was altogether at the mercy of the current, and the chance of swamping or striking a rock more than probable; this was in April, and from what information could be gleaned from the natives the most favourable time. From the appearance of what would become the river's bed in summer (now some thirty feet dry) the rapids must increase in danger and violence, inasmuch that even junks have to tranship their cargoes. On the return journey by junk, a line of soundings was obtained mid-channel, the depth of water in the gorges and above Ichang generally was found to be above twenty fathoms, rocky bottom. In one gorge forty-four fathoms were obtained. The various dangers are most abrupt, the lead giving no warning. No opportunity was lost of testing the speed of the current, although in the immediate vicinity of the rapids this had to be estimated, owing to the junk being tracked up close to the shore where the current's force was not so much felt. The river between Ichang and Yoh-chow is of similar nature as below Hankow, and quite as navigable for vessels of seven feet draught from the beginning of April to the end of September. Local report as to the fall of the river in these parts was so unsatisfactory, that although on the whole it tends to the conclusion that the river was in April at its lowest, still unless I spoke from actual observation, or better authority on the matter, I should feel much inclined to doubt this statement. This part of the river is subject to more changes than the river below Hankow, but nothing beyond what a pilot's experience could keep pace with. The general rule in navigating the river is to hug the steep bank, but the formation of the banks and difference of depth on either side of the ship as shown by the lead, are also of great assistance. Plans of the river in the vicinity of Ichang and Shaaze have been made, on scales of 2 and 3½ inches respectively."

Lieutenant and Commander Stokes, of Her Majesty's gunboat *Opossum*, was also so kind as to favour me with his views as expressed in his report to the admiral. I extract the following:—

"The great rise of water in the Upper Yang-tsze is, I suppose, caused by the melting of the snow on the mountains between China and Tibet. When the water is high it must be enormous in the gorges and confined portions of the river. I should certainly say that on the precipitous sides of the gorges the water must rise sixty to eighty feet, and I noticed the houses in the narrow parts were built very high up. We passed several rapids between Ichang and Kwei-chow-foo, and I consider in three of them the velocity of the current must have been eight to ten knots, very narrow, and appeared to be infested with rocks, with large boulders and rocks on both sides. I noticed most fearful eddies and whirlpools in the river also, before you came to and after you passed a rapid, which would, in case a vessel did not answer her helm at the very exact time, place her in a very critical position, and she would most likely be dashed to pieces against the numerous boulders and rocks that infest the river.

"It would not matter so much if the shore was of a mud nature, but being a place infested with large boulders on each side and very narrow channels, it would be a fearful risk for any steamer, and I would not like to cross these dangers in any ship that I commanded. It would not be so bad for a steamer proceeding up against the rapids, but the great difficulty would be in the downward course, on account of the velocity of the current. I inquired of several of the junk captains who trade up to Chung-king, if a vessel of my draught could cross; they all had the same opinion, that there was not sufficient water for her. I have been informed by some Chinese merchants at Ichang, that several of the large junks are lost in crossing the rapids by striking on rocks. It is my opinion, from Ichang to Kwei-chow the rapids are bad both during high and low water, and I should imagine they are worse during summer. I should think it would take nearly a year to survey them, and it would be a very perilous duty to perform. I did not observe any anchorage for a ship in the river, the junks when they wish to stop, either make fast to rocks, or drive piles into the shore, and make fast their ropes to them. The river is exceedingly tortuous in some places, and the width about eighty to a hundred yards. The high and precipitous mountains in the gorges, and the scenery on each side is a very imposing sight, especially after the low mud banks from Hankow to fifteen miles below Ichang."

I add the report of the gun-boat chief engineer, Mr. Ambler, to Lieutenant and Commander Stokes, on some fourteen tons of native coal from the Tung-moi coal district, about 200 miles above Ichang, which were burned in the gun-boat furnaces:—

"In appearance the coal greatly resembles good anthracite, but in the consumption of it, it requires large and roomy furnaces, and a slow combustion. There is no doubt it will prove a good and serviceable coal for steamers using the river, but for vessels with very small furnaces fitted to their boilers it is not so good."

The commercial information collected on the expedition was submitted by the Delegates of the Chamber of Commerce at Shanghae, in the form of a report to the members of that body, and has been published by them at Shanghae. On my return to England, I took an early opportunity of presenting a copy of this work to this Society, and those who wish to consult its pages will find it in the Society's library.

I had intended on my return from the up-river expedition to devote a few days to the exploration of the Poyang Lake, in company with Delegates of the Shanghae Chamber of Commerce; but a despatch from her Majesty's Minister, which

I received at Hankow, compelled me to abandon the idea. I, nevertheless, considered it desirable to wait over a steamer at Kiu-kiang to gather any further information that the residents might, since my last visit, have obtained on the commercial advantages to be gained by admitting steamers to certain places on he lake.

The lake is, at its lowest in winter, shallow and dry in many parts, with channels running in different directions. The water flows into it from the Yang-tsze, and with the river it rises in summer 30 or more feet, and is then, of course, navigable for the greater part of its expanse. Our desire was to ascertain the depth and courses of the main channels to the three places specially asked for as points of call, and to confirm by actual inspection their commercial importance.

Two of the places, Jao-chow-foo and Woo-ching, have been several times visited, and a survey made in 1866 by Lieutenant and Commander Kerr, of Her Majesty's gun-boat *Cockchafer*, of the northward channel, as far as Woo-ching, and of the eastward channel to within 26 miles of Jao-chow, has been published by the Admiralty. In 1868, these channels were again surveyed by W. Stuart, commanding His Imperial Chinese Majesty's steamer *Elfin*, who extended his examination of the eastern channel as far as Jao-chow itself; the survey has not yet been published. Of the channel to Jui-hung we know nothing.

Each of the three places named is at the mouth of two rivers. Their commercial recommendations have already been named by the Kiu-kiang Memorialists under date 1st May, 1867, and 2nd July, 1867, and by Mr. Consul Hughes, in his despatch of the 6th May, 1868. It has since been ascertained that Jao-chow might be made the port of shipment for some excellent coals that are worked on its lower river, 30 miles above the city and 9 below Lo-ping-hien, and it is even believed that steamers could ascend the river to the mines. Mr. Interpreter Cooper visited these mines in February last, and reported on them to Her Majesty's Minister at Peking. Mr. Holingworth of Messrs. Francis and Co., Kiu-kiang, was there more recently, and I append his observations which were published, in the 'Supreme Court and Consular Gazette,' Shanghae, of the 29th May, 1869.

Juihung is at the mouth of two rivers; the one leading eastward to Hokow, an outlet for the red-leaf black tea, and to Kuang-hsin-foo, noted for its paper manufactories, the other, southwards to Foo-chow-foo (Keang-se), whence native paper is also brought in large quantities. Juihung offers no mart, and is not even used as a place of transhipment by the natives. The boats from Hokow pass it, and go by a circuitous route past Nanchang, round to the great port of transhipment, Wooching. In summer, the land between the winter limit of the lake and Nanchang is all flooded, and boats can proceed by a nearer course from Juihung to Wooching, but they are never bold enough to ascend across the lake to Jaochow. As on the Yang-tsze so apparently on the lake, the Chinese boatmen dread the open waters, and avoid them by making use of creeks and side channels. Juihung has been named by the Memorialists simply on account of its position; and if a channel be found leading to it, it is not unlikely that steam would compel transhipment there, and thus save the goods the 70 extra miles to Wooching.

Wooching, in the south-west corner of the lake, about 45 miles from Kiu-kiang, is the great port of transhipment for goods that come from, and go to, the Yang-tsze. It is also at the confluence of two rivers, one communicating with E-ning-chow (Nanchang Prefecture), the chief black-tea district of Kiangse, and the other with the provincial capital, Nanchang. A better point could not be chosen for steam approach.

It is proposed to make the three places indicated points of call for steamers having their head-quarters at Kiu-kiang, or, in fact, to make them subsidiary to Kiu-kiang, so that the lake trade should flow through Kiu-kiang as at present, but with the greater despatch and security afforded by steam. This mode of transport, it is hoped, would defeat the local exactions, and by means of transit-passes, foreign goods would be laid down cheaper at, or near, the place of consumption, delays would be avoided in bringing out the produce of the country, and thus a fresh impetus would be given to foreign trade, while foreigners would be able to regain their share in its distribution.

That the grant of steam will facilitate and expedite the transport of goods, and make communication with places on the lake safer and surer there can be no doubt. It has been said, with great semblance of truth, that the lake offers as great an obstruction to intercommunication between places on different sides of its shores as does the Atlantic between Europe and America. Its strong currents, and the storms that without warning render its shallow waters into a boisterous sea of waves, make its navigation almost impracticable to the native junk, and cause very serious delays. At present, the better class of foreign goods, such as cottons, woollens, opium, and treasure, are transported at heavy cost overland, by means of wheel-barrows, to different parts of this province. The heavier goods, such as metals, sugar, seaweed, cotton, &c., on which the native merchant cannot afford cost of land-transit, he is obliged to convey by water; but much time is lost in waiting for fair wind.

At the same time it must not be forgotten that the land route enables the merchant to escape the taxes that await him at so many points on the water-ways. Boats downward-bound to Kiu-kiang sail or scull to Takootang, where they are bound to call for inspection by the tax-collectors. From Takootang they proceed in the same way to opposite Hookow, and there meeting a towpath, are easily tracked to Kiu-kiang, a distance of 15 miles. This leads me to the question of substituting Hookow-Hien as a Consular port in place of Kiu-kiang-foo. It was a startling suggestion made by one of the Memorialists, but has been ably answered. The answers appear to me distinctly to prove that the choice made of Kiu-kiang in the first instance as the port of the lake, was unquestionably correct. The opening of Hookow, even as a landing-place, would perhaps be a boon to the steam interest, but would be a death blow to Kiu-kiang.

The Kiu-kiang Memorialists, in their first Memorial, dated 1st May, 1867, ask to be allowed to run tug-boats on the Poyang Lake to help the junks in and out. River junks do not tow well and their boatmen have great objection to being in tow of a steamer, as we learnt to our cost on the up-river cruise. I am of opinion, consequently, that little advantage would be gained if this concession were made; whereas the right of steamers to call at certain points, with the right of residence of a conditional kind at those points, would no doubt give a fresh start to Kiu-kiang, and if properly managed might lead to a permanent prosperity of foreign residents at this port.

A comparative statement of the trade at Kiu-kiang from 1863 to 1868 shows no signs of decline in the commerce of the place, but what is vividly felt is that foreign trade is slipping from foreign hands. May we hope that the opening of the lake to steam will restore it with increased advantages to the foreign merchants.

I will conclude with a few remarks on the landing stations proposed on the Yang-tsze. My voyage in Her Majesty's ship *Salamis*, in company with his Excellency Vice-Admiral Sir H. Keppel, gave me an opportunity of making some inquiries as to the commercial advantages of most of the places suggested, and on my return from Szchuen, I was able to add to my former information. I will take the places in their order of position up the river.

The first in order is Kiang-yin Hien (in Chang-chow Prefecture, Keang-soo Province), on the south bank 94 miles by river from Shanghae, and 60 miles from Chinkiang. The opening of this place, with the right of residence is much desired. I have not visited this town, and have nothing to add to what is already known of the advantages it offers.

The second is Sien-neu-miao, up a broad canal on the north bank, and about 14 miles from Chinkiang, with which it is connected by a canal on the opposite bank. It is a market-town (or Chinshe in Yang-chow Prefecture—Keang-soo Province), which grew to considerable importance during the occupation of Chinkiang by the rebels, and numbers of Chinkiang merchants established themselves there; but since order has been restored, the merchants had been deserting the town and returning to their former positions at Chinkiang, which is more conveniently situated for the flow of trade. At present its chief trade is in timber, which comes down the river and gets distributed at Sien-neu-miao. It was only mentioned in one of the first memorials, when it was less known than it is now.

The third and fourth are Tai-ping-foo and Woo-hoo-hien, in the Province of Anhwuy, both situated on the south bank; the first 80 miles, and the second 95 miles from Chinkiang. I accompanied the Admiral on a walk through the city of Woo-hoo and its suburbs on the 12th March. The city is placed about 3 miles from the Yang-tsze, on the bank of a small river, with very rapid current. Within its walls ruin and desolation met our eyes on nearly all sides. Few houses had been rebuilt, and few were building. But the 3 miles of bank on either side of the little river were lined with substantial hongs, mostly new. The river was swarming with boats of various sizes, and there was every appearance of great commercial activity. It promises to be a thriving place. It is under the jurisdiction of Tai-ping-foo and the chief Mandarins (Taoutae and Chintae), of that prefectural city, which is still an almost hopeless mass of ruins, with only a few hundred inhabitants, have deserted it and taken up their residence at Woo-hoo. The highway from Tai-ping-foo to the river lies through Woo-hoo. Tai-ping-foo never seems to have been a business place, even before it was pillaged by the rebels. I fear little advantage would be gained by making a landing station there. We were much beset by crowds in our ramble through Woo-hoo, and the Admiral was pressed for time, I therefore had little opportunity to converse with the better class of natives.

Ta-tung (a market town or chinshe, in Tsing-yang-hien of Che-chow-foo, Province Anhwuy) on the south bank, 64 miles from Woo-hoo, and Nganking-foo (the capital of Anhwuy) on the north bank, 40 miles beyond Ta-tung, are the remaining two places of landing suggested by the Memorialists. The *Salamis* took the usual course, north of the island which shuts in Ta-tung, and we saw nothing of this proposed station.

Nganking-foo we skirted close to. It is the capital of Anhwuy and seems a large city, with a tolerably prosperous look; but there were few boats of any size about its banks, and few indications of trade.

A mercantile gentleman visited Woo-hoo and Ta-tung in January, and to him I am indebted for the following information :—

Woo-hoo is by far a more important port than Ta-tung, and is well adapted to become a commercial depôt of considerable importance. Its superiority, when compared with Ta-tung, is mainly owing to the excellence of its water communication with the interior, and its suitability to become a place of export for both green teas and silk. It also adjoins an extensive cotton-producing district, and its present trade is larger than of any other port on the river between Chinkiang and Kiu-kiang. A river with a depth of 5 to 6 feet of water in the winter, and 10 to 12 feet in the summer, connects Woo-hoo with the important City of Ning-kwo-foo, in southern Anhwuy, distant 50 miles. A branch of the same river runs inland over 80 miles, in a south-western direction, to Tai-ping-hien (of Tai-ping-foo), an extensive tea district. The Tai-ping tea districts are not situated in the vicinity of Tai-ping-foo, as the name would imply, but about 90 miles south-west of it, at Tai-ping-hien. They are much nearer to Ta-tung and Woo-hoo, being about 50 miles from the former, and 80 from the latter. This branch, which is only navigable in the summer, passes through Nan-ling-hien (of Ning-kwo-foo), where the cultivation of silk is carried on. The production of this article is not very large at present, owing to the devastations of the Tai-ping rebels; but it is steadily increasing every year, and is likely to become a trade of more importance before long. Boats carrying from 200 to 300 piculs of tea can come from Tai-ping-hien to Woo-hoo in the summer and autumn; but in the winter the creek is partially dry, and navigation of course rendered impossible. The silk districts of Nan-ling-hien are situated within 50 miles of Woo-hoo. Besides the watercourses leading to Ning-kwo-foo and Tai-ping-hien, there are two others communicating with Szan (a market town or chinshe in Kien-ping-hien of Kwang-tih-chow, Anhwuy Province) and Tung-po (also a chinshe in Kao-shun-hien of Kiang-ning-foo, Keang-soo Province). The Szan Canal is navigable for small boats in summer for nearly 100 miles, and passes through some silk-producing country; while that leading to Tung-po can be traversed by native craft of considerable size for about 70 miles. On the northern side of the Yang-tsze, a fine broad canal, navigable in summer for vessels drawing 10 to 12 feet of water, connects Woo-hoo with Sew-chow-foo (Anhwuy

Province), the chief mercantile epô for Central Anhwuy. In winter there is a minimum depth tof 4 feet of water in this canal, and its average width is over 200 yards.

Ta-tung is situated partly on the mainland and partly on an island on the Yang-tsze. It comprises two straggling villages; that on the island being the chief centre of trade at present. The place looks poor, and with but few signs of the busy traffic that are apparent at Woo-hoo. It is one of the calling-stations for salt-junks passing up and down the river, and its trade would seem to depend very materially upon this circumstance. Hemp or china grass is extensively grown to the south-west of Ta-tung; rice and cotton are also produced in the neighbourhood, but not to so large an extent as at Woo-hoo. It has only one navigable creek leading inland to Southern Anhwuy, and that for merely a distance of 12 miles. Ta-tung is nearer the Tai-ping tea districts than Woo-hoo, but the advantages of direct water communication possessed by the latter makes it the natural port of exit for all the green teas of that part of Anhwuy. Goods forwarded from Tai-ping-Hien to Ta-tung have to be conveyed overland for a distance of about 40 miles. Ta-tung is also connected by water with Sew-chow-foo on the north of the river, but the creek is said to be shallow and not much used.

On the opening of Woo-hoo as a landing-place opinions are unanimous, but there are some who think that the opening of Ta-tung and Nganking would be of no material benefit to foreign trade. Ta-tung is considered too near Woo-hoo, being only 64 miles distant; and if Woo-hoo is opened they believe Ta-tung will be quite unnecessary, and would not repay the expense; the steam interest would only benefit by it. Anking is also considered too near Kiu-kiang (distant 80 miles), and its trade not sufficiently promising to make it desirable to open it. Others, again, ask for as many centres of distribution as can be gained.

A sketch map, taken from Chinese sources, of the Yang-tsze from Chinkiang to Kiu-kiang, shows the water communication with the interior by means of small rivers and canals. From this Woo-hoo and Tai-ping certainly appear to have the widest connection by water. Chinese do not often make use of land transit for the conveyance of goods when water-ways are to be had; but many of these small streams are doubtless only available at certain seasons. The map corroborates the statements above offered in favour of Woo-hoo over Ta-tung, but these two places are some distance apart and in different Prefectures, and I do not see why both should not be opened. I think Tai-ping might be left out and the other three places, Woo-hoo, Ta-tung, and Anking, accepted as landing-places, with the *right of residence.*

X.—*The Irawady and its Sources.*   By Dr. J. ANDERSON.

Read, June 13, 1870.

I PROPOSE to restate all the information I have been able to collect regarding the sources of the Irawady, in the hope that by so doing I may throw some light on an interesting geographical problem. The little additional information I was able to gather on this subject, during my residence at Bhamaw, does not amount to much, but is worthy of record.

I shall first, however, reproduce all the facts which Wilcox brought to bear on this subject, along with those collected by his fellow-workers, Captains Burlton and Newfville; but as none of these observers had ever seen the main stream of the Irawady below the junction of its eastern and western branches, at lat. 26°, I shall reproduce the accounts that Hannay, Bayfield, Griffith, and Williams have given of it, as far north nearly as lat 25°, and the opinions they had formed as to its probable origin.

It may be as well to state, in the outset, that I am no disciple of the theory that the Sanpo is the Irawady, and I cannot see how it is possible, at the present day, in view of Turner's account of the Sanpo and the accurate observations made by Captain Montgomery's pundits, that any one could be found to be prepared to readvocate its claims. It appears to me, however, that Klaproth's hypothesis has done good service to the Irawady, in so far that it excited an interest in the discovery of its sources, and gave it that importance to which it is entitled by the enormous body of water which it carries to the sea. The very circumstance that so many able geographers have been found willing to pin their faith to the theory in question, seems to indicate that there must be some foundation for the opinion that the main stream has its source a long way to the north of the Khamti mountains. This, however, only by the way, for such evidence is of little practical value.

While Lieutenant Burlton was engaged on the survey of the Brahmaputra, in March, 1825, he considered that the eastern branch had its source in a high snowy range above the Brahmakund, and, from what the natives told him, he was inclined to believe that the *Seeres Lohit* or *Sri Lohit*, or Irawady, arose at the same place.

In June, 1825, Lieutenant Newfville learned that the ranges of the Brahmakund extended much further to the eastward than was at first supposed, and he was informed by some Khamtis that the Irawady took its source from the opposite side of the same mountain from which the Brahmakund rose, and he gave it as his opinion that this theory of the sources of the two streams was by far the most probable, and agreed more with the general accounts and geographical features of the country.

Captain Wilcox, on his visit to the Lári Gohains village, some distance up the Sudiya stream, learned that the Lama country, on the banks of the Brahmakund, was about 15 days distant, and the upper part of the Irawady about the same. On his return to Sudiya he met ambassadors from the tract beyond the Irawady in lat. 25° to 26°; Burmans and Shans, the latter from Mogoung, west of the Irawady, in lat. 25°, and the former from various parts of their own empire; and many Khamties from the *source of the Irawady.* (?) Taking advantage of these opportunities to investigate the connection of the Irawady with the Sanpo, all that he was able to establish was the existence of a large eastern branch of the former river.

Wilcox gained his first view of the supposed main stream of the Irawady from the hills which separate the Namlang, one of the affluents of the eastern branch of the Brahmaputra, from the plains of the Upper Irawady. The stream winds in a large plain spotted with light-green patches of cultivation and low grass jungle. On reaching its banks he states that he and Lieutenant Burlton were surprised to find but a small river, smaller even than they anticipated, *though aware of the proximity of its sources.* It was not more than 80 yards broad, and still fordable, though considerably swollen by the melting snows. The bed was of rounded stones, and, both above and below where they stood, they could see numerous shallow rapids, similar to those on the Dihing.

As to the general question of the origin of the Irawady, he proceeds to say he felt perfectly satisfied, *from the moment he made inquiries at Sudiya,* that Klaproth's theory that the waters of the Sanpo find an outlet through the channel of the Irawady, was untenable, and, now that he stood on the edge of the clear stream which he concluded to be the source of the great river, he could not help exulting at the successful termination of his toils and fatigues.

Before the two travellers a towering wall of mountains rose to the north, stretching from the west to the east, offering an awkward·impediment to the passage of a river in a cross direction, and they agreed on the spot that if Mr. Klaproth proved determined to make his Sanpo pass by Ava he must find a river for his purpose considerably removed towards China.

On the east and west of where they stood, about 27°26′, were peaks heaped on one another in the utmost irregularity of height and form, and at all distances. Their guide pointed out the direction of the two larger branches uniting to form the eastern branch of the Irawady—the Namkin, by which the Khamties distinguish the Irawady throughout its course to the sea. The mountain at the source of the western branch bore 315°, and of the other, which takes the name of the main stream, 345°. They could also perceive the snow to the westward, some continuing as far round to the south-west as 240°.

The elevation above the sea was found to be 1855 feet, and, on the theory that Bhamaw was 500 feet above the sea, which would be equivalent to a fall of the river of 8 inches each mile, there would remain 1300 feet of fall in the 350 miles between their position and Bhamaw, which he believed sufficiently accounted for the greater part of that distance being unnavigable *excepting for small canoes.*

The most important geographical information obtained by Wilcox was the existence of the eastern branch falling in at two days' journey about where the road turns off to Mogoung. This river had hitherto been a stumbling-block in reconciling the accounts of the Singphos and Burmans at Sudiya, for the latter were unacquainted with it from the fact that their route lay along to the south of it, by the Mogoung valley, while the former came from the eastward of the Hukoung valley. The Khamties, as well, appear to have been quite as ignorant about the eastern branch as any of the other tribes, and he states that they had no positive information about it.

This eastern branch, which no European eye has ever seen, and about which Wilcox professed he was unable to obtain any positive information, he calls the Suhmai Kha, Pougmai or Linmai Kha. It was described to him as rising in the northern mountains, at no great distance eastwards from the *heads of the Irawady,* and the objections to assigning it a very distant course are—First, its want of magnitude (for it is not described as larger than the Khamti branch); second, the direction of the high range (which would require it to break through the most elevated ground), in that quarter; and lastly, the want of room from the presence to the east of it of the Salween.

In an appendix to his account he again repeats what he had already said about having no positive statements to offer regarding the origin of the Suhmai Kha; but what is worthy of special note, he records that the Singphos generally were of opinion that it is somewhat larger than the western branch, though not materially; and he supposed it not at all improbable that it was the river which had been mentioned to him by an old man who had been a slave among the Lamas, as rising in the snowy mountains of the Khana Debas country, and flowing to the south near where the source of the eastern branch of the Deboug turns to the north-west.

These are all the facts which Wilcox from his own observation and research brought to bear on the question of the sources of this river. He may have contributed to disprove M. Klaproth's theory but he certainly did not discover the sources of the Irawady, as he seems to have thought.

We will now examine the estimate he had formed of the river whose sources he was thus locating, but whose main stream had never been seen by him; and, in connection with this subject, he asks himself the pertinent question, What is the magnitude of the Irawady compared with other rivers close at hand? I shall give his answer in nearly his own words, and with it before us we shall be able to judge whether his knowledge was sufficiently accurate to give his opinion on its probable source. much weight.

Speaking of the reported difficulty of stemming its current in the rainy season as a proof adduced by some of the great body of water which it sends down, he very justly remarks that such statements to those acquainted with the Ganges and Brahmaputra amounted to no more than that it resembled those rivers in the periodical difficulties of its navigation. He then, however, adverts to the fact that the Irawady is in one place contracted in breadth by its high banks to 400 yards, of which we have, he observes, no similar instance in the others; and, in view of such a circumstance, he could not consent to allow that the difficulty of stemming its current was a convincing argument of its superior importance. This comparison, however, and his deduction from it, would not have been made by one who had had any practical acquaintance with the character of the Irawady or of the country through which it flows, for the conditions of the river are entirely different from those which characterise the Ganges and Brahmaputra.

He reproduces Buchanan Hamilton's statement that during the dry months of January, February, March, and April, the waters of the Irawady subside into a stream that is barely navigable, and founding his deductions as to the magnitude of the river on this description, which is certainly apt to mislead one who had never visited the main stream, it is not to be wondered that he limited its source to the southern face of the mountains bounding the Khamti Plain to the north, in lat. 28°.

In his account he also enters into a comparison of the Brahmaputra and Irawady, with the object of combating Klaproth's theory regarding the course of the Sanpo, but it does not appear that he succeeded in proving that there was any very great disproportion between the size of the two rivers.

We shall now turn to the accounts of Hannay, Bayfield, and Griffiths, to give some idea of the true character of the Irawady about 60 miles below where it receives the branch Wilcox visited.

Colonel Hannay describes the Irawady in lat. 24° 56′ 53″, at the mouth of the Mogoung stream, as still a fine river flowing in a reach from the eastward, half a mile broad, at the rate of 2 miles an hour, and with a depth varying from 3 fathoms in the centre to 2 at the edge, and that it is not unnavigable to large boats is evidenced by the fact stated incidentally by Hannay, that at the town of Tsenbo, 10 miles below the mouth of the Mogoung River, they had to exchange their boats for others of a smaller description, better adapted for the navigation of a small tortuous stream like that of Mogoung. Even these, however, were not very small, for Colonel Hannay's required 25 men to paddle it.

In speaking of the first defile below Tsenbo, through which they had taken their large boats, he describes it as the most dangerous part of the Irawady, which I can fully verify from personal observation.

This portion of the river commences a few miles above Bhamo, and stretches to within 7 or 8 miles of Tsenbo. Between these two points it flows under high wooded banks formed by two parallel ranges. At the lower approach to the defile the channel is as much as 1000 yards wide, but as we proceed up it gradually narrows to 500, 200, 100, and even to 50 yards, according as the two ranges approach each other, again increasing in breadth as they recede, till at last, below Tsenbo it spreads out again into a noble river.

Considering all that portion to be defile in which the course of the river is defined by high hills, it may be stated to stretch over 25 miles in length. It must not, however, be imagined that the whole of this long stretch is equally difficult of navigation, for besides the so-called rapids and the narrowed portions, in which the bed of the stream is crossed by huge trap dykes, there are long deep lake-like reaches, in which there is hardly any perceptible current and no rocks.

The dangers mentioned by Hannay lie in that part of the defile where the channel is intersected by the greenstone beds. There the river has cut a passage for itself through the solid rock, in some places not more than 50 or 60 yards broad, but 10 to 12 fathoms in depth. The current is not so strong in the dry weather as to interfere materially with the passage of boats when they are kept under the lee of the greenstone beds; but we had telling evidences in the height of the high water-mark, which was about 25 feet above the level of the river in February, and in the shivered trunks and branches of large trees heaped in wild confusion among the rocks, that all navigation must be impracticable during the height of a flood, and that the body of water which pours through this narrow gorge must be enormous and of terrific power. Immediately below it there is a deep, still reach of black water, evidently of great depth, and so land-locked that it resembles a mountain lake.

Colonel Hannay, while at Hookoung, learned from Singphos, from the borders of China, that the Suhmai River rises in the mountains bordering the plain of Khamti to the north, and is enclosed in the last by the Goulang-sigong Mountains, which they considered the boundary between Burmah and China. This river was pronounced not to be navigable even for canoes. Several smaller streams were described as falling into it from the Shuedouggyee Hills to the west, and the name of Sitiong was given to the tract of country through which they flowed. Gold was very plentiful throughout the latter area, and, indeed, throughout the whole of the country between the eastern and western branches of the Irawady.

Griffiths states that Dr. Bayfield ascertained during his passage up, at a season when the waters were low, that in many places of the first defile no bottom was to be found at a depth of 45 fathoms.

Griffiths' own account of the Irawady above Bhamo is that it keeps up its magnificent character as far as he went to the mouth of the Mogoung River, where it is 900 to 1000 yards across, and he describes the appearance of its vast sheet of water as really grand. He observes that the general characters of the Irawady are very different from the Ganges and Brahmaputra, its waters being much more confined to one bed and comparatively seldom spread out. Generally speaking, it is deep, and the stream is not violent; and he states, what experience has proved to be perfectly accurate, that it affords every facility for navigation, although in one or two places troublesome shallows are met with. In the first defile the channel is occasionally impeded by rocks, but it is only in this part of the river that the navigation is attended with danger.

Further, in speaking of the tributaries of the Irawady between Mogoung and Ava, he remarks that they are exceedingly small, which tends to increase the astonishment with which one regards this magnificent river.

He hazards the opinion that its source will probably be found to be the Suhmai Kha, and points out the fact that the great body of water comes from the eastward, from between the Mogoung River and Bor Khamti, in which country Wilcox visited the Irawady, where it was found to be of no great size. No considerable branch finds its way from the westwards, neither are the hills which intervene between these points of such

height as to afford large supplies of water. On the whole, he
thought it probable that the Irawady is an outlet for some great
river which drains an extensive tract of country, for it appeared
to him that, if all its waters are poured in by mountain streams,
an expanse of country extensive beyond all analogy will be
required for the supply of such a vast body of water.

I attach great weight to this opinion, for it was formed by
Griffiths immediately after his visit to the Brahmaputra, and
because he was a man of thoroughly scientific habits and thought
—*très instruit, très zélé, et fort bon observateur*, as Mirbel observes.

Dr. Williams, in his book, entitled 'Through Burmah to
Western China,' gives it as his opinion that a river-steamer of
proper construction would have no difficulty in making her way
to the Tapeng, and for many miles beyond.

My comment on this is, that we proceeded to Bhamo in a
large steamer, drawing 4 feet of water, and experienced no
difficulty in the navigation, although the captain and all the
crew were Burmans, to whom the river above Mandalay was
entirely new.   This, too, happened in one of the months (January) in which Buchanan Hamilton stated the river to be hardly
navigable by native boats.

While at Bhamo I took the opportunity to make what I
can only characterise as a rush up to the first defile.   Our visit
was necessarily a hurried one, as our leader, Major Sladen, was
in the daily expectation of being able to make an immediate
advance ; so that if we had gone on for a thorough investigation
of the river above Bhamo, we should have certainly seriously
interfered with the progress of the expedition.

My visit, however, sufficed to convince me that the Tapeng
makes hardly any sensible difference in the general appearance
of the volume of this great river.

The Irawady, at the beginning of the first defile, about five
miles above Bhamo, is about 1000 yards across ; and its course
is defined by low wooded hills, which run close to its bank.
About two miles further on, the channel narrows to 500 yards,
and the hills become even closer and more abrupt over the
stream than before, and, about another mile beyond, a higher
range of hills from the south-west comes in behind the former
one, and both terminate on the bank as two headlands.   At
this point a ridge of rocks runs half across the bed, and, at this
season (February), they are eight feet above the water ; but the
river is so broad and deep, that I find myself speculating in my
notes, made on the spot, on the course a steamer would follow in
passing them.

The hills still continue on both sides, but they are highest on the west, and, as we proceed for four or five miles, the number of rocky points running out into the stream increases, and opposite to the village of Pivaw, about 20 miles above Bhamo, on the left hand, the channel has narrowed to about 150 yards, and here the first so-called rapids occur. The bank on which the village stands is about 80 feet high, and the country inland is undulating and runs up to low ranges of hills, a few miles to the north.

Leaving Pivaw, we proceeded about eight miles further up the defile, or Kyvakdweng, as it is called by the Burmese, still preserving the high wooded banks on either side. After we had gone about three miles above this, we came to a reach in which the river flows very sluggishly between two high conical hills, which so close in upon it, that one is puzzled to detect any outlet. The quiet motion of the water and its deep olive-black are suggestive of great depth. The breadth of this lake-like reach is about 250 yards, and its length about 1¼ mile; and passing on, we find it abruptly closing in at its northern end, and its channel broken up by numerous rocks which jut out boldly on either side into the stream, and in many cases approach each other so closely that the channel is reduced to 50 or 60 yards. The height of these rocks averages 30 feet, but many of them are not more than 15 to 20 feet. The current, although strong, did not interfere much with our progress.

There is a small isolated rock, on the right side of the channel, capped by a pagoda, and another little promontory, further on, with a similar structure. The first appears to be of great age, and its presence on this rocky island—well into the middle of the stream, and not higher, I should think, than 45 feet above it—gives us some indication as to the limit of the rise of the river, for the pagoda could not withstand the power of the current. It must be borne in mind, however, that the Irawady had not reached its lowest when I visited this spot.

This rocky reach stretches about a mile in a N.N.W. direction, and terminates abruptly above in an elbow, from which another reach stretches off in an E.N.E. course with a clear channel, overhung by the precipitous but grassy sides of high hills. This elbow is one of the most dangerous parts of the whole defile, owing to a number of large isolated rocks that stretch across it, exposed to about 20 feet or more, during February. Owing to the sudden bend, the current rushes between them with great violence, but not so much so as to prevent boats passing, and, indeed, while I stopped to admire this novel and grand bit of river scenery, round which I had taken my boat, I had the magnitude of the Irawady in the northern reach and of the picture *generally* brought out by some boats in the distance, that

had passed up before us, and which looked like specks a few
hundred yards off.

It was a matter of much regret that I could not afford time
to go beyond this point, but we learnt from the boatmen that
with what we had done and a glance up the reach ahead, we had
seen the whole of the defile.

The body of water which flows round this corner during
the rains must be very great, and its velocity and power tremendous; for all the rocks (greenstone) subjected to its influence
are rounded, and shine with an almost metallic glaze, produced
doubtless by the attrition of the flood.

The rocks are all *in situ*, and consist, as I have already mentioned, of greenstone dykes running across the stream from
E.N.E. to W.S.W., and through the hills on either side.   The
softer beds between them have been much denuded, and correspond to the shallow valleys and hollows of the hills.

It should be remembered that two other defiles occur on the
Irawady, one immediately below Bhamo, and the other about
40 miles above Mandalay.   Others may be said to exist below
Thayetmyo and at Prome, where the course of the river is
defined by high hills.   Throughout the whole of these the river
is of necessity restricted to a well-defined channel, and its breadth
depends entirely on the proximity or remoteness of the hills to
each other, so that its breadth is no indication whatever as
to the body of water which passes through these channels; but
the mere fact that the Irawady was contracted at one place to 400
yards, Wilcox considered sufficient evidence to make him doubt
the position which had been claimed for it by BuchananHamilton.

The following information was communicated to me by a
Khamti Shan, from the village of Khakhyo, three days' journey
by boat, below the junction of the main streams of the Irawady.
He professed to know the country well, as far as one day's
journey up the eastern river, to a village called Muanglow, where
his wife came from.

He informed me of his own accord that the river above the
first defile is exactly like what it is at Bhamo, and he stated
that if a steamer, as large as the one we had at the latter place,
could be got through the rocks of the first defile, that there
would be no difficulty in taking it as far as the village of Wyesoung, at the juncture of the two streams.

After the defile is passed, he described the mountains as
receding from the river, and, in speaking of the country about
his native village, he said that, in climbing a high tree, it
appeared a dead flat for miles around, with the mountains in
the distance.   Near the juncture of the two rivers, the hills
again close in, but not to the extent to form a gorge.

He described the eastern branch as the largest, and that it is navigable as far as the village of Muanglow, one day's journey from its mouth; but that, above that point, the channel becomes rocky and dangerous to navigate.

The country about Muanglow is inhabited by Kakhyens (Singphos), and the Khamti Shans, who are found along the Irawady as far south as Khakhyo.

Between the first defile and Khakhyo, the Kamsang stream from the east enters the Irawady, about one mile above the Mogoung River, and one day's journey by boat above the former. The Namthabet, another small river, flows into the Irawady from the east.

In a map of the north-eastern frontier of Burmah and Western China, compiled at the Surveyor-General's Office, Calcutta, in 1862, I find two rivers corresponding to the position of these. The southern one is named on the map *Shoomae*; but, at one place, along its course, I find the word *Mengzan* Khong, which has so close a resemblance to Namsang Khyoung, that I am inclined to regard it as the same. The name, however, is put in almost at right angles to the river, and evidently with some doubt, for it is difficult to say to what it refers. Any one familiar with the Burmese word " khyoung " would have had no hesitation in referring the name to the river, although the first part of the word is essentially Shan-Namsang, or the river Sang. To add khyoung to this, is an apparent tautology, but one which is in vogue among the Shan Burmese of Upper Burmah, as is instanced by the Nam-Tapeng-khyoung. I acknowledge the difficulty of reconciling the Meng with the Nam.

The name by which it goes on the map—Shoomae—is evidently Burmese, and should be written Shuaymai or Shuemai-khyoung. Now if we examine the orthography of Wilcox's eastern branch, we find that its name, as given by him, has a wonderful resemblance to that of this eastern khyoung or rivulet, the *Shuemai* opposite to Mogoung.

He writes it *Suhmai*, and gives it a Singpho origin, but it appears to me there can be little doubt that it is strongly attined to *Shue-mai* or golden-mai, which seems a plausible supposition, from the fact that it flows through a country rich in gold,* and from the additional circumstance that Pemberton, in his map of the Eastern Frontier of British India, gives the Singpho name as Zinmae, and writes the other Shuemae. It is certainly a remarkable coincidence that we should have two

---

* It seems probable that the mountain marked on the map lat. 25° 55′, long. 98° 50′, as Sine-shan, or snow mountain, should also be Shuay-shan.

rivers of the same name within 40 miles of each other, and both set down as eastern branches of the Irawady.

If two rivers of the same name do exist, I cannot avoid supposing, with the fact before me that Hannay made his inquiries at Mogoung, that his Shuaymai River was the small stream opposite, or nearly so to the Mogoung stream, and not the Shuaymai of Wilcox; for the Kakhyens (Singphos), from whom he derived his information, are stated to have come to Mogoung from the borders of China. If they travelled by the Bhamo route, it is highly improbable that they knew anything whatever about the affluents of the Irawady above Mogoung, and I think it therefore highly probable that they were some of the many Kakhyen traders who pass between these hills and Mogoung, by the Khakhyo and Wyemaw route. This would lead them across the Shuaymai (Namsang), and what could be more natural than that these ignorant hill-men, who have little or no acquaintance with the Burmese plain, when questioned about the Shuaymai, an eastern branch of the Irawady, should describe the small stream that they had so lately crossed.

They were quite correct in describing it as running from the Khamti Mountains, which is a very indefinite term, applicable to all the mountains in Upper Burmah on which the race is found; and as it extends as far south as Khakhyo, a short way to the north of the Mengzan, or Shuemai-khyoung, the mountains from which the stream rises might be accurately described as Khamti; and, as it flows from the south side of the Sine-shan Mountains, or our Shuay-shan, and through a country evidently rich in gold, it is easy to understand how the confusion arose between the two.

With Hannay's description of the Irawady above Mogoung before us, in which the river is stated to be a fine body of water half a mile broad and with a varying depth of 2 to 3 fathoms, the question suggests itself, that if the two main streams which go to form this splendid river are so insignificant as Hannay would lead us to suppose they are, where does all the water come from, for they do not supply it, and the Namsang and Namthabet can add but little? This consideration, and the foregoing facts, lead me to doubt the reliability of Hannay's information.

In the map a river corresponds to the Namthabet, but another is put in to the north of it of which I have no account.

According to my informant, the name of the eastern continuation of the Irawady is Kewhom, and he described it to me more as the upward prolongation of the Irawady than as a branch. Manila, or Muangla, which is placed on the maps at the junction of the two streams, he stated to be only half a day's journey by boat from Wyemaw ; whereas Wyesoung, which he places at the junction, is two days' journey above Muangla. Wyesoung is evidently the Maintsoung of the map, situated much on the position as that assigned to it by my informant.

The country above the first defile is stated to be very rich in cotton, rice, and oil-seed, cultivated by the Kukoos and Khamti-shans. The former raise the cotton, and after the rains it is taken, along with the other produce, in boats to Bhamo, where it is more highly valued than the cotton grown farther down the river. Mogmoung is another mart visited by these people from the upper Irawady.

The Shans, from Muangla, in the Sanda Valley, cross over by a hill-route to Khakhyo, from whence they proceed up the river in boats to the junction of the two streams, and for a day's journey up the eastern one, exchanging their salt for gold, which is said to be very plentiful.

I will now detail some information I obtained while at Momien on the course and extent of the rivers in Western Yunan, for the subject is closely connected with the sources of the Irawady.

Momien itself stands on the easterly branch of the Tapeng, the river through which Klaproth diverted the waters of the Sanpo to the Irawady. It is a small mountain-stream, about 20 yards broad at Momien, but during the rains it rises to about 8 feet in the Momien Valley, which is about 135 miles from the Irawady. Its sources lie in the Kananzan * range of mountains, about 10 miles to the east of Momien, and on the east side of which is the Shuelee.

The latter river was described to me as a comparatively small stream, rising in a range of hills about 40 or 50 miles to the north-east of Momien, and in the itineraries of the Burmese embassies I have since found it stated to be about 40 yards broad, and that it is spanned by an iron suspension-bridge, which was mentioned to me at Momien and described as the exact fellow of the iron bridge over the Taeping below Nanting, 25 miles below Momien, where the river is about the same breadth as the Shuelee, to the east of the latter city.

The Salween, which flows on the other side of the range of mountains defining the eastern watershed of the Shuelee, was described to me as a larger river than the latter, flowing in a narrow but comparatively level valley, like that of Sanda, and that there was a ferry over it, as is stated in the Burmese itineraries. It was further stated to be about the size of the Tapeng below Muangla, with which my informant was well

---

* This has a wonderful resemblance to Marco Polo's Karazar.

acquainted. It is improbable, therefore, that its sources can be more than 100 to 150 miles to the north of the latitude of Momien, if so much.

If we turn now to the Jesuits' map of the province of Yunan —the result of a survey by the Fathers Fidelli, Bonjour, and Regis, in the years 1714-15—and inquire into the data on which it is founded, it appears to me that the distribution of these rivers, as given in it, so agrees with the data I have collected that it is worthy of our full acceptance until other facts have been adduced to disprove its accuracy. Père Regis, one of the surveyors of Yunan, gives us the following account of how the survey was conducted. He informs us that " they omitted nothing for rendering their work perfect; that they visited all the places, even those of least consideration, throughout the province; that they examined the maps and histories of each city, made inquiries of the mandarins and their officers, as well as of the principal inhabitants, whose territories they passed through; and that by measuring as they advanced they still had measures ready to serve the triangles formed by such points as were to be fixed, and that they corrected their determinations by triangles by the meridian altitudes of the sun and polar stars." *

As a proof of the accuracy of this map, we have only to compare their distribution of the Tapeng to within 30 miles of the plains of Burmah with the result of the survey of our expedition.

If we take the position of Santa (Sanda), as triangulated by the Jesuits, and its relations to the branches of the Tapeng, we shall find that they agree in every way with the results of our observations, and that the most insignificant branch of the river about Sanda has been mapped with the most astonishing accuracy. The same remarks are also equally applicable to the river about Momien. Am I not, therefore, entitled to argue that, if such accuracy characterises their work in mapping out second-rate rivers, that a like accuracy will be found in their delineation of such important ones as the Salween and Cambodia?

They restrict the sources of the former to latitude 27° 10′ N., or thereabouts, and those of the latter to 27° 30′, or nearly so, which gives a course to the Salween of 640 miles from its sources to Moulmien, and to the Cambodia of 850 miles from its origin to the sea.

The Salween is a much less important river, however, than the Cambodia; and it may appear that the greater length of

_____

* De Halde, vol. i.

the latter, as indicated by the Jesuit maps, is not enough to account for its greater size. And, perhaps, it would be so if the Cambodia did not possess the immense reservoir that it does in the Lake of Jali, which stretches northwards from 25° 27′ N. lat. to about 26° 12′ N. lat., and from which this river derives a large body of water through the Hayquam River, which joins it at 24° 45′ N. lat.

The mention of the latter river gives me another opportunity of verifying the accuracy of the Jesuit Fathers; for I find it stated in the Burmese itineraries that the ambassadors, on their way to Pekin from Ava, after crossing the Cambodia by an iron bridge about 60 yards long, came to a branch of the Hâkyou, over which there was another iron bridge 40 yards in length; and that still further on they came to the Hâkyou itself, which flows from the Lake of Jali to join the Cambodia, and over which there is an iron bridge with a span of 42 yards.

These two streams, each 40 yards broad, must surely represent a large body of water; and with the knowledge that the Cambodia receives a supply of this magnitude from a lake fully 35 miles long, we do not expect it to have the same northerly extension that would be necessitated if this reservoir did not exist. It seems reasonable, therefore, to suppose that the three degrees of latitude in the Jesuit maps assigned to the main stream, above where it is joined by the Hayquam, is an area of sufficient extent to account for a river like that which the Cambodia appears to be above its junction with its Jali affluent.

The result of this argument, then, is to prove that the Tibetan rivers are in no way connected with either the Shuelee, Salween, or Cambodia. In the Jesuit map of Yunan they were certainly never connected with them; and it is only when we come to such maps as D'Anville's, which was drawn up for the express purpose of upholding a flimsy hypothesis, that we find them brought boldly down, through the most extraordinary series of windings, to their desired course.

With these facts before us, we are prepared to examine the position which Wilcox claimed for himself that he had discovered the sources of the Irawady.

After a careful consideration of all the statements advanced by him in his account of the survey of Assam and the neighbouring countries, I cannot avoid thinking that he came with a biassed judgment to the investigation of the sources of the Irawady, for he states that he felt perfectly satisfied as to the origin of the river before he left Sudiya. But, from the internal evidence of his paper, it is evident that he knew nothing of the main stream, and had never seen it. We are, therefore, fairly entitled to submit the evidence which he adduces for restricting its sources to the Khamti Mountains to a rigid criticism.

But, to appreciate his position, it must be borne in mind that he had set himself the task to demolish M. Klaproth, and no one had better facilities and information for doing so than this able explorer and geographer, and to my mind he was quite successful in his task; but in carefully reviewing his description of the question, it appears that in his desire to establish his position he was led unwittingly to depreciate the importance of the Irawady and to give it a restricted distribution, at utter variance with its magnitude.

The error was a likely one, for his whole acquaintance with the river was a few hours' observation of one of its streams between the 27th and 28th parallels of north latitude, to the east of Assam, and because what he learned of it beyond the spot on which he stood was derived solely from the Khamti Shans, who were, according to his own statement, little given to travel, and from Singphos from the eastward of Assam. He adduced no proofs, however, that the latter had ever been to the eastward of the eastern branch of the Irawady, which they made two days' journey above the Mogoung River, and, according to his own account, the former knew nothing of the river beyond the branch on which the villages were placed. Yet, notwithstanding all this, and the fact that the Singphos generally from the east of the western branch had informed him that the eastern one was the larger of the two, he adhered to the information which he had received at Sudiya that the western and smallest branch was the source of the river, and this on the authority of Khamti Shans, who knew nothing of the Irawady beyond their own river.

It is unnecessary to recapitulate what he has stated regarding the probable origin of the western or Khamti branch, his sources of the Irawady, for I agree with him, judging from the size of the stream where he met it, that its sources could not be far distant, but, at the same time, I am not disposed to limit its course on the grounds he advances, viz., the presence of a high range of mountains stretching to the east and west. We have only to look at the Dihong and the Brahmakund to find examples of how large rivers find their way through mountain ranges. I know no better example of this than the Tapeng piercing a mountain range, which in the distance one would never imagine it was possible for a river to flow through. So apparently unbroken by valleys are the Kakhyen Mountains, averaging 6000 feet, that I have found myself, at 4 miles from their base, speculating on the course of the Tapeng, and being puzzled to define it. The passage, therefore, of the Irawady through the mountains to the north would be no uncommon phenomenon, and indeed it seems to me to be necessitated in order to account for the volume of water which reaches the Khamti Plains about Wyesoung.

If Wilcox had only had the same practical acquaintance with the main stream of the Irawady as Griffiths, we should doubtless have had the testimony of both that the western stream was quite insufficient, even along with an eastern one of nearly the same dimensions, to account for the great body of water in the Irawady above Mogoung.

A glance at his description of the stream and of the weather during his visit to it will be sufficient to show that the only light he threw on the sources of the Irawady was to indicate that the weight of evidence pointed in the direction of the eastern branch as the great channel from whence that splendid river derives its supply from the highlands of Tibet, between the Yangtsekiang, the headwaters of the Cambodia and Salween, and the two eastern affluents of the Brahmaputra (Sanpo), the Dihong, and Brahmakund.

He says he was surprised to find but a small river, smaller even than he had anticipated, though aware of the proximity of its sources—a statement which has the sound of a foregone conclusion; but he goes on to describe it as 80 yards broad but still fordable, although considerably swollen by the melting snows. That this, however, was not the only cause of the rise of the river such as he describes, is evident from the frequent reference he makes to the very heavy rains he had experienced on the last eight days of his march, but which never occurred to him as the *vera causa* of the flood.

Now, with these facts before us, that the river during the height of a flood caused by the heavy rains and the melting snows was only 80 yards broad and fordable, the inference is forced upon us that it could be little more than a mountain rivulet during the dry weather.

Such, then, was Wilcox's supposed source of a river which 150 miles further down measured half-a-mile in breadth, with an average depth of from 2 to 3 fathoms, without receiving any notable stream on the way that would account for the unprecedented difference between the two points.

The conclusion, therefore, we arrive at is, that the eastern branch, as described by him, was only a small affluent of the main stream which flows down from the north-east, as described by my informant, and that the sources of the river, in all probability, lie considerably to the north of the so-called Khamti range of mountains, and that it thus becomes one of the Tibetan rivers; and as I have shown that the Jesuits, our only reliable authorities on the distribution of the Cambodia and Salween, restrict them to the 28th parallel of north latitude, it becomes probable that some of the Tibetan rivers flowing down from the north, in the direction of the Irawady, may be its upper sources, while the others may be branches of the Yangtsekiang, and that the Irawady drains part of that area between Lassa and Bathang, which has hitherto been apportioned to the Cambodia and Salween.

D'Anville was the first to connect the Tibetan rivers with the Cambodia and Salween, a course which was forced upon him from the circumstance that he believed the Sanpo to be the Irawady. Bringing the former river in the way he did to the west of Yunan, he considered he had provided an ample supply of water to account for the volume of the latter, and he had, therefore, to look to some other outlet for the drainage of that area of Tibet between Lassa and Bathang to the north of the supposed course of his Sanpo, and he hit upon the Salween and Cambodia as affording the means, and the unnatural and extraordinary course which he gave then has been perpetuated ever since in the maps of Klaproth, Dalrymple, and Berghaus, without a tittle of-evidence in its favour. Now that the Sanpo flows in its natural course to the Brahmaputra, as is now almost proved to a demonstration, it is to be hoped that the Irawady will not any longer be denied its due, as a river far surpassing the Salween and Cambodia in its northern distribution.

In conclusion, I may state that these remarks have been suggested by a note of Mr. Cooper's, in the 'Proceedings of the Royal Geographical Society' for June, 1869, in which he hazards the remarkable position that either the Sanpo or another large river to the east of it falls into a river called the Yarlong, which he supposes may be either the Brahmaputra or the Irawady; an amount of uncertainty which affords an ample field for conjecture, but certainly throws little light on the subject.

The French missionary, whose notes on the country were forwarded to the Society by Mr. Cooper, being almost certain that the Yar-kioute-tsanpo is the Irawady, he, of course, follows in the footsteps of D'Anville, considering the other rivers to be the Salween and Cambodia.

It is, however, extremely difficult to determine how much of the information he communicates is derived from personal observation and how much from maps. His remarks on the Lantsan-kiang and Loutse-kiang read as if they were almost exclusively due to the latter source; and it is to be observed that in describing the Yangtse-kiang, near which river I suppose him to be stationed, he states that it flows as it is drawn in the

LIBRARY
Leland Stanford, Jr.
UNIVERSITY

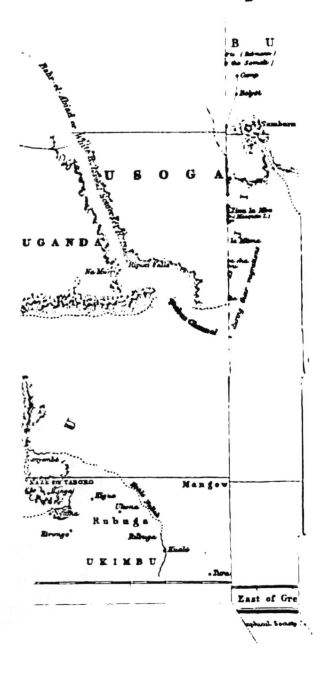

maps, a remark which he does not venture in the case of the former ones.

It may appear unreasonable, but the very circumstance that he states that he is almost certain that the Yar-kioute-tsanpo is the Irawady, would make me doubt that he was in a position to form a sound opinion on the course of the other Tibetan rivers.

I strongly suspect that Mr. Cooper's friend was J. Thomine Mazare, Vicar Apostolic of Tibet, who communicated similar views on the distribution of these rivers, including the Tsanpo, to Bishop Bigimdit, of Rangoon, in August, 1859, and on which Colonel Yule commented. I am led to this conclusion by the very strong similarity in the particulars of the notes communicated to Mr. Cooper to those published by the Bishop.

I may state that I have written to Captain Bowers, my fellow-traveller in the Yunan expedition, now stationed at Bhamo, pressing on him the importance of settling this question by a personal investigation of Wilcox's River where it joins the main stream of the Irawady, and that from the interest Captain Bowers takes in everything affecting the Irawady, I believe that we shall have an early solution of this problem.

XI.—*Routes of Native Caravans from the Coast to the interior of Eastern Africa, chiefly from information given by Sádi Bin Ahédi, a native of a district near Gázi, in Udigo, a little north of Zanzibar.* By T. WAKEFIELD, Missionary at Mombasa.

No. 1.—FROM TÁNGA TO LAKE NYÁNZA (OR NYANJA).

Tánga to Bwíti.—(A full day's march). [12 hours.]

At Bwiti there are many villages. The inhabitants are Wadígo and Wasegéju ; these occupy the *plain.* The summit of the hill is occupied by Wa-Teíta, and is laid out into plantations. The hill is probably about 500 or 600 feet high.

Bwíti to Dóngo Kúndu.—At Dongo Kundu there are villages, 2 or 3 miles to the south-west of camp. Agriculturist Wakwávi live there, and till the ground. The *soil* of this region (Dongo Kundu) is *red,* called by the Sawahilis—"*ngéu,*" and is used extensively by East Africans (having been mixed with oil), for lubricating their bodies. The redness of the ground commences at Bwíti, but is more extensive, and more deeply coloured, at Dongo Kundu ; the latter place taking its name from the nature of the soil,—Dongo, meaning *clay, earth,* or *soil* in the *plural,* (udongo, in the *singular*) ; and Kundu, *red.* [7 hours.]

Dóngo Kúndu to Mto wa Umba (River Umba).—This river takes its name from a district bordering on the sea-coast, near Vánga, and has its source in the Barámo Mountain. [12 hours.]

To *Mto wa Káti* (River Kati).—"Kati" means *between* or *central.* This river takes its rise in the Mshi Mountain, which lies about two hours to the south of the encamping place at "Mto wa Káti." [6 hours.]

To *Chwére Chwéré.* [6 hours.]

To *Barámo.*—The name "Barámo" is Kisámbá. There is a village here on the plain, at the base of Barámo Mountain, inhabited by Wasegúa. A chieftain lives here, whose name is Mulúgu, a son of Kimwéri. The summit of the mountain is populous, inhabited by Wasámbá. [6 hours.]

To *Gónja.*—(The name "Gonja" is Kipáré.) There are two villages of the Wazegúa, situated at the base of the mountain. Wa-Páré live on the summit, where they have *many* villages. [12 hours.]

To *Kisiwáni* (Páré district.) "Kisiwani" means *at the island.* At Lasitti there are no people. Wakwávi formerly lived there, but have long since abandoned the place. [8 hours.]

To *Sámè.*—The name "Sámè" is Kikwávi. A Kwávi chieftain and his people (who were very numerous), once lived here, but have long ago been displaced by the Masái. [8 hours.]

To *Mto wa Rúvu* (River Ruvu).—The word "Rúvu" is Kizegúa, and the river is called by the above name by the Wazegúa. This is the name by which it is known a few days in the interior; but on the *coast*, it is generally called "Pangáni," and is thus marked on our coast charts. It has probably received the latter designation from Pangáni, a village and district situated near its mouth, on the northern bank. [12 hours.]

To *Mihináni* (at the *hina trees*,—Muhína, sing., Mihína, pl.); the leaves of which being pounded, yield a reddish staining matter, with which the Sawahílis and others stain their finger-nails, &c. Probably the "*hinna*" mentioned by Lane, as used by the "Modern Egyptians." (Lawsonia inermis.) [6 hours.]

To *Arúsha.*—Here are villages and plantations. The people are called Wa-Arúsha. They are descendants of the Wakwávi, and were formerly their *subjects;* but the Wakwávi having left the country, and the Masai having taken possession of an adjacent region (Sigirári), the Wa-Arúsha are now recognized as the subjects of the latter. The people of this region have a few head of cattle, but that is all. As stated above, the Wa-Arúsha are agriculturists. [9 hours.]

It was from this place (Arúsha) that Baron Von Der Decken was forced to return, in his attempt to reach the Masai country.

To *Mikindúni*.—This locality takes its name from the Mikindu (brab-trees), which flourish here, and on the banks of many African streams. [9 hours.]

To *Mikuyúni* (at the sycamores), many of which grow in this locality. [8 hours.]

To *Arúsha wa Jú*, i.e. the Arusha *above*, or *higher up*, in contradistinction to the Arusha of the *lower region*, nearer the coast. The people of both districts are of the same nation, and intermarry. They combine, also, in their war expeditions. They frequently unite, and make a raid on the Chága country. This people (the Arusha wa Ju), in addition to their plantations, are also rich in *kina*. In this latter respect, they differ from their kinsmen nearer the coast. Arusha wa Ju is well stocked with villages. [12 hours.]

To *Kisóngo*.—The name "Kisongo," means *approach*. Sadi says it was given to this locality from the following circumstance: a very brave Mkwávi, in an engagement with a party of Masai, at this place, manifested astonishing dexterity in surprising the enemy, repeatedly coming upon them unawares. He was equally skilful in evading the weapons of the foe. Retreating for a little while, he was soon again upon the enemy, without his approach, or "kisongo" being observed! The word is Ki-Sawahili, and is derived from the verb *ku songa*, to approach, or draw near. The Masái acknowledged to the Wandoróbo, (a vassal race), that they were utterly unable to kill this man, that both spear and throwing-club had failed to strike him; when a young man (a Mdoróbo) asked them if they wished him to be killed. They replied in the affirmative. At the next encounter of the Wakwávi and Masái, the warrior mentioned above, again made his appearance, when the young Mdoróbo concealed himself in an adjacent wood, weapons in hand, and the Mkwávi bounding into the forest near his hidden foe, was shot with a poisoned arrow.

There are villages at Kisongo. The great chief of all the Masai lives at this place. About four years ago, the chief, whose name is Súvét (the Wa-Sawahili call him Subéti), having died, his son Batiyán succeeded to rule. Sadi says he is about sixty years of age. [7 hours.]

To the north-east of Kisongo, is the mountain *Méro*. It is situate two or three hours to the north of encamping-place at Arusha wa Ju.

To *Eét*.—This name is Kikwávi. There are Masai villages at Eet, but they are not permanent. There is a large mountain at Eet, called Kitúmbi. It is clothed with short grass, but has no forest. The summit is level, and the Masai live both at the base, and on the top. The villages here are permanent. [4 hours.]

X

To *Wángwa Angarúka.*—This name is a combination of Kisáwahili and corrupted Kikwávi. " Wángwa," in Kisáwahíli, means a *sandy plain,* and is equally applied to those which are covered with alluvial deposits from maritime rivers, and sandy plains in the interior. The Sawahilis pronounce the word angarúka, as above written, but the Wakwávi,—*gnarúka.* The *meaning* of this word I have forgotten. [8 hours.]

To *Mto wa Angarúka* (River Angarúka). [10 hours.]

To *Dóönyo Ngái* (Ngái Mountain). This mountain is very large. Sádi says that it is higher than *Kílima Njáro,* though not so massive. Its summit exhibits the same radiating and coruscant appearances as that of Kílima Njáro. Sádi says— " one moment it is yellow, like gold; the next, white, like silver; and again, black." This exactly agrees with the account given of the glittering or radiation of the summit of Kílima Njáro, by all the natives who have mentioned the subject to me. Sádi says that at night, numbers of " tawáfa" (lighted candles) ascend from the base of the mountain to the summit. *(Ignes Fatui.)* [9 hours.]

The " Dóenyo Ngái" is about one day's journey broad, situate about half an hour to the west of the camp.

To *Ngörót.*—There is a small stream at this place. [9 hours.]

To *Pínyínyi.*—This locality is called " Vinyínyi" by the Wakwávi, but by the Wasáwahili as above written. [8 hours.]

To *Ngúrumáni.*—Sádi says that the word " *ngúruman*" (which is Kikwávi), is synonymous with the Sáwahíli verb —" *ku líma,*" to cultivate or till; and that the Wakwávi who inhabit this region, cultivate beans, millet, sweet potatoes, &c. These are poor Wakwávi, who, having long since been robbed of their cattle by the Masái, were compelled to turn their attention to agricultural pursuits, as a means of obtaining a livelihood. There are also *Gallas* in the same condition as these Wakwávi of Ngúrumáni and other regions,— who, having been deprived of their cattle, have been reduced to poverty. A few of these are located near Takaúngn, in the district of Kaúma, learning the art of tillage, and adopting the settled habits of an agricultural life. Poverty must have pinched them sorely to have brought them to this, for the haughty Galla " cannot dig;" he regards the occupation as only fit for those less manly than himself. All agricultural races are despised by the Gallas, as mean people; and those of their nation who have been reduced, by circumstances, from (what *they* consider) the light and noble occupation of the fold, to the base toil of the field, are regarded as men disgraced. The men themselves *feel* the degradation, and no doubt cast many a sorrowful look back towards the congenial, pastoral life they have relinquished,—the daily wandering with the lowing herd among the " green pastures " of their beautiful and fertile plains, which was then a Paradise, and, like the *old* Paradise,

"———— a place of bliss, without drudgery or sorrow."

Sádi thinks the Wakwávi of Ngúrumáni number 1000 or 1500. They are disposed over the ground, a few miles apart, like the Wanyíka tribes. [10 hours.]

To *Utími.*—" Utimi " in Kikwávi, means *monkey*, and in this region there are immense numbers. From Sádi's account of the crowds he saw here, it is evidently a favourite habitat. Utimi is very populous with Masái, whose settlements are permanent. [6 hours.]

To *Mábokóni.*—This name is Sáwahíli, and means at the [place] of the hippopotami; singular—Bóko, pl. Mabóko. The affix " ni," is a locative particle, having the force of the English preposition *at* or *in.* The *diminutive* form is that generally used, when referring to the habitat of hippopotami, namely, Vibóko, sing. Kibóko. The form at the heading of this paragraph, may either be used to express the hugeness of the hippopotami of this region, or their immense numbers. [5 hours.]

To *Msíro.* [10 hours.]

To *Baráni.*—This word simply means—*in the country*, and is used on account of the district not having a specific name. [6 hours.]

To *Súllóíta.*—There are Wándoróbo villages at this place, but they are not permanent. However, there must be other settlements not far distant, for Sádi says, if a gun is fired off, the Wándoróbo hear the report, and are soon at the camp. [6 hours.]

To *Náívásha.*—This word is Kikwávi, and Sádi says means *sea.* This region has another name, *Bálíbáli*, meaning also *sea.* (There is a large *Lake* in this district, which is mentioned in the route " From Mombása to Dháicho.") [10 hours.]

To *Agnáta Elgék.*—(Pronounced Agnăt'elgék.) This name is Kikwávi. Sádi says " Agnata " means a tract of land covered only with very short grass, (moorland?) ; and elgék, firewood. This region is thickly dotted with withered trees, which are used for fuel, by the natives. [6 hours.]

To *Máu.* [10 hours.]

To *Lúmbwa.*—This region is called " Lúmbwa " by the Wasáwahíli, but by the *natives* (Wa-Lúmbwa and Wa-Nándi), *Kiskis.* This section is very populous, and the villages are those of a settled people. They are agriculturists, and cultivate beans of different kinds (*fiwi, kúndè*), millet, " wímbi " (a seed-bearing

x 2

plant, cultivated on and near the coast, and in many districts of East Africa, the seeds of which are pounded, and made into bread), &c., &c. They are also rich in cattle, on account of which, the Wakwávi of Ndára Serián (their south-eastern neighhours), frequently come upon them. The Wa-Lúmbwa engage them, but are unequal to a contest with them, for the Wakwávi are proverbially very brave; at times, however, they get the better of the fight, and drive off their assailants. Their weapons, like those of the Wakwávi, are spear and shield.

Their cattle, when brought from the pasturage, are put, for the night, into *houses*, and not left to sleep in *folds*, in the open air, like the cattle of the Masái, and of those of the natives near the coast. [10 hours.]

To *Kosóva.*—There are people here, and permanent settlements. The inhabitants are called Wa-Kosóva. They cultivate the soil, but have also plenty of cattle, which, Sádi says, are remarkably fine. Their weapons are spears and a shield. Each man carries four spears, three throwing-spears, with long shafts tapering to the end. The blades, which are short, are tipped with *poison.* The fourth spear, which is reserved for close combat, has a longer blade than those used for hurling, which is also poisoned, and the heel or end of the shaft armed with a long iron spike. These weapons are not only used for defence, but are also employed in hunting elephants and other animals. The Wa-Kosóva are very expert in hurling their light spears, which they can also throw to a great distance. Sádi says they go, quivering from the hand, and buzzing through the air, in an amazingly direct line to the mark. The *shields* these people carry are very large, larger even than those of the Masái. [6 hours.]

There is a small hill at Kosóva.

To *Káverónd.*—At Kaverond, and between it and Kosóva, there are villages. The people of this place are called Wa-Kaverond. They are the same as the Wa-Kosóva, only a different tribe or clan. The language is one. Like their neighbours at Kosóva, they have cattle, and also plantations. [4 hours.]

To *Máwka* (Móka).—This region is also inhabited. The people are called Wa-Mawka. They cultivate the soil, and also possess cattle, but not many. They are considered as the subjects of the Wa-Káverond and Wa-Nándi—their neighbours—a few hours to the north and south. [7 hours.]

To *Nándi.*—"Nandi" is also known by another name, that of *Ndéi Ngwálle.* [6 hours.]

To *Baráni* (camp). [5 hours.]

To *Gwáso Ngíshu.*—This name is Kikwávi, and means "*Cattle River*" ("Gwaso," River; and "Ngíshu," cattle). At this place there are Wakwávi villages, but they are not permanent settlements. [7 hours.]

To *Base of Nándi Mountain.*—Here is a mountain, large and wooded. It is called the *Nandi Mountain.* The Wa-Nandi (people of the district) live on the summit which they cultivate. They are also extensive cattle-owners, and, like the Wa-Lumbwa, secure their cattle at night in houses, as they fear the Masái and Wakwávi. On account of their fear of these wandering and insatiable robbers, their cattle are not taken to the pasturage until the sun is high in the heavens—about 10 A.M.—and they are brought back about 3 in the afternoon. Other prudent and cautionary measures are also adopted to protect and secure their coveted property. Outposts and scouts are daily sent to watch—a goodly distance off—in various directions, against a surprise, whilst the herdsmen attend to pasturing the cattle. The duty of scouting, or keeping watch, is performed alternately, and is daily changed. Thus, one day they are farming, and the next standing sentry!—an irksome necessity, arising from human cupidity!—the violent greed and covetousness of savages! [6 hours.]

To *Chamwáli.*—This region is very populous. The inhabitants are called Wa-Chamwáli. They cultivate the ground—growing millet, beans, bananas (the latter in large quantities), "wimbe" (a seed-bearing plant), &c. They have much cattle also. [4 hours.]

To *Kaverónd.* [6 hours.]

To *Baharíni.*—"Baharíni" is Kisáwahíli, and means *at the sea.* This is the terminus of Sadi's long journey from Tánga to the *Lake Nyánza,* which is so immense as to have led the natives (at least of the seaboard) to conceive of it as an *inland* " SEA ! " [4 hours.]

LAKE NYANZA.

Sádi calls it "*Nyánja,*" and is unacquainted with the above form. The first name he gave me was " bahari," sea ; and " *Báhari ya Ukára,*" the Sea of Ukara, Ukára being the name of the region on the eastern shore of the lake where the above route immediately terminated. The people of this region are also called *Wa-Ukára.* Sádi states that the lake is often designated " *Báhari ya Pili,*" the *Second Sea.* This latter expression I have frequently heard, but, at the time, thought it had reference to the Atlantic Ocean; and, when wishing myself to speak of the North Atlantic, I have used the above expression, " *the second Sea,*" thinking that they have heard of it. The term, however, refers to the *Lake.*

On asking Sádi why it was called a "*Sea*," he replied, was it not *like* a sea in its immensity? I asked him if it were a lake. He said he could not tell; but gave it as his opinion that it might, possibly (from its extraordinary length), be a huge river; "for," he remarked, "I have travelled sixty days (marches?) along the shore without perceiving any signs of its termination." Neither had the natives with whom he had conversed been able to give him any information about its northern or southern limit.

With regard to the *width* of the lake, Sádi was informed that it required six full days—from sunrise to sunset—to cross it in canoes, but that, if the men went right on, *day and night*, the journey was accomplished in *three* days.

Standing on the eastern shore, Sádi said he could descry nothing of land in a western direction, except the very faint outline of the summit of a mountain, far, far away, on the horizon.

Sádi states that the lake has a daily *tide*, and that its ebb and flow are as regular as that of the sea on the coast. That drifted foam, and other light matters, remain in lines upon the shore after the recession of the tidal waves. (Other natives, too, I believe, have made this statement.) The lake, says Sádi, has also its "mawímbi" (waves), but they are not very considerable. The water of the lake is *fresh* and sweet.

The eastern side of the lake, as far as Sádi's experience goes, is not mountainous, but rather level, and the shore sandy. There are a few detached mountains, as indicated on the map, but nothing more. From Ngóroínne to Ukára there are small, inconsiderable hills ("*vilíma vidógo 'dógo*"). There is a small *bay*, or bend in the shore, extending from Káverónd to Chamwáli.

With regard to the eastern shore, Sádi's bearings of his journey northward to Lake Baringo, cause the Nyánza to deflect north-westerly. He could not see the lake from any point of the journey; but from the summit of the Ligéyo Mountain it is clearly seen, appearing a long distance off, about two days' journey.

About eight or nine years ago, Sádi, arriving at the Nyanza, and observing a deep trench or channel on the shore, inquired what it meant, when the natives replied that a *large vessel* had recently been on the lake, and had anchored at that spot. The vessel has three masts, and another in the front (bowsprit?). The vessel had on it "ngúo nyíngi, neópè" (many white cloths) (sails). The visitors were described as "waópè sána," very *white*; they bought large quantities of eggs. They also purchased some *ivories*, but only *short* ones; *long tusks* they refused. Sádi thinks that it was about a month and a half after the departure of the visitors that he arrived at the lake.

THE WA-UKÁRA;

(*People of the Ukára region, Lake Nyánza.*)

Sádi describes them as a people most scrupulously clean in their habits. He was particularly struck with this feature of their character, and on several occasions referred to it. If they have been working in their plantations, they are particular, afterwards, in washing themselves and their clothes. They also keep their cooking-vessels and other utensils very clean. They are of average height. They cultivate the ground, growing maize, but not much; beans (of different kinds), millet, bananas, cassada, and sweet potatoes. Their agricultural implements are large iron hoes. The children use wooden ones. They also frequently employ themselves in fishing. With regard to their weapons, some carry bows and arrows; others, discarding the bow, are armed with spears, and carry large shields, like those of the Masái. Sádi says they are a peaceable people. Their clothing (like that of many of the interior tribes) is made of skins—goat-skins, sheep-skins, and those of wild animals taken in the chase. The apparel of the women are short kilts (of skin), which are anything but adequate to a decent appearance. The Ukára women present even a more nude appearance than those of the Wanyíka. The men do not wear ornaments, but the women wear a few, which are imported from the coast. The houses of these people are circular. The walls—which are high—are plastered inside and out with clay, surmounted with a conical roof, thatched with grass; resembling the huts of the Wa-Teita, and those represented in Captain Burton's book on the 'Lake Regions,' as built by the Wanyamwezi. With regard to language, the few specimens given below—which Sádi gave as the Ngóroínne dialect—show that the Wa-Ukára are of the pure African stock, *i. e.* as far as the affinity of language indicates it.

English.	Ki-Ukára (Ngóroínne dialect).	Kikámba.	Ki-Teíta.	Kinyíka (Kíbè).	Ki-Sáwahíli (Mombása).
Water	mátsi.	mánzi.	máchi.	mádzi.	máji.
Goat	chimbári.	mbúi.	mbúri.	mbúzi.	mbúzi.
Sheep	chignoóndu.	ilóodu.	gnóndi.	gnóozi.	kóndó.
Tree	múti.	múti.	múdi.	muhi.	míti.
House	chinyúmba.	nyúmba.	nyámba.	nyumba.	nyémba.
Fowl	chingúkú.	ngúku.	ngúkn.	kuku.	kúku.
Hen (*female fowl*)	chingúkú mkn.	..	..	..	..
Cock	chingúkú murámè (lit. *male fowl*).	ndzólólá.	jogóla.	dzogóla.	jógó; jímbi.
Cattle	gnómbè.	gnómbè.	gnómbè.	gnómbè.	gnómbè.
Dog	chítta.	jítta.	kóabi.	káro.	jíbwa; mbwá.
Fish	chíswi.	ikúyu.	nguísma.	swi.	samáki.
Man	mándu; pl. wánda.	mándu; pl. ándu.	múndu; pl. wandu.	mátu; pl. áta.	mtu; pl. wátn.
Woman	mdúmka.	mándu mdka.	mánda mdka.	mátu móobè.	mtámkè.
Child	chimwána.	kána.	mwána.	mwána.	mwána.
Grass..	nyáki.	nyíki.	nyási.	nyási.	nyási.
Hoe	ligémbè.	ye émbè.	igémbè.	jembè.	jémbè.

Sádi states that the Wá-Ukára are very numerous, and many of their settlements are large towns, containing about a thousand huts each. That they are about equal in size to the town of Mombása. This estimate, I imagine, is somewhat exaggerated.

The country (Ukára) is dotted with small hills, but the shore of the lake is level. The hills are not so large as those in Wanyika-land.

No. 2.   RETURN ROUTE

From Ukára (Nyánza) to Arúsha.

Búhari ya Ukára to Ngoroínne. The meaning of the word Ukára, Sádi is unacquainted with. Ngoroinne is **Kikwávi.** There are villages at Ngoroinne, and the inhabitants have cattle and plantations. [12 hours.]

To *Séro.* [10 hours.]

To *Kivdivái.* [12 hours.]

To *Ndá Sekéra.* [9 hours.] These are Kikwávi names, Sekéra meaning *Cowries.*

To *Sónjo.*—The name Sonjo is Kikwávi, and means "fíwi" (Kisáwahili), a large species of bean. The Wakwávi sometimes buy beans (fiwi), from the Wa-Sónjo. [4 hours.]

The settlements which are here are permanent. **The Wa**-sónjo cultivate the soil and keep cattle. The place is **populous.** Sádi says that the Wa-Sónjo are Wasegéju * immigrants, **who** left Shungwaya and came to this region, in which they **have** settled. (Shungwaya is a district between Goddóma **and** Kaúma (Wanyika-land); and Sádi states that it was the **ori**ginal home of all the Wasegéju.) To the south-east **is the** Sónjo Mountain. It is large, but not so large as Nándi **Mtu.** It is wooded from base to summit. To the south-west is **the** Kura Mtu, about as large as that of Sonjo.

To *Malámbo.*—Here are settled villages of the Wa-Malámbo, who, Sádi says are Masái, but "meskini" (poor). They have no cattle, but have plenty of goats. They also cultivate the soil. [3 hours.]

To *Sálè.*—Masái live at Sálè, and Sádi says that this **region** is the Masái country proper, "n'ti yáó *Rábisa.*" South-**east** of Sálè, is the mountain *Ngári.* [5 hours.]

To *Kíti.*—The name is Ki-Masái, "Kiti" meaning *little.* The mountain is lofty, but not very bulky. There is a luxuriant forest on its slope and summit; Wa-Ngúrumáni (some call them Wa-Utimi, and others, Wa-Bagbási, but Sádi says the

---

* **Wasegedshu,** History of the. *See* Krapf, in 'Church Miss. Intelligencer,' of 1849, p. 86.

first is the proper name) live on the top. There is water on the summit (a spring), which is perennial. The people cultivate the ground, and make trenches or channels, by which water is conducted in various directions to the plantations.

South of Ngári Kíti is the Mountain *Doënyo Sámbu* ("Sambu" is "Sahári" in Kisáwahili, a kind of cloth worn in Eastern Africa.) This mountain is very lofty, not inhabited. Sádi says that it is as high as Kilima Njáro. There is a forest on it; also rock on the base and summit. It stands about half an hour from their camping place at Ngúrumáni.

To *Másimáni.*—This word is Kisáwabili, and means "at the wells." (The *diminutive* form is generally used—namely, Visimáni, sing. Risimáni.) [6 hours from Sálè.]

To *Gíléi.*—Here is a very large mountain called Gilei. Masái live on the summit, where there is plenty of room for the pasturing of their cattle. Sádi's caravan encamped on the top, as there was no water at the base; but the ascent *tired* them. The summit is wooded, but the slope only covered with grass. [9 hours.]

To *Gwáso ná Ebór.*—(Gwaso na Ebór, means, literally, *white river.*) This is a region of *white sand.* Sádi says it sparkles (quartz?). There are Masái here. Six hours to the north of Gwasó na Ebór is a mountain, called *Méto* by the Masái, and *Malumbáto* by the Wa-Sáwahili. It is large, wooded on the top, with grassy slopes. [6 hours.]

To *N'daptúk.*—Masái live here permanently. They never migrate from Ndaptuk. There is a mountain here, called by the name of the locality. It is dotted with patches of jungle. Masái live on the top and at the base; the slopes are also encircled with settlements. [12 hours.]

To *Ngárè na Nyúki,* (ngáre, water; nyuki, red—red water).—The name is Ki-Masái. Here is a small stream, the surface of which is covered with a red dust. The *bed* of the stream is also formed of red mud. It is not deep. A man's arm easily reaches the bottom, and the least disturbance of the bed renders the water at that place a red mass. Though the stream is shallow, it never dries up. It comes from Kópia Ekópi. [12 hours.]

To *Ngárè na Eróbi.*—Ngárè na Eróhi, is Ki-Masái, and means *cold water.* There is a small stream here, the water of which is so intensely cold as to be unbearable. Sádi says if a man puts his foot into it, it makes the very bones ache! The whole country here, too (*i. e.* the ground), is so cold as to make the travellers put on shoes (sandals), if they commence the march in the morning. The cold, he says, comes from *Kilima Njaro.* This, and about fifty or sixty other streams, flow westward from

Kilima Njaro. Since writing the above, Sádi says there are more than sixty; some wander northward, and others south-ward, and others, again, easterly. [6 hours.]

To *Kirarágwa.* [3 hours.]

To *Sigirári.*—At Sígirári there are Masái villages, but not permanent. However, if the inhabitants migrate, they only remove to *Kiraragwa.* In course of time they remove again to Sígirári; spending their time between these two regions. Occasionally changing, no doubt, for the sake of pasture. [10 hours.]

To *Mibuyúni.* This name is Ki-Sáwahíli, being the plural of Mbúyu—the Baobab-tree. This region abounds with the Baobab (*Adansonia*). [9 hours.]

To *Márăgo ya T'émbo.*—This term is Ki-Mríma. Márăgo means camping-place; and Tembo, elephant. [8 hours.]

To *Arúsha.*—This is the end of the down-march, as Sádi falls into his old path at Arusha (see page 304). [5 hours.]

No. 3. From Mombása to Dhaicho.

To *Kimri.* (6 hours.]

To *Shimba.* Shimba is a long mountain, probably about 2000 feet high, which is occupied by various tribes of Wadígo and other Wanyika tribes. [6 hours.]

To *Mazóla.* There are two meanings to the word zola (which is Kidigo)—1. ku zóla ni ku fukúza, to expel or drive away; 2. ku zóla maíro, to run quickly. [9 hours.]

To *Gúrungáni.*—Gurúnga means, in Kisáwahílí, holes or pits in stones. N.N.W. of camp, at Gúrungáni is *Kilibássi*; at Ngu-rurungáni za Kilibássi (vizima via máwe, stone wells), there is always water. This locality is a camping-place for all caravans, on account of the perennial supply of water. The hill *Ukinga* is not so large as Kilibássi. [9 hours.]

To *Kisgáu* or *Kisigáu.*—Kisigáu is the Sáwahili name for Kádhiáro. The latter is a local name, and Sádi says is unknown to many Sáwahilis. [12 hours.]

To *Matátè.*—There is a hill at Matátè southward of the camp. A small stream also, about 1½ yard wide, comes from Teita. [9 hours.]

To *Búra Mountain.*—Búra is a large spreading hill. Wa-Teita inhabit the top, and cultivate the ground. Búra, Sádi says, is a Kidígo word (probably also Ki Teita), and means a fen or marsh, where water lies concealed amongst the sedge, and must be turned aside before the water can be discovered. There is such a marsh at the base of the Búra Mountain. [6 hours.]

To *Baráni* (Camp). [12 hours.]

To *Lanjóra.*—Lanjóra is a Kikwávi name. Wakwávi formerly lived here, but were driven away by the Masái. The remains of their villages are to be seen to this day. [6 hours.]

To *Tavéta.*—The latter is a Sáwahili corruption of a Kwávi word, which is *Ndovéta.* [4 hours.]

To *Uséri.*—This word is thus called by the Wasáwahili. By the Wakwávi it is called *M'su Géri.* Between Useri and the next stage (Kimangéla), the caravan crossed the River Tsávo. At this place it is only two or three yards wide. In the hot season the water merely covers the ankle. In the rainy season the river is not fordable. The caravan drivers fell a couple of trees, one on each side of the river, causing them to fall across the stream, thus constructing a bridge, by which the long line of travellers pass over. [12 hours.]

To *Kimangéla.*—Sádi says this is a *Chága* word, and that the Wakwávi call it *Kimangélia.* A little to the south of the next stage, the caravan crossed the Sabáki River, or, rather its bed, as, being the hot season, it was quite dry. Though they dug for water they failed to get any. In the rainy season the river is not fordable; but the water remains no great length of time. In a few days it is dry again. [5 hours.]

To *Léta Kotók.*—Sádi says this latter name means spring or fountain. The word Kotók being used to express the action of bubbling. There is a spring at this place. [12 hours.]

There is also a large lake at Léta Kotók—length, from station at Ribe to Makerúnge, or Ngú Hills (6 or 8 miles). Breadth, about a mile. It abounds with reeds and grass. There are about half a dozen small islands on it.

To *Baráni* (Camp). [6 hours.]

To *Mábarásha.*—This latter word is Ki-masái. There is a mountain called Mábarásha, about a quarter of an hour to the north of the camp, not very large. [6 hours.]

To *Ulasindio.*—This word was the name of a Kwavi chieftain, who formerly lived here. This region is called by the Wakwávi, *Lóbrabrána.* [6 hours.]

To *Lamwéa.*—Lamwéa was the name of a Kwavi chieftain, who gathered his followers about him, and determined not to leave the above region; but to live and die there. He did so. His followers, since the death of their chief, have long since migrated. [12 hours.]

Here is a large mountain called Lamwéa. On the northern slope is a jungle of Múrijón trees. Masái live on the sides, but not on the summit. It is less than an hour from the camp, bearing north-easterly.

To *Múrijón.*—Mírijon is the name of a tree very much

resembling, Sádi says, in appearance the clove-tree. It yields
the poison used by the Wandoróbos for their arrows. The
Wandoróbo women cook and prepare this poison for the men.
[6 hours.]

To *Kápté.* [4½ hours.]

To *Miviruni.*—Mvíru (pl. Mivíru) is the Ki-Sáwahili and Ki-
Nyika name of a large wild fruit tree. The fruit is about the
size of very small apples, and, when ripe, of a russet or brownish
colour, consisting of a thin rind, which is filled with fruit of a
globular form, pressed together, and about the size of marbles.
[12 hours.]

To *Mianzini.*—Míanzíni is Ki-Sáwahili; Mianzi meaning
bamboos or large reeds. At this place there are villages of the
Wandoróbos. Here (a little below Míanzíni) the River Tána is
crossed; which, Sádi says, passes mid-way through ("wa pasúa")
Kikúyu; and that its source is the Máu country. A number
of tributaries enter the Tána in the interior, four from Dhaicho,
and seven from Kikuyu. There are large numbers of crocodiles and hippopotami in the river. [12 hours.]

To *Naïvásha.*—At Náivásha there is a large lake—length,
from Jibána to Mombása (18 miles); breadth, from Ribe Station to Mombása (13 miles). Bearing N.N.E. and S.S.W. Large
numbers of hippopotami in it. The water on the north-eastern
side is fresh, whilst that on the opposite shore (north-western)
is salt. The water at the southern end is also salt. [12 hours.]

To *Vibokóni.*—Víbokóni is Ki-Sawahíli, and signifies the
"habitat of hippopotami." [9 hours.]

To *Agnáta Vús.*—Agnáta Vŭs is a Kikwávi name: *Agnáta,*
"country," or "wilderness;" and *Vús,* "mist," or "fog." This
is an excessively cold region, and is frequently buried in the
thick fog which descends from the Settíma Mountain; it is so
dense that a porter cannot see his fellow who is just before him.
Sádi says that the men link themselves together with their
cloths, to enable them to keep in the path. This foggy
mass continues until about 10 o'clock A.M., when it begins to
evaporate. This region is also a swamp, but underneath is a
formation of rock. [6 hours.]

To *Settíma.*—Here is a large mountain, called *Settíma.* Sádi,
speaking of it, says "it is a *Lima;*" a term denoting extraordinary size—mlima being the general term for mountain. Wakwávi make the summit a temporary dwelling-place. It is
clothed with grass, and is wooded on the eastern side. The
caravan slept on the top. If porters are tired, it is customary
to camp for the night at the base, and the following morning to
ascend it; but if not over-tired, they climb it, and camp on the
summit. [5½ hours.]

Three streamlets issue from this mountain, and, flowing south-easterly, ultimately coalesce and flow on in a single stream, called Ná-Erogwa, to the reedy fen or marsh near Mianzíni: "mianzi" meaning bamboos, which are here very numerous and as thick as a man's leg.

To *Agnáta Ndárè.* — This name, Sádi says, means goat pasturage: Ndárè meaning a herd of goats, and Agnata (báva, wilderness), as before remarked. [6 hours.]

To *Ndóro.* [9 hours.]

To *Dóěnyo Ebór.*—This name means "White Mountain." Sádi says it is very lofty, and the summit exceedingly white. He considers one day to be sufficient to go round it. It stands about an hour from where they encamped. Sádi states that the country at the base is called Vórè. The above Dóěnyo Ebór is the "*Kenia*" in Dr. Krapf's map.[*] [9 hours.]

To *Barani* (camp). [9 hours.]

To *M'sarára.*—Here is a large mountain, but less than Dóěnyo Ebór; about one hour to the east of camp. The summit is inhabited by people called by the Wakwávi "Liméro;" by the Wa-Sáwahili "Wa-Limeru" and "Wa-Méru;" and by the Wakámba "Embu." They are agriculturists. Lambúi, one of their chiefs, lives here. The whole country, from Dóěnyo Ebór to Dhaicho is called Liméro by the Wakwávi, but Méru by the Wa-Sawahili. To the north of Msarára, Sádi crossed the *Ozi River*, called by the Wakwávi and Masái, Gwaso (river) Limbárua (or Limbáruwa). This word is the name of a tree in the interior and near the coast; it is called Limbáruwa by the Wakwávi, and Mwáte by the Wa-Sawihili. At the place where Sádi crossed the river it was about 5 or 6 yards wide. Here is a broad and permanent bridge, which has been constructed by the Wakwávi, of trees, over which their cattle may cross. Wakwávi live on both banks. [6 hours.]

At other points the Ozi River is very wide—a hundred yards or more—receiving influents from both north and south, Dhaicho and Méru. The current, also, is strong; and, says Sádi, "were it not for the 'fungu' (sandbanks), which God has put into it, it would be utterly impassable."

To *Gwáso Nyíro.*—Gwáso (river), Nyuro (grey): Grey River it is called, on account of the river, at this place, being covered with a layer of *grey dust.* The Nyíro is a considerable river, and is perennial. Its lowest depth, *i. e.* in the hot season, reaches up to the knees. It flows from the Njémsi Mountain, and turns off northerly near the camping-place, and goes to

[*] *Vide* 'Travels,' &c., by Dr. Krapf, p. 360. "Oredoinio-eibor (White Mountain, Snow Mountain, the Kegnia of the Wakamba)."

Sambúru. There is also another stream of this name which rises in the Máu country, and thence flowing south loses itself in the nitre swamp near Ngúrumáni. To distinguish it from the above river, it is called "*Gwáso Nyíro*" ya Ngúrumáni. [12 hours.]

To *Lórián*. [8 hours.]

To *Dhaicho*.—Dhaicho is a long spreading hill, about as high as Búra (Teita), three or four days long and about one and a half broad (length, from station at Ribè to Wasin, 55 miles); bearing north-east and south-west, the northern part of the chain veering round again to the westward. [6 hours.]

THE WA-DHAICHO.

Inhabitants of the Dhaicho Hills are agriculturists. Mtama, mawěli, kimánga, ndíze, mbázi, muhógo, viázi, víkwa (like yams), miwa, fiwi, and mahúndi (*i. e.* banana, cassada, sweet potatoes, sugar-cane, beans, &c.), are amongst the things they cultivate; they have also cattle, goats, and fine sheep. They tenaciously cling to their house in the highlands, not daring to live in the plains below from fear of the Wakwávi. However they frequently come in contact with them. During the hot season, when the Wakwávi are in the habit of visiting those regions, the Dhaicho women descend from the hills, and carry to them fruit and vegetables for sale, or rather barter, obtaining from these nomadic pastorals flesh-meat in return. The fierce warrior, whose life is one constant foray on the flocks and herds of other tribes, can easily afford to' pay. On such occasions the women are not molested, but allowed to return. The Wakwávi also visit the Dhaicho people on the hills. On asking Sádi whether the Wakwávi do not kill or fight with the Dhaicho, he replied that, on their arrival, they propound terms of "peace and goodwill," and keep their word until just before they are ready to leave the country, when they attack them, plunder them of their cattle, and carry off captives. On the second visit they again propound peace and friendly relations, and keep their vows until the time of departure, when they again violate their covenant and display the "ruling passion" of their savage natures—a thirst for feud and plunder.

Sádi says, that now the Wakwávi and the Dhaicho people are not on friendly terms. The Wa-Dhaicho are respectable warriors, and are able to engage with the Wakwávi; but they are a peaceable people, nevertheless.

Their weapons: some carry bows and arrows, and others prefer the spear and shield. Their houses are circular, like those of the Ukára (Nyánza) people, but very large. With regard to their clothing, both men and women wear skins: the men short kilts, the women longer ones, and also a large skin to throw over the shoulders. There is some calico amongst them, which they have imported from the coast.

Ornaments: the men wear a little thick iron wire on the arms. The women wear iron wire round the neck and arms and legs. They also wear long ear-rings of thin brass wire. With regard to language, the following brief list of words, when compared with that given under the heading *Wa-Ukara*, will show that the Dhaicho people are genuine members of the African family:—

English.	Ki-Dhaicho.	English.	Ki-Dhaicho.
Water	ngárè (kikwávi).	Dog	kitte.
Goat	búri.	Fish	ngulúma.
Sheep	nöndu.	Man	múndu; *pl.* wándu.
Tree	mti.	Woman	múadu múka.
House	mjumba.	Child	chimwána.
Fowl	(no fowls).	Grass	uyági.
Cattle	gnömbè.	Hoe	ligémbe.

No. 4. FROM SÍGIRÁRI TO SAMBÚRU.

Sígirári to Kírarágwa.—(Both these names are Kikwávi.) [12 hours.]

To *Ngárè Na Eróbi.*—The meaning of this latter name in Kikwávi or Ki-Masái is *cold water.* There is water at this place, which, according to Sádi, fully justifies its name. It is so intensely cold that, if the natives drink it, they "endeavour to swallow it without it *touching their teeth.* [3 hours, *see* p. 313.]

To *Kópia Ekópi.*—This name is *old Kikwávi,* but Sádi does not know its meaning. [7 hours.]

To *Döenyo Erók.*—This name is Ki-Masái, and means *black mountain.* The Dóenyo Erok is very large. Like the Dóenyo Ngái, it is higher than the Kilima Njáro, though not so massive. Wándoróbo (not many) live constantly on the sides of this mountain, whilst there are some of those wild hunters on the summit itself. Elephants ascend this mountain, which is encircled with forests from base to summit, in which are large trees. The elephants frequent these woods for the sake of their abundant herbage. In bulk it is about the same as the Ngái mountain, but the former goes somewhat higher. Sádi says the Dóenyo Erók is "meúsi sáua sána!" (exceedingly *black*). 8 hours.]

To *Ngárè Rongéi.*—Rongéi is Ki-Masái, and means *narrow.* Ngárè means *water.* There is a very small streamlet here, or rill, about 5 or 6 inches * broad, but it never dries up. It comes from Chága, from the Kilima-njaro. [9 hours.]

To *Migungáni.*—This latter word is Kisáwahílí, and is the plural of mgúnga, a very large tree, producing immense thorns, 7 or 8 inches long. Elephants are fond of the succulent parts of this tree. [8 hours.]

To *Dóènyo Lamwéa.*—A Kwávi chieftain formerly lived here, whose name was Lamwea, after whom, no doubt, the region was called. The Doenyo Lamwea is a considerable mountain, as large as Nándi. It has rather a bare aspect—no wood upon it except on the north side, at the base, where there is a wood or jungle of the murijon-tree. There are rocks on the top. [10 hours.]

To *Mtangóni.*—Mtángo is a Kisáwahílí word, and means a desert—a dry waste, where no water whatever is to be obtained. The Masái and Wakwávi call this place *Garómi.* [12 hours.]

To *Mívirúni.*—Mivíru are a species of wild fruit-tree; Mívirúni *en route* to Dhaicho. [10 hours.]

To *M'to wa Gógnu Baghásè,.*—M'to wa Gógnu Baghásè, River Gógnu Baghásè (Kikwávi). Here is a small streamlet, about a yard wide, which never dries, coming from a mountain in Ku-kúyu. [6 hours.]

To *Náivasha.*—This word is Kikwávi, and Sádi says it means *salt water.* [12 hours, *see* routes 1 and 3.]

To *Mto wa Ngárè Motónyï.*—River Motónyi. Motónyi is the name of a species of bird, which make this region its habitat. The motónyi are numerous here. The "M'to" is a stream, about two yards wide, perennial, and comes from a district called Döndólè,† near the Njämsi Mountain. This region is perma-nently inhabited by Wándoróbo; they never leave it (*háwa tóki hápa Rabísa*). [7 hours.]

To *M'to wa Agnata Vûs.*—River Agnata Vus is perennial. It is only a stream, about a yard and half wide, and comes from the Séttima Mountain. [6 hours, *see* Route 3.]

To *Mlíma wa Séttima.*—A lofty mountain, a little to north of camp. After sleeping at base, the caravan climbed it the next morning, and afterwards descended the northern side. The eastern side is wooded; the other parts clothed well with grass. Wakwávi live on its sides and also at the base. In the hot season those living at the base ascend the mountain in search of pas-turage, and also on account of its cool temperature. The grass remains constantly fresh. [4 hours.]

To *Súbúgo Limárimárè.*—Here are *masima* (wells or pits), which have been dug by the Wakwávi to give drink to their cattle. [9 hours.]

---

* ? feet.   † Döndólè is a Kíndoróbo word.

To *M'to Migungáni.*—River Mígungání (Kisáwahílí), from mgúnga, a large tree, bearing immense thorns. The Mígungáni stream is about two yards wide, and reached to the loins. It never becomes quite dry. Water is left in places where the bed is deep. Flows out of the River Nyíro, near Suvúcha, and afterwards enters the Ozi. [5 hours.]

To *Suvúcha.* [6 hours.]

To *Másimá mikománi.*—This name is Kisáwahílí, meaning the wells or pits at the Mikoma—a species of fan-palm. [7 hours].

To *Mkwajúni.*—Mkwáju is Kisáwahílí, meaning tamarind. [10 hours.]

To *Mvóngonyáni.*—Mvongoènyáni (Kísawahílí), is a large tree, bearing long immense pods (calabash or monkey-bread tree). [9 hours.]

To *Ngárè Ndogéi.*—This name is Kikwávi; Ndogei, meaning the brab-tree, and Ngárè, water. There is a stream at this place, about four yards wide, and the Ndogei (brab) is growing on its banks. Sádi says it comes from Lórián, and that hippopotami and crocodiles are very numerous in it. [8 hours.]

To *M'swakíni.*—Mswakini is Kisáwahíli, meaning the tree from which Wa-Sáwahílis and others cut their tooth-sticks. [5 hours.]

To *Jiwéni.*—Jiweni (Kisáwahíli) means at the *stone.* Here is a small hill, on the top of which was a piece of rock, which, becoming split, half of it rolled down the slope of the hill to a distance of about a mile. Hence Jiwéni. [8 hours.]

To *Kitóni.*—This is a place of mud and puddles. With regard to the meaning of the word "kito," Sádi says it is used to express the bubbling sound made at the escape of air when releasing the foot from a hole into which it has sunk. Possibly the word is Ki-Mríma. [9 hours.]

To *Márăgo Khálfán.*—The word "marágo," thus accented, when *alone,* but on the ante-penultimate, when in combination, as above, is Ki-Mríma, and means camping-place. Khálfán is the name of an Arab, who was once taken very ill at this place, and has ever since been identified with that event. [6 hours].

To *Kítwa cha Ndóvu.*—This name, *The Elephant's Head,* is Kisáwahílí, and is thus called on account of the skull of an elephant being at this place. The tusks have, of course, long since been carried away. [9 hours.]

To *Túmbo la mlíma.*—Literally, the *belly of the mountain,* meaning its slope, or bulging part. [5 hours.]

To *Zíwa la Mbu.*—Mosquito Lake or Marsh (Kisáwahílí). The water reaches to the loins. [7 hours.]

To *Márăgo ya Fáu.*—"Fau" is Ki-Mríma, and means rhinoceros. Many rhinoceroses here. [5 hours.]

To *Ngulúvo Likária.*—This name is Kikwávi, and means red ochre or ruddle, which the Wakwávi mix with oil or grease (from the coast), and smear their heads, breasts, and arms with it. [6 hours.]

To *Mlíma wa Sambúru* (Sambúru Mountain).—Sádi says it presents a very forbidding aspect, cliffs stretching out, "arms" (horns) in all directions on the summit. It is very large, lofty, and massive. Uninhabitable. [9 hours.]

To *Bolyói.*—Bolyói is Kikwávi. Sádi says the Wakwávi dig large holes to obtain a species of *soil*, which *fattens their cattle.* The cattle know its purpose and voluntarily eat it. This soil, as well as the pit or excavations thus made, are called bolyói. [7 hours.]

To *Southern Sambúru.*—This is the *limit* of Sádi's journey from Sigirari. He did not really enter Sambúru, but only reached its southern frontier. He states that he saw the northern* end of the Sambúru Lake, which appeared about two days' off, and describes it as being—length, from station at Ribe to Kipínbui (128 miles); breadth, from Ribe to Gasi (37 miles); bearing, N.N.E., S.S.W.

WA-SAMBÚRU.

(*People of Sambúru.*)

The Sambúru people are subject to the Somalis, as the Wándoróbo are to the Masái. They are pastoral, and have much cattle. They do not cultivate the ground. Like the Gallas and Masái, they do not eat fish; consequently, their large and beautiful lake is lost upon them with regard to angling purposes. They have numerous horses and camels. They are hunters, and are said to hunt on horseback in a very singular manner; they tie a lot of spears together in two bundles, which they place one on each side of the horse. When approaching a place of game, they endeavour to conceal themselves by clasping the horse round the neck, with their heads underneath, and their feet resting in loops made for the purpose, which hang over the flanks of the horse. The horses are trained to go slowly towards the place of prey, and when near, the hunters very slowly turn themselves, until they have got on their horses' backs, when they give forth their shrill hunting-cry, and pursue the game until they come up to them, when they use their spears with great effect. The people of Bráva go to Sambúru for trading purposes, but the Sambúrus do not go to the coast.

► The Wa-Sambúru speak a dialect of Kikwávi; but though

* ? Southern end.

probably they and the Wakwávi have a common origin, they are by no means friendly, but fight whenever circumstances throw them together. They carry spear and shield, but no " *símè*" (native *sword*). They have, also, bows and arrows—the latter lubricated with a virulent poison, very strong. Those who do not carry spears have bows and arrows. The Wakwávi only carry spears, shield, and símè.

No. 5.—FROM LAKE NYANZA TO LAKE BARINGO.

We commence this route from *Gwáso Ngíshu*, to prevent the repetition of intermediate stages along the shore of Lake Nyanza. (See p. 308.)

To *Baráni* (Camp). [6 hours.]

To *Ligéyo.*—At this place is a *mountain* of the above name. It is not very lofty, but spreads out (nearly due north and south) for about two days' journey; also, about one and a quarter day broad. There is a little jungle about it, but not much. Pretty well clothed with grass. People called Wa-Ligeyo live on the top, where they also pasture their cattle; never coming down from the mountain, as they greatly fear their near neighbours— the Wakwávi of Gwáso Ngíshu. The summit is *very populous* —" nté nzima " (a whole country). The inhabitants have plenty of cattle, and also cultivate the ground. [8 hours.]

To *Mwisho wa Mlíma.*—*End of mountain* (Kisáwahílí), evidently no name for the locality. [9 hours.]

To *Dóěnyo Ebór.*—Kikwávi name, meaning *white mountain*. Sádi says it is but a hill, and not large. It has a *small white crown*—hence its name. [12 hours.]

To *Másimáni.*—At the *wells* or *pits* (Kisáwahílí). [9 hours.]

To *Mtánganyíko.*—Kisáwahílí, meaning the place of *mingling* or *mixture* (rendezvous). Being a damp, (rútuba) region, it affords pasturage for cattle during the hot season, yielding abundance of grass when it is scorched and withered elsewhere. Wakwávi of different tribes meet here to pasture their cattle, periodically, making the place quite a rendezvous. When the herds have cropped the grass the men again separate, returning in various directions to their homes. [12 hours.]

To *Agnata Láegŏb.* [8 hours.]

To *Lóbrubrána.* [10 hours.]

To *Lake Baríngo.*—When the caravan reached this point, they commenced to return, but by a different route from the one marked out above, fearing to go further northwards, as they were already very near the *Wa-Súku*, a fierce race, and much dreaded by the Wasáwihílí. [9 hours.]

LAKE BARÍNGO.

Length, from station at Ribe to Kipúmbui (128 miles); breadth, from the same to Wasín (56 miles); bearing: nearly north and south, but a little north-westerly and south-easterly. Sádi was told by the natives dwelling in the vicinity of the lake, that the meaning of the word "Baríngo" is *canoe*, possibly so called from its form resembling that of a canoe. There is an island—a small conical hill probably about a hundred feet high—in the lake, situated near the south end, and not far from the eastern shore. The slopes are clothed with vegetation, and the base is a circle of light-coloured sand (not so light as beach-coloured sand). Sádi says the island—which has no name, as far as he knows—is about as large as that of Mombása, and that it is appropriated as an asylum by the Njémsi people (*vide* below), who flee thither with their cattle from the marauding Wakwávi. It has a large village, containing about 120 huts, but closely packed. The water is shallow between the island and at the shores, reaching only up to the loins. He states that some of the Njémsis have taken up their permanent abode on the island, where they graze their cattle and cultivate the soil.

There are two rivers, both effluents of the Baríngo. Whether the northern is an effluent or not Sádi does not positively know, but gave it as his opinion that it is so. (The reason that so little is known of this river is on account of its flowing through the country of the savage Wa-Súku, of whom all parties seem to have a lively dread). One flows out of the lake at its northern end, and takes a north-westerly course, and the other flows out of the south end, continuing its course almost due south. Sádi calls them both by one name, that of *Nyarús;* but the southern one is frequently called by the Sáwahílis who have travelled in that region the *Jémsi* river, its source, and also course, for some distance, being in the Jémsi (or Njémsi) country. Sádi appears to know little of the northern stream, but conjectures that it enters the Nyanza lake to the northwards. Ulédi, one of our servants, who has been to the Baringo, says that he was informed that it was a considerable river, about 30 yards wide. The southern stream is narrower than the above, Ulédi says about 7 or 8 yards, and that it took them up to their necks when crossing it. This was in the hot season. He says, also, that a rivulet issues out of, or enters the lake (which, he cannot tell) directly opposite the island, on the eastern shore. He spoke of it as issuing from the lake, and then flowing south-easterly until lost amongst a group of lofty hills (higher than Shimba, south of Mombása), a short distance from Baringo (about a quarter of an hour). This rivulet is probably an influent, having its source in the above highlands, and thus a tributary to the lake. It was about two yards wide, and of shallow depth, only covering the ankles. He also crossed two other streamlets on the western side of the lake, flowing nearly parallel to each other, and the shore of the Baríngo, and only a short distance from it: one or both of them probably coming from the lofty hills he saw when in the Lugúmè country, which are situated to the north-west of the Baringo Lake. He says there are also large hills north-east of the lake, and that, in fact, the wholé is a mountainous or hilly region. Ngárè (or river) Davásh flows out of the Njémsi stream near Séro, and flowing due east, enters a region called "Kilíleóni," in the Wándoróbo country, and thence to that of the Masái. The Davash—though in Ki-Masái the word means broad—is but a narrow stream, being only a few yards across, and easily fordable. On both sides of the river (at Séro) the Wándoróbo have built their huts, the whole forming a permanent settlement, containing about a hundred dwellings.

The Njémsi Volcanic Mountain.—In the Njémsi country, to the south of the Baríngo, is a large volcanic mountain, with hot springs at the base.

There are 30 or 40 craters, not very large, and are all situate at the base of the mountain. From these craters large volumes of smoke are constantly issuing, resembling, Sádi says, "*mináro*" pillars or columns, like those from the funnels of steam-ships, the smoke is so abundant and dense as to obscure all objects in its vicinity, and beyond. The craters are constantly active, except at night, when they subside. The craters send out no fire, nor even stones, nothing but smoke. There are black stones at the base of the mountain. Sáwahílis sometimes pick them up and use them for gun-flints, but Sádi says they can only be used once or twice. They will not strike afterwards.

The mountain is somewhat of a cone in form, and is rocky and rugged from (túmbo la mlíma) the slope to the summit, here and there broken and jagged, and the rocks pointed and sharp. At the base there are hot springs, which are constantly boiling up or bubbling. The water is very hot; the fingers cannot be borne in it. If a little flesh-meat is put in, it is "done" immediately, quickly cooked. The water is in small but very numerous pits or pools ("Visima"). It wells over, but forms no stream, merely spreading over the ground. If some of the water is drawn very early in the morning, and put into a cool place, "it is not cold until about 3 p.m!" The caravan drank it cold, and also used it in cooking their food; and Sádi says it was very good (sweet).

THE PEOPLE OF BARÍNGO.

The races immediately dwelling about the Lake Baríngo are the Wa-Súku, the Wa-Ligéyo, and the Wa-Njémsi.

The Wa-Súku inhabit the region to the north-east and north-west of the lake, occupying both banks of the (northern) Nyarús River. Their country is called *Lugúmè* or *Súku-Lugúmè*, and is a region much feared, and consequently is evaded by caravans from the coast. Being surprised at Sádi knowing so little of the above stream, I asked him if he did not enquire about it from the lake people? when he replied that *Wa-Súku* were the only people who knew about it, and of them they were afraid; hence they turned back when they had reached the point marked on the map.

The Wa-Súku are feared on account of their ferocious and barbarous character. They are brave and daring, but guilty of many horrible and brutal deeds. They do not hesitate to give battle to the Masái. The latter pay them predatory visits and carry off their cattle, and they also go to the Masái country and lift their cattle in return. Despising sheep and goats, which they find with the Masái and others during their plundering excursions, they frequently spear them, and leave them on the spot; sometimes they content themselves by merely maiming them, by cutting off their tails, &c. They even spear dogs if they come in their way. If, during these raids, they capture a pregnant woman, they cut her open, take from the womb the unborn infant and cut it into pieces! Frequently they will cut off the hands of a captive warrior (Mkwávi, Masái, or any other), and then say to him, "go now, go your way, and how will you manage to eat?"

The Wa-Súku, though living near the lake, and on the banks of an ample stream, do not eat fish, though they abound in the water near them. They are agriculturists, and also pastoral. They cultivate the ground, and are also rich in cattle. Their weapons are light spears, which they throw with much dexterity and precision. But Sádi says, if near their enemy they do not hurl their spears. To catch a fleeing foe they throw, and to a good distance.

The *Wa-Ligéyo* live on the western side of the lake, but a good day's march off the shore.

The *Wa-Njémsi* live on the south-eastern and south-western shores of the lake, and on either bank of the Njémsi (or Nyarús) River; and their plantations stretch a considerable distance southwards along the stream. The people are agriculturists, but also possess cattle. They also employ themselves exceedingly in fishing. A species of large fish, of which they make great use, being abundant in the waters. Ulédi describes it as having a large head with a beard, and very long narrow body. He says the Wa-Njémsi prepare and preserve it in various ways —sometimes they split it and dry it in the sun, like the preserved shark and other kinds of fish, which are annually exported from Arabia to the east coast of Africa. At other times they cut it up, and submit it to a process of cooking, in order to extract the oil which it yields abundantly. This they store away in utensils.

Sádi says the Wa-Njémsi are the "ráya" (subjects) of the Wa-Súku. Both they and the Wa-Súku are said to be the descendants of the Masái. Reduced Wakwávi, namely those who have been robbed of their herds, have identified themselves with the Wa-Njémsi, and together with them cultivate the ground.

No. 6. DOWN-ROUTE.

Viz., from Lake Baringo to Ndára Sérián along the western bank of the River Njémsi.

1st day's march	..	..	..	..	..	5 hours.
2nd „ „	..	..	..	..	..	5 „
3rd „ „	..	..	..	..	..	8 „
4th „ „	..	to Siágnáu		..	..	7 „

This name is Kikwávi, and Sádi says, means "an open clear tract of land." He speaks of this locality as fully answering to its designation. He says—there are no trees, no thicket or bush of any description near the place, but that the whole region is a clear and open plain, furnishing no convenience whatever for a bivouac or defence in case of attack.

There are no thorns by which to make a fence for the camp, as is usual with the coast-caravans, when in the interior, not even a stick or bit of cow-dung to kindle a fire. For the night the camp is comfortless, and without protection, and the men, feeling their insecurity, sleep gun-in-hand.

5th day's march	.. ..			7 hours.
6th „ „	to the plantations of the *Wa-Njémsi*			6 hours.
7th „ „	to base of *Likámasía Mountain*		..	4 hours.

This is a large mountain; it is lofty, and spreads to a considerable extent. Its northern limit is at Ligéyo, and its southern termination, Subúgo. It is habited. The people, who are very numerous, are called Wa-Kamasía. They live on the summit, which is level, where they cultivate the soil and pasture their cattle. None of them live below.

Width of the mountain: about 3 hours or more. Length: 1¼ day. Especially lofty at the southern end.

8th day's march—from *Likámasía Mountain to Subúgo* 9 hours.
Subúgo to *Máu* .. .. 6 hours.
Máu is a large tract of country, embracing 5 or 6 days'
journey. The inhabitants—Wa-Máu—wander about a great
deal, hunting.

To *Northern frontier of Límbwa.* At Lúmbwa there are
about 1,000 villages. The people—Wa Lúmbwa—are agriculturists, and have much cattle. [8 hours.]

To *Southern frontier of Lúmbwa.* [9 hours.]

To *Líkunóno.* [4 hours.] The inhabitants—Wa-Líkunóno
—are blacksmiths. They do nothing else but forge—sometimes
implements of agriculture, and at others weapons of war—spears,
swords, axes, knives, hoes, &c. These things are purchased by
the Wakwávi and other tribes, with cattle, grain, honey, &c.

The villages of Líkunóno are not permanent.

To *Kítavéni* ... . 9 hours.
To *Kirísha* .. .. 12 hours.
To *Ndára Sérián* .. .. 10 hours.

Here is a mountain called *Keréto.* It is lofty, but has comparatively little length or breadth, about 1½ hours to the N.W.
of camp.

APPENDIX.

BY KEITH JOHNSTON, JUN.

Notes on the Map to Mr. Wakefield's Paper.

I. AUTHORITIES.

The sources whence the data for the map have been obtained, and to which
reference has been made in its construction, are:—

1. A manuscript map of the part of Eastern Africa which lies between the
Pangani and the Dana rivers on the coast, and the Victoria Nyanza in the interior,
drawn from personal and native information. By the Rev. Mr. Wakefield.

2. The notes on the caravan routes which are printed in this volume.

3. The results of the journey made by Baron Von der Decken to Mt. Kilima-
njaro, as contained in the first volume of his work, and in the accompanying map
by Mr. Hassenstein, 1870.

4. The journals of Dr. Krapf's routes to Usambara and Ukambani, preserved in
the volumes of the 'Church Missionary Intelligencer,' from 1849 to 1853, and
in Dr. Krapf's book of 'Travels in Eastern Africa.'

5. The map of Dr. Krapf's routes (on the basis of a preliminary reduction of
Von der Decken's positions), prepared by Mr. Hassenstein and published in
Petermann's Mittheilungen of 1864.

6. Dr. Krapf's manuscript map which accompanies his book of travels.

7. The journals and maps of Mr. Rebmann contained in the volumes of the
'Church Missionary Intelligencer,' above noted.

8. Dr. Kiepert's map of Krapf and Rebmann's discoveries (Zeitschrift für
Allgemeine Erdkunde, Berlin, 1860).

9. The manuscript known as the 'Mombas Mission Map,' prepared by the missionaries Erhardt and Rebmann in 1855.

10. Burton's report of a journey to Usambara, in the 'Royal Geographical Society's Journal' for 1858; 'Lake Regions of Central Africa,' 1859; and a paper on Lake Tanganyika, 1865.

11. Speke's narratives in 'Royal Geographical Society's Journal,' and elsewhere.

12. A map of Eastern Africa, by Léon des Avanchers, Missionaire Apostolique, and the accompanying paper in the 'Bulletin de la Soc. de Geog.' Paris, 1859.

13. Guillain; Documents sur l'Histoire, la Géographie et le commerce de l'Afrique Orientale, collected 1846 to 1848.

14. The narrative of the Rev. Mr. Wakefield's personal journey from Ribé to the Galla country, given in a pamphlet published by the United Free Churches Mission, and entitled 'Footprints in Eastern Africa.' London, 1866.

15. A portion of the narrative of a journey made by Messrs. Wakefield and New, in company, from Mombasa to Upokomo, on the Dana river; published in 'The United Free Churches Magazine' of 1866-67.[*]

16. A map of the routes of the missionary Richard Brenner in the region between the Dana and Juba rivers, contained in Petermann's Mittheilungen of 1868.

The Admiralty charts have been used as a basis for the delineation of the coast-line, but have been supplemented at several points by the more detailed surveys of recent travellers; the more important of these emendations are in the coast-line from Pangani river to Tanga and Wassin, from Burton and Speke's information; at Mombas, from the survey of the harbour by the brig 'Ducouëdic,' under M. Guillain; at Malindi and the chain of islands between the mouth of the Ozi and Wabushi rivers, from the sketches of the missionary Brenner.

II. THE CONSTRUCTION OF THE MAP.

Mr. Wakefield's manuscript drawing (the sole authority for the western portion of the map) has evidently been prepared by him independently of any of the existing materials, excepting only that he has placed on it the southern termination of Lake Nyanza in the form and position given to it by Captain Speke. In this respect the manuscript has great value, but, as is to be expected, it betrays a tendency to exaggerate the distances landward. Mount Kilima-njaro is shown on it nearly 30' further inland than its position astronomically determined by Von der Decken; the stations between this and the coast are correspondingly out of position, and westward the exaggeration appears to increase. Mount Kenia also is placed some 45' to westward of the position which is believed to be more nearly true. The distances from place to place on the reported routes may be to some extent checked by the time required to pass from one to the other, as given in Mr. Wakefield's notes; but for the relative bearings of these places, and the general direction of the routes, we are entirely dependent upon the manuscript map. That some confidence may be placed in these bearings is shown by the tolerable agreement of the positions, as laid down on the manuscript, with those determined astronomically. Beyond these known points, however, the adjustment of the reported routes admits of considerable latitude, and in order that a personal opinion may be formed by those who are competent to judge of the value of the new geography shown on the map, it has been thought advisable to give a statement of the manner in which each route has been laid down.

The two great landmarks in this region are the snowy mountains of Kilimanjaro and Kenia. The position of the former (Lat. 3° 5' s., Long. 37° 27' E.), as well as the geography of the whole of the country traversed by Baron Von der Decken from the coast, has been adopted from the large scale map which accompanies the first volume of his work, and may be considered as true. The real position of Mount Kenia is less certain. Its latitude can only be determined by the distances and bearings of Dr. Krapf's routes to its base in Ukambani. These

---

[*] It is to be regretted that though every effort has been made to recover the unpublished manuscript of this important journey, no success has been met with.

were first laid down by Dr. Kiepert, but the most elaborate reduction of these routes is that by Mr. Hassenstein, noted in the list of authorities (5).

Relatively to Kilima-njaro, on this map, Mount Kenia lies almost due north of it (2° E'). Since the production of the map of Dr. Krapf's routes (in 1864) a more critical examination of Von der Decken's astronomical observations has led to a change in the position of Kilima-njaro, as temporarily laid down, altering its longitude as much as 16' to westward. To maintain the same bearing from Kilima-njaro as before, the position of Kenia has also been moved westward, and now falls in Lat. 1° 16' s., Long. 37° 35' E., which is considered as not far from its true place. On the positions of these two points, the alterations which have been made in the bearings and distances shown on Mr. Wakefield's manuscript mainly depend.

ROUTE 1.

IN the earlier part of this route inland from Tanga, the positions of Bassano, Gonja, Kisiwani, and Arusha (at the base of Mount Kilima-njaro) are common to this and to Baron von der Decken's journey, during which their places were astronomically determined.

Beyond Arusha, the central part of the route, to as far as Naïvasha, has been laid down by the position of that lake, as given in the reduction of the route (3) from Mombasa to Dhaicho. From the altered Naïvasha to Baharini on the shore of the Nyanja, the bearings and distances of Wakefield's manuscripts have been accurately followed, since the distances from place to place there shown agree well with the average rate of travel, given by the portions of the routes which admit of correction by fixed points.

ROUTE 2.

From Baharini, as given by Route 1, the first portion of this route has been laid down to the position of the Gilei mountains, as shown on Mr. Wakefield's map between Ngorot and Pinyinyi in Route 1. The distances of this part, which appear to be considerably exaggerated on the manuscript, have been reduced to fall in with these terminal points, still the rate of travelling thus reduced agrees nearly with that of the route from Naïvasha to Baharini. Several points on this route, as well as on Route 1, are easily identifiable with the position on the route to Burgenei reported by Erhardt. The remaining portions from Gilei to Arusha have been entirely laid down by the positions on Route 1.

ROUTE 3.

From the coast at Mombasa to as far as Taveta (Dafeta), this route has been laid down with confidence, since several of the points along it are identical with positions ascertained by Von der Decken. The chief of these are Gurungani (Ngurungani), Kisigau (Kisigao or Kadhiaro), Matate and the Bura camp. The farther part of the route has been placed on the map from Decken's position of Taveta, and that of Mr. Kenia (Doenyo Ebor of Wakefield) previously noted, preserving the relative bearings of places between as given by Wakefield, but necessarily reducing the distances shown on his manuscript.

ROUTE 4.

This route starts from Sigariri, a position on the return route from the Nyanza (Route 2), and the direction of the former portion of it is determined by the position of Lamwea on Route 3. Here an alteration has been made on the rendering of the manuscript map, which gives two separate positions to Lamwea. This place is evidently identical in both routes from the descriptive notes. Beyond Lamwea to Settima Mountain, the middle portion of the route nearly coincides with Route 3 in this region, with the exception that between Lamwea and Miviruni, this route seems to take a more direct course than the other, as it also does between Miviruni and Naïvasha. The notes state that this route crosses the Settima Mountain (possibly by the same pass that is used in Route 3), but the manuscript map carries the route to westward of Settima. To correspond with the statement of the notes, the direction of the route has here been altered to pass over the mountain. Directly after descending the inward slope, this route diverges from Route 3, and goes northward. From this point the bearing of the route has been retained exactly as in the manuscript map. The distances, reduced in the same proportion as those of the corrected journey between Kilima-njaro and Kenia, bring the Samburu country, at the termination of the route, into nearly the same latitude as the Tsamburu of Rebmann, and of Lake Böö of Léon des Avanchers. There is no indication whatever in Wakefield's map or notes of the position of Samburu Lake (evidently Lake Böö), which, however, is said to rival Lake Baringo in extent. The statement in the notes that Sadi saw the *northern* end of Samburu Lake on arriving at the southern frontier of the territory of this name, is apparently a mistake. A lake of the extent described, with its northern end at the southern frontier of Samburu, and lying N.N.E. to S.S.W., besides being out of its proper country, would, if it be supposed to lie westward of the route to Samburu, overlap the Baringo, if to eastward, would occupy the space through which Mr. Wakefield has drawn an undescribed land route. The word *southern* has been read instead.

Route 5.

This route leads northward from the position of Gwaso Ngishu on Route 1. For the reason that there is no available means of checking the bearings of this journey, and since the distances shown are not in excess of the average reduced rate, it has been thought advisable to lay down this route exactly as it is shown in Mr. Wakefield's manuscript, to as far as the point where Lake Baringo is touched upon.

This is the point at which the traveller came upon the territory of the Wa-Suku, who inhabit the mountainous country lying round the northern part of the lake. Lake Baringo has been laid down according to the dimensions and bearing stated in the notes, agreeably to the above indications, and falls then into the place given to Bahari N'go by Captain Speke, and into that of the Baringo reported by Erhardt, in 36° east longitude. Krapf places the Baringo in 34° E.; Hanenstein in his map drawn from Krapf's reports has it in 38° E. On the map by Léon des Avanchers a district named Baharingo, with the note " près d'ici est un grand Lac," is placed to the south of Beségoujou (Wa-Suku?) and west of Lorian, agreeing with this later information.

Route 6.

The distances on this down route from the Baringo have also been preserved exactly as   Mr. Wakefield's manuscript, but if the bearings were retained this on te would cross over the waters of the lake, the dimensions given to it in the notes being adopted. To avoid this the direction of the route from Baringo has been altered so as to carry the path along the western coast of the lake as far as the Máu camp, south of Subúga. Thence to Kirisha, the second station from the termination of the route, the bearings have been retained exactly as on the manuscript. The position of the whole of the latter part of this route, and especially of the final station of Ndara Serian (Endarasereáni of Erhardt), is very uncertain. The notes give this place or district a south-easterly direction from the Lumbwa camp (on Route 1), but on the manuscript it is shown to south-west of this position, and near the point marked Sero. The latter position has been chosen for it on the map, since, to arrive at Ndara Serian, if it be supposed to lie south-eastward of Lmofoya, the river Njémsi must be crossed, and there is no mention of a crossing of this river in the notes on this route. Still this place, Ndara Serian, is the only one point on this route referred to in the notes, as being in any way connected with the other journies, and the uncertainty of this solitary point renders the construction of this particular route the least reliable of the whole.

The undescribed routes shown upon Mr. Wakefield's manuscripts have been laid down on the map, in the same bearing and relative position that he has given to them, but a distance correction, reducing their length by the average amount of exaggeration found in the other routes has been applied.

An estimate of the mean amount of error over the whole of the routes shown on the manuscript, gives an exaggeration in length in the proportion of 1·24 to 1 ; and the bearings throughout show a leaning in the direction to the west of north of 15° on an average.

III. THE RIVERS, STREAMS, AND LAKES.

The Hydrography of the new region described in the notes collected by Mr. Wakefield, and shown on his manuscript, presents numerous discrepancies. In the attempt to adjust these in a natural manner with the aid of former reports, various alterations have been made, especially in the direction in which certain rivers are stated to flow, the reasons for which seeming inversions are given in the following notes.

ROUTE 1.

In the route from Tanga, the rivers Umba¯ (rising in Baramo Mt.), Kati (rising in Msihi Mt.), and Chwere Chwere, are crossed in the above order, proceeding westward. Baramo Mt. is however furthest west, as is also the river Umba in Decken's route ; the Chwere Chwere (Chur-Chure of Decken) is given by him midway, and the Msihi Mt., on the east of the mountains of Baramo. It seems probable therefore that Mr. Wakefield's informant has transposed the order of arrival at these rivers, and that the name Kati should precede that of the Umba.

Ruvu River (Pangani). The position of the upper course of this river below Arusha has been drawn eastward of the previous representations to agree with the distances given in Route I., a portion of which passes along its banks.

Arusha Lake, identified (wrongly?) with Lake Ro by Erhardt, is shown by him to eastward of Meru mountains. Rebmann and Decken do not mention this lake, and neither does Mr. Wakefield's informant. Krapf, possibly following Erhardt's indication, places a lake Ro in this position. It has been indicated on the map, on Erhardt's authority, east of Meru Mountains, and north of the caravan route.

Augaruka Stream.—A stream at a place named Engorodo on Erhardt's map, is shown flowing round the base of Doenyo Engai, and falling into the salt swamp there indicated. This is probably the Augaruka. The stream at Ngorot is also apparently a feeder of the swamp.

Naïvasha Lake.—Neiwasa of Erhardt, Neiwascha of Krapf, and Neiwacha of Léon des Avanchers, has been laid down according to the dimensions given in the notes, and appears to have no outflowing river.

Gwaso Ngishu.—If this is a stream as the name implies, its course is most probably to the Nyanja.

Nyanja, Sea of Ukara, or *Second Sea.* The point of interest in connection with the great lake on whose shore Sadi stood at Baharini, is the question whether it is indeed the same lake which Captain Speke saw and named the Victoria Nyanza. It is observed that the reduction of the newly reported routes, made entirely without reference to the extent of this lake, places its eastern shores very nearly in the same position as that indicated for them indefinitely by Captain Speke, and that its supposed area is not materially altered. That the name of the lake here given should differ in some degree from that received by Captain Speke is of very little moment, but it is remarkable that not one single name of district, people or place * given in these new routes has any such remote resemblance to names reported by Speke and Burton, as to warrant an identification with any one of these. At page 275 of his 'Lake Regions,' Captain Burton says, "These races" (of the coast people on the eastern side of the Victoria Nyanza) "are successively from the south, the Washaki, at a distance of three marches, and their inland neighbours the Watatura" (lat. 2° 10' s. to 2° 20' s. on Speke's map) "then the Warudi, a wild tribe, rich in ivory, lying about a fortnight's distance; and beyond them the Wahumba or Wamasai." "Commercial transactions extend along the eastern shore as far as Thiri or Ut'hiri, a district between Urudi and Uhumba." The fortnight's distance from the south end of the lake should approach very near, if not actually to, the position given to Ukara by the new routes. It is possible that Thiri or Ut'hiri lies close on the south of the district of Ukara. Again, the names of the native states indicated by Captain Speke as lying between the north-east of the Nyanza and the Bahari 'Ngo, are in no degree similar to those of the people's districts named on the caravan routes which traverse these states.

In a paper on the Tanganyika Lake and on the Nyanza, published in the Journal of this Society of 1865, Burton says "the principal alterations which I would introduce into the map appended to Captain Speke's paper are as follows :—1. Draining Lake Tanganyika into the Luta N'zige; 2. Converting the Nyanza into a double lake, the northern part fed by rivers from the western highlands, and the southern by small streams from the south and south-east. The former in Captain Speke's book appears to be merely a broadening of the Kitangule River, and thus only can we explain the phenomena of six large outlets in 30 geographical miles.". . . . "Within a distance of 1° the map shows three first rate streams, viz.; the Mwerango, the Luajerri, and the Napoleon Channel issuing from the Nyanza. I believe this to be a physical impossibility." . . . . "In p. 130 of Captain Speke's Journal the petty chief Makaka assures Captain Speke that 'there were two lakes and not one'—unfortunately the hearer understood that the Bahari 'Ngo was alluded to."

That the arguments which Captain Burton used in recommending a division of the Nyanza had not a sufficient basis of proof to give them moment is shown by the acceptance of the lake as one sheet of water by the whole geographical world. Yet the only evidence that we have of the extension of the lake eastward of the meridian of its southern extremity is the statement given by Captain Speke that " A man who had been on the island of Ukerewe and had seen the broad expanse of the Victoria Nyanza beyond it told me that the lake was as broad on the eastern as on the western side though it could not be seen by us then " (from the Observatory Hill) " in consequence of Ukerewe standing in the way. He also said the lake was of indefinite length."

On the northern side also the evidence of eastward extension is limited to report (given on p. 330 of Royal Geographical Society Journal, 1863). " The Waganda confirmed the statements I had heard in Muanza" (south end of the lake) "regarding the extension of the lake to the eastward, where it was said there was as much water to the east of Observatory Hill as there was to the

---

* With the exception of that of the Wa-Masai, a general name for the people of the whole region east of the lake.

west; for the Waganda, who sometimes go to the Bahari Ngo for salt, said the strait leading into the 'Ngo lake was as far from the Ripon Falls as the mouth of the Katonga was in the opposite direction. They did not know the 'Ngo by name, but called it a salt lake, as they found salt there. No one in these regions knew of a river flowing into the 'Ngo, but all alike stated that one flowed out of it and joined the Nile, thereby making, as they called it, Usoga, an island."

Captain Burton's recommendation would seem to receive some slight support from the new information obtained by Mr. Wakefield. First, then, there is the (Arab?) name by which the lake is "often designated;" "Bahari ya Pili," the Second sea, which term we are expressly told has no relation to the Atlantic, and probably none to the Indian Ocean or the Baringo (which might be considered as the "first sea"), for Mr. Wakefield appears to have questioned his informant regarding the name, and Sadi who had seen the Baringo, would have referred the name "first sea" to this lake if he had heard it so termed. What then is the first sea? Surely it must be close to that which is named the "second."

Next, the width of the lake, if Sadi's information be correct, is too small for that of the Nyanza and Nyanja as one lake. Six full days' paddling from sunrise to sunset, or three days of continuous rowing, would scarcely suffice to transport a native canoe even to the shores of Uganda, on the north coast, distant 150 miles, much less to the western shore. Burton estimates the speed of rowing canoes on the Tanganyika at little more than 2 miles an hour, for long journeys, halts deducted; and even allowing that the canoe moved in a perfectly direct line at this speed for twelve hours in each day, the distance made could scarcely exceed 140 miles. It is more probable that a six days' journey would not reach so far as this, and the western shore is 250 miles distant.

Again, standing on the eastern shore, Sádi could descry nothing of land in a western direction, except the faint outline of the summit of a mountain on the horizon. But this mountain could not possibly be any part of the western shore of the Victoria Nyanza, or even the Mfumbiro Mt. (10,000 feet high), for this is more than 300 miles distant. The only suggestion which seems feasible, if the Nyanza and the lake at Baharini be considered as one expanse of water, is that this mountain summit, seen from the low eastern shore, is a high island rising in the midst of the lake; but such a feature could not well have been missed entirely by Captain Speke—he would either have seen or heard of it.

ROUTE 2.

Gwaso na Ebor, white river, sparkling like quartz, or possibly with chrystallized salts, is probably a branch or feeder of the Nitrate of Soda swamp, which appears to extend from Augaruka to Ngurumani northwards.

At *Ngare na Njuki* and *Ngare na Erobi* streams have been shown, as described in the notes, flowing from Doenyo Erok and from Kilima-njaro. These probably unite also to supply the great salt swamp.

ROUTE 3.

The information given respecting the part of this route which falls between Dafeta and Dhaicho, alters the representation of the upper course of several streams, which take their rise between Kilima-njaro and Kenia, as shown on previous maps.

From Dafeta the route appears to follow the valley of the *Lumi;* but the point at which the *Tzavo* is reached indicates that the Lumi cannot have such a northerly course as is shown on Von der Decken's map. The *Lake Tzavo* (? Luaya of Rebmann), through which the Tzavo is said by Krapf to flow, appears to lie eastward of the new route, since it cannot well lie at a higher level to be seen from the top of Julu Mt.

The *Sabaki* River, mentioned by Wakefield, and named *Yata* at its sources on his map, is evidently the Adi or Sabaki of Krapf. It appears to rise in the northern slope of Kilima-njaro, thus confirming the statement of Krapf that a branch of the Adi rises in Kilima-njaro (p. 233, ' Ch. Miss. Int.,' 1852).

Tiwa River, shown on the manuscript map as a tributary to the Sabaki, but not mentioned in the notes, is probably the Tiwa of Krapf, and it has been retained in the position given to it by that traveller.

The *lake* at Leta Kotok, mentioned in the notes but not shown in the manuscript, may lie to northward of the Sabaki River, since it is mentioned after it in turn; but there is no indication of whether it lies east or west of the route.

Tana River, the Dana, cannot well have its sources in the Mu country, as the notes describe it, for that district lies beyond the salt lake Naivasha, and that basin interferes between. Three streamlets, which rise in the Settima Mt., and uniting flow on to near Mianzini, appear to form the head stream of this river, and among the tributaries flowing to it from Dhaicho and Kikuyu, the Dika, Dida, Kingaji, and Ludi, spoken of by Krapf, are evidently four.

The *Ozi* River is represented on Mr. Wakefield's manuscript as rising in the Settima Mountain; but in his notes he describes it as being crossed to northward of Msarara, and this description would bring its course between this mountain and Dhaicho. Further he states that it receives tributaries from north and south, Dhaicho and Meru (Msarara) necessitating an alteration from the manuscript to agree with these statements. The source of the Ozi would then appear to be in the northern slope of Kenia, and this is confirmed by the report of the origin of this river from a lake in the north-east of Kenia, received by Dr. Krapf (p. 77, ' Ch. Miss. Int.,' 1852): " From this lake the Dana, the Tumbiri, and the Nsaraddi take their origin. The last-mentioned river goes to north-east to a much larger lake, called Baringo." Dr. Beke suggests (see p. 87, as above) that " Dr. Krapf has inadvertently transposed the two, named Tumbiri and Nsaraddi; . . . then his Tumbiri would correspond with the Tubirih of M. Werne." The Nsaraddi here mentioned is probably the Ozi.

Gwasi Nyiro, according to Sadi, " flows from the Njemsi Mountain, and turns off northerly near the camping-place" (west of Dhaicho) " and goes to Samburu."

Captain Speke, in vol. xxxiii. of this Journal, p. 322, says : " The Arabs, by their peculiar mode of expression, spoke of the flow of a river in a reverse manner to that in which we are accustomed to speak of the direction of a current." Captain Burton, in vol. xxxv. p. 4, says: " The African account of stream-direction is often diametrically opposed to fact; seldom the Arabs." Sadi, who is a M'swahili, and Uledi, another of Mr. Wakefield's informants, have occasionally used this reverse method of expression. One of the most palpable instances of this is given in the description of the rivulet, which is spoken of as " issuing from the lake (Baringo), and then flowing south-easterly until lost amongst a group of lofty hills." It seems probable that the direction of Gwaso Nyiro has been reported in this inverse manner, for, in the direction of Samburu, the country appears to rise, and the Njemsi Mountain, *from* which the river is said to flow, is apparently a single volcanic cone, which does not indicate any extensive elevation. We are told (Route 5) that " the springs from Njemsi do not form any stream, but merely spread over the (level ?) ground." The Gwaso Nyiro is spoken of again at Suvucha (Route 6). The water-parting of the whole of this region apparently lies in a line nearly continuing the direction of the summit peaks from Kilima-njaro to Kenia and Dhaicho, and therefore the more probable course of this river is *from* the direction of Samburu, or from northward round to westward to Suvucha and thence to Baringo Lake. A river thus indicated would correspond to the Tumbiri of Krapf.

The *second Gwaso Nyiro* seems to be accurately described by Sadi as a feeder of the swamp at Ngurmani, flowing from the Mu country. This stream probably has its origin in the hills indicated to southward of the "Mu camp," which lies westward of Naïvasha (on Route 1), since no mention is made of crossing it on that journey.

ROUTE 4.

Ngare Rongai, a stream coming from Chaga, is probably a seventh tributary to the great swamp at Ngurmani.

The *Gognu Baghase*, a stream mentioned in the notes, but not indicated in the manuscript, is evidently a tributary of Lake Naïvasha, and reaches the eastern shore at the place where the waters of the lake are reported to be fresh.

The course from Dondole, a district near Njemsi Mountain, given to the *Motonyi* River, is apparently a second case of inversion of the direction of flow, since it is much more probable that this stream issues from the heights which continue the water-parting south of Settima Mountain, and has its course towards the Njemsi and the valley of the Nyiro River, of which it would then be a tributary.

Agnata Vus, a stream from Settima Mt., must fall into the Motonyi.

Migungani River "flows out of the river Nyiro, near Suvacha, and afterwards enters the Ozi." Inversely this stream is a tributary of the Nyiro, flowing possibly from Kenia, in which mountain its source may be near that of the Ozi. Krapf (p. 37 of 'Ch. Miss. Int.') remarks that "rivers descending different slopes of the same watershed, from the contiguity of their sources, are often spoken of in native phraseology as having a common origin."

A second *Ngare Rongai*, not mentioned in the notes, is shown on the manuscript to westward of Kenia, and may possibly be the head stream of a tributary of the Migungani.

Ngare Ndogei, which crosses the route to Samburu, "comes from Lorian." Here again the direction of the current seems to be reversed. It is much more likely that the river comes *from* the "lofty hills" spoken of as rising on the east side of Baringo Lake, and flows *toward* Lorian, in which direction it would join the Gwaso Nyiro.

ROUTE 5.

The *Northern Nyarus*, "30 yards wide," in all probability flows out of Lake Baringo, and thus may possibly be identified with the Tubirih or White Nile of Werne, or the Asua of Speke. The Southern Nyarus (or Njemsi) we have identified with the Nyiro, coming from Sambúru.

ROUTE 6.

The *Njemsi*, spoken of in this route, is probably a separate river from the Southern Nyarus or Njemsi (a part of which flows through the Njemsi country), and is evidently not an affluent of the Baringo.

Its source and southward course, we are told, are in the Njemsi country, and it is again mentioned in connexion with Sero (Route II.), at which place a tributary—the Dawash—reaches it from the Kilileoni country eastward. This indicates that the direction of the main stream here is to westward or south-westward, possibly to the Nyanja. In this case it would coincide with the Sero River, described by Erhardt, flowing westward from near Ndara Serian.

There is no mention of an outflow to northward from Lake Sambúru, but such an outlet must exist. Léon des Avanchers shows a river flowing out of his Lake Bö to north-westward, "an affluent which the Somali say is the Nile," which may correspond to the Saubat; but the possession of the country round the lake by the Somali would rather indicate that it belonged to the opposite watershed,—to the Juba basin.

4. PHYSICAL FEATURES.

The broad physical features of the region under consideration are (1) the watershed to the Indian Ocean, marked by the Rufu, Dana, and Ozi Rivers; (2), a belt of continental drainage on the height of the plateau, containing the salt lake Naïvasha, and the great swamp at Ngurungani; (3), the inward slope to the Nile Valley beyond this, with the Nyanja and Baringo (fish-yielding Lakes of fresh water).

The Indian Ocean watershed has the general character of a gentle slope up to the high peaks which rise on the water-parting line, but is sharply divided into two sections of completely different aspect, by a remarkable line of landward sloping heights which extend for 250 miles from the western borders of Wanyika land diagonally to Mount Kenia. This ridge, of from 100 to 200 feet in height, appears to mark a great fault line, and would indicate that at one period a general sinking of the portion of this watershed which lies beyond it has taken place. It has this curious effect, that whilst the slopes from its edge to the coast receive the rains from the Indian Ocean, and have in consequence a fertile character, the lands immediately inland of the ridge, deprived of the rains because sheltered by it from the sea-breezes, are barren and desert.

As soon, however, as the land has again risen to a height sufficient to intercept the rain-bearing winds, there is a second belt of fertile country, in which the populous States of Pare, Chaga, Teita, and Kikuyu are found.

Similarly protected by the water-parting heights from the sea winds, but lying at a high elevation, the region of continental drainage has a rainfall which is very partially distributed. The higher grounds receive sufficient moisture to admit of plantations and agriculture, but the lower portions between are less fertile or even barren; and the salt lake Naïvasha shows that in its neighbourhood the rainfall and evaporation are nearly balanced, whilst the dried-up swamp behind the height of Kilima-njaro proves that there the evaporation is in excess.

The Njemsi Volcano in this region has a special interest, since if the report be true, it is the only one which is known to present any signs of activity in the African Continent. The information respecting it is confirmed by various independent reports. The missionary Erhardt, in the Memoir to his Map, says, "After passing the slope of Endara Seriani, there is a bare desert land having a rough strong soil, mixed with sulphur and interspersed with hot springs, which stretches as far as the neighbourhood of Burgenei" (Burkeneji of Wakefield).

Dr. Krapf also heard of it. "This morning, Kivoi" (Dr. Krapf's informant in Ukumbani) "made mention of a volcano which he placed in the Wakuafi country: to the north-west of the snow mountain Kenia. He called it a fire-mountain, of which the hunters were very much afraid." (Journal in 'Ch. Miss. Int.,' 1850.) In speaking of the country to north-west of Kenia, "Rumu," a native of a different locality, also knew of the fire-mountain of which Kivoi had first informed Dr. Krapf. "Rumu called it Kirima ja Jioki, or Mountain of Smoke. He stated that there is much water round it, and such miry ground that travellers cannot approach it." ('Ch. Miss. Int.,' 1852, p. 234).

5. THE LIMITS OF THE DIFFERENT PEOPLES.

From a combination of the information and reports given by the several
travellers named in the list of authorities, it has been possible to mark out
approximately the limits of the territories occupied by the various tribes of
this region.

A main line of division extends diagonally across the slope to the Indian
Ocean, and separates between the Galla races northward, and the completely
different tribes which lie to southward of it.

The non-Galla area, or the whole of the remaining portion of the map, may
be broadly subdivided between the hill peoples, or the settled inhabitants of
the fertile and isolated highlands, and the Masai and Wakwavi, the wandering
and raid-making hunters of the arid plains, the constant enemies of the high-
landers. The southern limit of the Gallas is clearly traceable from Dr. Krapf's
information along the line of mountains which protects Ukambani on the east,
and along the north of Wanyika Land, from Mr. Wakefield's personal knowledge. The caravans which proceed from the Wanyika inland are liable to be
pillaged on the direct route across the desert to Ukambani, now by the Wa-Galla,
now by Wa-Kwavi or Wa-Masai. The various isolated tribes and republics of
hill peoples on the water shed to the Indian Ocean appear to have some affinity
in race to one another. The more important of these states are Unikani,
Usambara, Pare, Ugono, Chaga, Teita, Ukambani, Limero, and Dhaicho, each
occupying a distinct elevation.

The region of continental drainage on the plateau in the great domain of
the Wa-Masai and Wa-Kwavi, two nomad peoples, hostile, but similar in habits
and completely intermixed, whose dwellings, in some parts of this area, are
permanent. The eastern limit of the country more constantly occupied by
these hunters is indicated on the map, and includes Kikuyu and the plain of
Kaptei, probably the most fertile country inhabited by them. The desert
which stretches from this plain to Unikani south-eastward is subject to their
temporary raids, as is also the country towards Samburu northward, and that
between the Baringo and Nyanja. On the north and west the furthest limit
of the Wakwavi is probably the edge of the lowland which seems to be defined
by the base of the mountains east of Baringo, and thence their boundary
might perhaps be drawn to Nandi and east of the country of Ukara to meet
the Wa-Humba (Masai), whose territory, according to the report received by
Burton, is probably to southward of Ukara, and not far from the Victoria Lake.
Southward the extent of the Masai is indefinite, but in his description of the
Wasegura people, whose country he traversed on the way to Tanganyika,
Burton remarks that the Wa-Kwavi are a sub-tribe of them.

On the slope to the Nile Valley, the most distinct hill tribes are those of
Nandi, Ukara, Ligeyo, and of Njemsi. Of the last named people it is remarked that they take refuge on being attacked by the Wa-Kwavi on an island
of the Baringo. These peoples are said to be subjects of the Wa-Suku, the
warlike people who occupy the mountainous country round the northern part
of the Baringo, and who apparently belong to a distinct race, probably Nilotic.

Samburu is evidently the limit of a new watershed, and if Sadi's information
is correct that the Wa-Samburu are subjects of the Somali, and speak a
dialect of Ki-Kwavi, there must exist an almost isolated southern portion of
this race inland, as there is of the Gallas on the coast slope, and the Somali,
Wa-Kwavi, Masai, and Wasagara are connected; somewhere in the Upper Juba
there must be a mingling of the Galla and Somali, elsewhere hostile.

* Léon des Avanchers says that the environs of the Lake Böö are inhabited by
Rendilé-Gallas, who are of a reddish colour, have long hair, and possess numerous
herds of cattle. Lake Böö is surrounded by conical mountains, the highest peaks
of which are snow-clad.

LIBRARY
Leland Stanford, Jr.
UNIVERSITY

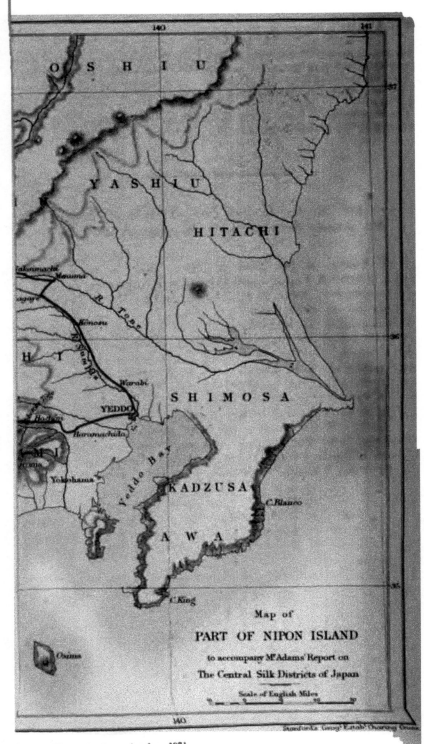

Map of

PART OF NIPON ISLAND

to accompany Mr Adams' Report on

The Central Silk Districts of Japan

Scale of English Miles

Murray, Albemarle Street, London, 1871.

On the south side of the Pangani River, the Wasegura people (Wasegua of Wakefield), met with by Burton and Speke in their journey to Fuga, appear to have advanced up the valley of the M'Komafi tributary of the Pangani, to as far as Gonja and Baramo, and the lower country between Pare and Usambara.

The Wandorobo, a vassal people to the Masai, are scattered in permanent villages over the central table-land.

XII.—*Report on the Central Silk Districts of Japan.* By Mr. ADAMS, Secretary to H. M. Legation in Japan.

Yeddo, Aug. 7, 1869.

I LEFT Yeddo on the 22nd of June, 1869, accompanied by Messrs. Davison, Piquet, and Brunat, silk-inspectors, belonging to three different firms in Yokohama, and by Mr. Wilkinson, of this legation, as interpreter.

We travelled on horseback, and, as has been usual in the expeditions into the interior of Japan, which have been taken from time to time of late years by members of the diplomatic body, we were attended by a mounted escort, consisting of ten yakunins, supplied by the Government. One or two of their number started before us in the morning, and gave warning of our approach to the officials of each post-town in the day's route. Much trouble was thus avoided with respect to the transport of our baggage and provisions, and on reaching our resting-place for the night we found the officials and the keepers of the "honjin," or hotels, prepared to receive us.

On the road we met with great civility, both from the retainers of Daimios through whose territories we passed, and from the yakunins of the post-towns and villages. Our escort were also uniformly attentive to our wants, and the Government had even inserted in the 'Official Gazette' a formal notice of our intended journey.

In order to show our route more clearly, I annex a tracing of a Japanese map.* The original map, though not strictly accurate, is sufficiently so for the present purpose. I also annex a table containing the names of a number of towns and villages through which we passed, and the distance between them in "ri," as nearly as they can well be computed from the do-chiu-ki, or road-books. A "ri" is calculated to be 2·442268 English miles, so that during our journey we rode from 280 to

---

* This refers to the map in the original Report; the map annexed to this paper has been constructed on the basis of the Admiralty Surveys.—[ED.]

290 miles. Owing to the practical knowledge of my companions, and to their acquaintance with Japanese merchants at each of the principal silk depôts, we experienced little delay in obtaining the information we desired, and we were thus enabled to cover a large extent of ground in the fortnight.

Before proceeding farther, I wish to state that I am indebted in a very great measure for the substance of this Report to my three companions, Messrs. Davison, Piquet, and Brunat, they having kindly placed at my disposal the copious and valuable notes which they had collected. Mr. Wilkinson acted as our interpreter, and I am sure that I am expressing the sentiments of the rest of the party by testifying to the ability and willingness with which he discharged his task.

The silk districts of Japan are confined to the principal island, and may be divided into three groups: the Northern, designated under the general name of Oshiu; the South-Western, including those of Echizen, Sodai, Mashita, &c.; and the Central (the object of our journey), which produces the Mayebashi, Shinshiu, and other varieties of hank silks, as well as the silks of Kôshiu and Hachôji.

We found the worms in general in the chrysalis state, and saw numbers of trays of cocoons baking in the sunny streets. Still, near the borders of Musashi and Jôshiu, and on the 30th of June at Kaminosuwa, in Shinshiu, we observed some late worms in the stage between the third and fourth casting of the skin. At Annaka, in Jôshiu (26th June), many were just ready to spin. At Uyeda (28th and 29th June), and other places in Shinshiu, and in parts of Kôshiu (first days of July), the moths were beginning to emerge from the cocoons, and to lay their eggs. Reeling was going on almost everywhere. In Musashi, Jôshiu, Shinshiu, and Sagami, is produced the greater part of the class of silk which is tied up in hanks, and sold under the name of Shinshiu and Mayebashi hanks. In Kôshiu the silk was formerly always made up in bundles; but since the decided preference shown by foreigners for the above two descriptions of hanks, much of the Kôshiu silk is made up in that form, instead of in bundles as before.

It has long been known that some regions of Japan are more favourable than others for the production of eggs. Mountainous districts, at a distance from the sea, appear to be the most propitious; and the Japanese rearers of silkworms, whom we questioned on the subject, invariably informed us that their seed came mostly from Shinshiu. None but the poorer class of peasants, as for example those in Kôshiu, rear silkworms hatched from their own eggs; so that in general the silk of the five provinces in question may be said to have a common origin, and the seed to be renewed year after year from Shinshiu. The difference in the quality manifestly results from the difference in climate and soil, and in the culture of the mulberry, the rearing of the worm, and the reeling of the silk.

Our route to Mayebashi lay through an extensive and well-cultivated plain, among the products of which are rice, both paddy and upland, grain, buckwheat, hemp, rape, and a great variety of beans and other vegetables. We first observed the mulberry some 12 ri from Yeddo, in the neighbourhood of Konosu; and we subsequently met with it almost universally, except upon certain high levels, where the Japanese had doubtless found by experience that the cultivation of its leaf, and the rearing of the worm, were rendered too uncertain by the frequent variations in the temperature. Near the borders of Musashi and Jôshiu the soil became more sandy and stony, and the country is traversed by a number of streams, mostly of little depth. We crossed the broad River Tone, the largest in the Kwantô, and then visited Mayebashi and Takasaki, the headquarters of the Jôshiu silk district. The former town is the seat of the Daimio Matsudaira Yamata no Kami, from whose retainers we experienced marked civility.

At Mayebashi we were conducted to a house where an office (Aratami Sho) has been recently established for the inspection of the silk of this Daimio's territory. We were told that there were fifteen merchants in the town, and seventeen in the country round about; that they formed a species of guild; and that they bought all the silk which was reeled in the territory. This silk, we were informed, was henceforward to be inspected at the office, where all foul hanks would be rejected, and a distinguishing mark put upon the remainder, which would then be despatched to Yokohama, to a newly-appointed agent of the name of Hikishimaya. After my return, when in Yokohama, I met this agent, and ascertained that he had already established himself there for the purpose above-mentioned.

After visiting Tagasaki we left the plain and entered a mountainous region, which we did not finally quit till we approached Hachôji on the 5th of July. As we rode up the valley to the Usui Tôge we observed a much larger quantity of the mulberry; the trees bordered many of the fields, and occupied the whole of some. They were larger than most of those which we had already seen, varying from 3 to 6 feet high, some being even considerably taller. The aspect of the country, with its sandy, stony soil, and its hilly character, reminded my French companions of that in which silkworms are reared in France.

After crossing the Usui Tôge, a high pass which divides Jôshiu from Shinshiu, we proceeded down a valley to Uyeda, one of the centres of the renowned silk district of Shinshiu. The town is situated in a large basin, which is bounded by hills of considerable height, some snowy summits appearing in the distance. The River Chikuma flows past it and falls into the Aka, one of the largest rivers in Japan, which runs into the sea at Neegata. The air in the whole of this high level was clear and bracing, and there was a healthier look about the inhabitants. Large tracts of land are devoted to the culture of the mulberry in the immediate neighbourhood of Uyeda.

In the little village of Nagase, which we passed soon after leaving Uyeda, all the silkworm cards are fabricated during the spring. The season was over for this year. Our route to Kôfu, the capital of Kôshiu, was by rough, stony roads, across the Wada Tôge to Kaminosuwa, where we received much attention from the retainers of Suwa Inaba no Kami, the Daimio of Takashima. Thence we continued through poorly cultivated valleys, which abounded in huge boulders, and where the broad, rocky beds of the streams, and strong stone breakwaters running out from the banks, testified to the violence of the torrents in rainy seasons.

Kôfu, the centre of the Kôshiu silk district, is situated in a large plain, surrounded by mountains. It is a town of some extent, and possesses one of the castles lately belonging to the Shôgun, but now held in trust for the Mikado by the Daimio Akidzuke Ukiô no Ske, President of the House of Representatives sitting in Yeddo. The plain is full of paddy; and when we were leaving it, and began to ascend again, we came upon a quantity of vines, trained on horizontal trellis-frames, which rested on poles at a height of 7 or 8 feet from the ground. Crystals are found in some of the surrounding hills.

The quality of Kôshiu silk has been found to vary considerably in different years, and even that of a single year is often of several qualities. The cause of this is manifestly to be sought in the climate. Mist covered the tops of the mountains during all the time we spent in this province—a contrast to the clear weather we had enjoyed at Shinshiu. Such a misty atmosphere in April and May, combined with the sudden returns of cold so common in this country, cannot but be prejudicial both to the culture of the mulberry and the rearing of the worm. The cocoons which we examined at Kôshiu were often of an inferior description; the villages are poorer, and the inhabitants have a less healthy appearance than those of the more favoured Shinshiu. We continued our journey through hilly country, and over two passes, till on 5th July we dropped down into the plain, and arrived at the well-known town of Hachôji, within the Treaty limits.

On the 6th July I returned to Yeddo, and my companions to Yokohama.*

TABLE OF ROUTE.

			Ri	chô
Nakasendô—				
June 22.	Yeddo to Warabi	.. ..	4	8
23.	Warabi to Kônosu	.. ..	6	22
24.	Kônosu to Kumagaye	.. ..	4	8
Cross road—				
June 25.	Kumagaye to Sakaimachi	.. ..	7	0
25.	Sakaimachi to Mayebashi	.. .. ..	6	18
Nakasendô—				
June 26.	Mayebashi to Takasaki	.. .. ..	3	0
26.	Takasaki to Annaka	.. ..	2	24
27.	Annaka to Oiwake ..	.. ..	9	24
Echigo road—				
June 28.	Oiwake to Uyeda ..	.. ..	8	0
29.	Uyeda to Oya ..	.. ..	1	18
Cross road—				
June 29.	Oya to Nagakubo	.. ..	3	18
Nakasendô—				
June 29.	Nagakubo to Wada ..	.. ..	2	0
30.	Wada to Shimonosuwa	.. ..	5	8
30.	Shimonosuwa to Kaminosuwa (Takashima) ..		1	18
Kôshiukaidô—				
July 1.	Kaminosuwa to Daignhara	.. .. ..	8	28
2.	Daigahara to Kôfu ..	.. .. ..	7	8
3.	Kôfu to Hanasaki	.. ..	10	6
4.	Hanasaki to Yose	.. ..	8	23
5.	Yose to Hachôji	.. ..	4	17
Cross road—				
July 6.	Hachôji to Haramachida ..	.. .. ..	5	0
6.	Haramachida to Yeddo ..	.. .. ..	8	0
			117	32

XIII.—*A Journey to the Western portion of the Celestial Range (Thian-Shan), or "Tsun-Lin" of the Ancient Chinese, from the Western Limits of the Trans-Ili region to Tashkend.* By N. SEVERTSOF. Translated from the Journal of the Russian Imperial Geographical Society, 1867, by ROBT. MICHELL, F.R.G.S.

THE earliest description of the Central Asian Uplands, the northern part of which, between the Chu and Syr-Daria, I surveyed in the year 1864, is to be found in the writings of the Buddhist monk Huen-Tszan, who journeyed through the whole of this region more than a thousand years ago (from 629 to 645 A.D.). I quote him here, because it is difficult to convey in a

_____

* In the original Report of Mr. Adams a detailed account of the mulberry-tree and silkworm follows the above sketch of his journey.—[ED.]

few words a more accurate idea of the district;—A proceeding from that country (the kingdom of Hogo) to the westward, one enters the Tsun-lin Mountains. They are situated in the centre of Djam-budvip; on the south they abut on a large snowy range; on the north they extend to the Warm Sea (lake Issyk-kul; on the west they reach the kingdom of Hogo; on the east that of U-sha. From east to west, and from north to south, these mountains stretch over several thousand *li*. Among them there are several steep summits. The valleys are gloomy, and full of precipices; masses of ice and snow are to be found in them at all seasons of the year; severe cold is experienced, and a strong wind prevails. Many onions (Tsun) grow in this country; hence the mountains are called 'Tsun-lin.' They are likewise so called, because the summits of the mountains are bluish; the word 'Tsun' in Chinese meaning also *blue*."

According to these boundaries of the Tsun-lin Mountains, it would appear that I visited the northern portion of this rugged region, between the Syr-Daria, and a line extending from Lake Issyk-kul to Mynbulak, or, more correctly, the northern boundary of the Tsun-lin.

Let us now examine the signification of the above description, which is in complete accordance with what I saw on the spot.

The Chinese traveller speaks of a mountainous region several hundred versts in length and breadth, in which are distributed several hundred peaks, not the slightest mention being made of mountain ranges. The valleys are deep, and the dark fissures are consequently clefts in a massive upheaved plateau, which is thickly dotted with peaks. Lastly, hollows with masses of ice and snow are spoken of; such is the depression of Rian-kul, situated to the northward of the sources of the Kashgar-Daria. The presence of snow in the hollows indicates the great height of the general upheaval, which is therefore, not without reason, called by the Orientals "the roof of the world."

In the northern part of the Tsun-lin surveyed by me, between the Chu and Talas, at the Arys and Upper Chirchik, mountain-ranges do apparently exist, as is indicated by the broader river-valleys which separate them. But a careful examination of their geology shows that these ranges are almost entirely an optical delusion, as they merge into the general plateau.

When this convex plateau was, at a distant geological period, an island in the middle of the sea, it must then have presented towards the north-west and west four narrow and deeply-indented inlets. These are the present valleys of the Talas, between the confluence of its upper waters and its passage through Cha-archa.

Mountain, of the Ferghanah, following the course of the Syr-Daria, between Andijan and Khodjend; of the Zeravshan at Samarkand and higher up to Varsaminar; and that of Badak-shan at the sources of the Oxus, or Amu-Daria.

These are the hollows that indent the Tsun-lin mountain-rise;* but I shall first describe its northern part, as I saw it in the order of my itinerary from Fort Vernoé to the south-west.

I shall consequently not describe what I saw in the steppe of the Siberian Kirghizes, and in the Alataù district. The ranges there, as far as I could observe them on my rather rapid journey through the country, will be mentioned in the proper place for the comparison of their characteristics. I will commence at once with Fort Kastek, from whence I started on the 5th (17th) May, 1864, with the detachment of General Cherniayef.

Kastek, as is well known, is situated at the northern base of the Trans-Ili-Alataù (Kungi-Alataù), as is Vernoé (Almaty), 80 versts (53½ miles) westward of the latter town, in 43° 8′ N. lat., and 93° 41′ 36″ E. long. of Ferro.

According to the measurement of Captain Golulef, the abso-lute height of Kastek is 3300 feet; consequently 1070 feet above Vernoé. It stands on a sloping descent, of a steppe character, from the mountains towards the Ili.

From Kastek I proceeded direct to the south, along the river of the same name. The point of issue of this rivulet out of the mountain defile occurs 12 versts to the southward of the fort. The chain of the Trans-Ili-Alataù has here no hilly promontory, as at Almaty; but this is replaced by an evenly sloping declivity, ascending from the River Ili to an elevation of nearly 4000 feet, as at Almaty. The ground, however, between the fort and the hills, though rather undulating, preserves the general appearance of a steppe, as the hillocks slope gently.

The transition from steppe to mountain vegetation is very sudden. The gorge opens into the mountains in the form of a deep and narrow crevice in the rocks, to which bushes occa-sionally cling. At the bottom of this fissure the Kastek rushes with great noise and rapidity, leaping over enormous loose rocks, and overturning those of smaller size, in the manner of all mountain torrents. The road in some places is not more

---

* This is the only general designation which appears to me to be correct; there are spots in the mountain district that receive distinct appellations, but no common name exists for the whole elevation. This mountain-rise is formed by the con-junction of two enormous ranges, the Thian-Shan and Himalayan. A separate Bolor, as a mountain system, does not exist. This is the name of a river and town, which European geographers have applied to the north-western part of the Himalayas. The mountains visited by me belong to the western Thian-Shan, but present in some parts the Himalayan extension of strata—north-west to south-east.

than 2 fathoms broad between the river and the sides of the gorge, and even then it is artificially widened, and constantly crosses from side to side. The rocks consist at first of hard limestone; their sombre colour adds to the gloomy character of the defile, which was still further increased, on the occasion of my first visit, by the summits of the mountains being hidden by rain-clouds lowly suspended. After a distance of about 3 versts the limestone is succeeded by crystalline formations, principally of granite of the most typical character, with large crystals of rose-coloured felspar, and smaller ones of quartz and mica; this is alternated on the northern side of the range by syenite and diorite. Without, however, any sharp division between either of the formations, higher up on the right bank dark limestone, and on the left granite, are the predominant formations. The summit of the pass consists of vertical strata of mica-schist, which on the southern slope are again succeeded by granite and porphyritic syenite. The granite below the southern slope changes to limestone, upheaved at an angle of about 60°. At the very base of the mountains the prevalent formation is again granite. The limestone thrown up by that rock dips towards the N. 65° E., as also does the limestone on the northern slope of the range. It does not present any traces of metamorphic action, but retains the same character at its point of contact with the granite, as it exhibits at a distance of 600 feet from it, that is to say, it is not crystalline. The interspersion of compound granitic minerals in the limestone at the point of junction of the strata, is only observable for a thickness of 3 or 4 inches. I noticed this also in the other chains of the western Thian-Shan system, as will be described further on.

The southern slopes of the chain are much more abrupt than the northern; the vegetation is scantier, except in the deepest hollows with rivulets running through them. The yellow tulips of the northern slope are superseded on the southern by those of a red, orange, and red streaked with yellow * colours; and they are, moreover, distinguished by a more marked colouring. Generally speaking, there is less verdure, but a greater variety of flowers, than on the northern slope.

On both the slopes the tulip ranges to the height of about 7000 feet. On the northern, however, as one ascends, the stems become shorter and the flowers smaller, while on the southern, although the stem decreases in height, the flowers remain large to the extreme limit of the line of growth. At the summit

---

* On both slopes I found variations of one and the same species of tulip ("djantyk"), together with every conceivable transition of form.

of the pass, at a height of 7500 feet, Alpine plants begin to appear. *

Another difference between the two slopes is, that the southern are less abrupt than the northern. In the former, ravines occur more frequently, and are drier and less rocky; and the lateral ravines intersect the slopes at more acute angles. Altogether, the southern slopes are less picturesque than the northern, and the vegetation upon them partakes more of the steppe character.

But on the northern slopes also the "tchi" (*Aira sp.*), a plant most characteristic of the Kirghiz steppe,† is found in the mountains, at the widening of the valley of the Kastek, at an elevation of 4300 feet. This widening commences about 5 versts from the northern base of the range, and continues for about 7 versts. The Kastek flows along the whole of this distance under the right rocky edge of the valley; while along the left bank, between the river and the granite crags, extends an undulating plain, from 200 to 300 fathoms in breadth, intersected by several rivulets, which run down from the escarped summits of the Sùok-tiubé into the Kastek. A series of eminences, in the form of steep and towering mounds of about 200 feet in height, extend here, running parallel to the Kastek. Opposite the issues of the transverse valleys which descend from the Sùok-tiubé, they are intersected by other ridges of similar form, but which extend in a direction perpendicular to the River Kastek.

All these mounds appear to me to be the moraines of ancient glaciers. They consist of blocks of granite, syenite, and diorite, of various sizes, many of them large; some are more or less rounded, and some have sharp edges.‡ The spaces between the boulders in these ancient moraines are filled with unstratified clay.

These were the first traces of moraines I observed in Central Asia. Fortunately they were very plain, and must have remained almost unchanged since the melting of the glaciers by which they were formed. They led me to identify as remains of similar moraines, which have subsequently been washed away, the numerous dispersed boulders I had previously seen in the neighbourhood of Almaty, but not lower down to-

---

* The plants are collected, but not yet classified, as I was obliged to leave my herbarium at Chemkend.

† Where, however, it serves to indicate the dry spots of fluvial valleys, and hollows with water at a small depth under ground.

‡ Many at their summits are worn by the wind, and are covered with a thin coating of "kaoline"—porcelain clay—produced by the action of the wind on the felspar of the granite and syenite.

wards the River Ili. It was the moraines of the Kastek that led me to seek for traces of the glacial period in the mountain ranges between the Chu and the Syr-Daria; and not in vain, for after having attentively studied the Kastek moraines, I was afterwards able to detect at first sight even less visible traces of ancient glaciers.

The view which presents itself to the sight from the summit of the pass, powerfully impresses the beholder with its desert grandeur. On the left, one mountain-mass rises up after another, each steeper and more rugged than the last, towards the Talgar, the highest summit of the Trans-Ili-Alataù, where two lines of ridges, running parallel with the principal range, become discernible; and it is evident to the eye that the southern ridge, split across by the Buam defile, through which flows the Chu, extends also in a westerly direction under the name of the Kirghiznin-Alataù (Alexandrofsky range). The valleys of both the Kebins, Great and Little, are also visible from this point. On the right the Chu, often forming a silvery network of branches, disappears in the boundless steppe; directly in front rises a gigantic wall of rock (the Kirghiznin-Alataù), tinged with blue by the mist. The mountain seemed to rest on a mass of dark clouds that had descended into its gorges. The summits of the Thian-Shan to the south of the Kutemaldy, which Mr. Veniukof saw from the Kara-Kupus Pass, were at this time, 8th (20th) May, hidden by clouds hanging at about the elevation of the Buam defile. All the innumerable mountains on this horizon of more than 200 versts are rocky and bare; and only at the Kebin, and in the Buam defile, is a narrow dark-green fringe of pine to be seen.

From the same range, though more to the westward, and nearer to Sùok-tiubé, Mr. Semenof, while proceeding to the Buam Pass in 1858, gazed on the valley of the Chu, and on the mountains of Kokand, as on coveted but forbidden fruit. In 1864 I also beheld these districts, but under more favourable circumstances. I took scientific possession of these long mysterious mountain-chains of Central Asia, and I felt rejoiced that it had fallen to my lot to continue the researches of the first scientific European who had visited the Thian-Shan.[*]

More eagerly still did I look toward the Kirghiznin-Alataù, with which my investigations were to commence; but for a whole week this enigmatical chain obstinately hid its head in clouds,

[*] Mr. Semenof, in 1867, descending from the mountain pass into the valley of the Chu, crossed this river at a point above the former Kokandian fort of Tokmak, and turned towards the south-east, proceeding up its course, and through the Buam defile, emerging at the foot of the Thian-Shan, at the western extremity of Issyk-kul.

increasing my impatience and intensifying my curiosity to a considerable degree. These dark clouds, rolling capriciously around the vast mountain-masses, metaphorically represented that impenetrable veil that had hitherto screened Central Asia from European investigation, and which had been so enticingly, though so partially, drawn aside in works of which Humboldt and Ritter availed themselves.

During this week I first became acquainted with the Sultu and Sary-bagysh tribes. They appeared to me to be less dirty than Messrs. Veniukof and Valikhanof describe them. To the accounts of these gentlemen I, who am but a poor ethnographer, have nothing to add, save perhaps that the formal visits of their " Manaps," or tribe-elders, to our camp were always accompanied by flute-playing and other music. The topography of their tribal encampments is but little known, and to this I intend, during my approaching second journey to the same region, to direct particular attention. The shape of their skulls is of the general Kirghiz character, being somewhat gable-shaped, towering towards the crown. Their features differ less than those of the Kirghiz-kasiaks; their faces are broad, angular, flat-nosed, and narrow-eyed, with high cheek-bones and scrubby beards.*

At last, on the 16th (28th) May, the curtain of clouds was drawn off from the Kirghiznin-Alataù, and its jagged snow-clad summits glistened in the clear blue sky. I proceeded towards them through the valley of the Issyk-kul, finding at the base of the range strata of red argillaceous sandstone, which afterwards merges into conglomerate. These strata extend w. 29° n., and their dip is 50° to 52°, not from the axis of the range, but towards it.

This proves the connexion of the ranges on both banks of the Chu. The sandstone strata at the base of the southern range are raised up by granite, which is denuded at the base of the northern range. The sandstone strata extend for about 2 versts beyond; the sides of the valley of Issyk-kul, to the height of several hundred feet, consist of unstratified clay, inter-mixed with boulders, and are frequently abrupt and precipitous; all the gentle slopes, as well as the narrow bottom of the valley, being covered with a luxuriant vegetation of various grasses and shrubs, which were then, in the middle of May, (O. S.), mostly in bloom. A blue unbroken carpet of "forget-me-nots" covered the slopes of the outlying hills of the Kirghiznin-Alataù. The

* In the Kirghiz features the skull and lower jaw are of almost equal breadth, so that the face is as nearly as possible square; the arch of the skull (*Arcus zygomaticus*) does not protrude much beyond the temples. In Kalmuck faces this arch is very curved, so that the face is rhomboidal in shape.

bushes commence at a height of about 4000 feet; 15 versts higher up the valley, at an elevation of about 5350 feet, the pine (*Picea Schrenkiana*) begins to grow. At this point we were obliged to return; and on the following day, crossing the spot where the granite is denuded, we entered the longitudinal valley extending between the outlying spurs and the main range of the Kirghiznin-Alataù. This valley is intersected by the River Nauruz and many of its affluents; its surface is undulating, and its rich black soil is covered with luxuriant pastures; but neither trees nor bushes grow upon it; its elevation is very nearly 5000 feet. This valley is remarkable for the plainly discernible moraines of ancient glaciers, similar to those of the Kastek already described; their summits protrude from the superincumbent deposits in regular lines of boulders, at the opening of the transverse valleys into the longitudinal one; the denudations at the rivers display unstratified clay with boulders, to a depth exceeding 80 feet. *Terminal* moraines occur here. The heights composed of unstratified clay, with boulders along the valley of the Issyk-kul, appear to me to be lateral moraines, and I consider the inclined strata of sandstone and conglomerate in the mountain promontory to belong to the Permian formation. The strata at the River Nauruz contain gypsum and rock-salt. The latter completely resembles lake-salt, being equally grey and muddy, and being mixed with black mould. In this part of the Kirghiznin-Alataù, we met with a great number of Kirghiz "aùls," of the Sultu tribe, particularly in the valley of he Issyk-aty.

On the 18th (30th) May, accompanied by Mr. Frehse, an officer of the Mining Engineers, we entered by the Ala-medyn rivulet the valley of the Chu, at Pishpek, and proceeded as far as Merké, along the northern base of the Kirghiznin-Alataù. This chain rises gradually from the Buam defile to the River Ala-archi, which is about 12 versts to the westward of Ala-medyn. Perpetual snow rests upon it opposite Tokmak, at the River Shamsi. The elevation of the highest peak at Ala-medyn is about 15,000 feet. Captain Protsenko, of the Etât Major, who visited these localities in 1863, told me that he saw glistening streaks, which appeared to him to be glaciers, between the mountain-snows on the Alarchi peaks, but as he saw them from a distance of 60 versts this statement requires confirmation by closer inspection. For my own part, I saw nothing of the kind. The presence in the Kirghiznin-Alataù of perpetual snow, and of peaks rising to a height of 13,000 or 14,000 feet, continues as far as the Karabalta River. The range gradually descends westwards towards Merké, near which place, at the sources of the Ourianda, it is not higher than 9200 feet.

and has there no peaks. Farther to the westward, between the sources of the Chanar and Makmala, the range again rises rapidly above the snow-line, that is, to above 13,000 feet; but to the westward of Makmala it descends again to Aulié-ata, where its western extremity, Tek-turmas, rises only 150 feet above the level of the Talas, and about 2600 feet above the level of the sea.

From Merké (height 2100 feet) we again made excursions into the mountains, proceeding first up the Merké River. Here broken and twisted strata of red argillaceous sandstone appear in the forelands; where they join the succeeding stratum of limestone, they are uplifted vertically. The limestone is dark, hard, and flinty, and contains many fossils, chiefly of the *Spirifer* group. In appearance, the latter strata seemed to me to belong to the mountain limestone, and consequently to the Carboniferous formation. However, I can only confidently assert that the limestone is Palæozoic; I cannot positively determine the formation.

The River Merké, flowing in a narrow fissure through the limestone, has a very uneven bed, strewn with small sharp stones, which make it very painful for horses. Frequently there is no road except along the bed of the torrent. We therefore left this ravine, and again ascended the mountains, up the Ourianda River, in the direction of Kyr-Djol Pass, which is the lowest elevation in this portion of the range.

At the two first Ourianda rivers, and at the sources of the third, the limestone forms hills, which, though rather declivitous, are not rocky, owing to which they can be easily traversed. They are covered with luxuriant grasses, and afford excellent pastures; high grasses, the mountain-poppy and the peony ascend to 7500 feet, that is, to the limit of snow in the ravines at the latter end of May (O. S.). In August (O. S.) this part of the range is quite free from snow; but at the end of May (O. S.) the pass itself is disencumbered of snow, and presents rich spreading grasses, with a number of flowers of Alpine habitus.

Trees (*Juniperus pseudosabina*) were only found near the second Ourianda, and they stood in small detached clumps, which contained also bushes of the black-currant (*Ribes sp.*). The third Ourianda flows through a narrow and deep fissure in the limestone, the walls of which rise almost perpendicularly to the height of 100 feet. The bottom of this abyss, as well as even the slightest projections of the rock, is thickly overgrown with bushes of the mountain-ash, black-currant, &c. The hawthorn also occurs below, and juniper trees a little higher up.

In some places the edges of the precipices form bare, smooth, perpendicular walls.

The sandstone headlands are neither rugged nor steep, and are clothed with grass; no bushes, however, are to be seen, excepting different varieties of the briar, which all bear a yellow flower. The grass on these outlying hills is also scantier than on the limestone elevations which are not rocky.

The sandstone formations are succeeded by a sharply-defined stratum of limestone, and above this the crest of the ridge is impregnated with crystalline formations. These extend uniformly from Tokmak to the end of the range at Aulié-ata; and by them, assisted by the sections surveyed, one could easily trace and mark down the geological formations of the northern declivity of the Kirghiznin-Alataù, which are distributed in regular order.

Entirely different is the formation of the southern declivity. Here, between the Talas and the chief crest, is a whole network of small ridges, which are still insufficiently explored and defined on the map, especially at the River Ken-Kala, an affluent of the Talas. Here, at the foot of the range, the oldest crystalline rocks prevail, intermingled with those of subsequent aqueous origin; their extension is not parallel with the main axis of the range, which runs E. 10° S., to W. 10° N., but at an acute angle from this axis E.N.E.—W.S.W., corresponding consequently with the prevalent direction of the Thian-Shan range.

The southern declivity had been already surveyed in August, after many other expeditions; but for the sake of giving a complete description of the Kirghiznin-Alataù, I will here embody all the accessible information regarding it. On Tek-Turmas Point, near Aulié-ata, geological formations are denuded, viz., some black sandstones dipping southwards, which are not met with in the northern slope. These strata form the undulating plain of the Tek-Turmas, above which, at a distance of only 7 versts to the eastward of Aulié-ata, rise frequent and bare masses of porphyritic syenite, consisting of an equal mixture of small crystals of rose-coloured felspar and hornblende, with dispersed and larger crystals of felspar. The black sandstone is upheaved by this syenite, and the former alternates beyond, at Kara-archa rivulet, with argillaceous schist.

Up the rivulet in the interior of the chain, the black sandstone, syenite, and micous and argillaceous schists succeed each other, but the alternation of the different strata occurs still more frequently than the scale of the map admitted of being shown. The syenite in some places merges into almost pure felspar, with a hardly perceptible mixture of crystals of hornblende. Some occasional thin layers of micous schist contain crystals of garnet (granite). There are also layers of micous schist which change into argillaceous schist, without coming into contact with pure layers of the latter. Generally speaking, the stratified formations are all metamorphic; the strata are very contorted, partly overturned, and partly standing in a vertical position, so that it is difficult to determine their dip and extension.

The valley of the Kara-archa, which falls into the Kainda, an affluent of the Talas, is overgrown with small birch-trees; the large ones have been cut down for building purposes at Aulié-ata. The rocks are covered by the *Juniperus pseudosabina*, which plant extends remarkably low, viz., to 3150 feet, the lowest limit of the birch. Transverse valleys frequently occur farther on, but up to the sources of the Kainda they are generally treeless. From the mouth of the Kara-archa up along the Kainda, the stratified formations at the foot of the Kirghiznin-Alataù are penetrated by those of a crystalline character, namely, by alternating syenites and diorites. The syenitic formations here lose their porphyry-like structure, and assume their typical character, resembling granite. Transitions between both forms of syenite are frequently observed even in small fragments of from 3 to 4 inches in size. Fragments of diorite are also perceptible in the syenite, showing that the former had penetrated into the latter.

The metamorphic formations penetrated by the syenites and diorites appear in the Cha-archa ridge, between the Kainda and the Talas, and, at the same time, alternating strata of black sandstone and of argillaceous and micous schists, which prevail at the mouth of the Kara-archa, also occur. The micous schists appear only in the eastern or most elevated portion of the range, which rises in a high steep wall, and joins the Kirghiznin-Alataù by means of the above-named crystalline formations (this being more particularly observable on the water-parting between the two affluents of the Talas, viz., the Kainda and Chaàldanyn-su). The latter rivulet flows across a transverse narrow fissure of the Cha-archa ridge. About 45 versts to the westward, the Talas breaks through another transverse fissure of the same ridge, but in an opposite direction, viz., from south to north.

The road from Kara-kol to Merké ascends along the Taldy-bulak rivulet, an affluent of the Kara-kol. Its valley is bare, as is also the opposite descent down the other Taldy-bulak,*

---

* Two rivers issuing opposite one another from opposite slopes always receive one and the same name from the Kirghises. Thus there are two Taldy-bulaks, two Kara-archas, two Mak-malas.

one of the upper sources of the Kara-kyspak. The valley of this northern Taldy-bulak commences in a crater-like expansion, covered with good pastures, out of which flows a small stream through a narrow gorge, between rocks of syenite and diorite. Of these rocks the whole of this range is composed.

The outlet of this gorge disclosed to view the longitudinal valley of the upper Kara-kyspak, along which flow, and in which afterwards also unite, its two upper sources, viz. the western Taldy-bulak, and the eastern, or the Kara-kyspak proper; the latter river being the longer, and having the more copious stream. The direction of the valley is from east to west, with a length of about 35 versts and a breadth between the ranges of about 8 versts; its surface is undulating, and a row of hillocks extends parallel with the longitudinal axis of the valley from east to west.

These hillocks consist of unstratified clay, intermixed with small boulders, and in them also I recognise moraines of ancient glaciers; particularly as the transverse valleys which run from the south into this longitudinal one evidently commence, as I observed in the Taldy-bulak valley, in craters or hollows,—a condition favourable for glaciers. These ancient moraines are covered with a somewhat scanty steppe vegetation, in which a *Ceratocarpus*,* called "ibelek," a pasturage grass, abounds.

After the confluence of the upper sources of the Kara-kyspak the river breaks through the northern chain of the Kirghiznin-Alataù by a narrow and almost impassable gorge. The road does not run through this gorge, but ascends to the third Eastern Ourianda: first by a narrow ledge along the southern slope of the northern chain; then trending through a small valley along the same slope it crosses, at an elevation of 8000 to 9000 feet, a high and rocky ridge, a spur of the northern chain, overgrown with juniper-trees, which afford cover to innumerable hares (*Lepus tolai*). Continuing along a ledge of this spur, it ascends the chief pass to the Ourianda, the descent to which northwards is already described.

The southern declivity of the northern chain presents the same syenites and diorites as exist in the valleys of both the Taldy-bulaks. From these central crystalline masses certain veins run into the limestone of the northern slope. The crystalline formations crop out through each other here, as along the whole southern slope of the Kirghiznin-Alataù. The veins of syenite run into the masses of diorite, and the syenite is itself impregnated with veins of granite; but besides these well-marked veins and fragments of broken formations, such as

---

* This species, doubtless, must be the Ceratocarpus arenarius. The "ibelek" is the same grass on which the Kirghizes feed their cattle in their winter pasturage.

diorite in syenite, diorite and syenite in granite, there are also gradual transitions of one crystalline formation into another.

The view southwards from the northern chain of the Kirghiznin-Alataù, at the Ourianda Pass, is magnificent. Immediately below, ledges of rock, overgrown with juniper-trees, hide the valley of the upper Kara-kyspak; beyond, rises the southern range of the Kirghiznin-Alataù, like a dark crenelated wall, as in the month of May (O. S.) it is already free from snow. Farther back still are seen the seemingly close and colossal snow-capped mountains of the Urtak-taù, glittering in the sun. They are particularly beautiful at the end of May (O. S.), when all that is seen of them from this mountain pass is covered with snow and contrasts distinctly with the black range which obstructs the view of their base, while their outline stands out in bold relief against the dark-blue sky. The snow in August rests only upon the topmost summits.

To the north, several snowless spurs of the Kirghiznin-Alataù are visible from this pass. They decline towards the Chu Steppe, and beyond them the boundless extent of this steppe blends with the heavens in the far horizon.

Such is the Kirghiznin-Alataù. Judging from what I saw of it, it consists of two principal ranges: of a southern, almost exclusively of crystalline formation; and of a northern range, for the most part of sedimentary origin. They are both intersected by rivulets between Aulié-ata and Merké; amongst others, the Ken-kol makes its way through the southern range, and, issuing from the longitudinal valleys, terminates its course at the syenitic hills at Tek-turma. Through the northern range, and issuing from the southern, rushes, as we have seen, the Kara-kyspak.

Let us now consider the localities at the base of the Kirghiznin-Alataù north and south.

Along the northern slope, between the sandy submontane region and the smooth steppe, there is another hilly zone of about 15 versts in breadth. Its soil is a sandy loam, mixed with boulders, chiefly of crystalline structure; and in this zone, between the rivers Djarsu and the north Kainda, flowing towards the Talas, there are visible moraines of ancient glaciers, similar to those existing in the longitudinal valley between Issyk-kul and Ala-medyn, which have already been described.* These mounds of drift have been principally formed

---

* In both these localities identification of the moraines is still more certain than on the Kastek, where the aggregation of granite boulders beneath rocks likewise granitic, might seem to favour the supposition of their having been so deposited by a landslip. Here, however, the granite boulders lie under sandstones and limestones, and the water from the annual melting of the present snows is powerless to remove them.

from the detritus washed down from the mountain-ranges. Their unstratified clay is interspersed with boulders; and, although they seem to me to be of glacial origin, they do not every where present the appearance of unmistakable moraines. They have probably been partly destroyed and washed away by the waters from the dissolving ancient glaciers. The greater portion of the pebbles in the beds of the mountain-torrents has been washed out of the moraines of the glacial period, and not directly from the rocks. Thus the Kara-kyspak at its upper course does not flow over crystalline formations, but upon an alluvial soil, containing, however, boulders of crystalline formation. Further on huge masses of limestone crop out; but there are few of them in the bed of the river after its issue from the mountains, nor are they more frequent in the beds of other rivers forcing their passage through the same sandstone, such as the Merké, Ourianda, &c.

The soil of this hilly zone, at the northern base of the Kirghiznin-Alataù, presents visible traces of its formation from crystalline elements, limestone, and sandstone. It consists of a light, sandy, calcareous, argillaceous loam, of which the proportions of clay and sand greatly vary.

Eastward of Merké, the loamy soil is chiefly of a yellowish colour; westward, it changes into a greenish marly mud, which is equally fertile under irrigation, but yields only a poor vegetation without such assistance. Opposite several mountain valleys there are tracts of black mould, but they are only found to the eastward of Merké.

From what I saw in the Issyk-kul valley, it appears to me that this black mould owes its origin to mountain forests, which for the most part are no longer in existence. They probably grew in a moister atmosphere than the present, during the glacial period and soon after it, as the glaciers here did not descend to the level of the sea which then covered the Kirghiz steppe, their lowest traces occurring at a height of more than 2000 feet above the Almatys and at the Kainda River.

The water-parting between the Chu and Talas, at the northern slope of the Kirghiznin-Alataù, is remarkable. It consists of a hardly-perceptible elevation of alluvial soils, intersected by hollows, of which some are transverse, sloping on both sides, and are filled in spring from the hollows running longitudinally with the elevation. This water-parting is situated between the Temdul rivulets, the affluents of the Chu and Djarsus, and the affluent of the Talas. The highest part of the range between Merké and Aulié-ata is to be found opposite this water-parting. Northwards, this low water-parting widens out

ed fields, irri-
e in the very
the Chemget

act of its rising
runs into the
ato the Talas in
ad to the Urtak-
water from the
f the mountains
om the moun-
ata. The first
ntains is a loose
ith a variety of
ick argillaceous
orge, are forts of
very mouth of
n of argillaceous

appears 8 versts

still in the sub-
imposed of con-
30° s.s.w. towards
of the Kara-bura

i-kene-kara-bura
than 30 versts
gillaceous schist
mestone. Both
The stream is
w-land; but at
Caragana sp.),
a and bearing
at elevations of
ot higher, and
ii, which latter,
crags covered
ura, in a hori-
o valley of the
elf flows direct
however, high;
bably by moun-

base of Tek-turmas ; after leaving which it proceeds
the north-west to Aulié-ata, where, skirting Tek-turr
proceeds in a direction towards north-east. Below A
the Talas emerges into the steppe. Through 8 ve
its course it is bordered by town-gardens, and beyond tl
banks are lined with reeds. Here, in its issue into the
the Talas preserves the character of a mountain river ; a:
number of limpid streams it runs rapidly over a pebb
crossed by innumerable fords.

High water, but of very changeable level, prevails fro'
to the middle of July (O. S.), owing to the melting of th
in the mountains where the Talas issues and receives
affluents. The water is lowest at the beginning of Sep
(O. S.). There is, again, an accumulation from the au
rains, which at elevations of more than 5000 to 6000 1
soon replaced by melting snow. I cannot say how far i
steppe the Talas retains its character of a mountain stre
lower course had not been surveyed up to the 1st (13th) J
1865.

. The Lake Kara-kul, into which this river disembo
merely a network of spreading pools among sand hillocks
at least, M. Patanin observed it t° be on his way from t
to Chalak-kurgan.

In volume of water the Talas is not much inferior
Chu. All the streams running towards it (and they a
merous) reach its bed ; whereas below Tokmak only two
run into the Chu, viz., the Ala-medyn and one other,
probability the Karagaty. This is probably owing to th
flowing in a single channel in its lower course, and not spr
out into a series of lagoons occasionally uniting with each
but, on the other hand, it is to be borne in mind that 50
below Aulié-ata this river enters upon the sands.
sands by the river, at the lower course of the Tal:
Kirghizes consider to be good winter grounds, and th
therefore occupied by their "aùls." As to the lower co:
the Talas, the Kirghizes say that it reaches Kara-kul in a
bed, but that it is very shallow, and is frequently silted u
sand.

The connexion of the Kirghiznin-Alataù with the Urt
noticeable, as has already been observed, at the Upper T
had previously explored at the end of June (O. S.), duri
excursion from Aulié-ata to the head of the Chirchik, alo
road to Namangan. This connexion is formed by the
archa ranges here described, and by the northern subm
region of the Urtak-taù.

The road leads from Aulié-ata up the right bank

Talas to the Cha-archa Mountains, along cultivated fields, irrigated by canals from the Talas which commence in the very gorge; it then passes through this gorge near the Chemget rivulet towards and along the River Kara-bura.

The Chemget rivulet is distinguished by the fact of its rising in a plain. It rises from several springs, and runs into the Cha-archa range, emptying itself through a cleft into the Talas in the midst of these mountains. Farther on, the road to the Urtaktaù winds along through fields irrigated with water from the Kara-bura, which gives its name to that part of the mountains where it passes. The issue of this river from the mountains is at a distance of 40 versts from Aulié-ata. The first geological formation that crops up in the mountains is a loose unstratified conglomerate of red clay, mixed with a variety of pebbles; then vertically-uplifted strata of black argillaceous schist. In the steppe, at the approach to the gorge, are forts of clay, erected by the Kara-Kirghizes. At the very mouth of the gorge, and before coming to the denudation of argillaceous schist, there is a spring and a small copse of old and large poplars (*Populus sp.*, as on the Talas, a medium between the common poplar and the ash). A similar copse appears 8 versts higher up, at the confluence of the two affluents of the Kara-bura.

Through all this extent the Kara-bura flows still in the submontane region of the range, which is also composed of conglomerate, but stratified; the strata dip about 30° s.s.w. towards the range. At the confluence of the sources of the Kara-bura argillaceous schist again crops up.

The road beyond proceeds up the Ketch-kene-kara-bura River, along the eastern summit, for more than 30 versts between bare and uniform rocks, in which argillaceous schist alternates with thin strata of dark-brown limestone. Both these formations are devoid of organic remains. The stream is very rapid; the valley exhibits patches of meadow-land; but at first no trees, and only small prickly shrubs (*Caragana sp.*), somewhat like what are seen on the Syr-Daria and bearing similar pink flowers. This plant grows chiefly at elevations of from 4000 to 5000 feet. It occurs lower, but not higher, and does not descend to the lowest limit of the birch, which latter, at a height of 5200 feet, extends first along the crags covered with detritus, from the left bank of the Kara-bura, in a horizontal line, for about 1 verst, and keeps to the valley of the river, which, higher up, is wooded. The river itself flows direct from south to north. The birch-trees are not, however, high; they are crooked and their tops are broken, probably by snow-slips.

We may observe that the lowest limit of the birch-tree here (5200 feet, *possibly* 5300) corresponds with the lowest limit of the fir (5300 feet) in the eastern Kirghiznin-Alataů.

In this comparatively wooded portion of its valley the Ketch-kene-kara-bura breaks in several places through narrow clefts, where the road overhangs precipices of 50 to 60 fathoms deep. Here the foliage of the trees below is thicker than elsewhere.

At an elevation of about 7000 feet are found the first indications of the *Juniperus pseudosabina*. Three versts farther, at a height of 7400 feet, is the junction of the two heads of the Ketch-kene-kara-bura River; and the road trends round the eastern one to the left towards the pass, following generally for 2 versts the bed of the torrent, which here, falling at an angle of 10° to 15°, leaps with terrible rapidity from stone to stone, between birches and flowering bushes. Here, however, the river is close to its source, and there is but little water; consequently, notwithstanding its rapidity, the stream is fordable.

On emerging from the gorge the wide basin of the Kara-bura opens before one, the converging wooded valleys of which river are separated by sloping passes.

Here only on the Kara-bura, at a height of 8000 feet, is found the mountain-ash;* but its highest limit corresponds with that of the birch and the high-trunked juniper at 8700 to 8800 feet. The former elevation was measured along the road; in the side valleys, which are more sheltered, the forest ascends somewhat higher.

The rise of the slope at the bottom of this basin is not more than 1000 feet in 5 versts; but it is steep towards the pass, viz., 1500 feet in 1½ verst. Here are Alpine pastures and creeping detached juniper-bushes; and here, by undulations and grassy declivities, close to the range itself, are easy passes over the heights of one valley to those of another; whereas lower down the valleys are separated by steep and hardly traversable or wholly impassable cliffs. Here, at a height of 9000 to 11,000 feet, are the summer pasturages of the Kara-Kirghizes; and when we travelled here at the end of June (O. S.) the pastures along the Kara-bura and its eastern source were all consumed to a height of nearly 10,000 feet. Moreover, the cattle are always driven through this locality from Aulié-ata to Namangan. The Kirghizes encamp on both sides of this road when they leave the route along the ridge of the mountains.

---

* In the Kirghiznin-Alataů, as we noticed, the juniper and the mountain-ash descend considerably lower.

Above 10,000 feet, we found on the 22nd June (4th July) that the snow had only just disappeared. There was as yet no vegetation; in the hollows there were large patches of snow. These patches lay there for several weeks longer, but the snow became porous and waterlogged and the intervening spaces were covered with alpine vegetation in full bloom.

On the same day the pass itself was free from snow to a height of about 10,500 feet. The ridge of this pass is saddle-shaped, rising at both ends into the region of perpetual snow; and here, as I remarked later, viz., in July and August (O. S.), the line of perpetual snow is at the height of about 12,000 feet.[*]

On the 23rd June (5th July) we observed from the ridge a herd of "teks," or mountain-goats (*Capra Siberica*, or another kind) on t_he slope. We did not not succeed in killing any. "Ullars" [†] (large-sized partridges, weighing from 10 lbs. to 15 lbs.) are found at the same altitude in the summer. The latter escape pursuit by running quickly up steep declivities, in order to rise and take wing beyond the reach of shot; so that the sport of shooting these birds is difficult and fatiguing. It is a remarkable fact that the "kikelik," or red-legged mountain partridge, which is common in all the mountains between the Chu and Syr-Daria, is not to be found on the mountain-ridges of the Urtak-taù; and it is also remarkable that on the Kara-bura alone I fell in with the dark-brown sea-sparrow of the Kuriles and Aleuts (*Cinclus Pallasii*); whilst on the southern slope of the same mountains, as well as in the Kirghiznin-Alataù, I saw only the "altayan," or white-bellied variety (*Cinclus leucogaster*) of this species of bird.

The valley of the Kara-kyspak is the most picturesque of all those that I saw in the Western Thian-Shan. The descent towards it from the pass is steep; and instead of running straight down, it proceeds along the side of the mountain, and the path leads on along the edge of an almost perpendicular cliff, which rises sheer from an abyss 900 to 1000 feet in depth. The summit of the pass is not more than 1½ fathom wide; and in proportion to its height above the abyss of 1000 feet, it may be compared to the blade of a knife. The descent follows a north-easterly direction, and does not lead to the issue of the Kara-kyspak, but to the confluence of its two sources. Farther down, the road proceeds along that river towards the south-east.

The valley of the Kara-kyspak, like that of the Kara-bura, first takes the form of a great hollow, divided by undulating

---

[*] Measured on the Urtuk-taù by angles from the Talas valley. The note of more accurate determination has, unfortunately, been lost.

[†] *Megaloperdrix* of Brandt. I shot only a couple of the *Megaloperdrix nigellii var.* on the *summits* of the Ourianda, in the Kirghiznin-Alataù.

ridges into several converging valleys. This hollow is walled up towards the west by a steep snowless declivity, along the face of which runs the descent from the Kara-bura Pass mentioned above. On all other sides rise snow-clad summits, nearly flat above the snow-line, but very steep below the level of snow. All these declivities, except the precipitous western side, are clothed with the soft fresh verdure of Alpine pastures. Spreading juniper-bushes are scattered everywhere, growing more especially in clefts. But in the lower part of the hollow, the Kara-kyspak, 100 fathoms below the confluence of its sources, where it still traverses the upland meadows, pours into a deep bed about 50 fathoms below the level of the meadow-land; after which it enters a regular gorge, through which it runs for about 17 versts, and receives on the right and left a number of streams from the lateral valleys. Thus, in the seventh verst of its course through the gorge, it becomes a torrent about 10 fathoms wide and 3 feet deep; rushing with a roar over the stones, between perpendicular walls. The lateral valleys are everywhere surrounded by snowy summits, and abundant springs pour their waters into the river over the walls of the gorge, which are more than 100 fathoms high, in frequent though narrow streams, resembling strings of silver between the rich green verdure of marginal brushwood. The river itself, swollen more and more by the streams of the gorge, and from torrents of snow-water, leaps and foams in one incessant cataract, between stupendous rocks several hundred feet in height, crowned by perpetual snow.

Snow-slips occur within 7 versts of the confluence of the sources of the Kara-kyspak, and they continue for about 10 versts further. The avalanches depend upon the form of the snow-clad summits above the gorge; their comparatively sloping sides (of 20° to 25°) terminating abruptly in almost perpendicular sides of crevices. Immense masses of snow accumulate in the hollows, and sliding down the declivities tumble into the gorge of the Kara-kyspak. This occurs in the spring, in April and May, according to the condition of the latest, or surface snow of these masses; which, by the end of June (O. S.) becomes water-logged. With the thaw it becomes pink, and gives birth to unilocular microscopic plants (*Protococcus nivalis*). These masses, tumbling annually, and always at exactly the same places, do not, however, obstruct the river, so as to dam it up until it again forces its way through; but, on the contrary, they form permanent bridges of snow, with regular semi-circular arches over the river; and there are no signs of the stream having been blocked up, or of the sides of these arches having been washed by the waters. It is everywhere easy to distinguish the yearly layers of snow; as each new avalanche descends upon the masses that were hurled down during the two preceding years, and which have become consolidated by the winter's frost succeeding the summer's thaw. Each of these bridges is sufficiently strong to bear the shock caused by the later falls; in a word, each avalanche falls upon a firm and permanent natural bridge of snow. These bridges almost always preserve the same dimensions; the addition of snow to the top being counter-balanced by the melting away of the inner part of the arch, and by the subsidence of its piers where they rest on strongly heated rocky ledges.

I always noticed three yearly layers of snow, either more or less; each being of the thickness of several fathoms, the bottom one being the thinnest. Consequently, the accumulation of each spring melts away in the course of two years.

There are in all seven bridges. The upper one is the smallest, and the two centre ones the largest, being each 400 to 500 fathoms across. The lower of these is at an elevation of 8650 feet, the lowest of all is 1200 feet below.

At the last but one of these bridges, where there is a bend in the gorge, the water has the appearance of being obstructed by steep and colossal snow-clad mountains, turret-shaped at their summits. This is the Namangau range on the left bank of the Chirchik; and its nearest snowy peaks are still 15 to 20 versts distant. Beyond this the gorge opens out into its lower hollow, and then again contracts. Here the Kara-kyspak passes under its last snow-bridge; and 3 versts lower down, enters the valley of the Chatkal, and merges into that river.

In the 20 versts of the course of the Kara-kyspak from the confluence of the two sources, to its mouth, the fall of the river equals 2500 feet; more by 125 feet to the verst than the fall of the Imatra which is not more than 30 feet to the verst, but its volume of water is less.

In the hollow last mentioned, and more especially on the right-hand side, the rocks retreat to nearly a verst from the river, which flows between hills of unstratified clay and boulders, several hundred feet in height, forming a slope between the rocks and the bottom of the hollow. These hills are covered with enormous parachute-shaped plants (*Umbellifera*, probably *Ferula* sp.), which shade also the southern submontane region of the Urtak-taù, composed of similar drift. The highest limit of those plants is about 500 feet above the level of the river in this locality, and therefore, about 8000 or 8100 feet above the sea-level. Here on the 20th June (11th July), I saw the ground covered with snow that had fallen during the previous night, while lower down by the river rain was falling.

Generally in the months of May and June (O. S.) the rains

are frequent at heights above 4000 to 5000 feet. They occur almost daily between four and seven in the afternoon: not so often in the morning or at night, and in June higher up the mountains than in May. Beginning from an elevation of 8000 feet, or a little below, during all the summer months it alternately rains and snows, but the snow soon melts. Above 9000 to 9500 feet rain never falls, but only snow; although here in summer the snow melts soon after it falls, or even in falling. This upper limit of rain is also the extreme limit of forest and of brushwood. Alpine grasses, and the spreading juniper alone are irrigated all the year round with snow-water only.

The rhododendron * which on the Alps and Caucasus grows abundantly above the upper limit of forest, and which in Siberia grows in the mountain forests, I did not fall in with either in the Kirghiznin-Alataû, or the Urtak-taù, at least not at these elevations. The upper limit of the birch at the Kara-kyspak is considerably lower than at the Kara-bura, viz., at 7450 to 8700 feet, which may be attributed to the fact that the entire valley of the Kara-kyspak winds between snow-clad mountains screening it toward the south-east and south-west. Three versts below the commencement of the forest, at a height of 7100 feet, the Kara-kyspak falls into the Chatkal; it is nowhere fordable in this later period of its course.

The valley of the Chatkal has less of the steppe character than that of the Talas. Copses of wood grow, not only at the sources of the Chatkal, but are sprinkled over the spaces between them, and near the hollows, moistened by springs. Even the lands along the river are also wooded. There are frequent patches producing birch, willow, and a variety of brushwood unknown to me, but amongst which I could distinguish the mountain-ash and black-currant bush. To these, at an elevation of 6350 feet, is added the (*Hippophae rhamnoides*), which rises to the great height of 2 fathoms; and the hawthorn, at an elevation of 6000 feet. The banks of the river here are steep, and are 10 to 15 fathoms in height; they form the first rise towards the mountains, the second consisting, as I have stated, of hills of drift; the unstratified clay and boulders of which appear to me to be of glacial origin.

The current of the Chatkal is extremely rapid throughout 30 versts. From the mouth of the Kara-kyspak to Chipash-Kurgan, it falls 750 feet;† there are no fords. The width of the principal current is 20 to 25 fathoms; its smaller side-

* There are no rhododendrons anywhere, either in the Thian-Shan or in the two Alataùs.

† The fall of the river from the mouth of the Kara-kyspak to Tashkend, is not less than 5000 feet throughout an extent of 270 versts.

channels are insignificant. The Chatkal or Upper Chirchik distinguishes itself from the many-armed Talas by the far greater concentration of its waters into one swift current. It has besides a greater volume of water, being fed by incomparably larger masses of snow, particularly in the Namangan range, as well as the Urtak-taù, the southern valleys of which do not, like those on the north, intersect a declivity generally snowless, but traverse the wide snow-plain forming the summit of the range; as is seen from a comparison of the Kara-Kyspak and the Kara-bura; and along the whole of the Chatkal, as on the Kara-kyspak, the perpetual snows upon the spurs, between the transverse valleys, commence at four or five versts from the bases of the mountains.

The current of the Chatkal, notwithstanding its velocity, is smooth and free from waves, and is, therefore, serviceable for floating down timber to Tashkend.

The Chatkal is formed by the Kara-kyspak and Kara-kuldja; the latter itself originating from the confluence in one valley of several torrents issuing from the mountains. These streams mostly rise in the short meridianal range or mountain-knot, covered with vast masses of snow, between the Urtak-taù and Namangan ranges. From both of these the Chatkal receives many affluents every 3 or 4 versts. The Namangan range is loftier than the Urtak-taù; and even the heights near its base ascend 15,000 or 16,000 feet. More than half its height is covered with snow. Taking the snow-line at the end of June (O. S.) at 11,000 feet (from a measurement of the snowless pass of the Urtak-taù), and the foot of the mountains as 6500 to 7500 high (about 500 feet above the level of the Chatkal), we obtain from the base to the snow-line a height of 3500 to 4500 feet.

The crest, however, of the Namangan range (i. e., the average height of the intervals between the peaks) is not higher than that of the Urtak-taù; but the peaks of the former rise higher than those of the latter.

Between these peaks are noticeable the cloven sides of the mountains, which are almost perpendicular; and on which for several thousand feet the snow does not rest. The peaks, which are either sharp-pointed or in the form of crenelated towers, with upper snow-covered platforms, and "counterforts" also snow-clad, are remarkably diverse and picturesque.

Within so grand a framework of snow-covered mountains, upon which the clouds are ever collecting, clinging to the crags, and then gradually dispersing, lies the beautiful valley of the Chatkal; and its loveliness is enhanced by the bright blue sky overhead. The river, clear and rapid, runs glittering along

through meadows of verdant green, sprinkled here and there
with copses. But it is this green verdure in the vicinity of
such masses of perpetual snow that leads me to suppose that
the valleys of the Chatkal must be subject to severe and prolonged winters.

The huts of the Kirghizes in the meadow lands by the river
ascend to the height of 6400 feet, consequently not much beyond
the limit of the *Hippophae rhamnoides.* Here are seen the enclosures for their cattle, formed of heaps of broken branches of
brushwood. There are no traces of permanent habitations, from
which we may conclude, that here also the Kirghizes pass the
winter in their summer felt "kibitkas;" as I afterwards observed in the western part of the Trans-lli-Alataù, near Suok-
tiubé. The Kirghizes, who encamp along the Chatkal for the
winter are of the Saru section of the Kara-Kirghizes.

I was able to descend the Chatkal 45 versts from the
mouth of the Kara-kyspak to a point 6000 feet above the sea-
level. Here it still flows in a south-westerly direction, but
a little further on it makes a decided bend towards the west.
Along its right bank runs only a spur of the Urtak-taù mountains. Between this spur and the principal chain there is
another mountain-torrent, at the source of which *verdigris*
has been discovered in the landslips. Mr. Frehse and myself
saw the ore, but not the spot where it was actually found.

From the Chatkal we returned to Aulié-ata by the same route,
along the Kara-kyspak and Kara-bura. We next examined the
valley of the Arys, and the northern slope of the Urtak-taù, to
the west of Aulié-ata; I exploring the Boraldai Mountains and
the elevated table-land, and Mr. Frehse the northern slope of
the Kara-taù.

From Aulié-ata we proceeded on the 7th July (19th July)
by the Chemkend road, which at a distance of 12 versts
from the town, intersects the River Asa, close to the southern
foot of the small chain called the Ulkun-burul Mountains,
appertaining to the Karataù system. About 5 versts from
Aulié-ata, and enclosed by mud walls, are the townlands
with small gardens scattered about. These are irrigated
from the Talas. Next comes an extremely flat and barren
watershed between the Talas and Asa; and a similar slope
towards the latter river, from which a few irrigation canals
are supplied, fringed with bright verdure, reeds and wild lucern
(*Medicago sp.*), or, in the Kirghiz vernacular, "djan-shike."

Between these canals is a marl steppe, with a gre parched
soil, covered by the poorest vegetation, and by withered wormwood and sickly bushes of the Ephedra. The valley of the Asa
is devoid of meadow-land, and is encumbered with boulders.

The river itself flows in many diverging streams of 8 to 10 fathoms in breadth, but of not more than 3 feet in depth. Judging from the river-bed, the waters of the Asa must, how-ever, increase considerably in the spring, and must then be very turbulent.

Immediately beyond the Asa is a flat elevation, as barren as the right bank of the river, strewed with small boulders from the Ulkun-burul, composed principally of hard limestone and cornelian; the latter lying in thick strata in the Ulkun-burul.

On the first slope from the Asa, and by the side of the first "aryk" (or canal) of this river, is a shady garden attached to an Aulié-ata summer residence which we found uninhabited. Surrounding it is a bare waste, over which reigns a death-like stillness, but the scenery is nevertheless grand, owing to the snowy heights of the Urtak-taù towering above the steppe. There is a perceptible decline in this steppe extending north-wards to the right, and stretching towards the small lake of Ak-kul, into which the Kuyuk rivulet discharges itself. Saline plants appear here, with the *Salicornia herbacea*, at a height of about 1700 feet, calculated by the measured heights of the Kuyuk and Biliù-kul mountain-chains.

The mountains of the Kuyuk are composed of schist, covered with a poor grass; there are hardly any bushes except the wild briar. The road passes into these mountains by the rocky defiles of the Kuyuk rivulet.

Having ascended through this defile, we entered upon a flat table-land, along which flows the River Tersa, afterwards falling into the Asa. The descent from the Kuyuk Pass to the Tersa is very short, viz., about 3 versts, and is extremely gentle. On the summit of the pass are several springs; these are the most southerly sources of the Myn-bulak district, celebrated in the ancient geography of Central Asia, and which Huen-Tszan places not far west of the Talas. The high mountains men-tioned by him, to which the district of Myn-bulak clings, are the Urtak-taù. This locality is celebrated from the fact that when Huen-Tszan visited it, in the seventh century after Christ, it was the site of the summer encampment of the Toorkish khans.[*] The Kirghizes, even now, consider Myn-bulak to be the best place for summer encampment between the Chu and Syr-Daria.

The large trees mentioned by Huen-Tszan no longer exist about Myn-bulak, but the summer there is still warm; the heat does not, however, exceed 25° R. There is good pasturage,

---

[*] *See* Humboldt's 'Central Asien, übers. v. Mahlman,' vol. i. part i.; chapters on the Thian-Shan and Bolor Mts.

with a dense and succulent herbage, and there are numerous clear springs.

From the southern slopes of the Urtak-taù, Myn-bulak, according to Huen-Tszan, is divided by the valley of the River Tersa; but the present district of Myn-bulak is separated even from the Tersa by a low ridge composed of schist. These strata, however, dipping from 25° to 30° N.W., and extending N.E., S.W., crop out in the bed of the Tersa.

The valley of the Tersa is rich in luxuriant meadows. In its course it has rather the character of a steppe river than of a mountain torrent, like most of the trans-Chu streams. It consists of lagoons of almost stagnant water, connected by rapid though noiseless currents.

The arms of the Talas are very insignificant, spreading out into small marshy pools. Waterfowl are plentiful on the Tersa, especially geese and snipe, while ducks are rather scarce. In the adjoining steppe are great quantities of sandgrouse* (*Pterocles arenarius*), and of bustards in the meadows along the river. In the winter I saw numerous traces of the wild-cat of the steppe (*Felis manul*), and of "corsaks" and other foxes. This river flows from the eastern declivity of the Kulan Mountains which belong to the Karataù chain, from one of the western ridges of which issues one of the sources of the Arys. The main sources of the latter river are, however, further south.

The Tersa does not possess many tributary streams. The principal ones flow from the Urtak-taù; the Ak-sai and Kok-sai from beneath the lofty peaks of the same name; but in its valley and even its bed are many springs.

The road from the Tersa to the sources of the Arys ascends obliquely a gentle slope of the Urtak-taù to the river. This slope is covered with drift, in which, however, I could detect no boulders; it is intersected by numerous ravines, streams and rivulets.

At the Chak-pak, which is the last tributary of the Tersa towards the west, perpendicularly upraised strata of sandstone crop out above the drift. These are of an unascertained palæozoic formation; they occur on the right bank of the rivulet, which, however, soon makes a bend towards the north-east and intersects this sandstone. The latter extends almost directly from north to south. The left bank of the Chak-pak is composed of limestone, which also forms the pass to the valley of the Arys. This is most probably mountain-limestone, as lower down the Arys sandstone and schist, which to all appearance is of the Carboniferous period, rest upon it. In these Mr. Frehse

---

* A large white bird found in the steppes of South and South-east Russia.

found indistinct impressions of calamites in the spurs of the Boroldai Mountains, forming the right side of the valley of the Arys at the Kulan-su rivulet. Seven versts below this rivulet on the Arys there occurs a reddish argillaceous sandstone, probably of the Permian formation.

The valley of the Arys, like that of the Tersa further east, separates the Kara-taù mountain system from that of the Urtaktaù. The watershed between the Arys and Chak-pak, the tributary of the Tersa, connects the Urtak-taù with the Kulan Mountains of the Kara-taù system. From the Urtak-taù this watershed extends in a hardly perceptible elevation, which descends gradually towards the north. At the sources of the Arys a steep mountain of limestone rises abruptly on this elevation, and beyond is a saddle-shaped ridge; while still further on are to be seen the numerous rocky summits of the Kulan, which increase in height as far as the source of the Tersa, and decline further still towards the north. This Kulan chain is the Mynbulak-tau on Humboldt's map, attached to his work on Central Asia; and it was considered by him to be the extreme northern continuation of the Bolor—of which more presently.

At the southern foot of the mountain with which the Kulun chain commences are several deep springs, giving rise to considerable streams, which, after a course of from 2 to 3 versts unite to form the Arys. The volume of water increases very rapidly, as the valley abounds with copious and fast-bubbling springs. The rivulets also that fall into the Arys, on the right from the Kara-taù, and on the left from the Urtak-taù, are very numerous. The latter chain declines in height towards the west, following the course of the river, and at Yaski-chù, about 45 versts from the sources of the Arys, terminates in a precipitous ridge not more than 2000 feet above the level of the river, which has here an absolute height of 1950 feet above the level of the sea. Farther west, however, of this projection there is still a low platform of limestone, separated from the former by a flat arid valley, one verst in breadth. The small patches of perpetual snow that still extend on the Urtak-taù west of the sources of the Arys, are no longer visible opposite the mouth of the Sary-bulak, at a distance of 20 versts above Yaski-chù. As to the mountains of the Kara-taù system that approach the Arys between its sources and Yaski-chù, they are simply the south-western slope of the Kulan Range, which terminates in a rather abrupt gradient at the River Arys. The deep valleys of the right tributaries of the Arys divide this slope into several seeming ridges, extending from north-east to south-west. Beyond these hills the principal chain of the Boroldai Mountains with their craggy heights, extends in the same direction between the river of the same name, falling into the Arys, and the River Bugun, flowing into the Chilik, an affluent of the Syr-Daria. The Boroldai Mountains abut upon the Kulan Range almost at a right angle. From their form, and from the yellowish tint of their crags, they seemed to me to consist of limestone.

The Kara-taû system is joined to the Western Thian-Shan, not only by means of the connexion between the Kulan and the Urtak-taû, described above, but also by the link between the Kuyuk and the Cha-archa, which, as already mentioned, is orographically connected with the Kirghiznin-Alataû, and geologically with the same and with the Urtak-taû.

The upper course of the Arys terminates at Yaski-chû, where it emerges from a mountain valley upon a perfectly level steppe, or rather upon a steppe which almost imperceptibly rises from the edges of the river-basin; these edges on all sides frequently forming a precipitous gradient of 6 or 7 fathoms in height, which, like the bottom of the valley, consists of drift. The river itself flows on in a deep bed. As far as Yaski-chû it is rapid, running over pebbles, and is of equal but not very great depth. In this locality it forms lagoons deeper than the river (about 1 fathom), but running more slowly. Lower down these lagoons are more numerous and still deeper, and their alternation with fords characterises the middle course of the river, to the mouth of the Badam; the depth of the fords here is, however, less than 3 feet. The bed of the river becomes more and more muddy, although occasionally it is sandy.

The middle course of the Arys differs from the upper, from the fact of there being fewer springs in the valley, and that its feeders, although larger, are not so numerous. These are: from the right, the Boroldai; from the left, the Mashat and Badam. Chemkend is situated on the latter. There are, however, several other rivulets flowing towards the Arys between the Mashat and Badam, without reaching it—their waters being diverted into irrigation canals. The lower course of the Arys extends 70 versts from the mouth of the Badam to the Syr-Daria. The current here is slow, and the depth from the ford at the embouchure of the Badam gradually increases, so that the Arys becomes navigable for vessels of 4 feet draught. Copses of "djida" (*Eleagnus angustifolia*), "turanta" (*Populus diversifolia*), and of thorn (*Caragana jubata*), cover the banks of the Arys for about 20 versts up the river from the Syr-Daria.

We arrived at Chemkend in the month of July and descended the Badam, crossing the Arys from the mouth of that river. Here I parted with Mr. Frehse, who proceeded to Turkestan, and thence, traversing the Kara-taû by the Turlan Pass, went on to Cholak, made an excursion to the coal-beds at the Kumyr-tas Stream, south of Cholak, and returned to Aulié-ata by way of the steppe, along the northern slope of the Kara-taú. My course was along the Bugun up to its sources, where I crossed the Kara-taú, and made my way back to Aulié-ata by the Bïliu-kul Lake and the Ulkun-burul Mountains.

I shall now describe these districts north of the Arys. Those to the south will be referred to further on, as I saw little of them in July, but examined them later in September. A few words, in the first place, regarding the valley of the Arys.

This valley is remarkably fertile, as well as the land on the left bank wherever water can be conducted; and this can be done everywhere in the valley without the smallest difficulty. By means of irrigation even the multifarious soils of the trans-Chu district, and of the Aralo-Caspian hollow—nay, those also which, without irrigation, do not produce a blade of grass, such as the grey clays of Khiva, or the red clays by the Ulkun-burul, near Aulié-ata—are made productive, and in some instances tolerably fertile. But the fertility of the valley of the Arys, with its luxuriant growth of lucern,* wheat, " djugara,"† maize, and "kunak," ‡ surpasses everything. The ears of the " kunak," instead of being, as at Aulié-ata, $1\frac{1}{2}$ inch in length and $\frac{1}{4}$ inch thick, attain a length of 3 inches and a thickness of 1 inch. The grain is not of the size of poppy seed, as at Aulié-ata, but nearly that of millet. The "djugara" reaches a height of $1\frac{1}{2}$ fathom, and the stalks are of the thickness of two fingers; they grow close to each other, stalk to stalk, so that their roots become entwined. The Indian wheat is almost equally large, although it does not grow so thickly. Wheat produces thirty-fold; the lucern, after three cuttings, grows up again nearly 3 feet high, and is prevented from bending down by its density, the stalks supporting each other, the outer ones alone bending down to the ground. The melons and water-melons are really superb.

The hay-fields an natural meadows are not sown, but are merely irrigated. This fertility is greatly owing to the properties of the soil, which is a rich dark loam, loose, and easily ploughed, and retaining moisture; it is composed of clay, the finest sand, lime and decomposed matter. It has been formed

---

* " Dïany-abke " (*Medicago* sp.), perhaps the known Chinese Mu-sui.

† *Holcus* sp., something like the sugar-cane. The stalks are not very sweet; the seed is used as grain, and the leaves are given to cattle, as well as the young stalks. The old stalks are used for fuel.

‡ A grass similar to *Alopecurus*. It is considered to be very good fodder for horses, especially if cut down while the seed is still unripe. The ripe seed falls out, and is too small for grain fodder. The Kokandians and Kirghizes, however, prepare it also as grain.

of granular particles of clay-schists and flinty limestones, blown down by the winds, the decomposed matter being probably the remains of forests which once covered the Urtak-taú, as we have had occasion to remark in allusion to similar soils between Tokmak and Merké. There are no woods along the Arys, excepting small artificial plantations of white willow, and partly of poplars in the valley. The soils on the left bank of the Arys, on the steppe uplands, and at the foot of the Urtak-taú, are also of good quality, and are quickly and easily covered with vegetation in the spring of the year, without irrigation. This, of course, becomes quite parched up in May, although it grows up in the condition of a passably good steppe hay-crop.

The last mountain declivities on the right bank, but only as far as Yaski-chù, are likewise covered with the same kind of herbage, growing in a similar soil. Further on, the steppe along the right bank becomes worse and worse, producing a scanty wormwood, and at last even "kali" between the mouths of the Boroldai and Badam, principally *Anabasis aphylla*; as yet there is no regular Salicornia. The highest limit of the "issegik" is here about 1550 feet above the level of the sea, near the mouth of the Boroldai; but it grows only on the lower decline of the steppe, close to the river itself, this decline being intersected by numerous waterless and precipitous ravines. The upper land, rising only 200 feet above the lower, is a rugged elevation, with a drifted sandy argillaceous soil, very dry and overgrown with scanty vegetation, consisting principally of *Festuca ovina*, interspersed with leguminous plants, particularly prickly *Alhagi camelorum*, an unwinding convolvulus, and small and meagre bushes of hawthorn (*Atraphaxis*). There is also a small proportion of wormwood. This elevation extends from the Boroldai Range, between the Boroldai and Arys rivers on the south, and the Bugun on the north. Similar steppe uplands rise also at the bases of the mountain ranges, and between all the affluents of the Arys. But to the north of this river each of these uplands assumes the form of several rows of rounded and gently sloping hillocks. Towards the south their form is that of elevated table-lands, intersected by deep and precipitous hollows, and they appear to me to have relation to the height of the mountains whence these deposits have been blown by the wind. To the south of the Arys especially the mountains are much higher, and the amount of deposit there is proportionately greater.

The valley of the Bugun into which we passed from the Arys, on our way back from the neighbourhood of Chemkend, is also occupied by meadows both natural and artificial, sprinkled with groups of willows, like the valley of the Arys. The cultivated plants are identical with those of the latter valley, but are not so fine in their growth, with the exception of the water-melons, which ripen earlier. The soil is not so dark, and particularly the irrigation is less effective, as the Bugun Rivulet has very little water; at 30 versts from its junction with the Chilik its bed is dry. Fifteen versts higher up the water stands in pools; but there even only where springs ooze up. These become much more numerous towards the mountains, where there is a tolerably full stream. The Chilik possesses a more copious current; it flows from the higher part of the Kara-taù, near the head waters of the Bugun. Between the Kulan and Boroldai ranges, with their spurs to the south, and the properly so called Kara-taù group to the north-west, there is, connecting them, an elevated plateau, not higher, however, than 3500 feet above the level of the sea, ploughed by deep hollows. These hollows are overgrown with bushes, principally hawthorn, which occurs here in the form of trees, with straight and thick trunks, reaching a height of 2½ fathoms, and attaining a thickness of 1½ foot. This plateau presents two slightly sloping sides —one towards E.N.E., the other towards W.S.W., both terminating in steep declines. The western slope is intersected by the Bugun, which first flows westward, then southward, and finally westward again, continuing in that direction after issuing from the mountains. At the foot of the western slope there is a sub-range of low hills of crystalline limestone, intersected by the Bugun. Dipping E.N.E. towards this mountain-range are alternating strata of crystalline limestone and clay schist; then between the base of the range and the bed of the Bugun, from north to south, non-crystalline limestone, a blackish limestone-conglomerate, and again a greyish crystalline unstratified limestone. The road here crosses to the left bank of the Bugun, and the upper surface of the plateau discloses nothing but black conglomerate, the strata of which dip towards the west and extend N.N.W., S.S.E.

The eastern declivity abounds with springs oozing from under the conglomerate which forms the summit of the pass. The waterline is not formed by it, but by a gradient above it intersected by gulleys following a westerly direction. The waterline itself forms at the foot of the conglomerate slope a plain of 2 versts in breadth, with waterless springs; at its eastern edge are gulleys, with bushes, chiefly of hawthorn, among which the roe (*Cervus capreolus Var. pygarga*) is not uncommon. In these gulleys are to be seen deposits of limestone upon metamorphic clay-schist, from which issue streams running through the foremost range to the north-east, the commencement of the Ketchkene Kara-taù (Little Kara-taù); two similar ranges, a smaller one to the north of a larger or chief range, with an intervening longitudinal valley,

Forty versts west of the Turlan Pass, as far as Suzak, the Kara-taù preserves its altitude. Here the *Juniperus pseudosabina* grows to the size of a large tree—large, that is, for this region. The trunks yield timber of 2 fathoms in length and about 10 inches in cross-section. I may at the same time observe that the dimensions of this tree become greater and greater towards the west. In the vicinity of the eastern extremity of the lake of Issyk-kul the trunks of this tree are very thick; but they lie along the ground, the branches alone stretching upwards. The erect trunks in the Buam defile astonished Mr. Semenof. At Ourianda, in the Kirghiznin-Alataù, these trunks occasionally yield timber 2 fathoms long; but it is generally of shorter lengths, not more than 2 or 3 yards, though still of considerable thickness. Further to the south-west, on the Kara-bura, the trunks are generally high, viz., from $1\frac{1}{2}$ to 2 and even to 3 fathoms; and at Chemkend, still further to the west, I saw timber of 3 fathoms in length of the same juniper, which would lead one to suppose that there are still larger specimens to be found. As this timber was from 8 to 10 inches in thickness in the bark, the tree must be 2 feet thick. These timbers were brought from the heights of the Badam; I was not there, so I have not seen the trees growing. But the tall common junipers that I saw have generally a trunk without any branches below half-way up; above this height the branches are numerous, short (one-fourth, or less, of the height of the tree), and of slight thickness; the top of the tree is generally broken off. The increasing size of the juniper towards the west in the Thian-Shan corresponds with Darwin's theory of the mutual competition of species of animal and vegetable life. As I did not see the juniper in the regions of the *Picea Shrenkiana*, I can imagine no other explanation. I will only observe that, in the lower woodlands on the mountains, its exclusion is not attributable to physical conditions, but to the presence of the fir and birch. On the Kara-archa it grows sporadically at an elevation of 3000 feet below the birch, and above it on the crags; and, what is remarkable at this extreme of the lowest limit, it lies along the ground, as it does at the uppermost limits. The best and thickest specimens that I saw were at an elevation of 7000 to 8000 feet.

Near Suzak the submontane region of the Kara-taù is covered with forests of saxaùl, which does not attain any great elevation. Farther westward the range is woodless, and gradually descends. Its heights have not been measured.

The Ketch-kene Kara-taù terminates between the Kumyr-Tas River and the Turlan Pass. At its highest part the Kara-taù is a single range; but to the west, from Turkestan, another range of not very high mountains stretches along the Kara-taù, parallel

to the main range, and on its south side only. The
declivity falls precipitately towards the steppe.

The steppe between the Kara-taù and the Syr-Daria i:
of cultivation only along the small streams, where there :
already cultivated. The further to the west that thi
extends the lower it inclines, and the more saline and
becomes. The best places to the north of the Arys ar
Arys-tandy Rivulet, between the Chilik and Turkestan
the south from the Turlan Pass. Further westward the
running from the Kara-taù are shorter and shorter, til
they are only springs. All these streams lose themselv(
steppe before reaching the Syr-Daria. Between them
Syr-Daria to the west of Yany-kurgan is a dense growth o
which at Suzak, on the northern side of the Kara-taù,
much further east to the base of the mountains; but beyo
castward, it again becomes scanty.

Between Cholak and Aulié-ata, the Ketch-kene Kara
scends towards the steppe in wide and rather steep g
furrowed by dry gulleys. These gradients are arid and
and consist of sedimentary formations. At their feet are
number of salines; and there are also salt lakes. The
issuing from the Ketch-kene Kara-taù are few, and flow
the gulleys of the outlying hills of drift, losing the
almost at their very outlet into the steppe. At Cho
Suzak only, are there any means of irrigation. At thes
there are cultivated fields and gardens; there is bru
such as willow, growing by the rivers; hawthorn occu
in the steppe, saxaùl and "djuzgun" grow at some
from the feet of the mountains, in the sands by the lower
of the Talas and Chu.

The neighbourhood of Biliu-kul is likewise saline;
salines *Salicornia herbacea* is found; but the lake itself is
fresh, though salt is cast up along its level banks, whicl
spring are inundated by its waters. The lake itself is
fringed with reeds, affording shelter to innumerable wil
and occasionally to tigers. There are large quantities
geese, herons, sui$_{pe}$ and pheasants; but very few wild-d

Biliu-kul is 1500 feet above the level of the se
length is about 20 versts, its breadth from north-v
south-east is 8 versts, and its form that of a paralle

Urtak-taù defile, but that a very considerable one, is the Tersa. Thus the Asa receives all the waters from the northern decline of the Urtak-taù, between the sources of the river Karabura, disemboguing into the Talas; and the sources of the Arys, flowing from a region of 45 versts in extent, of perpetual snow. But the waters being almost wholly derived from the melting snows, they subside in a very great measure towards the end of summer. The Arys is scarcely at all fed by perpetual snows, but almost entirely by those that fall during the winter, spring, and autumn; the water from which, after a thaw, being absorbed by the limestone, gives birth to countless numbers of perennial springs. The Arys has a far larger volume of water than the Asa, and sustains it in summer much more than the latter.

The Arys draws its principal tributaries also from the western Urtak-taù, chiefly from the range that separates itself from the Urtak-taù at the sources of the Mashat, within 30 versts to the s.s.w. of the place called Yaski-chù. This range proceeds in a south-westerly direction, terminating within 15 versts of the Chirchik, and 30 versts w.n.w. of Tashkend. From this range, within 20 versts south-west of the sources of the Mashat, flows likewise the Badam, the largest of the affluents of the Arys. Its course is 90 to 100 versts in length, if not more; its sources are believed to lie within 70 versts of Chemkend, which is a considerable exaggeration; and its embouchure is 45 versts from Chemkend; of the latter portion a survey has been made. In this calculation, the length of the valley is of course reckoned; not the course of the stream, which at every hundred fathoms makes several abrupt turns. At the sources of the Badam, striking off from the last-named range, directly to the west, is a not very high, but a precipitous, narrow, and lengthy spur, 45 versts in length, called the Kazy-kurt-ata; giving rise to the left tributaries of the Badam. Leaving the main range, it gradually descends, but again rises at its extremity, and ends in a steep double-crested peak, standing twice as high above the plain as the portion of the range immediately contiguous. Instead of rain, snow begins in September to descend upon the summit, and falls till November; it does not, however, lie upon it continuously, but frequently disappears altogether, thereby showing that the elevation is about 7000 feet. It is this elevation alone, and consequently not the range in which it terminates, that is known to the natives by the appellation of "Kazy-kurt-ata;" * which

---

* The solitary height of Kazy-kurt-ata, is considered, by the people of Chemkend and Sairam and the neighbouring Kirghizes, to be the identical mountain upon which Noah's ark rested.

unbroken, though tortuous, water-limit, of which tl
kurt was accounted the actual termination. But
the unbroken continuation of the mountain-ridge
direction, from the very sources of the Kara-bura,
western extremity of the Urtak-taù should be recko
the range between the Arys * and the Mashat, as a
scribed. As regards the Kazy-kurt; it separates the s
the Badam, and consequently those of the Arys, from tl
course of the Keles, which probably falls into the
between Tashkend and the Syr-Daria.† It is certain
Keles communicates with the Chilik by means of cana
the purpose of irrigating Tashkend and its neighbourhoo
the mountain properly called Kazy-kurt-ata, there is a f
tion separating the waters of the Arys and Chirchik ;
lofty peak standing alone, to the south-west of Chemk
to the south-east of the junction of the Badam with the

As regards the range stretching from the source of th
to the south-west towards the Chirchik, it separates t
from a considerable mountain stream; between which,
Chirchik there is another snowy range.

My geological map was drawn according to the views
from the valleys of the Chirchik, Keles, Badam, Mas
Arys, from the neighbourhoods of Chemkend and Ta
but only two ranges were drawn correctly ; one to t
of the Chatkal (Upper Chirchik), and the other to th
the Keles; they are both parallel to one another. Th
be between them some other parallel ranges and sever
tain streams of considerable dimensions, of which at
only the one nearest to the Chirchik is known to me.

---

* The more so, as its water-limit is of more importance than that of
kurt, i. e. that lying between the Syr-Daria and the steppe rivers. At t
of the Arys it passes from the Urtak-taù to the Kulan Kange.

† Up to the year 1864 the Keles was shown on all maps to disembo
diately into the Syr-Daria.

The steppes at the bases of the mountains just described, south of the Arys, are, so far as I could observe, all covered with an equally fertile soil, judging from their spontaneous vegetable productions. These consist of various thickly growing grasses of *Alhagi camelorum*, and many other varieties, chiefly of siliquose and many-coloured plants. There is no brushwood whatever. Cultivation on these steppes differs very greatly; as agriculture depends upon the local conditions respecting irrigation, as is the case in the whole district under consideration. It is more energetically pursued among the denser population between the rivers Arys, Badam and Mashat, and the mountains in which the two last rise; besides which, for purposes of irrigation, there are numerous other small streams between the Mashat and the Badam, fed from springs, and issuing from the hollows between the foot of the mountains and the Arys; such, for instance, as the Biuriudjar, a right affluent of the Badam, and the copious streams in these hollows, as, for instance, close to Chemkend, where from a single basin of spring-water there issues a stream in itself sufficient to irrigate all the gardens of the town, and also to work several mills in the neighbourhood.

There is again a fair system of irrigation, and accordingly a settled population, between the Badam and its left tributary the Sasyk, issuing from the Kazy-kurt. These districts about the Arys, Badam and Mashat, form the granary of the former Khanat of Tashkend; and from hence over and above the quantity retained for local requirements, supplies of corn are exported to Aulié-ata, Turkestan and Tashkend. In addition to what has been stated above as being grown in the valley of the Arys, cotton and sesamum are cultivated along the Badam and Mashat. Here, in the vicinity of Mankend, is the northern limit of cotton; but it can be grown even farther north, to judge from the experiments made by Mr. Kuznetsof at Almaty. Even at Gurief, at the mouth of the Ural, the seeds of cotton (though of a grassy kind, as it is generally in these parts) that were sown, grew up very well. But at Almaty, the capsules are smaller, and the stalk is shorter than at Chemkend. The Chemkend cotton, in its turn, is inferior in the size of its capsules, and the length of its stalk to that of Tashkend; which again is inferior to that of Bokhara. The Gurief cotton is, however, so bad, that its price in the market cannot repay the cost of cultivation. Even a grassy cotton requires a prolonged summer. At Chemkend it flowers in July and undeveloped capsules appear even then; but these ripen at the end of September, or more generally in October. At Tashkend, in the beginning of October I saw in one and the same field mature and unripe capsules, and even flowers. The cotton here outlived the frost of the (15th) 27th September, which killed the tendrils of the melons and water-melons. At Chemkend, however, one-third of the cotton perished in October before the capsules ripened; and at Almaty, probably one-half is destroyed by the frost; at Gurief, a still larger proportion. At Mankend, near Chemkend, I saw the gathered capsules; they are gradually picked as they arrive at maturity—an operation facilitated by the sub-division of the fields for the purpose of irrigation into squares of 1½ to 2 fathoms, separated by earth-mounds. Only the faded capsules are picked; and they are then left in the sun to burst.

These observations may serve toward the determination of the question of the acclimatization of grassy cotton in the south of Russia. I could not at the time have pursued the subject further in detail, without neglecting the real objects of my mission.

In the town-gardens between the Arys and the Badam, are grown grapes, peaches, apricots (ouriuk) and garden "djida." [*] Walnut and mulberry-trees also flourish; but I did not hear that the inhabitants of these parts attend to the rearing of the silk-worm, like the settled population of Kokan to the south of the Syr-Daria is known to do. It is therefore to be that if the silkworm is reared here at all, it is not done so very generally.

The outlines of the steppe south of the Arys, and gulleys, have already been dwelt upon. It may be said tion respecting these gulleys, that they are frequently of great width, *i.e.*, of 200 or 300 fathoms, and even of a verst broad. We may conclude from the fact of irrigation canals running from the gulleys, that from the Arys the steppe descends, not in one gradual slope, but by gradients. The rivers flow in deep valleys; and the affluents of the Arys are deeper than that river itself. At Mashat, both banks are high; in some places there are even overhanging cliffs standing 500 feet above the river; as for example, by the road from Aulié-ata to Chemkend. At Badam, the left margin of the valley is rugged and precipitous, but not higher than 100 feet. From this, towards the water-line from the Kazy-kurt, there is a very considerable slope extending 20 versts, with several abrupt gulleys. The chief valley here is that of the Sasyk, which river flows from the Kazy-kurt to the Badam, but reaches the latter only in spring. Both sides of the valley of

---

[*] *Elœagnus hortensis*, bearing a fruit of the size of the olive, of a yellowish-red colour. The wild "djida" (*Elœagnus angustifolia*) bears a smaller fruit, of a greyish-green colour even when ripe.

the **Arys,** as above stated, are abrupt, though of no great height.

The localities to the south of the Badam, I generally surveyed only from the road leading from Chemkend to Tashkend, in the beginning of October (O. S). Beyond the Sasyk, this road trends into the eastern outlying rocky mountains of the Kazy-kurt ; and leads out of them again along the banks of the affluent of the Keles. Here the steppe is rough and wavy ; it rises also in sloping hills, which fall, however, rather precipitously towards the Keles and its affluents. The Keles is 6 to 7 fathoms in breadth, and is shallow, being rarely deeper than $1\frac{1}{2}$ foot ; and at the utmost is $4\frac{1}{2}$ feet deep in its occasional pools. Its banks in the valley are abrupt and very low, viz., about 3 feet in height. The current is moderately rapid ; the bed consists of either mud or sand alternately. The waters of the Keles occasionally approaching the margin of the valley, sometimes lave the bases of abrupt crags about 10 fathoms in height, generally of sedimentary origin, and partly composed of the denudations of red sandstone mentioned above. The valley of the Keles is 2 versts wide, and displays numerous cornfields. .There is not so much lucern grown in this valley ; and as to cotton, I saw none at all. The soil is oozy, as on the Arys ; but less loamy ; there is also less irrigation, and the crops are inferior. Besides those in the valley of the Keles, there are fields also on its left bank which is considerably more depressed than its right, and is therefore capable of irrigation by means of canals cut from the river higher up. The steppe is, speaking generally, covered with a fertile soil, and by a tolerably dense growth of herbaceous plants ; but towards Tashkend it is less productive, owing to the predominant element of red clay.

Tashkend is built in the broad valley of the Chirchik, which is about 20 versts in breadth ; its northern side is not more than 25 to 30 fathoms above the bottom of the valley, but this valley is very steep. Tashkend is irrigated by means of canals cut from the Chirchik at Fort Niaz-bek.

The Chirchik flows within 8 versts of the town, which stands near the northern edge of the valley. I saw Tashkend from the outside, though from no great distance. . The nearest elevation (of which there are many in the broad valley, the whole of its surface being uneven), from whence I could overlook the interior of the town, stands at a distance of 100 fathoms from it. The town has the appearance of a large forest, with here and there a mud wall, and a flat mud roof visible amid the foliage of the trees. It is filled with gardens which screen the small houses from sight. There are two large buildings to be seen, and there are no monuments whatever.

The Mesjids, of which there are said to be a great number, according to some 50, to others 500, and even 2000, cannot at a distance be distinguished from the houses, and they do not rise above the garden poplars.

The town is enclosed by a mud wall, with barbets, surrounded by a deep trench, in some places filled with water from the irrigation canal, in some parts dry, as it is dug in uneven ground. Near the southern side of this enclosure, is a citadel, which at sight may be said to be 300 fathoms in length, and which is surrounded by another wall half-dilapidated, on the side of the town enclosure, which is here consequently stronger, and is kept in better repair; and the strong walls of the citadel overlooking the interior of the town are likewise carried on to the town enclosure. The space within the town walls is about equal to that of Moscow. The results of a survey show it to be about 12 versts in length from east to west, and 7 versts in breadth from north to south. Allowing, as I was assured was the case in 1858, that half the houses are there must be no less than 100,000 inhabitants. But there are many houses with gardens outside the walls of Tashkend; and those nearest the town are as close together as the courts with their houses and gardens are within the town itself. They form regular suburban villages, as, for instance, the one that lines the Niazbek road,* extending 4 versts; and the one on the Kokan road, stretching 5 versts.

Between these gardens and villas are fields of lucern and corn, but chiefly of cotton, sesame, and zedoary. I could see no madder; and even of the cultivation of the zedoary I speak only from accounts received from enquiry.

About Tashkend there is more cotton than anything else cultivated. Of cereals, rice is extensively grown; but far too little wheat for the requirements of the whole town. The latter is brought from the Keles, especially from the vicinity of Chemkend. The Tashkendians pay more attention to saleable products for exportation, and to their own manufactures. Mulberry-trees are common, and vines are abundant; as well as all those fruit-trees that occur in the towns between the Arys and Kazy-kurt; to these must, however, be added the fig-tree, which is not to be found in the latter places. It probably finds its most northern limit in the Trans-Chu region, in the neighbourhood of Tashkend, for at the beginning of October (O.S.) the fruit was not quite ripe, although then extremely sweet.

A few words concerning the towns of Kokan may he

---

* Niaz-bek is a fort on the Chirchik, within 15 versts south-east of Tashkend, commanding the canals which are conducted towards it from out of the Chirchik.

inappropriate. They differ from each other almost only in size, and perhaps also in the number of their gardens. To the north of the Arys, there are not so many gardens in the interior of the towns as in their outskirts; as, for example, at Aulié-ata and Azret.* In the last-named place there is hardly a single garden; they are all outside; and even those that I saw in 1858, under the outer side of the wall, have since been dug up. On the other hand, all thch towns to the south of the Arys present the appearance of aving large gardens; and their unsightly buildings are hid among the trees. These buildings require no description; they are all of mud, all look eastwards, are one-storied, and flat-roofed, and are without windows towards the streets. I will only remark that there is no communication between the rooms, which open upon the courtyard by several doors. Before the doors of each house is a common shed, supported by posts, beneath which the inmates are generally found sitting, if not reclining upon the embankment facing the street. The doors generally exhibit carved devices; the windows are trelissed with wood, but are without glass; in the winter they are pasted up with oilcd-paper; one aperture, where there is no regular stove, being left open for the escape of the smoke from the fire burning upon the mud floor. The stove is, properly speaking, a four-cornered chimney, with a wide opening at the bottom. When a fire is made in it, the least puff of wind fills the room with smoke. There are many recesses in the walls, which serve for cupboards.

I need not enumerate all these towns, as they are marked on good maps of Turkestan. Something might be said of their bazaars, their trade, and their monuments.† Of the latter, there are only two in all the towns put together, viz., the Holy Mesjids in Aulié-ata and at Azret. The last place I did not visit in 1858.

This article is already drawn out to a great length, I must therefore conclude the topographical portion of it; and, in con-formity with its physico-geographical character, I shall proceed to give the hearsay statements I obtained concerning the River Chirchik.

This river issues from the mountain 10 versts above Niaz-bek, through a very narrow and impassable defile. The road from Tashkend to its sources does not pass through this defile,

* Improperly called Turkestan, the ancient name of the countries between the Chu and Syr-Daria (Western Turkestan), and along the River Tarim (Eastern Chinese Turkestan).

† Official statistics relative to the towns of the Turkestan province are being now compiled; but, when I was there, those relative to Aulié-ata were alone completed.

irrigation canals, it is nowhere fordable. To the south of
kend the river forms swamps, which are partially sov..
rice. Lower down, there are again lagoons, through w...
river passes in a single stream, 120 feet wide without an...
and so disembogues into the Syr-Daria. The course ..
Chirchik, after issuing from the lagoons, is not more...
10 versts, if so much; but it flows rapidly.

In consequence of these lagoons, the Chirchik is na...
only below them, and therefore for a shorter extent tha...
Arys, which is navigable for steamers for at least 30 v...
if not more, from its mouth upwards. Yet the Chi...
is a more copious river than the Arys; and if a c
were to be cut through its flooded parts, overgrown with r...
the rapidity of the current would impede navigation highe...
if only to Tashkend. I judge of the rapidity of this st...
from the principal canal cut from it; the water in which,...
at Tashkend, rushes along like a mountain-torrent; and,...
the canal was constructed in very remote times, it has deep
its hard bed of limestone and sandstone.

Between the Lower Chirchik and the Syr-Daria tha...
another island, visible from the neighbourhood of Tashk...
not very elevated, but detached and intersected by val...
which are crescent-shaped and radiate from its centre. Bet...
this and the nearest spurs of the Western Thian-Shan, w...
lie south of the issue of the Chirchik through the defile,
saddle-shaped opening, through which passes the road
Tashkend to Kokan and Namangan. Another road
Tashkend to Kokan and Khodjend, passes round this high
from the west, along the Syr-Daria. The mountain valley
tween these roads are occupied by the Kurama tribe, belon...
according to some accounts, to the Uzbeg, and, accordin...
others, and to the result of my own inquiries, to the Ki...
race.

Let us now embrace, in a brief physico-geographical sum...
of the Trans-Chu region, all the particulars concerning it v
have just been obtained, so that the reader may himself see
far these observations serve to correct the too numerous
conceptions relative to Central Asia, and especially relati...
the mountainous country between the Chu and the We...

Himalayas, or Tsun-lin, in the recognised wide Chinese * sense of that appellation.

The first of these misconceptions arises in regard to the connexion of the Kara-taù with the Bolor. From lat. 34° to 45° N., between long. 71° and 69° E. from Paris, Humboldt supposes there to exist one unbroken chain, the Bolor, intersected by the Syr-Daria at about lat. 41° 40′, and long. 69½°; and that further south, at about lat. 40° 20′, this range is intersected by the Thian-Shan. Between the Thian-Shan and the Hindookush alone, it is called the Bolor; to the north of the Syr-Daria, it bears the name of "*Kozy-urt.*" The Thian-Shan not only intersects this range to the south of the Syr-Daria, but also connects itself with it in the north by its spurs. From the south bank of the Issyk-kul to the point of its intersection with the Bolor, the Thian-Shan proceeds in a south-westerly direction between lat. 42° and 40° N., and is called the Gakshal-taù and Terek-taù; westwards of that intersection it is known under the appellation of the Ak-taù or Asfera, and stretches away directly to the west along the parallel.

Along the Terek-taù, towards W.N.W., run several mountain-chains, of which little is known; on Humboldt's map these are, consequently, traced indistinctly, without any names being given to them, with the exception of the extreme northern one, which is called the Burul, and of the western chain, called the Khubaboi, through which the Talas forces its way. This chain stretches in the form of an arc, first towards north-west, then, beyond the issue the Talas, to the south-west. All these chains abut on the meridional Kazy-kurt, from which, however, diverge also the western spurs of the Kendyr-taù, Ala-taù and Kara-taù. The latter is the most northerly of all, being in lat. 45° N.

This construction Humboldt bases upon the evidence of Son-Yun and Huen-Tszan, and upon the continuation of the Tsun-lin (*i.e.*, according to him the Bolor, supposing the other term to be the name of one mountain erroneously applied to a whole range), to the Myn-bulak district.

I may add, that from the pass over the Kara-taù at Cholak, and up to the very Syr-Daria in the vicinity of Kokan, mountains are continually visible in the distance on the left side of the road; extremities of a mountain-range which are not meridional are also seen.

It is only these extremities of mountain-ranges, and their more easterly intersection by the road from Aulié-ata to

---

* Humboldt, as we know, applies the name of Tsun-lin only to an inferior portion of this extent, viz., to the convergence of the Kuen-lun with the Bolor, the limit of which he defines to be a mountain-knot south of Sary-kul (Sea of Victoria), at the sources of the Yarkend-Daria and the Amu-Daria.

Namangan, that are taken into account in the latest maps,
the continuous ranges visible along the road from Cholak
Kokan. Humboldt's meridional range between the Chu
Syr-Daria therefore disappears from these maps; and th
remain only the mountains running w.n.w from the Terek-
Thus, on Kiepert's map of Turan to Ritter's Geography
Asia, to indicate the absence of information, only detac
fragments of ranges are traced at the points of intersection
military routes; on the strength of which this portion of
map was constructed.

On the maps constructed on the basis of investigations m
by the Department of the General Staff of Western Sibe
these fragments are connected into long parallel ranges;
for example, on the first map of Kokan appended to the Jour
of the Geographical Society, and on the 4-sheet map of Cen
Asia.

In his article on the Bolor, Mr. Veniukof also rejects
northern prolongation of the Bolor beyond the Syr-Daria,
does not trace it on his map of Kokan,† which is nearer
truth than any of those previously published, even subsequen
to 1864, inasmuch as it indicated the prevailing direction of
mountain-ranges between the Chu and the Syr-Daria, to
from east to west, and not from south-east to north-west,
traced on other maps; ‡ and also as it shows their connex
with each other by means of short meridional chains in ma
places and under various latitudes. But in the western porti
to the west of the meridian of Aulié-ata, the direction of
mountains is only N.E., S.E., W.N.W. And yet it may be se
from my notes of travel given above, that Humboldt is h
correct. For, firstly, the Kulan Range occurs exactly in a li
with another range of the Bolor system, which separates
Kara-kul table-land from that of the Rian-kul, besides int
secting the Ak-taù, and then continuing toward the nor
to the Syr-Daria. The direction of these ranges is as
N.N.W. Secondly, there is a range of mountains extendi
between the Syr-Daria and the Kulan ranges, in the directi

---

* 'Memoirs of the Imperial Russian Geographical Society, 1861,' Book
page 160 :—" It is hardly possible to do otherwise than to detach the Kazy-k
and Boroldai from the system of the Bolor, and apply them to the offshoots of
Thian-Shan." This is true as regards the Kazy-kurt; but the Boroldai belo
to the system of the Kara-taù, the relation of which to the Thian-Shan and B
I will shortly explain.

† 'Memoirs of the Imperial Russian Geographical Society,' Book I. 1862.

‡ For instance, on the maps just alluded to, viz., Kiepert's, the 4-sheet map
Central Asia, published by the Geographical Society; Semenof's map in Pet
mann's 'Mittheilungen,' 18; on the map of the Orenburg region and south
portions of Central Asia, on the scale of 100 versts to the inch; on the map
Western Siberia, scale 50 versts to the inch.

of south-east, south-west, and nearing the meridional range. This is the Karjanyn-taù Range, running along the Keles between that river and the Chirchik. So that one who has not been an eye-witness, but has availed himself of indefinite Asiatic itineraries alone, may easily take this range to be a connecting link between the Kulan and the Bolor. But Humboldt must, of course, have been unaware that the direction of the Karjanyn-taù, with that of the Kulan Range, forms nearly a right angle.[*]

The more serious inaccuracy in his geographical deductions is the imaginary consecutiveness of the Northern Bolor, and even of the whole of that range generally. Arguing from the data we at present possess respecting the Central Asiatic mountain region (*i.e.*, the Tsun-lin of the ancient Chinese), the Bolor, in the sense of a distinct range, does not exist; and the mountains so called ought to be classed with the Himalayan system.

Each system consists of several ranges, following various directions, but all connected together. These ranges are distinguished by the directions in which they proceed, or, rather, by the direction in which their strata extend. The mountain chains appear to me, however, to be partly an artificial, schematistic arrangement, not quite in keeping with nature. Properly, those places on the map should be marked where there are actual ranges, that is, where there are upheaved strata preserving a uniform extension for a great distance; and also those places where the lines of the extension of these uplifted, contorted, and broken strata frequently cross each other, and where, consequently, many conterminous elevations of inconsiderable magnitudes, as regards their horizontal extension, unite in one vast contiguous mass of highlands, reft, however, by the narrow chasms of river valleys. Such an elevation the Tsun-lin appears to me to be, as it is formed by the intersection of the Thian-Shan system with elevations from the Himalayas. It would be more correct to represent it without any ridges, and to shade it more or less darkly, according to its height, with the river valleys, and with lines marking the direction and extent of the various strata; but for such a delineation, which would I believe be the more regular one, the geological observations have not been sufficiently numerous. No minute surveys have ever been ma e of the mountains to the south and south-west of the Chd.

I. Of the Western Thian-Shan, I am geologically acquainted with three ranges: the Western-Trans-Ili-Alataù, at the

---

[*] Among the 'Memoirs of the Imperial Geographical Society, 1862,' Book III., is a chart without any text, on which Mr. Veniukof's system of the Trans-Chui Range is connected with that of Humboldt.

Kastek, the Kirghiznin-Alataù, and the Urtak-ta
the points at which I saw them, neither of them
the appearance of independent ranges; that is, of   ges in
which upheaved sedimentary strata occur on both sides of the
crystalline axis which has upraised them. On the northern
declivity of the Kirghiznin-Alataù sedimentary strata pre-
ponderate; and on the southern slope crystalline formations,
alternating, however, with sedimentary. The same may be
said of the western portion of the Kungé-Alataù, between Tokmak and Kastek. In both these ranges crystalline
formations, granites and syenitcs, cropping up at the southernmost bases, connect each of these chains with the next towards

near the mouth of the Kara-kyspak, opening into the Chatkal.
The connexion is, as we have seen, exemplified by the sedimentary strata of the range; for example, some of these falli
away from a given range, and towards the nearest crystalli
formation.

The upheaved strata were observed to extend principally
three directions, viz. :—

1st, N.E., S.W., to which also refer directions to E.N.E., W.S.W.,
and N.N.E., S.S.W. The latter direction occurs only at one
place, the confluence of the sources of the Talas; the t
in the southern declivity of the Kirghiznin-Alataù, in the
Cha-archa Range, which geologically forms a part of it ; and in
both of the slopes of the Urtak-taù. In these directions
metamorphic schist is upheaved, with intermediate strata of
dark limestone and sandstone; in which, likewise, no organic
remains were discovered.

2nd, E., W., and E.S.E., W.N.W., sedimentary strata on the
northern slope of the Kirghiznin-Alataù; crystalline formations
on both of its ridges, and along the eastern summit of the Kara-
kyspak. In this direction there is an upheaval of sandstone,
exhibiting on the River Merké organic remains, chiefly of the
Spirifer, and occasionally *Productus*.

3rd, N.W., S.E., the least frequent, and only by the s
Such, in w. 30° N., E. 30° s., is the direction of the sands
with rock-salt, in the northern outlying mountains of
Kirghiznin-Alataù, between Ala-medyr and Aia-archa ; and of
the limestone in the western extremity of the Urtak-taù,
between the Arys and the Mashat. This extension in the
Thian-Shan is generally hardly distinguishable from the fore-

going, although here and there are traces of an independent upheaval.

It appears to me, from the nature of the upheaved formations, that the first of these three principal directions, viz., N.E., S.W., and the direction nearest to it, is the most ancient. The absence of organic remains does not, however, allow of the accurate determination of the geological period. I nowhere found limestones with spirifers resting on metamorphic and other alternating formations; but the relative extension of the strata in the Urtak-taù and the Kirghiznin-Alataù, and its general direction, seem to me to be favourable to the opinion that the upheavals north-east, south-west, happened earlier than those east, west, and E.S.E., W.N.W. The latter either abut on the extremities of the first, as on the northern slope of the Kirghiznin-Alataù, or intersect them, as on the water-parting of the Urtak-taù. Both of these systems of upheaval is illustrated by the valleys, previously clefts, parallel with and perpendicular to the upheavals. For instance, in a direction north-east, south-west, lies the valley of the upper Chirchik; in a line perpendicular to this direction are the valleys of the affluents of that river. In a direction east, west, lie the valley of the Talas, and the valleys of the affluents of the Kara-bura and Kara-kyspak; the valleys of the two last lie perpendicularly to that of the first-named river. The valleys of the Talas and Chirchik have been widened, but the others have preserved their original character of clefts.

The limestone with spirifers on the River Merké, upheaved in a direction from east to west, seemed to me to belong to the mountain-limestone formation (calcareous); but it may be Devonian. In Russia, however (for example, in the province of Tula), these two formations are not distinctly separated, but are intermixed, both lithologically and as respects the common *habitus* of their fossils—the difference is hardly perceptible to the uninitiated, except that there is one striking *Productus giganteus* of the mountain limestone, which does not exist in the neighbouring Devonian limestone. Neither did I find this fossil on the Merké, where there are small specimens of the *Productus*. Evidence of an independent upheaval of such strata as reddish sandstone and conglomerate, extending w. 30° N. by E. 30° S., is found, for example, in the outlying hills of the Kirghiznin-Alataù, on the Ouriandas, and between the Alamedyn and Issyk-ata, where these strata dip towards the limestone; but on the Ouriandas it is apparent that these sandstone strata are bent and undulating. There is a denudation of veins of granite in the bottom of the valley, at a hill formed of strata of this sandstone.

On the River Merké the strata of sandstone and [illegible] their point of contact are upheaved perpendicularly, [illegible] are entirely overturned. This dislocation is not [illegible] both sides of the cleft; on the right or eastern side [illegible] stone extends further to the south, which indicate [illegible] upheaval was very limited in extent, in a direction [illegible] s. 60° E., and that it occurred after the deposit of [illegible] and certainly not earlier than the Permian period.

This is confirmed by the coal-beds in the Kara-taü, whi[illegible] a direction from east to west, and by the coal-seams [illegible] between the mountain-ranges stretching in a similar [illegible] east and west, along both sides of the valley of the Riv[illegible]

Generally speaking, as has been already observed, [illegible] tion from east to west passes imperceptibly enough [illegible] direction of south-east, north-west, and connects, by [illegible] a complete series of intermediate rhumbs, even the p[illegible] opposing lines of elevation, north-east, south-west, and [illegible] east, north-west.

There are also, in the Thian-Shan system, traces [illegible] ridges almost meridional. I saw one of the kind as [illegible] range in the distance, running in the direction [illegible] w.s.w., at the sources of the Chirchik. I had not t[illegible] tunity of exploring it geologically; but the strata at [illegible] fluence of the two sources of the Talas, where sye[illegible] diorites alternate with the limestone and micaceous [illegible] laceous schists, have a similar extension, only not along [illegible] of the range at the upper source of the Chirchik, but [illegible] east. Consequently, such almost meridional ranges may [illegible] some degree of probability, be referred to the period [illegible] most remote upheavals north-east, south-west.

Corresponding with the direction of the strata along [illegible] Kara-bura and Kara-kyspak is the direction of the [illegible] striking off from the Urtak-taü to the south, parallel with [illegible] Chirchik. This direction bears E. 40° N., w. 40° s.

The Terek-taü, an offshoot of the Thian-Shan, between the sources of the Naryn and Kashgar-Daria, has generally the same direction; but here there are a great number of minor ranges running in other directions.

The northern range of the Kirghiznin-Alataü has a direction east, west, and E.S.E., W.N.W., the same as that of several short mountain chains of the southern range. Lateral valleys generally separate the range into the northern long range east to west, and into many southern short ranges, running in two directions, north-east, south-west, and south-east, north-west. Abutting on the northern range, at acute angles, the first-named direction predominates, and both are seen at the Upper Kara-k[illegible]

Eas to west is also the general direction of the Urtak-taù; but the water-parting itself is tortuous. Short ranges, with a direction E.S.E., W.N.W., and S.E., N.W., intersect the spaces between the more extended parallel chains north-east, south-west, and also invariably at the north-east extremities of the latter; so that across all these short intersecting ranges a direct line may be drawn east to west, which will be obviously the *central* direction of the range; while it will give an erroneous idea of the Urtak-taù, if drawn to characterise its configuration, without reference to the short mountain chains of which it is principally composed.

It may be seen by the map that the same directions exist in the western portion of the Thian-Shan, on the Susamir, at Son-kul, and at the upper course of the Chu River, which I have not visited. They also characterize generally the portion of the Thian-Shan system to the west of the meridian of Kute-maldy. This is a system of intersectional ranges; but, at the same time, in the Issyk-kul portion of the Thian-Shan and eastwards the predominant ranges are parallel, bearing E.N.E., W.S.W.

But there are intersectional ranges, having corresponding directions south-east, north-west, and north-east, south-west, even further to the south, in the upper course of the Oxus, as demonstrated in Veniukof's map.[*] All of this mountain region may, therefore, be considered as forming a whole: a fact observed, though not explained, by the Buddhist missionaries Son-Yun and Huen-Tszan. The determination by the latter of the boundaries of the Tsun-lin I am, from my personal observation, bound unconditionally to confirm.

This mountain region is divisible into two portions by a line drawn through lakes Kian-kul and Kara-kul. In both portions the directions of the intersectional ranges are alike, only in the northern half the preponderating direction is north-east, south-west, and in the southern south-east, north-west;[†] consequently,

---

[*] 'Bulletin of Russian Geographical Society, 1861-2.'

[†] This southern portion is then the Bolor in the strict sense, as it is accepted, for instance, by Kiepert. That it belongs to the Himalayan system has been stated by Veniukof ('Bulletin of Russian Geographical Society, 1861,' page 164). Apparently that marked distinction between the systems of the Bolor, Kuen-lun, Himalayas and Hindoo-kush, which is pointed out by Humboldt, does not actually exist. "The three first appear merged, as it were, into one common elevation, the axis of which stretches from north-west to south-east." According to the measurements of Schlagintweit ('Voyage,' Part II., Hypsometry), the highest range in this elevation is the central one of the Kara-korum, the Himalayas and the Kuen-Lun; the marginal ranges of the general elevation are lower, but have a direction from west to east. At Nepaul detached peaks of the Himalayas do rise above the Kara-korum, though the Himalayan Range does not exceed the latter in its general height. As far as the sources of the Indus, the Himalayan system, *i. e.* the elevations on both sides of the Kara-korum, presents, according to Cunning-

the northern portion is the Thian-Shan, and the southern the Himalayas.

Judging by the constant relations between the directions of ranges, and the general extension of their strata which I observed, one may form a general opinion as to the geological history of the whole of the Tsun-lin; notwithstanding that only a very small part of it, and that only in the Thian-Shan portion, has been geologically surveyed. This mountain district has been formed by an unintermitting and closely contiguous series of upheavals, each of which occupied no long period of time; indeed, many of these elevations running in a similar direction, were geologically contemporaneous, though, of course, they must not be considered as the result of a sudden simultaneous upheaval acting in different places. On the contrary, the process of upheaval was general, gradual and unremitting, although not equally so throughout the whole extent; as there were short and slowly-alternating directions of maximum pressure, and several of these during a given time. The unremitting nature of the process is demonstrated by the great variety of directions between the two marginal ones, intersecting each other at acute angles open to the north and south, N.E.E., S.S.W., and S.S.E., N.N.W.; the intermediate directions between between these two filling the obtuse angles opening towards the east and west. It may be supposed that by an uninterrupted process of upheaval the present Tsun-lin Mountains at first formed an archipelago of islands, and that these islands afterwards became joined together. As, however, their very traces are now hardly discernible, the bottom of the original strait probably became subsequently elevated above the original islands. But, judging from the distribution of the geological formations along the Kara-archa, Kara-bura and Karakyspak, it is probable that an unbroken continent was here formed by upheavals, north-east, south-west, prior to the deposit of limestone at Merké. Considerable upheavals occurred, after the deposit of limestone, in directions between east, west, and south-west, north-west, when the Tsun-lin attained its present elevations, and perhaps greater than exist at present; and

---

ham's map ('Ladak, Physical,' &c.) two different directions of mountain-chains, the principal ones being north-western and south-eastern ranges, lying north-east and south-west, which directions are similar to those in the Bolor Mountains. In the latter the north-western direction approaches nearer to the meridional than in the Himalayas; but on Kiepert's map, for example, this difference is far less considerable than on Veniukof's; yet the actual direction, *i. e.* its angle with the meridian, is not even determined: all that is known is that it lies north-west, south-east. Therefore the Bolor is not a distinct meridional range, but merely a north-western continuation of the Himalayas, or, more correctly, of the Himalayan branch of the Tsun-lin, which is a gigantic convexity, connecting, by means of gradual transitions, the system of the Thian-Shan with that of the Himalayas.

allowance should be made for the action of the wind on the débris, and subsequently of the water. Then came the formation of the gulfs, the present wide valleys of the Talas, of Ferghana at the Jaxartes, and that of Badakhshan at the upper course of the Oxus, as shewn by their directions from E.S.E. to W.N.W.; but their beds were raised above the level of the ancient sea, probably before the formation of the Permian strata, of which there are no denudations, at least in the valley of the Talas. The upheaval of the Permian formations on the northern and north-western limits of the Tsun-lin is inconsiderable, and denotes the termination of its general uplifting. There were still local upheavals towards east, west, and south-east, north-west, which at the northern base formed a long line of outlying mountains, and produced small ridges at the western base.

The geological periods subsequent to the Permian have left no traces whatever of their existence in the Tsun-lin. All that afterwards occurred was caused by the action of the wind and by the accretion of fluvial deposits.

The outlying hills of sedimentary formation, which, towards the glacial period, were raised above the level of the then existing Kirghiz Sea, as is proved by the large and scattered boulders near to the northern base of the Tsun-lin to a height of 2400 feet, may be attributed simply to the accumulation throughout the course of many geological periods—that is, of several millions of years—of detritus washed off from the highlands. At some future time some curious remains of continental organisms, which during early geological periods existed in the Tsun-lin, may be found in these deposits.

The observations respecting the traces of ancient glaciers made in the topographical portion of this paper may lead to the following general deductions:—

1st. The glaciers among the mountain ridges of the Tsun-lin surveyed by me have now disappeared, excepting, perhaps, the somewhat doubtful ones seen from afar by Captain Protsenko at the summits of the Ala-archa.

2nd. When they existed they did not descend below an elevation of 2500 feet. The lowest traces of moraines were observed on the northern slope of the Kirghiznin-Alataù, in the Makmala district; and evidently they were not mere accumulations of stones from landslips, as the boulders are crystalline, while the nearest hills (some, however, standing at a distance of 10 versts) are composed of sandstone. Lower than 2500 feet there are no boulders in the Thian-Shan. They are not found in the steppes, consequently they were not dispersed by icebergs; and either the sea, which at one time covered this steppe, did not attain to this elevation, or the glacier descend to the level of the then existing sea.

3rd. Large masses of non-stratified marl with bou discernible only at the bases of the mountains at presen with snow, or of those that approach the snow-line, are not less than 8000 feet in height, viz., the chair Thian-Shan system. There are no such formations mountains of the Kara-taù system—for instance, near tl Arys, or Bugun—but only fine gravel formed from local metamorphic formations, and clay.

4th. The highest limits of these ancient glaciers E in one place, viz., in the lower hollow of the Kar defile. The height of the snow-line during the glacia to judge from this locality, was about 8000 feet,—4 below the present snow-line. The perpetual snow wh covers the Khirgiznin-Alataù and Urtak-taù is not suf quantity to form glaciers, as it lies only on the sur detached peaks.

5th. The moraines are generally found welded toget one mass, exhibiting a wavy surface and being rent by which are sometimes, even at the present time, occasi rain and snow water. Some piles of boulders, marking tl of former moraines, rise just above the surface. This probability, to be attributed to the gradual washing the moraines, caused by the melting of the ancient and by the elevation of the snow-line. The various de the mountain valleys were also then washed away. Re these ancient glaciers and moraines, some may doubt these accumulations of boulders are not due to mount and to the subsequent action of water. But such . thesis is contradicted by the fact that the lowest limit boulders is 2500 feet. Mountain-streams washing throt and boulders, even at the present time, detach them as them a great distance; consequently, the volume o which was here greater than that of the present m streams, would have transported the boulders much be limit of 2500 feet, favoured, too, by a decline of 600 a 1000 feet in 20 versts between the mountains and t sent road from Tokmak to Aulié-ata. It therefore fell the force which set these boulders in motion was no but something that arrested them in their descent on a certain limit half-way down the mountain. Now, are thus propelled only by glaciers, the downward c which is checked by the thawing of their lower ends.

The formation of these ancient glaciers may be m

rectly associated with the former connexion of the Caspian Sea and the Sea of Aral with the Balkhash and the Arctic Ocean, the traces of which in the Kirghis steppe are mentioned by Humboldt.* I saw these traces in 1858, and farther east in 1864; but there is no occasion to enter here upon the hydrography of Eastern Europe and North-western Asia during the Glacial period.† I shall here only observe that the *average* annual temperature could, even then, have been little lower than at present; but the summer heats were not so great. Besides, the climate on the sea-coasts was evidently more moist than it is now; and this moisture, even with less frost than at present, was sufficient to produce upon the mountains masses of snow which are not now to be found.

There are no volcanic formations in the western portions of the Thian-Shan which I surveyed. From eastern sources Humboldt refers to evidences of volcanic action further south in the Ak-taù; but even these are doubtful. Fire may be produced in the mountains even by the ignition of the seams of coal, as well as of the carburetted hydrogen-gas filling the caverns of the seams. This conjecture is supported by the circumstance that Messrs. Bogoslovski and Lehmann discovered, on their journey to Bokhara, a burning seam of coal in the mountains at the Upper Zarafshan, a little to the south of the Ak-taù. Speaking generally of volcanic action in the Thian-Shan and the surrounding regions, the geological surveys hitherto made from Khan-tengri (east of Issyk-kul, near the sources of the Touba, Djirgalan, Tekes and Kegen) to the extreme western limits of the system, have given only negative results. To the east of Khan-tengri there are again seams of coal—for instance, at Kuldja, and perhaps also at Urumchi—the ignition of which is quite sufficient to create explosive gases. Whether the seams of coal were ignited at Urumchi by volcanic agency, or accidentally at their denudations, is a question that cannot be settled without close observation. It can only be said that the demonstrations in favour of volcanic action adduced by Humboldt are not sufficient proof of the volcanic origin of the Thian-Shan, excepting only as regards the lava, which, according to Chinese records, flowed from the Peshan Mountain during the sixth century. But a single crater—even if the fact of its exist-

* 'Central-Asien,' 1. Band, II. Theil, übers v. Mahlmann.
† I have only just commenced the study of the Glacial period in the Western Thian-Shan. A more minute explanation, plans, sections and views of moraines, on examination of the valleys where ancient glaciers existed, &c. &c., are yet to follow from another journey that I am about to undertake. I shall then be able to write more in detail on the Glacial period in Asia. Here I can, but for the first time, adduce proof that I am probably not mistaken in the discovery of his traces.

ence in an extensive mountain system extending, as the Thian-
Shan does, for 3000 versts can be proved—does not make
the whole of the range volcanic. Such a crater may belong to
a volcanic chain intersecting the Thian-Shan, and this chain
is, in reality, that of the Baikal.

Allusion may here be appropriately made to the relative
antiquity of the Thian-Shan, and to the elevation of the high
Gobi steppe from the upper course of the Onon through
Urga, and further to the south-west. This elevated plateau also
stretches from north-east to south-west. Humboldt considers it
to be the most ancient "upheaval" of Asia : of greater antiquity even than the great central Asiatic ranges. But the information communicated by Radde concerning its north-eastern
extremity, and relative to its salines and saline mud and lakes,
corresponds so closely with what I saw in the Aralo-Caspian
steppe, which was, geologically speaking, not very long ago a
sea-bottom, that it is difficult to admit the antiquity of the Gobi.
This steppe, like the high Sahara, was, in all probability, elevated during the latest geological period, while the formations
of the Thian-Shan hitherto examined are all palæozoic.

II.—Here I must also remark on the orography of the Kir-
ghizuin-Alataù, which I saw only on my way to the Thian-Shan.
The directions of the ranges in this system are identical with
those of the Thian-Shan, viz., north-east, south-west, and east,
west ; but the grouping of particular ranges is different, and
entirely the reverse of what I observed in the Urtak-taù and in
the ranges parallel with the Chirchik, especially to the west of
the main snowy range of the Alataù.

In the Alataù several elevations east to west, are intersected
by one north-east, south-west ; and in the Urtak-taù also, as Mr.
Semenof found in his exploration, the direction north-east, south-
west, is the latest. In the western Thian-Shan it is otherwise.
The sedimentary formations in the Alataù are exclusively
palæozoic. No metamorphic strata with organic remains, upheaved in a direction north-east, south-west, and capable of
serving to determine the antiquity of this upheaval, have as yet
been discovered. Carboniferous formations have been upheaved
in the small ranges bearing east by west of the Alataù system,
as well as in that of the Thian-Shan running in the same
direction.

The Alataù system may, therefore, be considered as having a
character of its own quite distinct from that of the Thian-Shan,
although having a geological connexion with it. From the
gradual deviation from one direction to the other (from north-
east, south-west, through east, west, to south-east, north-west),

and from the simultaneousness of their upheaval east, west, in the Thian-Shan and the Alataù, the sequence of geological revolutions in Central Asia may be thus represented:—

1st. Ancient upheavals north-west, south-west, in the Western Thian-Shan.

2nd. Their deviation to a direction east, west.

3rd. Upheavals east-west and south-east, north-west in the Western Thian-Shan and the Dzungarian-Alataù.

4th. Upheaval north-east, south-west, in the Alataù; the second of this direction.

5th. The Gobi elevation, also north-east, south-west, the third in this direction; and the formation of the Baikal volcanic zone.

It may be said by many that these deductions are premature, not being sufficiently grounded upon observation. I do not consider them myself to be finally decisive on the subject here treated of, in respect of the geology of Asia, but they will not be without use in the elucidation of the scientific questions that have yet to be solved. They are just as definitive as the incomplete observations upon which they are founded can allow, and no more.

III. The prevailing direction in the Kara-taù * system is south-east, north-west, *i.e.* the direction of the Himalayas. As mentioned above, it commences in a wide convexity, of which the axis — the Kulan range — bears S.S.E., N.N.W., which is an almost meridional direction. From the northern extremity of this plateau, a range with a general direction S.S.E., N.N.W., detaches itself. How the strata in this range extend, I am unable to say.

As regards the meridional portion of the Kara-taù, the two above-mentioned directions towards which the strata (the schist on the Tersa) extend, that towards north-east, south-west, is the most ancient; and that extending N.N.W., S.S.W., the more recent, having carboniferous and Permian formations. In the Ulkun-burul the latter are upheaved in a direction S.S.E., W.S.W., deviating almost to east, west. The latter direction of mountain ranges, with that of north-east, south-east, occurs also in the Western Kara-taù, affecting equally the carboniferous and more ancient strata, viz., schist, limestone and sandstone, which have consequently all been upheaved together.

It follows that the upheavals in the Kara-taù took place at the same geological epoch as those of a corresponding direction in the Thian-Shan, the difference being that the more ancient

---

* There is another range corresponding to the Kara-taù in its direction, viz., the lesser Kurdai Range on the right bank of the Chu, which is geologically known. I saw it only from a distance, and for this reason do not describe it.

upheaval was weaker and formed only an addition to tl
Shan in the flat elevation of the Upper Tersa, which w
raised at all independently of the nearer portions of tl
tañ. The later upheaval was considerably more powerful,
which formed out of the Kara-tañ small separate syst
direction towards the Himalayas.

Thus, in this respect, Humboldt was correct in conne
Kara-tañ with the Bolor, which, as has been said befo
north-western portion of the Himalayan system.

IV. I have already dwelt upon the relations of the
the Thian-Shan. As regards the period of upheava
north-western Himalaya (southern Tsun-lin), judging by
between its direction and the Kara-tañ, and by the i
strata of coal which Messrs. Bogoslofski and Lehmann
the upper Zaravshan; it may be argued that these u
occurred simultaneously with those in the Thian-S
similar direction; but it cannot be determined without
investigation whether the upheavals north-east, south-we
southern Tsun-lin were contemporaneous with those
Thian-Shan, or with the later ones in the Dzungariar
There is this circumstance in favour of the first conject
the ranges bearing in this direction are of more freque
rence, as those abutting on the Urtak-tañ parallel
Chirchik. The name of "Bolor" in the sense of a
mountain system, which I have denominated the Southe
lin, should, in my opinion, properly be excluded from
graphy of Asia, because it is really *not the name of a 1*
system, and in this sense is therefore an error. Bolo
name of a river, and of a town situated upon it; and is
according to Central Asiatic usage, the term for the on
tain from which the Bolor issues. They perpetuate an
giving this name to a mountain region which has for
ages borne another appellation, instead of one that is
and at the same time most appropriate, viz., "Tsun-lin
this name will be preserved in geography, although th
lin, as we have seen, does not constitute a complete ar
pendent mountain system, being formed by the weste
verging extremities of the Thian-Shan and the Hi
Both of these ranges, however, at their junction assu
common character as to their orography, somewhat disti
that of their more distant elevations, as is illustrated
dispersal of their peaks, and by the numerous short ran
detach themselves, and intersect each other.

The real orographical import of my observations
mountains between the Chu and the Syr-Daria, cur.
incomplete as they doubtless are, lies in the sufficien

explanation of the construction of the enigmatical Tsun-lin Mountains, and in the confirmation of the ideas of Huen-Tzsan, and of the Chinese generally, concerning them, viz. that they are not the mountain-knot of the Kuen-lun and Bolor, as Humboldt thought; but an extensive mountain region, formed by the meeting and blending of two distinct and colossal systems, those of the Thian-Shan and of the Himalayas.

The Kuen-lun and Bolor, as we have seen, do not form separate ranges, but both belong to the Himalayan system.

Speaking generally, Humboldt's *five* mountain systems of Central Asia, viz. those of the Altai, Thian-Shan, Kuen-Lun, Himalayas, and Bolor, are thus, according to more recent explorations on the spot, commencing with Cunningham's 'Ladak,' reduced to three—the Altai, Thian-Shan, and Himalayas.

Humboldt further considers that the principal and most fundamental feature in the orography of Asia is its long mountain ranges. To elevated table-lands he imparts only a secondary orographic significance (in contradistinction to Iran, where they prevail); and he repeatedly urges his view, chiefly on the ground of the concavity of Turkestan or Lop-Nor.

It is true that Central Asia is not one vast table-land; but if Schlagintweit's opinion as to the altitude of Kashgar and Yarkend* be correct, then the Lop-Nor concavity cannot be called a depression; and is only a concavity in comparison with the enormous heights skirting it, viz., the mountain systems of the Thian-Shan and the Himalayas.† Properly speaking, it is all a high table-land—an intermediate decline between Thibet and Siberia.

Humboldt's delineations of long ranges must evidently disappear from the orography of Central Asia, at least as constituting its principal feature, since the Thian-Shan and the Himalayan systems respectively represent a wide and continuous convexity, upon which rise numerous ranges subsidiary to the general convexity, and consequently of secondary orographic importance.

In the Thian-Shan this general convexity is of no great elevation; but towards the west, in the Thian-Shan portion of the Tsun-lin, it attains a height of 5000 feet at Issyk-kul, 10,500 feet an Son-kul,‡ and 7000 feet at the Upper Chirchik. To the east of the Issyk-kul table-land the Thian-Shan may perhaps appear in the form of one long ridge, but probably with a some-

* He estimates them at 3000 to 4000 feet above the level of the sea.
† The Kuen-lun alone abuts on the Lop-Nor level from the south; but this is only the northern margin of the elevation of Thibet, or of the Himalayan system.
‡ Where Captain Protsenko found ice even in June, though partially thawed.

what broad sub-montane belt on the north, the existence of which is not, however, positively ascertained.

But in the Himalayan system, commencing from Rian-kul and Kara-kul, the ridges scarcely rise above the general elevation of 14,000 to 17,000 feet, excepting in the case of isolated peaks. This elevation fails to be attained only in narrow valleys having the character of clefts.

The Altai also presents the appearance of a wide protuberance, studded with numerous ridges.

By referring thus critically to Humboldt's great work on Central Asia, the idea crosses my mind that the reader may imagine it to have been done with the object of enhancing the value of my own humble explorations. Still, I have no fear. I have opened my paper with a quotation from Tszan, and have accepted this quotation literally, as it to be understood; instead of using it for the purpose of mining the supposed continuity of the chain of the Bolor is not recognised as a range by Huen-Tszan), because reasons: firstly, the fact that I was there at the time of Gen. Cherniayef's expedition across the Chu; secondly, because of Lyell's geological theory of slow, gradual, and constant changes taking place upon the earth's surface, instead of that of rapid and general Plutonic convulsions simultaneously upheaving lengthy mountain chains, and of intermitting periods of repose when sedimentary deposits were formed.

Humboldt held the latter theory, which was in vogue during the period from 1820 to 1840. It rested to a very considerable extent upon his own geological observations in the Cordilleras; and it guided him when he wrote upon Central Asia, his Asiatic sources of information being meagre and unscientific. This geological theory must necessarily have misled even so scientific a genius when treating of mountains he had never seen.

My almost polemical tone refers to this antiquated theory only. It is now abandoned by geologists, but the theoretical systems of orography based upon it still survive, and continue to have weight only on account of Humboldt's authority, who was the founder of the present theory of orography and of physical geography generally.

It is, however, time to reconcile the geological theories of the present day, with respect to the processes which have caused the variations of the earth's surface, with orography. No theoretical construction of the general direction of long ranges is needed for the purpose; it is enough, avoiding these altogether, to be guided by the simple acquisition of facts.

Ritter's orographical deductions relative to Central Asia, contemporaneous with those of Humboldt, although independent of the geological teaching of any one, are for this reason not so complete as those of Humboldt; but, on the other hand, they will never become obsolete. They will be subject only to amplification, not to destruction. And yet they are not so suggestive, and do not equally stimulate fresh scientific exploration; and especially they do not afford the clue to new discoveries, like the ingenious, though often incorrect, geographical constructions of Humboldt; since those of the latter directly indicate the points particularly to be observed for the acquisition of correct scientific knowledge relative to an unknown country. Such is the character of Humboldt's genius. Even when erring, he renders services to science, for he makes the solution of the questions that he proposes easy even to an indifferent explorer, who has been placed for the purpose in much more favourable circumstances than he was himself.

The instruments used by me in these measurements were Greiner's hypsometer, indicating only temperatures $+92°$ Cent.,[*] and a barometer made by Brauer of Pulkova which was in a damaged condition when I received it from the Imperial Geographical Society; it was, however, very well repaired by Mr. Noak, a mechanist attached to the Physical Cabinet of the Academy of Science. The tube of this barometer got broken on the way from St. Petersburg at Fort Vernoe. At Aulié-ata a new one was cast out of some spare ones, and this was carefully tempered.

As regards the calculation of the heights, in the absence of corresponding observations, we might have calculated the altitude of the measured peaks, by reference to the barometrically determined situation of Fort Vernoë, and have obtained the absolute heights from these relative ones; but the deviations of the isolated observations from the mean heights, as indicated by the barometer, on the points under measurement would then, nevertheless, have remained unascertained. I therefore preferred, instead of making calculations, to transcribe from Schlagintweit's 'Travels in India and Central Asia' (p. 11, Hypsometry) the figures of absolute heights corresponding with the barometrical heights similar to my own. Then, also, when I found in Schlagintweit different figures of heights for identical indications of the barometer, or varying indications of the barometer for equal heights (which is unavoidable in determining the latter by separate observations), I took into consideration in my selections both wind and weather. And besides, from among Schlagintweit's measurements, I guided myself in preference by those

---

[*] *I. e.*, downwards from 101° to 92° Cent.; below, the mercury contracted into the globe.

TABLE OF HEIGHTS, MEASURED IN 1864,* BETWEEN THE CHU AND THE SYR-DARIA RIVER.

NAME OF PLACE AND OBSERVATION.	Boiling Point by Hypsometer.	Rise of Barometer.	Thermometer with Barometer.	12-13⅛° Barometer set to 13½° R.
		Eng. half lines.	Reaumur.	
I. TRANS-ILI-ALATAÚ.				
1. Fort Vernoé, 2nd (14th) May	98·69	..	..	552·08⎫
Ditto 3rd (15th) „	97·79	..	..	554·05⎭
Ditto by Golubef's determination	..	..	..	..
2. Kastek defile, opening at the lowest limit of ancient moraines, on the 7th (19th) May, 3 P.M., after rain. Hypsometer 95·65–95·69. A steppe grass "Chü "	95·67	..	..	512·76
3. Pass from the River Kastek to the Chu Valley, at the upper course of the Kara-Bulak, midday, 8th (20th) May, weather lowering, IV.	92·63	..	..	458·08
4. Kara-Bulak, at the foot of the mountains (Valley of Chu), 10th (22nd) May; morning, cirrhi	96·36	..	..	523·69
Ditto according to Ventükof ('Imperial Geographical Society,' 1861, IV.)	..	..	..	..

No. & Description						
7. Road from Tokmak to Merké, between River Chu and mountains, Kara-Balta Settlement, 21st May (2nd June), evening, clear, calm, cumuli and cirro-strati on the horizon	97·61	..	..	550·34	..	2,500
8. Fort Merké, 24th May (5th June), midday, clear, gentle, E.N.E. breeze, cumuli on the horizon	97·93	..	..	556·89	..	2,100
9. Lower limit of limestone in defile of River Merké, 25th May (6th June), evening, clear, cirrus. There is also the limit of yellow briar, generally of brushwood and mountain grasses. A spring here, 6·2 R.	96·83	..	..	535·03	..	3,300
10. Defile of river, 1st Ourianda, nearly lower limit of limestone,† within 6 versts (4 miles) from entrance, 27th May (8th June), 6 P.M., lowering, rain until half an hour before observation ..	95·74	..	..	514·11	..	4,200
11. Koi-Nar-Tas Pass, between 1st and 2nd Ouriandas, 28th May (9th June), 10 A.M., clear, cumuli	93·66	..	..	476·04	..	6,550
12. Summit of Koi-nar-Tas Mountain, 100 feet above the pass, composed of limestone	..	..	..	..	..	6,650
13. A spring of the same name, about 300 feet below the pass, 5·3 R.	..	..	..	..	..	6,200
14. Upper course of 2nd Ourianda, lower limit of snow in ravines, *Juniperus Sabina*, *Ribes sp*, growing, 28th May (9th June), midday, clear, calm	92·78	460·67	..	460·67	..	7,400
15. Sources of 2nd Ourianda, 200 feet still higher, estimated by sight	..	..	..	..	..	7,600

* The heights marked with an asterisk (*) were measured by Mr. Frehse; the others by myself.

† Here is also the lower limit of a kind of Mountain Liliacea, bearing yellow flowers, clustering thickly on the upper part of the stalk to 1¾ to 2¼ feet. There are similar plants, but with pink flowers, more to the east, in the Trans-Ili-Alataü and Khirgiz-Alataü. The highest limit is below 6000 feet.

Names of Places and Observations.		Eng. Imp. Feet.	Réaumur.	
16. Pass to Kara-Kystak; mercury of Hypsometer in globe, 26th May (9th June); barometrical measurement, 22nd August (3rd September) made in the evening, N.N.W., cloudy		430·7	9·8	481·05
17. Limit of mudstone and limestone on the 3rd Ouriande, 29th May (10th June)	95·81			806·04
18. Lower limit of junipers, mountain ash, and black current, about 200 feet higher				
19. Spring between 3rd Ouriande and Kara-Kystak in foreland of mudstone, temperature 7½ R., about				
20. Aulié-Ata. Depressed bank of Talas, 9th (21st) June, morning, clear	98·16			
21. Aulié-Ata, 3rd (15th) July, 2 p.m., clear		561·9 / 561·3	37·3 / 37·0	
Gentle wind, N.E. and N.W.				
4th (16th) July, 2 p.m.		548·3 / 552·1	38·3 / 38·2	
5th (17th) ,, ,, ,,				
,, ,, ,, ,, clouds dense from S.		564·0	38·1	
,, ,, ,, ,, clouds from S.				
Mean	4	552·6		

No.	Description					
24.	Talas, 5 versts above lower limit of forest in its valley, 18th (30th) August, 1 P.M.	3,500	523·07	24·0	524·2	..
25.	Talas, somewhat above the mouth of the Kenkol, 19th (31st) August, midday, clear, S.W. wind, gentle	4,100	515·61	..	..	95·82
26.	Talas, confluence of its upper courses, the Kara-Kol, with the Utch-Kosh-Sai, 20th August (1st September), clear, gentle, N.E.	4,500	508·84	..	..	95·46
27.	Tuldy-Bulak, a brook on the southern slope of the Range, between the Kara-Kol and the Kara-Kystak, Aconitum and gentians growing	8,150	448·85	11·8	448·7	..
28.	Pass from Tuldy-Bulak to the Kara-Kystak, 21st August (2nd September), midday, lowering, calm wind shifting from N.E. to S.W.	9,300	429·84	10·4	429·6	..
29.	Longitudinal valley at upper course of Kara-Kystak, with morning, confluence of its two sources, 22nd August (3rd September), 3 P.M., N.N.W., clear	5,300	495·34	18·0	526·0	..

Pass to the Ouriands, see No. 16.

III. USTARTAD AT THE UPPER COURSE OF THE ORMSERK (CHATKAL).

No.	Description					
30.	Chemget Spring (affluence of the Talas), at the southern base of the Cha-Archa Mountains 16th (28th) June, midday, clear, gentle, W.	8,000	540·55	..	..	96·99
31.	Summits of the Cha-Archa ridges, at the outbreak of the Talas, 1000 feet higher, estimated by sight	4,000	..	..	..	..
32.	Kara-Bura River, within 3 versts (2 miles) of its issue from the Urtak-tait Mountains, 17th (29th) June, morning, calm, clear	3,350	525·9	19·0	493·94	..
33.	Kara-Bura River, lower extremity of its defile, at issue from the mountains, 17th (29th) June, 10 A.M., calm, clear	3,400	524·39	20·5	526·41	96·54

TABLE OF HEIGHTS, MEASURED IN 1864, BETWEEN THE CHU AND THE SYR-DARIA RIVER

NAMES OF PLACES AND OBSERVATIONS.	Boiling Point by Hypsometer.	Rise of Barometer.	Thermometer with Barometer.
		Eng. half lines.	Reaumur.
34. Confluence of the two head waters of the Kara-Bura, 17th (29th) June, midday, clear, calm (the river here issues from the main range, and enters the sub-mountainous region of conglomerate)	..	517·58	21·0
35. Lower limit of Birch in the defile of the Eastern Kara-Bura River, 18th (30th) June, morning, calm, clear	94·83	491·51	16·3
36. First denudation of limestone from below, at the Eastern Kara-Bura, a little below the lower limit of Birch, 19th June (1st July), morning, clear	95·19	489·90	19·5
37. Confluence of upper sources of Eastern Kara-Bura; turn towards the pass across the water parting of the Talas and Chirchik, about 400 feet above the lower limit of Juniperus pseudo-sabina, 18th (30th) June, midday, cloudy	..	460·6	14·0
38. Mouth of Taity River into Eastern Kara-Bura, near lower limit of Birch and Willow, 21st June (3rd July), 4 P.M., lowering; a spring here, 6·4 R. .. .. .. little. Between (Locality higher than No. 36, though very Nos. 37 and 38 a spring, 4·2 R., near lower limit of	..	497·81	18·0

41. Extreme limit of spreading *Juniperus*, i.e. every kind of tree, and great accumulation of snow in gorges, 22nd June (4th July), evening, as on 28th June (10th July)..	‥	414·9	10·5	414·65	10,000
42. Pass from Kara-Bura to Kara-Kyspak River, i.e. from confluents of River Talas to those of the Chichik, no snow, 22nd June (4th July), evening, No. 10, lowering..	‥	408·0	7·3	408·56	10,500
43. Confluence of the two upper sources of the Kara-Kyspak, near the pass, 23rd July (4th August), 9 A.M., clear..	‥	423·8	18·7	423·34	9,500
44. Upper limit of leafed brushwood, at the Kara-Kyspak, 23rd July (4th August), 11 A.M., calm, clear; a spring here, 2·4 R.	‥	433·2	20·4	432·57	9,100
45. Upper Snow Slip at the Kara-Kyspak, 2 P.M., clear	‥‥	442·2	18·0	441·78	8,650
46. Lower Snow Slip at same place, between snow-covered mountains; upper limit of willow and birch near the southern base of the range, 23rd July (4th August), 6 P.M., clear, calm	‥	454·1	13·4	454·1	7,450
47. Valley of the Chatkal (Upper Chirchik) at the opening into it of the Kara-Kyspak, 24th July (5th August), 10 A.M., S.W., cloudy, rain alternating with sunshine..	‥	458·3	14·5	458·19	7,100
48. Upper limit of *Hippophae rhamnoides*, in the Chatkal Valley, near Chimash-Kourgan, 25th June (7th July,) midday, clear, S.W.	‥	478·4	16·0	478·14	6,350
49. Upper limit of Hawthorn (*Crataegus sp.*) at the Chatkal, 16 versts (10¾ miles), below Chimash-Kourgan, 25th June (7th July), 3 P.M., S.W., cloudy	‥	484·9	17·2	484·52	6,000
50. Spring at the Chatkal, at the month of the Chakmak, 10 versts (6¾ miles), below the month of the Kara-Kyspak, temperature 4·4 R, 27th June (9th July), 7 P.M., clear..	‥	464·7	18·2	464·8	7,000

* Measurements, Nos. 35, 36, and 38 were made over the extent of 1 verst (two-thirds of a mile). The order of the situations follows stream; the river accordingly runs as it were uphill. These observations may indicate the margin of inaccuracies from the figures relative to the weather. The winds in the defiles could not be noticed.

† Sic in orig.

TABLE OF HEIGHTS, MEASURED IN 1864, BETWEEN THE CHU AND THE SYR-DARIA RIVERS—continued.

Names of Places and Observations	Boiling Point by Hypsometer.	Rise of Barometer.	Thermometer with Barometer.	13–13½° Barometer set to 13½° R.	Eng. feet.
IV. KARATAU SYSTEM, VALLEYS OF THE TERSA, ARIT, AND BOROU.		Eng. half lines.	Reaumur.	Eng. half lines.	
51. Pass across the Kuyuk Mountains, at the sources of the Kuyuk Rivulet, 8th (20th July), 9 A.M., clear, calm	: :	587·8	20·4	536·53	8,090
52. Ford over the Tersa River (Tersa-Akhun), 8th (20th) July, 10 P.M., clear, calm, after gentle N.E.	: :	535·5	29·8	534·90	8,100
53. Kok-Sai Mountains, in the Urtak-tah range, from calculation of angle of elevation and distance by measurement, 10,150 feet above the Tersa valley-point, No. 52	: :	: :	: :	: :	18,590
54. Ak-Sai Mountain, in the Urtak-tah range, by same measurement, 11,725 feet higher than the same point	: :	: :	: :	: :	14,925
55. Pass from Chulpak, affluent of the Turm, to the Arys, about 500 feet higher than point No. 53	: :	: :	: :	: :	9,000
56. The Arys, as same ... July, 6 P.M., ...	: :				
57. 13th (25th July, 11 A.M., clear; 8 R. verts	: :	562·3	31·2	560·4	

Description					
58. The Arys, 15 versts (10 miles), above the mouth of the Boroldai, 14th (26th) July, 7·30 P.M., clear, calm; in daytime a varying wind, N.E. and N.W., near the level of the river	..	572·3	22·8	571·22	1,570
59. The Arys, at the mouth of the Badam, 25th July (6th August), morning, clear, on crest 50 feet above level of river..	..	575·0	28·0	573·31	1,500
60. Highest point on the Boroldai Mountains, near the sources of the Bugun, 27th July (8th August), strong N.W., clear, 10 A.M.	..	524·0	24·0	522·89	3,500
61. Town of Turkestan, at Mesjid of Hazret-Sultan, clear, hurricane (and consequently unreliable observations), 30th July (11th August)	99·53	589·87	..	589·87	400
62. Turlan Pass, summit of Karatau, between Turkestan and Chalak, 1st (13th) August	98·0	460·26	..	460·26	6,500
63. Coal at Kumyr-tas River, in longitudinal valley on northern slope of Karatau, 4th (16th August)	97·10	540·4	..	540·4	2,900
64. Biiliii-Kul 31st July (12th August), evening, calm, clear.. ..	..	574·0	27·0	573·26	1,800

8th (20th) J

I have the honour to report that the results of th
of the branch of the Expedition entrusted to my ch
been the discovery of hitherto unknown coal-beds ;
tainment of the extension and disposition of the coal f
in the Karataù ; the discovery of a whole system of be
dust by the affluents of the Tersa River, besides nume
of iron-ore. These discoveries were made in excursio
Karataù, to the Tersa River, and to the headwaters of th
during the month of May and in the beginning of Ju
I could not undertake any excursions before May on
the impossibility of obtaining a convoy, and the unset
of affairs consequent on the hostilities of the Bokhi
started on my first expedition on the 5th (17th) May
object of first examining the coal-beds already disc
the north of Chemkend, and then proceeding along th
Mounts in search for fresh layers, intending to con
explorations between Chemkend and Tashkend on the
of the road, where coal was said to exist. During m
sory stay in Chemkend, I was occupied in making a co
birds, insects, and plants. One ornithological collecti
1st (13th) May, comprised as many as 1600 specim
has now increased to about 2000, and, subsequent
journey of 1864, I have gathered a greater variety

During the winter and early in the spring, I succeeded in ascertaining the following particulars relative to the rearing of silk-worms, which were required by the Geographical Society :—

As an industry, the rearing of silk-worms in the valley of the Syr-Daria is pursued only in Hodjend. It is also pursued in Namangan, which is situated, however, by the river itself. It is general along all the southern affluents of the Syr-Daria, and particularly common in Margilan. The places here named are the three great centres of the rearing of silk-worms, although the practice is also followed to a large extent throughout Kokan. As to Andijan and Usk, I obtained no information. The northern limit of the regular rearing of silk-worms is the Namangan range, and its south-western extremity, the Kuraska hills which separate the Syr-Daria from the Chirchik. Thus the rearing of silk-worms is confined to the valley of the Ferghanah or the Khanat of Kokan Proper.

The mulberry-tree is largely grown in all the villages along the Chirchick, but more for the sake of its hard wood; it attains considerable dimensions, 30 feet in height and 18 inches in thickness. In Tashkend alone, some few of the natives rear silk-worms, but only in the way of experiment, although they are successful enough. The eggs are, however, dear; selling at 1000 tengas per lb. But in Tashkend they are rarely to be purchased; and, even in Namangan and Hodjend, they are preserved by the rearers for their own use, very few of the cocoons being reserved annually for sale. In Tashkend it is found that the expense attending the rearing of the worms is too great to admit of any competition with those of the Ferghanah, and that fruit-gardens are more profitable; it is also considered in Tashkend too great a trouble to gather the leaves of the immense mulberry-trees, of which there are nevertheless but a few in each garden, and to attend to the rearing of the worms. Therefore the small householders pay no regard to this occupation, and it is only a few of the wealthy merchants of Tashkend who rear silk-worms in their suburban gardens, keeping hired labourers to look after them; but owing to the want of proper supervision by the master himself, during the process of feeding these voracious worms, the Tashkend silk is worse and cheaper than that of Ferghanah, of Hodjend, &c., although it reimburses the expenses and gives even a small profit. I was told as to Tashkend rearers, that they were not indisposed to increase their establishments.

It is to be hoped that the recent capture of Hodjend will promote this, if only the Tashkendians have calculated on the encouragement of their own industry, by an unfair removal of all the silk-worms from Ferghanah and Bokhara during the war. There is yet a great deficiency in Tashkend of proper hands to do the work of unravelling the cocoons; there is no machinery for the purpose.

In Tashkend the rearing of silk-worms is a rarity, and is considered a novelty; and, although it has been carried on for ages, it is yet only experimentalized upon. As a branch of industry, it has been practised in Ferghanah from time immemorial.

Mulberry-trees are grown everywhere north of Tashkend, but they are not numerous, and there are very few in Auliéta. In a small degree the silk-worm is permanently reared in Kornak, near Turkestan. Experiments in this line have been made from time to time in Sairam, Mankend, and Karabulak, but in those places there is now no mention of the rearing of silk-worms.

The process of nursing the worm is very simple, and is commonly known in Europe. The worm is produced from the egg in a little pouch on its surface, before the leaves of the mulberry-tree are out, which is in April; later the worms are fed on the plucked leaves in-doors, and small twigs are supplied to them, among which they form their cocoons. Samples of white and yellowish cocoons, of unravelled silk, and of the eggs, besides other specimens of native produce, were sent by General Cherniozef to Moscow, with Hadji-Yunusof, a very intelligent Tashkendian.

Tashkendians say that even when transported during the summer heats, the eggs do not mature into worms on the road, but it is better to convey them in the spring and autumn; moderate frosts, of 10° R., do not destroy in them the germ of life. There is not the least sign here of a disease among the silk-worms (M. Meazza and his companions said the same of Bokhara). Both worms and eggs are perfectly healthy; the latter are very sound, and not at all delicate.

The common mulberry-worm is produced here. I did not find here either the *Bombyx mori* or any other variety of the wild silk-worm.

I turn now to my excursions and to their results, which apply practically to the region.

On leaving Chemkend I passed through Karabulak, and across the Arys River to the Karataù, proceeding then along the south-west base of those mountains. On the 19th May, at the Katurgan-Su Rivulet, I was joined by Captain Nikolski, of the Mining Engineers, who had arrived in Chemkend on the day after I left. Keeping to the south-west base of the mountains, we traversed the river Boroldoi, both the Buguns, the river Soyan, and the Arystamdy, determining the topographical and geognostic relations of the permian formations in the Karataù to the more ancient strata, which was the object of the excursion in this direction. On the 22nd May we reached the coal-beds by the rivers Batpak-Su and Isendy-Bulak, which were the first to be discovered in this region. After making a superficial examination of the disposition of the coal-layer, I parted from Captain Nikolski, who remained to make a more minute investigation into all the coal denudations in those parts, for the purpose of thoroughly studying the carboniferous formation of this region.

"I myself started for the rocky district of the Karataù, which had not yet been explored by any one, between the sources of the Babata River and the Turlan Pass, being accompanied by Skorniokof, the collector of the expedition, and by the dresser Shiliayef, both of whom very zealously occupied themselves in dressing the skins of birds and animals, and in preserving plants; while they exhibited great diligence and ability in following my directions in collecting specimens of ore and petrefaction.

Here I determined the position of a mineral vein, near the Turlan Pass, which M. Frehse had not visited. This vein is on the summit of one of the many parallel ridges compassing the Karataù, and occurs in a transverse cleft in the limestone; the vein runs from N. 40° E., and the limestone from W. 35° N. I afterwards found that all the mineral veins in the transverse clefts of the Karataù were similarly disposed. The ore is a mixture of *glittering lead* and other lead ore, with ochre of iron and lime. The *glittering lead* occurs either in lumps or in small crystals: the lead ore is cleansed by the Kirghizes at the scours of the mountain streams.

One exists also on the Ken-sara, in the rocks at the eastern extremity of the defile, where there is ochre, brown iron-stone, and *glittering lead*. A third mineral bed, not known even to the Kirghizes, we found by the Chulbar-Su; a thick vein of red *ochral* iron-stone, with sparkles of lead, protrudes in the rock.

Specimens of these ores were secured, and will be more minutely described by M. Nikolski, who is now fitting up a travelling laboratory in Chemkend. We likewise found in this part of Karataù layers of limestone, rich in lapidescences, in the Kandy-Mystaé locality, and at Usk-Tiubebas, by the sources of the Babata River. These layers, with the lapidescences, were discovered for the first time, and were traced by me throughout the whole extent of the Karataù, from Kandy-Mystaé towards the south-east, as far as to Boroldoi River; they are of great importance for the determination of the geological

construction of the Karataù Ranges. It appears prob
a first examination, and from a classification of some
lapidescences, which I have made, that these limest
their predominating *Productus* and *Spirifer*, belong to
of mountain limestone ; but in order accurately to
their antiquity, it is necessary to obtain defined spe
various shells and corals, which are guides to the dete
of all layers lying upon and under the limestone, o
primitive origin than the carboniferous strata.

Joining Captain Nikolski at Djar-Tiubé on the Babe
on the 24th of May, I learned that during my absen
examined five denudations of coal on the Isendy-B
three on the Batpak-su rivulet. In each of these places
only a thin layer of not more than six inches, and
abruptly. Mr. Tatarinof had found the same before
thick layer of coal in this locality, of which I reme
Frehse to have made mention, was not discovered.

On the 25th of May I inspected the coal-bed on th
Bulak, together with Captain Nikolski, who afterwards, a
to my directions, proceeded to Usk-tiubé-bas, to collect
specimens, whilst I ascended to the sources of the B
trace the connection between the coal leads there and
the Isendy-Bulak. On this occasion I passed at last
saddle-back of the Karataù, whence the Arystandy flo
south-west, and the Usk-tas to the north, between tl
mountains at the head waters of the Babata. Along the
this line of route I found denudations of a species of coa
tion such as occurs at the Isendy-Bulak ; viz. sandst
conglomerate. Denudations of pure coal I did not fin
came to the Tayan ; but even the conglomerate and sa
were here in great abundance, cropping out only here a
from under the thick masses of detritus ; the yare, howe
racteristic, so that even before I had seen the coal, I h
able, in 1864, to come to the conclusion from them th
was coal formation on the Bugun, where, in 1865, M. 'T
actually found it, after, however, seeing my geologic
From the Tayan to the Bugun I also traced a continu
of coal-formation. The collector Skorniokof discovered
thin streak of coal, extending for two fathoms, exposed
spring in the vicinity of the sources of the Sasyk.

On the Bugun I waited three days for Captain N
making in that interval small excursions in the neighb
of the denudation of coal found by M. Tatarinof. I h
up section-plans of all the Karataù formation on the
though not along the whole of its course through th
tains. Coal formations were also found here, resting

limestone conglomerate; closed in, between the Isendy-Bulak and the Usk-Tiubé-bas, by thick masses of deposit. Wild grape was found along the Bugun (afterwards also along the Boroldoi and Koturgan-Su), and a very good botanical and zoological collection was made here. Skorniokof picked up several petre-factions, while I geognostically determined the stratification of the bed in which they were found.[*] The defiles along the Bugun are wooded, as are also those in the mountains and by the head-waters of the Tayan, which nobody had visited before me. I was there too, for geological purposes. It may be here mentioned that the defiles along the Boroldoi and its affluents are equally wooded. The forests are all of the same kind; there are two kinds of ash and a tall hawthorn tree, the latter grows to a height of 18 feet, and is 8 inches thick; it is rarely higher, the old trees are mostly crooked, with forked or broken summits.

This timber is not fit for building purposes, but it will do for joiners' work, and for use at the arsenals; it may likewise be serviceable for coal mines, but the trunk of the hawthorn tree being short, and the copses lying in the narrow defiles, composed of one to three rows of trees, extending only a few versts, it is necessary to be economical in the use of timber in working the coal, and to utilize as much as possible the hard sandstone and limestone found with it. This is done in the copper mines in the Orenburg Steppes by the Takmara.

On the 31st of May Captain Nikolski returned from Usk-Tiubé-bas. We examined the denudations found by Tatarinof on the Bugun, after which Captain Nikolski left for Chem-kend. When he had gone I continued the investigation alone, and discovered perfectly new coal-beds on the Little Bugun and the Boroldoi Rivers.

Coal was found on the Little Bugun, which, by means of specimens picked up from the débris, we were enabled positively to ascertain was embedded in flaky, resinous, and inflammable schist containing, in a state of very good preservation, bended boughs and fruits of various plants, blossoms of ferns, and fruit of coniferous trees, as well as fish of the order *Ganoidia*. The coal was black and glossy.

The conditions in this structure, namely, the presence of the schist next to the coal, are the same as those of the best coal-beds in England and Belgium, but the denudations on the Bugun are everywhere hidden by the detritus.

By analogy with Tatarinof's, and with the Isendy-Bulak beds,

---

[*] Not having seen M. Tatarinof since his return from Bokhara, I cannot say whether I have supplemented anything to his observations on the Bugun, or merely confirmed them.

these denudations must be sought for at the height. of the frequent springs, on the slopes of the southern verge of the Little Bugun Valley, clearing the detritus away from the springs. Here and there are projections of sandstone of carboniferous formation, lying under the coal, and of finely stratified limestone next to the coal. A bed of conglomerate lies on the top of the limestone.

The same layers which are seen by the Little Bugun crop out by the River Malas, or the Northern Boroldoi, between two gorges formed by that stream. The lower stratum extends to the junction of that river with the Great Boroldoi. Here is an appearance of coal *subjected to the action of the wind*; it is flaky, and contains alternating and glassy veins, which owing to the windage have turned into an inflammable schist, not of a light-brown colour however; while the dark schistous clay is not combustible from the same cause. I tried to dig to the unwinded parts of the seam, but failed through lack of time. The similarity of the species in these denudations with that of the crumbling coal by the Little Bugun, inclines me to think that it is one single seam, cropping out at both rivers, or, rather, that there is here an entire system of seams, which is not a continuation of the larger by the Great Bugun, for the latter is not embedded in schist, but in a bluish stratified clay, as is the case by the Isendy-Bulak. I call this stratification by the little Bugun and Boroldoi a complete system, because three layers are visible by the Boroldoi, each being two feet thick; the general thickness of the denudations taken with the schist, where they are not covered with detritus, is 25 feet.

Judging from the height at which the Boroldoi denudation occurs—which is at the elevation of the spring by the Little Bugun, above the level of the seam on the Great Bugun—and seeing the similarity of the arrangement of the accompanying schist to the schist in the English coal-beds, it may be presumed that the thickness of the system of the coal-formations, taken with the schistous strata, is immense. To ascertain this exactly would require a concentration on the spot of all the mining operations of the Expedition; whereas the object of the Expedition is only to bring to certain knowledge the existence of mineral wealth in all parts of the entire region. But, although the thickest layer might not be worked upon, a seam might be selected which, with the means at hand, might be very advantageously examined into by the members of the Expedition.

There is no difficulty in bringing coal from the Little Bugun and Boroldoi in carts to the Syr-Daria. There is a sloping pass over the mountains, with a tolerably good cart-road between the rocky defiles of these rivers. The distance from the mouth of the Aryn is not more than about 54 miles, from Chemkend 40 miles. Timber, for mining purposes, can be obtained by the Boroldoi.

There is capital iron-ore in the Boroldoi defile. This is red iron-stone, occurring at a distance of 1 mile from the coal denudation in the limestone; and almost at the same distance from the coal there are immense projecting masses of brown iron-stone on a flat elevation, broken by a cleft in the mountains. Iron-works may conveniently be established at this spot, where coal and iron lie in close proximity to each other; the produce of these works would find a ready sale in Central Asia, while the establishment would, at the same time, be a source of great benefit to the Syr-Darian flotilla. A tramway could be laid from the coal-mines to the steamboat pier on the Syr-Daria. The expense of constructing this road would be recovered by the economy of the transport of coal. Instead of a load of 7 cwt., drawn by a horse along an ordinary road, each car on the tramway would contain 32 cwt. of coal. In no other localities do coal and iron lie so close together.

On my way back to Chemkend from the Boroldoi coal-seams I perceived numerous veins of red iron-stone at the sources of the Katurgan-su rivulet.

I returned to Chemkend on the 6th of June, where I found two topographers and a convoy of fifteen men, for whom I had applied to General Romanovski in the month of April. The services of the former were required for the purpose of mapping those parts in the Karataù where I had been, of tracing my line of march, and of marking down the newly-discovered beds of ore and coal, and the denudations with the embedded petrifactions.

I at once employed the topographers in the construction of a map of the Karataù, on a scale of 10 versts, from the surveys made for drawing up a geological plan. The details I had inserted in my diary, and the sketches of localities which I had made on the road enabled me to fill in the blanks on the map.

On the 10th of June, I despatched seven of the men, with their apparatus, to the sources of the Badam in the mountains, from whence they returned with a large collection of birds, insects, and plants, reporting also that they had hit upon iron-ore in the hills.

On that same day I myself started with Captain Nikolski, M. Osokin, a gold-seeker, and with two miners, towards Aùliéta. We pushed along the road—taking no convoy—as far as the

Tarsinsk picquet, where, according to arrangement, we found riding-horses.

The object of this excursion was to find out a gold-mine, of which I had received intelligence in 1864, through the eldest son of Tiuringildy, a Khirgiz Bi, of the Chemyr tribe. His name I have, unfortunately, forgotten, but I shall do my best to learn it, in order that he may obtain his merited reward through the application of his superiors.

He more particularly deserves an acknowledgment of his services, because his indications were not vague, like all the rumours about gold floating throughout the Asiatic Khanats; and amongst them the *on dits* in these parts concerning the existence of gold somewhere on the Tallel and Chirchik Rivers, where one has to search for it from 100 to 200 miles up and down their course. This Khirgiz so accurately described the spot where the gold was to be found that, although I had never been there before, I led my small party, without a guide, and without even inquiring the way, directly, and without fault, to the identical place. We found it by the Kukreü River, at its entrance into a rocky defile, in a schistous ridge of the Karataü system, which is also cloven by the Tersa. The pebble-stone at the spot and in the neighbourhood is indicative of gold, being composed of diorite, sienite, and ochreous quartz, of which latter there is an abundance. Moreover, the presence of a superstratum of detritus over the schist, which stands up on end slantingly across the valley, in a direction north-west to south-east, corresponding with the direction of the Karataü, appeared favourable for the formation of an auriferous bed. We selected a spot for working upon, but did not dig down to a gold vein; for, at a depth of about 6 feet, we came to water on a level with the rivulet. At a depth of 3 feet the welling of the water was so strong that the sides of the cutting gave way, and the pit was filled up with coarse-grained sand faster than the men could bale out the water which brought the sand in. We had, in the end, to abandon this cutting.

Gold was, however, found, though in small quantities, in some washings, beginning from a depth of 6 feet. The first three experiments in washing the still sands produced several grains of gold; but three other experiments in washing the *moving* sand did not succeed in bringing to light any of this metal.

From all these circumstances, and by analogy with all the Siberian mines of which I have read and where Osokin had worked, it may not be improbable that the auriferous layer is the deepest bed of detritus, the richest portion of this lying

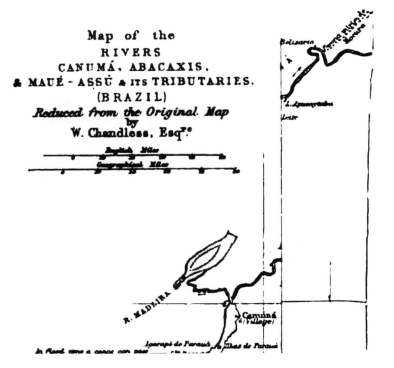

Map of the
RIVERS
CANUMÁ, ABACAXIS,
& MAUÉ - ASSÚ & ITS TRIBUTARIES.
(BRAZIL)
Reduced from the Original Map
by
W. Chandless, Esq.ʳᵉ

upon the vertical schist, of which the frequent projections have arrested the gold deposited on the top.

Having found gold, I considered it advisable not to lose time in digging any fresh pits, the object of the excursion—which was definitely to settle the question as to the existence of gold-dust in the Turkestan province—having been already gained. After this Report I shall proceed to fresh coal-fields between the sources of the Kelés and Chirchik Rivers.

Similar pebble-stone to that of the Kukreù was also found by other springs and rivulets feeding the Tersa from the left side below Chekpak, which leads me to believe that there exists there a complete system of unexplored gold-mines.

The proportion of gold to the sand in the detritus thrown away in the course of clearing the pit by the Kukreù is 7 grains in about 3 cwt., which is very promising.

In this excursion I determined geologically the extreme south-eastern limit of the Karataù range, and positively assured myself that it is perfectly distinct from the Thian-Shan system. Even the slight connexion between the two, which I spoke of with some doubt in a former paper, is an optical delusion.

On the Kukreù I made some observations which confirmed those I had previously made with regard to the traces of the glacial period in this region. There cannot now be a doubt that there was such a period there. Besides unmistakable evidences of ancient moraines, I observed coble-stone of the Thian-Shan on the schistous elevations of the Karataù system, like the coble-stone of the Alps or the Jura. These could only have slid down the glaciers which descended to 150 feet above the base of the lower schistous hills on which they rested. The glaciers in those parts must have descended to 2000 feet above sea-level.

XIV.—*Notes on the Rivers Maué-assú, Abacaxis, and Canumá; Amazons.* By W. CHANDLESS, Gold Medallist R.G.S.

THE rivers Maué-assú, Abacaxis, and Canumá all discharge themselves into what is now called the Paraná-mirim de Canumá, a side-channel that, leaving the river Madeira about 45 * miles (geographical) from its mouth, after a course of some 245 miles, enters the Amazons just below † the town of Villa Bella. About 25 miles below the mouth of the river Maué-

---

* According to the map of M. de Castelnau. It seemed to me rather less.
† Also just above the town, by a small branch known as the Furo de Limão, which is the usual canoe-route.

assú, the Paraná-mirim de Canumá (in this part often called P. de Maués) is joined by the Paraná-mirim de Ramos, a side channel of the Amazons; and below the junction, though the larger of the two, receives the name of the latter. Since my journey the whole channel from the Madeira to Villa Bella has, as well as the Upper Ramos (about 36 miles), been surveyed by Dr. Lisboa, a Brazilian civil engineer, whose map, no doubt, will be superior to mine; I have, therefore, inserted in my map the Paraná-mirim de Canumá, rather for the sake of present completeness, than as a thing of future value.

A glance at the map will show the great contrast between the Paraná-mirim and its influent rivers. The latter are of black water, but in the former the Madeira water usually predominates; consequently in its general features it is a white-water stream.

The three rivers are all much alike in general character and aspect. They may be said to have three phases. At a short distance from their narrow mouths, narrowed by the alluvial land of Madeira mud, they have long, wide, open, currentless reaches, often with a clear water-horizon, and with but a few small islands, which do not embarrass the view. In the second phase this is otherwise: the width from bank to bank is somewhat less than below, and between these is a labyrinth of islands and channels, perplexing enough not merely to map, but even to find one's way through. A back-water resembles a channel between an island and the coast, and probably was such before the head of the channel was silted up, and the island became a peninsula and the channel a backwater. Nor even by going in the middle are you always safe; for some of the islands also are skeletons, and enclose a back-water; the heads of two islands having joined together against you. Fortunately there is in this part of the river generally a faint current, perceptible enough not indeed to keep one right, but at any rate to prevent one's going far wrong. In time of full flood, however, when there is no current, there are, I am told, parts where a stranger cannot find his way. Finally, one reaches the third phase, where the river is in general one well-defined channel, with only here and there an island; with a width proportionate to the body of water that comes down, and with a fair current. In the Canumá River I did not reach the third phase. The transition from phase to phase is, for such things, rather abrupt.

Judging merely from appearances, without reference to any theory on the formation of the Amazon Valley, I should certainly consider the lower portions of these rivers to have been estuaries, swept through, and more or less excavated by strong tides (such as those of the river Guamá, near Pará), and now estuary channels, gradually being filled up by the formation of islands, and the joining of these to each other and to the shore. Who can think that these lower reaches are due to the present insignificant streams? Take especially the river Guarana-tuba, an eastern affluent of the Maué-assú; it is formed by the union of two small rivers, of which the Curuauhy is the larger, the bends of this will show how small it is; in fact, without a guide I was wholly unable to find its mouth, yet the Guarnua-tuba has a width of a mile or more.

The lòwer portion of the Tapajoz seems ,to me an example of the same fact on a much larger scale; and, but that the "white" (or muddy) water rivers have long since filled up with alluvial plain nearly the whole of their old estuaries, we should, I think, see the same there also. Why the one set of rivers are of clear dark water, and the other of muddy water, is another question, and one not easy to answer except hypothetically; just as in photography, " we know what *will* cause 'fogging,' but not (generally) what *has* caused it."

I will now speak briefly of each river separately. The Maué-assú (called Paranarý, above the mouth of the Amana) has a good many rapids, the lowest one almost exactly on the parallel of 5° s. The first group of ten may be passed up-stream in two days with a small canoe; after five days more another rapid is reached, and not very far from this another, and then is reached the "Salto Grande," which, however, is not a fall, but an incline with a very narrow channel and a furious current; all but this may be passed by water: of the river above the "Salto Grande" I could obtain no information. I went only to the fourth rapid of the first group.

The river Amana has a fine fall, about 30 feet high; the stream there narrowing (at the time I saw it) to about 25 yards. The rock is an extremely hard, light yellow sandstone, appa-rently more or less metamorphic; and close by on the right bank is much white quartz rock; but the lumps of rock in the river, 150 yards or so below the fall, are of common greyish purple sandstone, found, so to say, all over the Amazons Valley. There were also on the bank, both here and at some of the rapids of the Paranarý, large lumps and masses of a very friable white sandstone, just like that I saw at the Salto de S. Simão on the Tapajoz.

On the Paranarý, a few miles below the first or lowest rapid, is on the right bank a cave, where the water has undermined the rock some 10 or 12 feet for a distance of about 35 yards. The dark line of this, looking from a distance like the hull of a large canoe beside the bank, probably suggested its name— " Pedra *do* Barco." A man I met below told me that he had

been in this cave at low water, and that there were flowing
stone on the roof, which I imagined would be gypsum, forming
like those of the Mammoth Cave; but on arriving I found them
to be fossil shells, sometimes suspended by a mere thread of
stone. A good collection might be made in low-water times,
but now the water was very near the roof, and it was with difficulty that I could get a few thence, and some in much worse
condition from the outside edge. The strata are not far from
horizontal, but seem to incline a little eastwards. The lowest
bed, that forming the roof of the cave, is of a grey limestone,
probably impure, but effervescing pretty freely with lemon-
juice, and about 1½ foot thick. Above this is a similar bed,
divided from it by a shaley layer only 2 or 3 inches thick;
above is a pink rock, apparently quite different. From the two
chief beds project here and there what look like fossil roots in
situ. On the left bank of the river Amana, just below the
house "Frechal," and a few miles below the fall, I found this
same rock, easily recognised by the thickness of the beds, including the shale, and the "roots," but with no cave, consequently the fossil shells were in much worse preservation. The
beds were at about the same level with respect to the surface of
the water on the two rivers, and therefore must certainly have
a lower absolute level on the Amana. Among the fossils was, I
think, a *Spirifer*. One could hardly mistake this; and one
fossil, which I have compared at the Jermyn Street Museum,
seems to me a *Productus*. It may be well to add that Dr. Cou-
tinho found (as Professor Agassiz told me) Silurian fossils at one
of the lower rapids of the Tapajoz, in about lat. 4° 32' s. My
geological ignorance needs probably no explicit confession.

On the river Abacaxis I did not see these strata, and as I
travelled there in the time of low-water, and looked for them,
they could hardly have escaped my notice. Neither has the Abacaxis any rapids that deserve the name, though there are some
troublesome rocky currents; and the rock in and about them
is generally a flesh-coloured sandstone. What is most noticeable is the alternation, sometimes at short intervals, of cliffs of
white sand with the ordinary red clay cliffs. At the top of the
former the wood is generally very thin and low, sometimes mere
brush; but I nowhere saw open plain. In the higher part of
the river, above lat. 5° 35' s., the wood is everywhere lower, and
of a less tropical aspect than below; and the change comes rather
suddenly.

In the middle portion of the Abacaxis there is a bed of pebbles
and sand, about 4 feet thick, which much resembles some of the
diamond "formations" of Matto Grosso, but the quartz-pebbles
are perhaps more water-worn. I washed some of it in a cala-

bash, but, as might be expected, unsuccessfully; nor did I find undoubted stones of the diamond-formation; no "black beans," and—no gold. This bed is seen more or less for 70 or 80 miles.

The water of the Abacaxis, as of the other two rivers, is in the wide lower part clear and dark, but not coffee-colour, as that of the Rio Negro. Higher up however, that is above the Lago Grande, I found it continually less dark and with more sediment, till it resembled very closely the water of the Madeira or other white-water rivers at the same time of year, when low; and like that of a green, not a brown, tinge. But still further on, after rounding the great bend of the Abacaxis, I found the water again progressively clearer and darker, till brown as Rio Negro water, and so it continued as far as I went. It seemed to me that the change in the middle part was caused by the out-flow from the back-waters, in which mud-banks were being left bare; and the direction of the river favoured the action of the wind (N.E. or E.N.E.) in causing a wash, and it could not but suggest itself that the mud might perhaps absorb the brown colour of the water, and give it a greenish tinge. When the river is high, this cause would cease; and this may account why (as I have often noticed at the mouths of affluents of the Purús, &c.) the water of such rivers is much lighter when they are low, and darker in flood. That the dark colour of the water is due to vegetable matter is likely enough; but this proposition has often been joined with another more doubtful one, that the dark-water rivers come from lakes. That the lakes of the Amazons valley are almost universally of dark water is true; but this seems to be because the streams that feed them are such,—not that the water there grows darker, though, of course, it would become free from sediment. With trifling exceptions, generally of mere rivulets, all the streams flowing into the Abacaxis were of the same colour, approximately, as the main river; and one can hardly suppose them all to originate in lakes.

The Abacaxis, above the mouth of the Arapady is very small —sometimes the boughs from each side joined overhead—but still in a sense navigable, with a smaller canoe than mine and higher water, though fallen trees so greatly obstruct it that most of our time was spent in cutting through these; and on the last day, up-stream from morning to night, we made less than a mile. Copahiba-oil collectors go probably further, as we saw their old huts occasionally as far as we went. Tapirs are extremely numerous in and about the small stream, and very tame. Except in moments of hunger, I always regret killing these inoffensive creatures, that seem so enjoying their

bath. We killed one that must have had a recent encou[n]
with a panther; it was clawed all over, and one of its hind-l
dislocated at the knee-joint and the bone also broken, no[t]d
in its mad and successful rush to shake off the panther.
must soon have fallen a prey to another, had it not to us.

On the Canumá River, I went a much less distance than
either of the others. It has many rapids, some of them said
be difficult and dangerous to pass: probably none for at le
60 or 70 miles above the mouth of the Acáry, my farth
point. All agree that ague prevails on the upper river, fr
June to August inclusive, that is when the water is falling fi
and continues throughout the fall, but that the first rise e[r]
it; much as at New Orleans the first frost is considered to [a]
short yellow fever. But Manoel dos Santos Caldeira, an inte
gent man of colour settled on the lower river, told me that
the River Machado, a dark-water affluent of the Madeira on[y]
right, he and his party were completely prostrated by ague
February and March, when that river was rising; but for t
they could have made a fortune; "oil was so plentiful." On
River Madeira, too, it is always said that ague begins with
rise, and that during the fall there is absolutely none: t[his]
however, is not strictly true, at any rate among the rapids
three of my men had it there in June. On the River Mauó-a[ssú]
it is said that all who go above the rapids for the first ti
have ague badly, and perhaps on a second journey, but rar
afterwards; unless they have passed some years witho[ut]
journey up. The Abacaxis is considered the most healthy
the three rivers, but is much the least frequented.

Except on the Guaranatuba (an eastern affluent of the Ma
assú), where the Mauós live, the Indians of all these rivers
Mundurucús, a tribe so well-known and so often written of t[hat]
I need say little about them. Those on the Mauó-assú, bel
the rapids, are civilised, and live in families not as in tribe-l[ike]
and few under middle age are tattoed, excepting at Campina[s]
the settlement next below the rapids, the people of which (th[ree]
or four families) are from the plains above, as the name imp[lies]
Among them I found the pair whose photographs I had tak[en]
at Manáos. They welcomed me with apparent pleasure, a
gave me a supper of cutia (agoutí), which also was welco[me]
as I had been living for a week on salt fish. My hostess, t[he]
same of whom Mrs. Agassiz writes, "her expression is sw[eet]
and gentle," stood by laughing and talking pleasantly, a
doing the honours of her house with much grace. One of m[y]
men here, the last comer, was two or three months later m[ur]
dered as a magician by his more civilised compatriots bel[ow]
this nearly led to a raid in revenge by those above, who w

posed the "whites" would favour the untattooed Mundurucús: they were, however, persuaded to wait a while, and on hearing that the murderers were in gaol were satisfied, and pronounced the "whites" just.

The Abacaxis is all but uninhabited. The Mundurucús of the inland no longer have paths to the river; and the farthest house (in lat. 4° 47′ s.) had been recently abandoned. In three or four days travelling from the mouth we got beyond population; and from September 17th to November 10th did not see a soul. It is the most complete solitude I have been in. The Indians on the lower part, though so near the mouth are, being few, not much visited by traders; and consequently have much more freshness than those of the Maué-assú. Notably those of Jutahy: the women here were very pleasing in their manners, and their brightness and vivacity were unconscious and quite distinct from forwardness. It seemed to me to be the elasticity in the first freedom from woman's bondage of tribe-life, and probably may pass away; for those on the Maué-assú (except at Campineiros) had nothing of it: indeed generally they were dull and shy.

Mundurucú honesty is well known, yet I will mention one instance of it. These Jutahý people bought a few things of me, to be paid for, on my return, with tapioca and tobacco. As I came down the river, I met some of my debtors on an excursion a day's journey above Jutahý: unasked they told me that the pay was ready, and those at home had orders to pay for them, should I arrive during their absence: and on my arrival the next day, these brought me all that was due, and unasked also.

It was unfortunate that on my way up I could not obtain one of these Jutahý Indians as a guide, owing to their being then busied in their tobacco-gathering and manufacture. Consequently my map of the Abacaxis is very deficient in names, as I could identify with certainty but a few chief points. The Mundurucús seem to me rather weak in nomenclature, as the names Ouranahý and Arupadý occur as affluents both of the Abacaxis and Maué-assú. Igarapé Grande, Lago Grande are also frequent; but several of these I have not marked, believing them to have been given from ignorance of the true name.

The Indians on the Rio de Canumá, as far as I went, are much like those on the Maué-assú: higher up the tattooed sons of the Campinas are numerous.

On the Maué-assú tobacco and guaraná are planted pretty extensively, and on the Abacaxis on a small scale, and are reputed of good quality. On the Canumá River there is tobacco but not guaraná. Copahiba-oil seems to be the only natural product sought on the upper waters of all three rivers; but sarsaparilla and indiarubber are said to be met with on the Upper Maué-assú.

ASTRONOMICAL POINTS.

	Latitude S.	In Time.	In Space.
		M. M. S.	
On River Maué-assú.			
Maué (church)	3 23 55	3 00 50	
Mouth of River Guaranatuba (Ponta da Maloca)	3 41 16	..	
Mucaja-tuba (Indian village church)	3 54 4	3 49 56	
Sitio de Laranjal	4 18 46	..	
Mouth of River Amana	4 23 0	3 50 19	
Sitio de Namy*	4 32 0	3 51 3	
Campineiros (Indian village)	4 43 16	3 51 44	
1st rapid (Tambor)	5 0 45	3 53 6	
On River Amana.			
Sitio de Pindoval	4 31 50	3 50 2	
Salto*	4 44 20	3 49 55	
On River Guaranatuba.			
Igarapé de Quiynha (mouth)	3 45 30	3 49 3	
Mouth of River Arupady	4 13 36	3 44 15	
On River Abacaxis.			
Mouth of River Abacaxis (church at)	3 54 5	3 55 2	
Lago Grande (Munduruaú house) ..	4 24 40	..	
Barreira do Careca	4 55 6	3 53 55	
Taboleiro†	5 5 10	3 53 54	
Mouth of River Curauahy	5 20 16	3 55 0	
Remanso Grande (just below River Pupunha)	6 2 50	3 55 19	
Mouth of River Arupady	6 12 0	3 55 28	
On River de Canumá.			
Canumá (church)	4 2 17	3 56 15	
Igarapé de Prainha (mouth)	4 32 12	3 56 36	
Igarapé Jabuti-caá (mouth)†	4 56 0	3 57 8	
Sitio de Sucurujú	5 8 10	3 58 0	
Mouth of River Acary	5 16 45	3 58 36	

* These latitudes were determined by extra meridian observations, and are therefore probably worse approximations than the rest.

† On this "taboleiro" (or sand-bank on which turtles lay their eggs), was observed :—

			M. M. S.
1868—Nov. 3.	χ² Orionis Oc. D., giving long.	..	3 53 57·3
" "	" Oc. R. "	..	3 53 53·9

The Oc. R. was at the dark limb of the moon, and was probably therefore the better observation. A duplicate calculation was made of this only.

Other points fixed by observations are indicated on the map (as on my other maps) by a small cross (×). At all but about half-a-dozen the observations were both of latitude and longitude. With the exception of the above mentioned occultation, I obtained no absolute determination of longitude ; and the longitudes may be considered generally chronometrical with respect to Maués (church), and on each river with respect to its mouth : those on the river Maué-assú with two chronometers, and on the Abacaxis and R. de Canumá with one only. The longitude of Maués itself was determined chronometrically from Manáos on three occasions, with results : $-9'$ $11\cdot7''$, $-9'$ $12\cdot6''$, $-9'$ $11\cdot1''$, $-9'$ $5\cdot1''$, -9 $11\cdot3''$. The fifth was made long after the rest ; and from the first four the difference was taken as $-9'$ $10''$; and the longitude of Manáos being assumed as 4^h $0'$ $0''$, Maués has been placed in 3^h $50'$ $50''$. Perhaps $-9'$ $11\cdot7'$, obtained by omitting the fourth and including the 5th of the above results would be more exact.

On these rivers, and in fact on all my journeys, except that down the Tapajoz, the going of my chronometer has been tested by time-observations made at the same points both on the journey up and on the return ; and in no case have I relied upon a single determination with a merely assumed rate. On the Paraná-mirim de Canumá, not so to rely, instead of going straight to Manáos from the upper end, I returned to Maués, re-determining the differences of longitude. I may add that the chronometrical longitude of the "Taboleiro" on the Abacaxis agrees almost to one second with that given by the Oc. R., an agreement, however, the closeness of which is of course merely a coincidence.

Longitude of Manáos (Barra do Rio Negro).

At the end of my paper on the River Aquiry (Royal Geographical Society's Journal, vol. xxxvi. p. 126) I gave the results of three occultations observed by me at Manáos. In the second and third of these I have since discovered errors of $13\cdot8^s$ and of $3\cdot6^s$ respectively : the latter due to a slight miscalculation, the former to a strange error—my having taken from the 'Nautical Almanac' the H.P. and semidiameter of the moon (but not its R.A. or N.P.D.) for May 12th, instead of May 29th; the former day being that on which, in 1864, I had observed the first occultation of the three. These errors need throw no doubt over my longitudes from occultations and eclipses on the Purús, &c., as duplicate calculations were made of all but one [*] of

---

[*] ψ Ophiuci Oc. D. at Canotama (on the Purús), where Jupiter Oc. D. and Oc. R. were observed, and both calculated twice over.

these. In the present case, having recalculated the lo
from the first occultation (κ Cancri), I assumed the othei
as they agreed fairly with the first, to be correct ; b
observations led me to suspect them. I now give ag
longitudes from the three occultations: the first as bei
second and third corrected ; and also three others, of cc
twice calculated :—

					h. m.		
1864—May 12.	κ Cancri Oc. D., giving long.	..	4	0	2
1865—May 29.	a Cancri	„	„	..	4	0	0
1865—Dec. 21.	ν Aquarii	„	..	..	4	0	5
1865—Mar. 14.	λ Virginis Oc. R.	„	..	3	59	52
1868—July 22.	χ Leonis Oc. D.	„	..	3	59	54
1869—Feb. 4.	φ Ophiuci Oc. R.	„	..	4	0	0

All at the dark limb of the moon. It is to be notic
the fourth in this list (λ Virginis Oc. R.) gives the lowes
the divergence being on the side on which it would ha
caused by an error of observation. For this reason, th
was thought to have been observed with accuracy, I e
it before ; as, however, it agrees very nearly with
(χ Leonis), and its divergence from the other extreme :
lessened by the discovery of the above-mentioned errors,
thought fit now to insert it. If it be admitted the i
$3^h 59^m 59 \cdot 1^s$: without it $4^h 0^m 0 \cdot 4^s$.

From the eclipse of the sun of February 23rd, 18
served with a power of 300), Sig. Costa Azevedo, Chief
Boundary Commission, determined the longitude of M
be $3^h 59^m 58^s$. Probably this is much the best observati
has been made.

Appended are the results of such meteorological obse.
as I have made at Manáos in the intervals of travel. So
as these observations are over different years, the mean
whole (for each element) is of little value, but the s
results may, perhaps, be interesting. In particular ti
great diurnal variation of the barometer is noticeabl
observations of *maxima* and *minima* of this were, it is i
three months only ; but, during the other months, I hav
times noted and often looked at the barometer betwee
5 P.M., and always found it much lower than at 3 P.M. F
Raimondi (Royal Geographical Society's Journal, vol.
p. 426), on the River Apurimac, found a certainly ast
diurnal variation. The greatest I have observed at
is 0·202 ; at Pará it seems not to be very great. Po
increases towards the Andes.

Of the instruments (all by Casella), the barometer w
pared at Kew in 1863 and 1867 ; but the tube was b

the interval. The thermometers were all compared at Kew, except the air-thermometer used in January and February,

Mean Monthly Results of METEOROLOGICAL OBSERVATIONS made at Manáos, lat. 3° 8' s., long. 60° 0' w.; height above sea 120 feet (*i. e.*, 12 feet below the level of the church—cf. 'Royal Geographical Journal,' vol. xxxix. p. 309).

Year	Month	BAROMETER at 32° Fahr.						
		3 A.M.	9 A.M.	3 P.M.	9 P.M.	Mean.	Max.	Min.
1866	January (2 to 31) ..	29·824	29·923	29·784	29·853	29·846	29·995	29·733
1869	„ „ ..	·785	·81	·766	·811	·811	·931	·705
1866	February ..	·819	·90	·783	·825	·832	·974	·688
1869	„ „ ..	·793	·87	·767	·818	·816	·964	·726
1869	March (1 to 30) ..	·817	·95	·779	·841	·835	·977	·705
1868	April ..	·836	·927	·96	·870	·857	30·000	·759
1868	May ..	·831	·98	·82	·860	·848	29·970	·735
1868	June ..	·871	·98	·88	·91	·889	30·026	·748
1868	July ..	·850	·93	·814	·84	·865	29·980	·776
1869	October (2 to 31) ..	·790	·884	·749	·83	·809	·963	·682
1869	November ..	·766	·858	·723	·94	·785	·969	·655
1869	December (1 to 21) ..	·716	·811	·684	·741	·738	·858	·579
	Mean	29·809	29·897	29·771	29·835	29·828		

N.B. In obtaining the mean, the mean of the 2nd January (or February) results has been used as one quantity.

1866, which I had compared with a compared thermometer, and the self-registering thermometers (1868, 9), which were compared with the compared thermometer then in use for temperature of air.

The thermometers and hygrometer were hung during the day in an inner doorway, where there was usually a fresh breeze, and at night in an open verandah. I do not say the exposure was perfect, but I think it was good, and it was the best I could obtain. In the day the verandah was open to glare from the ground.

The *maxima* and *minima* are respectively the greatest and least heights observed in each month *at the regular hours of observation*. During three months' special observations (at short intervals) gave the following mean results:—

Year.	Month.	Oscillation.	A.M. Max. at		P.M. Min. at	
			H.	M.	H.	M.
1869	February	0ʰ·147	9	17	4	44
,,	March (1 to 30)..	0·157	9	37	4	44
,,	October (2 to 31)	0·154	9	8	4	26

The greatest observed height of the barometer (at 32°) was 30ʰ·029 on June 22nd, at 9ʰ 35ᵐ A.M.: the lowest was 29ʰ·556 on December 5th, at 5 P.M. The range, therefore, observed is 0ʰ·473: this includes the diurnal oscillation. The range at 2 A.M. was only 0ʰ·304.

I have on no day known the barometer *not* higher at 9 A.M. than at 3 A.M., or *not* lower at 3 P.M. than at 9 A.M.; and only on four days *not* higher at 9 P.M. than at 3 P.M.: but on more than twenty occasions it stood higher at 3 A.M. than at the previous 9 P.M: in February, 1866, even four times in five consecutive nights.

Note.—The slope of the three rivers (of which this paper treats) is evidently a very slight one, as shown by the current. I took with me only an aneroid, the indications of which, as it did not alter its zero during the journeys, may, probably, have been correct; but they give elevations so small, that I do not venture to send them. This may be due rather to meteorological causes. On the upper part of the Abacaxis we had frequent showers; but at the mouth of the river it did not rain at all during my absence. It is true that from this one would, *à priori*, expect on the upper river a (relatively) lower barometer; but *à priori* reasoning is often fallacious in these things. The change of weather seemed to take place in about lat. 5° s.

TEMPERATURE OF AIR, Fahr.

Year	Month	8 A.M.	9 A.M.	3 P.M.	9 P.M.	Mean	Mean Max.	Mean Min.	Absolute Max.	Absolute Min.
1866	January (2 to 31) ..	75·6	78·4	81·7	77·1	78·2				
1869	„ „ ..	74·8	79·4	83·4	76·8	78·6	85·1	74·2	90·2	72·3
1866	February	75·2	78·4	82·0	77·1	78·2				
1869	„	75·6	81·8	87·3	78·2	80·7	89·0	74·5	92·6	71·7
„	March (1 to 30) ..	75·2	79·4	83·4	77·2	78·8	85·5	74·2	90·4	71·7
1868	April	74·6	78·3	80·9	76·3	77·5	83·0	74·0	87·9	70·0
„	May	74·4	79·3	83·5	77·1	78·6	85·5	73·5	89·6	70·3
„	June	74·4	79·5	83·5	76·9	78·6	85·8	78·8	89·8	69·7
„	July	72·9	79·8	85·9	76·2	78·7	87·8	71·9	91·7	70·3
1869	October (2 to 31) ..	75·7	81·3	86·0	78·7	80·4	88·7	73·4	96·2	70·9
„	November	75·5	82·0	86·3	78·8	80·6	89·4	78·5	94·6	69·2
„	December (1 to 21)	75·6	81·4	85·8	78·3	80·3	88·5	73·5	94·6	71·3
	Mean	74·9	80·0	84·2	77·4	79·1				

FROM OBSERVATIONS of WET- and DRY-BULB THERMOMETERS by MR. GLAISHER'S TABLES.

VAPOUR-TENSION.

Year	Month	3 A.M.	9 A.M.	3 P.M.	9 P.M.	Mean	Max.	Min.
1869	January (2 to 31)	0·817	0·843	0·857	0·850	0·842	0·950	0·697
"	February	·818	·824	·803	·846	·823	·935	·690
1868	March (1 to 30)	·810	·853	·844	·842	·837	·947	·752
"	April	·810	·841	·849	·844	·836	·931	·735
"	May	·808	·851	·842	·855	·837	·930	·718
"	June	·782	·818	·785	·826	·803	·887	·683
"	July	·739	·781	·731	·778	·757	·846	·602
1869	October (2 to 31)	·788	·809	·770	·814	·795	·934	·566
"	November	·788	·826	·787	·823	·806	·942	·680
"	December (1 to 21)	·811	·857	·841	·849	·840	·980	·703
	Mean	0·796	0·830	0·811	0·833	0·818		

HUMIDITY (0 to 100).

Year	Month	3 A.M.	9 A.M.	3 P.M.	9 P.M.	Mean	Max.	Min.	Rain (inches). Total.	Rain (inches). Most in 24 hours.
1869	January (2 to 31)	93	85	76	91	86	96	52	7·306	3·566
"	February	91	77	64	87	80	94	50	4·549	1·673
1868	March (1 to 30)	93	85	75	90	86	97	59	6·966	1·441
"	April	94	87	81	92	89	96	67	10·136	2·238
"	May	93	86	76	91	86	98	61	3·079	0·560
"	June	91	82	69	88	83	94	59	2·779	1·067
"	July	90	77	60	84	78	91	49	1·831	0·574
1869	October (2 to 31)	90	77	63	85	79	94	43	3·553	1·431
"	November	91	76	64	85	79	94	49	5·547	1·424
"	December (1 to 21)	93	83	69	81	83	93	53	3·865	0·652
	Mean	92	81	70	86	83	..			

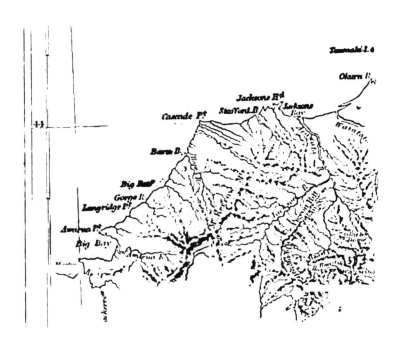

XV.—*Notes to accompany the Topographical Map of the Southern Alps, in the Province of Canterbury, New Zealand.* By JULIUS HAAST, PH. D., F.R.S.

IN a paper read before the Society on the 25th of January, 1864, I had the honour to submit some of the results of my researches into the Physical Geography of the Southern Alps of New Zealand; and in a second paper, printed in the Journal, vol. xxxvii., I described the principal passes and roads which lead through these mountain chains from coast to coast.

In the course of this year I have finished my topographical work in the interior of this province and the county of Westland, and prepared a map, the scale of which, 4 miles to 1 inch, has enabled me to give, with a greater degree of clearness than a smaller scale would have admitted, all the more remarkable features of these Alpine regions—a large extent of which had, previous to my explorations, never been trodden by the foot of man.

In presenting to the Royal Geographical Society a copy of this map, of which the original is in the possession of the Provincial Government of Canterbury, I think that I can pay no more appropriate tribute to the great utility and importance of an institution, by means of which geographical research is encouraged all over the world, and which lends a powerful helping hand to explorers, wherever they are found, irrespective of creed or nationality.

The lower and middle courses of the rivers were all laid down by the Canterbury Survey Department, the work of which generally ceased when the country was not available for pastoral purposes. From thence all the principal rivers on the east coast, and some on the west coast, were chained by myself and my assistants to their sources, generally issuing from glaciers.

In conducting this survey I repeatedly measured base lines, sometimes upon the glaciers themselves, to fix by triangulation the surrounding peaks and other peculiar features of the country.

On the west coast I tried to use the coast line as base line for fixing the orographical features, but found in several instances, when fixing prominent peaks in the Southern Alps proper, that the results did not correspond. This discrepancy was sometimes so considerable, that I was obliged to rely entirely upon the bearings obtained on the eastern side.

Since this map has been constructed I have been informed that the Colonial Marine Survey has found some serious errors both in latitude and longitude in that portion of the coast-line situated between Jackson's Bay and the mouth of the River Grey, which may account for the different results alluded to.

Owing to the rugged and precipitous character of the western side of the ranges, the difficulty of obtaining provisions, the matted and almost impenetrable nature of the forest vegetation, covering the lower portions of the mountain sides, the wild and impassable mountain torrents, as well as from want of time, I was unable to obtain such good and exhaustive sets of bearings as I had anticipated.

Thus I was only permitted to ascend a few of the western rivers to their glacier sources: in some other cases I obtained only a limited number of bearings, and in a few instances I had to fill up some portions of the map from eye-sketches; but nevertheless I may state my conviction that it will be found that none of the more important features have been overlooked by me, when future explorers in years to come may have more leisure at their command and less difficulties to contend with than I had during the eight years I devoted to researches into the geology and physical geography of this portion of New Zealand.

In the course of last autumn (March and April) I paid a second visit to Mount Cook and its immediate neighbourhood, principally with a view to collecting specimens illustrative of the natural history of the Southern Alps for the Canterbury Museum. During that journey I was accompanied by my friend Mr. Edward Seely, Government District Surveyor and an excellent amateur photographer, who, under great difficulties, took a number of very interesting photographs by means of dry plates along the principal glaciers, and of which he intends to present a set to your Society.

Some important changes have taken place at the terminal face of the Great Tasman Glacier since I visited it in 1862. The glacier has advanced down the valley in its central portion about half a mile, and has here also found an outlet—the one on its eastern side still remaining, the only one existing when I first visited it.

Formerly it was possible to walk along the terminal face of the whole glacier from west to east without meeting with any watercourse. Some changes in the appearance of the country have been brought about by the hand of man. The sheep-farmers have slowly crept up the river, in a great measure depriving the valleys and mountain sides of their magnificent subalpine vegetation by burning; thus changing the rich and varied tints of the graceful plants—both trees and shrubs—into a black mass of charred stumps and sticks now scantily covering the rocks. Even the southern slopes of the Mount Cook Range have not altogether escaped, the little grass-flat situated there—bounded on one side by the outlet of the Tasman and Murchison, on the other by that of the Hooker and Müller glaciers—has also been burnt, being now used as a ram-paddock.

I think that after this no other instance is needed to show that the Anglo-Saxon colonists of New Zealand are not deficient in enterprise and perseverance. There is, in fact, a fine race of mountaineers growing up in the interior of this island who in years to come, for strength and endurance, will fairly rival the hardy inhabitants of the European Alps.

Although I have, in the papers read to you, as well as in some others printed in the Journal of the Geological Society, given a general outline of the peculiar features of the Southern Alps, it may not be superfluous to offer a short *resumé* of their principal characteristics, in illustration of this topographical map.

In former publications I have indicated at some length the causes by which the present configuration of the Southern Alps has been brought about. I have endeavoured to show that the more plateau-like character of the ranges, before the great glacier period of New Zealand made its appearance, caused the accumulation of enormous snow-fields, and consequently the formation of gigantic glaciers. The latter have left their mark everywhere behind them, either by planing down the mountain sides, so that sharp or pyramidical peaks were formed, and by scooping out deep valleys or by carrying down enormous masses of *débris* to lower regions, either to throw up gigantic moraine accumulations, or to form extensive plains by the power of the huge torrents issuing from them, to fill up large tracts of country with boulders, shingle, sand, and ooze. And whilst the huge glaciers descended on the eastern side to 1500 feet, or even in one instance to 800 feet above the sea-level, at the more precipitous western side of the ranges they not only reached to the sea-level, but doubtless caused the formation of huge icebergs, strewing their moraine loads over the Pacific Ocean in comparatively low latitudes. The same difference in the relative position of the terminal face of the existing glaciers, the small remains of their gigantic predecessors, still exists.

Some of the west coast glaciers, such as the Francis Joseph and Prince Alfred Glacier, descend to such low positions as 700 feet above the sea-level, their terminal faces being close to a luxuriant forest vegetation, consisting of pines, arborescent ferns, and flowering shrubs, the remains of which, embedded in morainic accumulations now forming amongst them, would considerably puzzle the geologist of the future did he not possess the key to the explanation of such a phenomenon.

The glaciers of the eastern slopes, although being of much

larger proportions than those of the opposite side, nevertheless descend no lower than 2500 feet above the sea-level, partly owing to the fact of the ranges having a far more gradual slope and partly to the smaller amount of moisture which falls here.

The amount of rainfall on the western and eastern sides of the Southern Alps at once explains some of these principal characteristic features, the western having about four times more rain than the opposite side.

All the principal meteorological phenomena which are encountered in the European Alps, and which have been described and explained so differently, according to the point of view taken by each writer individually, also occur here, the norwester of New Zealand (equatorial current) being simply the Föhn of Switzerland or sirocco of Italy.*

As formerly pointed out, the snow-fields and glaciers of the Southern Alps, when compared with those of Europe, are of much larger dimensions, especially if we take the altitude of the mountains into due consideration. That they were formerly of still more gigantic proportions is, amongst other indications, well shown by the line of lakes on both sides of the southern Alps, and the enormous moraines surrounding them, which mark clearly the latest extension of the post-pliocene glaciers.

However, there are ample evidences that the glaciation of the country, anterior to this last well-marked event, has been on a still larger scale, by which even the front ranges have been affected to a considerable extent; in fact, the whole island having apparently been covered by one mass of snow and ice.

The continuation of the former glacier-course, of nearly the same breadth as the lake, and sometimes 20 to 25 miles long, showing the close relation between glacier and lake, as cause

---

* Discussions such as those going on for the last few years between Professor Dove of Berlin and some of the principal scientific men of Switzerland, would have been much simplified had those gentlemen been acquainted with all the characteristic features of our norwesters, which are in every respect identical with the phenomena described by Professor Dove, with whose writings I am best acquainted, and with whose conclusions I entirely concur, as being the characteristics of the föhn. In fact, his description of the föhn from its first setting in on the Italian side of the European Alps, its crossing and effects on the Swiss side is such, that if we change the word Italian for western, and Swiss for eastern, side, every inhabitant of this island who has travelled across it would consider it a faithful description of our norwesters as travelling from coast to coast. However, I may point out that occasionally our norwesters do not bring rain with them when crossing the height of land, having descended before they reached the southern Alps, thus becoming deprived of the principal portion of their moisture on the sea or on the low lands lying at the western foot of the ranges.

When these winds pass across the snowfields of the southern Alps, the cumulus clouds creeping up disappear as by enchantment, and the sky remains of a deep blue colour, but the wind sweeping down the valleys is very hot, and the rising of the glacier torrents shows at once its effect. A theory tracing them to the interior of Australia would be difficult to prove.

and effect, is very striking and suggestive, and for which only a mechanical theory can account; since, if Abysso-dynamic causes had been at work to effect such formations, we should be obliged to admit one of a different character for each particular case, owing to the different age, strike, dip, or character, of the rocks in which the glaciers have done their work.

There are, besides the passes I have described in a former communication, a few more crossing the Southern Alps, with which I have since become acquainted, and by which the ranges are divided into so many systems or *massifs*.

West of Arthur's Pass rises Mount Rolleston, partially covered with perpetual snow, from which a few glaciers of second order descend, and the highest point of which is about 7800 feet high. Further west, where the northern-source branch of the Waimakariri takes its rise, that mountain-chain which hitherto had preserved the character of a *massif*, breaks up into pointed peaks of less altitude, and with two low saddles between them, both leading into the head-waters of the Taipo, the chief tributary of the Teramakau (west coast). The lower of these passes—Harman's Pass, is 3980 feet high, or about 800 feet above the white river; but it would be of no use for ordinary traffic, as, along the bed of the Taipo, the mountain-sides are exceedingly rocky and precipitous, and the river-bed is very liable to be filled by avalanches from the western ranges.

South of this depression the Alps rise again to a more considerable altitude, forming a cluster of mountains, for which the native name Kaimatau has been preserved. In this *massif* the northern branches of the Rakaia, the main branches of the Waimakariri east, and the Arahura and Taipo (west), take their rise.

A well-defined depression, named Browning's Pass, from one of its discoverers, ends this portion of our Alps. The more remarkable features of the chains south of this pass I have previously described, and I therefore will only allude to another pass, which leads from near the sources of the Clarke River, a tributary of the River Haast, to the head-waters of the Makawiho (west coast), and which was discovered a few years ago by a miner, W. Doherty. This pass seems, according to his description, and the vegetation which he describes as growing there, to be about 4000 feet high. It is the same depression I observed in 1863 from Mount Brewster, and which separates the magnificent cluster of mountains south of it, of which Mount Hooker is the most conspicuous, from the northern ranges.

My main object being to present the topographical map to the Royal Geological Society, I shall only add a list of altitudes of the principal points obtained either by barometrical measurement, by the spirit-level, or, in a few instances, by the boiling-point of water, reserving a connected narrative of my journey for a future publication.

TABLE OF ALTITUDES.

River Waitaki.

	Feet.
Great Tasman Glacier, terminal face	2456
Junction of Hochstetter Glacier, ditto	4850
Murchison Glacier, ditto	3540
Müller Glacier, ditto	2578
Hooker Glacier, ditto	2091
Junction of Hooker with Tasman River	1296
„ „ Jollie	2014
Great Godley Glacier, terminal face	3593
Classen Glacier	3593
Junction of Grey with Gooley Glacier	4852
Separation Glacier between Mount Forbes and d'Archiac	4383
Macaulay Glacier, terminal face	4375
Junction of River Macaulay with River Godley	2611
Huxley Glacier, terminal face	5242
Faraday Glacier, ditto	4723
Richardson Glacier, ditto	4251
Selwyn Glacier, ditto	4311
Hourglass Glacier, ditto	3636
Junction of River Dobson with River Hopkins	2086
Junction of Holmes Creek with Hopkins	2190
End of terminal moraine, Ahuriri Valley	2464
Junction of Pukaki with Tekapo River	1547
„ „ Ohau with Waitaki	1475
„ „ Ahuriri with ditto	1168
„ „ Hakateramea with ditto	781
Lowest moraine accumulation in Waitaki Valley	716

Molyneux River.

	Feet.
Junction of Fish Creek with Makarora	1362
„ „ Blue River with ditto	1210
„ „ Wilkin with ditto	1068

Rangitata River.

	Feet.
Havelock Glacier, terminal face	3909
Forbes Glacier, ditto	3857
Clyde Glacier, ditto	3762
Tyndall Glacier, ditto	3950
Lawrence Glacier, ditto	4061
Junction of Forbes River with Havelock River	2871
„ „ M'Cay River with Clyde River	3269
„ „ Lawrence River with ditto	2284
„ „ Havelock River with ditto	2192
„ „ Potts' River with Rangitata	1762
„ „ Pudding stone Valley stream with ditto	1440
Where Rangitata enters Canterbury Plains	1150
Banks above it, uppermost terrace	1443

Ashburton River.

Ashburton Glacier, terminal face	**4882**
Junction of lower branches	2511
„ „ Clearwater Creek with Ashburton	1832
Two Brothers Rocks, where Ashburton enters Canterbury Plains	1326

Rakaia River.

Ramsey Glacier, terminal face	8354
Lyell Glacier, ditto	3568
Martius Glacier, ditto	4268
Junction of Whitecombe stream with Rakaia	2958
Hawker Glacier, forming Cameron River	4478
Junction of two glaciers, forming Hawker Glacier	5667
„ „ River Cameron with Rakaia	2052
Nerve Glacier, forming River Matthias	3786
Junction of two main branches of Matthias	2236
„ „ Matthias with Rakaia	1688
Stewart Glacier, terminal face	3584
Junction of two main branches of Stewart	3090
„ „ Stewart River with Wilberforce	2874
Camp Creek junction with ditto	3041
Junction of River Harper with River Avoca	2103
„ „ Western branch ditto	2531
Avoca Glacier, terminal face	4749
Junction of two source branches of Avoca	3416
„ „ River Harper with Wilberforce	1610
„ „ River Wilberforce with Rakaia	1857
„ „ River Acheron with ditto	1064
Gorge, where Rakaia enters Canterbury Plains	875
Canterbury Plains, above its uppermost terrace	1410

Waimakariri River.

Waimakariri Glacier, source of White River, terminal face ..	4162
Junction of two main source branches	2607
Ponds, sources of Northern branch	5141
Junction of Crow River with Waimakariri	2273
„ „ Bealey River with ditto	2065
„ „ Poulter River with ditto	1621
„ „ Esk River with ditto	1562
„ „ Porter River with ditto	1362
„ „ Kowai River with ditto	1003
Gorge Hill, in Canterbury Plains, river-bed	886

River Hurunui.

Eastern foot of Harper's Pass	2452
Junction of South Hurunui with Main River	1380

River Haast.

Brewster Glacier	4810
Junction of leading creek with River Haast	1510
„ „ River Wills with ditto	723
„ „ River Burke with ditto	892
„ „ River Clarke with ditto	248

River Waiau.

Francis Joseph Glacier, terminal face **705**

River Weheka.

Prince Alfred's Glacier, terminal face **702**

River Teramakau.

Western foot of Harper's Pass	**1781**
Junction of Otira with Teramakau	**719**
,, ,, Taipo with ditto	**350**
,, ,, Waimea with ditto	**148**

River Hokitika.

Sale glacier	**4183**
Hokitika River, where it enters West Coast Plains	**429**
Junction of Kokotahi with Hokitika	**178**

Lakes.

	Feet
Lake Hawea	1078
,, Wanaka	992
,, Pukaki	1717
,, Ohau	1837
,, Tekapo	2437
,, Alexandrina	2460
,, Acland	2205
,, Tripp	2228
,, Heron	2255
,, Browning	4616
,, Coleridge	1694
,, Selfe	1962
,, Ida	2304
,, Lyndon	2743
,, Taylor	1948
,, Summer	1702
,, Pearson	2065
,, Letitia	2079
,, Blackwater	2023
,, Poerua	345
,, Brunner	227
,, Kanieri	468
,, Hall	229

Passes.

	Feet
Harper's Pass	3008
Arthur's Pass	3013
Harman's Pass	3980
Browning's Pass	4752
Whitecombe Pass	4212
Haast's Pass	1716
Porter's Pass, leading into Waimakariri country	3026
Lake Lyndon Pass, from Rakaia into Waimakariri	2696
Burke's Pass, from Opihi to Lake Tekapo	2464
Walker's Pass, from Opihi to Waihi	720A

Tripp's Pass, from Oran to Opuka 2255
Pass between Cass and Godley Rivers, near Huxley Glacier 6565
Fraser's Pass, between Lake Pukaki and head of Lake Ohau 3992
Pass between Selwyn and Hawkins 1687
  "   "   Rubicon and sources of Kowai 3705
  "   "   Mount Somers and Mount Somers Range.. .. 3684
  "   "   sources of North and South Hireds 3025
  "   "   Ashburton and Rakaia, near Lake Heron .. 2290
  "   "   Ashburton and Rangitata, near Lake Tripp .. 2360
  "   "   Hurunui and Waitohi 4858

Mountains.

Mount Cook, highest summit 13,200
  "   Tasman, ditto 12,320
Sefton's Peak, in Moorhouse Range 11,690
Mount Hutt 7016
  "   Dobson, near Lake Tekapo 6271
Observation Mount, near Macaulay River 7862
Mount Sinclair 7022
Clenthills, highest point 4212
Ribbonwood Range 5662
Mount Harper 5216
Sugarloaf, Rangitata (Rochemont) 3268
  "   Lake Heron 3822
Mount Brewster, first peak 7200
  "   Torlesse 6136
Big Ben, Thirteen-mile Bush-range 5294

Miscellaneous.

Line of perpetual snow, south-eastern side Mount Cook .. 7800
  "   "   western ditto 6900
End of Fagus Forest in River Hopkins 3180
  "   "   River Dobson 3280
Limit of Fagus Forest on Mount Brewster 4320
  "   Alpine shrub vegetation, ditto 4920
  "   Fagus vegetation in Rakaia Valley 2430
  "   "   "   Wilberforce Valley 2360

XVI.—*On Surface Temperatures in the North Atlantic.* By
Admiral IRMINGER, Copenhagen, Corr. Mem. R.G.S.

SINCE I took the liberty of addressing a letter to you on the
21st of April last year, and which I see published in the 'Proceedings of the Royal Geographical Society,' vol. XIII. No. 3,
I have, in continuance with that letter, examined more
minutely the warmer streaks which are found on the surface of
the ocean, between Fair Island and Greenland, and I beg leave
to present you the results of it.

The drift or slow current in the North Atlantic Ocean,

coming from more southerly and more heated regions, gives this ocean a comparatively high temperature on its surface.

From Fair Island (about 59° 28' N. lat. and 1° 55' w. of Greenwich), and the direction towards Greenland, which usually is followed by the vessels belonging to the Royal Greenland Trade, between Copenhagen and our colonies in Greenland, the ocean has on its surface, until about 30° w. of Greenwich, and sometimes even not so far to the westward in the same season, a tolerably equal temperature, not varying more than 3°·6 to 5°·6 Fahrenheit.

In the annexed table I have noted seven voyages, made in the warmer as well as the colder seasons. They will give an idea of the temperatures of the Ocean.

This passage being but little frequented during winter, I do not possess any observations for voyages made in January and February; but, according to observations made at Thorshavn of the Faroe Isles, during the years 1846 and 1847, the mean temperature of the surface of the sea for December, January, February, and March, had only a difference of 2°·9 Fahr.,[*] and probably the temperature in the wide and open ocean being more constant than on the beach of the more enclosed Thorshavn, I think the difference from December to March (for which months voyages are noted in the Table) will not be very different from January and February.

Between the most westerly meridians mentioned in the table, and on approaching Greenland, the temperature fell quicker, which will be seen in the chart I sent you last year.

In order to examine the temperature of the ocean in about the same season of different years, I have quoted in the annexed Table, a voyage from May, 1844, and September and October, 1846, and compared these two voyages with two others, made in May, 1868, and October, 1867. On considering the different dates of the year, and the warming influence of the season, in which these voyages are made, it will be shown that the mean temperature accords tolerably between these voyages. The voyage in May, 1844, was made a little earlier in the year than the voyage in May, 1868, and gives a difference in the temperature of 1°·2 Fahr., and the voyage in September and October, 1843, was somewhat earlier than the voyage in October, 1867, which gives a difference of 0°·1 Fahr. If the said voyages had been made on the same dates of the year, the mean temperatures of the surface of the sea would undoubtedly have been still more corresponding.

---

[*] 'Havets Strömninger,' af Capt. C. Irminger. Archiv for Söndhond, 1852. Kiöbenhavn.

As the voyages quoted in the Table are made in the warmer as well as in the colder seasons, the mean temperature between Fair Island and the westerly meridians before cited—up to which the ocean, at the same season, did not vary more on its surface than from 3° to 5°·6 Fahr.—is not considerable for a mean of all voyages together, namely, 5°·89, as the mean temperature for March was, lowest, 45°·68, and highest, for July, 51°·57 Fahr. The greatest difference observed between the highest and the lowest temperatures was 10°·8, as the lowest temperature was 43°·7, on the 15th March, off Fair Island, and the highest 54°·5 on the 8th of July in 4° w. of Greenwich and 59¾° N. lat.

Besides the observations of the temperature on the surface of the ocean, I likewise have the temperatures of the air; and it is astonishing to notice how much equality there is on the open ocean between the temperatures of the sea and the air—except with gales from the northern quarters, during which the air generally is colder than the sea; but if the weather is only tolerable the difference is usually very insignificant.

The little difference of temperature which exists on the surface of this part of the ocean, can, on the contrary, not be attributed to the varying of the limits of the warmer water, which runs like streaks through this part of the ocean.

The Table shows the temperature of the warmest bands or streaks, whose eastern and western limits sometimes are found to be tolerably abruptly limited by the surrounding sea in which they have their course.

The Table shows where these streaks have been found, and I have indicated, as nearly as can be done, the extent of these warmer streaks from east to west in nautical miles.

In limiting the breadth of the warmer streaks, I have taken only the warmest belts; and it will be understood that the streaks would have become broader—particularly where they had not somewhat sharp limits—by allowing the temperatures observed on the voyages a greater scope.

To decide the longitude in which these streaks are to be found, in the different seasons, cannot be done (as the annexed Table will show, by comparing the voyages of May, 1844, and May, 1868, as well as the voyages made in September–October, 1846, and October, 1867), as the streaks indicate no great regularity, neither in their limits nor in their breadth from east to west, crossed by the vessels on their passages to and from Greenland.

The observations made elucidate, however, so much: that the warmer streaks are found on every voyage, and that usually two are to be found, one of which is met somewhat to the west of Fair Island, whereas the other is considerably more to the west

e NORTHERN ATLANTIC OCEAN.

Temperature at the most Western Place, Rubric 2.	The Warmest Temperature on the Surface.	Places with the Warmest Temperatures.		Extent of the Warmest Temperature in East and West.	The Coldest Temperature.	Greates Differen
		W. of Gr.	N. Lat.	Nautical Miles.		
44°·6, 23rd May.	47°·8 to 48°·9 47°·8 to 48°·9	5° to 9° 21° to 22½°	59¾° 59½°	120 40	44°·4 in 60° 30′ N. lat., 28° 30′ long. W. of Greenwich, 21st May.	40°·5
49°·3, th September.	52°·3 to 53°·1 52°·3 to 52°·7	Between 21° and 24° 8° and 9°	60° 59½°	80 25	49°·3 in 59½ N. lat., 30½ long. W. of Greenwich, 27th September.	3°·8
45°·5, 26th May.	49°·3 to 50°·0 49°·3 to 50°·0	6° and 7½° 14½ and 19°	60½° 59¾°	45 130	45°·5 in 58° N. lat., 32° long. W. of Greenwich, 26th May.	4°·5
48°·2, th October.	51°·1 to 51°·8 51°·1 to 51°·8	19½° and 20½° 2° to 5°	60½° 60°	30 90	48°·2 59° 18′ N. lat., 30° long. W. of Greenwich, 9th October.	3°·6
44°·4, 18th March.	47°·8 to 48°·9	22° and 24¾°	59°	80	43°·7, 15th March, close to Fair Island.	5°·2
50°·0, 28th July.	52°·3 to 54°·5 52°·3 to 53°·4	3° to 5° 26° to 30°	59¾° 60½°	60 120	48°·9 60° 50′ N. lat., 7° 15′ long. W. of Greenwich, 10th July.	5°·6
44°·4, h November.	47°·8 to 48°·4 47°·8 47°·8	20° to 28° 12¾° to 13½° 2° to 5°	57° 59¾° 60½°	250 12 90	44°·4 56° N. lat., 33° long. W. of Greenwich, 26th November.	4°·0

gth are nautical miles = 60 on 1 degree of the meridian. The longitudes are from Greenwich

—C. Irminger.

in the ocean, and sometimes even more westerly than the meridian of the south-westernmost land of Iceland, Cape Reikianæs, in 22° 50′ w. of Greenwich. Likewise it will be observed by the Table that these two streaks have about the same temperature.

The Gulf Stream is so well known that I shall not enlarge upon it, but only say, that not only its limits are very variable in the different seasons, but likewise the breadth of the stream, which is stated by many observations given by Major James Rennell and others.

That the Gulf Stream, after having passed the Bank of Newfoundland, from whence only a branch goes to the north-east, could spread, or cover, the Atlantic, in the whole breadth, where an equal and comparatively high temperature is found on the surface, I think less probable, as the warm water of this stream is not to be found in greater depths, and the volume of the heated water is scarcely so considerable, as many believe it; besides this, it is regarded as a fact that the water of the Gulf Stream is but little inclined to mix with the waters of the surrounding ocean.

It must be well remembered that the equal high temperature in the Atlantic Ocean on the route from Fair Island to Cape Farewell extends until between 30° and 40° w. of Greenwich. Fair Island is in nearly 2° w. of Greenwich, and even when these 2° are subtracted from the longitude of the western meridians, indicated in the Table, where the nearly equal and high temperature is found on the surface, it will give at least 30° of longitude, or a distance of more than 900 nautical miles of the ocean from Fair Island towards west. The many thousands of square miles in the North Atlantic, which thus are found warmed, I think must be ascribed to the drift to the north of the great wide Atlantic from about 40° N. lat., and as a branch of the Gulf-stream, following the drift of the Atlantic in a northerly direction, I think it highly probable that *only the warmest streaks*, which always are crossed by the vessels passing the Atlantic between Fair Island and Greenland, can be admitted to be branches of, or be connected with, the Gulf-stream itself, which, by constantly succeeding confluence of the warmer water from the stream, in this manner maintains a higher temperature than the surrounding ocean. Why these streaks are met with at times more easterly or westerly I have stated in my last letter.

The above-mentioned warmer streaks may be followed much farther to the north; and as for those which find their way between Iceland and Norway, they are met with even up in the icy sea, which, according to my opinion, is proved by the discoveries of Parry, Scoresby, and so many other distinguished navigators.

Between 62° N. lat. and the south coast of Iceland, and 18° and 23° W. of Greenwich, nearly the longitude of Cape Reikianæs, south-west cape of Iceland, the current has been found proceeding in a north-westerly direction, and on the west coast of Iceland the current is to the north. This drift of the North Atlantic, as well as the warmer streaks noted in the table to the westward of about 18° W. of Greenwich, wash the south and west coasts of Iceland, and continue proceeding between Iceland and Greenland, until stopped by the Arctic current coming from the sea around Spitsbergen.

The high temperatures observed on the surface near this part of Iceland prove sufficiently the presence of a current coming from more heated regions; and as a proof—shown by the description of the currents and icedrifts near Iceland, which I took the liberty of sending last year—I shall only mention that 8°·4 Reaum. = 50°·9 Fahr. is marked E.S.E. from Cape Reikianæs, 8°·8 Reaum. = 51°·8 Fahr., also some miles north-west of Snefelsjákul, in 65° N. lat., and 7°·6 Reaum. = 49°·1 Fahr. in nearly 66° N. lat., north-west of Patriksfiord, close to the limits of the Arctic current, in which, but 30 miles farther north, the surface of the sea was found only 0°·2 Reaum. = 32°·4 Fahr.

By the seven voyages marked in the Table, *two* warmer streaks have thus always been found in that part of the ocean I have described, with the exception of the voyage of the brig *Etna*, where a third narrow streak was crossed in about 13° W. Greenwich, and on the voyage in March, 1869, for which I have only marked in the Table the most western streak; then from Fair Island, until 15¾° W. Greenwich, the surface was frequently colder than the mean of the voyage 45°·68. Still 45°·9 to 46°·6 was found between 7½° and 11¼° W. Greenwich, and though this being higher than the mean temperature, I found the difference so inconsiderable, that I did not think it right to mark this as a warm streak in the Table.

How far west some isolated warmer streaks may be found now and then, I can quote the following:—May 6th, 1869, Captain Bang, brig *Constance*, on her home-passage from Greenland, was beating against contrary winds between 53½° and 54° N. lat. and 40° and 41° W. of Greenwich. From the coast of Greenland, where no ice had been in sight, the temperature gradually had risen from 33°·6 to 41° and 42° Fahr., when that of the sea pretty suddenly rose to 45°·5. With this rising of the temperature in the surface the thermometer was frequently used, and the temperature rose even to 46°·6, and in this warmth of the surface the brig sailed 16 nautic miles true N.N.W. Likewise the temperature of the air rose from 41° to 45° to 45°·5 during the brig's being in this warmer streak. A few miles more east the temperatures of the sea and the air fell again to $42°·6$ to $42°$; on $39\frac{3}{4}°$ w. Greenwich the thermometer marked in the surface again $45°·5$, but fell very soon to $42°·6$. The surface of the sea was afterwards between $41°$ and $43°·2$, and first in $58°$ N. lat. and $30\frac{1}{4}°$ w. of Greenwich the temperatures of the surface of the sea and the air rose to $45°·5$, and shortly after to $47°·8$ to $48°·9$.

After all that I have explained in the above, it seems to me that probability speaks for admitting that the warmer streaks noted in the Northern Atlantic, which are crossed on every voyage between Fair Island and Greenland, are branches of the Gulf-stream. The westernmost warmer streak connects undoubtedly with that part of the Gulf-stream which passes nearest the banks of Newfoundland, while the streak more to the east probably has followed that part of the Gulf-stream which, according to the opinion of Rennell and others, proceeds in a direction towards Europe, after which it bends still more to the north, and thus gets a more eastwardly run, passing nearer Fair Island.

The mild winter-climate which is found on the western coasts of Europe, I suppose, can however not be ascribed to the Gulf-stream *alone*, but chiefly to the great Atlantic Ocean, over which, particularly during the colder season, a comparatively high temperature of the air is found, which, with the reigning south-westerly and westerly. winds, is carried to the coasts of Europe.

Copenhagen, February 5th, 1870.

XVII.—*Topographical Sketch of the Zarafshan Valley.* By Mr. A. FEDCHENKO, Professor in Moscow University.

(From the Russian,—communicated by ROBERT MICHELL, F.R.G.S.)

THE portion of the Zarafshan Valley occupied by the Russians, includes only one-fourth part of the course of that river. The river terminates within the limits of the Khanat of Bokhara. This part of its course has been described by several travellers. The upper course is far less known. Lehmann in 1841 proceeded up the Zarafshan Valley, and his is the only account we have relating to the head waters. But even he got only as far as the Fan Rivulet, and by a long way failed to reach the sources of the Zarafshan. At that time, and long after, the lands along the upper course of the Zarafshan, as also along the right bank of the Oxus, belonged to the Emir of Bokhara. But as the power of that potentate declined before the aggressions of Russia, the captures of Ura-Tiubé and Jizakh involved

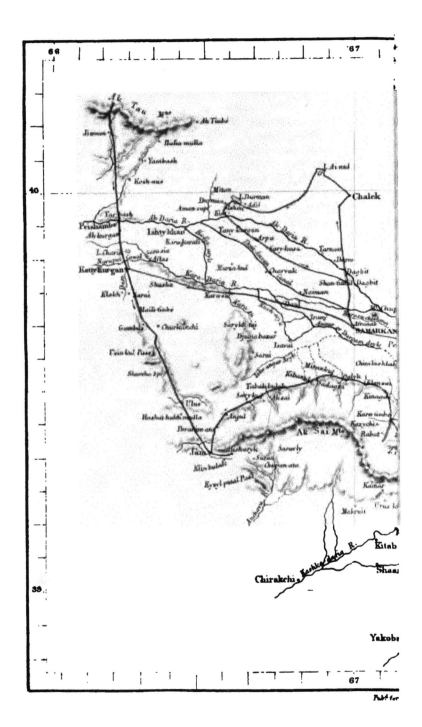

the defection of the Bekships in the mountain regions. Macha and Maghian became then the centres of independent principalities at the head of the Zarafshan Valley. The town of Macha is, according to accounts given by the natives, situated in the same meridian as Kokan from which it is not more than about 27 miles distant. The sources of the Zarafshan are said to be somewhat still farther to the east.

It is asserted by the natives that the river first flows under the name of the Macha-daria. "There are no sources of the Zarafshan-daria," said the inhabitants of Pianjakent to Mr. Fedchenko; "there are the rivers *Macha-daria, Fan-su, Kshtut-su* and *Maghian-su*, which join together, and so form the Zarafshan-daria;" "the Zarafshan," they added, "is so called by the people of Miankal, situated on its banks at its mid-course, but to the inhabitants of the hill-district it is not known by that name." * Similar statements were made in answer to Mr. Fedchenko's inquiries about the sources of the Oxus,—"Five rivers," they replied," "flowing from the east combine in forming the Piandj-daria (meaning five rivers) which is called the Amu-daria lower down." As Mr. Fedchenko did not himself penetrate farther east than the Fan River, he found it utterly impossible to learn from the natives which was the largest stream of those they mentioned; whereabouts they sprang from, and what was their general direction. The people were not even positive about the names of those rivers. It may, however, be reasonably considered that the river rising east of Macha is the main head-stream of the Zarafshan, for it runs almost in a direct line from east to west. It is only in the Khanat of Bokhara that the river first makes a bend towards the southwest, and finally flows towards the south. All the above named affluents run into the Zarafshan from the south. From the north the streamlets which approach it are for the most part diverted for irrigation purposes into canals. Thus there are a great many more valleys and gorges in the mountains on the south, separating the water systems of the Oxus and the Zarafshan, than in those on the north side of the valley, between the latter and the Jaxartes. This frequency of defiles on the northern slopes is characteristic of the whole of the Thian-Shan. Baron Osten-Sacken has made the same observation with reference to the Thian-Shan in the meridian of Issyk-kul.† On the

---

* The Zarafshan has been subsequently traced to its source by a Russian military detachment, led by General Abramof. The valley is said by the Russians to be closed at the top; the Zarafshan was found to issue from a stupendous glacier, extending from 32 to 37 miles up the valley. The same detachment reached the Iskander-kul. This survey was performed in the month of June, 1870.

† See 'Mém. de l'Acad. de Science de St. Pétersbourg.' S. vii. T. xiv. No. 4, p. 5.

north side of the valley, that is, in the southern slopes of the range along the right bank of the Zarafshan, there are only a few small defiles, such as the Jora, Mindanaù, and the Vichy, whilst opening upon the basin of the Zarafshan from the south there are several very wide and branching valleys, occupied by the streams above named, and with their feeders. These valleys are so very spacious that the entire independent Bekship of Maghian is located within one of them, though it is true that this principality is insignificant and poor.

The first of these valleys is that of the Fan, which appertains to the Macha Bek-ship; its wild and lovely scenery is described by natives in strains of extravagant enthusiasm. Of all the lateral valleys this is the most interesting, and it is known to us through Mr. Lehmann, who visited it with the mining-engineers sent by the Russian Government, at the request of the Emir of Bokhara, in search of gold.

Lehmann relates that following up the course of the Zarafshan, he and his party reached Varsaminar, where they crossed the river, and entered the Fan valley; here they passed a fort, called Sarvada, and reached a burning mountain. At this place coal was found, and Lehmann explains the phenomenon of the burning-mountain by attributing it to the ignition of the carboniferous strata. Among the people of those parts this mountain is famed for a variety of minerals procured from it. Thus, in the bazaar at Samarkand Mr. Fedchenko saw *sal-ammoniac*, alum, and a certain substance unknown to him, called *Zak-Sia*, used in black paint.

From the burning mountain Lehmann turned away to the east, reaching the village of Fan. From here, in consequence of some misunderstanding with their guide, the travellers had to turn back, journeying westwards along the Pasriut rivulet, and leaving the Alpine lake of Iskander-Kul on the left. Lehmann discovered only one lake in those parts, but every one referred to by Mr. Fedchenko assured the latter there were two—one called Alaùdin-kul, which, judging by his map and description, was the one Lehmann saw; and the other Kulikalan, a large lake, otherwise styled Iskander-kul, in honour of Alexander the Great, who, of course, was never there.

The lake Iskander-kul lies in the mountains separating the Zarafshan from the Oxus. It must be well known to the inhabitants of that country, since there is a road from Varsaminar to Hissar which passes by it. The lake is situated at the great height of 7000 to 8000 feet, and is surrounded by high mountains, from which some small streams pour into it. The road from the lake to Hissar is described as being very difficult; the natives affirm that the watershed can be traversed only on foot, for which reason travellers dispose of their horses in the village at the foot of the mountains, and procure fresh animals on the other side. In the winter the lake is frozen. From the description of the lake given by the natives, its length may be said to be about 8 miles, and its width about 7; its shape is oval, narrowing to the north, where a stream issues from it; this stream is small, and is not the main source of the Fan. This will in a measure serve to explain why Mr. Lehmann and his companion did not suspect that they were within only 17 miles of an alpine lake. A great many stories are related about this lake by the natives; for instance, they say that Adam-oba, *i.e.*, water people, live in it. What can this mean? It is hardly possible that there are seals in the lake.

We now pass on to the defiles farther west, within the Bekship of Maghian. In the month of June, when it was seen by Mr. Fedchenko, the Maghian river was deep and most remarkably rapid. It is spanned by a bridge in the village Sudjana, where it flows through a hilly country, but at Kostarach village it breaks impetuously through the rocks and falls in a series of cascades.

The defiles are visible from a great distance, going from Pianjakent to Dashty-kazy. The mountains on the left side of the river stand 10 miles off at one point, and slantingly approach the bank; the intervening space being occupied by undulating hills. The Mazar-taù, a massive range, approximatively 12,000 feet high, rises between the two valleys of Maghian and Kshtut. On the left side, by the Kostarach village, two snowy peaks, called the "Shin" raise their hoary heads to a great height, and beyond them, in the cast, are visible the stupendous massy heights of Kshtut, with their numerous snow-clad crests. A peculiar beauty is lent to the grandeur of the scene by a confused distribution of various flowers. The foreland of undulations terminates by the Zarafshan in precipitous masses of rosecoloured clay, several hundred feet thick. Farther away in the Mazar-taù, is discernible a white streak of bare limestone; above this, like a broad, bright green ribbon, lies a zone of brushwood, then another barren streak, and over these tower the snow-capped heights of the mountains.

The Maghian bekship includes also Farap-Kurgan, situated at the sources of the Hujaman-Su, a river watering the principality of Shahr-i-Subz. The pass from Maghian into the valley of the Hujaman-Su is said to be very low, so that Maghian must be at a very great height, probably at an elevation of about 6000 feet. Farap-Kurgan is a small town with only one village in its vicinity, which is called Musa-Bazar.

The mountains stretching in a parallel line with the course of

the Zarafshan, along the northern side of that river, terminate
a little to the west of Pianjakent, where the valley suddenly
widens. On the southern side this valley is skirted by a continuation of the Mazar-taù, bearing the name here of the
Shahr-i-Subz mountains. On the northern side there are several
ranges, the Godun-taù, a small oblong-shaped range, stretching
directly north of Samarkand within a distance of 24 miles ; and
the Ak-taù, in a line with the former, running from near Ak-
tiubé village, parallel with the Zarafshan, and 20 miles off these
two mountain chains are linked by a series of tolerably high
hills called Karadal. Although broken, and not very high, these
mountains may be taken as the northern limits of the Zarafshan valley ; at all events they constitute the water-parting of
the Zarafshan. To the north of these mountains stretches a
range called the Karacha-taù, between which and the Godun and
Ak-taù lies an independent valley, or rather an elevated plain.
Mr. Fedchenko conceives that this plateau, beginning in the
east in the narrow Sanzar defile, widens out gradually, and so
extends between the two mountain ranges, although the
southern ranges of the Godun and Ak-taù are far from
being unbroken in their continuity ; and, although they are
for the most part lower than the Karacha-taù, yet the Zarafshan
does not receive a single affluent from the plateau. The only
river rising in that plateau first called the Sanzar, and then the
Djelanuty, runs through a narrow and deep fissure in the
Karacha-taù, and flows through the fields at Jisakh, which
are within the basin of the Jaxartes. The other parts of the
plateau are irrigated by streams descending from the mountains
in the north ; they are occasioned by the melting of the winter
snow, but the quantity of water supplied by them is said to be
quite enough to enable the inhabitants of that country to
pursue agriculture successfully.

The Zarafshan, issuing from the narrow part of the valley,
enters upon a flat country, and flows between less abrupt
banks. As far as the Jarty-tiubé village, it is walled in on
both sides by strata of conglomerate ; here, however, it runs
between sloping banks of clay. The conglomerate consisting of
rounded boulders of varying sizes, occupy the valley from Dashty-
kazy, the extreme eastern point of the Russians, to Jarty-
tiubé. There the Zarafshan rushes with great impetuosity in a
very narrow and deep bed. Above Pianjakent the river is not
fordable in the summer, but bridges are laid across where there
are villages on both sides. Within the Russian limits there are
two such bridges ; one half-way between the Jora and Mindanaù,
and another near the Gusar village. This last bridge is particularly remarkable. The Zarafshan here widens out into a

large pool; the banks are composed of comglomerate, but the velocity of the current is such that it has broken them, and detached a number of heavy masses of conglomerate, washed into a variety of shapes. The predominating form of these is, however, that of a pillar 30 or 35 feet high. From this large pool the Zarafshan forces its way in two precipitous channels through the conglomerate. These two precipices are spanned by little bridges, resting on the immense natural buttress formed by the island between them. One is a wooden bridge, the other is made of brick. It is difficult to conceive to oneself the fragile and shaky nature of these constructions, of which the many component parts give way under one's feet. The construction of the bridges is simply this. Two trunks of the juniper tree are laid across the chasm, and over these, crossways, are laid logs of wood. No rider ever ventures to cross the bridge on horseback, he invariably leads his horse over by the bridle. Although the width of the bridge is even less than 6 feet, there is no guard on either side. The stone bridge inspires not more confidence than the other; the supports upon which it is built are of wood, and are very much bent and broken from the weight of the arch. It bears an inscription, with the year 1233 (A.D. 1816), but what this signifies is unknown, as there is not a single person in Samarkand who is able to decipher it.

Where the conglomerate ceases to prevail, there one finds the first of a system of very large canals, or "aryks" dug for the irrigation of the southern parts of the valley. The lands here are much more elevated than those of the north, and have a precipitous fall to the Zarafshan River. The streamlets flowing from the Shahr-i-Subz Mountains are all absorbed into the canals serving to irrigate the fields at the very base of the mountains; but for the supply of water to the whole of the southern portion of the valley it has been necessary to divert streams at great elevations. Thus the "aryks" are constructed on very large scales. The most remarkable one is the Dargam, re-christened "Angar;" it is 47 miles long. In the formation of this watercourse the inhabitants have very cleverly availed themselves of natural gullies and ravines, although it must, nevertheless, have required great exertion to execute the work. This watercourse near Samarkand—that is, at the midcourse of the Zarafshan—has still the appearance of a tolerably large river. The Nurapai is another gigantic "aryk," which has been mistaken for a natural current. The head of this canal is in the vicinity of the Aflas village, 4 miles from Katty-Kurgan, which it supplies with water passing into the Bokhara territories.

At Chupanata Hill, about 5 miles from Samarkand, the waters of the Zarafshan run off into two channels: the one on

the north called the Ak-daria, and the one on the south the
Kara-daria. After separating from each other to a distance of
10 to 12 miles, these branches re-unite near Khatyrchi, at the
Russo-Bokhara frontier. Thus the Zarafshan forms an island,
which is divided into the two districts of Afarinken and Pai-
shambé, the richest and most populous part of the entire valley.
The excellent quality of the soil and the abundance of water
make this island strikingly productive. But, though it lies
between two branches of the river, the water-supply of this
island is derived only from one side—that is, from the Kara-
daria—the Ak-daria serving to irrigate the fields extending
on the northern side.

The principal "aryks," however, watering the northern parts
of the valley are conducted from greater elevations; from
heights equal almost to those from which the southern canals
are dug. In this manner irrigation is supplied to localities far
away from the Zarafshan, like Tash kupriuk, or the fort named
the "Stone Bridge," situated at a distance of 20 miles. This is
watered by a canal from the river.

We find, then, that to describe the valley it can very conveniently be divided into *Northern, Southern,* and *Insular.*

The country to the north of the Zarafshan is pure steppe—at
least that is clearly its character wherever agriculture has not
altered its physiognomy. But agriculture has spread over a
considerable portion of this tract; the road from Tash-kupriuk
to Samarkand, a distance of 20 miles, passes almost entirely by
gardens and cultivated fields. The great volume of water diverted from the Zarafshan abundantly satisfies the thirsty
grounds over this extent of country, which cedes only to the
Insular district in point of fertility.

The town of Chalek is situated in the Northern district. It
was once the residence of Omar Bek, the fanatic, who, by his
incursions, occasioned so much annoyance and trouble to the
Russians when their advanced post was at Jizakh. Chalek
stands on a line of road from Bokhara to Jizakh, which
avoids Samarkand. This is the shorter road; but there is
another reason for its being preferred to the Samarkand road
by caravans. Owing to the melting of the snows at the upper
sources of the Zarafshan, that river fills to a great depth at
midsummer. The velocity of the stream is so great that it is
impossible to cross over in boats; and sometimes by day it is
not fordable near Samarkand, where there is a shallow. In such
case the people avail themselves of the early morning, before
the snow-water from the mountains reaches Samarkand. In
1869 the river was so much swollen that all communication
between the two banks was suspended for several days. Bazar

prices went up considerably, as most of the supplies come from the northern districts. There are, on the southern bank of the river, signs of an attempt having once been made to overcome the difficulty of the passage across the river by building a bridge. This evidence consists of two stone arches at the foot of the Chupanata, which, apparently, were intended to serve for a bridge. When these were built, and whether the bridge was completed and afterwards carried away by the river, or the work impeded through want of technical knowledge, there are no means of ascertaining. Mr. Fedchenko observes that a plan of a bridge has now been drawn, and leaves it to be concluded that the project is one of the Russian Government.

The tract of land under cultivation diminishes greatly to the west of Chalek, where, in the same proportion the zone of virgin steppe country opens out wider; and here also is a corresponding change in the form of life and in the nature of the occupations of the inhabitants. In the purely agricultural districts the population is a fully settled one, but towards the steppes it is semi-nomadic, cattle-breeding, on a large scale, being allied with agricultural pursuits. In Central Asia, cattle-breeding and farming do not go hand-in-hand. Where much attention is paid to the soil, and where, consequently, the field and garden yield abundant produce, there cattle-breeding is very little in vogue; but where the cultivated zone merges with steppe pastures, and where there is a scarcity of water, one finds immense herds of oxen, sheep, and horses. Here the people live in villages only during the winter; in the summer-time they are away with their tents, camping in the trackless steppes. In some cases a portion of these villagers remain in their permanent dwellings; in most, however, they all leave for the plains. It is sometimes difficult to say, on lighting on an abandoned "ulus," whether the place is a ruin or serves yet for habitation. In most cases, even when in "residence," these semi-nomads live in their tents in the court-yards, while their mud houses are reserved for the shelter of their beasts.

There are four lakes in the steppe district between Mitan and Chalek; it is therefore called the Djurt-kul (Four Lakes) district. The lakes are called Chibisht, Airaùchi or Ai, Bigisht, and Durman *kuls*. They all lie in a depression between the Ak-daria and the Karadal hills. They are below the level of the Ak-daria, as is evidenced by the canals leading into them almost at straight angles from that river, and by another canal conducting the waters of the lakes towards the Karadal hills. Mr. Fedchenko saw Airaùchi and Durman kuls in the month of August. The latter is the largest of the four: it was covered with reeds over an extent of 3 to 4 miles. The clear space in the centre of the lake is not more than 180 feet long by 120 wide, the depth being about 1½ fathom. The water is fresh, turbid, and has the smell of sulphuretted hydrogen. There is a spring in the middle of the lake. The inhabitants of Hoji-Kishlak village pay the government 72*l.* a year for the right of cutting down the reeds. The other lakes yield a revenue of 30*l.* These reeds are used for roofing houses and for fuel; mats are also made of them. Innumerable quantities of leeches are found in the lakes, which have been proved to answer perfectly well in the hospital at Samarkand. They sell for about 1*s.* 6*d.* the hundred.

West of Kitaù the population is grouped under the Ak-taù Mountains. The gardens of these people, in the small villages of Ak-tiubé, Koshaùs, Jisman, and some others, are watered by small rills running from the mountains. The Ak-taù range has a hilly foreland of raised schist. The chain itself is composed of white marble, whence, probably, the name, Ak, meaning white, and taù, mountain. The highest point of the Jisman defile has an absolute height of 4076 feet. But the most elevated peaks, though not very much above the Jisman-taù, are to be found within the Bokhara limits, where this range extends far to the west, under the name of the Nurataù.

The Jisman rivulet forms the boundary line between the Russian territories and Bokhara.

The steppe, stretching away from the foreland of the Ak-taù to the Ak-daria, is nearly all sown with spring wheat, and the fields are never irrigated. The seeds are sown in February (O.S.), in the rainy season, so that the wheat, favoured by the little rain that continues to fall in March, and even partly in April, ripens by the beginning of June (O.S.). But there is a great difference in the heights attained by the wheat sown in the autumn, which is distinguished as the " irrigated wheat," and that sown in the spring which is called the " rain-watered." On seeing a field of wheat in the Steppe, after having passed through the Peishambé wheat-fields, Mr. Fedchenko was surprised, and asked if it would ever ripen, and when? He was told that it would arrive at maturity as early as the tall wheat growing in the irrigated fields. The short stalks compensate, too, for their diminutive stature by bearing a superior grain.

There is not an atom of land in the insular district which is not turned to account. It is all under the most careful cultivation. The landscape presented by each island is a multiplicity of fields sown with cotton, wheat, barley, rice, millet, and lucern, divided by hedge-rows of trees. These fields are sprinkled over with villages surrounded by gardens, and are irrigated by means of numerous "aryks" of large and small dimensions.

It may not be out of place here to give a more complete idea of the system of irrigation adopted in Central Asia. The insular district, for instance, in the valley of the Zarafshan is irrigated by means of water conduits leading from the Kara-daria. The valley slopes considerably, though smoothly towards the west, rendering the process of irrigation easy. A canal is simply dug, and one side of it is made to project into the bed of the river from which the water is to be drawn. When a great body of water is required then a large weir is constructed. Thus at the parting of the Zarafshan, at Chupanata, a weir is built across the Ak-daria to force the greater bulk of the river into the Kara-daria. The importance of the weir is very great, for the greater part of Bokhara depends upon it for its supply of water. The Ak-daria and the Kara-daria reunite on the Russo-Bokhara frontier, but, as mentioned before, a large quantity of water from the Kara-daria is drawn into the large "aryk" of Nurapai, near Katty-Kurgan. This "aryk" serves to irrigate the majority of the fields of the Zièddin Bek-ship. There is, nevertheless, only enough water conveyed through this "aryk" when the Kara-daria is full, and therefore the duty of constructing and regulating the weir on the Ak-daria is not imposed on the natives in its vicinity, but on the inhabitants of Katty-Kurgan and Zièddin. The weir is so badly built that it is re-constructed every year, and has to be frequently repaired. There are special regulations bearing on this work, and the 1000 men who are each year required to execute it are brought to Chupanata from a distance of 64 miles. Notwithstanding this, however, Katty-Kurgan and Zièddin were last year (1869) deprived of water for three consecutive days.

A little consideration given to this matter will convince any one, says Mr. Fedchenko, that the rest of Bokhara is not at the mercy of the Russians, through the fact of their tenure of the weir on the Ak-daria.

The water of the Zarafshan serves not alone to irrigate, it also fertilises the soil it passes over. Rapid as it is throughout a course of 134 miles in a mountain-valley, it brings down such a quantity of earthy particles that it is quite muddy. This slime deposited on the fields enriches them very much. Manure and marsh mud are used to a very small extent.

Dagbit, Yany-kurgan, and Peishambé, are the chief places in the insular district. Dagbit, within 9 miles north of Samarkand, is notorious for a great fair held in its vicinity on the banks of the Ak-daria. The inhabitants of five circuits assemble at this fair, and it is said that more business is done in it than at the

Samarkand bazaar. The Dagbit fair is held twice a week, like that in Samarkand. The fair in Chalek occurs once a week. In this manner there are fairs at different places on every day in the week, and many traders, particularly Jews, proceed from one bazaar to the other.

The feature in Dagbit is a mesjid, with the tomb of the Saint Mokhsum-Asam, who is believed to have died four hundred years ago. The mesjid is a low and long room, with two rows of columns, between which the ceiling is composed of a series of twelve cupolas. A high gallery, with paintings of various figures, runs along the outside of this structure. The impression produced by it on the people is so great that, in answer to Mr. Fedchenko's inquiry as to when and by whom it was built, they replied, "How could man have built such a place? God built it!" By some others the architect was, however, named, and his tomb at the entrance to the grave-yard where the saint lies was pointed out. The builder was one Jelengtash, to whom Samarkand is indebted for two splendid " medresses" (colleges). There are two remarkable palanquins preserved in the mesjid, which are quite Chinese in appearance; they are called "tokhtaravans," and were used more than a century ago by Musa-Khan and Isban-Khan, descendants of the Saint Mokhsum-Asam.

Yany-kurgan is a very small town, enclosed, like all Central Asiatic towns, within a low mud wall. It is the residence of an "Amilakdar," the chief of the Afarinken "tiumen," otherwise district.

Peishambé is situated not far from Katty-kurgan. There is a citadel in the place which was at one time occupied by a Bek, and by Bokharian troops. In the south portion of the valley the most conspicuous place is, of course, Samarkand, the capital of Timur. It is the largest town occupied by the Russians in Bokhara, and contrasts favourably with other Central Asiatic towns; such, for instance, as Tashkend, which, from a bird's-eye view, is a scene of flat roofs. In Samarkand the many stately edifices rise above the level of low dwelling-houses.

The town of Samarkand is surrounded by a thick clay wall, with six gateways. The gardens are outside the walls. On the eastern and southern sides of the town these gardens are particularly extensive. There is a cemetery on the northern side, and a desolate spot, which is supposed to be the ruined site of the ancient city. To the east of the town there is a place colonised by Persians who were once slaves in the khanat.

The town is chiefly indebted to its bazaar for its popularity and animation. In the centre of the place where the bazaar is held stands a stone building, called the "Char-Su." From the rooms within this building five corridors radiate to what were formerly covered avenues of the bazaar. These were, however, destroyed by General Kaufmann after the sudden attack made on the Russian garrison in 1867, after which the bazaar was rebuilt. The streets are now wide, and the shops are better than they were. The monotonous character of the town, composed mostly of mud-houses, with their backs turned to the streets, is relieved by the mesjids and their green ponds, by a variety of diminutive bazaars at the crossings and town gates, and also by old grave-yards.*

The citadel, in which was the Emir's palace, and where his "Sarbazes" were located, is at the western extremity of the town. Now, of course, it is occupied solely by the Russians.

The gardens around Samarkand, as well as the ravines, give a picturesque appearance to the place, and present many charming landscapes.

The valley of the Kara-Su-Chishma is particularly lovely. A large canal runs through it, and as the fall of the ground is very great the water runs rapidly, setting in motion a considerable number of mills. On the northern side the valley is closed in by a wall of schistous formation, which forms the extremity of the Chupanata mountains. The " aryks " here, as in other places, are cut through rock, and the river tumbles in a series of romantic falls.

Besides Samarkand, the towns on the left bank of the Zarafshan are Piandjakent and Katty-Kurgan. Pendjkend was in earlier times the capital of a distinct Bek-ship. After the capture of Samarkand by the Russians, it first declared its independence of Bokhara and then succumbed voluntarily to Russia. It is a small town, with a proportionately small bazaar. The inhabitants occupy themselves with weaving and with agriculture; they grow wheat, barley, rice, &c., in fields by the Zarafshan. In the month of May, Mr. Fedchenko saw them already harvesting their barley. Fruit ripens here about a week later than at Samarkand, owing to the superior elevation of this locality—about 3393 feet—which is more than 1200 feet above Samarkand.

Villages, cornfields, and gardens line the entire road from Pianjakent to Samarkand. The zone under cultivation stretches

* The number of edifices, &c., in Samarkand is as follows:—Shops, 1846; caravanserais for storage, 7; caravanserais with shops, 11: Indian caravanserais, 9; bath-houses (Hummums), 7; mesjids, 86; medresses (Colleges), 23.

Since the Russian occupation 25 mesjids have been demolished—18 in the citadel and 7 in the town—so that, previously, the number of mesjids alone was 111, or 134 with the medresses, which are of the same religious character, being devoted to religious teaching; this gave a proportion of 1 mesjid to every 200 of the inhabitants, who numbered 30,000.

almost to the foreland of the Shahr-i-Subz mountains.—Below
Samarkand there is very little irrigation, but the greater
portion of the steppe is sown with spring wheat, which requires
none. The absence of cultivation here is owing to the absence
of water. The Zarafshan canals, after a course of from 30 to
50 miles, contain here very little water; and to the west of the
Kara-Su the steppe is undulating and considerably elevated, so
that water cannot be transmitted through it. The irrigated
zone must, however, have extended farther in former days, for
one can yet see the traces of the Iské-angar "aryk" passing
along the very bases of the undulations. The Shahr-i-Subz
mountains give rise to several small streams, which irrigate
only the fields around the little settlements at their base.
These mountains, rising 7000 feet, completely wall off the
Shahr-i-Subz Bek-ship from the Russian possessions. From
Samarkand there is only one pass over them, the Kara-tinbé,
and this is practicable on horseback: the path lies through a
narrow gorge traversed by the river Kara-Su. The other road
from the Russian side lies through Djani; this also passes
through a defile, but is suitable for wheeled carriages.

Mr. Fedchenko entered both these defiles, but could form a
conception of the Shahr-i-Subz valley on the other side only, after
ascending to the top of the Aksai mountain, which is 6986 feet
high. This ascent, and a survey of the mountain district of
Oalyk, enabled Mr. Fedchenko to project the little map of the
Shahr-i-Subz valley which is incorporated with the one attached
to this paper. The two principal towns of that Bek-ship—shown
on the map—were determined instrumentally. The valley is
bounded on the north by the Shahr-i-Subz mountains; on the
east, by a chain of mountains stretching from Maghian, at first
directly to the south, and then turning westwards. This chain
is much higher than that forming the northern boundary, and
is covered with perpetual snows. It gives rise to several streams:
the principal one—the Hujaman-Su—flows by Farap, and on
emerging upon a level country it receives the name of Kashka-
daria. This river runs by Kitab to the Bokharian town of
Chirakchi, and passing by Karshi discharges itself into a small
lake.

The capital towns of the Shahr-i-Subz Bek-ship are Shaar
(meaning *town*) and Kitab. The population is centred in these
and in a few villages by the Kashka-daria and its affluents.
Some of the villages of the Bek-ship stand on the northern
slopes of the Shahr-i-Subz mountains and by the Anchava river.
Chirakchi and Yakobak belong to the Emir of Bokhara.

A few words may here be added respecting the northern slope
of the Shahr-i-Subz mountains, which is now claimed by Russia.

Some of the defiles, like those of Jam, Aksai, Oalyk, Kara-tiubé, and Urgut, which are long and deep, penetrate from the steppes almost to the main range. The sides are nearly bare, verdure appearing only in those places where there is trickling water. These springs magnify into rivers, or rather water-falls, and the inhabitants turn them into account for irrigation purposes, and the size of the villages here is quite proportionate to the number of such sources of benefaction. Urgut is the largest place in this district; it is a town with 5000 inhabitants, and was once an independent Bek-ship. The other places are of inconsiderable dimensions.

Where there is a scarcity of water, or where the ground is too steep to allow of its cultivation, the people pasture flocks. But besides the flocks of the native inhabitants of this country, sheep from Khulum and Balkh—from beyond the Oxus—are driven to these pastures. Owing to the absence of forest, these mountains have a most dreary and melancholy aspect. They contain no minerals whatever, excepting limestone, which is burned in great quantities in the village of Oalyk.

NOTE ON THE MAP OF THE ZARAFSHAN VALLEY ATTAHCED
TO MR. FEDCHENKO'S PAPER.

THIS map is based on surveys, made during Mr. Fedchenko's explorations in the valley of the Zarafshan, by Lieut. Kutzei and Mr. Novasëlof, a topographer. They mapped all the southern portion of the valley and the country extending along the road to the Ak-taù Mountains. This was done instrumentally, with the aid of the plane-table, on a scale of 5 versts to the inch. The bases of 5 versts were measured in the neighbourhoods of Samarkand, Katty-kurgan, and Karatiubé. The island lying between the Kara-daria and Ak-daria is inserted in the map, according to a survey of it conducted under the personal super-intendence of General Abramof.

No satisfactory survey having previously been made of the northern part of the valley, it is left in blank on the accompanying map; Mr. Fedchenko's route and the lakes on the northern side being, however, approximately traced upon it.

The Shahr-i-Subz valley is given from a sketch; the towns of Shar and Kitab having been alone determined by notches from the summit of the Aksai-taù and from the Oalyk Mountains.

The delineation of the Bek-ship of Magian, and of the Valley of the Fan River, are based on inquiries; but the main direction of the mountains was determined on the road from Pianjakent to Dashty-kazy.

The latitude and longitude of Samarkand are taken from Mr. Struve's determinations.

Annexed is a Table of the elevations determined during the Expedition into the Zarafshan Valley in 1869.

The absolute elevation of Samarkand was determined after a three months' observation (from the 14th (26th) January to the 13th (25th) April, according to Hauss's formula).

The height of Chupanata was determined by simultaneous observations at its base and summit. The elevation of Katty-kurgan was arrived at after thirteen observations, while corresponding observations were being made in Samarkand.

The elevations to the north of Katty-kurgan were ascertained by means of simultaneous observations.

The barometer left in Samarkand having been damaged, no corresponding observations were made of the remaining heights.

With reference to the elevation of Katty-kurgan, shown in the following table, Mr. Fedchenko says that, although he does not know by what method Burnes made his determinations, he believes that Burnes's showing of 1200 feet as the elevation of Bokhara is incorrect. He thinks that is too high a figure. From Samarkand to Yar-basha—a distance of 60 versts (40 miles)—he says there is a fall of 1000 feet in the valley, and the elevation of Yar-basha is only 1260 feet; so that, admitting the decline beyond is not so great, it is his opinion that the elevation of Bokhara is not so great as given by Burnes.[*]

TABLE OF ELEVATIONS.

	Number of Observations.	Name of Place.	Elevation in English Feet.
1	270	Samarkand	2154
2	1	Chupanata Mountain	9639
3	3	Karasu	1667
4	13	Katty-Kurgan	1366
5	2	Yar-bash	1260
6	4	Jizman	2040
7	1	Base of Jizman Mountain	2974
8	1	Summit of ditto ditto	4076
9	3	Ulus	1780
10	3	Jam	2047
11	1	Pass of Kizil-Kutal	9153
12	5	Aksai Village	2732
13	1	Aksai Mountain	6986
14	3	Oalyk	2850
15	3	Kara-tiubé	2900
16	3	Hojaduk	3205
17	2	Urgut	3715
18	4	Huz	3561
19	1	Kulbasy Mountain	7118
20	8	Pianjakent	3206
21	2	Jora Village	4085

[*] 'Travels in Bokhara,' new edition, vol. iii. p. 187. London, 1839.

LIBRARY
Leland Stanford, Jr.
UNIVERSITY

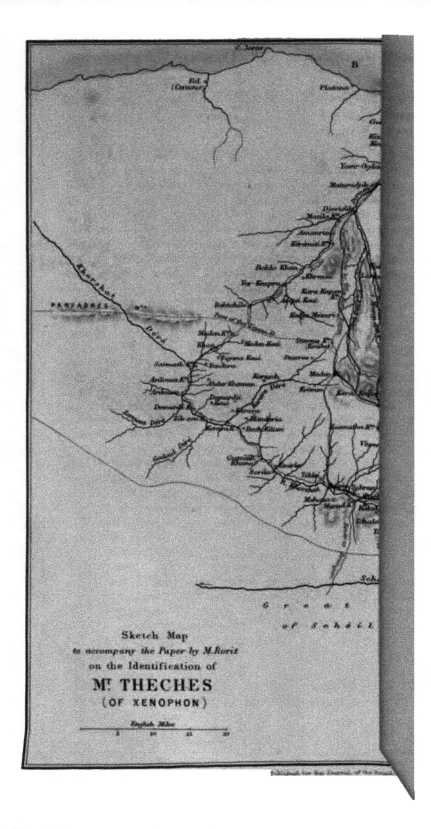

Sketch Map
to accompany the Paper by M. Rorit
on the Identification of
M. THECHES
(OF XENOPHON)

English Miles

XVIII.—*Identification of the Mount Théchés of Xenophon.* By M. P. Rorit, Chief Engineer at Trebizond. Translated by T. K. Lynch, Esq., F.R.G.S.

No point of the route followed by the Ten Thousand has been traced or determined with certainty, from their entry into the mountains of the Carduchi till they reached Trebizond. There exists even a doubt whether Trebizond is the Trapezus where the Greeks reached the Euxine Sea. Few travellers have cared to make researches as to the road taken by the Ten Thousand. Kinneir is perhaps the only one who has given this subject a serious thought. He makes the Greeks wander about in the plains of Kars, and return towards Ispir. Kinneir is right, and Gymnias is found to be in the neighbourhood of Ispir.

Arrian, in his 'Periplus,' visited, near Trebizond, the monument raised to the memory of the Ten Thousand "upon the mountain from whose summit the Greeks caught sight of the sea." Arrian does not appear to have left the coast. The monument he mentions must have been situated upon the Boztépé* of Trebizond, at an elevation of 200 mètres above the level of the sea that flows at its foot, whilst, according to the 'Anabasis,' Mount Théchés is distant five days' march from Trapezus. No one has either recognised or visited Mount Théchés since the passage of the Ten Thousand. The Greek colonists on the banks of the Euxine only commenced penetrating into the interior of the country after the time of Alexander. Mount Théchés is situated at the head of the valley of Khorshat, a river of Gumish Khana that flows into the sea at a distance of some miles to the east of Tireboli. After having consulted several works of modern travellers and the geography of Strabo, I perceived that one must keep to Xenophon, and it is by the very letter of the 'Anabasis' that I arrived at this discovery of the sacred mountain. Let us then return to Trapezus.

The Greeks went by land in three days from Trapezus to Cerasus, which, consequently, cannot be Kerassund, as it cannot be reached under eight days. Might Trapezus, then, be Eski-Trebizond (old Trebizond), situated more than a degree to the east of Trebizond? But, then, to reach Cotyora one must calculate a march of more than three days. Eight stages, and then two days—twenty days would have been necessary, for the windings of the shore are very tortuous, thereby exceedingly lengthening the route. Cerasus must have been,

---

* Boz means in Turkish *whitish grey*, and Tépé means a hill. Boztépé is a common name for hills, and that rising behind Trebizond, over which the road leads to the interior, is so called.—[T. K. L.]

then, at Fol, between Trebizond and Tireboli. The country of
Massyneques was evidently the environs of Tireboli, for chesnut
woods abound there, as do cherry trees. At Fol, the country is
high, and deeply indented. The Chalybes were evidently
between Kerassund and Ordou (Cotyora), "the country of
Tibarenes being much more level." Here one discovers the slag
of forges, which the inhabitants presented to me as coal. Trebizond is evidently Trapezus.

The summit of the mountain chain that extends along the
Black Sea (Mounts Paryadres and Moschiques) is generally
distant 70 to 80 kilomètres from it. Facing Trebizond is Kara
Kapan, the Kolat-dagh, the Mountain of the Colques—Mount
Theches being three days further, and must be found in the
environs of Balahar or Baiburt, in the high spurs of the Kara
Kapan or of the Kop Dagh (Mounts Scydisses). The Greeks
caught the first glimpse of the sea through a hollow of the
Moschiques Mountains, or through the valley of Khorshat.
Let us lay this down only hypothetically, until we find the river
of the Macrones.

I need scarcely say that I had made myself perfectly
acquainted with the country. I had travelled from Erzerum
by the Kara-Kapan, Vesernik, Baiburt, and the Kodja Pounar,
returning by the Kop Dagh and the valley of Gumish Khanah.
The river of Macrones flowed on the right, that is to say, from
the north or cast; so that the river on the left, into which it
flowed, could only flow in a westerly or southerly direction.
These characteristics can only be met with in the river of
Gumish Khanah, and the river of the Macrones could only be
that of the Vezernik. That one on the left into which it flows
is much smaller; it is in the prolongation of the valley of
Khorshat, flowing from Kaledjik, Tarkanas, and Mount Théchés.
The two rivers unite between Sabrau Khan and Mourad Khan,
19 kilomètres up from Gumish-hanah and 6 kilomètres from
Tekké. There is a bridge over the Khorshat at Sabran Khan.
At 500 mètres down from the bridge at the meeting of these
two rivers (which of them is Khorshat?), exactly where the
Greeks passed, one sees a square enclosure which was an odjiak
of the Sultan Murad's, according to Hadji Khalfah Orlah, or
Ourlah Gumish Khanah. The river comes from Ghévanis, where
there is a fortress. The country, filled with forests, is an odjiak
of the Sultan Murad. This part of the country still retains the
name of Murad Khan. As to the forests, they have disappeared.
The river of Macrones is in summer merely a rivulet of some
mètres wide, quite shallow, flowing through beds of gravel 100
mètres wide, full of thorny brushwood, just as one finds everywhere in the same country growing along the river. The banks

were bordered by low but closely-interlaced brushwood. Though the river was not wide, the Macrones were separated from it by the brushwood, which fact explains why the stones they threw at the Greeks did not hit them.

"At the right a very steep mountain." This mountain is at a distance of 5 kilomètres higher up. It is crowned with very considerable ruins, attributed to the Genoese. At the foot lies the village of Kaledjik, or Kalé (Mudirlik of Ghevanis). The present road is a narrow path cut in the rock, at about 10 mètres above the river. On the other side of the river there is a land-slip which confines it and makes it a narrow ravine.

According to the text of Xenophon, the very steep mountain would be near the river of the Macrones: it is actually at a distance of 5 kilomètres. This discrepancy is to be explained—Xenophon noted his march at the stations where he encamped. At the same place the army encountered two difficulties, the very steep hill and the river of the Macrones.

The river of the Macrones is distant 25 kilomètres from Mount Théchés, and about 30 kilomètres from the mountain of the Colques. The river of the Macrones once settled, the circle in which one must find Mount Théchés is narrowed. It cannot be more than a few hours from Murad Khan in the chain of Vavough-Dagh, which separates the basins of the Khorshat and Tchoruk. Armed and convinced on these given points, I again went up the valley of Gumish Khanah; I visited the highest summits of the chain of Vavoug-Dagh; I found here and there little pyramids of loose stones which took the form of a semicircular shelter, the work of shepherds. I had lost all hope, and was about to take myself off further towards Baibout, when, to satisfy my conscience, I decided to scale a group of hills isolated towards the south, that is to say, in a direction which seemed to remove me from any possibility of seeing the sea. I scaled a central summit which commanded the rest, and found there some heaps of stones, but they were nothing more than dykes of porphyritic rock, which had loosened themselves from their bed and rolled away (not in a direct line) on to the level, which spreading out towards the south formed the vast plain of Scheilan-déré.

I then directed my steps towards another summit at 1200 or 1500 mètres towards the north. I arrived at the first summit, on which I found a mound of stones. By a slight curvature of the ground at 150 or 200 mètres further on, I arrived at another summit; it was there I had found the mound raised by the Ten Thousand. To the north-west, right ahead, plunged deeply the valley of Khorshat, at the bottom of which it struck me that I saw the sea. To the north a very low dip of the mountains of Moschiques, overhanging Vesernik, was covered with mist. To the south-west I thought I saw a large and remote valley, at the extreme end of the plain of Scheilan-déré. Was that the valley of the Euphrates below Erzinjan towards Malatia? or the valley of Kizil Irmack towards Sivas? ·I had that day no instrument to take the direction.

I wrote in my diary, "On the 5th September, 1867, at ten o'clock in the morning, I arrived at the summit of Mount Théchés." I breakfasted, and then poured out libations on the mound in memory of the Ten Thousand.

The mound was 10 mètres in diameter, and 80 centimètres in height, in the centre; it crowned the whole summit of the elevation, and is formed of large rubble stones collected immediately below on the sides of the mound, where there is a round bank of porphyritic stone, these rubble stones had flat level faces, but are not so rectilinear as if they had been sculptured, for they are without doubt crude.

In the middle of the mound exists a semicircular screen like the half kerbstone of a well, a mètre and a half in diameter, facing south. The interior of this heap of rough stones is occupied by a wild currant-bush, a *Ribes*, if I am not mistaken; round the mound have been erected, evidently with the stones of the mound, 6 (or 7) cylindrical pyramids or truncated cones of a mètre and 20 centimètres to 1½ mètre in diameter, and a half to 2 mètres in height, by which one can conclude that the mound itself had a conical height of many mètres.

In the little heap of stones of the neighbouring mound there exists also a hut of the same material, partly destroyed. The Greeks apparently occupied the two hills, and the space separating them.

One finds on the top of the surrounding mountains as well as here, small pyramids of rough stone; at a distance one would take them for men. Who could have erected these? Evidently shepherds in imitation of those of the Greeks. One can obtain no information from the inhabitants; their history and traditions do not extend beyond the memory of their old men; all their ruins are attributed to the Genoese, and all their old fortresses are those of Kiz-kalessi.

Turning over the stones of the mound, I made a slight search and found there common broken pieces of a red and black pottery, such as is still in use in the country, and such as one finds in the mound of Balahor. At the bottom is a brown earth, the remains, perhaps, of the shields of osier and leather of the Scythians. I filled up again the breach I had made with the stones I had taken out. The red potteries were among the stones, and the black in the brown earth.

The northern and western slopes of Mount Théchés still preserve the remains of forests of pine trees; the summit is barren, covered by coarse grass, and with a low artichoke plant, which they call jubarbe: I there gathered a bouquet of yellow "immortelles."

Mount Théchés is situated at about 500 mètres above the summit of Vavoug-Dagh, which by measurement is 1900 mètres above the Black Sea. The highest points of Kara-Kapan must attain 3000 mètres, and the surrounding peaks of the mountains of Paryadres and Moschiques 2000 mètres.

I delayed the publication of my discovery, because I had no time to reduce into form the present mémoires, and to map the chart in explanation thereof. I had no fear that the honour of my discovery would be taken from me, because Mount Théchés is in a position upon which one is continually tempted to turn one's back, and the commentaries made upon the retreat of the Ten Thousand serve only to baffle one in the research.

I have studied during one year, ancient and modern works, treating of the history and geography of the country, without being able to draw therefrom the least positive information. It is by a general knowledge of the country, the account given by Xenophon, a reasonable conviction, persistency, and good legs that I have arrived at Mount Théchés.

From the three roads which lead from Trebizond to Baiburt, one can see Mount Théchés from the plain of Nive and Varzahan, from the peak of Vavoug-Dagh, and as far as Kaderak. From the bridge of Balahar one sees it lifting its summit over the little valley, between the Scheilan-déré and the River of Kaderak. The Greeks ought to have seen it on the fourth day of their march from Gymnias or Ispir.

If we follow the Winter Road which they are trying at the present moment to render serviceable for carriage traffic, on leaving the stone bridge of Gumish-Kanah, we find—

At 13 kilomètres, 200 mètres			Tekke.
„ 19	„	.. .. ..	River of the Macrones.
	(Vezerik) 1290 mètres above the level of the sea.		
At 24 kilomètres, 500 mètres			Kaledjik (the very steep mountain).
„ 31	„	800 „	Guelchid Khan.
„ 34	„	500 „	the river bifurcates.

The largest arm trends towards the East Tarkanas, and Mount Théchés, which is not discernible; the other arm turns towards the north, and then almost immediately to the east, through a gorge of rocks in order to reach Vavoug-Khan at 37 kilomètres from Gumish-Khanah. At 1800 mètres further, or Vavoug-Khan, one attains the summit of the mountain or the peak of Vavoug-Dagh, a height of 1900 mètres above the Black Sea.

On the route from·Erzerum, by Gumish, and desc
a few steps to the right, and south about 6 kilomètre
comes at once to the discovery of Mount Théchés,
forms a ridge towards the north, at the foot of whic
perceives the village of Tchartchi.

From the peak of Vavoug-Dagh one ought to distin
with an eye-glass, the pyramids which surround the s
and almost the brushwood which occupies the centre.

The Greeks directed their march towards the sea.
guide informed them of a village where they encamped ;'
the Tarkanas of the present day. Although situated in the
of the Korshat, Tarkanas was Scythian. Vavoug-Dagh
rotund mountain; the Arabs had free access round it a
had over all the country of the Scythians, and to
Erzerum, on the side of the basin of the Tcharuk-Su, the
of the Vavoug-Dagh have not more than 150 mètres of ele

Below Tarkanas, as also below Vizernik, the valleys b
rocky gorges, which were then covered with wood, inhabi
they are at present, and separating the Scythians fro
Macrones. This explains the limits of these last at N
Khan, at the river of the Macrones, who inhabit a
uneven country, where the use of carts was not possible.
Macrones, and the Scythians of the present day, are sti
distinct types.

From Tarkanas the Greeks, descending a rocky and
valley, passed to the foot of the Mountain of Kaledjik
steep"—arrived at the river of the Macrones, over whi
passed, and encamped there among the Macrones, w
nished them with cattle and provisions. The Greeks
changed their direction when they learnt that Trapezus
the north. The Macrones, in order to remove them from
territory, conducted them toward Vizernik, over which
encamped. The third day they scaled the mountain, ove
the Colques, and encamped in the villages, abounding in
of provisions and much honey, which makes one del
One still finds in all these mountains a honey which
produce of the flower of the *Rhododendron Ponticum.*

In two days they descended rapidly and light-hearted to
the sea. Xenophon says nothing of that part of the
which he reckons as 7 parasangs. That is too little.
mountain of·the Colques is 60 to 70 kilomètres from Treb
Admitting that, after having scaled the mountain, they
have advanced for some hours in the yailas of to-day, whe
villages were found, there would still remain more th
parasangs, but one can fancy that Xenophon (all diffi
being at an end) found the distance short.

Where were those villages? In the yailas of Koulat, Kodja-Mezari, the Upper Meiran-anadéré, or of Galian-déré, which are the superb prairies of the mountains. Did the Greeks follow the present route of Vezernik by Maden-Khan, Koulat, Kodja-Mezari, by the Larkana-déré or the Meiran-anadéré, to come out at Djevislik? Did they pass the Kazecli? In the latter case they would not have descended the Galian-déré, which is as steep as a ladder, between Amborlo and Galian Keupru. From Karagumse they would have followed a ridge of the mountain, which is practicable at the present day, coming out at the sea at Campos. But no, they must have passed at Djevislik? a very old place. The route of Kara Kapan must be the most ancient in the country between the Black Sea and Persia, and must have been frequented since the earliest period.

I am not perfectly acquainted with the summits of the Kara Kapan. I know the three to the south and the one to the north, towards Tach Keuprn. By the lines punctuated in red I have indicated the different directions which the Greeks might have followed. Did the Greeks really cross the mountain of the Colques by one of the clefts which separate the high summits of the Kara Kapan, as does the actual route at present? In that case they would have to re-descend from 700 to 800 mètres in order to reach the yailas where the villages were to be found. We fall altogether in doubt should we read, "The Greeks having arrived at the top encamped in several villages." One of two things: either the Greeks went over the mountain of the Colques by a very low pass, which led them, without descending too much in the yailas—which are at the sources of Meiran-anadéré, or they came down the Korshat, towards Tekké, in order to reach Kromm, the river of which is in a box-sided gorge. In scaling the right flank, inaccessible from the valley, one arrives, by an easy walk, in the yailas of Koulat. A survey of the Kara Kapan, and of the routes from Kromm to Tekké, would clearly settle all that.

The Greeks did not pass by Tach Keupru. There is between that point and Kazecli Khan a spur of mountain which is not impracticable, but where the Colques would easily have arrested the army by rolling great stones from the mountain.

The Greeks encamped at Trebizond, between Boztepe and the old town above Meidan, on a sloping plateau which had no houses 30 years ago. This plateau, separated from the sea by an irregular declivity, which, covered with wood, answers well to the narrative of Xenophon. The neighbourhood of Platana answers only to the narrative of the expedition amongst the Drilles, whose capital ought to be found above towards Aska, 5 hours from the sea above Platana. Does not Pliny report that the yailas of the mountains of Asia Minor were studded with villages? In the yailas of Kera Gueul Dagh, between Kerassund and Karahissar, I found, last year, traces of ancient villages, which, as they explained to me, existed up to about 50 years ago, but were abandoned after the destruction of the forests which sheltered them.

There is no doubt but that the mountains of Asia Minor have been everywhere covered with forests; their destruction is the work of those who adopt a pastoral life, which is the enemy of progress and civilisation, the source of polygamy, the hotbed of idleness, carelessness, want of foresight, and Oriental fatalism.

The nakedness of Arabia and of the vast tracts of Asia to the north and to the west, the sterility which extends like an oil spot over Persia, cannot be traced to any other cause than the pastoral life of the inhabitants. The people adopting it are locusts; they destroy all woodland and vegetation, modifying thereby even the climate,—from whence the necessity of emigration. Had the invasions of barbarians any other cause? A study of the question in this sense would, perhaps, give us a key to the great migrations of mankind. These lazy nomads destroy the forests by the axe and by fire, their goats achieve the rest down to the very roots; the goat is a scorching-iron which one must destroy—they lay bare the very rocks. They are in Greece as numerous as stones. The destruction of the ancient Greek towns of Asia, and the filling up of their ports with sand, had no other cause than the cutting down of their mountain forests.

After this digression, which the present state of destruction of the forests of Asia Minor has inspired, I will say yet a few words on the retreat of the Ten Thousand. The pile of stones on Mount Théchés is, without doubt, the only witness left by the Ten Thousand on their march from Cunaxa to Trebizond, with this exception, perhaps, that in the numerous battles in which they may have been engaged, they should have left some arms or military acoutrements in places which have not been subsequently overrun by the Greek and Roman armies. I think, nevertheless, that the points where they encamped, where they fought, where they crossed the rivers, and the town of Gymnias, can be identified, and the march of the Ten Thousand traced with some degree of exactitude.

I only know the routes which lead from Trebizond to Erzerum, and I have not been further than the latter town.

With the map of Kiepert in my hand if I traversed the mountains of the Carduchi, the Centrite, the Teleboas, and across the plain, I should arrive at castles surrounded by numerous villages, "where there was wine of an excellent bouquet, all sorts of vegetables and raisins;" we must discover this locality—I think it is Mouch. It had a mountain both in front and in the rear. From such a locality the Greeks leave early in the morning as quietly as possible, fight Tiribzze, and encamp; then they make three stages in the desert, along the Euphrates, which they pass, towards Melas Gerd.

Further down one must also pass the Binquel. According to Kinneir, the Euphrates is here as large, as deep, and as rapid as the Tigris at Mosoul. After making three stages in the plain, the advanced guard arrived at a village where there was a fort, and a fountain near the fort. The rear guard repulsed the enemy, who threw themselves into a valley. The rest of the army passed the night outside. The soldiers bent their steps towards dark spots, where the snow had melted; those were springs. The next day they rejoined Chirisophe, and found villages underground, where there was a Comarque, &c.

This spot ought to be easily identified; the country and its customs have not changed since the earliest antiquity; perhaps, in the direction of Boscheg (Boschich), Teranos Kara-Kilissey, Mollah-Osman, on the road from Erzerum to Bayazid. One follows that road towards Erzerum, in order to pass the mountain; but by that route one has only 100 kilomètres as far as (Keupru-Keui) Kopru-Koi, where one crosses the Phasis (Araxes), and the Greeks marched three stages, 35 parasangs, more to reach the Phasis: one cannot allow that they passed further down, since it was only a plethrum wide. Most likely the guide led the Greeks wrong, to where there were no villages; they were directed towards the west, across a very high and very difficult mountain, in the snowy season. But afterwards they took the direction towards the north, which in their position was that towards the Euxine Sea. Did they believe that they had found the Phasis, where ships terminated their voyages? They would have advanced along that river the Phasian-su, and seeing the mistake that it went eastwards, they would have turned towards Kars.

Following the route from Erzerum to Kars, after a march of ten parasangs from Kopru Koi, we reach the mountain Soghauli-Dagh. Let us get over that mountain and we are in the plain of Toaques. Thirty parasangs more and we arrive at a pass very difficult to cross. The details which Xenophon gives regarding that place ought to enable us to recognise it. It is perhaps Kars.

The Greeks saw before them a high chain of mountains which stood between them and the Caucasus (I do not know if from

Kars one can see the Caucasus). They discovered that they were too much to the east, and that they had passed the extreme curve of the Euxine Sea; they then turned towards the west, and found the country of the Chalybes, who had long spears; these are the Kurds. About Kars they had fortified places, and the villages of the country are still situated on points very difficult of attack.

After a march of 50 parasangs, in seven stages, the Greeks arrived at the Harpasus, 5 plethrums in breadth. It is the river Olti. Spring was approaching, the snow commenced to melt, and the rivers became swollen. The Greeks could not see the sea by the valley of Tchorouk-su because of the chain of the mountains of Artwin (or Ardween).

From Kars to the river of Olti is not 50 parasangs, but one must admit that the march of the Ten Thousand was lengthened by the attacks of the Chalybes, "the most warlike of the people through whom they passed. The Greeks made 20 parasangs, in four stages, through the country of the Scythians, in a plain well studded with villages. We are in the Thortum. The Greeks of mountainous Greece would consider it as an open, plain country. Its inclines are not very steep. Vehicles traversed it, as also all the round mountains of those countries, which are of an *infrajurassique* formation. The road which leads from Ispir is not difficult. We are at Gymnias. Was Gymnias to the right or to the left of Tchorouk-su? There had been a bridge of some kind near that town, "large, rich, and populous." The Greeks passed the river, which otherwise they would have had to pass higher up, towards Baiburt, in order to reach Mount Théchés. I hope this year to visit the country between Rizet, on the Black Sea, and Ispir, and repass the mountain towards Baiburt.

Let us continue. The Greeks might in one day have reached the summit of the Moschiques Mountains, and have seen the sea at their feet, but the Governor of the country directed them towards the country of his enemies, which they understood when the guide enjoined on them to burn and destroy everything. They remounted the left bank of the Tchorouk-su, went along the plain of Kart, passed the Balahor, and arrived at Mount Théchés.

The village of Djinnis, situated on the left bank of the Euphrates of Erzerum, at 40 kilomètres from that town, on the route from Erzingan, cannot be Gymnias. In fact, from Phasis to Gymnias the Greeks made a march of nineteen days, when they could have gone to Djinnis in four days, by Erzerum. One cannot suppose that they went towards Kars and returned to the sources of the Phasis. Besides, they would have passed

Hassan-Kalé, where there are springs of hot water at Elidja, (12 kilomètres in straight line from Erzerum), over the very source, which has 100° of heat—rising from the earth, in a volume as thick as a man's body, and forming a kind of lake, in which, at all seasons, clouds of water-fowl disport themselves. One must pass the very source, having on the left a mountain, and on the right marshes, 25 kilomètres long. That source could not have risen since that time, for it has formed there quite a hill of travertine of a very resisting nature.

From Djinnis one must pass the Euphrates, which is 30 mètres wide in the low season, across Kop-Dagh and the gorge of Kop-Keni, where a few men could arrest an army. One must pass by the Valley Saman-Su, after having travelled along the Euphrates. Is it possible that Xenophon would have said nothing about that? Djinnis and Ispir are both exactly five days from Mount Théchés.

If the Mahometans were not iconoclasts, I would propose to erect a statue to Xenophon, on the place of Charki-Meidan of Trebizond, which should take the name of "The Place Xenophon, or of the Ten Thousand."

INDEX

TO

VOLUME THE FORTIETH.

END OF VOLUME XL.

LONDON. PRINTED BY WILLIAM CLOWES AND SONS, STAMFORD STREET, AND CHARING CROSS.

Lightning Source UK Ltd.
Milton Keynes UK
UKHW021819281118
333125UK00009B/395/P